Systems Biology

Edited by
Robert A. Meyers

Related Titles

Meyers, R. A. (ed.)

Encyclopedia of Molecular Cell Biology and Molecular Medicine

Online version:
www.meyers-emcbmm.com

Klipp, E., Liebermeister, W., Wierling, C., Kowald, A., Lehrach, H., Herwig, R.

Systems Biology

A Textbook

ISBN: 978-3-527-31874-2

Fu, P., Panke, S. (eds.)

Systems Biology and Synthetic Biology

ISBN: 978-0-471-76778-7

Dehmer, M., Emmert-Streib, F. (eds.)

Analysis of Complex Networks

From Biology to Linguistics

ISBN: 978-3-527-32345-6

Junker, B. H., Schreiber, F.

Analysis of Biological Networks

ISBN: 978-0-470-04144-4

Mandoiu, I., Zelikovsky, A.

Bioinformatics Algorithms

Techniques and Applications

ISBN: 978-0-470-09773-1

Systems Biology

Advances in Molecular Biology and Medicine

Edited by
Robert A. Meyers

WILEY-BLACKWELL

The Editor

Dr. Robert A. Meyers
Editor in Chief
RAMTECH Limited
122, Escalle Lane
Larkspur, CA 94939
USA

Cover

Network-based analysis of DNA microarray data, with kind permission by Mikael Benson (for more information see Chapter 18, Figure 5).

Library of Congress Card No.: applied for

British Library Cataloguing-in-Publication Data
A catalogue record for this book is available from the British Library.

Bibliographic information published by the Deutsche Nationalbibliothek
The Deutsche Nationalbibliothek lists this publication in the Deutsche Nationalbibliografie; detailed bibliographic data are available on the Internet at <http://dnb.d-nb.de>.

Wiley-Blackwell is an imprint of John Wiley & Sons, formed by the merger of Wiley's global Scientific, Technical, and Medical business with Blackwell Publishing.

Composition Laserwords Private Limited, Chennai, India

Printing and Binding Strauss GmbH, Mörlenbach

Cover Design Adam Design, Weinheim

Printed in the Federal Republic of Germany
Printed on acid-free paper

Print ISBN: 978-3-527-32607-5

Contents

Preface and Commentary

Our project is based on the recognition that a true and complete understanding of biology and indeed, human disease must go beyond genomics. Genes and their transcripts and translation products are of course crucial, but in fact these molecules never act alone; they rather operate in tremendously complex and noisy networks. Consider it is now estimated that the human genome encodes about 20 000 to 32 000 distinct proteins, while the expression and alternative splicing of mRNAs indicates that humans may be able to produce 10^6 different proteins. Hence the need for the emerging discipline of Systems Biology. Thus, Systems Biology is a recently established field in life sciences that aims at promoting a global, top-down, mathematical, systems understanding of living matter through the integration of various scientific domains.

Our compendium is written for university undergraduates, graduate students, faculty and investigators at research institutes and is a team effort which began with overall guidance from our Board, which includes 11 Nobel Prize winners. Each article and author selection was then validated by several reviewers from major university research centers and each manuscript was then reviewed by peers from other universities. There is a glossary of terms with definitions provided at the beginning of each article for students and the articles average close to 30 print pages – which provides significantly more depth for advanced researchers than other systems biology reviews.

Our treatment consists of 22 articles or chapters, and begins with an overview of Systems Biology which the reader, at any level, can use as a roadmap to the content of our book. Then, the detailed articles are organized into five sections, the first of which is: the *Biological Basis of Systems Biology* (the chapters range from developmental biology to the Interactome and protein abundance variation); the section on *Systems Biology of Evolution* includes a chapter by Nobel Laureate and Board Member, Werner Arber; the *Modeling of Biological Systems* section includes chapters ranging from computer simulation methodology of specific systems such as the brain and cell migration as well as protein modeling. Considering that there are now about 21 000 distinct proteins encoded directly by the human genome while the expression and alternative splicing of mRNAs indicates that humans may be able to produce 10^6 different proteins, and we can only surmise the function of nearly half of these proteins – a systems approach is clearly needed. This section then includes top down mathematical modeling methods involving chaos, fractals, dynamics of biomolecular networks and informatics; the *Systems Biology in Medicine and Disease* section includes chapters on inferring networks for disease (termed the diseasome) and culminates in a chapter on systems aspects in personalized medicine for prediction and prevention. Our concluding section is *Systems Biology of Organisms* which includes plant systems biology as well as two chapters from scientists at the J. Craig Venter Institute, one on microbiomes which encompasses metagenomics as well as the human gut microbiome which has recently been discovered to contribute to viral infection cycles, and the other on synthetic biology which describes methods and the recent creation of a synthetic cell.

Our team hopes that you, the reader, will benefit from our hard work – finding the content useful in your research and education. We wish to thank our Managing Editor, Sarah Mellor as well as our Executive Editor, Gregor Cicchetti for both their advice and hard work in the course of this project.

Larkspur, California, March 2012

Robert A. Meyers
RAMTECH Limited

List of Contributors

Mario Albrecht
Max Planck Institute for Informatics
Department of Computational Biology
and Applied Algorithmics
Campus E1 4
66123 Saarbrücken
Germany

Mikkel A. Algire
J. Craig Venter Institute
Synthetic Biology
9704 Medical Center Drive
Rockville, MD 20850
USA

Miguel Antonio Aon
Johns Hopkins University
Department of Medicine
720 Rutland Avenue
Baltimore, MD 21205
USA

Werner Arber
University of Basel
Department of Microbiology
Biozentrum
Klingelbergstrasse 70
4056 Basel
Switzerland

Mikael Benson
University of Linköping
The Center for Individualized Medicine
Linköping
58185
Sweden

Tilo Beyer
Institute for Molecular and Clinical
Immunology
Otto-von-Guericke University
Leipziger Str. 44
39120 Magdeburg
Germany

Robert R. Bies
Indiana University School of Medicine
Department of Clinical Pharmacology
Department of Medicine
1001 W. 10th Street
W.D. Myers Bldg W7125
Indianapolis, IN 46202
USA

Melanie Boerries
Freiburg Institute for Advanced
Studies – LifeNet
School of Life Sciences
Albertstraße 19
79104 Freiburg
Germany

Ian David Lockhart Bogle
University College London
CoMPLEX – Centre for Mathematics and Physics in the Life Sciences and Experimental Biology
Gower Street
London WC1E 6BT
UK

Hauke Busch
Freiburg Institute for Advanced Studies – LifeNet
School of Life Sciences
Albertstraße 19
79104 Freiburg
Germany

Dong-Yeon Cho
National Institute of Health (NIH)
National Center of Biotechnology Information
NLM, 8600 Rockville Pike
Bethesda, MD 20894
USA

Valentina Corradi
Department of Biological Sciences and Institute for Biocomplexity and Informatics
2500 University Dr. NW
Calgary, AB T21N4
Canada

Sonia Cortassa
Johns Hopkins University
Department of Medicine
720 Rutland Avenue
Baltimore, MD 21205
USA

Emanuele Cozzo
University of Zaragoza
Institute for Biocomputation and Physics of Complex Systems (BIFI)
50018 Zaragoza
Spain

Pawan K. Dhar
Symbiosis International University
School of Biomedical Sciences
Pune 411 042
India

Francisco S. Domingues
Max Planck Institute for Informatics
Department of Computational Biology and Applied Algorithmics
Campus E1 4
66123 Saarbrücken
Germany

Frank John Dye
Western Connecticut State University
Biological and Environmental Sciences
181 White Street
Danbury, CT 06810
USA

Roland Eils
German Cancer Research Institute
Im Neuenheimer Feld 280
69120 Heidelberg
Germany

Alisdair R. Fernie
Max Planck Institut für Molekulare Pflanzenphysiologie
Am Mühlenberg 1
14476 Potsdam-Golm
Germany

Antony Finkelstein
University College London
CoMPLEX – Centre for Mathematics and Physics in the Life Sciences and Experimental Biology
Gower Street
London WC1E 6BT
UK

Karl J. Friston
University College London
Wellcome Trust Centre for Neuroimaging
Institute of Neurology
12 Queen Square
London WC1N 3BG
UK

Michele S. Garfinkel
J. Craig Venter Institute
Synthetic Biology
9704 Medical Center Drive
Rockville, MD 20850
USA

Aaron David Goldman
Princeton University
Department of Ecology and Evolutionary Biology
106A Guyot Hall
Princeton, NJ 08542
USA

and

NASA Astrobiology Institute
Center for Ribosomal Origins and Evolution at Georgia Institute of Technology
NASA Postdoctoral Program
225 North Ave. NW
Atlanta, GA 30332
USA

Greco Hernández
National Institute for Cancer
Division of Basic Research
Av. San Fernando No. 22, Col. Sección XVI, Tlalpan
C. P. 14080 Mexico City
Mexico

and

McGill University
Department of Biology and Developmental Biology Research Initiative, Bellini Life Sciences
3649, Sir-William-Osler Street
Montréal, H3A 1B1
Canada

Jeremy A. Horst
University of California San Francisco
School of Pediatric Dentistry
Department of Orofacial Sciences
707 Parnassus Ave
San Francisco, CA 94143
USA

and

University of Washington
Department of Microbiology
960 Republican Street
Seattle, WA 98109
USA

Ling-Hong Hung
University of Washington
Department of Microbiology
960 Republican Street
Seattle, WA 98109
USA

Rajiv Jalan
University College London
CoMPLEX – Centre for Mathematics and Physics in the Life Sciences and Experimental Biology
Gower Street
London WC1E 6BT
UK

Konstantinos Krampis
J. Craig Venter Institute
9704 Medical Center Drive
Rockville, MD 20850
USA

Michael A. Langston
University of Tennessee
Department of Electrical Engineering and Computer Science
203 Claxton Complex
1122 Volunteer Boulevard
Knoxville, TN 37996-3450
USA

Thomas Lengauer
Max Planck Institute for Informatics
Department of Computational Biology and Applied Algorithmics
Campus E1 4
66123 Saarbrücken
Germany

David Lloyd
Cardiff University
Microbiology (BIOSI 1)
Cardiff
Wales CF10 3AT
UK

Gabriele Angelo Losa
Institute of Scientific Interdisciplinary Studies
v.F. Rusca 1
6600 Locarno
Switzerland

Ramana Madupu
J. Craig Venter Institute
9704 Medical Center Drive
Rockville, MD 20850
USA

Michael Meyer-Hermann
Helmholtz Centre for Infection Research
Department of Systems Immunology
Inhoffenstr. 7
38124 Braunschweig
Germany

and

Bio Center for Life Science
University of Technology Braunschweig
Spielmannstr. 7
38106 Braunschweig
Germany

Jason Miller
J. Craig Venter Institute
9704 Medical Center Drive
Rockville, MD 20850
USA

Michael G. Montague
J. Craig Venter Institute
Synthetic Biology
9704 Medical Center Drive
Rockville, MD 20850
USA

Yamir Moreno
University of Zaragoza
Institute for Biocomputation and Physics of Complex Systems (BIFI)
50018 Zaragoza
Spain

and

University of Zaragoza
Department of Theoretical Physics
Faculty of Sciences
50009 Zaragoza
Spain

Yoichi Nakayama
Keio University
Institute for Advanced Biosciences
Tsuruoka
Yamagata 997-0035
Japan

Karen E. Nelson
J. Craig Venter Institute
9704 Medical Center Drive
Rockville, MD 20850
USA

Sonia Osorio
Max Planck Institut für Molekulare Pflanzenphysiologie
Am Mühlenberg 1
14476 Potsdam-Golm
Germany

Teresa M. Przytycka
National Institute of Health (NIH)
National Center of Biotechnology Information
NLM, 8600 Rockville Pike
Bethesda, MD 20894
USA

Yu-Hui Rogers
J. Craig Venter Institute
9704 Medical Center Drive
Rockville, MD 20850
USA

Doug Rusch
Center for Genomics and Bioinformatics at Indiana University
Bloomington
Jordan Hall A302D 1001 E. 3rd Street
Bloomington, IN 47405-3700
USA

Ram Samudrala
University of Washington
Department of Microbiology
960 Republican Street
Seattle, WA 98109
USA

Joaquín Sanz
University of Zaragoza
Institute for Biocomputation and Physics of Complex Systems (BIFI)
50018 Zaragoza
Spain

Jeffrey H. Schwartz
University of Pittsburgh
Departments of Anthropology and History and Philosophy of Science
3302 Posvar Hall
University of Pittsburgh
Pittsburgh, PA 15260
USA

Robert Seymour
University College London
CoMPLEX – Centre for Mathematics and Physics in the Life Sciences and Experimental Biology
Gower Street
London WC1E 6BT
UK

Elizabeth Shephard
University College London
CoMPLEX – Centre for Mathematics and Physics in the Life Sciences and Experimental Biology
Gower Street
London WC1E 6BT
UK

Klaas E. Stephan
University of Zurich
Laboratory for Social and Neural Systems Research
Department of Economics
Bluemlisalpstr. 10
8006 Zurich
Switzerland

and

University College London
Wellcome Trust Centre for Neuroimaging
Institute of Neurology
12 Queen Square
London WC1N 3BG
UK

Thomas Sumner
University College London
CoMPLEX – Centre for Mathematics and Physics in the Life Sciences and Experimental Biology
Gower Street
London WC1E 6BT
UK

Kouichi Takahashi
Keio University
Institute for Advanced Biosciences
Tsuruoka
Yamagata 997-0035
Japan

Gritta Tettweiler
McGill University
Department of Biology and Developmental Biology Research Initiative, Bellini Life Sciences
3649, Sir-William-Osler Street
Montréal, H3A 1B1
Canada

D. Peter Tieleman
Department of Biological Sciences and Institute for Biocomplexity and Informatics
2500 University Dr. NW
Calgary, AB T21N4
Canada

Masaru Tomita
Keio University
Institute for Advanced Biosciences
Tsuruoka
Yamagata 997-0035
Japan

John Urquhart
AARDEX Group, S.A.
Ave de la Gare, 29
1950 Sion
Switzerland

and

University of California
San Francisco Medical Center
Department of Biomedical Engineering and Therapeutic Sciences
505 Parnassus Ave
San Francisco, CA 94122
USA

Vincent VanBuren
Texas A&M Health Science Center
College of Medicine
Systems Biology and Translational Medicine
702 SW HK Dodgen Loop
Medical Research Building
Temple, TX 76504
USA

Sanjay Vashee
J. Craig Venter Institute
Synthetic Biology
9704 Medical Center Drive
Rockville, MD 20850
USA

Anne Warner
University College London
CoMPLEX – Centre for Mathematics and Physics in the Life Sciences and Experimental Biology
Gower Street
London WC1E 6BT
UK

Alette M. Wessels
Indiana University School of Medicine
Department of Clinical Pharmacology
Department of Medicine
1001 W. 10th Street
W.D. Myers Bldg W7125
Indianapolis, IN 46202
USA

Marian R. Zlomislic
Department of Biological Sciences
and Institute for Biocomplexity and
Informatics
2500 University Dr. NW
Calgary, AB T21N4
Canada

Part I
Biological Basis of Systems Biology

1
Systems Biology

Melanie Boerries[1], *Roland Eils*[2], *and Hauke Busch*[1]
[1]*Freiburg Institute for Advanced Studies – LifeNet, School of Life Sciences, Albertstraße 19, 79104 Freiburg, Germany*
[2]*German Cancer Research Institute, Im Neuenheimer Feld 280, 69120 Heidelberg, Germany*

Systems Biology: Advances in Molecular Biology and Medicine, First Edition. Edited by Robert A. Meyers.

Keywords

Systems biology
A new field of biology that studies the functional structure and dynamics of intercellular and intracellular networks with the help signal- and systems-oriented methods.

Synthetic biology
Studies life as networks of biological objects such as DNA proteins RNA and metabolites.

Network biology
Studies the static organization of life as networks made up of biological entities such as DNA proteins RNA or metabolites.

System
A set of interacting parts functioning as a whole and distinguishable from its surroundings by identifiable boundaries.

Systems theory
This denotes the cross-disciplinary investigation of the abstract organization of systems irrespective of their substance type or spatiotemporal scale of existence. The goal is the study of emerging properties that arise from the interconnectedness of the individual parts making up the system.

Robustness
The robustness of biological systems denotes the maintenance of specific system functionalities in the presence of fluctuations or change in environmental parameters.

Control
Control is defined as the response action taken by a system to counteract parameter changes to maintain system functions at a certain predefined level.

Modularity
A design concept of complex systems to integrate simpler self-contained functional building-blocks into the framework of one larger system.

Model
The concept of representing causal relationships from real systems in the language of mathematics.

Systems Biology is a new field of biology, which places the theoretical foundations of systems analysis of living matter into the context of modern high-throughput quantitative experimental data, mathematics, and *in silico* simulations. The aim is to analyze the organization and to gain engineering-control of metabolic and genetic pathways. The ultimate goal is to gain an "holistic" view of the complex workings of life. The need for a system level understanding of biology is reviewed in this chapter, and comments are provided on the current scientific progress in this field. The current and future directions of experimental design strategies and theoretical approaches are also highlighted.

1 Introduction

Systems Biology is a recently established field in life sciences that aims at promoting a global systems understanding of living matter through the integration of various scientific domains (see Refs [1–3] for special journal sections and Refs [4–7] for textbooks on the topic).

The considerate attention that Systems Biology receives is due to the fact that it currently causes a paradigmatic shift in many areas of biological research. Modern molecular biology has been mostly a descriptive science, devoting insight to small, isolated compartments of a system as a whole – for example, by investigating the influence of individual proteins within the behavior of a whole cell. Thus, the study of the interconnected nature of cellular processes has long been avoided in favor of a reductionist approach. On the one hand this is due to the sheer amount of new challenges that come about when tackling complex systems, whereas on the other hand it has been the common leitmotif in other natural sciences, such as physics, to shed light mostly on well-defined and controllable systems. Those systems are then either small and isolated, or large and homogeneous, so that they can be tackled by applying the laws of statistics. Novel challenges, however, lie in the description of dynamical, mesoscopic, open, spatiotemporally extended, nonlinear systems, operating far away from thermodynamic equilibrium, which are the most important type to understand, as these are the systems that support life.

Although reductionist approaches have been successful in elucidating key processes and key factors of many fundamentally important biological processes, contemporary science is now realizing the importance of wholeness by studying problems of organization. Emergent phenomena arise from the interaction of various units or modules, which are neither resolvable nor understandable through the study of local events or the respective parts in isolation. Hence, traditional reductionist models and methods of cell and molecular biology are not very well suited, and can be incomplete, misleading, or even completely wrong.

Historically, Jan Christiaan Smuts was among the first to formulate a theory of the whole that was hoped to fill the gap between science and philosophy. In his book *Holism and Evolution*, which was published in 1926 [8], Smuts argued that Nature consisted of discrete objects, or "wholes," that are not entirely resolvable into their respective parts. The wholes and parts mutually depend on each other in their functionality, thus forming one organic, unified web of relations, which comprises matter, life, and mind, and which cannot be accounted for by a reductionistic analysis. Smuts saw his idea confirmed in evolution, regarding Holism as the active driving force towards more perfect wholes or species.

The theoretical foundations of systems engineering were laid some 60 years ago, when the concept of systems theory in biology was proposed during the 1940s by the biologist Ludwig von Bertalanffy [9]. The proposal was further developed during the 1950s by Ross Ashby [10], as a counter-movement against reductionism in science. In the sense of holism, von Bertalanffy emphasized the need for a study of the informational organization within real, open systems. The assembly of such inter-related elements then comprise a unified whole, which in turn can show new emergent properties.

In 1948, the mathematician Norbert Wiener established the field of *cybernetics* [11] as the science of communication and control of systems in regard to their environment. Cybernetics is closely related to systems theory, using the same concepts of information, control, or feedback. However, whereas the former focuses on systems function for providing regular and reproducible behavior, the latter deals more with system structure. Even so, both terms are often used in conjunction, for both structure and function cannot be understood as separate entities.

Today, biology embarks on systems thinking in two different ways. One way is to regard Systems Biology as a new way toward integrating information from different organizational levels, starting from DNA to proteins via signaling pathways to functional modules, into the context of a holistic organizational view [12]. The primary goal of the second view on Systems Biology is to establish a conceptual framework and working methodologies for the augmentation of knowledge on biological phenomena by combining systems theory and molecular biology: "Systems Biology is not a collection of facts but a way of thinking" [13]. This view has already been shared during the late 1960s by Mesarović, who predicted that Systems Biology would be an established field of science as soon as "... biologists start asking the right questions" [14]. Put differently, biologists need not recast facts already known from molecular biology in a different language, but they need to ask questions based on system-theoretic concepts [15].

Both approaches share the extensive need for high-quality, quantitative biological data obtained through extensive experimental measurements, and this is the reason why the use of systems theory in biology has gained momentum during recent years. New techniques can provide the necessary amount of quantitative data for the establishment of appropriate holistic models of cellular processes. Eventually, those new experimental techniques will lay the foundation for the integration of mathematics, engineering, physics, and computer science into biology, to permit an understanding of the range of complex biological regulatory systems at multiple hierarchical and spatiotemporal levels of cellular organization [16].

2
What Is Systems Understanding?

The word "system" derives from the Greek συστημα, and is composed of the prefix syn, which means "together," and the root of histanai, meaning "cause to stand." A *system* is defined as "... the assembly or set of inter-related elements comprising a unified whole that is distinct from its environment," and can be hierarchically organized and made up of other subsystems or modules, which allows the construction of a complex entity from simpler units. For example, organelles such as mitochondria constitute distinct subsystems within the organization of a cell. The subdivision of natural entities into systems is an abstract construct. Systems *per se* do not really exist in reality; rather, they are defined as a set of elements interacting over time and space.

Systems theory denotes the transdisciplinary investigation of the abstract organization of phenomena, independent of their substance, type, or spatiotemporal scale of existence [17]. The goal of systems theory is to study emerging properties arising from the interconnectedness and complexity of relationships between parts. Such theory argues that however complex or diverse a system is, there are always different types of organizational structures present, which can be represented as a network of information flow. Because these concepts and principles remain the same across different scientific disciplines such as biology, physics, or engineering, systems theory can provide a basis for their unification. The systems view distinguishes itself from the more traditional analytic approach by emphasizing the concepts of system–environment boundaries, signal input–output relationships, signal and information processing, system states, control, and hierarchies. Albeit systems theory is valid for all system types, it usually focuses on complex, adaptive, self-regulating systems which are termed "*cybernetic*."

Elegant, simple, and globally valid models are rare in biology as compared to other fields of science. Few examples exist where a function can be attributed to the workings of a single small molecule or few proteins, as in the case of hemoglobin for the transport of gases in the bloodstream, or bacterial chemotaxis [18].

In general, many genes and proteins are involved in cellular responses to external stimuli. In general, biology follows a reductionist approach by investigating small, isolated parts of a cell, tissue, or organism; typically, biology tries to deduce biological phenomena from molecular behavior, which often results in a simplistic "one gene for one function" approach. However, since genetic analysis has shown that the genotypes of different species are mostly identical, it would appear that it is the signal processing stage(s) on the way from the genome to the phenotype – in other words, an ever more elaborate regulation of gene expression [19] – which carries the subtle particularities in the respective genetic codes.

As a consequence, biological phenomena should be explained within the vocabulary of system theory, such as amplification, control, adaptation, sensitivity, autoregulation, and error correction, taking an holistic view of the system under consideration [20]. In short, Systems Biology is required to uncover the laws of the whole that cannot be inferred by delving deeper into the details.

The systematic investigation of biological matter comprises the understanding and control of system structure and dynamics in the sense of systems theory and cybernetics, respectively [1].

System structure denotes the identification of the static connection topology and regulatory relationships within the network of genes, proteins, and other small molecules that constitute the signal transduction and metabolic pathways, as well as the physical structure of organisms or cells. Experimental techniques to elucidate the cells' global system structure include for example DNA microarrays [21], deep sequencing [22], and protein–protein interaction screens via yeast-two-hybrid [23] or split-ubiquitin approaches [24]. System dynamics refers to the qualitative and quantitative evolution of the above network over time. Dynamics include the temporal variation of molecule concentrations, as well as the structure of the network itself. Examples of experimental techniques to study cellular system dynamics include fluorescent imaging techniques to monitor molecular dynamics and interactions on the level of the individual cell [25]. Moreover, mass spectrometry or transcriptomics can be used to investigate the collective behavior of the proteome or gene expression in cells over time.

3
Why Are Biological Systems Different?

While biological systems must still be based on the laws of physics, chemistry, and thermodynamics [26], biology also incorporates the notion of organismal function. This notion represents the need for survival and reproduction, as well as the possibility to evolve in and adapt to changing environments, and it is this inherent purpose that distinguishes biology significantly from all other natural sciences [27]. Moreover, there is no distinct separation of information storage and regulatory units. Genes, for example, can regulate their own expression by gene splicing [28].

Classical physics views the emergence of every effect to be determined by a cause residing in the past. Biological systems, on the contrary, are teleonomic – that is, they are oriented towards a state in the future [17]. It was the insight of cybernetics that purposeful activity can be described through the use of circular mechanisms, where the effect equals the cause. The simplest example of a circular logic is a feedback loop, in which the output of the system is fed into its input again (see Sect. 3.2).

Having these prerequisites in mind, the concepts of systems theory and cybernetics should be utilized, if there is a desire to establish successful formal mathematical models in biology. The introduction of circular causalities has far-reaching consequences on the general design and global properties of biological systems, such as robustness, complexity, and control, all of which are discussed in detail in this chapter.

3.1
Biological Complexity

In order to comprehend the challenges in Systems Biology, it is important first to understand the origin of the complexity encountered in Nature. For once, the complexity within a system does not necessarily come about as a consequence of the number of its component. In physics, the macroscopic state or the dynamics of objects are often well described by a few parameters or simple mathematical equations. For example, although the temperature of an object comes about due to the thermal agitation of its individual atoms, the average energy of the atoms can be ascribed to a single, macroscopically measurable, quantity.

A complex system is one in which the laws that describe its behavior are qualitatively different from those that govern its units, such that new features emerge when moving from one temporal, spatial, or organizational scale to another. The science of complexity is about revealing the principles that govern the ways in which these new properties appear [29]. Thus, it is rather the organization of the system into irreducible, heterogeneous parts that are highly structured and hierarchically organized on various spatiotemporal scales, which makes a system become complex or, in other words, "complicated" [30].

The notion of complexity is not well defined amongst the various science disciplines. Information theory usually characterizes complexity as the amount of information needed to optimally predict the behavior (or state) of the system based on entropy measures [31]. Adami uses sequence complexity of biological genomes to define the amount of information stored about the environment [32]. Another viable approach toward defining the complexity of biological systems may be the identification of the topological structure at a higher level of large-scale organization in terms of hierarchically organized networks [33].

Complex systems can show the emergence of ordered macroscopic behavior, termed *self-organization*. While biological systems are ideal candidates to demonstrate self-organization, the applicability of its principles to elucidate biological phenomena is, to date, rather limited. This is due mainly to the fact that biological systems require a re-definition of complexity that is quite different from that in physics. Biological systems are heterogeneous, modular, highly structured on multiple irreducible spatiotemporal scales, and also self-dissimilar, with each entity usually having several functional and regulatory properties. Nevertheless, the individual components "... interact selectively and nonlinearly to produce coherent rather than complex behavior" [34].

One explanation for the emergence of self-organization is the so-called "slaving-principle." Under certain conditions the global, macroscopic behavior of a system is governed by a few, slowly evolving state variables, which "slave" all other dynamics. In this way, the relevant degrees of freedom are largely reduced, allowing the system to find its own structure – that is, to self-organize [35]. One use of the slaving principle is the analytical description of lasers, explaining the spontaneous, synchronous light emission by the atoms in the lasing medium. Another framework for the explanation of emergent phenomena in multibody systems is the notion of self-organized criticality (SOC) [36], which was acclaimed to explain the frequent occurrence of long-tail distributions in many natural phenomena. SOC denotes the ability of open and dissipative systems to display critical – that is, scale-invariant – behavior even in the absence of external pressures. However, whilst this theory did not live up to its promises of being universally applicable, it is applied in earthquake, forest fire, or avalanche models.

Carlson and Doyle introduced the theory of highly optimized tolerance (HOT) [37], which accounts for the intrinsic design of biological and engineered systems. Such theory reflects the behavior of such high-throughput, high-density systems, which are faced with limited resources. These systems show a high tolerance (i.e., robustness) against environmental parameter fluctuations, with the robustness being achieved at the cost of a high degree of complexity through the addition of control units to the system. Resource constraints then call for an optimized trade-off between fault tolerance towards frequently experienced perturbations and fragility to rare, yet possible, events. Thus, there exists a "conservation law of robustness" [38].

The complex design of the living cell is often compared to the make-up of computers or today's commercial aircraft. These man-made machines have a bewildering complexity, and no man alone understands all the parts and their interplay in complete detail. Many of the strongly interacting and irreducible complex functions are there to automate, control, or back-up operations.

These definitions elucidate the main differences between systems describable by SOC or HOT; the former theory relates to scale-invariant and self-similar features, while the latter deals with self-dissimilar structures and sensitivities (dis-)appearing on each scale of observation, such that the degrees of freedom in these systems are irreducibly many.

3.2 Global Properties of Biological Systems

Typically, systems thinking seeks universal properties for biological systems, linking the emergence of complexity to general design features. Such features can be used as a guideline for abstract mathematical modeling by examining the principles that are commonly shared between diverse species. These properties, as discussed below, show how biology and engineering converge on a systems level view.

3.2.1 Robustness of Biological Systems

The fault tolerance or robustness of biological systems denotes the maintenance of specific system functionalities, even in the presence of fluctuations or a change in environmental parameters [39, 40]. In biology, this refers to the concept of homeostasis and the stability of developmental control. Robustness is achieved through the incorporation of regulatory control loops into a system, thus shielding or buffering the desired system functionality from environmental influences. Robustness is a relative system property: in order to maintain a certain equilibrium, other properties must evolve and adapt. There is, therefore, a need to define which cellular functions change, and which resist as a reaction towards a disturbance.

Most of the "complicatedness" encountered in biological systems is a direct consequence of the implementation of control schemes, rather than the core function itself. In particular, if robustness is to be achieved for a wide range of disturbances, the control must have an equally increasing variety, as stated in the "Law of Requisite Variety" [10]. Consequently, complexity serves simplicity in the sense that the complicated control schemes are hidden underneath the simple, yet reliable, output [41].

Robustness in biological systems is usually achieved by various strategies, including redundant or functionally overlapping regulatory pathways [42], feedback loops to regulate signal responses [43], or checkpoints within the cell [44, 45]. All of these control elements add to the total number of components regulating cellular signaling elements, while keeping the number of phenotypic expression levels low [20]. Cellular checkpoints play important roles, such as in mitosis [45, 46], in the cell cycle [47], and during the early stages of embryo development [48]. Bacterial chemotaxis is an intensively studied example of a robust behavior due to multiple feedback control [49, 50]. Examples of genetic buffering to achieve robustness are shown by the fact that only about 20% of genes in budding yeast are essential for viability [51]. To a certain extent, genetic buffering can safeguard against environmental changes or genetic defects. The simplest bacteria (e.g., *Mycoplasma pneumonia*) may have only a couple of hundred genes yet, despite their unexpectedly complex transcriptome organization (which includes antisense transcripts, alternative transcripts, and multiple regulators per gene [52]) they survive only within a narrow band of environmental parameters. In contrast, the bacterium *Escherichia coli* possesses about 3000 genes, and is able to survive under a variety of environmental conditions [30] by activating additional control schemes under environmental stress.

A further consequence of robustness is that dynamical behavior in biological systems is coupled less to the parameters themselves, and more to the overall system structure. Previously, Von Dassow [53] showed that robustness was the simplifying criterion for determining the correct topology of the segment polarity network, producing a highly robust patterning over a large range of parameter variation.

Robustness in complex systems comes at the price of some fragility due to tiny, yet rare, events. Indeed, a small rearrangement of some cellular signal pathways can often lead to a spectacular failure – that is, impaired cell death. For example, although cancer may be caused through an accumulation of mutations in the genetic code over the human life span [54], the occurrence of cancer during the reproductive years of a human is rather rare. It appears that Nature can cope well with this trade-off for the benefit of robust and reliable "normal" functioning.

3.2.2 System Adaptation and Control

Adaptation is the ability of a system to accommodate for varying external input stimuli or disturbances in order to gain and maintain the correct and optimized output; this is usually achieved with the help of feedback control. A well-studied example of adaptation using integral feedback control is chemotaxis, in which bacteria adapt their movement not to the absolute level of pheromones but rather to the chemical gradient, only [18, 50, 55]. Another example is the activation of heat shock proteins in *E. coli* under temperature stress [56], which function as chaperones for correct protein folding and the prevention of unwanted protein aggregation.

Control is a recurrent design encountered in natural systems, which effectively increases robustness, with the goal of keeping certain values within predefined physiological limits. It is possible to distinguish between feed-forward and feed-back system controls; the former is an open-loop sequence of predefined actions triggered by a certain stimulus, working dependably only within strict ranges of input stimuli, yet simple in design [57], while the latter employs a closed-loop design, which feeds part of the output signal back into the system input. Depending on the sign of the feed-back, positive feed-back (or autocatalysis) will amplify the output signal, enhancing reliable sensitivity toward cellular decision derived from noisy input signals [58], whereas negative feedback usually stabilizes the output around some desired value by opposing any changes caused by disturbances. For example, a combination of feed-back control schemes can be used to stabilize the receptor presentation of a cell to increase its sensitivity toward a broad range of external ligand concentrations; this was demonstrated in the case of erythropoietin receptor signaling [59].

Although both types of control are successfully employed in intracellular signaling [60], both also bear fragilities. Autocatalysis can lead to self-sustained, unrestricted signal amplification, as observed in uncontrolled tumor growth [61], while ultra-stable homeostasis regulation can cause large, possibly disturbing, transient signals in response to external fluctuations [38].

Most often, biological systems use combinations of open- and closed-loop designs, basing their control action on the absolute, accumulative or differential value of the input stimulus, in this way balancing both sensitivity and stability.

3.2.3 **Modules and Protocols**

Modules are subsystems characterized as mostly self-contained entities with fixed interfaces for external communication. They are evolutionarily detached, possessing their own identity, and have many internal – but only few external – links for information and matter exchange [27].

Modularity seems to be an important concept in biology, probably stemming from the evolutionary pressure for optimal flexibility and the low chance of damage spread. Spatially distinct entities, such as organelles in the cell or metabolic units, appear to be a recurrent scheme found throughout various organisms. The gene regulatory network in cells seems to be organized in modules [62]. Interestingly, modular design is a commonly found principle in modern engineering, as it enables the independent development and testing of units before their integration into a common system. The benefits comprise savings in developmental and maintenance costs, as well as the prospect of graceful degradation rather than catastrophic failure, as errors are usually restricted to a single module.

If the modular design of biological systems is a governing principle, this opens up new possibilities for the simplification of *in vivo* and *in silico* experiments. Modules would provide fixed levels of detail and size, which are easily abstracted once their functions are known. Moreover, with the help of these core modules it would be possible to build ever-more complex models, without the need for segmentation of every level of detail [62].

Modules communicate by using *protocols*; these are fixed, commonly agreed-on rules that standardize communication with their respective surroundings. In addition, protocols ensure error correction, cellular coordination, and also evolvability through the possibility of adding new functions to a particular module. Protocols have been shown to be an efficient means for ensuring the hierarchical organization of complex systems through their integration of different layers, thus reducing the costs of information transmission [63]. For example, negative feed-back is a powerful module for the establishment of homeostasis. Similarly, gene regulation, membrane potentials, or signal transduction pathways can all be regarded as protocols which are utilized by different biological modules, such as the DNA, ion channels or kinases, and phosphatases [38]. Elaborate feed-back control protocols have been robustly established for the spatiotemporal development of various species [60, 64]. An extensive genome-wide analysis of prokaryotic cell-cycle progression has revealed a hierarchical control structure with three to four master regulatory proteins acting in a coordinated fashion [65].

Interestingly, the use of protocols on a cellular level has striking similarities with the way that modern computers communicate via the internet. The internet employs many protocols for persistent communication that can be ordered into a general hierarchy, starting from the data link via the network and transport to the application layer. Popular examples of each hierarchy are the reverse address resolution protocol (RARP) for the IP (internet protocol) look-up and search of the communication partner, the IP for the routing for the appropriate subnet of the data, the TCP (transmission control protocol) for the secure, error-free data exchange and, finally, the application layer such as the HTTP (hypertext transfer protocol) or FTP (file transfer protocol) for retrieving a web page or downloading files. The fact that these protocols have been used for decades, irrespective of the rapid and ongoing evolution of computer hardware and software, highlights their importance.

4
Systems Biology Modeling

The word "model" derives from the Latin *modus* – that is, "manner" or "measure," and refers to the concept of representing causal relationships from real systems in the language of mathematics. This mapping often involves simplifications of the original systems, with the hope of gaining predictive power on experimental results and explaining functional design principles. Life is an emergent property stemming from the interaction among molecules, and cannot be reduced to the individual properties of the molecules. Modeling tries to infer and predict the relations between molecules in terms of causation – that is, it tries to establish explanatory relationships of the spatiotemporal changes of matter [66].

Modeling and simulation have become indispensable tools for gaining insights into natural systems. In biology, they help to bridge the gap between theory and experiment. Often, biologists are faced with the dilemma that experiments do not provide sufficient data for theoretical interpretation, while at the same time, clues for new experiments are missing as a consequence of lacking hypotheses [67].

Experimental results require correct mathematical interpretation, and model hypotheses require experimental proofs [68]. The process of knowledge generation is iterative in nature, and consists of two feedback loops (see Fig. 1). Here, the experimental part employs high-throughput techniques to obtain quantitative data

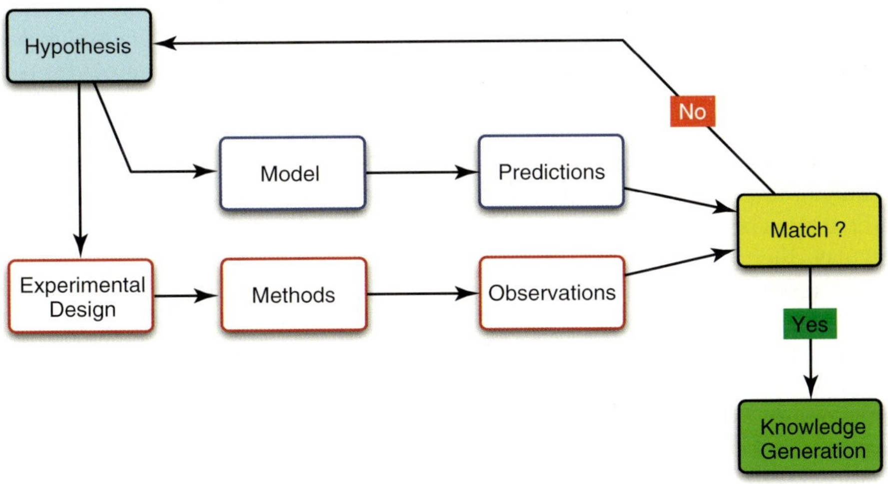

Fig. 1 Schematic to the experiment–modeling loop for knowledge generation in Systems Biology.

relating to the system dynamics; the data acquired are then compared with predictions obtained from the mathematical model. In the case of failure – that is, in the case of diverging results from model and experiment – this process must be repeated by adjusting the old hypothesis and thus reconciling the model with the new results from the experiment.

As an example of this cycle of model building, experiment, and model refinement, Ideker *et al.* [69] have recently developed an integrated approach to construct, test, confirm and refine the simulation of the yeast galactose utilization network, through the combined numerical simulation and analysis of networks with systematic experimental perturbation and global measurements.

The major guideline for optimal modeling should be the concept of Occam's razor, which states that models should be void of any redundant information. Yet today, this paradigm needs also to be viewed from the opposite aspect – models must not be oversimplified, so as to miss the essential clues of functionality of real life systems. In particular, it is important to note that modeling in biology must establish different concepts from those in physics. The paradigm of nonlinear science is that even simple systems can lead to a dynamically rich behavior. Despite the beauty of the simplicity of this idea, however, it must be realized that living systems regulate their dynamics differently – that is, through a complex make-up of interconnected regulatory functions, a fact which is often neglected or ignored [70]. Models in biology are usually heuristic; they arise embedded in the process of biological experiments, and are coupled tightly to them. Information is, for example, deduced from a perturbation analysis of the experimental system, and thus contains assumptions regarding the causality and the passage of time [70].

In general, experimental data must be comprehensive with respect to four aspects [57]:

- **Factor**: the need to capture the behavior of all important target factors, such as genes and proteins that play decisive roles in the experimental system under consideration.
- **Item**: this refers to the simultaneous measurement of the necessary sets of variables that are required for reliable hypothesis building, such as transcription level, molecule concentration, or spatial location.
- **Time and space**: this refers to the need for a sufficiently high sampling rate of the experimental data to obtain a reasonable resolution of the spatio-temporal dynamics.
- **Repetition**: experiments need to be repeated to obtain a statistically reliable estimate for the biological variability and other sources of error induced by the experimental set-up.

The construction of a valid working model of a biological system from experimental data can be approached from two directions:

- **Bottom-up modeling**: This is based on the integration of established biological knowledge on the dynamics of the relevant biological components of the system or regulatory network under consideration. This attempt is useful when most of the reaction partners are known and their interaction dynamics are understood. The research goal is the establishment of an accurate computer simulation that allows for: (1) the analysis of the system dynamics;

(2) the scan of parameter ranges that are unattainable in experiments; and (3) the prediction of unknown functionality or interactions [57]. Attempts at bottom-up modeling include the λ-phage decision circuit [71] or the data-driven simulation of a cancer cell [72].
- **Top-down modeling**: This approach attempts to apply statistical analyses to data from high-throughput experiments derived, for example, from DNA microarrays. Data mining techniques search for clusters of coexpressed genes as a consequence of cell state or an external perturbation, such as the knockout or overexpression of certain genes.

The working assumption for these methods is that coexpressed genes also share common relationships with respect to other biological processes. For example, Müller *et al.* [73] showed, from clustering analysis, how pluripotent stem cells are under tight control by specific molecular networks across species.

The result of a cluster analysis is the construction of an interaction network of genes or proteins, the topology of which can hint at biological reasonable organization principles [33, 74, 75]. While top-down modeling is well in line with the need for holistic approaches, it has been criticized for violating individuality and locality in the cell. Although cellular stimulus–response patterns are highly coordinated, these patterns emerge from individual protein–protein interactions, which cannot be deduced from the high-throughput data. Hence, the question of gaining new biological knowledge on individual genes or proteins from this approach often remains open.

Although knowledge-based bottom-up approaches of modeling possess a certain appeal to biologists, successful modeling needs to overcome the gaps in understanding. Most protein–protein interactions remain unknown, and the underlying physics of interacting molecules require better attention. Until now, the influence of high molecule concentrations in the subcompartments of a cell environment on reaction rates is largely unknown, but the spatial distribution of reactants and their molecular crowding may well affect the reactions and their rates [76, 77]; an example of this is when explaining the symmetry breaking process of mitosis [78]. One particular challenge arises when trying to identify the relevant components, as this is very difficult due to the vast number of combinations of active molecules. Hence, it remains unclear as to how this approach can be scaled up to large networks on the cellular or even tissue level. Indeed, this situation presently poses major difficulties for detailed mathematical modeling in biology.

Regardless of the approach taken, there is a need to define the level of model abstraction, complexity, and spatiotemporal scale of the system under investigation. Cellular processes have characteristic time scales that range from milliseconds for individual protein–protein interactions to minutes for phosphorylation events, up to hours, days, and years for changes in gene transcription, cell growth, and gene mutations, respectively. This has important consequences for the appropriate choice of model detail level. Processes that occur much faster than the time scale of observation can be assumed to be instantaneous, while slow processes can be assumed to be quasi-static. As a consequence, the level of abstraction leads to certain dynamics being modeled and simulated in detail, whereas for other parts the details can be neglected – for example, due to low sensitivity towards specific parameter values. Recently, Busch *et al.* [79] have used time-scale separation between fast protein signaling and slow transcription dynamics to infer a dynamic decision network for hepatocyte growth factor-induced migration of keratinocytes. On the time-scale of observation, all protein concentrations were assumed to be in quasi-equilibrium, such that it sufficed to focus on the change in gene regulation as the decisive element of regulation.

The major goal of theory and model simulations is the reverse engineering of biological systems, which is "... the process of analyzing a system to identify its components and their interrelationships and create representations of the system in another form or at a higher level of abstraction" [80]. Despite major efforts in this field [81], biological systems still pose a major challenge towards reverse engineering, due to the hidden complexities, inherent robustness, and possibly suboptimal design of functional units. Circular causality makes it difficult to distinguish between cause and effect from biological data, and all of this imposes ambiguities when attempting to deduce the correct biological "wiring diagram" from the experimental data. It is hoped that systems thinking in biology, with its concepts of robustness, hierarchy and modularity, in addition to the necessary protocols, will provide detailed bottom-up models with testable hypotheses for model discrimination [39].

4.1 Network Biology

Network biology is the study of the static organization of life as networks of biological entities such as DNA, proteins, RNA, and metabolites [33, 82].

The relationship between these entities is depicted graphically, the result being a set of nodes connected by edges. Usually, the nodes represent the state variables of the system (such as the molecule concentrations), while the connections define the interaction between the nodes. The networks are characterized by their connectivity, path length, and clustering distributions. Connectivity indicates how many neighbors each node has on average, while the path length denotes the average separation between arbitrarily chosen nodes, and the clustering coefficient is a measure of the grouping tendency of the nodes.

Network biology attempts to discover universal design and organization principles, which govern the functioning and evolution of intercellular and intracellular networks [83, 84]. Biological networks appear to share certain topologies that are best described as scale-free networks; these possess few highly connected nodes, termed *hubs,* while many nodes share only a few connections. Hence, the connectivity distribution follows a power-law: $P(k) \propto k^n$, where k and $P(k)$ denote the connectivity and distribution, respectively, and n is the connectivity exponent. These networks appear if the new nodes have a preferential attachment to already highly connected hubs [85, 86], although, interestingly, this type of complex network also emerges in other aspects of life and society, such as the worldwide web and social interaction webs [87].

Network architecture and task are closely linked. By comparing the topology of the regulatory network in *E. coli* with the call graph of the kernel of the Linux computer operating system, Yan *et al.* showed that both networks show a hierarchical layout, despite having fundamentally different design principles [88]. The *E. coli* network is optimized toward robustness, with few global regulators at the top and many downstream targets, whereas the

Linux kernel is designed for code efficiency and the re-use of software modules at the cost of robustness, and has many regulators controlling a small set of highly connected generic functions. By using high-throughput methods, it is possible to draw ever more-detailed interaction webs of protein–protein, metabolite, and gene transcription networks. Experimental results obtained with yeast [89] and *E. coli* [90] seem to support the view that both metabolic and genomic regulatory networks show a hierarchical organization with few recurrent subnetworks, termed *motifs*, which hint at the existence of elementary regulatory units [5, 74]. While modules are discrete functional units that are semi-detached from the whole system, motifs comprise a set of genes or metabolites that form recurring, significant patterns of interconnections, that are inseparable from the remainder of the system [87]. Subsequently, Ravasz *et al.* showed the metabolic networks of 43 distinct organisms to have a modular organization that was interconnected in a hierarchical manner – a system-level cellular organization that might be generic in nature [91].

The need to investigate the interconnections of genes–proteins and proteins–proteins stems from the fact that the genome of various species is quite similar, despite it having been argued that the evolution of ever more-complex species is correlated with an increased functional connectivity between a constant number of genes. Motifs might provide a means to increase the interconnectivity of existing proteins, in this way creating new functionalities. As noted by Ravasz *et al*:

> *"This is likely one of several reasons that the apparent complexity of organisms can increase so markedly without a corresponding increase in gene number. An attribute of proteins encoded by the human genome is that they have a richer assembly of domains than do their counterparts in invertebrates or yeast, and indeed the assortment of domains into novel combinations is likely an important aspect of genome divergence" [92].*

Today, network analysis has opened up new avenues in the global analysis of diseases, and their mutual connections. Such analysis allows for the identification of the genetic basis of disease [93], and can also reveal novel gene association overlapping across common human disorders, which in turn helps to unravel the general patterns of human diseases that are not clear from studies of the individual conditions [94].

4.2 Dynamic Network Models

Despite the intuitive appeal of network biology for the structure identification of biological systems, it is limited in the sense that it does not include the temporal dynamics of the system. Life is an emergent property of the interaction of cellular proteins. Hence, the dynamic interplay occurring via chemical reactions and changing cellular protein numbers is essential to the cell function, and constitutes the essence of many modeling approaches in Systems Biology. Boolean networks are an abstract, yet relevant, approach towards the inclusion of a temporal evolution of large-scale networks, as they offer a qualitative modeling approaches to build and analyze simplified, but still rigorous, dynamical models. Boolean networks are used for the elucidation of large-scale dynamic protein signaling networks, where it is assumed that the detailed inclusion of, perhaps, concentration gradients or the stochastic effects of protein concentrations, can be neglected in favor of including a large number of players [95]. Each node of a Boolean network can assume an ON/OFF value. Then, the state of each node at the next time step $(t+1)$ is deduced deterministically from a logical Boolean function (i.e., AND, OR, NAND...), based on its current state and external input [96]. Boolean networks have been used successfully for the analysis of protein and gene regulatory networks [97], for the modeling of the epidermal growth factor receptor (EGFR)/ErbB signaling pathways [98], and apoptosis [99].

4.3 Reaction–Diffusion Models

Biochemical reactions can be considered as the most fundamental processes in cells, wherein the concentrations of the reacting species change subject to the other molecule species involved within the respective reactions. Ordinary differential equations/partial differential equations (ODEs/PDEs) are a natural choice for the mathematical description of system dynamics with continuous system states in time and space, respectively [6, 100]. They are, therefore, ideally suited to describe changes in concentrations in biochemical reactions, being the most widespread formalism to model systems throughout the various scientific domains. In biology, they are widely used to describe the time and space course of molecular concentrations. In this case, differential equations relate the rate of change of a variable (protein substrate) to the current state of other variables (reactant), wherein the interaction between the various reactants is modeled through functional and differential relations. There is a vast literature available on modeling biochemical reactions based on differential equations, especially in the context of metabolic processes (cf. Ref. [100] and references therein).

In general, it is possible to distinguish between purely time-dependent ODEs, and more general PDEs, which additionally include spatial dimensions. A PDE describing the spatiotemporal evolution of a system has the general form:

$$\frac{\partial \psi_i(\mathbf{r},t)}{\partial t} = \underbrace{f\left[\psi_i(\mathbf{r},t)\right]}_{\text{temporal evolution}} + \underbrace{\nabla \mathbf{D}_i(\mathbf{r},t)\nabla \psi_i(\mathbf{r},t)}_{\text{spatial Diffusion}}, \tag{1}$$

where $\psi_i(\mathbf{r},t)$ denotes the respective system state variables of the various molecules labeled by the subscript i. The above equation reduces to an ODE in the absence of spatial diffusion, that is, $\mathbf{D}_i(\mathbf{r},t)=0$. The term $f[\psi_i(\mathbf{r},\ t)]$ denotes the respective synthesis rates, depending on the various concentrations of $\psi_i(\mathbf{r},\ t)$ and possibly external signals. It is often comprised of the various, mostly nonlinear, interaction functions between the system reactants based on the "law of mass action," which leads to, for example, Michaelis–Menten-like enzymatic degradation or the Hill-type cooperative activation [6, 100]. The organization of interacting molecules within a cell implies the concept of pathways, in which information processing in the cell is organized. In terms of mathematics, such a biochemical network is then represented as a system of coupled differential equations, as shown above. It is those nonlinear interactions that are essential for biological systems to show nontrivial behavior such as multistability, hysteresis, or oscillations [43].

A stability analysis of the system of coupled equations is usually carried out to unravel the qualitative behavior of steady-state solutions and their stability, as well as the occurrence of periodic solutions in space and time [101]. As a matter of fact, the dynamic behavior of ODE systems depends heavily on the reaction parameters of the underlying chemical reactions. Therefore, the estimation and identification of parameters from experimental data [102, 103], as well as the optimal experiment design to yield a maximal amount of new biological knowledge [104, 105], is a field of active research in Systems Biology. Due to the nonlinear interaction terms $f[\psi_i(\mathbf{r}, t)]$, an analytical solution of the differential equations is usually impossible, and solutions must be found via numerical integration. A variety of numerical integration algorithms for ODEs and PDEs can be found in the literature [106]. In addition, various computer software packages have been developed specifically for the simulation and bifurcation analysis of nonlinear systems, such as GEPASI [107], DBSolve [108], Cell Designer [109], or a Matlab extension such as SB toolbox [110] or Potters wheel [111].

The great success of modeling is its predictive power. From model simulations, it is possible to obtain a better insight into the regulatory logic, and it is also possible to perform experiments in the computer that otherwise were impossible. A good example of this is the mitogen-activated protein cascade model, as originally proposed by Huang and Ferrell [112]. Here, the pathway is highly conserved and implicated in various biological processes, conducting signals from the membrane to the nucleus. It is composed of three kinases that sequentially phosphorylate and activate each other. By converting and simulating the rate equations into a system of coupled ODEs, Huang and Ferrell showed that this particular pathway architecture could be used to convert a graded response into a switch-like robust output, appropriate for mediating processes such as mitogenesis or cell fate induction.

The spatial aspect of the equation enters the description due to the inclusion of a diffusion term, where the diffusion term $\mathbf{D}_i(\mathbf{r}, t)$ depends on space and time in general. Alan Turing was the first to point out such reaction–diffusion systems as a possible explanation for morphogenesis in natural systems, when he showed – theoretically – that a system of reacting and diffusing chemicals can evolve spontaneously to a spatially heterogeneous state as a response to an infinitesimal small forcing [113]. These pioneering studies led to a new branch of research, and many so-called "Turing systems" have been proposed (though not finally proven) to account for pattern formation in developmental biology. Examples include an explanation of the pattern formation on snail houses [114], the modeling of *Drosophila* embryogenesis [115], or the coating patterns of animals [100]. The importance of spatiotemporal inhomogeneities in molecular concentrations and molecular crowding in signal transduction pathways has also recently been acknowledged [77, 116]. As a consequence, cellular functions may also employ active transport mechanisms to sustain reliable cellular processes [117, 118].

The above ODE formalism fails in the case of low molecule concentrations [119], as the discrete and random nature of the individual, elementary reactions between molecules then becomes non-negligible, showing an impact on a macroscopic scale as fluctuations in the molecule concentration over time [120]. This chemical noise becomes most significant in the regulation of gene expression, which is usually accompanied with low copy numbers of mRNA transcripts and the genes themselves [121]. Stochastic effects in cellular pathways have been attributed to cause phenotypic diversity in isogenetic populations of cells [122], or to play a major role in the lysis-lysogeny decision circle of the λ-phage [123].

4.4 Holism versus Reductionism: The Global Dynamics of Networks

Currently, most Systems Biology approaches follow either a top-down or a bottom-up approach. However, while both have their unique motivation, they each also hold certain criticisms. Typically, a top-down approach allows for the holistic, unbiased view of cellular events, albeit at the expense of limiting detail level, thus violating the individuality and locality that are important in the understanding of cellular processes. The bottom-up approach aims at a detailed mechanistic and causal understanding of biochemical networks, but it is not yet been determined how such studies of "isolate" signaling pathways can be scaled up to the cellular or even tissue and whole organism levels. As an interim solution, Huang and Ingber [20, 124] have proposed the study of cellular behavior on the level of global network dynamics to reveal any higher-order, collective behavior of the interacting genes and proteins. These authors have described global network organization in terms of a state space, spanned by the expression levels of the whole genome, as well as attractors providing a mathematical and molecular basis for an epigenetic landscape. Genome-wide expression levels cannot take up arbitrary values; instead, they are tightly coupled to respective cell states – that is, to the various cell types and distinct, stable phenotypic states in a multicellular organism. Such stable network configurations are referred to as *attractor states*. Huang and Ingber have argued that the attractors naturally capture the essential properties of cell behavior, such as the mutual exclusivity of cell fates, robustness, and all-or-none transitions in response to a large variety of signals. As a proof of principle, it was shown that HL60 cells are capable of following different routes into the same differentiated state. Subsequently, upon the application of two different stimuli, separate transcriptome response transients were created which, nevertheless, settled onto the same gene expression pattern after several days [125].

This formal framework on the orchestrated role of cellular gene network dynamics could serve as a potential explanation for stem cell differentiation and the reprogramming of differentiated back into pluripotent stem cells [126], or even cancer progression. The accumulation of genetic mutations over time would distort the attractor landscape such that it would eventually lead to an altered cellular response within the same cellular context [124]. While the above suggestions are intuitive and appealing, their value must still be proven with respect to knowledge gain on the level of individual genes and proteins as potential targets for cell control and intervention.

4.5 Modeling Resources and Standards

The development of a standardized computer infrastructure is of utmost importance to manage the increasing amount of knowledge relating to biological systems. This supports the effective utilization of resources and the exchange of models, ideas, and data. An integrative software tool for Systems Biology should comprise the following features for proper systems understanding [67]. From a mathematical point of view, these tools need to support different simulation algorithms, such as deterministic and stochastic ODE and PDE solvers, and should include analysis algorithms for parameter estimation and model discrimination.

From an experimental point of view, a package should support a standardized modeling language, while in terms of software the simulation package should run independently from the computer platform, preferably on a computer grid or cluster environment.

A number of simulation packages are currently under active development, most of them freely available. As a consequence of the complexity encountered in biological systems, there is no integrative software package presently capable of handling all phenomena in signal transduction, metabolic pathways, or spatiotemporal simulations. Some of the currently most versatile packages among many are ECell [127], Virtual Cell [128], Cellware [129], Cell Designer [109], or Smart Cell as simulation packages for spatiotemporal simulation [130].

Model building is a complex enterprise, and is usually accomplished in collaboration with scientists from different research institutions. Serving the need for seamless information exchange of computer-based models, progress has been made to define standardized biological "wiring diagrams" [131] together with a common description language, SBGM [132]. Moreover, common standard mark-up languages based on the XML format help in the unique definition of biological entities, pathways, and events for the rapid exchange of models between experimentalists and theoreticians. The most common of these are CellML [133], BioPAX [134], and the Systems Biology Markup Language, SBML [135]. The Systems Biology workbench extends the above approach by providing a standardized application interface for researchers to exchange not only model data and results, but also the simulation tools themselves [136]. Another challenge lies in the inherent modeling of the stochasticity of inner cell processes.

The SSA (stochastic simulation algorithm) from Gillespie is computationally infeasible with an increasing number of molecules, and the computational power for exact stochastic simulations will be immense. Consequently, different algorithms have been developed, such as to reduce the number of random variables for simulation [137] or to allow larger time steps to be taken as a justifiable error in the respective reaction probabilities [138]. Several program packages for the simulation of stochastic molecular dynamics exist, including StochSim [139] or Stocks [140].

Regardless of the modeling approach of biological systems employed, there is always a need for parameter estimates of at least a few constituents of the model. Because of the sheer amount of parameters required for successful modeling, and the huge number of experiments already performed, the need to organize experimental data has brought about several publicly available databases of molecular properties, interactions, and pathways [141–143]. These provide an invaluable infrastructure for future modeling efforts, enabling the modeler to begin simulations from a certain degree of abstraction [144].

Spatiotemporal modeling will additionally require information concerning the physical structure of its model constituents, as obtained using microscopy. For these needs, the development of an

Open Microscope Environment, which is currently being built as a joint effort among various research institutions, will provide a unified data format and database environment for consistently annotating, storing, and retrieving five-dimensional microscope images (four dimensions in space and time, with additional color information) and exchanging them between research institutions [145, 146].

5 Future Prospects of Systems Biology

The systematic generation and analysis of quantitative experimental data slowly, but definitely, is turning biology into branch of science that is close to engineering. Once the "language of the genes" – that is, their syntax and their semantics – is decoded [147], then theoretical knowledge and experimental expertise will suffice to draft and create synthetic model cells or even organisms from scratch, on which new drugs and cures can be tested *in silico* before their possible assembly *in vivo* [148].

There is indeed the prospect that Systems Biology will change medical practice by allowing the prediction of new combinatorial and/or personalized drugs for diseases that currently are regarded as severe, including Alzheimer's, diabetes, human immune deficiency virus (HIV), or cancer [149]. Indeed, Systems Biology is set to change today's medicine from being responsive to being predictive, preventive, personalized, and participatory; this situation is often referred to as *"P4 medicine"* [150]. For example, viewing cancer from a systemic level as a robust system might provide physicians with a framework for future anticancer strategies [61]. As a consequence, anticancer therapies might be developed from mathematical modeling to first identify the peculiarities of an heterogeneous tumor cell population, and then try to predict and control cell activity by detecting and using specific fragile points for each tumor cell type [151]. As a first success in this line, novel, computationally predicted anticancer drugs targeting the ErbB family of receptor tyrosine kinases have been developed from *in silico* models of signaling pathways, and are currently undergoing clinical trials [152, 153].

5.1 Synthetic Biology

The aim of Synthetic Biology is to (re-)design new or already existing biological parts, devices, and systems with the help of mathematical modeling and engineering approaches; in other words, Synthetic Biology is the technological counterpart of Systems Biology. Yet, progress in Synthetic Biology goes hand in hand with current progress in cellular and molecular biology, as well as genetics and associated fields of engineering and computer sciences. It is hoped that the combination of this knowledge will enable the creation of essentially artificial systems, by employing biological design principles with new combinations of building modules from existing (sub)cellular systems [154, 155]. In fact, the first steps in the creation of artificial genomes and their implantation into host organisms has already achieved a degree of success [156].

Synthetic Biology departs from a component-based approach by viewing a living system as a programmable entity that is composed of interacting modules, each having particular functions and which exchange their information via protocols. The results of current research have suggested that these modules are (possibly) limited in number, albeit their specific tasks are diverse. Synthetic Biology implies a scientific agenda on a higher level of abstraction towards identifying and categorizing the various module types within the cell and across organisms, and investigating their interactions on the basis of the modular rather than the molecular interaction. From an engineering point of view, this means that modules can be taken out of their current evolutionary context and assembled differently, in the sense of versatile building blocks.

Ideally, Synthetic Biology starts with mathematically inspired designs, such as a system of coupled differential equations with desired dynamic properties, which are then translated into the biological and chemical realities as promoters, enzymes, or metabolites within a cell. Along this line of thought, the workings of switches and bistable systems of stabilizing control loops have been systematically recast in terms of chemical reaction schemes [148, 157]. Different authors have described the fundamental principles of building logic circuits into the language of gene regulatory networks. For example, Tyson *et al.* presented a systematic overview of designs for biological control systems such as switches, sniffers or buzzers, and combined their mathematical description with experimental findings [43]. Likewise, Hasty *et al.* have reviewed the possibilities of constructing gene circuits which serve various functions such as autoregulation, repression, or logical gates [157]. In a first *de novo* design study of a gene regulatory network, Guet *et al.* systematically analyzed the phenotypic behavior of different parameter and topology combinations in a genetic regulatory network [158]. These authors constructed various logic gates, such as NAND, NOR, and NOT IF, through the combination of three nodes together with five promoters. As an important result, it was found that not only the parameter values but also the network topology was important in order to determine unambiguously the computational function of the systems.

Recently, efforts have been undertaken to establish a Biological Information System (BIS) which extends genomic databases with quantitative mechanistic knowledge [70]. The BioBricks foundation (http://bbf.openwetware.org/), founded by engineers and scientists from the Massachusetts Institute of Technology (MIT), Harvard and the University of California, San Francisco (UCSF), have set up a free, publicly available repository of standardized biological parts for the common re-use and combination to design *de novo* functions in living organisms. Bio-Bricks constitute promoters, proteins, RNA-coding sequences, or transcriptional terminators; physically, they are DNA sequences stored on a circular plasmid distributed by the Registry of Biological Parts (http://www.partsregistry.org). By using such data in an integrated form, it will become possible to build refined synthetic systems by checking on key molecules and replacing complex pathways with effective reaction parameters, thus defining biologically meaningful reaction subsets from the large amount of possible reactions.

5.2
Conclusions: Where Are We?

Although, today, Systems Biology is still at the start of a new branch of science, the road ahead is clearly marked and commonly agreed on by many people. Yet, the problems associated with all of the processes lie in the details.

To date, most experimental biosciences are method-driven, and pay much attention to detail and the production, in extreme cases, of large amounts of data which are out of context. Theoretical sciences, on the other hand, are principle-driven; they neglect important details and thus come up with theorems and models that are "out of this world." As a consequence, research groups must learn to focus on the important areas of their respective systems under investigation, while also determining which part to measure in detail and which part to neglect:

- *Biologists* need to adopt abstract thinking, and to trust in the language of mathematics and necessary simplification.
- *Physicists* must learn to abolish oversimplified thinking [159] and become used to the analysis of strongly interacting systems with many degrees of freedom.
- *Engineers* must gain a deeper understanding of their systems under investigation beyond numerical solutions, for example, in terms of process optimization.

The necessary unification of these various branches of science will, for the sake of the advancement of Systems Biology, make communication skills among scientists ever more important. Previous years of "isolated" research have diversified the scientific vocabulary and methods, making it sometimes very difficult to mediate and discuss interdisciplinary goals and methods. Moreover, all scientists must learn to place their own research into a much wider frame of interdisciplinary research and research teams.

Despite this increased complexity of the systems under study, one of the greatest challenges in Systems Biology is to bridge the gap between detailed kinetic protein signaling models (of up to 100 different molecules) and large-scale -omics approaches, thereby providing – simultaneously – data on thousands of genes or proteins. At present, the means by which the detailed models can be scaled up to include more variables and to bridge several time scales (from minutes to hours or even days) is largely unknown. In contrast, it is very difficult to move from the statistical analysis of -omics data to the prediction of individual protein interactions. Yet, taken together, despite not being able to bridge this modeling gap, there is a clear need for an intuitive understanding of the relevant elements of the system under study. Vilar *et al.* [160] have argued, using the example of the lac operon, that "... even in the 'postgenomic era,' (modeling) will still rely more on good intuition and skills of quantitative biologists than on the sheer power of computers."

Today, it is strongly believed that Systems Biology will promote a unification of the sciences, especially if the scientific community continues to work "holistically" in a joint effort of biology, mathematics, physics, and engineering. Only then might a new generation of scientists with an interdisciplinary training emerge, whose everyday business will comprise not only working in the laboratory but also performing data analysis and model simulations, in addition to exchanging their data and results freely through publicly available databases. With the establishment of new experimental techniques and mathematical tools, and asking the right questions, it is clear that the systems approach to biology will become a major success. Indeed, it will change not only modern life sciences, but also views on life itself.

References

1 Kitano, H. (2002) Systems biology: a brief overview. *Science*, **295**, 1662–1664.

2 Special Issue (2002) Systems biology. *Science*, **295**.

3 Special Issue (2002) Systems biology. *Nature Insight*, **420**.

4 Szallasi, Z., Stelling, J., Periwal, V. (2006) *System Modeling in Cell Biology*, The MIT Press.

5 Alon, U. (2007) *An Introduction to Systems Biology: Design Principles of Biological Circuits*, Chapman & Hall/CRC.

6 Klipp, E., Liebermeister, W., Wierling, C., Kowald, A., Lehrach, H., Herwig, R. (2009) *Systems Biology*, Wiley-Blackwell.

7 Fu, P., Panke, S. (Eds) (2009) *Systems Biology and Synthetic Biology*, AIChE-Wiley Press.

8 Smuts, J.C. (1926) *Holism and Evolution*, Macmillan & Co. Ltd, London.

9 Bertalanffy, L. (1973) *General Systems Theory*, Penguin, Harmondsworth.

10 Ashby W.R. (1956) *Introduction to Cybernetics*, Chapman & Hall, London. http://pespmc1.vub.ac.be/books/IntroCyb.pdf (accessed 8 April 2011).

11 Wiener, N. (1948) *Cybernetics: Control and Communications in the Animal and the Machines*, The MIT Press, Cambridge.

12 Ideker, T., Galitski, T., Hood, L. (2001) A new approach to decoding life: systems biology. *Annu. Rev. Genomics Hum. Genet.*, **2**, 343–372.

13 Wolkenhauer, O. (2001) Systems biology: the reincarnation of systems theory applied in biology? *Brief. Bioinform.*, **2** (3), 258–270.

14 Mesarović, M.D. (1968) System Theory and Biology – View of a Theoretician, in: Mesarović, M.D. (Ed.) *Systems Theory and Biology*, Springer, New York, pp. 59–87.

15 Wolkenhauer, O., Hofmeyr, J.-H.S. (2007) An abstract cell model that describes the self-organization of cell function in living systems. *J. Theor. Biol.*, **246** (3), 461–476.

16 Westerhoff, H.V., Palsson, B.O. (2004) The evolution of molecular biology into systems biology. *Nat. Biotechnol.*, **22** (10), 1249–1252.

17 Pittendrigh, C.S. (1958) Adaptation, natural selection, and behavior, in: Roe, A., Simpson, G.G. (Eds) *Behavior and Evolution*, Yale University Press, New Haven, pp. 390–416.

18 Baker, M.D., Wolanin, P.M., Stock, J.B. (2006) Signal transduction in bacterial chemotaxis. *BioEssays*, **28** (1), 9–22.

19 Levine, M., Tjian, R. (2003) Transcription regulation and animal diversity. *Nature*, **424**, 147–151.

20 Huang, S. (2004) Back to the biology in systems biology: what can we learn from biomolecular networks? *Brief. Funct. Genomic. Proteomic.*, **2** (4), 279–297.

21 Dufva, M. (Ed.) (2009) *DNA Microarrays for Biomedical Research*, Methods in Molecular Biology, Vol. 529, Humana Press.

22 Hawkins, R.D., Hon, G.C., Ren, B. (2010) Next-generation genomics: an integrative approach. *Nat. Rev. Genet.*, **11** (7), 476–486.

23 Fields, S., Song, O. (1989) A novel genetic system to detect protein-protein interactions. *Nature*, **340** (6230), 245–246.

24 Johnsson, N., Varshavsky, A. (1994) Split ubiquitin as a sensor of protein interactions in vivo. *Proc. Natl Acad. Sci. USA*, **91** (22), 10340–10344.

25 Spiller, D.G., Wood, C.D., Rand, D.A., White, M.R.H. (2010) Measurement of single-cell dynamics. *Nature*, **465** (7299), 736–745.

26 Demirel, Y., Sandler, S.I. (2002) Thermodynamics and bioenergetics. *Biophys. Chem.*, **97**, 87–111.

27 Hartwell, L.H., Hopfield, J.J., Leibler, S., Murray, A.W. (1999) From molecular to modular cell biology. *Nature*, **420**, C47–C52.

28 Malygin, A.A., Parakhnevitch, N.M., Ivanov, A.V., Eperon, I.C., Karpova, G.G. (2007) Human ribosomal protein s13 regulates expression of its own gene at the splicing step by a feedback mechanism. *Nucleic Acids Res.*, **35** (19), 6414–6423.

29 Vicsek, T. (2002) Complexity: the bigger picture. *Nature*, **418**, 131.

30 Carlson, J.M., Doyle, J. (2002) Complexity and robustness. *Proc. Natl Acad. Sci. USA*, **99** (Suppl. 1), 2538–2545.

31 Grassberger, P. (1986) Toward a quantitative theory of self-generated complexity. *Int. J. Theor. Phys.*, **25**, 907–928.

32 Adami, C. (2002) What is complexity? *BioEssays*, **24**, 1085–1094.
33 Barabási, A.-L., Oltvai, Z.N. (2004) Network biology: understanding the cell's functional organization. *Nat. Gen.*, **5**, 101–114.
34 Kitano, H. (2002) Computational systems biology. *Nature*, **420**, 206–210.
35 Haken, H. (1987) *Advanced Synergetics*, Springer, Berlin.
36 Bak, P., Tang, C., Wiesenfeld, K. (1987) Self-organized criticality: an explanation of the 1/f noise. *Phys. Rev. Lett.*, **59**, 381–384.
37 Carlson, J.M., Doyle, J. (1999) Highly optimized tolerance: a mechanism for power laws in designed systems. *Phys. Rev. E*, **60** (2), 1412–1427.
38 Csete, M.E., Doyle, J.C. (2002) Reverse engineering of biological complexity. *Science*, **295**, 1664–1669.
39 Stelling, J., Sauer, U., Szallasi, Z., Doyle, F.J., Doyle, J. (2004) Robustness of cellular functions. *Cell*, **118**, 675–685.
40 Kitano, H. (2007) Towards a theory of biological robustness. *Mol. Syst. Biol.*, **3**, 137.
41 Lauffenburger, D.A. (2000) Cell signaling pathways as control modules: complexity for simplicity? *Proc. Natl Acad. Sci. USA*, **97** (10), 5031–5033.
42 Gu, X. (2003) Evolution of duplicate genes versus genetic robustness against null mutations. *Trends Genet.*, **19**, 354–356.
43 Tyson, J.J., Chen, K.C., Novak, B. (2003) Sniffers, buzzers, toggles and blinkers: dynamics of regulatory and signaling pathways in the cell. *Curr. Opin. Cell Biol.*, **15**, 221–231.
44 Novak, B., Tyson, J.J., Gyorffy, B., Csikasz-Nagy, A. (2007) Irreversible cell cycle transitions are due to systems-level feedback. *Nat. Cell Biol.*, **9** (7), 724–728.
45 Lpez-Avils, S., Kapuy, O., Novk, B., Uhlmann, F. (2009) Irreversibility of mitotic exit is the consequence of systems-level feedback. *Nature*, **459** (7246), 592–595.
46 Clarke, D.J., Giménez-Abián, J.F. (2000) Checkpoints controlling mitosis. *BioEssays*, **22** (4), 351–363.
47 Kearns, W.G., Liu, J.M. (2001) Cell cycle checkpoint genes and aneuploidy: a short review. *Curr. Genomics*, **2** (2), 171–180.
48 Fulka, J., Tesařík, J., Loi, P., Moor, R.M. (2000) Manipulating the human embryo: cell cycle checkpoint controls. *Cloning*, **2** (1), 1–7.
49 Barkai, N., Leibler, S. (1997) Robustness in simple biochemical networks. *Nature*, **387**, 913–917.
50 Alon, U., Surette, M.G., Barkai, N., Leibler, S. (1999) Robustness in bacterial chemotaxis. *Nature*, **397**, 168–171.
51 Hartman, J.L., Garvik, B., Hartwell, L. (2001) Principles for the buffering of genetic variation. *Science*, **291**, 1001–1004.
52 Gell, M., Noort, V., Yus, E., Chen, W.-H., Leigh-Bell, J., Michalodimitrakis, K., Yamada, T., Arumugam, M., Doerks, T., Khner, S., Rode, M., Suyama, M., Schmidt, S., Gavin, A.-C., Bork, P., Serrano, L. (2009) Transcriptome complexity in a genome-reduced bacterium. *Science*, **326** (5957), 1268–1271.
53 von Dassow, G., Meir, E., Munro, E.M., Odell, G. (2000) The segment polarity network is a robust developmental module. *Nature*, **406**, 188–192.
54 Hanahan, D., Weinberg, R.A. (2000) The hallmarks of cancer. *Cell*, **100** (1), 57–70.
55 Yi, T.-M., Huang, Y., Simon, M.I., Doyle, J.C. (2004) Robust perfect adaptation in bacterial chemotaxis through integral feedback control. *Proc. Natl Acad. Sci. USA*, **97** (9), 4649–4653.
56 Guisbert, E., Yura, T., Rhodius, V.A., Gross, C.A. (2008) Convergence of molecular, modeling, and systems approaches for an understanding of the *Escherichia coli* heat shock response. *Microbiol. Mol. Biol. Rev.*, **72** (3), 545–554.
57 Kitano, H. (Ed.) (2001) *Foundations of Systems Biology*, The MIT Press, Cambridge, pp. 1–36.
58 Angeli, D., Ferrell, J.E., Sontag, E.D. (2004) Detection of multistability, bifurcations, and hysteresis in a large class of biological positive-feedback systems. *Proc. Natl Acad. Sci. USA*, **101**, 1822–1827.
59 Becker, V., Schilling, M., Bachmann, J., Baumann, U., Raue, A., Maiwald, T., Timmer, J., Klingmller, U. (2010) Covering

a broad dynamic range: information processing at the erythropoietin receptor. *Science*, **328** (5984), 1404–1408.
60 Freeman, M. (2000) Feedback control of intercellular signalling in development. *Nature*, **408**, 313–319.
61 Kitano, H. (2004) Cancer as a robust system: implications for anticancer therapy. *Nat. Rev.*, **4**, 227–235.
62 Carter, G.W., Rush, C.G., Uygun, F., Sakhanenko, N.A., Galas, D.J., Galitski, T. (2010) A systems-biology approach to modular genetic complexity. *Chaos*, **20** (2), 026102.
63 Guimerà, R., Arenas, A., Díaz-Guilera, A. (2001) Communication and optimal hierarchical networks. *Physica A*, **299**, 247–252.
64 Davidson, E.H., Rast, J.P., Oliveri, P., Ransick, A., Calestani, C., Yuh, C.-H., Minokawa, T., Amore, G., Hinman, V., Arenas-Mena, C., Otim, O., Brown, C.T., Livi, C.B., Lee, P.Y., Revilla, R., Rust, A.G., Pan, Z., Schilstra, M.J., Clarke, P.J.C., Arnone, M.I., Rowen, L., Cameron, R.A., McClay, D.R., Hood, L., Bolouri, H. (2002) A genomic regulatory network for development. *Science*, **295**, 1669–1678.
65 Laub, M.T., McAdams, H.H., Feldblum, T., Fraser, C.M., Shapiro, L. (2000) Global analysis of the genetic network controlling a bacterial cell cycle. *Science*, **290**, 2144–2148.
66 Pearl, J. (2000) *Causality*, Cambridge University Press.
67 Dhar, P.K., Zhu, H., Mishra, S.K. (2004) Computational approach to systems biology: from fraction to integration and beyond. *IEEE Trans. Nanobiosci.*, **3** (3), 144–152.
68 Crampin, E.J., Schnell, S., McSharry, P.E. (2004) Mathematical and computational techniques to deduce complex biochemical reaction mechanisms. *Prog. Biophys. Mol. Biol.*, **86** (1), 77–112.
69 Ideker, T., Thorsson, V., Ranish, J.A., Christmas, R., Buhler, J., Eng, J.K., Bumgarner, R., Goodlett, D.R., Aebersold, R., Hood, L. (2001) Integrated genomic and proteomic analyses of a systematically perturbed metabolic network. *Science*, **292**, 929–934.
70 Endy, D., Brent, R. (2001) Modelling cellular behavior. *Nature*, **409**, 391–395.
71 McAdams, H., Shapiro, L. (1995) Circuit simulations of genetic networks. *Science*, **269**, 650–656.
72 Christopher, R., Dhiman, A., Fox, J., Gendelman, R., Haberichter, T., Kagle, D., Spizz, G., Khalil, I.G., Hill, C. (2004) Data-driven computer simulation of human cancer cell. *Ann. N. Y. Acad. Sci.*, **1020**, 132–153.
73 Müller, F.-J., Laurent, L.C., Kostka, D., Ulitsky, I., Williams, R., Lu, C., Park, I.-H., Rao, M.S., Shamir, R., Schwartz, P.H., Schmidt, N.O., Loring, J.F. (2008) Regulatory networks define phenotypic classes of human stem cell lines. *Nature*, **455** (7211), 401–405.
74 Jeong, H., Tombor, B., Albert, R., Oltvai, Z.N., Barabási, A.-L. (2000) The large-scale organization of metabolic networks. *Nature*, **407**, 651–654.
75 Jeong, H., Mason, S.P., Barabási, A.-L., Oltvai, Z.N. (2001) Lethality and centrality in protein networks. *Nature*, **411**, 41–42.
76 Elowitz, M.B., Surette, M.G., Wolf, P.-E., Stock, J.B., Leibler, S.J. (1999) Protein mobility in the cytoplasm of *Escherichia coli*. *J. Bacteriol.*, **181**, 197–203.
77 Schnell, S., Turner, T.E. (2004) Reaction kinetics in intracellular environments with macromolecular crowding: simulations and rate laws. *Prog. Biophys. Mol. Biol.*, **85**, 235–260.
78 Howard, M., Rutenberg, A.D. (2003) Pattern formation inside bacteria: fluctuations due to the low copy number of proteins. *Phys. Rev. Lett.*, **90** (1–4), 128102.
79 Busch, H., Camacho, D., Rogon, Z., Breuhahn, K., Angel, P., Eils, R., Szabowski, A. (2008) Gene network dynamics controlling keratinocyte migration. *Mol. Syst. Biol.*, **4**, 199.
80 Chikofsky, E.J., Cross, J.H. (1990) Reverse engineering and design recovery: A taxonomy. *IEEE Softw.*, **7**, 13–17.
81 He, F., Balling, R., Zeng, A.-P. (2009) Reverse engineering and verification of gene networks: principles, assumptions, and limitations of present methods and future perspectives. *J. Biotechnol.*, **144** (3), 190–203.
82 Strogatz, S.H. (2001) Exploring complex networks. *Nature*, **410**, 268–276.
83 Sharom, J.R., Bellows, D.S., Tyers, M. (2004) From large networks to small

molecules. *Curr. Opin. Chem. Biol.*, **8**, 81–90.

84 Alm, E., Arkin, A.P. (2003) Biological networks. *Curr. Opin. Struct. Biol.*, **13**, 193–202.

85 Barabási, A.-L., Albert, R. (1999) Emergence of scaling in random networks. *Science*, **286**, 509–512.

86 Keller, E.F. (2005) Revisiting "scale-free" networks. *BioEssays*, **27** (10), 1060–1068.

87 Milo, R., Shen-Orr, S., Itzkovitz, S., Kashtan, N., Chklovskii, D., Alon, U. (2002) Network motifs: Simple building blocks of complex networks. *Science*, **298**, 824–827.

88 Yan, K.K., Fang, G., Bhardwaj, N., R.P., Alexander, Gerstein, M. (2010) Comparing genomes to computer operating systems in terms of the topology and evolution of their regulatory control networks. *Proc. Natl Acad. Sci. USA*, **107** (20), 9186–9191.

89 Lee, T.I., Rinaldi, N.J., Robert, F., Odom, D.T., Bar-Joseph, Z., Gerber, G.K., Hannett, N.M., Harbison, C.T., Thompson, C.M., Simon, I., Zeitlinger, J., Jennings, E.G., Murray, H.L., Gordon, D.B., Ren, B., Wyrick, J.J., Tagne, J.-B., Volkert, T.L., Fraenkel, E., Gifford, D.K., Young, R.A. (2002) Transcriptional regulatory networks in *Saccharomyces cerevisiae*. *Science*, **298**, 799–804.

90 Almaas, E., Kovács, B., Vicsek, T., Oltvai, Z.N., Barabási, A.-L. (2004) Global organization of metabolic fluxes in the bacterium *Escherichia coli*. *Nature*, **427**, 839–843.

91 Ravasz, E., Somera, A.L., Mongru, D.A., Oltvai, N., Barabási, A.-L. (2002) Hierarchical organization of modularity in metabolic networks. *Science*, **297**, 1551–1555.

92 Pawson, T., Nash, P. (2003) Assembly of cell regulatory systems through protein interaction domains. *Science*, **300**, 445–452.

93 de la Fuente, A. (2010) From differential expression to differential networking identification of dysfunctional regulatory networks in diseases. *Trends Genet.*, **26**, 326–333.

94 Goh, K.-I., Cusick, M.E., Valle, D., Childs, B., Vidal, M., Barabsi, A.-L. (2007) The human disease network. *Proc. Natl Acad. Sci. USA*, **104** (21), 8685–8690.

95 Saez-Rodriguez, J., Alexopoulos, L.G., Epperlein, J., Samaga, R., Lauffenburger, D.A., Klamt, S., Sorger, P.K. (2009) Discrete logic modeling as a means to link protein signalling networks with functional analysis of mammalian signal transduction. *Mol. Syst. Biol.*, **5**, 331.

96 Jong, H. (2002) Modeling and simulation of generic regulatory systems: a literature review. *J. Comput. Biol.*, **9** (1), 67–103.

97 Faur, A., Thieffry, D. (2009) Logical modelling of cell cycle control in eukaryotes: a comparative study. *Mol. Biosyst.*, **5** (12), 1569–1581.

98 Samaga, R., Saez-Rodriguez, J., Alexopoulos, L.G., Sorger, P.K., Klamt, S. (2009) The logic of egfr/erbb signaling: theoretical properties and analysis of high-throughput data. *PLoS Comput. Biol.*, **5** (8), e1000438.

99 Schlatter, R., Schmich, K., Vizcarra, I.A., Scheurich, P., Sauter, T., Borner, C., Ederer, M., Merfort, I., Sawodny, O. (2009) On/off and beyond – a Boolean model of apoptosis. *PLoS Comput. Biol.*, **5** (12), e1000595.

100 Murray, J.D. (1993) *Mathematical Biology*, Springer, Heidelberg.

101 Cross, M.C., Hohenberg, P.C. (1993) Pattern formation outside of equilibrium. *Rev. Mod. Phys.*, **65** (3), 851–1112.

102 Moles, C.G., Mendes, P., Banga, J.R. (2003) Parameter estimation in biochemical pathways: A comparison of global optimization methods. *Genome Res.*, **13**, 2467–2474.

103 Hengl, S., Kreutz, C., Timmer, J., Maiwald, T. (2007) Data-based identifiability analysis of non-linear dynamical models. *Bioinformatics*, **23**, 2612–2618.

104 Bandara, S., Schlder, J.P., Eils, R., Bock, H.G., Meyer, T. (2009) Optimal experimental design for parameter estimation of a cell signaling model. *PLoS Comput. Biol.*, **5** (11), e1000558.

105 Skanda, D., Lebiedz, D. (2010) An optimal experimental design approach to model discrimination in dynamic biochemical systems. *Bioinformatics*, **26** (7), 939–945.

106 Press, W.H., Teukolsky, S.A., Vetterling, W.T., Flannery, B.P. (1993) *Numerical Recipes in C*, 2nd edn, Cambridge University Press, New York.

107 Mendes, P. (1993) Gepasi: a software package for modelling the dynamics, steady states and control of biochemical and other systems. *CABIOS*, **9** (5), 563–571.

108 Goryanin, I., Hodgman, T.C., Selkov, E. (1999) Mathematical simulation and

analysis of cellular metabolism and regulation. *Bioinformatics*, **15** (9), 749–758.

109 Funahashi, A., Matsuoka, Y., Jouraku, A., Morohashi, M., Kikuchi, N., Kitano, H. (2008) Celldesigner 3.5: a versatile modeling tool for biochemical networks. *Proc. IEEE*, **96** (8), 1254–1265.

110 Schmidt, H., Jirstrand, M. (2006) Systems biology toolbox for matlab: a computational platform for research in systems biology. *Bioinformatics*, **22** (4), 514–515.

111 Maiwald, T., Timmer, J. (2008) Dynamical modeling and multi-experiment fitting with Potters wheel. *Bioinformatics*, **24** (18), 2037–2043.

112 Huang, C.Y., Ferrell, J.E. (1996) Ultrasensitivity in the mitogen-activated protein kinase cascade. *Proc. Natl Acad. Sci. USA*, **93** (19), 10078–10083.

113 Turing, A.M. (1952) The chemical basis of morphogenesis. *Phil. Trans. R. Soc. London B*, **237**, 37–72.

114 Meinhardt, H. (2008) Models of biological pattern formation: from elementary steps to the organization of embryonic axes. *Curr. Top. Dev. Biol.*, **81**, 1–63.

115 Kauffman, S.A. (1993) *The Origins of Order: Self-Organization and Selection in Evolution*, Oxford University Press, New York.

116 Volz, D., Eigel, M., Athale, C., Bastian, P., Hermann, H., Kappel, C., Eils, R. (2005) *Spatial Modeling and Simulation of Diffusion in Nuclei of Living Cells*, CMSB 2004, Lecture Notes in Computer Science, Springer, Heidelberg, pp. 161–171.

117 Kholodenko, B.N. (2003) Four-dimensional organization of protein kinase signaling cascades: the roles of diffusion, endocytosis and molecular motors. *J. Exp. Biol.*, **206**, 2073–2082.

118 Slepchenko, B.M., Schaff, J.C., Carson, J.H., Loew, L.M. (2002) Computational cell biology: Spatiotemporal simulation of cellular events. *Ann. Rev. Biophys. Biomol. Struct.*, **31**, 423–441.

119 van Kampen, N.G. (1992) *Stochastic Processes in Physics and Chemistry*, Elsevier Science Publishers B.V., Amsterdam.

120 Kaern, M., Elston, T.C., Blake, W.J., Collins, J.J. (2005) Stochasticity in gene expression: from theories to phenotypes. *Nat. Rev. Genet.*, **6** (6), 451–464.

121 Golding, I., Paulsson, J., Zawilski, S.M., Cox, E.C. (2005) Real-time kinetics of gene activity in individual bacteria. *Cell*, **123** (6), 1025–1036.

122 Elowitz, M., Levine, A., Siggla, W., Swain, P. (2002) Stochastic gene expression in a single cell. *Science*, **297**, 1183–1186.

123 Arkin, A., Ross, J., McAdams, H.H. (1998) Stochastic kinetic analysis of developmental pathway bifurcation in phage lambda-infected *Escherichia coli* cells. *Genetics*, **149** (4), 1633–1648.

124 Huang, S., Ingber, D.E. (2007) A non-genetic basis for cancer progression and metastasis: self-organizing attractors in cell regulatory networks. *Breast Dis.*, **26**, 27–54.

125 Huang, S., Eichler, G., Bar-Yam, Y., Ingber, D.E. (2005) Cell fates as high dimensional attractor states of a complex gene regulatory network. *Phys. Rev. Lett.*, **94** (12), 128701.

126 Huang, S. (2009) Reprogramming cell fates: reconciling rarity with robustness. *BioEssays*, **31** (5), 546–560.

127 Tomita, M., Hashimoto, K., Takahashi, K., Shimizu, T., Matsuzaki, Y., Miyoshi, F., Saito, K., Tanida, S., Yugi, K., Venter, J.C., Hutchison, C.M. (1999) Ecell: software environment for whole cell simulation. *Bioinformatics*, **15** (5), 72–74.

128 Slepchenko, B.M., Schaff, J.C., Macara, I.G., Loew, L.M. (2003) Quantitative cell biology with the virtual cell. *Trends Cell Biol.*, **13**, 570–576.

129 Dhar, P., Meng, T.C., Somani, S., Ye, L., Sairam, A., Chitre, M., Hao, Z., Sakharkar, K. (2004) Cellware – multi-algorithmic software for computational systems biology. *Bioinformatics*, **20**, 1319–1321.

130 Ander, M., Beltrao, P., Ventura, B.D., Ferkinghoff-Borg, J., Foglierini, M., Kaplan, A., Lemerle, C., Tomas-Oliveira, I., Serrano, L. (2004) SmartCell, a framework to simulate cellular processes that combines stochastic approximation with diffusion and localisation: analysis of simple networks. *Syst. Biol.*, **1**, 129–138.

131 Kitano, H., Funahashi, A., Matsuoka, Y., Oda, K. (2005) Using process diagrams for the graphical representation of biological networks. *Nat. Biotechnol.*, **23** (8), 961–966.

132 Novre, N.L., Hucka, M., Mi, H., Moodie, S., Schreiber, F., Sorokin, A., Demir, E., Wegner, K., Aladjem, M.I., Wimalaratne, S.M., Bergman, F.T., Gauges, R., Ghazal, P., Kawaji, H., Li, L., Matsuoka, Y.,

Villger, A., Boyd, S.E., Calzone, L., Courtot, M., Dogrusoz, U., Freeman, T.C., Funahashi, A., Ghosh, S., Jouraku, A., Kim, S., Kolpakov, F., Luna, A., Sahle, S., Schmidt, E., Watterson, S., Wu, G., Goryanin, I., Kell, D.B., Sander, C., Sauro, H., Snoep, J.L., Kohn, K., Kitano, H. (2009) The systems biology graphical notation. *Nat. Biotechnol.*, **27** (8), 735–741.

133 Lloyd, C.M., Halstead, M.D.B., Nielsen, P.F. (2004) CellML: its future, present and past. *Prog. Biophys. Mol. Biol.*, **85**, 433–450.

134 BioPAX Homepage Internet (2010), http://www.biopax.org/ (accessed 8 April 2011).

135 Hucka, M., Finney, A., Sauro, H.M., Bolouri, H., Doyle, J.C., Kitano, H., Arkin, A.P., Bornstein, B.J., Bray, D., Cornish-Bowden, A., Cuellar, A.A., Dronov, S., Gilles, E.D., Ginkel, M., Gor, V., Goryanin, I.I., Hedley, W.J., Hodgman, T.C., Hofmeyr, J.-H., Hunter, P.J., Juty, N.S., Kasberger, J.L., Kremling, A., Kummer, U., Le Novre, N., Loew, L.M., Lucio, D., Mendes, P., Minch, E., Mjolsness, E.D., Nakayama, Y., Nelson, M.R., Nielsen, P.F., Sakurada, T., Schaff, J.C., Shapiro, B.E., Shimizu, T.S., Spence, H.D., Stelling, J., Takahashi, K., Tomita, M., Wagner, J., Wang, J. (2003) The systems biology markup language (SBML): a medium for representation and exchange of biochemical network models. *Bioinformatics*, **19**, 524–531.

136 Hucka, M., Finney, A., Sauro, H.M., Bolouri, H., Doyle, J., Kitano, H. (2002) The erato systems biology workbench: enabling interaction and exchange between software tools for computational biology. *Pac. Symp. Biocomput.*, 450–461.

137 Gibson, M.A., Bruck, J. (2000) Efficient exact stochastic simulation of chemical systems with many species and many channels. *J. Phys. Chem.*, **104**, 1876–1889.

138 Gillespie, D.T. (2001) Approximate accelerated stochastic simulation of chemically reacting systems. *J. Chem. Phys.*, **115**, 1716–1733.

139 Morton-Firth, C.J. (1998) Stochastic simulation of cell signalling pathways. PhD thesis, University of Cambridge.

140 Kierzek, A.M. (2002) Stocks: stochastic kinetic simulations of biochemical systems with Gillespie algorithm. *Bioinformatics*, **18** (3), 470–481.

141 Lindvall, J.M., Emelie, K., Blomberg, M., Smith, C.I.E. (2003) In silico tools for signal transduction research. *Brief. Bioinform.*, **4**, 315–324.

142 Jensen, L.J., Kuhn, M., Stark, M., Chaffron, S., Creevey, C., Muller, J., Doerks, T., Julien, P., Roth, A., Simonovic, M., Bork, P., Mering, C. (2009) String 8 – a global view on proteins and their functional interactions in 630 organisms. *Nucleic Acids Res.*, **37** (Database issue), D412–D416.

143 Kanehisa, M., Goto, S., Furumichi, M., Tanabe, M., Hirakawa, M. (2010) Kegg for representation and analysis of molecular networks involving diseases and drugs. *Nucleic Acids Res.*, **38** (Database issue), D355–D360.

144 Arkin, A.P. (2001) Synthetic cell biology. *Curr. Opin. Biotechnol.*, **12** (6), 638–644.

145 Swedlow, J.R., Goldberg, I., Brauner, E., Sorger, P.K. (2003) Informatics and quantitative analysis in biological imaging. *Science*, **300**, 100–102.

146 Goldberg, I., Allan, C., Burel, J.-M., Creager, D., Falconi, A., Hochheiser, H., Johnston, J., Mellen, J., Sorger, P.K., Swedlow, J.R. (2005) The Open Microscopy Environment (OME) Data Model and XML File: Open Tools for Informatics and Quantitative Analysis in Biological Imaging. *Genome Biol.*, **6**, R47.

147 Searls, D.B. (2002) The language of the genes. *Nature*, **420**, 211–217.

148 François, P., Hakim, V. (2004) Design of genetic networks with specified functions by evolution in silico. *Proc. Natl Acad. Sci. USA*, **101** (2), 580–585.

149 Butcher, E.C., Berg, E.L., Kunkel, E.J. (2004) Systems biology in drug discovery. *Nat. Biotechnol.*, **22** (10), 1253–1259.

150 Hood, L. (2008) A personal journey of discovery: developing technology and changing biology. *Annu. Rev. Anal. Chem. (Palo Alto CA)*, **1**, 1–43.

151 Fitzgerald, J.B., Schoeberl, B., Nielsen, U.B., Sorger, P.K. (2006) Systems biology and combination therapy in the quest for clinical efficacy. *Nat. Chem. Biol.*, **2** (9), 458–466.

152 Schoeberl, B., Pace, E.A., Fitzgerald, J.B., Harms, B.D., Xu, L., Nie, L., Linggi, B., Kalra, A., Paragas, V., Bukhalid, R.,

Grantcharova, V., Kohli, N., West, K.A., Leszczyniecka, M., Feldhaus, M.J., Kudla, A.J., Nielsen, U.B. (2009) Therapeutically targeting erbb3: a key node in ligand-induced activation of the erbb receptor-pi3k axis. *Sci. Signal.*, **2** (77), ra31.

153 Hendriks, B.S. (2010) Functional pathway pharmacology: chemical tools, pathway knowledge and mechanistic model-based interpretation of experimental data. *Curr. Opin. Chem. Biol.*, **14** (4), 489–497.

154 Purnick, P.E.M., Weiss, R. (2009) The second wave of synthetic biology: from modules to systems. *Nat. Rev. Mol. Cell. Biol.*, **10** (6), 410–422.

155 Kiel, C., Yus, E., Serrano, L. (2010) Engineering signal transduction pathways. *Cell*, **140** (1), 33–47.

156 Gibson, D.G., Glass, J.I., Lartigue, C., Noskov, V.N., Chuang, R.-Y., Algire, M.A., Benders, G.A., Montague, M.G., Ma, L., Moodie, M.M., Merryman, C., Vashee, S., Krishnakumar, R., Assad-Garcia, N., Andrews-Pfannkoch, C., Denisova, E.A., Young, L., Qi, Z.-Q., Segall-Shapiro, T.H., Calvey, C.H., Parmar, P.P., Hutchison, C.A., Smith, H.O., Venter, J.C. (2010) Creation of a bacterial cell controlled by a chemically synthesized genome. *Science*, **329** (5987), 52–56.

157 Hasty, J., McMillen, D., Collins, J.J. (2002) Engineered gene circuits. *Nature*, **420**, 224–230.

158 Guet, C.C., Elowitz, M.B., Hsing, W., Leibler, S. (2002) Combinatorial synthesis of genetic networks. *Science*, **296**, 1466–1470.

159 May, R.M. (2004) Uses and abuses of mathematics in biology. *Science*, **303**, 790–793.

160 Vilar, J.M.G., Kueh, H.Y., Barkai, N., Leibler, S. (2002) Mechanisms of noise-resistance in genetic oscillators. *Proc. Natl Acad. Sci. USA*, **99**, 5988–5992.

2
Developmental Cell Biology

Frank John Dye
Western Connecticut State University, Biological and Environmental Sciences, 181 White Street Danbury, CT 06810, USA

Systems Biology: Advances in Molecular Biology and Medicine, First Edition. Edited by Robert A. Meyers.

Keywords

Cell cycle
A series of phases designated G_1, S, G_2, and M, that a cell passes through sequentially between the time it is formed by cell division and the time it itself divides.

Chimera
An organism consisting of parts derived from more than one pair of parents. The chimera of Greek mythology consisted of a lion's head, a goat's body, and a serpent's tail. Chimeras may be made in the laboratory by, in the case of mammals, removing the zona pellucida of each embryo and pushing the sticky embryos together; these merge together to form a single chimeric embryo.

Embryonic stem (ES) cells
Stem cells derived from the inner cell mass of the blastocyst stage embryo.

Growth factors
Extracellular polypeptide signal molecules that can stimulate a cell to grow or proliferate.

Induced pluripotent stem cells (iPSCs)
Adult differentiated cells that have been converted to cells with the pluripotency of embryonic stem cells, generally accomplished by the activation of specific transcription factors.

***In-vitro* fertilization (IVF)**
Human/mammalian fertilization occurring under artificial conditions outside of the body.

Intracellular signal transduction
Cascades of chemical reactions which are used by cells to convey information from the cell surface to the cell interior (e.g., the nucleus), where the information is acted upon. A cell can make use of multiple and interacting intracellular signal transduction pathways.

microRNAs (miRNAs)
microRNAs are post-transcriptional regulators that bind to complementary sequences of target messenger RNA transcripts (mRNAs), usually resulting in gene silencing.

Organogenesis
Organ formation.

Pluripotent cell
A cell that is capable of giving rise to many of the cell types of an organism.

Receptors
Proteins that bind other molecules or ions and, as a result of the binding, influence cellular activity.

Stem cell
A cell that undergoes mitotic cell division to give rise to the same type of cell. At some point, stem cells leave the pool of mitotically dividing cells to begin a process of cell differentiation.

Transcription factors
Proteins that attach to DNA at a gene regulatory site and, by so doing, influence the rate of transcription of a specific gene.

Developmental cell biology is concerned with the consequences of what cells do in the context of developing organisms. Developmental biologists are also concerned with events before fertilization, such as gametogenesis, and after birth (or its equivalent in nonmammalian species), such as metamorphosis and regeneration. The range of species in the purview of the developmental biologist covers, essentially, the entire living world – animals, plants, protists, fungi, and even prokaryotes. In the creation of a human being, cell proliferation is required to create the trillions of cells of which the body is composed; cell differentiation provides the more than 200 different types of cells identified to date in the human body; while morphogenesis shapes the body form that is characteristic of the human species. The present understanding of the molecular and cellular mechanisms that underlie these phenomena has progressed at an ever-increasing pace, and is testimony to the army of cell and developmental biologists worldwide. To understand the regulation of the cell cycle, apoptosis, cell differentiation, pattern formation, and cell movement (to mention only a few such phenomena), to understand the integration of their underlying molecular and cellular mechanisms, and to understand the developmental consequences of their operation, makes the early twenty-first century an exciting time to be a developmental cell biologist. It is also a hopeful time for those people suffering the ravages of dreadful diseases, such as cancer, and for those awaiting the promise of regenerative medicine.

1 Historical Perspective

Cell biology involves the investigation of the molecular basis of how cells work, for example, how cells move, how organelles are moved within cells, and how genes are regulated. Developmental biology is concerned with understanding how these processes are integrated to make the organism; for example, how cells move collectively to accomplish gastrulation; how cells are polarized as part of cell differentiation; and how cell communication affects the origin of the different cell types within close proximity to each other so as to create an organ composed of functionally integrated tissues. As the twenty-first century moves into its second decade, the boundaries between biochemistry, molecular biology, genetics, cell biology, and developmental biology are beginning to be diminished.

1.1 Origins of Cell Biology

The cell theory of Schwann and Schleiden (1838/1839) is a scientific milestone of the nineteenth century. Subsequently, Schwann and Muller (Johannes Muller, German physiologist and comparative anatomist) recognized the implications of the cell theory for human pathology, while Rudolph Virchow laid the foundation for cellular pathology during the mid-nineteenth century, replacing the long-lived "humoral pathology" concept of disease. The realization that organisms are composed of cells, and the development of improved microscopy, made the studies of early developmental biology possible. During the last quarter of the nineteenth century, the fundamental details of mitosis and meiosis were elucidated at the level of the light microscope.

1.2 Origins of Developmental Biology

The nineteenth century – a century which was pregnant with advances in cytology (cell biology) and embryology (developmental biology) – initially witnessed the laying of the foundations of descriptive and comparative embryology, and ultimately provided the beginnings of experimental embryology.

Karl Ernst von Baer (1792–1876), an Estonian naturalist and pioneer embryologist, is considered by some to be the father of modern embryology. He was the first to describe the mammalian egg (in 1827), and became the first to trace the egg to the embryo; indeed, his *Uber Entwickelungsgeschichte der Tiere* (1828) became a standard text of embryology. Also regarded as the father of comparative embryology, von Baer was the first to make embryology truly comparative and, together with Pander (Christian Pander, Russian zoologist, regarded as a founder of the science of embryology), he formulated the germ layer concept as a structural fact for vertebrate embryology.

Wilhelm Roux (1850–1924), a German embryologist, is considered to be a pioneer in (and by some, the founder of) experimental embryology. On the basis of his experimental results, Roux concluded that eggs are self-differentiating rather than being driven by external conditions

1.3 Relationship between Cell and Developmental Biology

By the dawn of the twentieth century, cell biology, and developmental biology – although not generally referred to as such at the time – were burgeoning disciplines of biology. Early in the twentieth century, Edmund Beecher Wilson (1856–1939), an American zoologist who specialized in cytology and embryology, wrote *The Cell in Development and Heredity*, third edition (1925). This book is correctly considered to be a classic in, at least, the fields of cytology (cell biology) and embryology (developmental biology).

According to Gilbert [1], the experiments of Hans Spemann (1869–1941; a German zoologist and 1935 recipient of the Nobel Prize in Physiology or Medicine for his discovery of the organizer effect in embryonic development) and his students framed the questions that experimental embryologists continued to ask for most of the twentieth century.

In the editorial of the article, "Where Cell Biology and Developmental Biology Meet," in the inaugural issue of the journal *Developmental Cell* [vol. **1** (1): 1, (2001)], Siegel and Sweet proposed that it really did seem as though we were beginning to understand how cells function, both individually and as parts of multicellular organisms, and that there was a growing interaction and overlap between cell biology and developmental biology.

During the second half of the twentieth century, the merging of cell and developmental biology was reflected in the titles of books and journals (e.g., *Annual Review of Cell Biology* became *Annual Review of Cell and Developmental Biology*, *In Vitro* became *In Vitro Cell and Developmental Biology* and, in 2001, Cell Press established a new journal, *Developmental Cell*).

Just as one endeavored to explain cellular phenomena at the molecular level, during the twentieth century it became increasingly clear that development needed to be explained at the cellular level. Attempts to explain the mechanisms and control of cell proliferation, cytodifferentiation, embryo patterning, and morphogenesis necessitated a molecular understanding (see Ref. [2]).

2 Cell Activities Underlying Development

2.1 Intracellular Signal Transduction

Developmentally important cellular activities include cell division and differential gene expression; in a multicellular organism, these cellular activities are under the control of the cell's environment, including signals from other cells. Although some of these signals (e.g., steroid hormones) are nonpolar and are able to pass through the plasma membrane of the cell, other signals are polar (e.g., the hormone epinephrine), and attach to receptors in the plasma membrane. In order to convey information carried by the signal from the cell surface to the interior of the cell (e.g., the nucleus), where the information is acted upon, the cell makes use of cascades of chemical reactions that constitute the intracellular signal transduction pathways.

2.1.1 Receptors

Receptor proteins are capable of binding other molecules or ions such that, as a result of the binding, they can influence cellular activity; examples include insulin receptors at the cell surface and progesterone receptors in the cytoplasm. A *morphogen* is a substance that can direct the differentiation of cells since, along a gradient of morphogenetic substances, cells will respond differently at different morphogen concentrations. A *morphogen receptor gradient* is a gradient of those molecules that recognize the morphogen; the expression of the *Brachyury* and *goosecoid* genes has been correlated with the number of activin receptors on each cell that binds activin.

2.1.2 Intracellular Signal Transduction

Signal transduction is the conversion of a signal from one form to another form; for example, converting a cAMP signal to a kinase signal as an intracellular signal transduction pathway progresses into the interior of a cell.

The receptor tyrosine kinase (RTK) signal transduction pathway is an intracellular signal transduction pathway that is important in development, and was one of the first pathways to unite various areas of developmental biology. In the migrating neural crest cells of humans and mice, the RTK pathway is important in activating the microphthalmia transcription factor (Mitf) to produce pigment cells.

Homologous signal transduction pathways are composed of homologous proteins arranged in a homologous manner. Although homologous pathways form the basic infrastructure of development, the targets of these pathways may differ among organisms. For example, the dorsal-cactus pathway used in *Drosophila* for specifying dorsal–ventral polarity is also used by the mammalian immune system to activate inflammatory proteins. When homologous developmental pathways are used for the same function in both protostomes and deuterostomes, they are said to have "deep homology." Homologous signal transduction pathways illustrate the conservation of molecular mechanisms across taxons.

2.1.3 Transcription Factors

A *transcription factor* is a molecule that attaches to DNA at a gene regulatory site, and by so doing influences the rate of transcription of a specific gene. A transcription factor is also a protein that regulates the transcription of genes, often, but not exclusively, by binding to cis-regulatory elements (promoters and enhancers). Transcription factors exist as two general types: (i) basal transcription factors (e.g., TFIID, TFIIA, etc.), which attach to gene promoters; and (ii) cell-specific transcription factors (e.g., Pax6, Mitf, etc.), which bind to gene enhancers. Trans-regulatory factors are usually transcription factors. Four major families of transcription factor have been identified, based on their DNA-binding motifs: homeodomain; basic helix-loop-helix (bHLH); basic leucine zipper (bZip); and zinc finger.

The trans-activating domain of a transcription factor is concerned with activating or suppressing the gene's transcription. A *transcription initiation complex* is, collectively, the RNA polymerase and associated basal transcription factors that attach to a gene promoter. A *battery* is a group of genes regulated by the same transcription factor.

Zygotic genes are those genes expressed by the embryonic genome; that is, the embryo's genes as opposed to maternal genes expressed in the oocyte by the maternal genome. In early *Drosophila* development, most of the zygotic genes, which are first activated along the anteroposterior and dorsoventral axes set up by maternal genes, encode transcription factors, which then activate more zygotic genes. *Combinatorial regulation* is the control of gene transcription by two or more transcription factors; the spatial patterns of gene expression are often delimited by the combined action of transcription factors.

Historically, the bicoid protein gradient provided the first reliable evidence for the existence of the morphogen gradients that had been postulated to control pattern formation. The bicoid protein, which is a transcription factor that acts as a morphogen, switches on certain zygotic genes at different threshold concentrations, thus initiating a new pattern of gene expression along the axis. The bicoid protein is a member of the homeodomain family of transcriptional activators, and activates the *hunchback* gene by binding to regulatory sites within the promoter region.

2.2 Cell Signaling

Signaling between and among cells is necessary for the order (i.e., the organism) to emerge from the potential chaos, which is multicellularity. Signaling between cells of the anterior pituitary gland and the gonads orchestrates the development of gametes, and signaling between cells may result in the type of self-inflicted cell death known as *apoptosis*.

2.2.1 Juxtacrine Signaling

Juxtacrine signaling is a mode of cell–cell communication in which signaling molecules are retained on the surface of the signaling cell and interact with receptor proteins on adjacent cell surfaces. An example is the interaction between the Bride of Sevenless protein and its receptor Sevenless, in fly eye photoreceptor specification (see Sections 2.2.2, 2.2.3, and 2.2.4).

2.2.2 Autocrine Signaling

Autocrine signaling is a mode of cell–cell communication in which signaling molecules (autocrine factors) attach to receptors on the same cell that produced them. An example is the explosive proliferation of placental cytotrophoblast cells in response to platelet-derived growth factor (PDGF), which these cells themselves produce.

2.2.3 Paracrine Signaling

Paracrine signaling is a mode of cell–cell communication in which signaling molecules (paracrine factors) act as local mediators, and only affect cells in the immediate environment of the signaling cell. Paracrine factors are the protein molecules used in paracrine signaling; *these factors are the inducing factors of the classical experimental embryologists.* Most paracrine factors fall into one of four major families: Hedgehog; Wnt; transforming growth factor-β (TGFβ); and fibroblast growth factor (FGF).

2.2.4 Endocrine Signaling

Endocrine signaling is a mode of cell–cell communication in which signaling molecules (endocrine factors) are released into the circulatory system and may affect cells (target cells) that are some distance from the signaling cell; an example is the effect of anterior pituitary gland hormones on the gonads. Endocrine factors are molecules (hormones) that function in endocrine signaling; examples include estrogens, testosterone, progesterone, and polypeptide hormones such as the gonadotropins. Endocrine mimics (endocrine disruptors) are exogenous substances that act like hormones in the endocrine system and disrupt the functions of endogenous hormones (e.g., DDT, PCBs, bisphenol A, and phthalates).

2.2.5 Growth Factors

Growth factors are extracellular polypeptide signaling molecules that can regulate proliferation, differentiation, cell survival, and cell death. The FGFs constitute a family of protein growth factors which were first identified as mitogens for fibroblasts in tissue culture, and which stimulate the proliferation of many cell types, inhibit the differentiation of various types of stem cell, and act as inductive signals in embryonic development. They may also play a key role in the induction of the ventral mesoderm in *Xenopus* embryos, where FGF is secreted by the vegetal cells. Apoptosis may be initiated by a withdrawal of growth factors from the cell, or by an active response to a signal.

2.3 Cell–Cell Interactions

Cells may interact one with another in a variety of ways. For example, in addition to them interacting through cell signaling of the juxtacrine, autocrine, paracrine, and endocrine varieties, they may make physical contact with each other through a variety of cell junctions, referred to as "tight junctions," "gap junctions," and "adherens junctions/desmosomes."

2.3.1 Cell Junctions

Cell junctions are specialized regions of contact between the cells, that generally fall into three categories: tight junctions; adherens junctions; and gap junctions:

- *Tight junctions* are specialized contacts that are formed between cells, thereby establishing partitions between the isolated compartments of the body. During the compaction of early mammalian embryos, tight junctions are formed between the cells of the trophoblast; these cells, with their tight junctions, seal the blastocyst cavity off from the embryo's environment.
- *Gap junctions* are specialized contacts that are formed between cells, so as to establish a cytoplasmic continuity between the cells. Cells with these junctions rapidly communicate with each other. During the compaction of early mammalian embryos, gap junctions form between the cells of the inner cell mass (ICM).
- *Adherens junctions/desmosomes* are a type of cell junction that join one epithelial cell to another and provide structural integrity to an epithelium; subsequently, hemidesmosomes join the epithelial cells to the basal lamina, which is a specialized extracellular matrix (ECM) on which epithelial cells reside.

2.3.2 Cell Signals

A pathway in the context of cell–cell signaling consists of the components required for the sending, receiving, and transduction of a signal, including one or more ligands, membrane-associated receptors, intracellular signal transducers, and (depending on the type of pathway) transcription factors.

2.4 Cell–Matrix Interaction

During morphogenesis, growing, quiescent, and dying cells may all be found within the same microenvironment. The

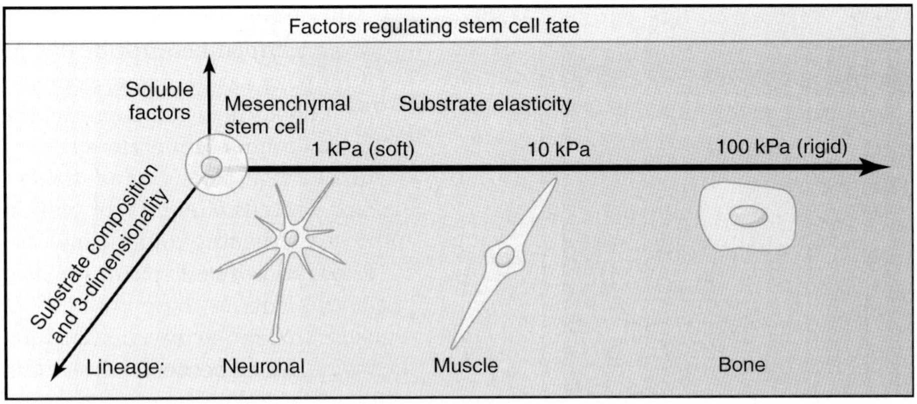

Fig. 1 Effect of substrate elasticity on stem cell fate. Reproduced with permission from Ref. [4]; © 2006, Springer-Verlag GmbH.

normal stem cell biology also depends on the availability of the stem cell niche. The ECM serves as a component of the cell's microenvironment; in fact, it has been shown that the tension exerted by a stem cell's ECM may determine the nature of its differentiation. Engler *et al.* [3] examined the effect of matrix elasticity on the differentiation of human mesenchymal stem cells, and showed that soft matrices would favor the differentiation of mesenchymal stem cells into neuronal-like cells. In contrast, a moderate elasticity would promote a myogenic differentiation, while a rigid matrix stimulated osteogenic differentiation (see Fig. 1).

2.4.1 **Integrin Signaling**

Cells reside in a protein network – the ECM – which they secrete into the extracellular space and which exerts a profound control over cells. The effects of the matrix are primarily mediated by integrins, a family of cell-surface receptors that attach cells to the matrix and mediate the mechanical and chemical signals emanating from the matrix. The integrins span the plasma membrane, bind the fibronectin of the ECM, and provide anchorage sites for the actin microfilaments of the cytoskeleton; that is, they integrate the extracellular and intracellular scaffolds. Most integrins are capable of recognizing several ECM proteins, while the individual ECM proteins (e.g., fibronectin, laminins, collagens, and vitronectin) may bind to several integrins. The integrins are able to signal through the plasma membrane in either direction; an extracellular binding activity of the integrins is regulated from the inside of the cell, whereas binding of the ECM elicits signals that are transmitted into the cell.

In order to survive, adherent cells must be anchored to an appropriate ECM; indeed, depending partly on the signals from the matrix, they can either proliferate or exit the cell cycle and become differentiated. Such an anchorage requirement is lost in neoplastic cells.

The cytoplasmic domains of integrins always lack enzymatic activity; hence, they are able to transduce signals by associating with adaptor proteins that connect the integrin to the cytoskeleton, cytoplasmic kinases, and transmembrane growth factor receptors. Integrin signaling and the assembly of the cytoskeleton are intimately linked. As integrins bind to the ECM, they become clustered within the plane of the plasma membrane and associate with a cytoskeletal and signaling complex that promotes the assembly of actin filaments. The reorganization of actin filaments into larger stress fibers, in turn, causes further integrin clustering, thus enhancing the matrix binding and organization by integrins in a positive feedback system. As a result, ECM proteins, integrins and cytoskeletal proteins will assemble into aggregates on each side of the membrane. Well-developed aggregates that are detectable using immunofluorescence microscopy are referred to as *focal adhesions* and *ECM contacts*. Thus, integrins can serve as integrators of the ECM and the cytoskeleton, the property for which they were named.

2.4.2 **Cell Movement**

Cell shape changes and cell movements underlie the morphogenetic movements that create the form of the developing organism. Examples of the burgeoning literature on this topic are available from Keller [5], Adler [6], and Hall [7].

Polarized cell movements shape the major features of the vertebrate body plan during development. The head-to-tail body axis of vertebrates is elongated in embryonic stages by "convergent extension" tissue movements, during which the cells intercalate between one another transverse to the elongating body axis to form a narrower, longer, array. Recent discoveries have shown that these polarized cell movements are controlled by homologs of genes that control the polarity of epithelial cells in the developing wing and eye of *Drosophila*.

The actin cytoskeleton mediates a variety of essential biological functions in all eukaryotic cells. In addition to providing a structural framework around which the cell shape and polarity are defined, its dynamic properties provide the driving force for cells to move and to divide. An understanding of the biochemical mechanisms that control the organization of actin is, therefore, a major goal of current cell biology, with implications for both health and disease. In this respect, members of the Rho family of small guanosine triphosphatases (GTPases) have emerged as key regulators of the actin cytoskeleton; furthermore, through their interaction with multiple target proteins, they ensure a coordinated control of other cellular activities, such as gene transcription and adhesion. Recent observations have suggested that members of the Rho GTPase family are key regulatory molecules that link the surface receptors to the organization of the actin cytoskeleton. The JAK/STAT pathway is a central component of the signal transduction network that controls cell proliferation, fate, and movement.

3
Cell Differentiation

With the field of cell differentiation having generated vast amounts of information, two models of cell differentiation will be considered here: (i) the role of retinoblastoma protein (pRB) in differentiation; and (ii) a model of how retinoic acid (RA) and FGF9 act antagonistically to determine germ cell fate.

The active role of pRB in differentiation has been studied in a number of cell lines inducible for differentiation. In each instance, an early event in differentiation was the dephosphorylation of pRB, which correlated with cell cycle arrest in G_0/G_1, which is a prerequisite to enter the differentiation pathway. During the differentiation of muscle cells, pRB accumulates in the nucleus and forms complexes with muscle-forming transcription factors such as MyoD and myogenin, thus preventing pRB rephosphorylation and locking the cell in the differentiated state. However, the inactivation of pRB reverses the differentiated phenotype and allows cells to re-enter the cell cycle.

Ajioka *et al.* [8] have described a new mouse model of retinoblastoma, that bears on the relationship between cell differentiation and the cell cycle. These authors have shown that retinoblastoma is not driven by an uncontrolled expansion of retinal progenitor cells, but rather is the result of cell cycle re-entry and the expansion of differentiated horizontal interneurons in the retina (see Fig. 2).

Recent studies conducted by Bowles *et al.* [10] have provided details regarding how the sex determination of mammalian germ cells occurs during fetal development, and depends on signals from gonadal somatic cells. Previously, it had been established that RA would trigger ovarian germ cells to enter meiosis and thereby commit to oogenesis; in contrast, in the developing testis the enzyme CYP26B1 would degrade RA such that the germ cells would not be induced to enter meiosis. Bowles *et al.*, using both in vitro and in vivo models, showed that FGF9 produced in the fetal testis acted directly on the germ cells to inhibit meiosis; in addition, FGF9 maintained the expression of pluripotency-related genes and up-regulated markers associated with male germ cell fate. Based on these results, it was concluded that two independent and mutually antagonistic pathways involving RA and FGF9 would act in concert to determine mammalian germ cell sexual fate commitment. This supported a model in which the mitosis/meiosis switch would be robustly controlled by both positive and negative regulatory factors (see Fig. 3).

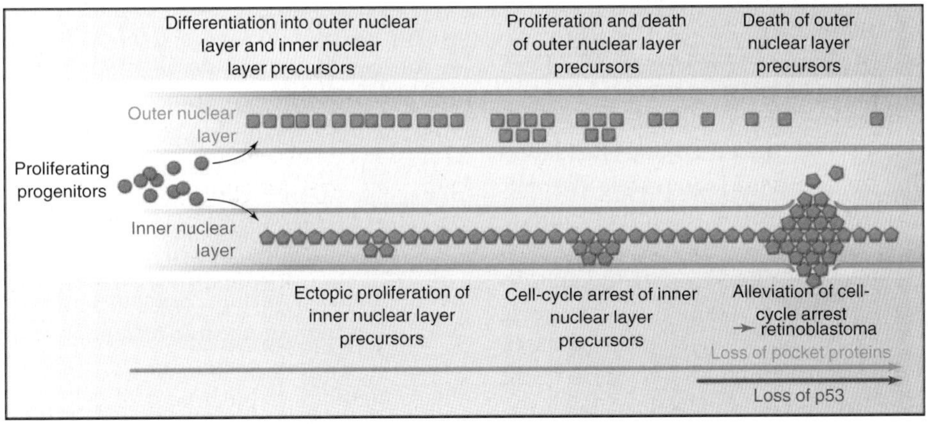

Fig. 2 Retinoblastoma development in mice. The loss of pocket proteins allows proliferating progenitor cells in the retina to differentiate into outer and inner nuclear layer precursor cells. These cells continue to proliferate at least for a while, but then certain retinal cell types (rods, cones, ganglion, and bipolar cells) die, whereas other retinal cells (amacrine, horizontal, and Muller glia cells) survive but become arrested in the cell cycle. A second event that suppresses the p53 pathway may alleviate cell-cycle arrest, leading to proliferation of differentiated cells. Reproduced with permission from Ref. [9]; © 2007, Springer-Verlag GmbH.

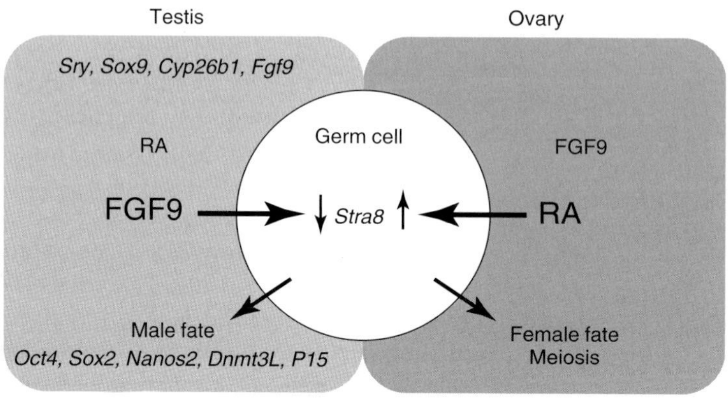

Fig. 3 Retinoic acid (RA) and fibroblast growth factor 9 (FGF9) act antagonistically to determine germ cell fate. The germ cell sexual fate is determined by two signaling molecules produced by the somatic cells of the gonad, FGF9, and RA. *Cyp26b1* and *Fgf9* are highly expressed in the testis, but are downregulated in the ovary. Since CYP26B1 degrades endogenous RA, levels of RA are low in the testis while FGF9 levels are high. In the ovary, RA is not degraded and FGF9 levels are low. Both RA and FGF9 act directly on germ cells (one white cell shown) to upregulate *Stra8* (RA) or to prevent its upregulation (FGF9). *Stra8* expression in gonadal germ cells is essential for entry into meiosis, by an unknown mechanism. FGF9 acts directly on germ cells to antagonize *Stra8* expression, maintain expression of pluripotency markers, *Oct4* and *Sox2*, and to induce male germ cell fate markers, *Nanos2*, *Dnmt3L*, and *P15*. Reproduced with permission from Ref. [10]; © 2010, Springer-Verlag GmbH.

4 The Cell Cycle and Development

The relationship between the cell cycle and cell development is of great interest to developmental cell biologists (also see Bock *et al.* [11]). Since its initial elucidation during the 1950s, the canonical cell cycle, G_1, S, G_2, and M, has undergone intensive examination by cell biologists. Indeed, it has long been known that the cell cycle of embryos undergoing cleavage is not classical, but rather is an abbreviated cycle consisting of S alternating with M. This embryonic (pre-gastrulation) cell cycle, which uses maternal gene products and does not provide time (G_1) for cell growth, results in cleavage of the zygote into an ever-increasing number of blastomeres of progressively decreasing size; in other words, multicellularity is achieved at the expense of cell size. The midblastula transition of early *Xenopus* development has been used to study the conversion of S/M to $G_1/S/G_2/M$. Changes in the cell cycle and its regulation during development represent fundamental topics when considering the relationship between the cell cycle and development.

Xenopus oocytes and early embryos may serve as excellent systems to study the relationship between the cell cycle and development. Typically, *Xenopus* embryos bracketing the midblastula transition provide a developing system for the study of the sequential establishment of cell cycle checkpoints and the connection of signaling pathways to the regulation of cell cycle progression.

The extent to which the cell cycle can be altered, to serve the changing needs of the developing organism, is provided by studying *Drosophila* development. In

Drosophila, the embryonic cells undergo minimal growth between divisions; rather, the key event during this developmental phase is *patterning*. As the development progresses to the larval phase, however, the developmental objective changes, and so too does the nature of the cell cycle. The major objective of *Drosophila* larval tissues is cell growth rather than cell proliferation; hence, the larval cells undergo repeated rounds of endoreplication, an effort that may be designed to increase the total gene dosage in a given cell in an attempt to support the drastically increased cell size. The imaginal disc cells in the developing larva must undergo both growth and patterning, and exhibit a "classical" $G_1/S/G_2/M$ cell cycle, with a growth-sensitive cell cycle checkpoint at the G_1/S boundary and a patterning checkpoint at the G_2/M boundary. The developmental program thereby neatly adjusts the cell cycle to promote patterning without growth, growth without patterning, or growth and patterning simultaneously, as each is required for correct development.

5 Organogenesis

Differentiated groups of cells constitute the tissues and organs (also see Arnone [12]), with each organ having a characteristic structure and function that emerges during development of the embryo and so must be studied on an individual basis. The practical applications of a general understanding of organogenesis will include the repairing and/or replacement of patients' compromised organs, most likely starting with the patient's own stem cells.

The process of organogenesis has a prerequisite step, namely the specification of spatially defined regulatory domains that promote the differentiation programs. This commonplace mechanism also enables the recognition and definition of regulatory fields as discrete territories of specific gene activities.

Selector genes control the formation and identity of the various fields. Field-specific genes are a special class of selector genes that have the unique property of directing the formation of complex specialized structures, such as organs. Examples of such organ-identity genes include: *Pax6/eyeless*, which is required for eye formation in *Drosophila* imaginal discs, and *Pit 1*, which, together with *Gata2*, controls pituitary differentiation.

Recently, Fang *et al.* [13] carried out a genome-wide expression analysis of embryonic development, and reported transcriptome profiles of human early embryos covering development during the first third of organogenesis. These authors identified two major categories of genes, displaying gradually reduced or gradually increased expression patterns across this developmental window. The decreasing group appeared to include stemness-specific and differentiation-specific genes that were important for the initiation of organogenesis, whereas the increasing group appeared to be largely differentiation-related and indicative of diverse organ formation (see Fig. 4). Based on these findings, a putative molecular network was devised that may provide a framework for the regulation of early human organogenesis.

6 Stem Cells

A stem cell is a cell that undergoes mitotic cell division to give rise to the same type of cell. At some point, the stem cells leave the

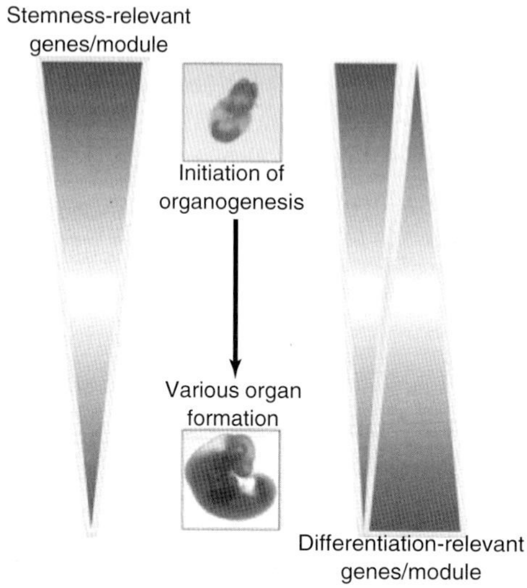

Fig. 4 Transcriptome analysis of early organogenesis in human embryos. Reproduced with permission from Ref. [13]; © 2010, Springer-Verlag GmbH.

pool of mitotically dividing cells to begin a process of cell differentiation. Adult stem cells are, in effect, an embryonic population of cells, continually producing cells that can undergo further development within an adult organism. The path of differentiation that a stem cell descendant enters depends on the molecular milieu in which it resides. It is noteworthy that most stem cells are relatively quiescent.

The term *pluripotent* describes a cell that is capable of giving rise to several of the cell types of an organism; examples include the pluripotent epidermal stem cell and the pluripotent hematopoietic stem cell. Lineage-restricted stem cells are cells that can produce only one type of cell in addition to renewing itself; an example is the burst-forming unit, erythroid (BFU-E). Embryonic germ (EG) cells are primordial germ cell (PGC)-derived cells; these pluripotent stem cells may be produced in culture by treating PGCs with stem cell factor, leukemia inhibition factor, and basic FGF2. Embryonic stem (ES) cells are stem cells derived from the embryo; for example, they may be derived from normal mouse inner mass cells cultured *in vitro*.

A *meristem* is a part of a plant that is characterized by dividing cells; these regions of plants contain stem cell populations that produce cells, some of which go on to differentiate into plant tissues and some of which constantly renew the stem cell population.

The induced pluripotent stem cell (iPSC) is a type of pluripotent stem cell that is artificially derived from a non-pluripotent cell (typically an adult somatic cell) by inducing a "forced" expression of certain genes. The recent creation of iPSCs raises the question of whether the genomes of these cells and those of human ES cells have similar pluripotent states. Guenther *et al.* [14] have suggested that a comparison of the global chromatin structure and the gene expression programs of these two cell types would provide a robust means to assess whether the genomes of these cells have similar pluripotent states. Subsequently, the data reported by Guenther *et al.* showed that: (i) genome-wide maps of nucleosomes with certain histone modifications indicate that there is little difference between ES cells and iPSCs with respect to this criterion; and (ii) gene expression

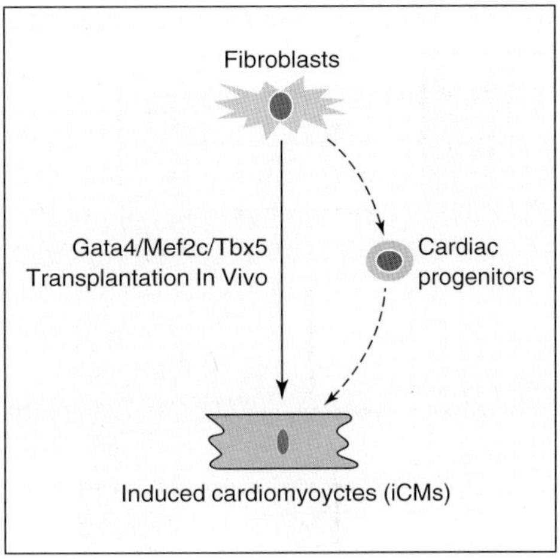

Fig. 5 Direct reprogramming of fibroblasts into functional cardiomyocytes. Reproduced with permission from Ref. [15]; © 2010, Springer-Verlag GmbH.

profiles confirmed that the transcriptional programs of ES cells and iPSCs showed very few consistent differences. This led to the conclusion that these variations did not serve to distinguish ES cells from iPSCs.

The reprogramming of fibroblasts to iPSCs raises the possibility that a somatic cell could be reprogrammed to an alternative differentiated fate, without first becoming a stem/progenitor cell. A large pool of fibroblasts exists in the postnatal heart, yet no single "master regulator" of direct cardiac reprogramming has been identified. Masaki Ieda *et al.* [15] have reported that a combination of three developmental transcription factors (i.e., Gata4, Mef2c, and Tbx5) rapidly and efficiently reprogrammed postnatal cardiac or dermal fibroblasts directly into differentiated cardiomyocyte-like cells. The induced cardiomyocytes expressed cardiac-specific markers, had a global gene expression profile similar to that of cardiomyocytes, and contracted spontaneously. Fibroblasts transplanted into mouse hearts at one day after transduction of the three factors also differentiated into cardiomyocyte-like cells. Based on their findings, Masaki Ieda *et al.* proposed that functional cardiomyocytes could be directly reprogrammed from differentiated somatic cells by defined factors (see Fig. 5). Indeed, they speculated that the reprogramming of endogenous or explanted fibroblasts might provide a source of cardiomyocytes for regenerative approaches.

7 Chimeras

A *chimera* is an organism that consists of parts derived from more than one pair of parents; for example, the original chimera of Greek mythology had a lion's head, a goat's body, and a serpent's tail! In the laboratory, chimeras may be produced – in the case of mammals – by removing the zona pellucida of each embryo and pushing the sticky embryos together. The latter then merge to form a single chimeric embryo which, if transferred into the uterus of a maternal host, may develop into a chimeric organism. As an example, such constructs have been created with different strains of mice; a plant that contains genetically distinct cell types is also referred to as a chimera.

Subsequently, when ES cells derived from the embryo (e.g., from normal mouse inner mass cells cultured *in vitro*) were used to create chimeras, this allowed an engineered gene with a known mutation to be introduced into ES cells, such that any mice that were then bred would have the same mutation in all of their cells.

Beatrice Mintz (b. 1921), an American developmental biologist, is especially well known for developing mouse chimeras. In one of her experiments, early embryos consisting of only a few cells were removed from pregnant mice and placed in close contact with similar cells of genetically unrelated embryos to form a composite that was then implanted in a mouse uterus to create a cellular mosaic. This technique enabled the tissue site of specific genetic diseases to be traced.

A recent report [16] tells of the creation of viable rat-mouse chimeras that developed to term and became fully functional adults, the initial aim being to demonstrate that xenogeneic organ complementation could be achieved. For this, fluorescently labeled

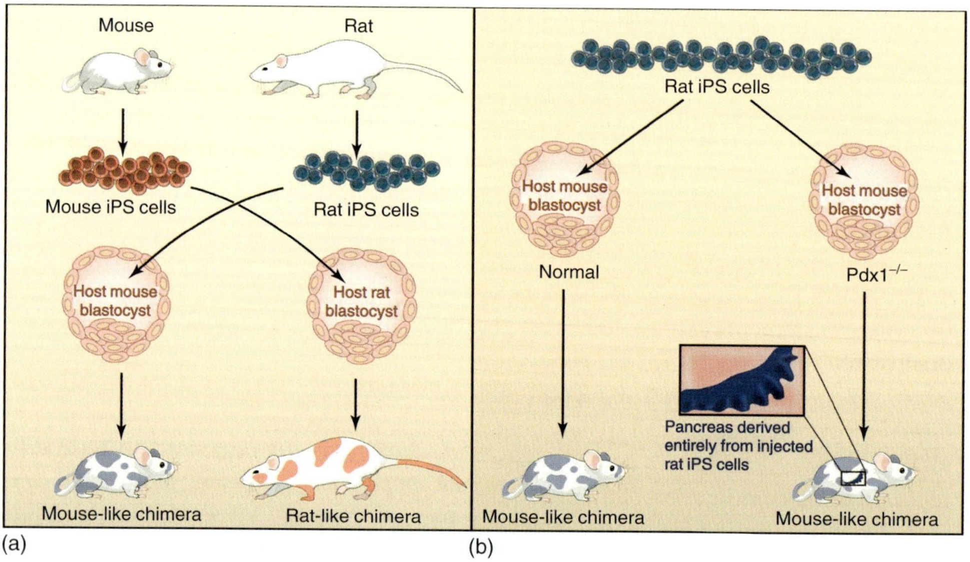

Fig. 6 The generation of rat–mouse chimeras. (a) Induced pluripotent stem (iPS) cells were derived from adult mouse and rat cells and labeled with different fluorescent proteins. Rat (blue) and mouse (red) iPS cells were injected into reciprocal blastocysts (mouse into rat, and vice versa) to produce intergeneric chimeras. From these blastocysts, several chimeras were born and some survived to adulthood. The contribution of injected donor stem cells was observed throughout the body of the host. The size and morphology of the newborn and adult chimeras was determined by the host blastocyst; (b) Fluorescently labeled rat iPS cells (blue) were injected into normal mouse blastocysts (left) or blastocysts lacking the *Pdx1* gene (right), which encodes the transcription factor pancreatic and duodenal homeobox 1 that is required for pancreas development. Chimeras derived from normal or Pdx1-deficient mouse blastocysts showed an extensive contribution of rat cells to all tissues. However, in the Pdx1-deficient chimeras, the entire pancreas was derived from donor rat cells (inset, blue) and was fully functional, including the production of insulin by β islet cells. Reproduced with permission from Ref. [17]; © 2010, Springer-Verlag GmbH.

mouse or rat iPSCs were injected into rat or mouse blastocysts, which were then returned to blastocyst-compatible (to avoid rejection) pseudopregnant females. The donor cells (i.e., the injected iPSCs) were shown to have made major contributions to the tissues of the host. Moreover, it was found that the host blastocyst would control the size of the chimera, as well as imposing additional morphogenetic regulations, as exemplified by the presence or absence of a gallbladder. Ultimately, it was shown that cells derived from rat iPSCs were able to completely rescue a genetic deficiency of the host mouse blastocyst, resulting in a normal functional pancreas, and confirming that xenogeneic organ complementation was indeed achievable. On reviewing these investigations Solter [17] considered that, although xenogeneic organ complementation would be unlikely to serve as a viable strategy for regenerative medicine, the elegant studies of Kobayashi and coworkers would prove invaluable to research groups seeking to understand the biology of stem cells and mammalian development (see Fig. 6).

8 microRNAs (miRNAs)

microRNAs (miRNAs) are post-transcriptional regulators that bind to complementary sequences of target messenger RNA (mRNA) transcripts, usually resulting in gene silencing. The miRNAs are short RNA molecules that, on average, are only 22 nucleotides long. Over 1000 miRNAs may be encoded by the human genome, they may target about 60% of mammalian genes, they are abundant in many human cell types, and each miRNA can repress hundreds of mRNAs. The miRNAs are well conserved in eukaryotic organisms, and are thought to be a vital and evolutionarily ancient component of genetic regulation.

9 *In vitro* Fertilization

An excellent example of the merging of cell biology and developmental biology is the technique of in-vitro fertilization (IVF), which it is estimated has resulted in the birth of four million babies worldwide, beginning with the birth of Louise Brown in 1978. A pioneer in this field is Robert Edwards (b. 1925), who in 2010 received the Nobel Prize in Physiology and Medicine, for the development of human IVF therapy. In order to successfully carry out human IVF, it was first necessary to understand the cell and developmental biology of human gametes, zygotes, and early embryos *in vivo*, and then to apply this knowledge to the in-vitro conditions that allowed for the success of the technique.

References

1 Gilbert, S.F. (2010) *Developmental Biology*, 9th edn, Sinauer Associates, Inc., Sunderland.
2 Dye, F.J. (2004) Developmental Cell Biology, *Encyclopedia of Molecular Cell Biology and Molecular Medicine*, Vol. **3**, 2nd edn, Wiley-VCH Verlag GmbH.
3 Engler, A.J., Sen, S., Sweeney, H.L. (2006) Matrix elasticity directs stem cell lineage specification. *Cell*, **126** (4), 677–689.
4 Even-Ram, S., Artym, V., Yamada, K.M. (2006) Matrix control of stem cell fate. *Cell*, **126** (4), 645–647.
5 Keller, R. (2002) Shaping the vertebrate body plan by polarized embryonic cell movements. *Science*, **298**, 1950–1954.
6 Adler, P.N. (2002) Planar signaling and morphogenesis in *Drosophila*. *Dev. Cell*, **2** (5), 525–535.

7 Hall, A. (1998) Rho GTPases and the actin cytoskeleton. *Science*, **279**, 509–514.

8 Ajioka, I., Martins, R.A.P., Bayazitov, I.T., Donovan, S., Johnson, D.A., Frase, S., Cicero, S., Boyd, K., Zakharenko. S.S., Dyer, M.A. (2007) Differentiated horizontal interneurons clonally expand to form metastatic retinoblastoma in mice. *Cell*, **131** (2), 378–390.

9 Te Riele, H. (2007) Retinoblastoma teaches a new lesson. *Cell*, **131** (2), 227–229.

10 Bowles, J., Feng, C.-W., Spiller, C., Davidson, T.-L., Jackson, A., Koopman, P. (2010) FGF9 suppresses meiosis and promotes male germ cell fate in mice. *Dev. Cell*, **19** (3), 440–449.

11 Bock, G., Cardew, G., Goode, J.A. (Eds) (2001) *The Cell Cycle and Development*, Novartis Foundation Symposium 237, John Wiley & Sons, Ltd, West Sussex.

12 Arnone, M.I. (2002) Bringing order to organogenesis. *Nat. Genet.*, **30** (4), 348–350.

13 Fang, H., Yang, Y., Li, C., Fu, S., Yang, Z., Jin, G., Wang, K., Zhang, J., Jin, Y. (2010) Transcriptome analysis of early organogenesis in human embryos. *Dev. Cell*, **19** (1), 174–184.

14 Guenther, M.G., Frampton, G.M., Soldner, F., Hockemeyer, D., Mitalipova, M., Jaenisch, R., Young, R.A. (2010) Chromatin structure and gene expression programs of human embryonic and induced pluripotent stem cells. *Cell Stem Cell*, **7** (2), 249–257.

15 Ieda, M., Fu, J.-D., Delgado-Olguin, P., Vedantham, V., Hayashi, Y., Bruneau, B.G., Srivastava, D. (2010) Direct reprogramming of fibroblasts into functional cardiomyocytes by defined factors. *Cell*, **142** (3), 375–386.

16 Kobayashi, T., Yamaguchi, T., Hamanaka, S., Kato-Itoh, M., Yamazaki, Y., Ibata, M., Sato, H., Lee, Y.-S., Usui, J.-I., Knisely, A.S., Hirabayashi, M., Nakauchi, H. (2010) Generation of rat pancreas in mouse by interspecific blastocyst injection of pluripotent stem cells. *Cell*, **142** (5), 787–799.

17 Solter, D. (2010) Viable rat-mouse chimeras: where do we go from here? *Cell*, **142** (5), 676–678.

3 Principles and Applications of Embryogenomics

Vincent VanBuren
Texas A & M Health Science Center College of Medicine, Systems Biology and Translational Medicine, 702 SW HK Dodgen Loop, Medical Research Building, Temple, TX 76504, USA

Systems Biology: Advances in Molecular Biology and Medicine, First Edition. Edited by Robert A. Meyers.

Keywords

BAC
Bacterial artificial chromosome.

cDNA
Complementary DNA; a DNA that corresponds to all or part of a gene transcript.

DIGE (or 2-D DIGE)
Difference gel electrophoresis (or two-dimensional difference gel electrophoresis): a type of electrophoresis where two protein samples are run together in the same 2-D gel, such that each sample is labeled with a different fluorescent dye.

DNA Microarray
Sometimes also called a "*gene chip*"; spots of DNA are placed in a 2-D array on a surface, typically either a glass surface or a nylon surface, so that a nucleotide sequence labeled for detection may be hybridized to the DNA on the surface, and then quantified.

Embryo
After fertilization of an ovum, the resultant zygote and subsequent stages until birth are collectively considered "embryonic." In human and other larger animals, the later stage embryo is called a *fetus*.

Genome
The entire nucleotide sequence of an organism's chromosomes.

Genomics
A field of study that addresses biological questions at the scale of the genome or transcriptome.

Hybridization
The process of annealing complementary nucleotide sequences, as performed in a variety of techniques, including DNA microarrays, SAGE, ISH, and RT-PCR.

ISH
In situ hybridization, a technique used to characterize the localization of mRNA in cells and tissues by visualizing (using microscopy) labeled probes that hybridize with RNA in the cell.

Next-generation sequencing
New sequencing technologies, such as 454, Solexa, and SOLiD, that use non-Sanger methods of sequencing that achieve a rate of data acquisition that is hundreds of times what a Sanger-type capillary sequencer can produce.

PCR
Polymerase chain reaction, a method using known probes flanking a nucleotide sequence of interest and cycles of polymerase action to amplify, or make many copies of, that sequence of interest.

Q-PCR
Real-time, quantitative PCR (Q-PCR); a variation of PCR that scales the amplified nucleotide sequence of interest to the original frequency of that sequence, thereby allowing an accurate measure of the original abundance.

SAGE
Serial analysis of gene expression; a technique that estimates the amount of mRNA expression by ligating small pieces, or tags, of the entire set of expressed transcripts, determining the sequence of this ligated sequence, and counting the occurrence of each tag.

Transcriptome
The collection of sequences and abundances of all transcripts, or RNAs, uniquely collected from a cell type, tissue, organ, or whole organism.

Embryogenomics, or developmental genomics, is defined as a systematic analysis of embryo development with large-scale approaches. This systematic analysis permits consideration of the system as a whole by keeping whole-genome sequences and a complete gene catalog in perspective. By merging genomics and developmental biology, embryogenomics contributes to both disciplines: large-scale methodologies and genome-wide concepts are contributed to developmental biology, and the materials and knowledge of genes specifically expressed in embryos are contributed to genomics. In this chapter the history and recent strides in this effort are reviewed, and discussions included of the genome projects, complementary DNA (cDNA) library construction, and large-scale experiments and analysis for model organisms that have the greatest impact on the field of embryology. Emphasis will be placed on progress in mouse embryogenomics, as this provides the basis of discussion for what is arguably the most accessible model organism with a developmental plan that has high similarity to that of humans. Finally, the status of current technologies as they relate to embryogenomics, the obstacles to overcome, and the future aims for this new field are discussed. Large-scale approaches to developmental biology require some special considerations in this discussion, as compared to approaches using samples from adult organisms. This is due to the relatively small size of tissue samples and the temporal-spatial migration and proliferation of cells in embryonic tissues of interest.

1 Introduction

One of the fundamental questions in developmental biology is how the expression of individual genes is regulated in specific cell types, and how the differential expression and action of genes drive the development of complex organisms from simpler forms (e.g., fertilized eggs in animals) [1]. During the 1960s and early 1970s, hybridization kinetic analyses of early embryos and many different cell types was used to examine global transcriptomes, and this established that different groups of genes are indeed actively transcribed in each cell type [2, 3]. The majority of these RNA molecules, however, could not be identified at that time. The advent of molecular cloning during the mid-1970s made it possible to isolate and characterize individual genes that play crucial roles in the development of organisms [4, 5]. This "one-gene-at-a-time" approach has become the standard of molecular developmental biology, and forms the foundation of the current understanding of development. When using such an approach, however, it is difficult to provide a global view of genome-wide gene actions. Simultaneous expression analyses of hundreds of genes during development were started during the early 1980s with hybridization measurements from complementary DNA (cDNA) clones spotted on membranes. The extension of such analyses to entire genomes began to be formulated during the early 1990s.

The earlier approaches both global aspects of gene expression, as in the 1960s to the 1970s, and individual gene aspects of gene expression since the 1970s. However, owing to the much larger scale of whole-transcriptome/whole-genome studies, the development of new technologies was inevitable. The Human Genome Project, which was started in the late 1980s revolutionized molecular biology by developing and incorporating large-scale technologies into biology [6]. Massive handling and the analysis of DNA clones, aided by robotics and informatics, are for the first time being brought into biology. As a consequence, embryogenomics has emerged at the interface between genomics and developmental biology.

Genomics offers three main conceptual frameworks to developmental biology:

- Large-scale approaches offer investigators the ability to screen a large number of candidate molecules simultaneously, to determine which candidates are of interest for further study of the phenomenon of interest. Large-scale efforts in molecular biology offer a cost-effective method of providing comprehensive resources and data sets. The development of permanent, reusable resources (e.g., cDNA clone archives), once and for all establish a strong infrastructure, thus increasing the efficiency of development research. These reusable resources for large-scale approaches offer a huge advantage, and have increased the pace of research over previous techniques that allowed only a very limited subset of molecules to be screened in a single experiment. Paradoxically, this advantage also brings the problem of having more data available from one experiment than would previously have been obtained from hundreds or even thousands of experiments prior to the advent of technologies such as DNA microarrays. With this tidal wave of data comes the need to improve analytical and data-visualization techniques, so that the results can be fully analyzed and understood.
- While the cost-effectiveness of comprehensive coverage provides a central motivation for large-scale efforts, an ongoing pursuit in the use of large-scale approaches in molecular biology is motivated in part by the desire to move from using large-scale endeavors as merely screening tools (data) to synthesizing these results with those obtained with other techniques, so as to build sophisticated models (knowledge). Embryogenomics poses questions at a large scale compared to how questions are framed with traditional molecular biology techniques. Integrative approaches to large-scale questions have the best hope of providing cohesive answers. While questions aimed at a larger scale have the merits of efficiency and a broadened perspective, the real advantage of embryogenomics is the ability to begin considering the combinatorial actions of molecules in the specification of developmental processes. This interface of embryogenomics with systems biology approaches allows a re-evaluation of the one-gene-at-a-time paradigm to development research.
- The evolutionary aspects of developmental biology will be explored through comparative genomics. The completion or draft completion of many projects to sequence entire genomes has ushered in the so-called “postgenomic era for

molecular biology." Comparisons of the genome sequences of different species have provided opportunities to identify orthologous genes and conserved regulatory sequences. The emerging view is that molecular mechanisms executed by orthologous genes operate similarly in different species.

Genomics approaches also benefit from the interface with embryogenomics because of the experimental knowledge and materials offered by embryogenomics approaches. While available genome sequences play a central role in large-scale approaches (Fig. 1), much of the present understanding of transcript expression and gene structure within the genome derives from cDNA collections, where transcripts or partial transcripts may be assembled into consensuses and mapped to the genome. The construction of these cDNA libraries is normally performed with adult tissues because they are easier to obtain. Consequently, cDNAs uniquely expressed in embryonic tissues will not be represented in these libraries, and the inference of gene structure from transcripts could not be achieved for these embryo-specific sequences. Efforts to build libraries from embryonic tissues have contributed to the ability of genomics approaches to identify gene structures, both for the genes actually represented by these libraries and for others through computational predictions using these unique transcripts as part of their training set.

As the above definition of embryogenomics is somewhat broad, the following discussions will be focused primarily on the organisms and large-scale approaches that are most aptly suited for investigation of development at the level of the transcriptome.

2 Approaches

2.1 Overview

While large-scale methods offer many insights previously inaccessible by traditional experiments, many problems remain for those conducting investigations by using these methods, including questions of how to design experiments, how to analyze the data, and how to achieve reproducibility between different experiments and investigators.

2.1.1 Experimental Design and the Null Hypothesis Test

A correct experimental design is a critical aspect of any experiment, and the scale of the experiment magnifies the importance of design. Although the analysis of microarray data has, historically, been the subject of much discussion and controversy, several currently popular techniques of analysis may be effective if the experiments are correctly designed. Some difficulties may arise in selecting a design strategy, because no one design is suitable for all types of experiment, or for all practical considerations, such as the availability of resources. Some of the design problems that might arise include: the biological questions being asked (or which comparisons are the most important to consider); whether to use a universal reference sample; the number of hybridizations that can be performed; sample availability; technical variability; biological variability; and the desired compatibility with future experiments.

Although many standard practices have been adopted for the analysis of microarray data to assess differential gene expression, there is a source of confusion as to how

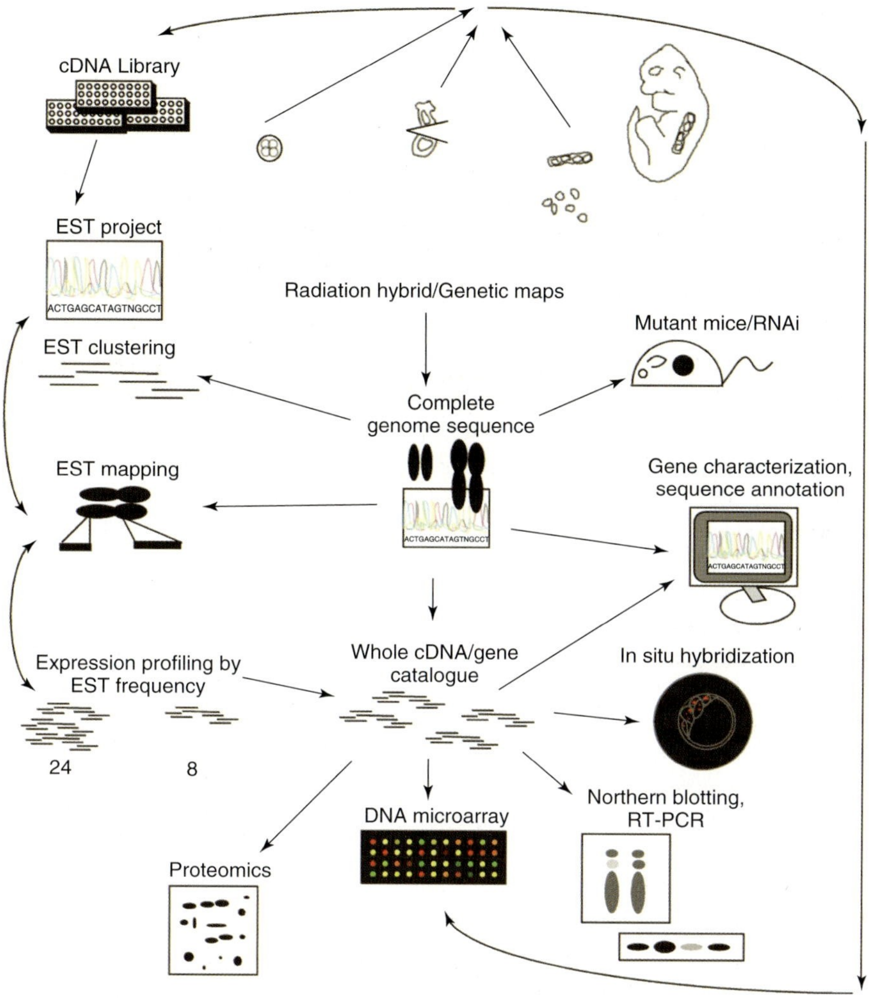

Fig. 1 An approximation of how knowledge gained from technologies used in embryogenomics supports the use of other technologies. Just as the genome provides a central framework for the biology of an organism, the sequence of a genome provides a resource of central importance to the field of embryogenomics.

microarray results should be interpreted, namely the null hypothesis statistical test. In this case, a typical experiment compares untreated samples used as "controls" with samples that have undergone some experimental treatment. Occasionally, the investigators may infer that a list of "significant" genes from this process are the genes that changed in response to the treatment, whereas the other genes are judged to have remained unchanged. Yet, this is epistemologically wrong. As Cohen and Abelson have pointed out [7, 8], and others before them, there are two problems with this line of thinking: (i) it should be predicted that there will always be a difference between samples when the null hypothesis test is used; and (ii) that a significant result in which the null hypothesis test is rejected merely provides the direction (sign) of the difference. Thus, in the context of microarray experiments (or similar high-throughput experiments), it is better to think of the null hypothesis test as producing a result where the answer given is that the gene expression for a particular gene was higher in the treated sample, was lower in the treated sample, or "was too close to call." In the latter case, the null hypothesis is "retained" rather than "accepted." For example, if an experiment is performed with an array of 40 000 gene features, and the null hypothesis is rejected for 1000 (thus provisionally giving the sign of the difference for those genes), the reason that such a result is obtained is because these genes had a combination of effect size and a low variance that produced an appreciable deviation from the null hypothesis. It does not indicate that the other 39 000 genes did not change; only that, given their relatively lower effect size or higher variance (or a combination thereof), it is impossible to make a determination regarding the direction of the difference.

Whilst, at present, there is no ready replacement for the null hypothesis test, Cohen has offered some suggestions for augmenting its use that include the employment of confidence intervals, using data visualization techniques, and performing a meta-analysis. These ideas are explored further in Sect. 2.6.

2.1.2 Gene Coverage and Analysis

A more complete coverage of genes by microarrays can be expected to provide a better foundation for global assessments and the reconstruction of biochemical pathways. One obstacle here is that transcripts may be of low abundance, and thus be missed when seeking genes that are significantly differentially expressed and measurable above background intensities. Without addressing the problem of measuring transcripts with low abundance, a complete assessment of the transcriptome cannot be achieved.

Subsequently, new technologies have been developed to increase gene coverage, while other technologies are emerging to increase the sensitivity for genes expressed with low abundance. These include high-density microarrays with exon-level coverage, and new non-Sanger sequencing technologies that can be employed for measuring gene expression, respectively.

2.1.3 Sensitivity and Specificity

The amplification of target sequences continues to be used in instances where target samples are rare, such as with mammalian oocytes. Previously, some concern has been expressed that amplification may skew the apparent abundance of transcripts, especially when multiple rounds of amplification are used. However, many research groups have addressed this problem, and with only a few rounds of linear amplification the reproducibility was found to be good.

The sensitivity of microarray measurements remains one of the greatest technical concerns when acquiring comprehensive data. One way of approaching the problem of low-abundance transcripts with measured intensities that are not significantly different from background levels, would be to use a better probe design that would provide a means of increasing the measured signal intensity without corrupting the data. A knowledge of how the probe qualities might affect the measurements would also provide the ability to grapple with the problem of deriving absolute abundance estimates from these measures. The current practice with expression microarrays generally yields information only about the *relative* changes in expression for the same gene between two different samples. To reconstruct genetic/biochemical networks from transcript data, however, a knowledge of the *absolute* abundance of transcripts would be invaluable. Although several attempts at making a conversion from measured fluorescence intensity have produced useful results, these early efforts have not been applied widely to the diverse commercial and custom platforms used throughout the scientific community.

2.1.4 Spatial Resolution

The application of DNA microarrays can also address the question of *where* transcription occurs, to the extent that tissues or cells of interest may be microdissected or otherwise separated from other tissues and cells before examination with a microarray. A combination of laser capture microscopy and microarray analysis appears to offer a solution to this problem, although in order to obtain a sufficient spatial resolution a prohibitively large number of arrays must be used. This requires the use of *in-situ* hybridization (ISH) techniques, as described below.

2.1.5 Complementary DNA (cDNA) Projects

Expressed sequence tags (ESTs) are short sequence reads from either the 5′ or 3′ end of cDNA. The construction of cDNA libraries, and their large-scale analysis, forms part of the core of embryogenomics [9–13]. The ability to build a gene catalog from the fundamental gene identification and gene structure information provided by EST projects is a prerequisite for most of the expression profiling techniques discussed in Sect. 2.2.

The ESTs provide gene signatures that help to define gene structure by alignment with the genome, and they also allow for an evaluation of the frequency of randomly selected cDNA clones within a library. The cDNA clones from which ESTs are derived are useful for functional studies of a particular gene; alternatively, they may be used as part of a collection in large-scale efforts, such as spotting on cDNA microarrays or in large-scale ISH. Early efforts at cDNA library construction and EST sequencing typically resulted in clones with relatively short inserts (ca. 1.5 kb average insert size) with relatively short EST sequence reads (100–500 bp). These truncated sequences are useful as unique tags for functional studies, and also for defining gene boundaries. Nonetheless, the demand for more complete sequences and gene definitions have driven the development of technologies to increase clone insert length and sequence read length, and there is also much interest in collecting cDNA clones with full-length inserts. Although several cDNA projects were started during the early 1990s, many of these early efforts resulted in the construction of libraries from adult tissues, and resulted in the assembly of gene indices that mainly represented adult organisms. As expression of some genes is likely to be specific to certain developmental stages or tissues, and this practice of constructing cDNA libraries from adult tissues results in the exclusion of uniquely expressed developmental genes from gene indices.

2.2 Large-Scale Analysis of Gene Expression at the Transcriptome Level

Profiling the expression of genes at a large scale provides valuable insights into which genes are important to, or have unique expression patterns in, a particular cell type, or which genes have enriched or depleted expression in response to a given cue.

Several issues, all of which concern the complexity of biochemical and gene regulatory networks *in vivo* and *in vitro*, have confounded efforts to analyze expression data and to reconstruct gene regulatory networks from that analysis:

- Transcript expression and protein activity sometimes do not correlate.
- Although information may be readily gathered about how a single gene varies expression between samples, the finding of absolute expression values, so that the expression levels of different genes may be compared, is not yet a standard practice.
- Comparisons of different cell types may lead to erroneous conclusions because of heterogeneity of some sample collections and volumetric differences between cell types.
- Even when the experimental conditions are well described (see Minimal Information About a Microarray Experiment; MIAME standard compliance, below), it is not yet clear how the results from different experimental platforms and methods may be compared.

The use of genomics approaches such as microarrays forces questions to be asked, or hypotheses to be framed, at a different scale or purpose. While genomics approaches may not always provide consistent answers or yield the biological "truth," the same is true for traditional biological methods, which have in the past certainly provided some mistaken conclusions. Just as integrative approaches are used to independently verify traditional experiments, the same is necessary for microarrays and other high-throughput approaches, and this should not be of any surprise. Conversely, these high-throughput approaches are also not a panacea for biological discovery, but instead represent a useful tool to pose some very broad questions that cannot otherwise be asked.

In the following sections, some of the primary techniques used to determine expression profiles without spatial resolution will be discussed, followed by an outline of recent efforts at large-scale ISH. Although the latter approach offers a spatial resolution of the transcripts, most approaches have a poor quantitative resolution. Hence, ISH will complement the quantitative techniques of expression profiling.

2.2.1 EST Frequency

Standard (not normalized or subtracted) EST libraries are thought to be random samplings of the transcriptome from the cell type, or of the tissue from which it was derived. As such, it has been proposed by Meyerowitz that EST frequencies correlate with the relative abundance of each transcript in the sample [14]. The EST sequences are obtained from cDNA libraries, which are transcript-encoding sequences or sequence fragments that have been cloned into plasmids and then cultured in a biological host (Fig. 2a). The primary consideration for the construction of cDNA libraries is to identify and analyze gene structures, to make these cDNA clone sequences available for spotting onto cDNA microarrays, or to make these clones available for downstream functional studies. The availability of EST sequences for the analysis of EST frequencies is an additional benefit of the cDNA projects, which have the primary goal of providing sequenced cDNA clones as reagents for molecular biology investigations [15–18].

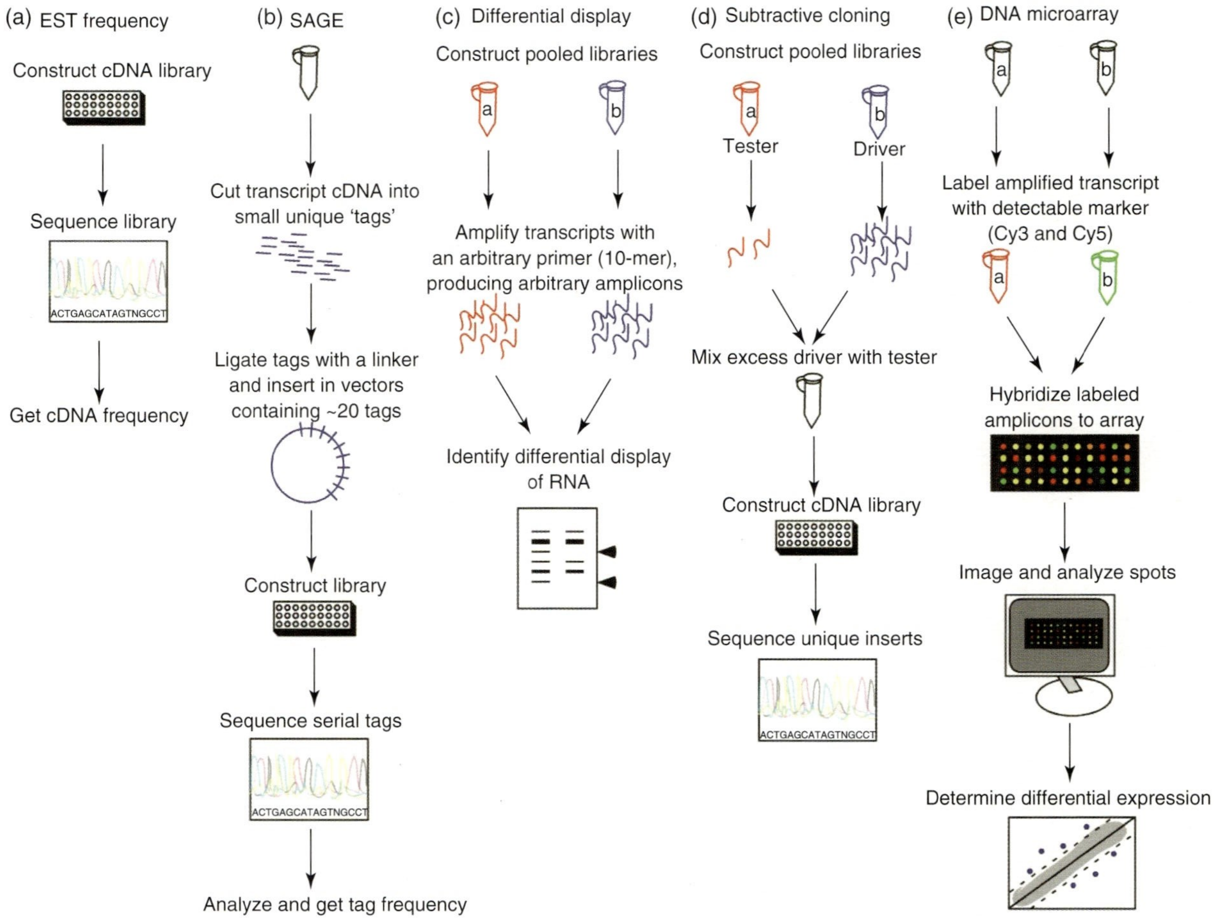

Fig. 2 An overview of some methodologies important to embryogenomics research.

2.2.2 Subtractive Cloning

One strategy for cloning transcripts that are uniquely expressed in one of a pair of transcriptomes is to use "subtractive cloning," which removes transcripts common to the pair [19]. Initially, the investigator will select the transcriptome in which novel transcripts are being sought (this is termed the *tester*), after which the transcriptome will be chosen that will be used to subtract non-unique clones (this is termed the *driver*). By using modern polymerase chain reaction (PCR) methods, double-stranded cDNA from the cells or tissues of interest is created, linkers are ligated to the ends of the cDNA, and the cDNA pools are amplified using PCR. After amplification, the linkers are removed and unique linkers are ligated to the tester pool; the tester pool is then hybridized to a large excess of the driver. Tester–tester hybrids (where the cDNA is unique to the tester pool) are then amplified using PCR (Fig. 2c). The so-called representational difference analysis (RDA) of cDNA employs two rounds of hybridization, since tester–tester hybrids in the first phase of hybridization may arise as a result of random annealing events and thus may not necessarily be unique to the tester pool.

2.2.3 Differential Display

An alternative PCR technique that may be used to identify unique transcripts is that of differential display [19], or "RNA fingerprinting," which amplifies the transcriptome by first using reverse-transcriptase and an oligo-dT primer (or another arbitrary primer) to produce cDNA from the transcriptome; the arbitrary transcript fragments are then amplified using a poly-dT primer and another arbitrary primer (usually a 10-mer). Any differences in transcriptomes may be evaluated by electrophoresis of the two amplified pools (Fig. 2d).

2.2.4 DNA Microarrays

During recent years, DNA microarray technology has led the way in the development of applications and knowledge-seeking strategies in large-scale biology. Typically, microarrays provide a "snapshot" of gene expression for thousands of transcripts, in simultaneous fashion.

Currently, three types of microarray are used: (i) cDNAs spotted onto either a glass slide or a nylon membrane; (ii) 25-mer oligonucleotides synthesized by the lithographic technique (Affymetrix); and (iii) 50–80-mer oligonucleotides synthesized *in situ* (e.g., Agilent) or spotted onto either a glass slide or a nylon membrane. The transcripts are labeled with Cy3 and Cy5 flourophores for two-channel glass arrays, and with the ^{32}P radioisotope for one-channel nylon membranes. When transcripts are collected from rare sample tissues, such as preimplantation embryos, they are often linear-amplified using an *in vitro* RNA polymerase reaction and then hybridized to the microarray. The intensity of the spot is used as a means of quantifying the transcript abundance relative to the abundance of the same transcript from a different sample. This may be measured in the second channel on the same two-channel array, or from a different array when using one-channel arrays (Fig. 2e).

2.2.5 Next-Generation Sequencing

The early genome sequencing techniques used with multicellular eukaryotes employed large sequencing centers involving hundreds of Sanger-type capillary sequencers, which in turn required numerous operators [20]. Yet today, new non-Sanger-type sequencing methods have been developed that can produce sequence data at rates that are many hundred-fold higher than a Sanger-type capillary sequencer [20]. Although these newer technologies cannot yet produce read lengths equivalent to Sanger-type capillary sequencers, new breakthroughs continue to increase the read lengths attainable. These technologies, which include 454, Solexa, and SOLiD, allow for the rapid sequencing of small genomes with a low abundance of low-complexity regions. This makes the use of sequencing for expression profiling more appealing, and may be combined with serial analysis of gene expression (SAGE) (see below) in emerging technologies for rapid expression profiling.

2.2.6 SAGE

SAGE, which offers an alternate strategy for quantifying gene expression [19, 21], is essentially an accelerated version of determining the EST frequency from an EST sequencing project. In this case, expressed sequences are cut with restriction enzymes to produce small unique "tags" that represent each expressed sequence, and which are then ligated together with linker sequences into vectors containing approximately 20 tags, separated by the linker sequences. The sequencing of these vectors allows the frequency of the expressed sequences to be determined in a high-throughput capacity, when compared to normal EST sequencing efforts (Fig. 2b). Although many plasmid clones must be sequenced for SAGE, these costly sequencing efforts produce only the expression profiles, and not the reagents to be used for other purposes, as in EST projects.

2.2.7 Large-Scale ISH

One of the key strategies when investigating developmental processes is to examine the differential distribution of molecules as development progresses. In accomplishing this feat with regards to transcripts, ISH represents the most direct means of visualizing the tissue-based or subcellular localization of transcripts. Consequently, several high-throughput ISH efforts produced image databases that have been of tremendous value to the scientific community. Indeed, such efforts will undoubtedly be accelerated in the near future, especially with the recent development of robots and other technologies capable of streamlining this laborious technique [22].

The technique of ISH involves the hybridization of detectable probes to the transcripts of fixed tissues, in order to localize gene expression. The technique also permits the characterization of gene expression that is complementary to the quantification of transcript abundance provided by DNA microarrays. As many of the tissues used in the microarray studies are heterogeneous, being composed of several cell types, a cell-type-specific enriched expression may be missed in microarray studies, due mainly to dilution within the heterogeneous tissue. In addition, some tissues are difficult or even (at least at present) impossible to separate by microdissection for applications in array experiments. Hence, the ability to localize expression patterns to particular cell types should provide many insights that are currently inaccessible to array experiments. Moreover, as the cellular compartmentalization of transcripts is also generally not accessible to microarray studies, ISH may also offer a means of exploring expression patterns at a higher resolution (this often provides insights that are invaluable for the elucidation of gene function). Probes with detectable tags attached may be created either by the synthesis of an oligonucleotide from a known sequence, or by using

PCR-amplified cDNA sequences of the target sequences.

New gene expression atlases of human chromosome 21 orthologs in mouse at various developmental stages have been described in two recent reports [23, 24]. In the first report, an ISH robot and an automated microscope were used to analyze approximately 6500 sections, thereby generating an atlas which covered embryonic day (E) 9.5, E10.5, E14.5, and adult tissues [23]. Of the 178 confirmed human genes on chromosome 21, expression was recorded for 98% of the 161 mouse orthologs. In the second report (from the HSA21 expression map initiative), the mapping of 158 mouse homologs of human chromosome 21 confirmed the presence of genes (258 human chromosome 21 genes, 168 identified mouse cognates) [24]. In the latter case, emphasis was placed on mapping whole-mount E9.5 embryos, although a subset of clones was mapped at other stages.

Each of these ISH mapping efforts produced a freely available, high-quality online database for retrieval of the ISH images and the supporting data and methods. Moreover, with many new gene patterns of expression having been discovered in each project, these studies have clearly provided important resources for future studies on any of the genes with transcripts mapped in these investigations.

2.3 Large-Scale Analysis of Gene Expression at the Proteome Level

2.3.1 2-D Gels and DIGE

Proteomics – the large-scale study of proteins – began in 1975, when the introduction of two-dimensional (2-D) electrophoresis allowed for the separation and quantification of proteins by isoelectric focusing (based on pI) in the first dimension, and by size in the second dimension. Typically, 2-D gels have been used for the creation of extensive databases, for cataloging spots, and for measuring the differential expression of proteins. Whereas 2-D gels have become recognized as providing the "classical" definition of proteomics, large-scale yeast two-hybrid experiments, and the protein identification and detection of post-translational modification using mass spectrometry (MS) have since also become members of this illustrious field of analysis [25]. Today, a number of computational techniques have emerged for deciphering metabolic networks, including a combination of yeast two-hybrid data with microarray studies to help define network topology. Additionally, efforts to create "protein-chips" that are analogous to DNA microarray have been targeted at supplanting the utility of 2-D gels.

Recently, the use of 2-D gels has been greatly improved by the development of difference gel electrophoresis (DIGE) [26]. In the past, a major problem encountered with traditional 2-D gels was that the many sources of variance as to how the two gels would run made it difficult to make comparisons between spots on any two gels, and this greatly limited their use. In the case of DIGE, however, the aim is to remove those sources of variance by running two samples on the same gel; this is achieved by labeling each sample with a different fluorescent dye (commonly Cy5 and Cy3).

2.3.2 Mass Spectrometry

One very promising new technique for monitoring gene expression involves the application of matrix-assisted laser desorption ionization mass spectrometry

(MALDI MS) to tissue samples that have been either laser capture microdissected (LCM) or contact blotted. In this way, it is possible to characterize the spatial distribution of a large number of proteins in tissue slices of about 12 μm thickness. Unlike immunohistochemistry and fluorescence microscopy, which permit only one or a few measurements of proteins to be made in the tissues or cells, MALDI MS provides a much larger scale of analysis, with hundreds of protein measurements being made simultaneously. In particular, the use of MALDI MS with contact-blotted tissues has offered a spatial resolution of approximately 50 μm, producing striking results in sections of mouse brain and human glioblastomas.

2.4 Development and Evolution: Comparative Genomics

Recently, it has become clear that many organisms share a similar mechanism of development with, in many cases, those genes that are conserved among species conveying a similar function. For example, following transfection, some transcription factors from one species may function in the cells of other species.

One approach to identifying genes and regulatory sequences that are essential for the execution of a developmental plan is to compare the genomes of closely related species, and to seek any highly conserved sequences. The converse would be to seek genes that are uniquely expressed in a developmental profile, and this may yield insights into how that profile generates the unique features of its developmental plan. The two forerunners for this type of analysis were the comparisons between *Drosophila* and *Anopheles* [27], and between *Mus musculus* and *Homo sapiens* [28–30]. Until now, comparative genomics approaches have mainly been performed at the global level, without any published accounts detailing the comparison of genes with important developmental roles between genomes. The analysis of the current draft genomes, with a focus on embryology, has some beginnings in the supplemental materials accompanying the published comparative genomics of the *Drosophila* and *Anopheles* genomes, though little else can be found in the literature that approaches this problem. This scarcity of "comparative embryogenomics" approaches is most likely a consequence of the present poor understanding of large regulatory networks for the specification of development. It is difficult to make comprehensive comparisons of the molecular details of development in different organisms without the corresponding network topologies to use as a conceptual framework of comparison.

It has been proposed by Meyerowitz that a comparison of plant and animal developmental processes offers the only true comparative studies of development, as studies between animal species compare only developmental plans that are similar by descent [14]. For example, segmental identity is established in *Drosophila* by spatially specific transcriptional activation of the *Hox* master regulatory genes. Whilst vertebrates exhibit similar processes in the developing brain, *Arabidopsis* has master regulatory genes, mainly of the MADS family [these are named after the founding family members Mcm1, Agamous, Deficiens, and serum response factor (SRF)], but these sequences have no similarity to the *HOX* genes. A second pattern specification in animals, the dorsal–ventral specification, has a logical analog in plants, although most of the animal genes involved in this specification are absent from plants. Gurken, the receptor tyrosine kinases, Ras protein, and Pipe, all of which are implemented in the dorsal–ventral specification in *Drosophila*, have no homologs in plants. Adaxial–abaxial axes in the leaves of *Arabidopsis*, the final example of pattern formation given by Meyerowitz, involves an activation of the YABBY (named after the Australian fresh water crayfish, "yabby," as a link to Crabs claw, the founding member of the gene family) family of transcription factors by the GARP (glycoprotein A repetitions predominant) family of transcription factors. No animal genomes are known to encode members of the YABBY or GARP protein families, which again demonstrates a lack of molecular similarity between patterning in these two kingdoms. Plants and animals have a very different molecular specification for pattern formation, although the logic of that patterning has a striking similarity. This relationship presents an opportunity for "true" comparative developmental studies, where the focus of study will be how similar logical patterning is achieved with a different molecular toolset.

Now that complete genomes for human and mouse are available, comparative genomics approaches should yield valuable insights into development. The mouse and human genomes have 217 syntenic blocks that range from 303 kb to 64.9 Mb in size, and which are composed of 90.2% of the human genome, and 93.3% of the mouse genome. While much comparative analysis was completed with the publication of the mouse genome, no special attention was given to any of the developmental processes in this analysis.

2.5 Functional Genomics/Large-Scale Manipulation of Expression

2.5.1 Mutagenesis

The availability of mutant model organisms underpins much of the present understanding of development and disease. In the case of the mouse, less than 5% of the genes have been mutated, although several large-scale mutagenesis projects are currently under way [31]. Most of these mutations were created as targeted mutagenesis by homologous recombination in embryonic stem (ES) cells. Whereas, targeted mutations offer a precise selection of the gene to mutate, the standard strategy causes the creation of a null allele. In fact, it is often preferable to generate point mutations that may have a range of effects, as this provides more knowledge about protein function. Chemical mutagenesis strategies have the advantage that they allow for phenotype screening, which may be preferable to targeting a specific gene; moreover, they can also produce point mutations that have a spectrum of effects on protein function.

Exposure to X-rays, to chlorambucil (CHL), and to ethylnitrosurea (ENU) all lead to the production of random genomic mutations. Typically, X-irradiation produces large mutations and has a low efficiency, while CHL creates large deletions but is more efficient at creating mutations. The advantage of ENU is that it creates point mutations by causing the ethylation of either O or N, with the most frequent base pair changes being A/T → T/A (44%), A/T → G/C (38%), and G/C → A/T (8%), most of which are missense mutations (64%). The general idea of using ENU is to affect the early spermatogonial cells of male mice, which subsequently will produce mutated sperm for the remainder of their lives (following a period of sterility).

2.5.2 Gene Trapping

Gene trapping is a technique that permits reverse genetics (i.e., the determination of what phenotype results from a particular gene) and also provides a mechanism that does not require additional steps to isolate the gene responsible for the phenotype of interest. The overall procedure creates a high-throughput random disruption of genes, with each disrupted gene being simultaneously captured in a cassette with a reporter driven by the endogenous promoter; this cassette may be tagged for an easy identification of the gene sequence insert. This also allows for any phenotype that results from a disruption to be quickly and easily associated with a particular gene. The principal method of gene trapping is to randomly insert a promoterless construct with reporter, selectable marker, and splice site(s) into the genome. The correct integration and splicing of this construct will result in the creation of a fusion protein with one or more exons of the trapped locus; ideally, the native transcript will be truncated and so will be nonfunctional.

Constructs with a splice acceptor (SA) or with both a SA and a splice donor (SD), are the most commonly used gene trap vectors. The murine engrailed-2 and the adenovirus late major transcript SA are both widely used, while the human *BCL-2* gene intron 2/exon 3 SA has been introduced more recently. One potential problem here is that typical reporters are only detectable if the insertion is in-frame, but with gene trapping the insertion of the trapping vector is random. Reporter activity is enhanced by the insertion of an internal ribosomal entry site (IRES) between the SA and the reporter in the gene trap vector. The IRES allows an independent translation of the reporter, such that reporter detection does not require in-frame gene trapping when the gene trap vector contains an IRES [31].

2.5.3 RNAi/siRNA

RNA interference (RNAi) is one of the most exciting technologies to be introduced into molecular biology during recent years [32, 33]. RNAi takes advantage of the cellular mechanisms that carry out the destruction of specified transcripts, and may provide a loss-of-function phenotype, thus providing insights and motivation for further analysis. Before the application of RNAi, gene knockouts and knockdowns could take many months or even years to achieve useful results. Nowadays, the large-scale transient inactivation of individual genes (e.g., in *Caenorhabditis elegans*) can be achieved for thousands of genes (separately) in only a few months (see Sect. 3.1.3).

Briefly, the mechanism of RNAi begins in the cell with the introduction of double-stranded RNA (dsRNA), which mediates interference with the target of interest. Short interfering RNAs (siRNAs), that are 21–23 nt in length, are produced by the cutting action of the enzyme Dicer on dsRNA. The siRNAs are incorporated into the RNAi-induced silencing complex (RISC), a multisubunit protein complex that is able to unwind the siRNA so as to facilitate interaction with the target transcript. The interaction of siRNAs with targets leads to target cleavage by DICER and the liberation of new siRNA as a result; the remaining mRNA target is then degraded by exonucleases. It also seems likely that siRNA can act as a primer for an RNA-dependent RNA polymerase (RdRP), which would allow for the amplification of siRNA sequences. The positive feedback loop provided by an RdRP might explain how a relatively small amount of siRNA can cause the degradation of a larger amount of target, and why RNAi is inheritable in worms.

Although RNAi may be applied to many eukaryotic systems, among the metazoans *C. elegans* is particularly well suited to this approach, as these worms may be fed bacteria transfected with the dsRNA and used as an RNAi stimulus.

2.6 Computational Approaches

The increased scale of experiments, and the subsequent increase in the flow of data into public repositories, has fueled the increasing importance of computational approaches for storing, searching, and analyzing data.

2.6.1 Databases

Databases are essential for storing, organizing, and accessing the large amount of data produced by embryogenomic experiments. For microarray expression data, and for sequence data for DNA, RNA, and proteins, the United States National Center for Biotechnology Information provides an important collection of centralized databases that are easily accessible from any part of the world via the Internet [34]. Likewise, Ensembl provides a similar function at the European Molecular Biology Laboratory [35]. Independent research groups have constructed numerous other Internet-based specialty databases in an effort to complement these centralized efforts by addressing the requirements of certain scientific communities. Many of these efforts are defined by a particular model organism (e.g., WormBase, Mouse Genome Informatics, and the Zebrafish Model Organism Database [36–39]), while others are defined by a group of closely related disease states. Other efforts may reflect a breadth of coverage similar to NCBI or Ensembl, but may focus on a database functionality not covered by those central databases.

2.6.2 Meta-Analyses and Reproducibility

Although many laboratories have produced quality microarray experiments with acceptable reproducibility, the reproducibility of experiments, and of the conclusions drawn between different laboratories, remains a major challenge to the scientific community. Even when research groups conduct experiments to answer the same questions with microarrays of the same construction, the results and conclusions may be very different. The formulation of guidelines for data sharing in the MIAME standard has been an important step towards realizing the goal of reproducibility. Here, "Minimal" is something of a misnomer, as the MIAME standard requires the reporting of all aspects of the microarray experiment, including detailed information regarding the platform, sample collection, and all hybridizations. The standardization of microarray data analysis would also be of benefit to the scientific community, by promoting a common understanding of techniques used, and by allowing for a greater ease of comparing results from different groups. This will be created with time, as the best techniques emerge from the numerous available ideas.

A formal meta-analysis represents a means of combining statistical results from multiple, highly similar studies. One important reason for formalizing these comparisons is that the naïve approach to comparing the results of two microarray experiments directly, for example, does not yield results that are useful for inference.

Two similar microarray experiments may produce very different lists of "significant" genes. As discussed in Sect. 2.1.1, nothing useful can be said about genes for which the null hypothesis is retained. Neither can it be asserted that the expression of such genes is the same for the control and experimental treatment – only that it is "too close to call" the sign of the difference. In other words, if a gene is significantly different on a list produced by one experiment, but absent from a list produced by another experiment, this should not be interpreted as a conflicting result.

2.6.3 Data Visualization

Data visualization often negates the need for a statistical analysis. The visualization of data can often make patterns in the data obvious, whereas such patterns may be indiscernible from a detailed list of the same data. Given the large quantities of data presently streaming into public databases,

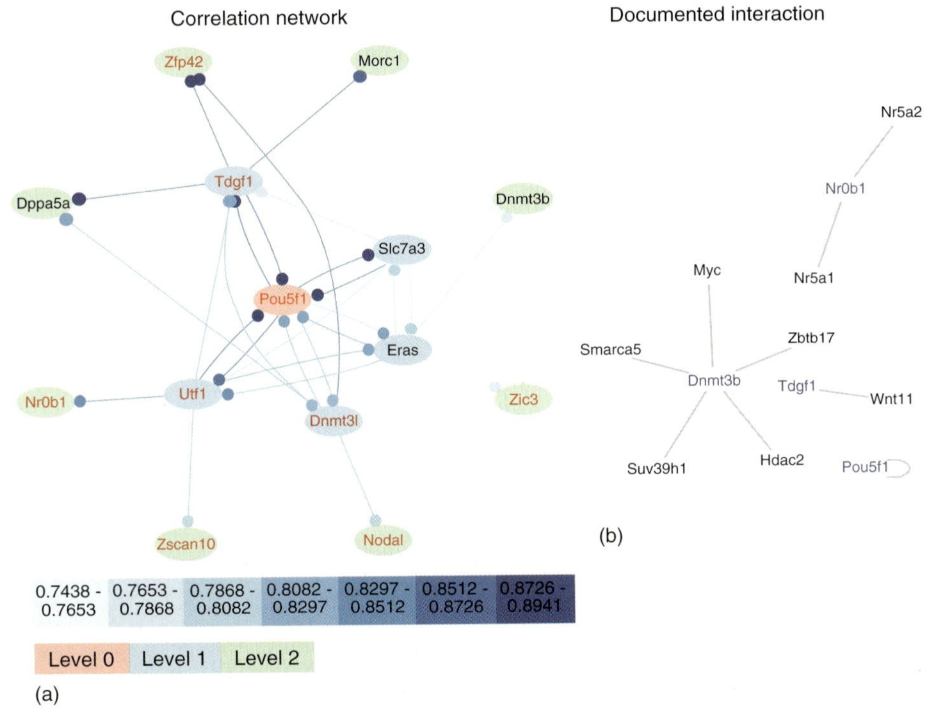

Fig. 3 (a) A representative correlation network drawn from data for more than 2000 microarrays using StarNet (http://vanburenlab.medicine.tamhsc.edu/starnet2.html), and documented interactions involving genes in the correlation network [40, 41]. These networks are the result of a query for the gene Pou5f1 (also known as *Oct4*), where the five highest magnitude correlations are shown, and the network is propagated to two levels of correlations radiating from Pou5f1. The scale bar shows the magnitude of correlation for the edges in the graph; (b) The graph of known interactions is drawn by searching against a database of known interactions for every gene in the correlation network. Genes from the correlation network are shown in blue text in the interaction graph. Visualizations such as this facilitate the construction of new hypotheses for the functional associations between genes.

it is evident that novel visualizations of these data will be an important aspect of emerging efforts to acquire systems level knowledge of transcriptional regulatory networks and metabolic networks.

An integration of a meta-analysis of microarray data with a powerful visualization of database queries is exemplified by StarNet, which draws correlation networks from numerous microarray experiments [40, 41]. StarNet includes coverage of 10 important model organisms, and includes selected developmental data cohorts for mouse, human, rat, and *Drosophila*. All together, over 12 000 arrays are analyzed in StarNet, with over 3600 for human and over 2000 for mouse. An example result from a StarNet query for *Pou5f1*, a gene that is important in the maintenance of pluripotent stems cells, and one of four genes used recently in reprogramming non-stem cells to become stem cells, is shown in Fig. 3. StarNet draws the highest magnitude correlations for a query gene, where the number of top-ranking correlates and the number of levels to propagate from the gene of interest may be specified in the query (the default parameters draw five connections and propagate two levels from the query gene). The interpretation of the resulting correlation network is supported by searching a database of known interactions for all genes in the correlation network, and drawing a separate graph of those results (Fig. 3a), as well as by performing an enrichment test of Gene Ontology terms in the correlation network, relative to the complete set of genes represented on the array platform from which the correlation data were derived (Table 1). This latter approach allows the investigator provisionally to annotate the biological relevance of the group of genes in the correlation network.

2.6.4 Discovery of Transcriptional Regulatory Networks

In addition to providing research groups with powerful data visualization tools, it is a desirable aim of computational approaches to automatically discover regulatory relationships from the data. This is an ambitious aim, and care must be taken not to supplant the investigator's expertise and judgment, but rather to extend it by judiciously automating certain aspects of inference. Methods for achieving this include Bayesian methods of inference, as well as numerous algorithms for inferring functional relationships, using either a single type of data or integrated data from different types of data. These methods may combine rule-based methods with statistical procedures.

3
Model Organisms for Embryogenomics

Model organisms are chosen in order to optimize their suitability for a particular type of investigation and the use of knowledge acquired during the study [42]. The term "suitability" means the accessibility of the organism's physiology to investigation, including considerations such as gestation period, physical properties of the body tissues of interest, and overall effort required in maintenance of the organism. Utility of the knowledge gained refers to the closeness of the model organism's genetics and physiology to the organism (or more generally, organisms) being modeled (e.g., *Homo sapiens*). At this point, the various model organisms studied in embryology using genomics approaches [43] will be discussed, the advantages and disadvantages of the various model

Tab. 1 Gene Ontology (GO) enrichment test for the representative correlation network for Pou5f1 shown in Fig. 3. StarNet automatically performs this test upon retrieval of a specified correlation network [40, 41].

GO term	*GO ID*	*Raw p-value*	*Bonferroni corrected p-value*	*Symbols (Entrez IDs)*
Negative regulation of cell differentiation	GO:0045596	6.48E−07	7.52E−05	Pou5f1 (18999), Nr0b1 (11614), Nodal (18119)
Positive regulation of transcription from RNA polymerase II promoter	GO:0045944	1.08E−06	1.25E−04	Pou5f1 (18999), Tdgf1 (21667), Utf1 (22286), Zic3 (22773), Nodal (18119)
Nuclear heterochromatin	GO:0005720	1.23E−05	1.43E−03	Dnmt3l (54427), Dnmt3b (13436)
Stem cell maintenance	GO:0019827	1.88E−05	2.18E−03	Pou5f1 (18999), Nodal (18119)
Trophectodermal cell differentiation	GO:0001829	3.10E−05	3.60E−03	Pou5f1 (18999), Nodal (18119)
Nucleus	GO:0005634	3.51E−05	4.07E−03	Pou5f1 (18999), Tdgf1 (21667), Utf1 (22286), Dnmt3l (54427), Zfp42 (22702), Morc1 (17450), Nr0b1 (11614), Zscan10 (332221), Dnmt3b (13436), Zic3 (22773)
Heart looping	GO:0001947	3.58E−05	4.15E−03	Zic3 (22773), Nodal (18119)
In utero embryonic development	GO:0001701	5.82E−05	6.75E−03	Tdgf1 (21667), Dnmt3l (54427), Nodal (18119)
Determination of left/right symmetry	GO:0007368	1.90E−04	2.20E−02	Zic3 (22773), Nodal (18119)
Negative regulation of transcription from RNA polymerase II promoter	GO:0000122	2.01E−04	2.33E−02	Pou5f1 (18999), Nr0b1 (11614), Nodal (18119)
Cell fate commitment	GO:0045165	3.05E−04	3.54E−02	Pou5f1 (18999), Nodal (18119)
Transcription	GO:0006350	3.13E−04	3.63E−02	Pou5f1 (18999), Utf1 (22286), Zfp42 (22702), Nr0b1 (11614), Zscan10 (332221), Zic3 (22773)

organisms will be considered, and the current status of genomics approaches to developmental questions reviewed. As each organism discussed here has a rich history as a model for developmental studies, the discussions will necessarily be cursory, with only the salient features of each organism and their place in the study of development in general and in embryogenomics highlighted.

3.1 Non-Mammalian Animals

3.1.1 *Drosophila melanogaster*

The developmental stages of metamorphosis in the fruit fly *Drosophila melanogaster* (hereafter *Drosophila*) are the first developmental plans to be studied with microarray analysis [44–46]. *Drosophila* has a small genome, an extremely short life cycle, and is relatively inexpensive to maintain, which makes this organism appealing for use in scientific investigations. *Drosophila* has been the most powerful system used for genetic analysis, with its ease of mutant screening and the large archives of such mutants. Completion of the *Drosophila* genome has increased the value of this model species further. There is a large body of experimental knowledge about development in *Drosophila*, and over 125 000 cDNA clones from *Drosophila* have been derived from the embryonic stages. The systematic large-scale ISH of 2179 genes has also been reported recently.

3.1.2 *Anopheles gambiae*

The genome sequencing of the malaria mosquito, *Anopheles gambiae* (hereafter *Anopheles*) was published in 2002, thus providing the opportunity to compare the genomes of the two dipteran species *Drosophila* and *Anopheles* [47]. Prior to the availability of this genome sequence, interest in *Anopheles* was primarily focused on the prevention and treatment of malaria. Although there has been interest in the life cycle of *Anopheles*, this has mainly been with the goal of a better understanding of the corresponding life cycle of *Plasmodium falciparum* (the malaria-causing organism that invades erythrocytes), and there are no cDNA libraries with sequences in dbEST (database of expressed sequence tags) that report derivation from *Anopheles* embryonic stages.

A comparative genomics analysis of *Anopheles* and *Drosophila* appeared in a companion report to the *Anopheles* draft genome report [27]. Although most of this analysis emphasized a global comparison, the supplementary materials for this work included a simple analysis of some developmentally important genes. Flybase (*Drosophila*) was queried with keywords related to embryos and development to identify 314 *Drosophila* genes (after manual corrections), and the orthologs in *Anopheles* were identified. This information was placed in a table according to the *Drosophila* gene name, and the number of *Anopheles* orthologs was listed. This procedure showed that most of the characterized *Drosophila* genes listed that are important in development had orthologs in *Anopheles* (88%), whereas approximately 12% of these developmentally important genes appeared to be absent in *Anopheles*. This analysis offers some clues about genes that may drive important differences in the developmental plan of *Drosophila*. The absence of a reciprocal analysis underscores that there is much less gene information available for *Anopheles*.

3.1.3 *Caenorhabditis elegans*

Caenorhabditis elegans has been of profound importance for developmental studies, primarily because it has been possible to follow the developmental lineage and migration of every cell in the organism, from one cell through adulthood.

In addition to the usefulness of *C. elegans* in small-scale functional studies of development, this organism has proved useful beyond its anticipated advantages as a model organism because of the application of RNAi. *C. elegans* may be fed bacteria carrying RNAi vectors, and these vectors can cause the destruction of targeted native transcripts in the worm. Kamath *et al.* have recently performed a comprehensive functional analysis of *C. elegans* using RNAi. By their count, there are 19 427 predicted genes in *C. elegans*, and they inhibited transcript function for 16 757 (about 86%) of this total. A total of 1528 single-gene mutant phenotypes was identified, where a mutant phenotype was defined as reproducible sterility, embryonic or larval lethality, slow postembryonic growth, or any of a number of specifically defined postembryonic defects. These studies also established that there is high reproducibility of RNAi techniques, regardless of methods, and showed there to be a false positive rate of <1%. Kamath *et al.* claimed to have identified more than 70% of all embryonic lethal phenotypes in *C. elegans*. Their efforts have produced both an intellectual contribution and a tremendous resource for the scientific community.

3.1.4 *Xenopus laevis*

Xenopus laevis has been of great value to embryology in the areas of early fate decisions and morphogenesis (with detailed fate maps constructed for blastula and from gastrula through organogenesis), patterning of the vertebrate body plan, and early organogenesis. Moreover, several functional genomics studies have focused on *X. laevis* development [3, 48–54]. When females are injected with human chorionic gonadotropin, they ovulate and lay hundreds of oocytes that may be manipulated or subjected to *in-vitro* fertilization. The early embryos are quite large, and are thus suitable for microinjections and dissections; the embryos are also transparent, which makes the effects of experimental conditions easy to visualize. Recently, *Xenopus tropicalis* has been identified as a model system that retains the advantages of *X. laevis*, but has the additional benefits of a diploid genome that is suitable for genetics study and a shorter generation time, making this species more accessible to multigenerational approaches [55].

3.1.5 *Brachydanio rerio*

The zebrafish *Brachydanio rerio* (hereafter *Danio*) has been a cornerstone model organism for developmental processes, mainly because *Danio* is extremely hardy and easy to raise, a large number of *Danio* mutants have been created by mutagenesis and, most importantly, *Danio* embryos are transparent, thereby facilitating their study by noninvasive observation [42, 48, 56]. Large-scale ISH has been carried out to identify genes specific to the embryonic stages in *Danio*. Currently, more than 34 000 cDNA clones derived from *Danio* embryos are available. Recently, the first cDNA microarray constructed from *Danio* cDNA libraries, including a library derived from three-day-old embryonic heart, has been used to study development and hypoxia [48, 57].

3.1.6 *Strongylocentrotus purpurratus*

The long history of the sea urchin as a developmental model has its beginnings with the discovery by Theodore Boveri that chromosomes contain information for controlling the development of embryos. Of the developmental plans in the metazoans, only the sea urchin endomesoderm specification has been reconstructed from existing experimental knowledge. This tremendous effort combined many years of acquired experimental knowledge with computer simulation to construct what Davidson *et al.* have described as a provisional model for the specification of the endomesoderm in *Strongylocentrotus purpurratus* [58].

3.2 Mammalian

Studies of mammalian systems are expected to offer the most relevant insights into human developmental processes. Here, the initial discussion relates to the use of mammalian model organisms, by examining human, mouse, and rat as systems as a basis for inquiries into developmental questions. As moral and practical considerations severely restrict studies with human tissues, the emphasis of the discussion will relate to the nonhuman mammals, *Mus musculus* and *Rattus norvegicus*, with most attention on *M. musculus*, the preeminent model organism that closely matches human developmental patterns.

3.2.1 *Homo sapiens*

Mammalian species are of the greatest interest to embryonic investigations, because the traditional view is that these species will provide the most relevant model for humans. Ethical concerns limit the availability of human embryonic tissues, and essentially forbid any *in vivo* embryological experimentation. Yet, despite these limitations several large-scale experimental studies have been performed using human embryo tissues, and there is great interest in studies with stem cells derived from human. The human draft genome sequence was first made available in 2001, and the completed human genome was announced in March 2003 [28, 59]. The recent availability of the mouse genome will greatly contribute to an understanding of the regulatory elements and conserved genes important to development, by comparison with the human sequence [29, 30, 60].

3.2.2 *Mus musculus*

The house mouse *M. musculus* is the forerunner of model organisms for use in studies of development and pluripotency. Similarity to human development, a short generation time and a relatively low cost of maintenance each contribute to the notion that *M. musculus* is the optimal model organism for relevant and rapid large-scale developmental investigations [31]. Of the mouse, rat, and human, only mouse has cDNA libraries derived from preimplantation stage embryos. Transcript expression specific to these early embryonic stages may be overlooked in rat and human because of the difficulty in working with and acquiring these rare tissues. Another advantage of the mouse as a model is that the largest number of full-length cDNA clones has been assembled and sequenced for mouse [10, 13, 15–17, 31, 61, 62].

In response to the under-representation of embryonic tissues in cDNA collections, many new libraries have been created from these rare tissues in mouse. The availability of over 115 000 cDNA clones derived from preimplantation tissues in mouse offer a compelling reason to choose a mouse model for study of early development. Sequencing and subsequent analysis has identified many novel or rare transcripts from early embryonic stages.

Given multiple EST sequences from these libraries, an effort to assemble collections of representative transcripts culminated in the construction of the NIA Mouse 15K and NIA Mouse 7.4K cDNA collections [9, 63]. Each of these collections has been distributed worldwide to more than 100 cDNA microarray facilities, and these clones are used in both spotting cDNA microarrays and in downstream functional studies.

The human and mouse genomes together represent valuable tools for comparative studies, revealing important *cis* regulatory elements, definitions of gene structures, and key genes for shared developmental processes. Comparisons of the high-quality mouse and human genomes underscore the usefulness of comparative studies. For example, the Sox family of transcription factors is important to many developmental processes in both mouse and human, and includes the *Sry* gene on the Y chromosome, which specifies testis determination. A comparative analysis of the human and mouse genomes has determined the precise number and ortholog identity of the members of Sox family of transcription factors, thus resolving the nomenclature for this important family of genes [64].

The utility of dsRNA for RNAi in mouse oocytes and preimplantation embryos has been demonstrated [65]. It was shown that dsRNA could be used to prevent gene expression of a GFP (green fluorescent protein) transgene (MmGFP) in heterozygotes where the transgene was parentally derived. The 2-cell expression of the GFP transgene was prevented by the injection of dsRNA MmGFP in the one-cell zygote. The results of these studies confirmed that a strong, specific depletion of target transcripts could be achieved from the 4- to 6-cell stage through the blastocyst stage (it was later shown that hairpin RNA could achieve similar results). For these promising techniques, the main problem is that the injection of the RNAi vector is required in each instance. This difficult procedure currently limits the extent of the application of RNAi in these early embryonic tissues, and poses a challenge to any present notion of making this procedure large scale.

3.2.3 *Rattus norvegicus* and Others

The first genome assembly of *Rattus norvegicus* (rat), was announced by the Rat Genome Sequencing Consortium on 19th June 2003. The rat genome is slightly larger than the mouse genome (2.57 Gb compared to 2.5 Gb). At the time of writing this chapter, details of the assembly, as well as of the comparative genomics of rat with other species, await future publication. The rat is used primarily as a model organism for cardiovascular and/or physiological research, though the availability of the genome sequence may spark interest in comparative genomics approaches towards a better understanding of the common features of genetic programs of development among mammals. Similarly, genome projects for other mammals or vertebrates may also spark interest in large-scale comparative studies of development.

3.3 Plants

Plant embryology enters this discussion because, although plant and animal development and physiology are quite distinct at the molecular level, many of the experimental procedures and analyses share common techniques and strategies for the elucidation of developmental pathways [14]. There is also the additional motivation that comparative studies between plant and animal development may yield evolutionary insights and provide clues about the convergence of some aspects of their developmental plans (see Sect. 2.4).

3.3.1 *Arabidopsis thaliana*

Completion of genome sequencing for the thale cress *A. thaliana* has strengthened the utility of this plant model for comparative approaches to development [66]. *Arabidopsis* has long been of interest for genomics and developmental studies, because of its small genome and short germination and development cycle [66–68]. Among plants, *Arabidopsis* is the canonical species for large-scale studies.

3.3.2 *Zea mays* and Others

Crop plants such as *Zea mays* (corn) and *Oryza sativa* (rice) have been extensively studied because of their importance as food sources for much of the world [68–70]. Genome projects for each of these organisms are currently progressing, and will contribute to the comparative genomics approaches between plants and the more general comparisons of development between kingdoms.

3.4 Suitability of Approaches for Particular Model Organisms Applied to the Study of Development

The traditional view has been that mammalian systems are closest to the human in genome organization and developmental plan; hence, these systems are of the greatest interest as model species. The biggest trade-off here is that viviparity makes mammalian systems much more difficult to work with than egg-laying organisms. In addition, the increased cost and effort involved in working with mammalian systems provides an incentive to work with other organisms that are more suitable for some large-scale efforts. For example, the nematode *C. elegans* has for many years been an apt choice for developmental studies, including large-scale mutagenesis. More recently, however, *C. elegans* was the first metazoan to be studied extensively, by using RNAi to selectively render transcripts nonfunctional [71].

The primary benefits of the mouse as a developmental model are the closeness of the developmental plan to that of the human, and the availability of a large number of cDNA clones throughout the preimplantation stages and later embryo development. By comparison, *C. elegans* offers a rich base of experimental knowledge, a relatively easy application of RNAi gene silencing, a large number of available mutants, and easy maintenance; *Xenopus* offers a replenishable source of oocytes and a large oocyte (ca. 1 mm) that may be collected and manipulated relatively easily; and *Danios* offers a rich base of experimental knowledge, a large number of available mutants, a translucent embryo for easy visualization of development, and an easy maintenance.

4 Conclusions

Embryogenomics presents an exciting new area that continues to grow and attract great interest from the scientific community. Although various difficulties have been encountered with the currently available hardware and analytical technologies, these issues are being constantly and aggressively addressed in an effort to acquire an effective systems biology approach that will not only yield accurate predictions to guide the present understanding of development, but also provide a fertile ground for future investigations.

References

1 Ko, M.S. (2001) Embryogenomics: developmental biology meets genomics. *Trends Biotechnol.*, **19** (12), 511–518.

2 Young, B.D., Harrison, P.R., Gilmour, R.S., Birnie, G.D., Hell, A., Humphries, S., Paul, J. (1974) Kinetic studies of gene frequency. II. Complexity of globin complementary DNA and its hybridization characteristics. *J. Mol. Biol.*, **84** (4), 555–568.

3 Purdom, I., Williamson, R., Birnstiel, M. (1972) Kinetic analysis of the base sequence heterogeneity of RNA molecules by RNA–DNA hybridization. *Results Probl. Cell Differ.*, **3**, 25–36.

4 Marx, J.L. (1976) Molecular cloning: powerful tool for studying genes. *Science*, **191** (4232), 1160–1162.

5 Hershfield, V., Boyer, H.W., Yanofsky, C., Lovett, M.A., Helinski, D.R. (1974) Plasmid ColEl as a molecular vehicle for cloning and amplification of DNA. *Proc. Natl Acad. Sci. USA*, **71** (9), 3455–3459.

6 Lander, E.S., Linton, L.M., Birren, B., Nusbaum, C., Zody, M.C., Baldwin, J., Devon, K., Dewar, K., Doyle, M., FitzHugh, W., Funke, R., Gage, D., Harris, K., Heaford, A., Howland, J., Kann, L., Lehoczky, J., LeVine, R., McEwan, P., McKernan, K., Meldrim, J., Mesirov, J.P., Miranda, C., Morris, W., Naylor, J., Raymond, C., Rosetti, M., Santos, R., Sheridan, A., Sougnez, C., Stange-Thomann, N., Stojanovic, N., Subramanian, A., Wyman, D., Rogers, J., Sulston, J., Ainscough, R., Beck, S., Bentley, D., Burton, J., Clee, C., Carter, N., Coulson, A., Deadman, R., Deloukas, P., Dunham, A., Dunham, I., Durbin, R., French, L., Grafham, D., Gregory, S., Hubbard, T., Humphray, S., Hunt, A., Jones, M., Lloyd, C., McMurray, A., Matthews, L., Mercer, S., Milne, S., Mullikin, J.C., Mungall, A., Plumb, R., Ross, M., Shownkeen, R., Sims, S., Waterston, R.H., Wilson, R.K., Hillier, L.W., McPherson, J.D., Marra, M.A., Mardis, E.R., Fulton, L.A., Chinwalla, A.T., Pepin, K.H., Gish, W.R., Chissoe, S.L., Wendl, M.C., Delehaunty, K.D., Miner, T.L., Delehaunty, A., Kramer, J.B., Cook, L.L., Fulton, R.S., Johnson, D.L., Minx, P.J., Clifton, S.W., Hawkins, T., Branscomb, E., Predki, P., Richardson, P., Wenning, S., Slezak, T., Doggett, N., Cheng, J.F., Olsen, A., Lucas, S., Elkin, C., Uberbacher, E., Frazier, M., Gibbs, R.A., Muzny, D.M., Scherer, S.E., Bouck, J.B., Sodergren, E.J., Worley, K.C., Rives, C.M., Gorrell, J.H., Metzker, M.L., Naylor, S.L., Kucherlapati, R.S., Nelson, D.L., Weinstock, G.M., Sakaki, Y., Fujiyama, A., Hattori, M., Yada, T., Toyoda, A., Itoh, T., Kawagoe, C., Watanabe, H., Totoki, Y., Taylor, T., Weissenbach, J., Heilig, R., Saurin, W., Artiguenave, F., Brottier, P., Bruls, T., Pelletier, E., Robert, C., Wincker, P., Smith, D.R., Doucette-Stamm, L., Rubenfield, M., Weinstock, K., Lee, H.M., Dubois, J., Rosenthal, A., Platzer, M., Nyakatura, G., Taudien, S., Rump, A., Yang, H., Yu, J., Wang, J., Huang, G., Gu, J., Hood, L., Rowen, L., Madan, A., Qin, S., Davis, R.W., Federspiel, N.A., Abola, A.P., Proctor, M.J., Myers, R.M., Schmutz, J., Dickson, M., Grimwood, J., Cox, D.R., Olson, M.V., Kaul, R., Shimizu, N., Kawasaki, K., Minoshima, S., Evans, G.A., Athanasiou, M., Schultz, R., Roe, B.A., Chen, F., Pan, H., Ramser, J., Lehrach, H., Reinhardt, R., McCombie, W.R., de la Bastide, M., Dedhia, N., Blocker, H., Hornischer, K., Nordsiek, G., Agarwala, R., Aravind, L., Bailey, J.A., Bateman, A., Batzoglou, S., Birney, E., Bork, P., Brown, D.G., Burge, C.B., Cerutti, L., Chen, H.C., Church, D., Clamp, M., Copley, R.R., Doerks, T., Eddy, S.R., Eichler, E.E., Furey, T.S., Galagan, J., Gilbert, J.G., Harmon, C., Hayashizaki, Y., Haussler, D., Hermjakob, H., Hokamp, K., Jang, W., Johnson, L.S., Jones, T.A., Kasif, S., Kaspryzk, A., Kennedy, S., Kent, W.J., Kitts, P., Koonin, E.V., Korf, I., Kulp, D., Lancet, D., Lowe, T.M., McLysaght, A., Mikkelsen, T., Moran, J.V., Mulder, N., Pollara, V.J., Ponting, C.P., Schuler, G., Schultz, J., Slater, G., Smit, A.F., Stupka, E., Szustakowski, J., Thierry-Mieg, D., Thierry-Mieg, J., Wagner, L., Wallis, J., Wheeler, R., Williams, A., Wolf, Y.I., Wolfe, K.H., Yang, S.P., Yeh, R.F., Collins, F., Guyer, M.S., Peterson, J., Felsenfeld, A., Wetterstrand, K.A., Patrinos, A., Morgan, M.J., de Jong, P., Catanese, J.J., Osoegawa, K., Shizuya, H., Choi, S., Chen, Y.J. (2001) Initial sequencing and analysis of the human genome. *Nature*, **409** (6822), 860–921.

7 Abelson, R.P. (1995) *Statistics as Principled Argument*, Psychology Press, New York, p. 221.
8 Cohen, J. (1994) The Earth is Round (p <. 05). *Am. Psychol.*, **49** (12), 997–1003.
9 VanBuren, V., Piao, Y., Dudekula, D.B., Qian, Y., Carter, M.G., Martin, P.R., Stagg, C.A., Bassey, U.C., Aiba, K., Hamatani, T., Kargul, G.J., Luo, A.G., Kelso, J., Hide, W., Ko, M.S. (2002) Assembly, verification, and initial annotation of the NIA mouse 7.4K cDNA clone set. *Genome. Res.*, **12** (12), 1999–2003.
10 Strausberg, R.L., Feingold, E.A., Grouse, L.H., Derge, J.G., Klausner, R.D., Collins, F.S., Wagner, L., Shenmen, C.M., Schuler, G.D., Altschul, S.F., Zeeberg, B., Buetow, K.H., Schaefer, C.F., Bhat, N.K., Hopkins, R.F., Jordan, H., Moore, T., Max, S.I., Wang, J., Hsieh, F., Diatchenko, L., Marusina, K., Farmer, A.A., Rubin, G.M., Hong, L., Stapleton, M., Soares, M.B., Bonaldo, M.F., Casavant, T.L., Scheetz, T.E., Brownstein, M.J., Usdin, T.B., Toshiyuki, S., Carninci, P., Prange, C., Raha, S.S., Loquellano, N.A., Peters, G.J., Abramson, R.D., Mullahy, S.J., Bosak, S.A., McEwan, P.J., McKernan, K.J., Malek, J.A., Gunaratne, P.H., Richards, S., Worley, K.C., Hale, S., Garcia, A.M., Gay, L.J., Hulyk, S.W., Villalon, D.K., Muzny, D.M., Sodergren, E.J., Lu, X., Gibbs, R.A., Fahey, J., Helton, E., Ketteman, M., Madan, A., Rodrigues, S., Sanchez, A., Whiting, M., Young, A.C., Shevchenko, Y., Bouffard, G.G., Blakesley, R.W., Touchman, J.W., Green, E.D., Dickson, M.C., Rodriguez, A.C., Grimwood, J., Schmutz, J., Myers, R.M., Butterfield, Y.S., Krzywinski, M.I., Skalska, U., Smailus, D.E., Schnerch, A., Schein, J.E., Jones, S.J., Marra, M.A. (2002) Generation and initial analysis of more than 15,000 full-length human and mouse cDNA sequences. *Proc. Natl Acad. Sci. USA*, **99** (26), 16899–16903.
11 Stapleton, M., Liao, G., Brokstein, P., Hong, L., Carninci, P., Shiraki, T., Hayashizaki, Y., Champe, M., Pacleb, J., Wan, K., Yu, C., Carlson, J., George, R., Celniker, S., Rubin, G.M. (2002) The *Drosophila* gene collection: identification of putative full-length cDNAs for 70% of *D. melanogaster* genes. *Genome. Res.*, **12** (8), 1294–1300.
12 Adams, M.D., Soares, M.B., Kerlavage, A.R., Fields, C., Venter, J.C. (1993) Rapid cDNA sequencing (expressed sequence tags) from a directionally cloned human infant brain cDNA library. *Nat. Genet.*, **4** (4), 373–380.
13 Ko, M.S. (1990) An 'equalized cDNA library' by the reassociation of short double-stranded cDNAs. *Nucleic Acids Res.*, **18** (19), 5705–5711.
14 Meyerowitz, E.M. (2002) Plants compared to animals: the broadest comparative study of development. *Science*, **295** (5559), 1482–1485.
15 Okazaki, Y., Furuno, M., Kasukawa, T., Adachi, J., Bono, H., Kondo, S., Nikaido, I., Osato, N., Saito, R., Suzuki, H., Yamanaka, I., Kiyosawa, H., Yagi, K., Tomaru, Y., Hasegawa, Y., Nogami, A., Schonbach, C., Gojobori, T., Baldarelli, R., Hill, D.P., Bult, C., Hume, D.A., Quackenbush, J., Schriml, L.M., Kanapin, A., Matsuda, H., Batalov, S., Beisel, K.W., Blake, J.A., Bradt, D., Brusic, V., Chothia, C., Corbani, L.E., Cousins, S., Dalla, E., Dragani, T.A., Fletcher, C.F., Forrest, A., Frazer, K.S., Gaasterland, T., Gariboldi, M., Gissi, C., Godzik, A., Gough, J., Grimmond, S., Gustincich, S., Hirokawa, N., Jackson, I.J., Jarvis, E.D., Kanai, A., Kawaji, H., Kawasawa, Y., Kedzierski, R.M., King, B.L., Konagaya, A., Kurochkin, I.V., Lee, Y., Lenhard, B., Lyons, P.A., Maglott, D.R., Maltais, L., Marchionni, L., McKenzie, L., Miki, H., Nagashima, T., Numata, K., Okido, T., Pavan, W.J., Pertea, G., Pesole, G., Petrovsky, N., Pillai, R., Pontius, J.U., Qi, D., Ramachandran, S., Ravasi, T., Reed, J.C., Reed, D.J., Reid, J., Ring, B.Z., Ringwald, M., Sandelin, A., Schneider, C., Semple, C.A., Setou, M., Shimada, K., Sultana, R., Takenaka, Y., Taylor, M.S., Teasdale, R.D., Tomita, M., Verardo, R., Wagner, L., Wahlestedt, C., Wang, Y., Watanabe, Y., Wells, C., Wilming, L.G., Wynshaw-Boris, A., Yanagisawa, M., Yang, I., Yang, L., Yuan, Z., Zavolan, M., Zhu, Y., Zimmer, A., Carninci, P., Hayatsu, N., Hirozane-Kishikawa, T., Konno, H., Nakamura, M., Sakazume, N., Sato, K., Shiraki, T., Waki, K., Kawai, J., Aizawa, K., Arakawa, T., Fukuda, S., Hara, A., Hashizume, W., Imotani, K., Ishii, Y., Itoh, M., Kagawa, I., Miyazaki, A., Sakai, K., Sasaki, D., Shibata, K., Shinagawa, A., Yasunishi, A., Yoshino, M., Waterston, R., Lander, E.S., Rogers, J., Birney, E., Hayashizaki, Y. (2002) Analysis of the mouse

transcriptome based on functional annotation of 60,770 full-length cDNAs. *Nature*, **420** (6915), 563–573.

16 Tanaka, T.S., Jaradat, S.A., Lim, M.K., Kargul, G.J., Wang, X., Grahovac, M.J., Pantano, S., Sano, Y., Piao, Y., Nagaraja, R., Doi, H., Wood, W.H. III, Becker, K.G., Ko, M.S. (2000) Genome-wide expression profiling of mid-gestation placenta and embryo using a 15,000 mouse developmental cDNA microarray. *Proc. Natl Acad. Sci. USA*, **97** (16), 9127–9132.

17 Ko, M.S., Kitchen, J.R., Wang, X., Threat, T.A., Hasegawa, A., Sun, T., Grahovac, M.J., Kargul, G.J., Lim, M.K., Cui, Y., Sano, Y., Tanaka, T., Liang, Y., Mason, S., Paonessa, P.D., Sauls, A.D., DePalma, G.E., Sharara, R., Rowe, L.B., Eppig, J., Morrell, C., Doi, H. (2000) Large-scale cDNA analysis reveals phased gene expression patterns during preimplantation mouse development. *Development*, **127** (8), 1737–1749.

18 Lennon, G.G., Lehrach, H. (1991) Hybridization analyses of arrayed cDNA libraries. *Trends Genet.*, **7** (10), 314–317.

19 Carulli, J.P., Artinger, M., Swain, P.M., Root, C.D., Chee, L., Tulig, C., Guerin, J., Osborne, M., Stein, G., Lian, J., Lomedico, P.T. (1998) High throughput analysis of differential gene expression. *J. Cell. Biochem. Suppl.*, **30-31**, 286–296.

20 Reis-Filho, J.S. (2009) Next-generation sequencing. *Breast Cancer Res.*, **11** (Suppl. 3), S12.

21 Menssen, A., Hermeking, H. (2002) Characterization of the c-MYC-regulated transcriptome by SAGE: identification and analysis of c-MYC target genes. *Proc. Natl Acad. Sci. USA*, **99** (9), 6274–6279.

22 Komiya, T., Tanigawa, Y., Hirohashi, S. (1997) A large-scale in situ hybridization system using an equalized cDNA library. *Anal. Biochem.*, **254** (1), 23–30.

23 Reymond, A., Marigo, V., Yaylaoglu, M.B., Leoni, A., Ucla, C., Scamuffa, N., Caccioppoli, C., Dermitzakis, E.T., Lyle, R., Banfi, S., Eichele, G., Antonarakis, S.E., Ballabio, A. (2002) Human chromosome 21 gene expression atlas in the mouse. *Nature*, **420** (6915), 582–586.

24 Gitton, Y., Dahmane, N., Baik, S., Ruiz i Altaba, A., Neidhardt, L., Scholze, M., Herrmann, B.G., Kahlem, P., Benkahla, A., Schrinner, S., Yildirimman, R., Herwig, R., Lehrach, H., Yaspo, M.L. (2002) A gene expression map of human chromosome 21 orthologues in the mouse. *Nature*, **420** (6915), 586–590.

25 Pandey, A., Mann, M. (2000) Proteomics to study genes and genomes. *Nature*, **405** (6788), 837–846.

26 Liley, K.S., Friedman, D.B. (2004) All about DIGE: quantification technology for differential-display 2D-gel proteomics. *Expert Rev. Proteomics*, **97**, 248–257.

27 Zdobnov, E.M., von Mering, C., Letunic, I., Torrents, D., Suyama, M., Copley, R.R., Christophides, G.K., Thomasova, D., Holt, R.A., Subramanian, G.M., Mueller, H.M., Dimopoulos, G., Law, J.H., Wells, M.A., Birney, E., Charlab, R., Halpern, A.L., Kokoza, E., Kraft, C.L., Lai, Z., Lewis, S., Louis, C., Barillas-Mury, C., Nusskern, D., Rubin, G.M., Salzberg, S.L., Sutton, G.G., Topalis, P., Wides, R., Wincker, P., Yandell, M., Collins, F.H., Ribeiro, J., Gelbart, W.M., Kafatos, F.C., Bork, P. (2002) Comparative genome and proteome analysis of *Anopheles gambiae* and *Drosophila melanogaster*. *Science*, **298** (5591), 149–159.

28 Saluja, S.K., Kohane, I. (2003) Localization and characterization of mouse-human alignments within the human genome. Does evolutionary conservation suggest functional importance? Proceedings, AMIA Annual Symposium, 18–23 November 2003, Vancouver, British Columbia, p. 994.

29 Pennacchio, L.A., Baroukh, N., Rubin, E.M. (2003) Human-mouse comparative genomics: successes and failures to reveal functional regions of the human genome. *Cold Spring Harbor Symp. Quant. Biol.*, **68**, 303–309.

30 Mural, R.J., Adams, M.D., Myers, E.W., Smith, H.O., Miklos, G.L., Wides, R., Halpern, A., Li, P.W., Sutton, G.G., Nadeau, J., Salzberg, S.L., Holt, R.A., Kodira, C.D., Lu, F., Chen, L., Deng, Z., Evangelista, C.C., Gan, W., Heiman, T.J., Li, J., Li, Z., Merkulov, G.V., Milshina, N.V., Naik, A.K., Qi, R., Shue, B.C., Wang, A., Wang, J., Wang, X., Yan, X., Ye, J., Yooseph, S., Zhao, Q., Zheng, L., Zhu, S.C., Biddick, K., Bolanos, R., Delcher, A.L., Dew, I.M., Fasulo, D., Flanigan, M.J., Huson, D.H., Kravitz, S.A., Miller, J.R., Mobarry, C.M., Reinert, K., Remington, K.A., Zhang, Q., Zheng, X.H., Nusskern, D.R., Lai, Z., Lei,

Y., Zhong, W., Yao, A., Guan, P., Ji, R.R., Gu, Z., Wang, Z.Y., Zhong, F., Xiao, C., Chiang, C.C., Yandell, M., Wortman, J.R., Amanatides, P.G., Hladun, S.L., Pratts, E.C., Johnson, J.E., Dodson, K.L., Woodford, K.J., Evans, C.A., Gropman, B., Rusch, D.B., Venter, E., Wang, M., Smith, T.J., Houck, J.T., Tompkins, D.E., Haynes, C., Jacob, D., Chin, S.H., Allen, D.R., Dahlke, C.E., Sanders, R., Li, K., Liu, X., Levitsky, A.A., Majoros, W.H., Chen, Q., Xia, A.C., Lopez, J.R., Donnelly, M.T., Newman, M.H., Glodek, A., Kraft, C.L., Nodell, M., Ali, F., An, H.J., Baldwin-Pitts, D., Beeson, K.Y., Cai, S., Carnes, M., Carver, A., Caulk, P.M., Center, A., Chen, Y.H., Cheng, M.L., Coyne, M.D., Crowder, M., Danaher, S., Davenport, L.B., Desilets, R., Dietz, S.M., Doup, L., Dullaghan, P., Ferriera, S., Fosler, C.R., Gire, H.C., Gluecksmann, A., Gocayne, J.D., Gray, J., Hart, B., Haynes, J., Hoover, J., Howland, T., Ibegwam, C., Jalali, M., Johns, D., Kline, L., Ma, D.S., MacCawley, S., Magoon, A., Mann, F., May, D., McIntosh, T.C., Mehta, S., Moy, L., Moy, M.C., Murphy, B.J., Murphy, S.D., Nelson, K.A., Nuri, Z., Parker, K.A., Prudhomme, A.C., Puri, V.N., Qureshi, H., Raley, J.C., Reardon, M.S., Regier, M.A., Rogers, Y.H., Romblad, D.L., Schutz, J., Scott, J.L., Scott, R., Sitter, C.D., Smallwood, M., Sprague, A.C., Stewart, E., Strong, R.V., Suh, E., Sylvester, K., Thomas, R., Tint, N.N., Tsonis, C., Wang, G., Williams, M.S., Williams, S.M., Windsor, S.M., Wolfe, K., Wu, M.M., Zaveri, J., Chaturvedi, K., Gabrielian, A.E., Ke, Z., Sun, J., Subramanian, G., Venter, J.C., Pfannkoch, C.M., Barnstead, M., Stephenson, L.D. (2002) A comparison of whole-genome shotgun-derived mouse chromosome 16 and the human genome. *Science*, **296** (5573), 1661–1671.

31 Perkins, A.S. (2002) Functional genomics in the mouse. *Funct. Integr. Genomics.*, **2** (3), 81–91.

32 Schutze, N. (2004) siRNA technology. *Mol. Cell. Endocrinol.*, **213** (2), 115–119.

33 Kumar, L.D., Clarke, A.R. (2007) Gene manipulation through the use of small interfering RNA (siRNA): from *in vitro* to *in vivo* applications. *Adv. Drug Deliv. Rev.*, **59** (2-3), 87–100.

34 Edgar, R., Domrachev, M., Lash, A.E. (2002) Gene Expression Omnibus: NCBI gene expression and hybridization array data repository. *Nucleic Acids Res.*, **30** (1), 207–210.

35 Parkinson, H., Kapushesky, M., Shojatalab, M., Abeygunawardena, N., Coulson, R., Farne, A., Holloway, E., Kolesnykov, N., Lilja, P., Lukk, M., Mani, R., Rayner, T., Sharma, A., William, E., Sarkans, U., Brazma, A. (2007) ArrayExpress – a public database of microarray experiments and gene expression profiles. *Nucleic Acids Res.*, **35** (Database issue), D747–D750.

36 Shaw, D.R. (2009) Searching the mouse genome informatics (MGI) resources for information on mouse biology from genotype to phenotype. *Curr. Protoc. Bioinformatics*, Suppl. 5, 1.7.1–1.7.13.

37 Sprague, J., Clements, D., Conlin, T., Edwards, P., Frazer, K., Schaper, K., Segerdell, E., Song, P., Sprunger, B., Westerfield, M. (2003) The Zebrafish information network (ZFIN): the zebrafish model organism database. *Nucleic Acids Res.*, **31** (1), 241–243.

38 Bradford, Y., Conlin, T., Dunn, N., Fashena, D., Frazer, K., Howe, D.G., Knight, J., Mani, P., Martin, R., Moxon, S.A., Paddock, H., Pich, C., Ramachandran, S., Ruef, B.J., Ruzicka, L., Bauer-Schaper, H., Schaper, K., Shao, X., Singer, A., Sprague, J., Sprunger, B., Van Slyke, C., Westerfield, M. (2011) ZFIN: enhancements and updates to the Zebrafish model organism database. *Nucleic Acids Res.*, **39** (Database issue), D822–D829.

39 Harris, T.W., Antoshechkin, I., Bieri, T., Blasiar, D., Chan, J., Chen, W.J., De La Cruz, N., Davis, P., Duesbury, M., Fang, R., Fernandes, J., Han, M., Kishore, R., Lee, R., Muller, H.M., Nakamura, C., Ozersky, P., Petcherski, A., Rangarajan, A., Rogers, A., Schindelman, G., Schwarz, E.M., Tuli, M.A., Van Auken, K., Wang, D., Wang, X., Williams, G., Yook, K., Durbin, R., Stein, L.D., Spieth, J., Sternberg, P.W. (2010) WormBase: a comprehensive resource for nematode research. *Nucleic Acids Res.*, **38** (Database issue), D463–D467.

40 Jupiter, D.C., VanBuren, V. (2008) A visual data mining tool that facilitates reconstruction of transcription regulatory networks. *PLoS ONE*, **3** (3), e1717.

41 Jupiter, D., Chen, H., VanBuren, V. (2009) STARNET 2: a web-based tool for accelerating discovery of gene regulatory networks using microarray co-expression data. *BMC Bioinf.*, **10**, 332.

42 Anderson, K.V., Ingham, P.W. (2003) The transformation of the model organism: a decade of developmental genetics. *Nat. Genet.*, **33**, 285–293.

43 Reinke, V., White, K.P. (2002) Developmental genomic approaches in model organisms. *Annu. Rev. Genomics Hum. Genet.*, **3**, 153–178.

44 White, K.P., Rifkin, S.A., Hurban, P., Hogness, D.S. (1999) Microarray analysis of *Drosophila* development during metamorphosis. *Science*, **286** (5447), 2179–2184.

45 Tomancak, P., Beaton, A., Weiszmann, R., Kwan, E., Shu, S., Lewis, S.E., Richards, S., Ashburner, M., Hartenstein, V., Celniker, S.E., Rubin, G.M. (2002) Systematic determination of patterns of gene expression during *Drosophila* embryogenesis. *Genome Biol.*, **3** (12), RESEARCH0088.

46 Arbeitman, M.N., Furlong, E.E., Imam, F., Johnson, E., Null, B.H., Baker, B.S., Krasnow, M.A., Scott, M.P., Davis, R.W., White, K.P. (2002) Gene expression during the life cycle of *Drosophila melanogaster*. *Science*, **297** (5590), 2270–2275.

47 Holt, R.A., Subramanian, G.M., Halpern, A., Sutton, G.G., Charlab, R., Nusskern, D.R., Wincker, P., Clark, A.G., Ribeiro, J.M., Wides, R., Salzberg, S.L., Loftus, B., Yandell, M., Majoros, W.H., Rusch, D.B., Lai, Z., Kraft, C.L., Abril, J.F., Anthouard, V., Arensburger, P., Atkinson, P.W., Baden, H., de Berardinis, V., Baldwin, D., Benes, V., Biedler, J., Blass, C., Bolanos, R., Boscus, D., Barnstead, M., Cai, S., Center, A., Chaturverdi, K., Christophides, G.K., Chrystal, M.A., Clamp, M., Cravchik, A., Curwen, V., Dana, A., Delcher, A., Dew, I., Evans, C.A., Flanigan, M., Grundschober-Freimoser, A., Friedli, L., Gu, Z., Guan, P., Guigo, R., Hillenmeyer, M.E., Hladun, S.L., Hogan, J.R., Hong, Y.S., Hoover, J., Jaillon, O., Ke, Z., Kodira, C., Kokoza, E., Koutsos, A., Letunic, I., Levitsky, A., Liang, Y., Lin, J.J., Lobo, N.F., Lopez, J.R., Malek, J.A., McIntosh, T.C., Meister, S., Miller, J., Mobarry, C., Mongin, E., Murphy, S.D., O'Brochta, D.A., Pfannkoch, C., Qi, R., Regier, M.A., Remington, K., Shao, H., Sharakhova, M.V., Sitter, C.D., Shetty, J., Smith, T.J., Strong, R., Sun, J., Thomasova, D., Ton, L.Q., Topalis, P., Tu, Z., Unger, M.F., Walenz, B., Wang, A., Wang, J., Wang, M., Wang, X., Woodford, K.J., Wortman, J.R., Wu, M., Yao, A., Zdobnov, E.M., Zhang, H., Zhao, Q., Zhao, S., Zhu, S.C., Zhimulev, I., Coluzzi, M., della Torre, A., Roth, C.W., Louis, C., Kalush, F., Mural, R.J., Myers, E.W., Adams, M.D., Smith, H.O., Broder, S., Gardner, M.J., Fraser, C.M., Birney, E., Bork, P., Brey, P.T., Venter, J.C., Weissenbach, J., Kafatos, F.C., Collins, F.H., Hoffman, S.L. (2002) The genome sequence of the malaria mosquito *Anopheles gambiae*. *Science*, **298** (5591), 129–149.

48 Lohr, J.L., Yost, H.J. (2000) Vertebrate model systems in the study of early heart development: *Xenopus* and zebrafish. *Am. J. Med. Genet.*, **97** (4), 248–257.

49 Bogi, C., Levy, G., Lutz, I., Kloas, W. (2002) Functional genomics and sexual differentiation in amphibians. *Comp. Biochem. Physiol. B Biochem. Mol. Biol.*, **133** (4), 559–570.

50 Grammer, T.C., Liu, K.J., Mariani, F.V., Harland, R.M. (2000) Use of large-scale expression cloning screens in the *Xenopus laevis* tadpole to identify gene function. *Dev. Biol.*, **228** (2), 197–210.

51 Pollet, N., Schmidt, H.A., Gawantka, V., Niehrs, C., Vingron, M. (2000) In silico analysis of gene expression patterns during early development of *Xenopus laevis*. Proceedings, Pacific Symposium on Biocomputing, 4–9 January 2000, Island of Oahu, Hawaii, pp. 443–454.

52 Sargent, T.D., Dawid, I.B. (1983) Differential gene expression in the gastrula of *Xenopus laevis*. *Science*, **222** (4620), 135–139.

53 Stancheva, I., El-Maarri, O., Walter, J., Niveleau, A., Meehan, R.R. (2002) DNA methylation at promoter regions regulates the timing of gene activation in *Xenopus laevis* embryos. *Dev. Biol.*, **243** (1), 155–165.

54 Tanaka, T.S., Kunath, T., Kimber, W.L., Jaradat, S.A., Stagg, C.A., Usuda, M., Yokota, T., Niwa, H., Rossant, J., Ko, M.S. (2002) Gene expression profiling of embryo-derived stem cells reveals candidate genes associated with pluripotency and lineage specificity. *Genome. Res.*, **12** (12), 1921–1928.

55 Hirsch, N., Zimmerman, L.B., Grainger, R.M. (2002) *Xenopus*, the next generation: X.

tropicalis genetics and genomics. *Dev. Dyn.*, **225** (4), 422–433.

56 Fishman, M.C. (2001) Genomics. Zebrafish – the canonical vertebrate. *Science*, **294** (5545), 1290–1291.

57 Kudoh, T., Tsang, M., Hukriede, N.A., Chen, X., Dedekian, M., Clarke, C.J., Kiang, A., Schultz, S., Epstein, J.A., Toyama, R., Dawid, I.B. (2001) A gene expression screen in zebrafish embryogenesis. *Genome. Res.*, **11** (12), 1979–1987.

58 Davidson, E.H., Rast, J.P., Oliveri, P., Ransick, A., Calestani, C., Yuh, C.H., Minokawa, T., Amore, G., Hinman, V., Arenas-Mena, C., Otim, O., Brown, C.T., Livi, C.B., Lee, P.Y., Revilla, R., Schilstra, M.J., Clarke, P.J., Rust, A.G., Pan, Z., Arnone, M.I., Rowen, L., Cameron, R.A., McClay, D.R., Hood, L., Bolouri, H. (2002) A provisional regulatory gene network for specification of endomesoderm in the sea urchin embryo. *Dev. Biol.*, **246** (1), 162–190.

59 Venter, J.C., Adams, M.D., Myers, E.W., Li, P.W., Mural, R.J., Sutton, G.G., Smith, H.O., Yandell, M., Evans, C.A., Holt, R.A., Gocayne, J.D., Amanatides, P., Ballew, R.M., Huson, D.H., Wortman, J.R., Zhang, Q., Kodira, C.D., Zheng, X.H., Chen, L., Skupski, M., Subramanian, G., Thomas, P.D., Zhang, J., Gabor-Miklos, G.L., Nelson, C., Broder, S., Clark, A.G., Nadeau, J., McKusick, V.A., Zinder, N., Levine, A.J., Roberts, R.J., Simon, M., Slayman, C., Hunkapiller, M., Bolanos, R., Delcher, A., Dew, I., Fasulo, D., Flanigan, M., Florea, L., Halpern, A., Hannenhalli, S., Kravitz, S., Levy, S., Mobarry, C., Reinert, K., Remington, K., Abu-Threideh, J., Beasley, E., Biddick, K., Bonazzi, V., Brandon, R., Cargill, M., Chandramouliswaran, I., Charlab, R., Chaturvedi, K., Deng, Z., Di Francesco, V., Dunn, P., Eilbeck, K., Evangelista, C., Gabrielian, A.E., Gan, W., Ge, W., Gong, F., Gu, Z., Guan, P., Heiman, T.J., Higgins, M.E., Ji, R.R., Ke, Z., Ketchum, K.A., Lai, Z., Lei, Y., Li, Z., Li, J., Liang, Y., Lin, X., Lu, F., Merkulov, G.V., Milshina, N., Moore, H.M., Naik, A.K., Narayan, V.A., Neelam, B., Nusskern, D., Rusch, D.B., Salzberg, S., Shao, W., Shue, B., Sun, J., Wang, Z., Wang, A., Wang, X., Wang, J., Wei, M., Wides, R., Xiao, C., Yan, C., Yao, A., Ye, J., Zhan, M., Zhang, W., Zhang, H., Zhao, Q., Zheng, L., Zhong, F., Zhong, W., Zhu, S., Zhao, S., Gilbert, D., Baumhueter, S., Spier, G., Carter, C., Cravchik, A., Woodage, T., Ali, F., An, H., Awe, A., Baldwin, D., Baden, H., Barnstead, M., Barrow, I., Beeson, K., Busam, D., Carver, A., Center, A., Cheng, M.L., Curry, L., Danaher, S., Davenport, L., Desilets, R., Dietz, S., Dodson, K., Doup, L., Ferriera, S., Garg, N., Gluecksmann, A., Hart, B., Haynes, J., Haynes, C., Heiner, C., Hladun, S., Hostin, D., Houck, J., Howland, T., Ibegwam, C., Johnson, J., Kalush, F., Kline, L., Koduru, S., Love, A., Mann, F., May, D., McCawley, S., McIntosh, T., McMullen, I., Moy, M., Moy, L., Murphy, B., Nelson, K., Pfannkoch, C., Pratts, E., Puri, V., Qureshi, H., Reardon, M., Rodriguez, R., Rogers, Y.H., Romblad, D., Ruhfel, B., Scott, R., Sitter, C., Smallwood, M., Stewart, E., Strong, R., Suh, E., Thomas, R., Tint, N.N., Tse, S., Vech, C., Wang, G., Wetter, J., Williams, S., Williams, M., Windsor, S., Winn-Deen, E., Wolfe, K., Zaveri, J., Zaveri, K., Abril, J.F., Guigo, R., Campbell, M.J., Sjolander, K.V., Karlak, B., Kejariwal, A., Mi, H., Lazareva, B., Hatton, T., Narechania, A., Diemer, K., Muruganujan, A., Guo, N., Sato, S., Bafna, V., Istrail, S., Lippert, R., Schwartz, R., Walenz, B., Yooseph, S., Allen, D., Basu, A., Baxendale, J., Blick, L., Caminha, M., Carnes-Stine, J., Caulk, P., Chiang, Y.H., Coyne, M., Dahlke, C., Mays, A., Dombroski, M., Donnelly, M., Ely, D., Esparham, S., Fosler, C., Gire, H., Glanowski, S., Glasser, K., Glodek, A., Gorokhov, M., Graham, K., Gropman, B., Harris, M., Heil, J., Henderson, S., Hoover, J., Jennings, D., Jordan, C., Jordan, J., Kasha, J., Kagan, L., Kraft, C., Levitsky, A., Lewis, M., Liu, X., Lopez, J., Ma, D., Majoros, W., McDaniel, J., Murphy, S., Newman, M., Nguyen, T., Nguyen, N., Nodell, M., Pan, S., Peck, J., Peterson, M., Rowe, W., Sanders, R., Scott, J., Simpson, M., Smith, T., Sprague, A., Stockwell, T., Turner, R., Venter, E., Wang, M., Wen, M., Wu, D., Wu, M., Xia, A., Zandieh, A., Zhu, X. (2001) The sequence of the human genome. *Science*, **291** (5507), 1304–1351.

60 Waterston, R.H., Lindblad-Toh, K., Birney, E., Rogers, J., Abril, J.F., Agarwal, P., Agarwala, R., Ainscough, R., Alexandersson, M., An, P., Antonarakis, S.E., Attwood,

J., Baertsch, R., Bailey, J., Barlow, K., Beck, S., Berry, E., Birren, B., Bloom, T., Bork, P., Botcherby, M., Bray, N., Brent, M.R., Brown, D.G., Brown, S.D., Bult, C., Burton, J., Butler, J., Campbell, R.D., Carninci, P., Cawley, S., Chiaromonte, F., Chinwalla, A.T., Church, D.M., Clamp, M., Clee, C., Collins, F.S., Cook, L.L., Copley, R.R., Coulson, A., Couronne, O., Cuff, J., Curwen, V., Cutts, T., Daly, M., David, R., Davies, J., Delehaunty, K.D., Deri, J., Dermitzakis, E.T., Dewey, C., Dickens, N.J., Diekhans, M., Dodge, S., Dubchak, I., Dunn, D.M., Eddy, S.R., Elnitski, L., Emes, R.D., Eswara, P., Eyras, E., Felsenfeld, A., Fewell, G.A., Flicek, P., Foley, K., Frankel, W.N., Fulton, L.A., Fulton, R.S., Furey, T.S., Gage, D., Gibbs, R.A., Glusman, G., Gnerre, S., Goldman, N., Goodstadt, L., Grafham, D., Graves, T.A., Green, E.D., Gregory, S., Guigo, R., Guyer, M., Hardison, R.C., Haussler, D., Hayashizaki, Y., Hillier, L.W., Hinrichs, A., Hlavina, W., Holzer, T., Hsu, F., Hua, A., Hubbard, T., Hunt, A., Jackson, I., Jaffe, D.B., Johnson, L.S., Jones, M., Jones, T.A., Joy, A., Kamal, M., Karlsson, E.K., Karolchik, D., Kasprzyk, A., Kawai, J., Keibler, E., Kells, C., Kent, W.J., Kirby, A., Kolbe, D.L., Korf, I., Kucherlapati, R.S., Kulbokas, E.J., Kulp, D., Landers, T., Leger, J.P., Leonard, S., Letunic, I., Levine, R., Li, J., Li, M., Lloyd, C., Lucas, S., Ma, B., Maglott, D.R., Mardis, E.R., Matthews, L., Mauceli, E., Mayer, J.H., McCarthy, M., McCombie, W.R., McLaren, S., McLay, K., McPherson, J.D., Meldrim, J., Meredith, B., Mesirov, J.P., Miller, W., Miner, T.L., Mongin, E., Montgomery, K.T., Morgan, M., Mott, R., Mullikin, J.C., Muzny, D.M., Nash, W.E., Nelson, J.O., Nhan, M.N., Nicol, R., Ning, Z., Nusbaum, C., O'Connor, M.J., Okazaki, Y., Oliver, K., Overton-Larty, E., Pachter, L., Parra, G., Pepin, K.H., Peterson, J., Pevzner, P., Plumb, R., Pohl, C.S., Poliakov, A., Ponce, T.C., Ponting, C.P., Potter, S., Quail, M., Reymond, A., Roe, B.A., Roskin, K.M., Rubin, E.M., Rust, A.G., Santos, R., Sapojnikov, V., Schultz, B., Schultz, J., Schwartz, M.S., Schwartz, S., Scott, C., Seaman, S., Searle, S., Sharpe, T., Sheridan, A., Shownkeen, R., Sims, S., Singer, J.B., Slater, G., Smit, A., Smith, D.R., Spencer, B., Stabenau, A., Stange-Thomann, N., Sugnet, C., Suyama, M., Tesler, G., Thompson, J., Torrents, D., Trevaskis, E., Tromp, J., Ucla, C., Ureta-Vidal, A., Vinson, J.P., Von Niederhausern, A.C., Wade, C.M., Wall, M., Weber, R.J., Weiss, R.B., Wendl, M.C., West, A.P., Wetterstrand, K., Wheeler, R., Whelan, S., Wierzbowski, J., Willey, D., Williams, S., Wilson, R.K., Winter, E., Worley, K.C., Wyman, D., Yang, S., Yang, S.P., Zdobnov, E.M., Zody, M.C., Lander, E.S. (2002) Initial sequencing and comparative analysis of the mouse genome. *Nature*, **420** (6915), 520–562.

61 Sasaki, N., Nagaoka, S., Itoh, M., Izawa, M., Konno, H., Carninci, P., Yoshiki, A., Kusakabe, M., Moriuchi, T., Muramatsu, M., Okazaki, Y., Hayashizaki, Y. (1998) Characterization of gene expression in mouse blastocyst using single-pass sequencing of 3995 clones. *Genomics*, **49** (2), 167–179.

62 Neidhardt, L., Gasca, S., Wertz, K., Obermayr, F., Worpenberg, S., Lehrach, H., Herrmann, B.G. (2000) Large-scale screen for genes controlling mammalian embryogenesis, using high-throughput gene expression analysis in mouse embryos. *Mech. Dev.*, **98** (1-2), 77–94.

63 Kargul, G.J., Dudekula, D.B., Qian, Y., Lim, M.K., Jaradat, S.A., Tanaka, T.S., Carter, M.G., Ko, M.S. (2001) Verification and initial annotation of the NIA mouse 15K cDNA clone set. *Nat. Genet.*, **28** (1), 17–18.

64 Schepers, G.E., Teasdale, R.D., Koopman, P. (2002) Twenty pairs of sox: extent, homology, and nomenclature of the mouse and human sox transcription factor gene families. *Dev. Cell*, **3** (2), 167–170.

65 Svoboda, P., Stein, P., Schultz, R.M. (2001) RNAi in mouse oocytes and preimplantation embryos: effectiveness of hairpin dsRNA. *Biochem. Biophys. Res. Commun.*, **287** (5), 1099–1104.

66 Boyes, D.C., Zayed, A.M., Ascenzi, R., McCaskill, A.J., Hoffman, N.E., Davis, K.R., Gorlach, J. (2001) Growth stage-based phenotypic analysis of *Arabidopsis*: a model for high throughput functional genomics in plants. *Plant Cell*, **13** (7), 1499–1510.

67 Finkelstein, D., Ewing, R., Gollub, J., Sterky, F., Cherry, J.M., Somerville, S. (2002) Microarray data quality analysis: lessons from the AFGC project. Arabidopsis Functional

Genomics Consortium. *Plant Mol. Biol.*, **48** (1,2), 119–131.

68 Somerville, C., Somerville, S. (1999) Plant functional genomics. *Science*, **285** (5426), 380–383.

69 Tao, Q., Chang, Y.L., Wang, J., Chen, H., Islam-Faridi, M.N., Scheuring, C., Wang, B., Stelly, D.M., Zhang, H.B. (2001) Bacterial artificial chromosome-based physical map of the rice genome constructed by restriction fingerprint analysis. *Genetics*, **158** (4), 1711–1724.

70 Woo, Y.M., Hu, D.W., Larkins, B.A., Jung, R. (2001) Genomics analysis of genes expressed in maize endosperm identifies novel seed proteins and clarifies patterns of zein gene expression. *Plant Cell*, **13** (10), 2297–2317.

71 Kamath, R.S., Fraser, A.G., Dong, Y., Poulin, G., Durbin, R., Gotta, M., Kanapin, A., Le Bot, N., Moreno, S., Sohrmann, M., Welchman, D.P., Zipperlen, P., Ahringer, J. (2003) Systematic functional analysis of the *Caenorhabditis elegans* genome using RNAi. *Nature*, **421** (6920), 231–237.

4
Interactome

Teresa M. Przytycka and Dong-Yeon Cho
National Institute of Health (NIH), National Center of Biotechnology Information, NLM, 8600 Rockville Pike, Bethesda, MD 20894, USA

Systems Biology: Advances in Molecular Biology and Medicine, First Edition. Edited by Robert A. Meyers.

Keywords

Coevolution
Correlated evolutionary changes in the genetic composition of biological objects, such as genes. Typically, it is assumed that the change in one object is triggered by the genetic change in a second object (although within the context of this chapter the causality assumption is not made).

Coexpression
Analogous spatial or temporal expression patterns of two or more genes. Typically, coexpression can be identified by calculating correlation coefficients and assessing these correlations through statistical tests.

Graph/network
A representation of a binary relation between a set of objects. The objects are called vertices (or nodes). The relation between two vertices is represented by an edge (or link). Specifically, a graph of which the edges are directed is termed a directed graph. Directed graphs can be used to represent an asymmetric relation, such as gene regulation.

Gene regulatory network
A directed network of genes where one gene (a transcription factor) regulates the expression of another gene (target gene).

Homologs and orthologs
Homologs are genes which originated from a common ancestor. Two homologs found in different species are termed orthologs if their least common ancestor in the gene evolutionary tree corresponds to a speciation event. Orthologs typically perform closely related roles in the corresponding organisms.

Protein–protein interaction network
A network where the nodes of while denote proteins and links that represent the binding of proteins to carry out their biological function.

Transcription factor
A protein which binds to specific DNA sequences to regulate the transcription or expression of genes. By controlling the access of RNA polymerase to the genes, transcription factors can either activate or repress the expression (the transfer of genetic information from DNA to mRNA) of the corresponding gene.

Yeast two-hybrid (Y2H)
A yeast-based biological technique used to test protein interactions. One of tested proteins is fused to a DNA-binding domain and the other to an activation domain. If the two proteins interact a reporter gene is expressed.

The interactome of a cell consists of all the molecular interactions that occur within that cell. Among the diverse types of molecular interaction, genome-wide protein–protein interactions are the most broadly studied. In this chapter, attention is focused on genome-scale methods to infer and to study protein–protein interactions. Following a brief introduction to the subject, the experimental and computational techniques used to uncover interactions between proteins are outlined. In the context of computational approaches, methods are also discussed for predicting interactions between protein domains. Later, the basic topological properties of protein interaction networks are described, such as node degree distribution, modularity, and network motifs. The motivation and approaches for comparing biological networks are then detailed, followed by a brief introduction to protein and domain interaction databases. Finally, some sample applications are provided that take advantage of interactome data, and details are proposed for the possible future direction of interactome-related research.

1 Introduction

Within a cell, diverse biomolecules work together in a coordinated fashion to provide specific cellular functions. This coordinated action is achieved, in large part, by a variety of intermolecular interactions including protein–protein interactions, protein–DNA interactions, RNA interactions, and many others. Here, attention is focused mainly on the interactions between proteins, including those corresponding to physical interactions (here, protein–protein binding) as well as to more abstract "functional" interactions between them. However, other interaction types will also be briefly mentioned.

Biomolecular interactions are conveniently represented as networks (graphs), where the nodes (vertices) represent molecules, and the links (edges) represent the interactions between them. Depending on the type of interaction, the corresponding edge might be directed, or not. For example, a binding of two proteins is typically represented by an undirected edge, while an interaction between a transcription factor and a gene, the expression of which is regulated by the given transcription factor, is usually represented by a directed edge. A representation of the interactome as a network/graph not only provides a convenient visualization but also enables the use of graph theoretical concepts and tools in the study of

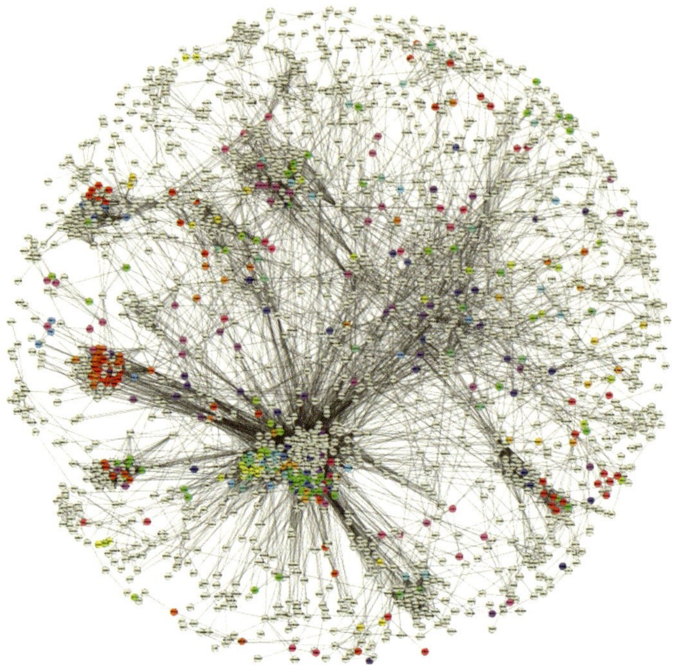

Fig. 1 Yeast protein–protein interaction network drawn using Cytoscape [1] constructed from based on complex purification experiments ([2, 3]) with statistical scores from Ref. [4]. Modules (proteins that are simultaneously present in several protein complexes) are color-coded.

biological networks. For example, the Cytoscape suite has recently emerged as a leading tool for visualization and network analysis, and also allows for the development of third-party tools (plugins) that take advantage of its functionality [1]. An example of yeast protein–protein interaction network visualized, with the help of Cytoscape, is shown in Fig. 1.

2
Experimental Techniques for Detecting Protein Interactions

The two main technologies that are used in high-throughput protein–protein interaction detection experiments are the yeast two-hybrid (Y2H) assay, and protein complex detection: co-immunopreci pitation (co-IP) and coaffinity puryfication followed by mass spectrometry (caIP/MS). These two techniques are vastly different, with each having its own strengths and limitations.

In the Y2H experiment, as pioneered by Fields and Song [5], one of the tested proteins – the so-called "bait" – is fused with a DNA-binding domain (usually GAL4 or LexA), while the second protein (the "prey") is fused with a transcriptional activation domain for a transcription factor that can activate expression of a reporter gene (e.g., β-galactosidase). Both chimeras are then expressed in a yeast cell and, if they interact, their interaction prompts expression of the reporter gene. The two most important properties of this experimental technique to be borne in mind are:

- It detects binary interactions only, and thus might miss interacting proteins that require additional proteins, such as scaffold proteins or other members of a protein complex, to facilitate the interaction.
- It uncovers the *potential* for interaction; that is, whether or not the two proteins actually interact in a cell depends upon spatial, temporal, and contextual constraints such as the cell cycle phase, stress, and the presence (or lack) of a particular nutrient.

Currently, high-throughput Y2H interaction networks (maps) are available for many organisms, including *Saccharomyces cerevisiae* [6–9], *Caenorhabditis elegans* [10, 11], *Drosophila melanogaster* [12], and humans [13, 14].

The second key technology used to obtain high-throughput interaction maps is that of co-complex identification including tandem affinity purification (TAP) followed by mass spectrometry (TAP/MS) [15–17]. In contrast to the Y2H procedure, this approach reveals one-to-many interactions in a particular experimental condition. Specifically, a bait protein is tested for interaction with all other proteins (preys) expressed in the given condition. This is achieved by allowing a complex formation of the bait protein with other proteins in the cell, and then retrieving and purifying the corresponding complexes and identifying any co-complexed proteins using MS. The retrieval of the corresponding complexes requires either antibodies for the bait protein, or tagging of the bait protein with a peptide for which antibodies are available. Consequently, it should be borne in mind that:

- As the experiment recovers whole complexes, the bait protein does not necessarily interact directly with all proteins in the complex.
- Transient interactions are often difficult to capture using this approach. As with the Y2H experiments, a number of high-throughput co-complexed interactions have been recognized [2, 3, 18–20].

The two above-described experimental methods are, in many ways, complementary and have unique strengths and limitations [21–23]. Hence, the results of both types of experiment are commonly combined into one protein–protein interaction network. While such a combined network provides a more complete interactome, it is important to bear in mind that, if treated individually, the two types of experiment used to obtain such a network define two rather different networks in terms of their biological and topological properties (e.g., Refs [9, 24]).

3 Computational Prediction of Protein Interactions

Experimental procedures detecting protein–protein interactions are complemented by computational approaches. The latter methods incorporate a variety of techniques that can be subdivided into three categories: evolutionary-based approaches; statistical methods; and machine learning techniques. Evolutionary-based approaches for predicting protein interactions typically explore the idea that interacting proteins are subject to common evolutionary constraints. Such constraints can impact the spatial organization of interacting genes in the genome, the position in the protein–protein interaction network, and/or the amino-acid sequence. While some of these approaches are designed to predict physical interactions, many do not attempt to distinguish between physical and functional interactions, as both types of interaction might be subject to similar evolutionary constraints.

3.1 Interaction Prediction from the Gene Patterns Across Genomes

3.1.1 Gene Fusion

The gene fusion method is an evolution-based approach for predicting physical interactions. The main idea follows from the observation that if a pair of proteins (A and B) present in one organism are, in another organism, fused together into a single protein, then these two proteins are likely to interact (Fig. 2a) [25, 26]. A natural explanation for this observation is that if A and B interact, then bringing them together in the fused protein will facilitate their interaction. Marcotte *et al.* coined the term *"the Rosetta Stone protein"* for the fused protein and, by using this approach, were able to identify 6809 such putative protein–protein interactions in *Escherichia coli,* and 45 502 in yeast (Fig. 2a). In a larger study, Enright and Ouziunis uncovered 7224 component and 2365 unique composite proteins across 24 species [27].

3.1.2 Gene Order

The whole-genome sequencing of large numbers of genomes allows observations to be made of the patterns of genome organization and evolution [28]. Within the context of bacterial and archaeal genomes it has been observed that, whilst in general the ordering of genes along genomes is not well conserved between species, the order of genes encoding interacting proteins does tend to be conserved [29, 30]. Conversely, proteins encoded by gene pairs with a conserved gene order often interact physically. Based on these observations, Dandekar *et al.* proposed a method for predicting interacting proteins [30] as being proteins encoded by genes with a conserved gene order (Fig. 2b). It should be noted that this so-called "gene order method" and its variants [31–34] are most suitable in the context of bacterial and archaeal genomes, where groups of genes are organized into operon structures.

3.1.3 Phylogenetic Profiling

The phylogenic profiling method is based on the premise that functional interactions are conserved across a range of species. Consistent with this assumption, there should be a correlation between patterns of the presence and absence in various genomes of functionally interacting genes. Such a presence/absence pattern is referred to as the *phylogenic profile* of a gene [35–38]. More formally, the phylogenic profile of a gene within a set of n reference genomes is a vector of length n, where the i-th element of the vector is set to 1 if the given gene is present in the i-th genome, and to 0 otherwise. Similarities between phylogenic profiles can be measured using metrics such as Hamming distance, correlation coefficient, or mutual information. The presence/absence of a gene within a genome can be also quantified using probability scores instead of binary values. Similarities between phylogenetic profiles can be used to predict functional linkage between proteins (Fig. 2c) [35, 36]. The grouping of genes with common evolutionary patterns also allows not only

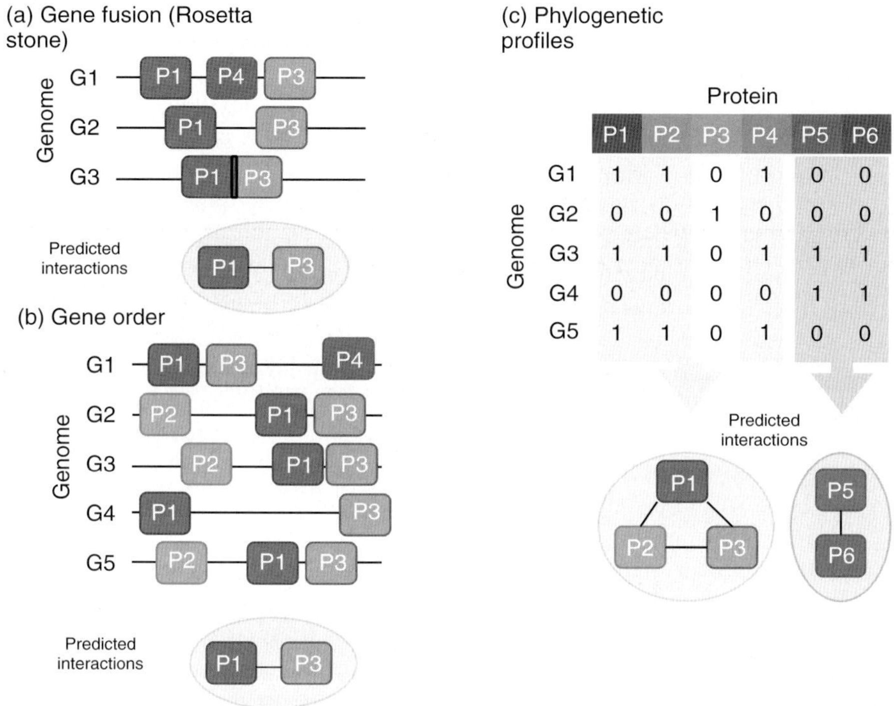

	P1	P2	P3	P4	P5	P6
G1	1	1	0	1	0	0
G2	0	0	1	0	0	0
G3	1	1	0	1	1	1
G4	0	0	0	0	1	1
G5	1	1	0	1	0	0

Fig. 2 Prediction of protein interaction from gene pattern in the genomes. (a) The "Rosetta stone" method that predicts interaction between proteins P1 and P2, based on the fact that they are fused together into one protein in genome 3; (b) The gene order method predicts interaction between P1 and P3, based on the conservation of their order; (c) The phylogenic profile method predicts (functional) interaction, based on similar patterns of presence and absence in genomes G1–G5.

the prediction of functional associations for genes with unknown functions, but also the discovery of any previously uncharacterized cellular pathways and functional network modules [39–54]. Moreover, it has been shown that the reliability of the phylogenic profile method depends on the selection of the reference genomes [53, 55].

Bowers *et al.* extended the phylogenic profiling method so that it could take into consideration three proteins simultaneously. In this case, instances were sought in which the combined logical patterns embodied by two proteins determined the behavior of a third protein [56].

Whilst the phylogenic profile predicts functional associations rather than physical interactions, an important drawback of the method is that it cannot be used to predict interactions between proteins that are present in all (or nearly all) reference genomes.

3.2 Predicting Interaction from Sequence Coevolution

Whereas, investigations conducted so far have centered on approaches that utilize patterns of gene organization within genomes, the "mirror tree" method (as discussed here) takes the next step and zooms in on the details of sequence features. This approach is based on the premise that sequences of interacting proteins are expected to coevolve to maintain their interactions [57–60]. Motivated by this supposition, the mirror tree method predicts protein–protein interactions by assessing the extent of agreement between evolutionary trees or, more precisely, between the distance matrices used to infer such trees, as illustrated in Fig. 3 [57, 59–67].

In its simplest form, the mirror tree method assesses the coevolution of two proteins by the correlation between the distance matrices constructed individually for the sets of sequences orthologous to each of the two proteins. That is, given two proteins A and B, it considers *n* sequences orthologous with A and the same number of sequences orthologous with B deriving from the same set of species. Subsequently, for each pair (i,j) of orthologs of A (respectively B), the method estimates the evolutionary distance $A_{i,j}$ (respectively $B_{i,j}$) between them. The degree of coevolution between the two families of orthologs is then assessed by computing the correlation coefficient between the distance matrices (see Fig. 3).

As described above, the mirror tree method measures the correlation between rates of evolutionary change (rates of divergence). Two reasons have been identified for which such correlation might occur: (i) a common speciation history; and (ii) common evolutionary constraints imposed by physical and/or functional interactions. Thus, one of the main challenges related to the mirror tree method is to separate these two sources of correlation, and several techniques have recently been developed to address this problem,

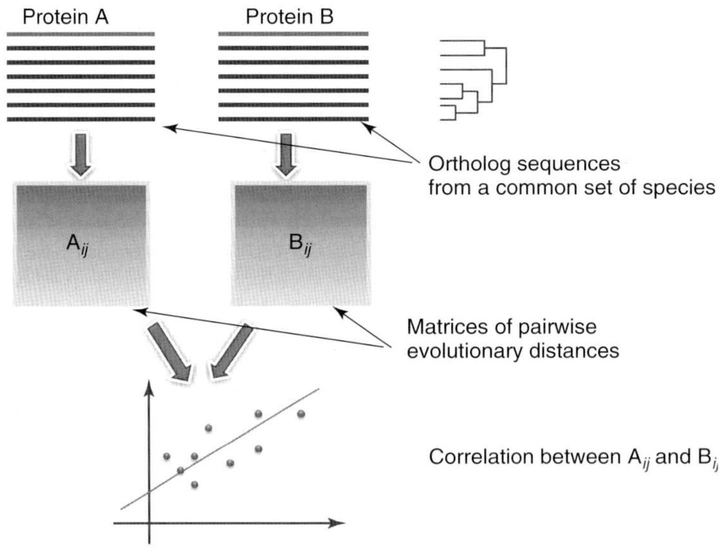

Fig. 3 The basic variant of the mirror tree method. Coevolution between proteins A and B is assessed by comparing the evolutionary rates of a family of sequences orthologous to A, and the corresponding family of sequences orthologous to B. Subsequent variants of the method attempt to account for the common speciation history (illustrated by the evolutionary tree on the right) of both families.

with marked improvements in accuracy when predicting interactions [64, 65, 68]. In the methods developed by Pazos *et al.* and Sato *et al.*, the estimated organism divergence rates were first subtracted from the combined coevolution signal. Subsequently, Kann *et al.* further improved the mirror tree performance by restricting the coevolution analysis to the more conserved regions in the protein domain sequences while disregarding any highly divergent regions. The latter regions are likely to be diverged by neutral evolution, and would not be expected to contain a functional coevolution signal.

It is natural to hypothesize that the driving mechanisms of sequence coevolution derive from compensating mutations, where a mutation in one binding partner is compensated by a complementary mutation in another partner, so as to maintain amino acid interactions. Yet, it has been shown that compensating mutations are not the only – perhaps not even the dominant – contributors to the correlation of evolutionary rates [69, 70]. Rather, the coevolution signal is more likely a composite of many other factors, such the coexpression of interacting proteins, similar codon usage, and interaction with other proteins in a complex. This suggests that it might be more meaningful to consider the coevolution (or coadaptation) of proteins within a broader, network-level context. Subsequently, this proposal was explored by Juan *et al.*, who replaced the vector of evolutionary distances by the vector of coevolutionary correlations between all proteins within a genome and, in this way, obtained a reliable interaction network of *E. coli* [71].

Currently, several other variants of the mirror tree method are available. For example, Tiller and Charlebois replaced a simple matrix correlation by a more sophisticated search for the most similar common subtree, that can also include paralogs [72]. Another variant of the mirror tree approach has also been used to predict interaction specificity. In this case, given two families of proteins which are known to interact, the objective is to establish a mapping by defining interaction partners between the members of one family and members of the other family [62, 73, 74]. Finally, Jothi *et al.* used the mirror tree method to identify interacting domains within interacting proteins [61].

3.3 Domain Interactions

A large proportion of prokaryotic proteins, as well as most eukaryotic proteins, are composed of more than one domain [75]. Protein interaction typically involves binding between two or more specific domains; indeed, the domain composition of two proteins can be used to predict interactions between them [76–84]. Conversely, knowledge of a protein interaction network can be used to infer interacting domains [85–90].

The concept of using protein–protein interaction networks to predict domain–domain interactions was first explored by Sprinzak and Margalit, who proposed a simple statistical approach that was referred to as the *Association Method* [82]. The idea here was to score each domain pair by the log ratio of the frequency of occurrences in interacting proteins to the expected frequency of independent occurrences of the two domains [82]. That is, if P_i is the observed frequency of domain i in the interaction network, and P_{ij} is the observed frequency of domain pair (i, j) in the interacting protein pairs, then the Association Score $(\mathrm{AS}_{ij}) = \log \frac{P_{ij}}{P_i P_j}$. The interacting domain pairs are then predicted as those with the highest AS-value.

Following investigations conducted by Sprinzak and Margalit, several related methods have been proposed (for a review, see Ref. [86]). For example, Deng *et al.* developed a maximum likelihood approach to estimate the probability of domain–domain interactions [81] in which the idea was to estimate, for each domain pair, the probability of an interaction between domains, so that the likelihood of the interaction network would be maximized. An elegant feature of this approach was the explicit modeling of errors in the high-throughput data that constituted the protein interaction network.

Whilst these early domain interaction prediction methods were successful in identifying domains that would interact in a constitutive manner, they were challenged in situations where a domain pair (i, j) interacted in the context of some protein pairs but, at the same time, there are also many proteins that contained i and j, respectively, but did not interact. In an attempt to discover such context-specific interactions, Riley *et al.* introduced the so-called domain pair exclusion analysis (DPEA) [91], which was a clever utilization of the maximum likelihood approach. Notably, the likelihood score of a network can be viewed as a measure of how well the probabilities assigned to putative domain interactions could explain the network. Thus, if the domain pair (i, j) were to mediate protein–protein interactions in a context-specific way, then the exclusion of such a domain pair as a possible interacting pair should reduce the likelihood score of the network. Consistent with this premise, DPEA can be used to predict interactions between domains by monitoring the fall in the likelihood score when a particular domain pair was not allowed to interact. Moreover, this approach could be used to detect interacting domain pairs that had been missed by previous approaches. More recently, Wang *et al.* [90] improved this concept further by suggesting the use of a scoring method that would account more fully for the context in which the interaction occurred. In order to achieve this, rather than globally disallowing all interactions between two specific domains, as had been proposed by Riley *et al.*, only interactions in the context of a specific pair of interacting proteins were disallowed.

The idea of recovering interacting domains by examining how well the potential domain contacts could explain the protein interaction network, also formed the basis of a method proposed by Guimares and colleagues [85]. Building on the assumption that protein interactions had evolved in a most parsimonious way, these authors proposed the *Parsimonious explanation* method, which identified the smallest weighted set of domain interactions that could "explain" the protein interaction network. In other words, the method located the smallest set of domains such that, if they were considered to interact, then each interacting protein pair would contain at least one interacting domain pair. This model was formalized as an optimization problem, and solved with a linear programming procedure. In this case, the variables of the linear program represented the potential domain contacts derived from the protein interaction network, while the constraints were defined as protein–protein interactions (edges). The construction is illustrated schematically in Fig. 4.

So far, the methods used to predict interacting domains have utilized only protein interaction–interaction networks and the domain composition of each protein. Yet, this basic information can be enriched in many ways [90, 92]. For example, by analogy to the Rosetta Stone approach (see Sect. 3.1.1), it has been observed that two domains that co-occur in one protein chain are also likely to mediate interactions between different proteins. The incorporation of such additional information has the potential of improving domain interaction predictions [93, 94].

One formidable challenge that is faced by the methods used to predict interacting domains is an evaluation of the method's quality. Typically, such an evaluation would be based on a knowledge of the domain interactions derived from crystal structures, although unfortunately these data would be biased towards particular types of interaction, thus limiting the ability to test the methods to the full [94].

3.4 Coexpression Networks

Starting from the assumption that proteins from the same protein complex are likely to be coexpressed, expression data have been utilized to predict new and to validate previously known protein–protein interactions [95–101]. Coexpression networks are usually inferred by computing the Pearson correlation coefficients or mutual information between every pair of gene expression profiles, across a variety of experimental conditions. It has been shown that genes with similar expression patterns across a set of samples tend to be functionally related [102]. Consequently, the coexpression data are frequently combined with other types of data in order to predict protein–protein interactions and to create functional networks. Such networks, built using expression information, have been constructed for variety of organisms [95–97, 103–107].

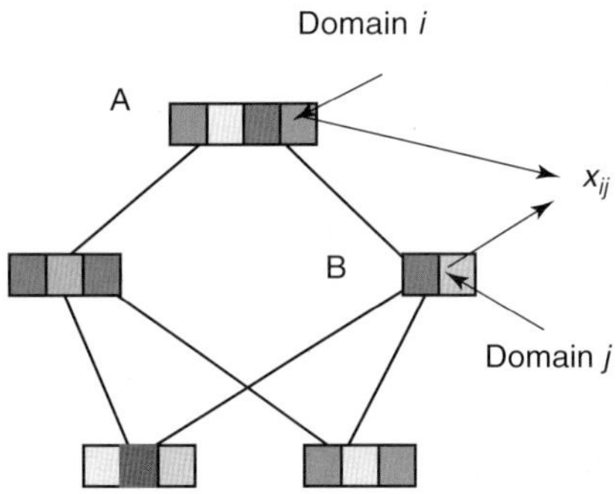

Fig. 4 Domain–domain interaction inference from protein interaction networks via a linear programming construction. In this toy network, the rectangles correspond to proteins and the colored squares to domains. For each domain pair (i,j) there is a variable taking real values between 0 and 1. For each pair (A,B) of interacting proteins A and B, there is one constraint $\Sigma_{i \in A j \in B} x_{ij} \geq 1$ where the summation is over all domain pairs (i,j), where i is a domain of protein A and j is a domain in protein B. Thus, the constraint for each interacting protein pair enforces that the values of the variables representing the potentially interacting domain pairs add up to at least 1.0. Following the parsimony principle, the objective function aims to minimize the overall sum of the variables.

Expression profiles are often used to infer regulatory networks. The underlying assumption explored in gene regulatory network reconstruction programs, such as ARACNE [108], is that any change in the expression of a transcription factor should be mirrored by change in the expression of genes regulated by a given transcription factor. Coexpression alone does not provide information on the direction of regulatory relationship; however, expression data can also be used to construct Bayesian networks that can represent the conditional dependence of expression levels (for details of a primer on Bayesian network analysis utilizing expression data, see Ref. [109]; for a recent review, see Ref. [110]). A related approach to orient the edges was proposed by Schadt *et al.* [111, 112].

4 Exploring the Topology of the Interactome

Graph theory provides a unifying language to describe the relations within complex systems, and during recent years has played an increasingly important role in understanding biological systems. The process enables the use of graph-theoretical tools and concepts to interrogate the properties of interaction networks. Currently, several packages are available for visualizing, modeling, and analyzing various types of network, including the popular Cytoscape package [1, 113].

4.1 Global Properties

Studies of protein interaction networks from several model organisms have revealed that such networks often have interesting topological properties. One of the most celebrated such properties is a particular distribution of node degrees (the number of immediate network neighbors for a given node). Specifically, it has been argued that for these networks the degree distribution is consistent with a *power law* [114–116]; in other words, it has been suggested that interaction networks are scale-free. Formally, in a scale-free network the fraction $P(k)$ of nodes in the network having degree k is proportional to $k^{-\gamma}$; that is, $P(k) \sim k^{-\gamma}$, where γ is a constant the value of which is typically in the range $2 < \gamma < 3$. While the accuracy of the supposition that the protein interaction networks are scale-free has been questioned [117–119], the main property of protein interaction networks consistent with a scale-free network – a small number of highly connected nodes (hubs) and a large number of weakly connected nodes – is generally not disputed.

4.2 Network Centrality and Protein Essentiality

Network centrality is a measure of the topological prominence of a node within a network. However, it is unclear whether network centrality is related to protein function. Several approaches exist for the measurement of node centrality, emphasizing the different aspects of network topology [120]. In the context of biological networks, the most studied centrality indices are *degree centrality* and *betweenness centrality*, although many other indices have been also considered [24, 121, 122]. Degree centrality evaluates the node's centrality by the number of its immediate neighbors in the network (Fig. 5), with nodes having a high degree being referred to as *hubs*. In contrast, in the shortest-path betweenness centrality, the node's centrality value is proportional to the fraction of shortest paths between all pairs

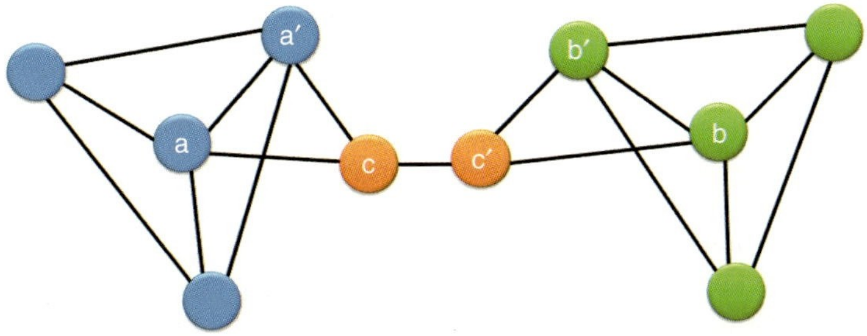

Fig. 5 Illustration of centrality measures. Vertices a, a′, b, and b′ tie for the highest degree centrality of 4. The orange vertices c and c′ have the highest betweenness centrality, since every shortest path between the green component and the blue component must pass through these vertices.

of nodes that pass through a given node (Fig. 5). Thus, betweenness centrality measures whether a node might be central for the information flow within the network. Alternatively, by modeling the interactive as an electric circuit, the node's centrality can be measured by the amount of current passing through it. Specifically, for any pair of nodes the current in the network can be computed under the assumption that a given pair of nodes is taken as the source and the sink. The current flow centrality for an individual node is the sum of currents that pass through this node [24, 123–127].

Subsequently, in an influential report Jeong *et al.* [114] showed that a protein interaction network of *S. cerevisiae*, high-degree nodes contained more essential proteins than would be expected by chance. (In that study, a gene was termed "essential" if its deletion prevented yeast growth under optimal laboratory conditions.) Similarly, nodes with high betweenness centrality (bottlenecks) have been correlated with gene essentiality [128]. The correlation between vertex degree and essentiality has been studied extensively [24, 129–131]; interestingly, however, this relation has not been clearly observed in Y2H networks [9, 24, 130].

It has been proposed that an enrichment of essential proteins among high-degree nodes is an implication of the central role that hubs play in mediating interactions among other proteins. Indeed, the removal of hubs disrupts the connectivity of the network (as indicated by the network diameter or the size of the largest connected component) to a greater degree than does the removal of an equivalent number of random nodes [114, 132, 133]. However, the lack of any clear relation between essentiality and vertex degree in the Y2H-derived networks, as well as computational evidence [24], suggests that the correlation between vertex degree is related to an enrichment in the essential proteins of large dense subnetworks, typically corresponding to complexes. This observation was further confirmed by Wang *et al.*, who also showed that the enrichment of complexes with essential proteins was increased with complex size [134]. In contrast, in yeast binary interactions Y2H, Yu *et al.* identified a relationship between the essentiality of a protein and the number of cellular processes in which it participates.

By combining vertex degree information with expression data, Han *et al.* were able to distinguish two groups of hubs that they referred to as "*party*" hubs and "*date*" hubs [135]. In the case of a party hub, the expression of the hub node is correlated with that of its neighbors, which suggests that all of the interactions may take place simultaneously, or under similar conditions. In contrast, in the case of a date hub, the correlation between the expression of the hub and its neighbor is, on average, low. Thus, these two types of hub are proposed to play distinct roles in the interactome, with party hubs being members of protein complexes or functional modules, and date hubs corresponding to global regulators that possibly link various functional modules.

4.3 Network Modules

In their landmark report, Hartwell *et al.* proposed that functional modules served as a critical level of biological organization [136]. Thus, they defined a functional module as an entity, composed of many types of interacting molecule, the functions of which were separable from those of other modules. Currently, the modularity of biological systems is a widely accepted phenomenon and, indeed, by analyzing an early yeast protein–protein interaction network, Schwikowski *et al.* observed that proteins of known function and cellular location tended to cluster together. Subsequently, the genome-scale reconstruction of biological networks that were enabled by current technologies provided the context for identifying such modules. There is, however, no unique way in which the functional modules could be mathematically defined. Computationally, most methods will seek densely connected subgraphs or clusters by using a variety of heuristics ranging from growing modules from seed clusters, to clustering based on graph-theoretical distance measures, to Markov Chain Monte Carlo (MCMC) clustering approaches [137–154]. Additionally, gene expression information can be utilized to obtain more reliable modules [155–157].

One characteristic property that modules would be expected to satisfy naturally, is that the molecules within the module are more strongly connected between themselves, than are molecules within the module to those outside. This intuition can be formalized by the following concept of modularity. Given a partition of nodes of network with m edges (links) into groups $C_1, C_2, \ldots, C_r$, the modularity of such a partition can be defined as:

$$Q = \frac{1}{2m} \sum_{i,j} \left(A_{ij} - \frac{k_i k_j}{2m} \right) C_{ij}$$

where A_{ij} equals 1 if i and j are connected in the network, and zero otherwise; and C_{ij} is equal to 1 if i and j are in the same group, and zero otherwise. Thus, the defined Q takes values of between -1 and 1, where a positive value of Q indicates that the number of edges within groups is higher than would be expected by chance. Following such (or related) a definition, some methods have identified modules as partitions into "communities" that maximize modularity [150, 158].

Most module-finding algorithms assign each node to at most one module although, in practice, biological modules can overlap and/or form modular hierarchies. Furthermore, a given component may belong to a different module at a different time. Thus, some of the more recently developed approaches have focused on identifying overlapping modules [139, 159], their hierarchy [160], or their dynamics along activity pathways [161].

Given the wealth and diversity of the module-finding algorithms, it is important to provide some means of establishing the biological relevance of uncovered modules. The most commonly applied strategy is to evaluate how well the various methods perform in uncovering either known complexes [162] or potentially overlapping functional modules [163], and/or how well they are conserved through evolution [164]. In a recent evaluation of module-finding algorithms, Song and Singh noted that the performances of various algorithms in uncovering functional modules could differ substantially when run on the same network. Moreover, their relative performances changed, depending on the topological characteristics of the network under consideration. Taken together, these results indicated that there is, at present, no single best approach to this problem.

4.4 Network Motifs and Related Concepts

In Sect. 4.1, it was observed that the vertices in a protein–protein interaction network tend to have a characteristic vertex degree distribution, manifested by a small number of high-degree nodes and a large number of nodes with a very small degree. As a natural extension of node degree, the distribution of small subnetworks can be considered as triangles, squares, and so on. Along this line, Milo *et al.* defined network motifs as subgraphs that occurred in a network much more often than would be expected by chance [165]. The "by chance" occurrence in a given type of network is usually estimated by constructing a set of random networks with the same basic properties as the tested network; for example, the same degree distribution. Subsequently, in a landmark report Milo *et al.* showed that the various networks were characterized by an overrepresentation of certain network motifs. By focusing on directed networks, Milo *et al.* showed – among other findings – an overrepresentation within the gene regulatory network of feed-forward loops and bi-fan network motifs (Fig. 6a). Network motifs have been shown to support specific regulatory functions [166–170]. In the context of undirected protein–protein interaction network, Przulj *et al.* studied the distribution of small induced subgraphs (graphlets) [171]. As a somewhat related concept, but taking advantage of protein functional annotations, Banks *et al.* introduced various network schemas to describe patterns of labeled subgraphs [172]. A network schema consisted of descriptions of proteins (e.g., their molecular functions or putative domains), along with the desired topology and types of interaction (e.g., physical, phosphorylation, or regulatory) (Fig. 6b). In addition to searching for matches to particular network schemas, it is also possible to infer which network schemas are frequent and which are overrepresented in networks [173], and thereby to uncover any general recurring patterns that might underlie a range of biological processes.

5 Comparing Protein–Protein Interaction Networks

The interaction prediction methods discussed in Sect. 3.3 were based on the assumption that, if two proteins interact in one organism, then the orthologs of these two proteins in another organism would also be very likely to interact. With this in mind, Walhout *et al.* coined the term *interologs* to describe such orthologous

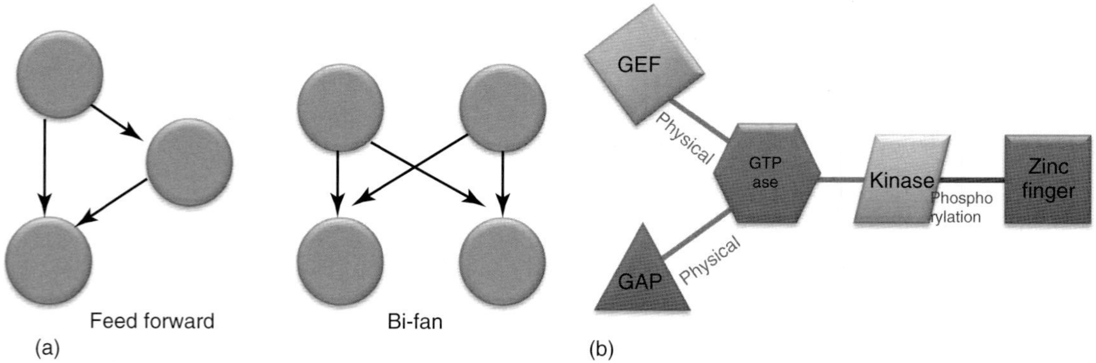

Fig. 6 (a) Examples of network motif in regulatory network motifs: feed-forward and bi-fan motifs. The arrows indicate the direction of regulation; (b) Example of a network schema associated with signaling. The nodes are labeled with specific feature description, such as GTPase and proteins kinase, while the edges are labeled with the interaction type.

pairs of interacting proteins [174]. Indeed, it has been shown that a protein interaction map generated in one species can be used to predict interactions in another species [174–176].

The idea of transferring interaction annotation between organisms can be extended further by comparing whole interaction networks between different organisms. Such network comparison allows a number of fundamental biological questions related to the evolution of protein interaction networks to be addressed, and predictions to be made regarding new functional information relating to proteins and interactions that are at present poorly characterized [177]. Just as sequence alignment represents the cornerstone of sequence comparison, network comparison demands methods for the alignment of biological networks. Indeed, the alignment of interaction networks from different organisms allows the discovery of evolutionarily conserved pathways and functional orthologs [178–182].

From an algorithmic-theoretical perspective, network alignment is a difficult problem which, in the most general formulation, is reduced to the finding of a maximal common subnetwork of two (or more) networks. This classical problem in graph theory is known to be NP-complete and thus difficult to solve computationally [183], and it is unlikely therefore that a fast algorithm to solve the problem in the full generality actually exists. Over the years, however, a number of insightful and efficient algorithms have been proposed that have taken advantage of the various specific properties of biomolecular networks [180, 184–191]. These algorithms allow for global network alignment and local alignments, as well as identifying the alignments of a subnetwork within a larger network.

With these tools at hand, it is possible to search for conserved network regions such as conserved protein complexes and pathways, to identify proteins despite a lack of sequence similarity, to perform the same function in the network, or even to use network similarities to infer evolutionary trees [192, 193].

6 Databases of Protein and Domain Interactions

During recent years, as the amount of data relating to protein and domain interactions has steadily increased, various databases and public repositories have been constructed to enable a sharing of the knowledge, and the support of subsequent studies. Recent reviews of these databases can be found in Refs [194–196], and some representative examples of such databases are provided in this section. The database of interacting proteins (DIP) catalogs experimentally determined protein–protein interactions which are obtained from various resources, including the literature, the Protein Data Bank (PDB), and high-throughput experiments [197, 198]. Both, IntAct [199] and BioGRID [200] contain the details of not only protein–protein interactions but also of other types of interaction, such as protein–small molecule interactions and genetic interactions. MINT [201] annotates each interaction with a score which ranges from 0 to 1 for quantifying the interaction support. As none of the current databases can provide the complete information regarding interactions for all species, some research groups have attempted to extract and unify interaction data from different repositories;

APID [202] and PINA [203] are two representatives of such meta-databases. Furthermore, some databases (e.g., STRING [54, 204, 205] and I2D [206]) include protein–protein interactions predicted by computational approaches. Finally, some dedicated protein interaction data focusing on specific model organisms have been incorporated as a part of the organism-related resources such as FlyBase [207] for *D. melanogaster*, and the *Saccharomyces* Genome Database (SGD) [208] for yeast.

The analysis of domain interactions often provides important insights into both the role and the mechanism of an interaction. The databases 3did, [209], iPFAM [210], and PIBASE [211] allow for exploring the details of domain interactions by studying the three-dimensional structures of interacting domains extracted from the PDB [212]. The Conserved Binding Mode (CBM) database [213] categorizes interacting domains by the Conserved Domain Database (CDD) family type and interaction mode. DOMINE contains both known and predicted domain interactions obtained from two three-dimensional, structure-based databases (iPFAM and 3DIA), and eight different computational approaches [214].

7
Applications

A knowledge of protein interactions can provide important clues concerning the functioning of cells and organisms. In this section, several examples are provided of how interaction networks can be explored to empower biomedical research. Clearly, the examples listed below represent only a small sample of the diverse applications of bimolecular networks.

7.1
Predicting Protein Function

The observation that the majority of interactions occurs between proteins with a common functional assignment [215] has paved the way towards several approaches for predicting protein function, based on the location of the protein in the network in relation to functionally annotated proteins. Sharan *et al.* have subdivided the emerging methods into two types, namely direct and module-assisted schemes [216]. In the direct annotation schemes, individual links in the network are used for inferring the functions of proteins, whereas module-based methods first detect modules of interconnected proteins and then assign protein functions based on the functional annotation of other proteins in the module. Thus, the key step of module-based methods is to use a module-finding approach (as discussed in Sect. 4.3).

At this point, attention will be focused on the direct annotation approaches. The pioneering method developed by Schwikowski *et al.* predicted the biological process of a non-annotated protein by considering its neighboring interactions, and assigning to this protein the annotations that were most frequent among the neighbors [215]. This strategy worked very well for biological networks with a high proportion of annotated proteins, where un-annotated proteins had many annotated neighboring proteins [217]. Subsequently, Hishigaki *et al.* extended this neighbor-derived annotation method by considering all of the proteins within a particular radius, rather than just the direct neighbors of given proteins [218]. Later, Nabieva *et al.* argued that during the transfer of functional annotation from more distant neighbors, account should be taken not only of the distance to the annotated proteins, but also of the topology of the network. This view was utilized in the "functional flow" algorithm described by these authors [217].

The integration of network topology with other types of data, such as gene expression profile, domain context, and text mining may further inform functional annotation [216, 219, 220]. Kourmpetis *et al.* proposed the use of a Markov random field analysis for integrating protein interaction networks with multiple data sources [221].

7.2
Application to Human Diseases

Today, it is increasingly recognized that complex diseases should be studied from the perspective of dysregulated pathways and processes, rather than of individual genes. Indeed, during recent years the availability of genome scale protein–protein interaction and other interaction maps has made it possible to begin such systems-level investigations of human diseases. By following this principle, Chuang *et al.* proposed a network-based method for the classification of breast cancer metastasis [222], the main idea being to combine gene expression profiles with network/pathway information and thus to seek disease-altered subnetworks. The underlying assumption here is that disease-related perturbations, manifested by gene expression changes, would propagate over the interaction network and lead to clusters of perturbed nodes. Importantly, whilst some abnormally expressed genes might differ between disease cases, many clusters would be expected to remain common. Such a perspective has proven to be helpful for disease classification [223, 224], in the identification of disease dysregulated pathways [225, 226], and for identifying disease-associated genes [227–233].

The results of recent studies have also begun to connect disease-perturbed networks with genetic variations. By integrating expression and genotypic data from an intercross population, Cheng *et al.* were able to identify a liver and adipose macrophage-enriched subnetwork that was associated with metabolic disease traits [234]. More recently, Kim *et al.* proposed a method to trace the propagation of the effects of copy number variations through interaction network and the identification of biological pathways affected by such perturbation. This employs gene expression profiles, copy number variation information, and diverse interaction networks [124, 126, 127].

8
Looking Ahead: Towards the Dynamic Interactome

The availability of a genome-scale interactome has provided the possibility of asking general questions with regards to the organization of biological systems, and of informational flow within such systems. Whilst in this chapter attention has been focused on protein–protein interaction networks, for practical applications – including some of those described here – integrative networks combining the various types of interaction are increasingly being used. Today, a growing number of tools are available that allow the interrogation of existing networks to generate testable hypotheses, despite the majority of current studies still treating molecular networks as "hardwired" structures. Clearly, real biological networks are characteristically dynamic – a property on which their functioning is heavily dependent. Recently, Przytycka *et al.* categorized interactome dynamics as being spatial, temporal, and contextual [235]. Hopefully, the second decade of interactome studies will incorporate a shift from static to dynamic network analyses, thereby providing a deeper understanding of molecular systems.

Acknowledgments

These studies were supported by the Intramural Program of the National Institutes of Health, National Library of Medicine.

References

1 Shannon, P., Markiel, A., Ozier, O., Baliga, N.S., Wang, J.T., Ramage, D., Amin, N., Schwikowski, B., Ideker, T. (2003) Cytoscape: a software environment for integrated models of biomolecular interaction networks. *Genome Res.*, **13** (11), 2498–2504.

2 Gavin, A.C., Aloy, P., Grandi, P., Krause, R., Boesche, M., Marzioch, M., Rau, C., Jensen, L.J., Bastuck, S., Dumpelfeld, B., Edelmann, A., Heurtier, M.A., Hoffman, V., Hoefert, C., Klein, K., Hudak, M., Michon, A.M., Schelder, M., Schirle, M., Remor, M., Rudi, T., Hooper, S., Bauer, A., Bouwmeester, T., Casari, G., Drewes, G., Neubauer, G., Rick, J.M., Kuster, B., Bork, P., Russell, R.B., Superti-Furga, G. (2006) Proteome survey reveals modularity of the yeast cell machinery. *Nature*, **440** (7084), 631–636.

3 Krogan, N.J., Cagney, G., Yu, H., Zhong, G., Guo, X., Ignatchenko, A., Li, J., Pu, S., Datta, N., Tikuisis, A.P. (2006) Global landscape of protein complexes in the yeast *Saccharomyces cerevisiae*. *Nature*, **440** (7084), 637–643.

4 Collins, S.R., Kemmeren, P., Zhao, X.C., Greenblatt, J.F., Spencer, F., Holstege, F.C., Weissman, J.S., Krogan, N.J. (2007) Toward a comprehensive atlas of the physical interactome of *Saccharomyces cerevisiae*. *Mol. Cell. Proteomics*, **6** (3), 439–450.

5 Fields, S., Song, O. (1989) A novel genetic system to detect protein-protein interactions. *Nature*, **340** (6230), 245–246.

6 Fromont-Racine, M., Rain, J.C., Legrain, P. (1997) Toward a functional analysis of the yeast genome through exhaustive two-hybrid screens. *Nat. Genet.*, **16** (3), 277–282.

7 Ito, T., Chiba, T., Ozawa, R., Yoshida, M., Hattori, M., Sakaki, Y. (2001) A comprehensive two-hybrid analysis to explore the yeast protein interactome. *Proc. Natl Acad. Sci. USA*, **98** (8), 4569–4574.

8 Uetz, P., Giot, L., Cagney, G., Mansfield, T.A., Judson, R.S., Knight, J.R., Lockshon, D., Narayan, V., Srinivasan, M., Pochart, P., Qureshi-Emili, A., Li, Y., Godwin, B., Conover, D., Kalbfleisch, T., Vijayadamodar, G., Yang, M., Johnston, M., Fields, S., Rothberg, J.M. (2000) A comprehensive analysis of protein-protein interactions in *Saccharomyces cerevisiae*. *Nature*, **403** (6770), 623–627.

9 Yu, H., Braun, P., Yildirim, M.A., Lemmens, I., Venkatesan, K., Sahalie, J., Hirozane-Kishikawa, T., Gebreab, F., Li, N., Simonis, N., Hao, T., Rual, J.F., Dricot, A., Vazquez, A., Murray, R.R., Simon, C., Tardivo, L., Tam, S., Svrzikapa, N., Fan, C., de Smet, A.S., Motyl, A., Hudson, M.E., Park, J., Xin, X., Cusick, M.E., Moore, T., Boone, C., Snyder, M., Roth, F.P., Barabasi, A.L., Tavernier, J., Hill, D.E., Vidal, M. (2008) High-quality binary protein interaction map of the yeast interactome network. *Science*, **322** (5898), 104–110.

10 Reboul, J., Vaglio, P., Rual, J.F., Lamesch, P., Martinez, M., Armstrong, C.M., Li, S., Jacotot, L., Bertin, N., Janky, R., Moore, T., Hudson, J.R., Jr, Hartley, J.L., Brasch, M.A., Vandenhaute, J., Boulton, S., Endress, G.A., Jenna, S., Chevet, E., Papasotiropoulos, V., Tolias, P.P., Ptacek, J., Snyder, M., Huang, R., Chance, M.R., Lee, H., Doucette-Stamm, L., Hill, D.E., Vidal, M. (2003) *C. elegans* ORFeome version 1.1: experimental verification of the genome annotation and resource for proteome-scale protein expression. *Nat. Genet.*, **34** (1), 35–41.

11 Li, S., Armstrong, C.M., Bertin, N., Ge, H., Milstein, S., Boxem, M., Vidalain, P.O., Han, J.D., Chesneau, A., Hao, T., Goldberg,

D.S., Li, N., Martinez, M., Rual, J.F., Lamesch, P., Xu, L., Tewari, M., Wong, S.L., Zhang, L.V., Berriz, G.F., Jacotot, L., Vaglio, P., Reboul, J., Hirozane-Kishikawa, T., Li, Q., Gabel, H.W., Elewa, A., Baumgartner, B., Rose, D.J., Yu, H., Bosak, S., Sequerra, R., Fraser, A., Mango, S.E., Saxton, W.M., Strome, S., Van Den Heuvel, S., Piano, F., Vandenhaute, J., Sardet, C., Gerstein, M., Doucette-Stamm, L., Gunsalus, K.C., Harper, J.W., Cusick, M.E., Roth, F.P., Hill, D.E., Vidal, M. (2004) A map of the interactome network of the metazoan *C. elegans*. *Science*, **303** (5657), 540–543.

12 Giot, L., Bader, J.S., Brouwer, C., Chaudhuri, A., Kuang, B., Li, Y., Hao, Y.L., Ooi, C.E., Godwin, B., Vitols, E., Vijayadamodar, G., Pochart, P., Machineni, H., Welsh, M., Kong, Y., Zerhusen, B., Malcolm, R., Varrone, Z., Collis, A., Minto, M., Burgess, S., McDaniel, L., Stimpson, E., Spriggs, F., Williams, J., Neurath, K., Ioime, N., Agee, M., Voss, E., Furtak, K., Renzulli, R., Aanensen, N., Carrolla, S., Bickelhaupt, E., Lazovatsky, Y., DaSilva, A., Zhong, J., Stanyon, C.A., Finley, R.L., Jr, White, K.P., Braverman, M., Jarvie, T., Gold, S., Leach, M., Knight, J., Shimkets, R.A., McKenna, M.P., Chant, J., Rothberg, J.M. (2003) A protein interaction map of *Drosophila melanogaster*. *Science*, **302** (5651), 1727–1736.

13 Rual, J.F., Venkatesan, K., Hao, T., Hirozane-Kishikawa, T., Dricot, A., Li, N., Berriz, G.F., Gibbons, F.D., Dreze, M., Ayivi-Guedehoussou, N., Klitgord, N., Simon, C., Boxem, M., Milstein, S., Rosenberg, J., Goldberg, D.S., Zhang, L.V., Wong, S.L., Franklin, G., Li, S., Albala, J.S., Lim, J., Fraughton, C., Llamosas, E., Cevik, S., Bex, C., Lamesch, P., Sikorski, R.S., Vandenhaute, J., Zoghbi, H.Y., Smolyar, A., Bosak, S., Sequerra, R., Doucette-Stamm, L., Cusick, M.E., Hill, D.E., Roth, F.P., Vidal, M. (2005) Towards a proteome-scale map of the human protein–protein interaction network. *Nature*, **437** (7062), 1173–1178.

14 Stelzl, U., Worm, U., Lalowski, M., Haenig, C., Brembeck, F.H., Goehler, H., Stroedicke, M., Zenkner, M., Schoenherr, A., Koeppen, S., Timm, J., Mintzlaff, S., Abraham, C., Bock, N., Kietzmann, S., Goedde, A., Toksoz, E., Droege, A., Krobitsch, S., Korn, B., Birchmeier, W., Lehrach, H., Wanker, E.E. (2005) A human protein–protein interaction network: a resource for annotating the proteome. *Cell*, **122** (6), 957–968.

15 Puig, O., Caspary, F., Rigaut, G., Rutz, B., Bouveret, E., Bragado-Nilsson, E., Wilm, M., Seraphin, B. (2001) The tandem affinity purification (TAP) method: A general procedure of protein complex purification. *Methods*, **24** (3), 218–229.

16 Rigaut, G., Shevchenko, A., Rutz, B., Wilm, M., Mann, M., Seraphin, B. (1999) A generic protein purification method for protein complex characterization and proteome exploration. *Nat. Biotechnol.*, **17** (10), 1030–1032.

17 Collins, M.O., Choudhary, J.S. (2008) Mapping multiprotein complexes by affinity purification and mass spectrometry. *Curr. Opin. Biotechnol.*, **19** (4), 324–330.

18 Gavin, A.C., Bosche, M., Krause, R., Grandi, P., Marzioch, M., Bauer, A., Schultz, J., Rick, J.M., Michon, A.M., Cruciat, C.M. (2002) Functional organization of the yeast proteome by systematic analysis of protein complexes. *Nature*, **415** (6868), 141–147.

19 Ho, Y., Gruhler, A., Heilbut, A., Bader, G.D., Moore, L., Adams, S.L., Millar, A., Taylor, P., Bennett, K., Boutilier, K., Yang, L., Wolting, C., Donaldson, I., Schandorff, S., Shewnarane, J., Vo, M., Taggart, J., Goudreault, M., Muskat, B., Alfarano, C., Dewar, D., Lin, Z., Michalickova, K., Willems, A.R., Sassi, H., Nielsen, P.A., Rasmussen, K.J., Andersen, J.R., Johansen, L.E., Hansen, L.H., Jespersen, H., Podtelejnikov, A., Nielsen, E., Crawford, J., Poulsen, V., Sorensen, B.D., Matthiesen, J., Hendrickson, R.C., Gleeson, F., Pawson, T., Moran, M.F., Durocher, D., Mann, M., Hogue, C.W., Figeys, D., Tyers, M. (2002) Systematic identification of protein complexes in *Saccharomyces cerevisiae* by mass spectrometry. *Nature*, **415** (6868), 180–183.

20 Butland, G., Peregrin-Alvarez, J.M., Li, J., Yang, W., Yang, X., Canadien, V., Starostine, A., Richards, D., Beattie, B., Krogan, N. (2005) Interaction network containing conserved and essential protein

complexes in *Escherichia coli*. *Nature*, **433** (7025), 531–537.

21 Deane, C.M., Salwinski, L., Xenarios, I., Eisenberg, D. (2002) Protein interactions: two methods for assessment of the reliability of high throughput observations. *Mol. Cell. Proteomics*, **1** (5), 349–356.

22 Deng, M., Sun, F., Chen, T. (2003) Assessment of the reliability of protein-protein interactions and protein function prediction. *Pac. Symp. Biocomput.*, 140–151.

23 Goll, J., Uetz, P. (2006) The elusive yeast interactome. *Genome Biol.*, **7** (6), 214.1–214.6.

24 Zotenko, E., Mestre, J., O'Leary, D.P., Przytycka, T.M. (2008) Why do hubs in the yeast protein interaction network tend to be essential: reexamining the connection between the network topology and essentiality. *PLoS Comput. Biol.*, **4** (8), e1000140.

25 Marcotte, E.M., Pellegrini, M., Ng, H.L., Rice, D.W., Yeates, T.O., Eisenberg, D. (1999) Detecting protein function and protein-protein interactions from genome sequences. *Science*, **285** (5428), 751–753.

26 Enright, A.J., Iliopoulos, I., Kyrpides, N.C., Ouzounis, C.A. (1999) Protein interaction maps for complete genomes based on gene fusion events. *Nature*, **402** (6757), 86–90.

27 Enright, A.J., Ouzounis, C.A. (2001) Functional associations of proteins in entire genomes by means of exhaustive detection of gene fusions. *Genome Biol.*, **2** (9), R esearch 0034 .1–Research 0034.7.

28 Koonin, E.V., Wolf, Y.I. (2008) Genomics of bacteria and archaea: the emerging dynamic view of the prokaryotic world. *Nucleic Acids Res.*, **36** (21), 6688–6719.

29 Mushegian, A.R., Koonin, E.V. (1996) Gene order is not conserved in bacterial evolution. *Trends Genet.*, **12** (8), 289–290.

30 Dandekar, T., Snel, B., Huynen, M., Bork, P. (1998) Conservation of gene order: a fingerprint of proteins that physically interact. *Trends Biochem. Sci.*, **23** (9), 324–328.

31 Overbeek, R., Fonstein, M., D'Souza, M., Pusch, G.D., Maltsev, N. (1999) The use of gene clusters to infer functional coupling. *Proc. Natl Acad. Sci. USA*, **96** (6), 2896–2901.

32 Tamames, J., Casari, G., Ouzounis, C., Valencia, A. (1997) Conserved clusters of functionally related genes in two bacterial genomes. *J. Mol. Evol.*, **44** (1), 66–73.

33 Huynen, M.A., Snel, B. (2000) Gene and context: integrative approaches to genome analysis. *Adv. Protein Chem.*, **54**, 345–379.

34 Snel, B., Lehmann, G., Bork, P., Huynen, M.A. (2000) STRING: a web-server to retrieve and display the repeatedly occurring neighbourhood of a gene. *Nucleic Acids Res.*, **28** (18), 3442–3444.

35 Gaasterland, T., Ragan, M.A. (1998) Microbial genescapes: phyletic and functional patterns of ORF distribution among prokaryotes. *Microb. Comp. Genomics*, **3** (4), 199–217.

36 Pellegrini, M., Marcotte, E.M., Thompson, M.J., Eisenberg, D., Yeates, T.O. (1999) Assigning protein functions by comparative genome analysis: protein phylogenetic profiles. *Proc. Natl Acad. Sci. USA*, **96** (8), 4285–4288.

37 Tatusov, R.L., Koonin, E.V., Lipman, D.J. (1997) A genomic perspective on protein families. *Science*, **278** (5338), 631–637.

38 Huynen, M.A., Bork, P. (1998) Measuring genome evolution. *Proc. Natl Acad. Sci. USA*, **95** (11), 5849–5856.

39 Zheng, Y., Roberts, R.J., Kasif, S. (2002) Genomic functional annotation using coevolution profiles of gene clusters. *Genome Biol.*, **3** (11), Research 0060 .1–Research 0060.9.

40 De Las Rivas, J., Lozano, J.J., Ortiz, A.R. (2002) Comparative analysis of chloroplast genomes: functional annotation, genome-based phylogeny, and deduced evolutionary patterns. *Genome Res.*, **12** (4), 567–583.

41 Levesque, M., Shasha, D., Kim, W., Surette, M.G., Benfey, P.N. (2003) Trait-to-gene: a computational method for predicting the function of uncharacterized genes. *Curr. Biol.*, **13** (2), 129–133.

42 Martin, M.J., Herrero, J., Mateos, A., Dopazo, J. (2003) Comparing bacterial genomes through conservation profiles. *Genome Res.*, **13** (5), 991–998.

43 Peregrin-Alvarez, J.M., Tsoka, S., Ouzounis, C.A. (2003) The phylogenetic extent of metabolic enzymes and pathways. *Genome Res.*, **13** (3), 422–427.

44 Wu, H., Mao, F., Su, Z., Olman, V., Xu, Y. (2005) Prediction of functional modules based on gene distributions in microbial genomes. *Genome Inform.*, **16** (2), 247–259.

45 Gutierrez, R.A., Larson, M.D., Wilkerson, C. (2004) The plant-specific database. Classification of *Arabidopsis* proteins based on their phylogenetic profile. *Plant Physiol.*, **135** (4), 1888–1892.

46 Mikkelsen, T.S., Galagan, J.E., Mesirov, J.P. (2005) Improving genome annotations using phylogenetic profile anomaly detection. *Bioinformatics*, **21** (4), 464–470.

47 Wu, J., Hu, Z., DeLisi, C. (2006) Gene annotation and network inference by phylogenetic profiling. *BMC Bioinformatics*, **7**, 80.

48 Date, S.V., Marcotte, E.M. (2003) Discovery of uncharacterized cellular systems by genome-wide analysis of functional linkages. *Nat. Biotechnol.*, **21** (9), 1055–1062.

49 Glazko, G.V., Mushegian, A.R. (2004) Detection of evolutionarily stable fragments of cellular pathways by hierarchical clustering of phyletic patterns. *Genome Biol.*, **5** (5), R32.

50 Marcotte, E.M., Xenarios, I., van Der Bliek, A.M., Eisenberg, D. (2000) Localizing proteins in the cell from their phylogenetic profiles. *Proc. Natl Acad. Sci. USA*, **97** (22), 12115–12120.

51 Snel, B., Huynen, M.A. (2004) Quantifying modularity in the evolution of biomolecular systems. *Genome Res.*, **14** (3), 391–397.

52 Yamada, T., Kanehisa, M., Goto, S. (2006) Extraction of phylogenetic network modules from the metabolic network. *BMC Bioinformatics*, **7**, 130.

53 Jothi, R., Przytycka, T., Aravind, L. (2007) Discovering functional linkages and uncharacterized cellular pathways using phylogenetic profile comparisons: a comprehensive assessment. *BMC Bioinformatics*, **8** (1), 173.

54 von Mering, C., Zdobnov, E.M., Tsoka, S., Ciccarelli, F.D., Pereira-Leal, J.B., Ouzounis, C.A., Bork, P. (2003) Genome evolution reveals biochemical networks and functional modules. *Proc. Natl Acad. Sci. USA*, **100** (26), 15428–15433.

55 Sun, J., Xu, J., Liu, Z., Liu, Q., Zhao, A., Shi, T., Li, Y. (2005) Refined phylogenetic profiles method for predicting protein–protein interactions. *Bioinformatics*, **21** (16), 3409–3415.

56 Bowers, P.M., Cokus, S.J., Eisenberg, D., Yeates, T.O. (2004) Use of logic relationships to decipher protein network organization. *Science*, **306** (5705), 2246–2249.

57 Goh, C.S., Bogan, A.A., Joachimiak, M., Walther, D., Cohen, F.E. (2000) Co-evolution of proteins with their interaction partners. *J. Mol. Biol.*, **299** (2), 283–293.

58 Pazos, F., Helmer-Citterich, M., Ausiello, G., Valencia, A. (1997) Correlated mutations contain information about protein-protein interaction. *J. Mol. Biol.*, **271** (4), 511–523.

59 Pazos, F., Valencia, A. (2001) Similarity of phylogenetic trees as indicator of protein-protein interaction. *Protein Eng.*, **14** (9), 609–614.

60 Goh, C.S., Cohen, F.E. (2002) Co-evolutionary analysis reveals insights into protein–protein interactions. *J. Mol. Biol.*, **324** (1), 177–192.

61 Jothi, R., Cherukuri, P.F., Tasneem, A., Przytycka, T.M. (2006) Co-evolutionary analysis of domains in interacting proteins reveals insights into domain–domain interactions mediating protein–protein interactions. *J. Mol. Biol.*, **362** (4), 861–875.

62 Gertz, J., Elfond, G., Shustrova, A., Weisinger, M., Pellegrini, M., Cokus, S., Rothschild, B. (2003) Inferring protein interactions from phylogenetic distance matrices. *Bioinformatics*, **19** (16), 2039–2045.

63 Pazos, F., Valencia, A. (2002) In silico two-hybrid system for the selection of physically interacting protein pairs. *Proteins*, **47** (2), 219–227.

64 Sato, T., Yamanishi, Y., Kanehisa, M., Toh, H. (2005) The inference of protein-protein interactions by co-evolutionary analysis is improved by excluding the information about the phylogenetic relationships. *Bioinformatics*, **21** (17), 3482–3489.

65 Pazos, F., Ranea, J.A., Juan, D., Sternberg, M.J. (2005) Assessing protein co-evolution in the context of the tree of life assists in the prediction of the interactome. *J. Mol. Biol.*, **352** (4), 1002–1015.

66 Tan, S.H., Zhang, Z., Ng, S.K. (2004) ADVICE: automated detection and validation of interaction by co-evolution. *Nucleic Acids Res.*, **32** (Web Server issue), W69–W72.

67 Craig, R.A., Liao, L. (2007) Phylogenetic tree information aids supervised learning for predicting protein-protein interaction based on distance matrices. *BMC Bioinformatics*, **8**, 6.
68 Kann, M.G., Jothi, R., Cherukuri, P.F., Przytycka, T.M. (2007) Predicting protein domain interactions from coevolution of conserved regions. *Proteins*, **67** (4), 811–820.
69 Hakes, L., Lovell, S.C., Oliver, S.G., Robertson, D.L. (2007) Specificity in protein interactions and its relationship with sequence diversity and coevolution. *Proc. Natl Acad. Sci. USA*, **104** (19), 7999–8004.
70 Kann, M.G., Shoemaker, B.A., Panchenko, A.R., Przytycka, T.M. (2009) Correlated evolution of interacting proteins: looking behind the mirrortree. *J. Mol. Biol.*, **385** (1), 91–98.
71 Juan, D., Pazos, F., Valencia, A. (2008) High-confidence prediction of global interactomes based on genome-wide coevolutionary networks. *Proc. Natl Acad. Sci. USA*, **105** (3), 934–939.
72 Tillier, E.R., Charlebois, R.L. (2009) The human protein coevolution network. *Genome Res.*, **19** (10), 1861–1871.
73 Jothi, R., Kann, M.G., Przytycka, T.M. (2005) Predicting protein-protein interaction by searching evolutionary tree automorphism space. *Bioinformatics*, **21** (Suppl. 1), i241–i250.
74 Ramani, A.K., Marcotte, E.M. (2003) Exploiting the co-evolution of interacting proteins to discover interaction specificity. *J. Mol. Biol.*, **327** (1), 273–284.
75 Apic, G., Gough, J., Teichmann, S.A. (2001) An insight into domain combinations. *Bioinformatics*, **17** (Suppl. 1), S83–S89.
76 Huang, C., Morcos, F., Kanaan, S.P., Wuchty, S., Chen, D.Z., Izaguirre, J.A. (2007) Predicting protein-protein interactions from protein domains using a set cover approach. *IEEE/ACM Trans. Comput. Biol. Bioinform.*, **4** (1), 78–87.
77 Kim, W.K., Park, J., Suh, J.K. (2002) Large scale statistical prediction of protein-protein interaction by potentially interacting domain (PID) pair. *Genome Inform.*, **13**, 42–50.
78 Nguyen, T.P., Ho, T.B. (2006) Discovering signal transduction networks using signaling domain-domain interactions. *Genome Inform.*, **17** (2), 35–45.
79 Singhal, M., Resat, H. (2007) A domain-based approach to predict protein-protein interactions. *BMC Bioinformatics*, **8**, 199.
80 Wojcik, J., Schachter, V. (2001) Protein–protein interaction map inference using interacting domain profile pairs. *Bioinformatics*, **17** (Suppl. 1), S296–S305.
81 Deng, M., Mehta, S., Sun, F., Chen, T. (2002) Inferring domain-domain interactions from protein–protein interactions. *Genome Res.*, **12** (10), 1540–1548.
82 Sprinzak, E., Margalit, H. (2001) Correlated sequence-signatures as markers of protein–protein interaction. *J. Mol. Biol.*, **311** (4), 681–692.
83 Wuchty, S. (2006) Topology and weights in a protein domain interaction network – a novel way to predict protein interactions. *BMC Genomics*, **7**, 122.
84 Liu, M., Chen, X.W., Jothi, R. (2009) Knowledge-guided inference of domain–domain interactions from incomplete protein–protein interaction networks. *Bioinformatics*, **25** (19), 2492–2499.
85 Guimaraes, K.S., Jothi, R., Zotenko, E., Przytycka, T.M. (2006) Predicting domain–domain interactions using a parsimony approach. *Genome Biol.*, **7** (11), R104.
86 Guimaraes, K.S., Przytycka, T.M. (2008) Protein Interaction Network Based Prediction of Domain-Domain and Domain-Peptide Interactions, in: Panchenko, A.R., Przytycka, T.M. (Eds) *Protein-Protein Interactions and Networks*, Springer-Verlag, London, pp. 83–98.
87 Ng, S.K., Zhang, Z., Tan, S.H. (2003) Integrative approach for computationally inferring protein domain interactions. *Bioinformatics*, **19** (8), 923–929.
88 Kluger, Y., Basri, R., Chang, J.T., Gerstein, M. (2003) Spectral biclustering of microarray data: coclustering genes and conditions. *Genome Res.*, **13** (4), 703–716.
89 Zhao, X.M., Chen, L., Aihara, K. (2010) A discriminative approach for identifying domain–domain interactions from protein-protein interactions. *Proteins*, **78** (5), 1243–1253.

90 Wang, H., Segal, E., Ben-Hur, A., Li, Q.R., Vidal, M., Koller, D. (2007) InSite: a computational method for identifying protein–protein interaction binding sites on a proteome-wide scale. *Genome Biol.*, **8** (9), R192.

91 Riley, R., Lee, C., Sabatti, C., Eisenberg, D. (2005) Inferring protein domain interactions from databases of interacting proteins. *Genome Biol.*, **6** (10), R89.

92 Lee, H., Deng, M., Sun, F., Chen, T. (2006) An integrated approach to the prediction of domain-domain interactions. *BMC Bioinformatics*, **7**, 269.

93 Wang, R.S., Wang, Y., Wu, L.Y., Zhang, X.S., Chen, L. (2007) Analysis on multi-domain cooperation for predicting protein-protein interactions. *BMC Bioinformatics*, **8**, 391.

94 Guimaraes, K.S., Przytycka, T.M. (2008) Interrogating domain-domain interactions with parsimony based approaches. *BMC Bioinformatics*, **9**, 171.

95 Jansen, R., Yu, H., Greenbaum, D., Kluger, Y., Krogan, N.J., Chung, S., Emili, A., Snyder, M., Greenblatt, J.F., Gerstein, M. (2003) A Bayesian networks approach for predicting protein–protein interactions from genomic data. *Science*, **302** (5644), 449–453.

96 Rhodes, D.R., Tomlins, S.A., Varambally, S., Mahavisno, V., Barrette, T., Kalyana-Sundaram, S., Ghosh, D., Pandey, A., Chinnaiyan, A.M. (2005) Probabilistic model of the human protein–protein interaction network. *Nat. Biotechnol.*, **23** (8), 951–959.

97 Tirosh, I., Barkai, N. (2005) Computational verification of protein-protein interactions by orthologous co-expression. *BMC Bioinformatics*, **6**, 40.

98 Soong, T.T., Wrzeszczynski, K.O., Rost, B. (2008) Physical protein–protein interactions predicted from microarrays. *Bioinformatics*, **24** (22), 2608–2614.

99 Jansen, R., Greenbaum, D., Gerstein, M. (2002) Relating whole-genome expression data with protein–protein interactions. *Genome Res.*, **12** (1), 37–46.

100 Kemmeren, P., van Berkum, N.L., Vilo, J., Bijma, T., Donders, R., Brazma, A., Holstege, F.C. (2002) Protein interaction verification and functional annotation by integrated analysis of genome-scale data. *Mol. Cell*, **9** (5), 1133–1143.

101 von Mering, C., Krause, R., Snel, B., Cornell, M., Oliver, S.G., Fields, S., Bork, P. (2002) Comparative assessment of large-scale data sets of protein–protein interactions. *Nature*, **417** (6887), 399–403.

102 Eisen, M.B., Spellman, P.T., Brown, P.O., Botstein, D. (1998) Cluster analysis and display of genome-wide expression patterns. *Proc. Natl Acad. Sci. USA*, **95** (25), 14863–14868.

103 Marcotte, E.M., Pellegrini, M., Thompson, M.J., Yeates, T.O., Eisenberg, D. (1999) A combined algorithm for genome-wide prediction of protein function. *Nature*, **402** (6757), 83–86.

104 Greenbaum, D., Luscombe, N.M., Jansen, R., Qian, J., Gerstein, M. (2001) Interrelating different types of genomic data, from proteome to secretome: 'oming in on function'. *Genome Res.*, **11** (9), 1463–1468.

105 Ramani, A.K., Li, Z., Hart, G.T., Carlson, M.W., Boutz, D.R., Marcotte, E.M. (2008) A map of human protein interactions derived from co-expression of human mRNAs and their orthologs. *Mol. Syst. Biol.*, **4** (180), 1–16.

106 Costello, J.C., Dalkilic, M.M., Beason, S.M., Gehlhausen, J.R., Patwardhan, R., Middha, S., Eads, B.D., Andrews, J.R. (2009) Gene networks in *Drosophila melanogaster*: integrating experimental data to predict gene function. *Genome Biol.*, **10** (9), R97.

107 Lee, I., Date, S.V., Adai, A.T., Marcotte, E.M. (2004) A probabilistic functional network of yeast genes. *Science*, **306** (5701), 1555–1558.

108 Margolin, A.A., Nemenman, I., Basso, K., Wiggins, C., Stolovitzky, G., Dalla Favera, R., Califano, A. (2006) ARACNE: an algorithm for the reconstruction of gene regulatory networks in a mammalian cellular context. *BMC Bioinformatics*, **7** (Suppl. 1), S7.

109 Pe'er, D. (2005) Bayesian network analysis of signaling networks: a primer. *Sci. STKE*, **2005** (281), l4.

110 Alterovitz, G., Liu, J., Afkhami, E., Ramoni, M.F. (2007) Bayesian methods for proteomics. *Proteomics*, **7** (16), 2843–2855.

111 Schadt, E.E. (2005) Exploiting naturally occurring DNA variation and molecular profiling data to dissect disease and drug response traits. *Curr. Opin. Biotechnol.*, **16** (6), 647–654.

112 Sieberts, S.K., Schadt, E.E. (2007) Moving toward a system genetics view of disease. *Mamm. Genome*, **18** (6–7), 389–401.

113 Cline, M.S., Smoot, M., Cerami, E., Kuchinsky, A., Landys, N., Workman, C., Christmas, R., Avila-Campilo, I., Creech, M., Gross, B., Hanspers, K., Isserlin, R., Kelley, R., Killcoyne, S., Lotia, S., Maere, S., Morris, J., Ono, K., Pavlovic, V., Pico, A.R., Vailaya, A., Wang, P.L., Adler, A., Conklin, B.R., Hood, L., Kuiper, M., Sander, C., Schmulevich, I., Schwikowski, B., Warner, G.J., Ideker, T., Bader, G.D. (2007) Integration of biological networks and gene expression data using Cytoscape. *Nat. Protoc.*, **2** (10), 2366–2382.

114 Jeong, H., Mason, S.P., Barabasi, A.L., Oltvai, Z.N. (2001) Lethality and centrality in protein networks. *Nature*, **411** (6833), 41–42.

115 Wagner, A. (2001) The yeast protein interaction network evolves rapidly and contains few redundant duplicate genes. *Mol. Biol. Evol.*, **18** (7), 1283–1292.

116 Yook, S.H., Oltvai, Z.N., Barabasi, A.L. (2004) Functional and topological characterization of protein interaction networks. *Proteomics*, **4** (4), 928–942.

117 Przulj, N., Corneil, D.G., Jurisica, I. (2004) Modeling interactome: scale-free or geometric? *Bioinformatics*, **20** (18), 3508–3515.

118 Han, J.D., Dupuy, D., Bertin, N., Cusick, M.E., Vidal, M. (2005) Effect of sampling on topology predictions of protein-protein interaction networks. *Nat. Biotechnol.*, **23** (7), 839–844.

119 Tanaka, R., Yi, T.M., Doyle, J. (2005) Some protein interaction data do not exhibit power law statistics. *FEBS Lett.*, **579** (23), 5140–5144.

120 Brandes, U., Erlebach, T. (Eds) (2005) *Network Analysis: Methodological Foundations*, Springer, Berlin, New York.

121 Zielinski, R., Przytycki, P.F., Zheng, J., Zhang, D., Przytycka, T.M., Capala, J. (2009) The crosstalk between EGF, IGF, and insulin cell signaling pathways – computational and experimental analysis. *BMC Syst. Biol.*, **3**, 88.

122 Lohmann, G., Margulies, D.S., Horstmann, A., Pleger, B., Lepsien, J., Goldhahn, D., Schloegl, H., Stumvoll, M., Villringer, A., Turner, R. (2010) Eigenvector centrality mapping for analyzing connectivity patterns in FMRI data of the human brain. *PLoS ONE*, **5** (4), e10232.

123 Missiuro, P.V., Liu, K., Zou, L., Ross, B.C., Zhao, G., Liu, J.S., Ge, H. (2009) Information flow analysis of interactome networks. *PLoS Comput. Biol.*, **5** (4), e1000350.

124 Kim, Y.-A., Wuchty, S., Przytycka, T.M. (2010) Simultaneous identification of causal genes and dysregulated pathways in complex diseases. *Lect. Notes Bioinformatics*, **6044** (RECOMB 2010), 263–281.

125 Suthram, S., Beyer, A., Karp, R.M., Eldar, Y., Ideker, T. (2008) eQED: an efficient method for interpreting eQTL associations using protein networks. *Mol. Syst. Biol.*, **4**, 162.

126 Kim, Y.-A., Przytycki, J.H., Wuchty, S., Przytycka, T.M. (2011) Modeling information flow in molecular networks. *Phys. Biol.*, **8** (3), 035012.

127 Kim, Y.-A., Wuchty, S., Przytycka, T.M. (2011) Identifying causal genes and dysregulated pathways in complex diseases. *PLoS Comput. Biol.*, **7** (3), e1001095.

128 Yu, H., Kim, P.M., Sprecher, E., Trifonov, V., Gerstein, M. (2007) The importance of bottlenecks in protein networks: correlation with gene essentiality and expression dynamics. *PLoS Comput. Biol.*, **3** (4), e59.

129 He, X., Zhang, J. (2006) Why do hubs tend to be essential in protein networks? *PLoS Genet.*, **2** (6), e88.

130 Batada, N.N., Hurst, L.D., Tyers, M. (2006) Evolutionary and physiological importance of hub proteins. *PLoS Comput. Biol.*, **2** (7), e88.

131 Hahn, M.W., Kern, A.D. (2005) Comparative genomics of centrality and essentiality in three eukaryotic protein-interaction networks. *Mol. Biol. Evol.*, **22** (4), 803–806.

132 Albert, R., Jeong, H., Barabasi, A.L. (2000) Error and attack tolerance of complex networks. *Nature*, **406** (6794), 378–382.

133 Yu, H., Greenbaum, D., Xin Lu, H., Zhu, X., Gerstein, M. (2004) Genomic analysis of essentiality within protein networks. *Trends Genet.*, **20** (6), 227–231.

134 Wang, H., Kakaradov, B., Collins, S.R., Karotki, L., Fiedler, D., Shales, M., Shokat, K.M., Walther, T.C., Krogan, N.J., Koller, D. (2009) A complex-based reconstruction of the *Saccharomyces cerevisiae* interactome. *Mol. Cell. Proteomics*, **8** (6), 1361–1381.

135 Han, J.D., Bertin, N., Hao, T., Goldberg, D.S., Berriz, G.F., Zhang, L.V., Dupuy, D., Walhout, A.J., Cusick, M.E., Roth, F.P., Vidal, M. (2004) Evidence for dynamically organized modularity in the yeast protein–protein interaction network. *Nature*, **430** (6995), 88–93.

136 Hartwell, L.H., Hopfield, J.J., Leibler, S., Murray, A.W. (1999) From molecular to modular cell biology. *Nature*, **402** (6761, Suppl.), C47–C52.

137 Bader, G.D., Hogue, C.W. (2003) An automated method for finding molecular complexes in large protein interaction networks. *BMC Bioinformatics*, **4**, 2.

138 Spirin, V., Mirny, L.A. (2003) Protein complexes and functional modules in molecular networks. *Proc. Natl Acad. Sci. USA*, **100** (21), 12123–12128.

139 Adamcsek, B.Z., Palla, G., Farkas, I.S.J., Derãnyi, I., Vicsek, T.S. (2006) CFinder: locating cliques and overlapping modules in biological networks. *Bioinformatics*, **22** (8), 1021–1023.

140 Altaf-Ul-Amin, M., Shinbo, Y., Mihara, K., Kurokawa, K., Kanaya, S. (2006) Development and implementation of an algorithm for detection of protein complexes in large interaction networks. *BMC Bioinformatics*, **7**, 207.

141 Rives, A.W., Galitski, T. (2003) Modular organization of cellular networks. *Proc. Natl Acad. Sci. USA*, **100** (3), 1128–1133.

142 Arnau, V., Mars, S., Marín, I. (2005) Iterative cluster analysis of protein interaction data. *Bioinformatics*, **21** (3), 364–378.

143 Bader, J.S. (2003) Greedily building protein networks with confidence. *Bioinformatics*, **19** (15), 1869–1874.

144 Asthana, S., King, O.D., Gibbons, F.D., Roth, F.P. (2004) Predicting protein complex membership using probabilistic network reliability. *Genome Res.*, **14** (6), 1170–1175.

145 Brun, C., Chevenet, F., Martin, D., Wojcik, J., Guenoche, A., Jacq, B. (2003) Functional classification of proteins for the prediction of cellular function from a protein-protein interaction network. *Genome Biol.*, **5** (1), R6.

146 King, A.D., Przulj, N., Jurisica, I. (2004) Protein complex prediction via cost-based clustering. *Bioinformatics*, **20** (17), 3013–3020.

147 Luo, F., Yang, Y., Chen, C.F., Chang, R., Zhou, J., Scheuermann, R.H. (2007) Modular organization of protein interaction networks. *Bioinformatics*, **23** (2), 207–214.

148 Pereira-Leal, J.B., Enright, A.J., Ouzounis, C.A. (2004) Detection of functional modules from protein interaction networks. *Proteins*, **54** (1), 49–57.

149 Wang, C., Ding, C., Yang, Q., Holbrook, S.R. (2007) Consistent dissection of the protein interaction network by combining global and local metrics. *Genome Biol.*, **8** (12), R271.

150 Newman, M.E. (2006) Modularity and community structure in networks. *Proc. Natl Acad. Sci. USA*, **103** (23), 8577–8582.

151 Dunn, R., Dudbridge, F., Sanderson, C.M. (2005) The use of edge-betweenness clustering to investigate biological function in protein interaction networks. *BMC Bioinformatics*, **6**, 39.

152 Qi, Y., Balem, F., Faloutsos, C., Klein-Seetharaman, J., Bar-Joseph, Z. (2008) Protein complex identification by supervised graph local clustering. *Bioinformatics*, **24** (13), i250–i258.

153 Navlakha, S., Schatz, M.C., Kingsford, C. (2009) Revealing biological modules via graph summarization. *J. Comput. Biol.*, **16** (2), 253–264.

154 Jiang, P., Singh, M. (2010) SPICi: a fast clustering algorithm for large biological networks. *Bioinformatics*, **26** (8), 1105–1111.

155 Chen, J., Yuan, B. (2006) Detecting functional modules in the yeast protein–protein interaction network. *Bioinformatics*, **22** (18), 2283–2290.

156 Maraziotis, I.A., Dimitrakopoulou, K., Bezerianos, A. (2007) Growing functional modules from a seed protein via integration of protein interaction and gene expression data. *BMC Bioinformatics*, **8**, 408.

157 Feng, J., Jiang, R., Jiang, T. (2011) A max-flow based approach to the identification

of protein complexes using protein interaction and microarray data. *IEEE/ACM Trans. Comput. Biol. Bioinform.*, **8** (3), 62 1–634.

158 Rosvall, M., Bergstrom, C.T. (2007) An information-theoretic framework for resolving community structure in complex networks. *Proc. Natl Acad. Sci. USA*, **104** (18), 7327–7331.

159 Palla, G., Derenyi, I., Farkas, I., Vicsek, T. (2005) Uncovering the overlapping community structure of complex networks in nature and society. *Nature*, **435** (7043), 814–818.

160 Sales-Pardo, M., Guimera, R., Moreira, A.A., Amaral, L.A. (2007) Extracting the hierarchical organization of complex systems. *Proc. Natl Acad. Sci. USA*, **104** (39), 15224–15229.

161 Zotenko, E., Guimaraes, K.S., Jothi, R., Przytycka, T.M. (2006) Decomposition of overlapping protein complexes: a graph theoretical method for analyzing static and dynamic protein associations. *Algorithms Mol. Biol.*, **1** (1), 7.

162 Brohee, S., van Helden, J. (2006) Evaluation of clustering algorithms for protein-protein interaction networks. *BMC Bioinformatics*, **7**, 488.

163 Song, J., Singh, M. (2009) How and when should interactome-derived clusters be used to predict functional modules and protein function? *Bioinformatics*, **25** (23), 3143–3150.

164 Poyatos, J.F., Hurst, L.D. (2004) How biologically relevant are interaction-based modules in protein networks? *Genome Biol.*, **5** (11), R93.

165 Milo, R., Shen-Orr, S., Itzkovitz, S., Kashtan, N., Chklovskii, D., Alon, U. (2002) Network motifs: simple building blocks of complex networks. *Science*, **298** (5594), 824–827.

166 Goentoro, L., Shoval, O., Kirschner, M.W., Alon, U. (2009) The incoherent feedforward loop can provide fold-change detection in gene regulation. *Mol. Cell*, **36** (5), 894–899.

167 Rosenfeld, N., Young, J.W., Alon, U., Swain, P.S., Elowitz, M.B. (2007) Accurate prediction of gene feedback circuit behavior from component properties. *Mol. Syst. Biol.*, **3**, 143.

168 Alon, U. (2007) Network motifs: theory and experimental approaches. *Nat. Rev. Genet.*, **8** (6), 450–461.

169 Mangan, S., Zaslaver, A., Alon, U. (2003) The coherent feedforward loop serves as a sign-sensitive delay element in transcription networks. *J. Mol. Biol.*, **334** (2), 197–204.

170 Mangan, S., Alon, U. (2003) Structure and function of the feed-forward loop network motif. *Proc. Natl Acad. Sci. USA*, **100** (21), 11980–11985.

171 Przulj, N., Corneil, D.G., Jurisica, I. (2006) Efficient estimation of graphlet frequency distributions in protein–protein interaction networks. *Bioinformatics*, **22** (8), 974–980.

172 Banks, E., Nabieva, E., Chazelle, B., Singh, M. (2008) Organization of physical interactomes as uncovered by network schemas. *PLoS Comput. Biol.*, **4** (10), e1000203.

173 Banks, E., Nabieva, E., Peterson, R., Singh, M. (2008) NetGrep: fast network schema searches in interactomes. *Genome Biol.*, **9** (9), R138.

174 Walhout, A.J.M., Sordella, R., Lu, X., Hartley, J.L., Temple, G.F., Brasch, M.A., Thierry-Mieg, N., Vidal, M. (2000) Protein interaction mapping in *C. elegans* using proteins involved in vulval development. *Science*, **287** (5450), 116–122.

175 Yu, H., Luscombe, N.M., Lu, H.X., Zhu, X., Xia, Y., Han, J.D., Bertin, N., Chung, S., Vidal, M., Gerstein, M. (2004) Annotation transfer between genomes: protein–protein interologs and protein–DNA regulogs. *Genome Res.*, **14** (6), 1107–1118.

176 Matthews, L.R., Vaglio, P., Reboul, J., Ge, H., Davis, B.P., Garrels, J., Vincent, S., Vidal, M. (2001) Identification of potential interaction networks using sequence-based searches for conserved protein–protein interactions or "interologs". *Genome Res.*, **11** (12), 2120–2126.

177 Sharan, R., Ideker, T. (2006) Modeling cellular machinery through biological network comparison. *Nat. Biotechnol.*, **24** (4), 427–433.

178 Kelley, B.P., Yuan, B., Lewitter, F., Sharan, R., Stockwell, B.R., Ideker, T. (2004) PathBLAST: a tool for alignment of protein interaction networks. *Nucleic Acids Res.*, **32** (Web Server issue), W83–W88.

179 Kelley, B.P., Sharan, R., Karp, R.M., Sittler, T., Root, D.E., Stockwell, B.R., Ideker, T. (2003) Conserved pathways within bacteria

and yeast as revealed by global protein network alignment. *Proc. Natl Acad. Sci. USA*, **100** (20), 11394–11399.

180 Sharan, R., Suthram, S., Kelley, R.M., Kuhn, T., McCuine, S., Uetz, P., Sittler, T., Karp, R.M., Ideker, T. (2005) Conserved patterns of protein interaction in multiple species. *Proc. Natl Acad. Sci. USA*, **102** (6), 1974–1979.

181 Bandyopadhyay, S., Sharan, R., Ideker, T. (2006) Systematic identification of functional orthologs based on protein network comparison. *Genome Res.*, **16** (3), 428–435.

182 Ogata, H., Fujibuchi, W., Goto, S., Kanehisa, M. (2000) A heuristic graph comparison algorithm and its application to detect functionally related enzyme clusters. *Nucleic Acids Res.*, **28** (20), 4021–4028.

183 Gary, M.R., Johnson, D.S. (1979) *Computers and Intractability a Guide to the Theory of NP-Completeness*, WH. Freeman and Co.

184 Flannick, J., Novak, A., Srinivasan, B.S., McAdams, H.H., Batzoglou, S. (2006) Graemlin: general and robust alignment of multiple large interaction networks. *Genome Res.*, **16** (9), 1169–1181.

185 Koyuturk, M., Kim, Y., Topkara, U., Subramaniam, S., Szpankowski, W., Grama, A. (2006) Pairwise alignment of protein interaction networks. *J. Comput. Biol.*, **13** (2), 182–199.

186 Berg, J., Lassig, M. (2006) Cross-species analysis of biological networks by Bayesian alignment. *Proc. Natl Acad. Sci. USA*, **103** (29), 10967–10972.

187 Kalaev, M., Bafna, V., Sharan, R. (2009) Fast and accurate alignment of multiple protein networks. *J. Comput. Biol.*, **16** (8), 989–999.

188 Singh, R., Xu, J., Berger, B. (2008) Global alignment of multiple protein interaction networks with application to functional orthology detection. *Proc. Natl Acad. Sci. USA*, **105** (35), 12763–12768.

189 Pinter, R.Y., Rokhlenko, O., Yeger-Lotem, E., Ziv-Ukelson, M. (2005) Alignment of metabolic pathways. *Bioinformatics*, **21** (16), 3401–3408.

190 Cheng, Q., Harrison, R., Zelikovsky, A. (2009) MetNetAligner: a web service tool for metabolic network alignments. *Bioinformatics*, **25** (15), 1989–1990.

191 Dost, B., Shlomi, T., Gupta, N., Ruppin, E., Bafna, V., Sharan, R. (2008) QNet: a tool for querying protein interaction networks. *J. Comput. Biol.*, **15** (7), 913–925.

192 Kuchaiev, O., Milenkovic, T., Memisevic, V., Hayes, W., Przulj, N. (2010) Topological network alignment uncovers biological function and phylogeny. *J. R. Soc. Interface*, **7** (50), 1341–1354.

193 Erten, S., Li, X., Bebek, G., Li, J., Koyuturk, M. (2009) Phylogenetic analysis of modularity in protein interaction networks. *BMC Bioinformatics*, **10**, 333.

194 Shoemaker, B.A., Panchenko, A.R. (2007) Deciphering protein-protein interactions. Part I. Experimental techniques and databases. *PLoS Comput. Biol.*, **3** (3), e42.

195 Lehne, B., Schlitt, T. (2009) Protein–protein interaction databases: keeping up with growing interactomes. *Hum. Genomics*, **3** (3), 291–297.

196 De Las Rivas, J. Fontanillo, C. (2010) Protein–protein interactions essentials: key concepts to building and analyzing interactome networks. *PLoS Comput. Biol.*, **6** (6), e1000807.

197 Salwinski, L., Miller, C.S., Smith, A.J., Pettit, F.K., Bowie, J.U., Eisenberg, D. (2004) The database of interacting proteins: 2004 update. *Nucleic Acids Res.*, **32** (Database issue), D449–D451.

198 Xenarios, I., Rice, D.W., Salwinski, L., Baron, M.K., Marcotte, E.M., Eisenberg, D. (2000) DIP: the database of interacting proteins. *Nucleic Acids Res.*, **28** (1), 289–291.

199 Aranda, B., Achuthan, P., Alam-Faruque, Y., Armean, I., Bridge, A., Derow, C., Feuermann, M., Ghanbarian, A.T., Kerrien, S., Khadake, J., Kerssemakers, J., Leroy, C., Menden, M., Michaut, M., Montecchi-Palazzi, L., Neuhauser, S.N., Orchard, S., Perreau, V., Roechert, B., van Eijk, K., Hermjakob, H. (2010) The IntAct molecular interaction database in 2010. *Nucleic Acids Res.*, **38** (Database issue), D525–D531.

200 Stark, C., Breitkreutz, B.J., Reguly, T., Boucher, L., Breitkreutz, A., Tyers, M. (2006) BioGRID: a general repository for interaction datasets. *Nucleic Acids Res.*, **34** (Database issue), D535–D539.

201 Ceol, A., Chatr Aryamontri, A., Licata, L., Peluso, D., Briganti, L., Perfetto, L., Castagnoli, L., Cesareni, G. (2010) MINT, the molecular interaction database: 2009

update. *Nucleic Acids Res.*, **38** (Database issue), D532–D539.

202 Prieto, C., De Las Rivas, J. (2006) APID: Agile protein interaction data analyzer. *Nucleic Acids Res.*, **34** (Web Server issue), W298–W302.

203 Wu, J., Vallenius, T., Ovaska, K., Westermarck, J., Makela, T.P., Hautaniemi, S. (2009) Integrated network analysis platform for protein-protein interactions. *Nat. Methods*, **6** (1), 75–77.

204 Jensen, L.J., Kuhn, M., Stark, M., Chaffron, S., Creevey, C., Muller, J., Doerks, T., Julien, P., Roth, A., Simonovic, M., Bork, P., von Mering, C. (2009) STRING 8 – a global view on proteins and their functional interactions in 630 organisms. *Nucleic Acids Res.*, **37** (Database issue), D412–D416.

205 von Mering, C., Huynen, M., Jaeggi, D., Schmidt, S., Bork, P., Snel, B. (2003) STRING: a database of predicted functional associations between proteins. *Nucleic Acids Res.*, **31**, 258–261.

206 Brown, K.R., Jurisica, I. (2007) Unequal evolutionary conservation of human protein interactions in interologous networks. *Genome Biol.*, **8** (5), R95.

207 FlyBase Consortium (2003) The FlyBase database of the *Drosophila* genome projects and community literature. *Nucleic Acids Res.*, **31** (1), 172–175.

208 Cherry, J.M., Ball, C., Weng, S., Juvik, G., Schmidt, R., Adler, C., Dunn, B., Dwight, S., Riles, L., Mortimer, R.K., Botstein, D. (1997) Genetic and physical maps of *Saccharomyces cerevisiae. Nature*, **387** (6632, Suppl.), 67–73.

209 Stein, A., Panjkovich, A., Aloy, P. (2009) 3did update: domain-domain and peptide-mediated interactions of known 3D structure. *Nucleic Acids Res.*, **37** (Database issue), D300–D304.

210 Finn, R.D., Marshall, M., Bateman, A. (2005) iPfam: visualization of protein-protein interactions in PDB at domain and amino acid resolutions. *Bioinformatics*, **21** (3), 410–412.

211 Davis, F.P., Sali, A. (2005) PIBASE: a comprehensive database of structurally defined protein interfaces. *Bioinformatics*, **21** (9), 1901–1907.

212 Berman, H.M., Westbrook, J., Feng, Z., Gilliland, G., Bhat, T.N., Weissig, H., Shindyalov, I.N., Bourne, P.E. (2000) The protein data bank. *Nucleic Acids Res.*, **28** (1), 235–242.

213 Shoemaker, B.A., Panchenko, A.R., Bryant, S.H. (2006) Finding biologically relevant protein domain interactions: conserved binding mode analysis. *Protein Sci.*, **15** (2), 352–361.

214 Raghavachari, B., Tasneem, A., Przytycka, T.M., Jothi, R. (2008) DOMINE: a database of protein domain interactions. *Nucleic Acids Res.*, **36** (Database issue), D656–D661.

215 Schwikowski, B., Uetz, P., Fields, S. (2000) A network of protein-protein interactions in yeast. *Nat. Biotechnol.*, **18** (12), 1257–1261.

216 Sharan, R., Ulitsky, I., Shamir, R. (2007) Network-based prediction of protein function. *Mol. Syst. Biol.*, **3**, 88.

217 Nabieva, E., Jim, K., Agarwal, A., Chazelle, B., Singh, M. (2005) Whole-proteome prediction of protein function via graph-theoretic analysis of interaction maps. *Bioinformatics*, **21** (Suppl. 1), i302–i310.

218 Hishigaki, H., Nakai, K., Ono, T., Tanigami, A., Takagi, T. (2001) Assessment of prediction accuracy of protein function from protein–protein interaction data. *Yeast*, **18** (6), 523–531.

219 Zhang, S., Chen, H., Liu, K., Sun, Z. (2009) Inferring protein function by domain context similarities in protein-protein interaction networks. *BMC Bioinformatics*, **10**, 395.

220 Jaeger, S., Gaudan, S., Leser, U., Rebholz-Schuhmann, D. (2008) Integrating protein–protein interactions and text mining for protein function prediction. *BMC Bioinformatics*, **9** (Suppl. 8), S2.

221 Kourmpetis, Y.A., van Dijk, A.D., Bink, M.C., van Ham, R.C., ter Braak, C.J. (2010) Bayesian Markov Random Field analysis for protein function prediction based on network data. *PLoS ONE*, **5** (2), e9293.

222 Chuang, H.Y., Lee, E., Liu, Y.T., Lee, D., Ideker, T. (2007) Network-based classification of breast cancer metastasis. *Mol. Syst. Biol.*, **3**, 140.

223 Lee, E., Chuang, H.Y., Kim, J.W., Ideker, T., Lee, D. (2008) Inferring pathway activity toward precise disease

classification. *PLoS Comput. Biol.*, **4** (11), e1000217.

224 Daemen, A., Signoretto, M., Gevaert, O., Suykens, J.A., De Moor, B. (2010) Improved microarray-based decision support with graph encoded interactome data. *PLoS ONE*, **5** (4), e10225.

225 Ulitsky, I., Karp, R., Shamir, R. (2008) Detecting disease-specific dysregulated pathways via analysis of clinical expression profiles. *Res. Comput. Mol. Biol.*

226 Mani, K.M., Lefebvre, C., Wang, K., Lim, W.K., Basso, K., Dalla-Favera, R., Califano, A. (2008) A systems biology approach to prediction of oncogenes and molecular perturbation targets in B-cell lymphomas. *Mol. Syst. Biol.*, **4**, 169.

227 Navlakha, S., Kingsford, C. (2010) The power of protein interaction networks for associating genes with diseases. *Bioinformatics*, **26** (8), 1057–1063.

228 Kohler, S., Bauer, S., Horn, D., Robinson, P.N. (2008) Walking the interactome for priorization of candidate genes. *Am. J. Hum. Genet.*, **82**, 949–958.

229 Vanunu, O., Sharan, R. (2008) A propagation-based algorithm for inferring gene-disease associations. German Conference on Bioinformatics, Dresden, pp. 54–62.

230 Wu, X., Jiang, R., Zhang, M.Q., Li, S. (2008) Network-based global inference of human disease. *Mol. Syst. Biol.*, **4**, 189.

231 Karni, S., Soreq, H., Sharan, R. (2009) A network-based method for predicting disease-causing genes. *J. Comput. Biol.*, **16** (2), 181–189.

232 Vanunu, O., Magger, O., Ruppin, E., Shlomi, T., Sharan, R. (2010) Associating genes and protein complexes with disease via network propagation. *PLoS Comput. Biol.*, **6** (1), e1000641.

233 Lee, E., Jung, H., Radivojac, P., Kim, J.W., Lee, D. (2009) Analysis of AML genes in dysregulated molecular networks. *BMC Bioinformatics*, **10** (Suppl. 9), S2.

234 Chen, Y., Zhu, J., Lum, P.Y., Yang, X., Pinto, S., MacNeil, D.J., Zhang, C., Lamb, J., Edwards, S., Sieberts, S.K., Leonardson, A., Castellini, L.W., Wang, S., Champy, M.F., Zhang, B., Emilsson, V., Doss, S., Ghazalpour, A., Horvath, S., Drake, T.A., Lusis, A.J., Schadt, E.E. (2008) Variations in DNA elucidate molecular networks that cause disease. *Nature*, **452** (7186), 429–435.

235 Przytycka, T.M., Singh, M., Slonim, D.K. (2010) Toward the dynamic interactome: it's about time. *Brief Bioinform.*, **11** (1), 15–29.

5
Protein Abundance Variation

Greco Hernández[1,2] *and Gritta Tettweiler*[2]
[1] *National Institute for Cancer, Division of Basic Research, Av. San Fernando No. 22, Col. Sección XVI, Tlalpan, C. P. 14080, Mexico City, Mexico*
[2] *McGill University, Department of Biology and Developmental Biology Research Initiative, Bellini Life Sciences, 3649, Sir-William-Osler Street, Montréal, H3A 1B1, Canada*

Systems Biology: Advances in Molecular Biology and Medicine, First Edition. Edited by Robert A. Meyers.

Keywords

Aptamers
Short strands of oligonucleotides species that have been engineered through repeated rounds of *in vitro* selection. Aptamers are designed to bind specifically to a given target molecule, such as proteins or nucleic acids.

Chaperones
Proteins that assist the noncovalent folding or unfolding of other proteins. Some chaperones fold newly made proteins as they exit from the ribosome. Chaperones prevent both newly synthesized polypeptide chains and assembled subunits from aggregating into nonfunctional structures. As protein folding is severely affected by heat and other stresses, many chaperones are heat shock proteins, and act to repair the damage caused by misfolding.

Proteases
Enzymes that catalyze the cleavage of proteins by hydrolyzing the peptide bonds that link amino acids together in a polypeptide chain. Proteases are classified into six broad groups, according to the amino acid residue recognized as performing the cleavage. The groups are serine proteases, threonine proteases, cysteine proteases, aspartate proteases, metalloproteases, and glutamic acid proteases.

Proteasome
Large protein complexes containing proteases complexes that degrade damaged proteins by proteolysis. The proteasome is a cylindrical complex containing a "core" of four stacked rings, each composed of seven individual proteins. The inner two rings contain six protease active sites that are oriented toward the interior surface of the rings, so that the target protein must enter the central pore before it is degraded. The outer two rings form an entrance through which proteins enter the barrel.

***cis*-Acting elements**
Sequence-specific control elements in DNA or RNA which regulate expression within the same message.

***trans*-Acting factors**
DNA- or RNA-binding proteins and miRNAs which control expression distant from the message they regulate.

In all forms of life, the amount of cellular protein present at any given time is tightly regulated by a combination of mechanisms of gene expression, namely transcription, mRNA degradation, translation, and protein degradation. A variety of events can trigger changes in protein abundance throughout the cell cycle by affecting the different steps of gene expression. Although the mechanisms that regulate gene expression in prokaryotes and eukaryotes are well known, the way in which protein abundance is controlled at each regulatory step remains unclear. The quantitation of protein levels is essential to understand the different aspects of gene expression, while a solid knowledge of the mechanisms and causes that affect protein levels should enable the development of new technologies, whether for human needs or environmental care. In the past, considerable effort has been invested in quantifying protein abundances in different situations such as stress, disease progression, or changing environmental conditions. More recently, significant advances have been made in the methods used to quantify protein abundance on the genome scale, in specific cells at certain time points, and/or under different environmental conditions. In this chapter, the different mechanisms that control protein abundance variation in prokaryotes and eukaryotes – that is, the biochemical features of mRNAs and proteins, as well as external factors such as environmental changes, stress, diseases, or developmental stages – are reviewed. Details are also be provided of the most relevant methodologies for the qualitative and quantitative analyses of protein abundance.

1 Introduction

Proteins are the ultimate result of the expression of protein-encoding genes. In prokaryotes, the transcription of genes is coupled to translation of the nascent messenger RNA (mRNA), whereas in eukaryotes the transcription of genes into mRNAs is compartmentalized in the nucleus, after which the mature mRNAs are exported to the cytoplasm. Having reached the cytoplasm, the mRNAs may be translated, stored for later use, or degraded. The amount of a protein present in a specific cell at any given time is a dynamic issue throughout the cell cycle, and is determined by the sum of the different steps of gene expression, namely transcription, mRNA degradation, translation, and protein degradation.

Protein abundance is known to change due to a variety of causes, and in this respect the cellular processes and mechanisms used to regulate gene expression in prokaryotes and eukaryotes have been extensively investigated. Nevertheless, the way in which protein abundance is controlled in different cells, and under specific extracellular conditions, at the transcriptional, post-transcriptional, translational and post-translational levels, remains a central question in modern biology. Moreover, an ability to monitor such protein levels is essential if the different aspects of gene expression that are encountered in various situations such as disease, development, stress, or changing environmental conditions are to be understood. It is only by knowing the mechanisms and causes that affect protein abundance, and having an ability to measure such changes, that it may become possible to develop technologies that could be applied to combat infectious agents of both humankind and crops, or to maintain the environment. In particular, bearing in mind the importance of monitoring protein abundance in diagnostics and in determining the targets of drugs, considerable effort has been expended not only in identifying the factors that affect gene expression and variations in protein abundance, but also in providing the ability to measure such variation in prokaryotes and eukaryotes.

During the past few years, the number of powerful techniques used to monitor variations in protein abundance under different biological conditions has witnessed a major increase. Details of the most recent high-throughput and bioinformatics-based technologies, which have been developed to measure global or particular protein abundances, in addition to systematic quantitative analyses conducted to correlate transcription and translation efficiencies with protein abundances, are available in Refs [1–25]. More recently, methods have also been developed for the analysis of collective protein compositions and the abundance of whole microbial communities [26, 27].

In this chapter, a review is provided of the intrinsic and extrinsic causes that control protein abundance variations in prokaryotic and eukaryotic organisms; these are applicable to the biochemical features of mRNAs and proteins, as well as to various factors that are external to cells such as environmental changes, stress, disease, or developmental stages. Details of the most prominent differences in gene regulation among prokaryotes and eukaryotes are listed in Table 1. Details are also provided of the most relevant techniques used for the qualitative and quantitative analyses of protein abundance.

2 Biochemical Aspects Affecting Protein Abundance in Prokaryotes

A major part of the versatility of prokaryotes to adapt resides in their ability to rapidly fine-tune their gene expression in order to respond to an external stimulus. Such an ability is due mostly to the fact that, in prokaryotes, transcription and translation are coupled and both processes occur almost simultaneously (Fig. 1). It is generally accepted that the prokaryotic regulation of gene expression is exerted primarily at the transcriptional level, and secondarily at the level of translation [28–34]. The results of previous genome-wide studies have shown there to be only a weak correlation between the absolute mRNA levels and protein abundances. This indicates that different levels of regulation act to determine the abundance of each protein synthesized from a given mRNA, and that groups of mRNAs

Tab. 1 Overview of most prominent differences of gene regulation in prokaryotes in eukaryotes.

	Prokaryotes	*Eukaryotes*
Compartmentalization	No nuclear membrane: coupled transcription and translation	Nuclear membrane: spatial and temporal separation of transcription and translation
Transcription	Single RNA polymerase	Different RNA polymerase for different RNAs
RNA processing	No spliceosomal pathway	Splicing of coding mRNAs
RNA-based mRNA expression control	Riboswitches, small RNAs	microRNAs
Ribosome recruitment	Shine–Dalgarno-containing mRNAs, Shine–Dalgarno-less mRNAs, ribosomal protein S1, leaderless mRNAs	5′ cap structure or IRES
Ribosomes	50S + 30S	40S + 60S

IRES, internal ribosome entry site.

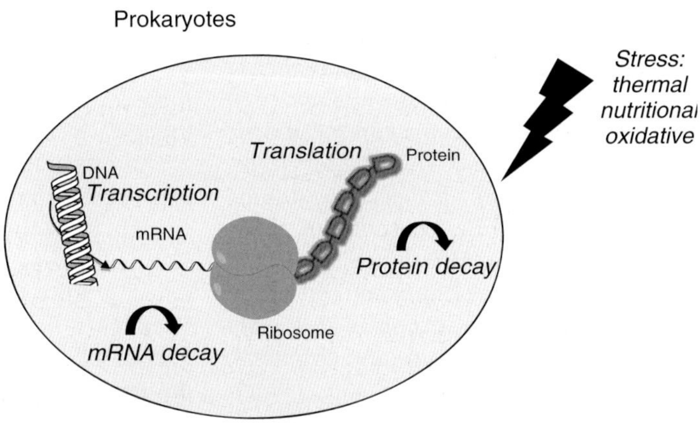

Fig. 1 Cellular mechanisms and extrinsic causes affecting protein abundance in prokaryotes. Highlighted in black are the most relevant mechanisms of gene expression regulating protein abundance, as well as the external causes affecting protein abundance.

can be regulated in concert in order so as to keep physiologically tightly related proteins separate from other groups of mRNAs [29, 35–37]. The main mechanisms that affect protein abundance in prokaryotes are described in the following subsections.

2.1
Transcription Rate

Transcription is the process of transcribing DNA into different RNAs, namely mRNAs, transfer RNAs (tRNAs), ribosomal RNAs (rRNAs), and small RNAs (sRNAs). Each RNA polymerase associates with one of the several sigma factors, a prokaryotic transcription initiation factor (IF), that enables the specific binding of RNA polymerase to a target gene promoter [30–32]. Different sigma factors are activated in response to different transcriptional cues. Typically, transcription consists of five phases: (i) preinitiation; (ii) initiation; (iii) promoter clearance; (iv) elongation; and (v) termination. The promoter is a *cis*-element that regulates the transcription rate of a gene and which is located in the immediate vicinity upstream of the transcription start site. A typical prokaryotic promoter comprises two relatively conserved hexanucleotide sequences with the consensus sequences TTGACA at position −35 and TATAAT at position −10 relative to the transcription start site. During preinitiation, the sigma factors associate with the RNA polymerase to enable recognition of the −35 and −10 sequences of the selected core promoter. The binding of the RNA polymerase induces a separation of the DNA strands and the creation of a transcription "bubble" (based on its appearance under the electron microscope) between positions −10 and +2. Before successful transcription occurs during promoter clearance, the RNA polymerase often slips, which gives rise to abortive transcripts. The stability of RNA polymerization is greatly increased when the transcript reaches 23 base pairs in length, after which elongation will be continued until transcription termination. During termination, the mRNA dissociates from its template DNA, either by binding of the Rho cofactor the RNA polymerase or by hairpin structures in the DNA (transcriptional terminators; Rho-independent termination). Because the supply of both sigma factors and free RNA polymerase in a cell are limiting, transcriptional regulation takes place mainly during pre-initiation and promoter clearance [30–33].

Different genes are transcribed at different levels, and the frequency at which a given gene is transcribed depends on several regulatory mechanisms. One of the most prominent mechanisms of regulation is the binding of different sigma subunits to RNA polymerase; this in turn confers the binding specificity of this complex to a selected promoter [30–34, 38]. Furthermore, transcriptional factors (proteins that function either to repress or to activate transcription on specific DNA targets) also add another level of regulation to the transcription process. The binding and release of the repressors and the activators are themselves often controlled by cofactors; these encompass a wide variety of molecules, ranging from small ions, nucleotides, covalently attached phosphate moieties and sugars to transcriptional factors [30–34, 38].

The promoter activity is also regulated by several sequence-dependent properties or structural modifications of the DNA. The stability of the promoter region during the initial separation of the DNA strands is strongly dictated by the sequence of this region. During this step, the RNA polymerase locally separates the two DNA strands around the −10 region; this is a thermodynamically unfavorable process that takes place without the input of external energy. Subsequently, the G–C-rich sequences are more difficult to separate than A–T-rich sequences. Another sequence-dependent property is the occurrence of unusual DNA structures in the promoters, such as DNA curvatures; these may be an inherent property of a DNA molecule or they may be induced by protein binding. As the promoter DNA wraps around the RNA polymerase during the transcription start, DNA bending can modulate the binding of transcription factors. Modifications to the structure of the DNA, either by methylation or by altering the chromosome's organization, can also lead to a regulation of the transcription rate. However, chromosome organization provides a more global control of transcription than the aforementioned mechanisms [30–34].

2.2 mRNA Decay

The amount of RNA available for translation is greatly affected by RNA decay, a process which exerts an RNA quality control to degrade any aberrant mRNAs and represents a means of adjusting the mRNA availability to the cell's requirements. The degradation of mRNAs is mediated by a large protein complex – the RNA degradosome – and involves the coordinated activities of both endoribonucleases and exoribonucleases. Genome-wide studies conducted in different prokaryotes have identified half-lives of between 1.5 and 10 min for all mRNAs. It has also been shown that, while mRNA abundance is not a predictor for its stability, in the case of some organisms there is a correlation between mRNA stability and functional categories; notably, the "housekeeping" gene mRNAs are more stable than those involved in stress adaptation. The main structural features of the mRNAs (which can provide them with protection against nucleases and thus enhance their stability) are 5′-end-triphosphorylated nucleotides, stem-loops near to 5′, or secondary structures in 3′- untranslated regions (UTR)s [37, 39–42]. Variations in mRNA sequence may also influence transcript decay [43]. When a ribosome is recruited to the 5′-UTR of an mRNA and translation begins, the mRNA stability will be greatly increased.

2.3 Translation Rate

In most prokaryotes, the ribosomal subunit binds to the ribosome-binding site (RBS; this is also known as the Shine–Dalgarno; SD) sequence. This sequence is a purine-rich region which is located at around 10 nt upstream of the start codon and is complementary to the 3′ end of the 16S rRNA located inside the ribosome. A second type of translation occurs in "leaderless" mRNAs (these are mRNAs that lack a 5′-UTR and hence have no SD sequence), in which the start codon itself serves as the signal for the ribosome to start translation. Leaderless genes have been shown to be widespread in a variety of different bacteria, and are most abundant in the Archaea. Translation initiation can also be mediated exclusively by the ribosomal protein S1 (RPS1), which is a component of the 30S ribosomal subunit. RPS1 interacts with a 5′-UTR of an mRNA, thus initiating translation efficiently, regardless of the presence of a RBS [44–48].

In general, initiation is the limiting step in prokaryotic translation, as the initial anchoring of ribosomes onto the mRNAs strongly determines the global translation efficiency. Thus, RBS and the start codon play a major role in translation efficiency. Secondary structures in the RBS and in the sequence surrounding the start codon have a strong effect on the translation efficiency [36, 37, 46, 48, 49], while translation initiation is also regulated by different noncoding RNAs, including "riboswitches" and sRNAs [37, 43, 50, 51]. The riboswitches, which are RNA secondary structures composed of an aptamer, bind to nucleotides, sugars, or amino acids, and can control the cytoplasmic concentrations of these latter compounds by affecting the translation expression of enzymes related to their metabolism. This elicits the formation of stem–loop structures containing the RBS, which in turn inhibits the initiation of translation. The sRNAs match the mRNAs either in UTRs, the RBS, or the coding region, causing a blockade of translation and leading to RNA degradation of the target transcript. Finally, some ribosomal proteins, when present in excess within the cell, may also bind their own mRNA and act as translation repressors [37, 50, 51].

Several mechanisms are known to influence the elongation phase of mRNAs. Codon usage, based on the degeneracy of the genetic code and on the abundance of specific tRNAs, is one such mechanism [37, 52–55]. For example, GGU, GGA, GGG, and GGC each encode the amino acid glycine, but their respective tRNAs are not found in equal proportions within a cell. Genes rich in codons that are recognized by abundant tRNAs are more efficiently translated. The ribosomes pause on rare codons and wait for diffusion of the tRNA; this causes a decrease in the translation rate and ultimately leads to mRNA destabilization. This phenomenon, which is referred to as *codon usage bias*, has a major impact on the translation efficiency and, as a consequence, on the protein abundance in all forms of life. A notable correlation has been identified between the expression level of a gene and the degree of its codon bias. It has been reported that the use of abundant codons in an mRNA can increase the abundance of a protein by over 1000-fold [37, 52–55].

The phase of bacterial growth may also affect the translation efficiency. For example, entry into stationary phase leads to dimerization of the 70S ribosomal complexes, followed by binding of the ribosome modulation factor (RMF), thus forming translationally inactive 100S particles. When the cells re-enter log phase, however, the inactive 100S particles dissociate, and this leads to a rapid degradation of the RMF and the reactivation of translation [37, 56, 57].

Recently, a strict relationship was shown to exist between the translation rate and the number, length, and order of the genes in an operon. Increasing the distance from the start of a gene to the end of the operon (known as the *"transcription distance"*) provides more time for translation to occur during transcription, which in turn results in an increased expression [58].

2.4 Protein Stability

The amount of a given protein in a cell is also dictated by the balance between the rates of protein synthesis and degradation. Protein degradation is a crucial biological process aimed at discarding nonfunctional proteins, and provides a quick means of achieving adequate protein levels in response to environmental changes [37, 59, 60]. Protein quality control mechanisms consist of proteases and chaperones acting in concert to monitor the function and correct folding of proteins; these systems remove denatured, misfolded, or aggregated proteins, either by refolding or degradation. In general, the chaperones recognize hydrophobic stretches of polypeptides that become surface-exposed as a consequence of either misfolding or unfolding. These systems assist proteins to regain their native, functional structure, whilst if the refolding process fails any irreversibly damaged polypeptides are removed by proteases [61–63]. The proteolytic core particle of large protease complexes contain the 20S proteasome – a large, cylinder-shaped protease of approximately 700 kDa that is present in all kingdoms of life. In eubacteria, another proteasome-related protein-degradation machinery has been identified, termed HslV. In this case, the core particles and proteasome-related systems generate peptide fragments with a typical length of between eight and 15 amino acid residues that are then further degraded into single amino acids to be reused by the cell for *de novo* protein synthesis. An example of this type of protease are the so-called "tricorn" proteases, which exist in some archaea and eubacteria. Further degradation of the short peptides is performed by tricorn-interacting proteases, such as the aminopeptidases F1, F2, and F3 [37, 59, 60].

3
Extracellular Causes Influencing Protein Abundance in Prokaryotes

From an evolutionary standpoint, bacteria have developed immense capabilities to rapidly modulate their gene expression according to changing environmental conditions [29, 56, 61, 64–67]. Many agents of physical, chemical, or biological nature, as well as environmental conditions, cause different degrees of cell stress, which affect general gene expression and, hence, the abundance of most proteins. Upon sensing adverse conditions, cells are able to adapt to their new milieu by rapidly downregulating the expression of their housekeeping genes and activating specific anti-stress genes. As protein denaturation is a common outcome of cell stress, many anti-stress genes encode chaperones, chaperonins, and protease systems [61–63]. Three of the most important and widely studied stress responses in prokaryotes are described in the following subsections.

3.1
Nutritional Stress

Nutritional limitation elicits a strong response in bacteria that involves the modulation of expression of several genes. This is mainly triggered by the intracellular accumulation of two small molecules – the nucleotides guanosine tetraphosphate (ppGpp) and guanosine pentaphosphate (pppGpp) – which are rapidly synthesized in response to amino acid starvation, and the deprivation of phosphorus, iron, glucose, or fatty acids. Such deprivation triggers the rapid synthesis of ppGpp and pppGpp, which in turn causes a modulation of the expression of several genes. In this way, ppGpp and pppGpp can modulate the expression of genes that are involved in helping the bacteria to adapt to nutritional scarcity [56, 68, 69]. The cellular levels of both ppGpp and pppGpp are maintained by the products of the genes *relA*, *spot*, and *relV*. ppGpp and pppGpp each shut down growth and promote cell survival by triggering adaptive responses at different levels. At the transcriptional level, ppGpp and pppGpp each produce profound changes that include the transcriptional repression of rRNA and tRNA genes, and also the induction of genes required for amino acid synthesis and transport. pppGpp and ppGpp can each also initiate the global reprogramming of transcription by switching the use of general sigma factors for alternative, stress-related sigma factors [56, 68, 69]. At the translational level, pppGpp and ppGpp induce RMF synthesis, thus forming the translationally inactive 100S particles [56, 70].

Moreover, pppGpp and ppGpp may inhibit the activity of several translation factors, including the elongation factor thermounstable (EF-Tu), elongation factor G (EF-G), and initiation factor 2 (IF2) [56, 68, 69]. Global protein synthesis is also inhibited via the RMF under other adverse conditions, including heat, acidic, and osmotic stresses [56, 70].

3.2 Thermal Stress

Although, upon heat stress, global transcription and protein synthesis are immediately arrested, the genes for more than 20 heat shock proteins (HSPs) are, in contrast, rapidly transcribed and translated. Whilst many other genes may be either directly or indirectly induced by heat shock, the most conspicuous – and hence the most characterized – are the HSPs genes and their products. The synthesis of these proteins is usually regulated by a single transcription factor, which in *Escherichia coli* is σ^{32}. Directly induced HSPs genes are those encoding for the main cellular chaperone machineries of GroEL and Hsp70 (also known as DnaK), Hsp60, ATP-dependent proteases of Lon, HslUV, Clp, and FtsH, the components of the chaperonin complex Hsp40 (also known as DnaJ) and GrpE, Hsp60, the proteases DegP, DegS, and DegQ, and some other proteins involved in protein folding, refolding, quality control, and degradation. The fact that many HSPs are conserved across prokaryotes indicates that they have fundamental physiological functions. The HSPs play crucial roles not only in the rescue or removal of proteins damaged by heat-shock stress, but also in the response to ionic stress, as well as in the routine folding of proteins under normal growth conditions [28, 29, 61, 66, 67, 71].

Two crucial control elements in Gram-negative bacteria are quickly transcribed when heat shock begins; that is, the alternative heat shock σ^{32} and σ^{24} factors that regulate transcription of the major HSPs genes. Both sigma factors are further regulated at the translation and post-translational levels. Whereas a cognate small RNA regulates the translation of σ^{24} mRNA, the translation of σ^{32} mRNA is regulated by structural changes of its own mRNA, while the phosphorylation of σ^{32} modulates its activity [29, 61, 66, 67, 71].

3.3 Oxidative Stress

Reactive oxygen species (ROS) are known to cause damage to DNA, lipid membranes, and proteins, and are lethal when present in excess. Two major transcriptional regulators – OxyR and SoxRS – are known to regulate bacterial response to oxidative stress. OxyR transcription is activated by high levels of hydrogen peroxide, after which the activated OxyR induces the transcription of a set of antioxidant genes, including *katG* (hydroperoxidase I), *ahpCF* (alkylhydroperoxidase), *dps* (encoding a DNA binding protein), *gorA* (glutathione reductase), *grxA* (glutaredoxin I), and *oxyS* (a regulatory RNA). Superoxide-generating compounds activate the transcription of the transcription factor SoxR which, when oxidized, then induces expression of the gene *soxS* (that encodes a transcription factor), which in turn directly activates the transcription of *sodA* (manganese superoxide dismutase), *fpr* (ferredoxin/flavodoxin-NADP$^+$ reductase), *zwf* (glucose 6-phosphate dehydrogenase), *fumC* (fumarase C), *nfo* (endonuclease IV), *acnA* (aconitase A), and *micF* (encoding a regulatory RNA), all of which are involved in the response to oxidative stress [29, 61, 65, 72]

The toxicities of both hydrogen peroxide and superoxide are exacerbated by iron which, in its reduced form, enhances oxygen toxicity by converting the less-reactive hydrogen peroxide to the more-reactive oxygen species, hydroxyl radical and ferryl iron. Since an increase in intracellular free iron leads to oxidative stress, the bacteria possess a strict regulatory control of iron intake from the milieu, whereby iron metabolism is coordinately regulated with the defenses against oxidative stress. OxyR and SoxRS also each control the expression of Fur, the global repressor of ferric ion intake and a regulator of iron homeostasis in bacteria. Fur and Fur-like proteins control the transcription of genes involved in iron acquisition, and also coordinate iron metabolism with defense against oxidative stress. Besides triggering the synthesis of both Fur and redox-sensitive proteins, oxidative stress also activates the expression of Hsp33, which is the major class of redox-regulated chaperone found in bacteria. Hsp33 employs redox-sensitive cysteines, the oxidative stress-induced disulfide bond formation of which assists in the partial unfolding of the chaperone and that, in turn, leads to the exposure of a high-affinity binding site for unfolded proteins. In contrast, during iron scarcity, Fur becomes inactivated such that all genes involved in Fe^{3+} and Fe^{2+} intake are derepressed. Fur also allows the bacteria to adapt to adverse environmental conditions caused by a scarcity of iron; this includes the regulation of purine biosynthesis and the ability to control not only motility but also the activity of several sigma transcription factors [29, 61, 65, 72].

The results of recently conducted genome-wide experiments have shown that more than half of the genes that are indispensable for growth at 47 °C (which is a critical high temperature) are sensitive to H_2O_2 at normal temperatures, thus confirming that the mechanism of thermotolerance partially overlaps with that of oxidative stress resistance.

4
Biochemical Aspects Affecting Protein Abundance in Eukaryotes

In contrast to prokaryotes, transcription and translation in eukaryotes are spatially separated by the nuclear membrane. Within the nucleus, the DNA is transcribed into a pre-mRNA, which is further processed into a mature mRNA that is transported to the cytoplasm. At this point the ribosomes bind to the mRNA and translate the message into a polypeptide chain (Fig. 2). These processes can be divided into multiple steps, each of which are highly regulated and thus contribute to the abundance of proteins.

4.1
Transcription Rate

In eukaryotes, three different types of RNA polymerase catalyze transcription into different types of RNA: RNA polymerase I transcribes rRNA (except 5S rRNA, which is transcribed by RNA polymerase III); RNA polymerase II transcribes pre-mRNAs, sn-RNAs, sno-RNAs and other noncoding RNAs; and RNA polymerase III transcribes 5S rRNA, tRNAs, and snRNAs.

In eukaryotes, transcription follows the same five phases as in prokaryotes, namely pre-initiation, initiation, promoter clearance, elongation, and termination. The regulation of transcription is different in eukaryotes, notably because during the

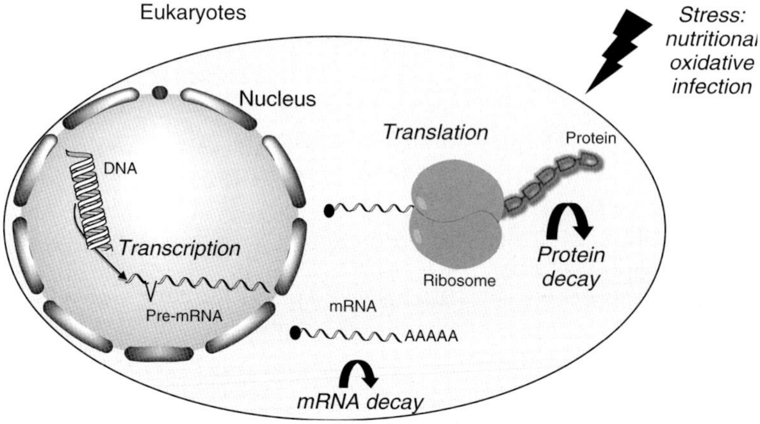

Fig. 2 Cellular mechanisms and extrinsic causes affecting protein abundance in eukaryotes. Highlighted in black are the most relevant mechanisms of gene expression regulating protein abundance, as well as the external causes affecting protein abundance.

pre-initiation phase the RNA polymerase binds a promoter sequence, which is found at position −30, −75, and − 90 upstream from the transcription start site. Within this sequence is located a core promoter, the so-called TATA box, of consensus sequence TATAAA [73–75]. Unlike prokaryotes, the initiation step in eukaryotes requires additional transcription factors to mediate the binding of RNA polymerase to the TATA box. However, similar to prokaryotes the *eukaryotic termination* of transcription is defined as the completion of RNA synthesis and the release of the RNA polymerase from the DNA strand. In contrast to prokaryotes, however, the termination process in eukaryotes links polyadenylation with cleavage of the RNA at the 3′ end. The RNA polymerase carries other factors, which participate in a cleavage/polyadenylation complex that recognizes a polyadenylation signal of consensus sequence AAUAAA in the nascent RNA, cleaves the RNA, and generates a poly(A) tail by the addition of adenine nucleotides at the 3′ end of the transcript. The poly(A)-binding proteins (PABPs) then bind to the poly(A) sequence, so as to protect the newly synthesized mRNA from degradation [73–76].

4.2 Alternative Splicing

In eukaryotes, most protein-coding genes are interrupted by intervening sequences known as *introns*. When the pre-mRNAs have been transcribed, the RNA undergoes a modification during which the introns are removed and the exons[1] are joined. The mRNA must undergo this so-called *"splicing"* process before it can be used to produce the correct protein. Typically, eukaryotic splicing is

1) An exon is a nucleic acid sequence that is represented in the mature form of an RNA molecule, either after portions of a precursor RNA (introns) have been removed by *cis*-splicing, or when two or more precursor RNA molecules have been ligated by *trans*-splicing. Depending on the context, the exon may refer to the sequence in either the DNA or its RNA transcript.

catalyzed by the "spliceosome", which is a complex of small nuclear ribonucleoproteins (snRNPs). In many cases, the splicing process of a single gene can create a diversity of different proteins by: (i) varying the exon composition of the same pre-mRNA; (ii) extending or skipping exons; or (iii) retaining the introns. This phenomenon is referred to as *alternative splicing*.

Alternative splicing may serve as a major regulator of protein abundance, mainly because the different mRNA isoforms can often alter the gene-coding capacity or exhibit a distinct RNA stability. Moreover, they may also add or remove regulatory elements and protein domains that can alter the protein's half-life. Alternative splicing is a major contributor to protein diversity in metazoans, with a general correlation having been recognized between organismal complexity and the proportion of genes that are alternatively spliced. Thus, organisms with more tissues and cell types tend to undergo more alternative splicing (up to 60% of all human gene products may undergo alternative splicing). Genetic switches based on alternative splicing have also been identified as important in diverse cellular and developmental processes, including apoptosis, sex determination, axon guidance, and cell contraction [77–79].

4.3 mRNA Features Regulating Protein Abundance

Eukaryotic mRNAs contain distinct properties which allow for maturation of the transcribed pre-mRNA and subsequently promote stability, nuclear export, and the activation of protein synthesis. At the 5′ end of the mRNA, the capping process takes place cotranscriptionally, such that the modified guanine nucleotide m^7Gppp – referred to as the *"cap"* – is added shortly after the initiation of transcription. The cap consists of a terminal 7-methylguanosine that is connected through a triphosphate bond to the first transcribed nucleotide. The correct addition of the cap regulates mRNA splicing and nuclear export, as well as providing protection against degradation and promoting the initiation of translation [80, 81].

At their 3′ end, the eukaryotic mRNAs carry a poly(A) tail (polyadenylation forms part of the termination step of transcription). In this case, a cleavage/polyadenylation specificity factor (CPSF) binds to the transcript such that, with assistance from other specificity factors, cleavage occurs upstream of the factor's binding site. Following cleavage, multiple adenine nucleotides are then added to the 3′ end, catalyzed by polyadenylate polymerases. The newly synthesized poly(A) tail can then serve as a binding site for the PABPs, which in turn ensures its export from the nucleus, protects against degradation, and promotes the initiation of translation [82–84].

4.4 mRNA Stability

The stability of the mRNA is a key feature that contributes to the ensuing protein abundance, and elements in both the coding and UTRs of a eukaryotic mRNA will regulate its turnover. The above-mentioned 5′ cap and 3′ poly(A) sequence increase the mRNA's half-life by preventing its degradation. Typically, the 5′ cap structure is inaccessible to the 5′ exonucleases, as it functionally resembles a 3′ end. In addition, the nuclear and cytoplasmic cap-binding proteins protect the mRNA against the decapping enzymes, which compete for binding to the cap. At the 3′ end, the poly(A) tail brings PABP to the mRNA, so as to prevent degradation by the exosome [47, 80, 81, 85, 86]. The PABP may also interact with the cap-binding complex eukaryotic initiation factor (eIF) 4F, which is formed by the translation initiation factors eIF4E, eIF4G, and eIF4A. This results in a circularization of the mRNA, thereby increasing the efficiency of protein synthesis [84, 87–89].

mRNA turnover is strongly regulated at the mRNA 3′-UTR, and most *cis*-acting elements that control mRNA turnover and translation are located in the 3′-UTR. The AU-rich elements (AREs), natural antisense RNAs, cytoplasmic polyadenylation element (CPE) and microRNAs are the most studied of such elements. Whereas, the AREs and microRNAs are crucial regulators of mRNA stability, natural antisense RNAs block translation, while the CPE regulates the translation rate by controlling the mRNA poly(A) tail length. Moreover, most *cis*-acting elements that drive the localization of mRNAs – a process that often helps to drive the spatially-asymmetric translation of specific proteins – are found in the 3′-UTR [1, 20, 90–98].

While even relatively stable mRNAs can produce little to no protein, any relatively unstable messages are efficiently translated and result in an abundant protein production. With regards to protein abundance, the regulation of mRNA stability and the translation rates are connected processes, with the outcome being influenced by an interplay of multiple pathways of these two events [14, 87, 99, 100].

4.5 Translation Rate

Translational control plays a major role in determining protein abundance. Indeed, it has been shown in recent analyses that two-thirds of protein abundance variation can be attributed to post-transcriptional processes such as mRNA sequence lengths, amino acid properties, codon usage, and secondary structures in the UTRs [1, 2, 6, 87]. A global quantification of gene expression in mammalian cells showed that the cellular abundance of proteins is controlled predominantly at the level of translation [14]. Eukaryotic protein synthesis can be divided into three steps – initiation, elongation, and termination – at each of which there exists the possibility of regulating the translation rate [100].

Although the overall speed of translation is determined by the rates of initiation, elongation, and termination, the main point of regulation appears to be the initiation (which in eukaryotes is also the limiting step). In eukaryotic mRNAs the SD sequence does not exist; rather, the ribosomes are recruited to the start codon with assistance from the eIFs. Most eukaryotic mRNAs are translated via a cap-dependent mechanism, such that the mRNA is recruited to the ribosome through recognition of the 5′ cap structure by the cap-binding protein eIF4E as part of the eIF4F complex. In turn, the eIF4F recruits the 43S pre-initiation complex, which consists of eIF3 and eIF1A bound to a 40S ribosomal subunit and the ternary complex eIF2-GTP-Met-tRNAi. The 40S subunit then scans along the 5′-UTR in a 5′ → 3′ direction until it reaches an initiation codon. For some viral and cellular mRNAs, 5′-UTR recognition by the 40S ribosomal subunit occurs without the involvement of eIF4E; it is instead driven by an RNA structure located in *cis* within the message itself. This structure, which is referred to as an internal ribosome entry site (IRES), is located in the proximity of the start codon. The eIFs are released when they reach the start codon, and this allows the 40S and 60S ribosomal subunits to join together to form an 80S initiation complex. The elongation of a polypeptide chain is then started [88, 100–102]. The regulation of translation initiation is mainly achieved by both the regulation of eIF4E activity and the formation of a ternary complex. One crucial point of regulation here is the interaction of eIF4E with eIF4E-binding proteins (4E-BPs). When bound to eIF4G, ribosome recruitment occurs, although other 4E-BPs can act as translational repressors by competing with eIF4G for eIF4E binding. The 4E-BPs play a crucial role in regulating protein abundance variation in development, cell differentiation, or in response to external stimuli [88, 101, 103].

It has been shown that the sequence surrounding the start codon is an important structural element in mRNA that has a major influence on the translation rate. The specific sequence GCC(A/G)CCAUGG – which is known as the "*Kozak consensus sequence*" – was found to be located around the initiator codon of vertebrate mRNAs [104, 105], while A/G in position −3 and G in position +4 (relative to the A of the AUG start codon) are regarded, functionally, as the most important bases [105, 106]. A comprehensive analysis of genomes from many species belonging to various taxonomic groups showed that, although the Kozak consensus is in fact the optimal sequence in vertebrate mRNAs, the preferred nucleotide sequence flanking the start codon may vary considerably among different taxonomic groups, with the A/G at position −3, the A/C at position −2, and the C at position +5 being the most common nucleotides across the eukaryotes [47, 107, 108]. In addition to the sequence surrounding the start codon, the formation of a secondary structure along the 5′-UTR also influences the efficiency of the scanning process during the initiation step, and in turn determines the translation rate of an mRNA. A tight secondary structure will reduce the translation efficiency, especially if it is located near to the translation starting codon [87, 107, 109, 110]. Finally, as noted for prokaryotes, codon usage has a major effect on the elongation speed of translation [87].

4.6 Protein Stability

In addition to transcription and translation efficiencies and mRNA stability, protein stability represents another mechanism that is important for controlling cellular protein abundance. In eukaryotic cells, nonfunctional proteins are discarded via two major pathways – the ubiquitin–proteasome pathway and lysosomal proteolysis.

Ubiquitination is the formation of a covalent bond between the carboxy terminus of ubiquitin and a lysine residue of the target protein, catalyzed by the E3 ligase. This leads to the formation of a multiubiquitin chain, which serves as a marker for rapid proteolysis in the proteasome. The different forms of ubiquitination, and the length of the polyubiquitin chain(s), may have diverse effects on the target protein [59, 60, 111].

In the case of lysosomal proteolysis, proteins can be incorporated into particular organelles that contain various enzymes, among which are included the proteases. A mechanism that allows the uptake of proteins for degradation in lysosomes is that of *autophagy* which, at its basal level, allows not only the turnover of proteins and protein aggregates but also the elimination of damaged/old organelles, thereby ensuring cellular homeostasis. On encountering environmental stress, however, the autophagy level may be dramatically increased so as to allow adaption and survival. Unfortunately, the downside of this response mechanism may be that overly excessive autophagy would result in cell death [60, 112, 113].

5 Other Factors Influencing Protein Abundance in Eukaryotes

In eukaryotic cells, under unfavorable conditions, protein synthesis is the most prominent mechanism to be employed for regulating protein abundance. The evolution of regulatory mechanisms at the translational level has endowed eukaryotes with the potential for a spatial control of protein production, an efficacy in tissues lacking transcription, and a rapidity and reversibility of responses to environmental challenges without invoking transcription, mRNA processing, or transport [88, 91, 103, 114]. The regulation of translation initiation allows for the most efficient control of protein abundance before interrupting the pathway midstream, or dealing with the consequence of faulty or undesired polypeptides. The various factors that influence protein abundance via the translation machinery are discussed in the following subsections.

5.1 Environmental Stress

Environmental factors such as temperature changes, starvation, or hypoxia can each influence overall protein synthesis at each level of regulation, as noted above. If, for example, the nutrients are limited then survival may only be achieved by shifting the energy supply away from the highly energy-consuming process of protein synthesis. In order to ensure an adequate stress response, several signaling cascades are critical, and one important molecule in this respect is the evolutionarily conserved target-of-rapamycin (TOR), a kinase that serves as a checkpoint for downstream effects on translation [88, 115, 116]. Sufficient nutrients (among other signals) convey a positive signal to TOR, whereas starvation (and other signals) influences TOR signaling in a negative fashion.

TOR exists in two complexes: (i) TORC1, which includes TOR, GβL, and Raptor; and (ii) TORC2, which includes TOR, GβL, and Rictor [88, 115, 116]. TORC2 phosphorylates Akt, which in turn is an upstream component of the TOR signaling pathway, while TORC1 phosphorylates the ribosomal protein S6 kinase (S6K) and the translational inhibitor, 4E-BP. The phosphorylation of S6K leads to its activation, and the subsequent phosphorylation of eIF4B, S6, and eukaryotic elongation factor (eEF) 2K [88, 115, 116]. These events result in enhanced rates of translation. The phosphorylation of 4E-BP leads to a decreased affinity to eIF4E, thereby allowing binding to eIF4G that results in increases in eIF4F formation, ribosome recruitment and, ultimately, in protein synthesis rates. Under conditions of starvation, TOR signaling is inactive, thereby repressing cap-dependent translation [88, 115, 116].

The downregulation of translation rates under conditions of nutritional stress may be very beneficial, however. For example, dietary restriction may extend the life span of various organisms [117, 118] and, indeed, recent analyses in the fruit fly *Drosophila melanogaster* and the worm *Caenorhabditis elegans* have shown dietary restriction to result in an overall reduction in the number of polysomes which, in turn, decreases the overall relative translation rate [119].

5.2 Infection

On contracting infection, the overall level of protein synthesis will be downregulated, while the translation of selected mRNAs will be enhanced. Microbial infections affect protein homeostasis through a variety of pathways, while the regulation of protein synthesis in both a time- and tissue-dependent manner represents just one way in which a host organism will respond to an infection [103, 114].

The translation of viral proteins is dependent on the host cells' translational machinery. Indeed, several viruses have been shown capable of circumventing the cap-dependent translation initiation, allowing a preferential synthesis of viral proteins by recruiting ribosomes directly to an IRES [120–122]. Other mechanisms may affect the translation initiation factors directly, however. For example, infection with vesicular stomatitis virus (VSV) results in the dephosphorylation of 4E-BP, which is then free to bind eIF4E, preventing eIF4F formation and inhibiting the overall translation [123, 124]. In contrast, other viruses may induce the hyperphosphorylation of 4E-BP, which would result in an increased translation of not only cellular but also viral mRNAs [125].

5.3 Development

Spatial and temporal variations in protein abundance are particularly important during the development of multicellular organisms, when translational control plays a crucial role [103, 126, 127]. During development, protein gradients are established in many metazoans via the translation of localized mRNAs; this pattern formation is especially important in early embryogenesis, when transcription is typically silenced. Two key factors in establishing the embryonic anterior–posterior gradient are known as Caudal and Hunchback. At the anterior, the translation of *caudal* mRNA must be repressed, and this is achieved by binding of the eIF4E homologous protein (4EHP) at the 5′ cap structure, which in turn interacts with Bicoid bound to the 3′-UTR. At the posterior, 4EHP binds to a multiprotein complex at the 3′-UTR of *hunchback* mRNA [128].

6 Techniques Used to Measure Protein Abundance

Currently, in Systems Biology, evidence from several fields is typically combined to address the question at hand, with information from developmental biology perhaps being combined with established methods from the areas of physics, chemistry, and mathematics. A typical example of this would be the blending of "classic" methods such as one-dimensional gel electrophoresis with Western blotting which, today, is regarded simply as a "stepping-stone" to the provision of conclusive evidence relating to protein abundance. Today, access to computational data mining and rapidly improving algorithms allow measurements to be made at the genome-wide scale, with predictions for whole dynamic networks also possible. Recently, the details have been described of new technologies that allow transcription or translation efficiencies to be correlated with protein abundance, as well as the high-throughput measurement of protein abundances in either diverse organisms or microbial communities [1–27]. Some of the most recently developed – and most popular – methods employed are outlined in the following subsections.

6.1 Correlation between mRNA Abundance and Protein Abundance

One common approach employed to understand variations in protein abundance is focused on the transcriptional output from genomes. This involves the use of microarray chips, which allow the simultaneous measurement of gene expression by a large number of genes [5, 129] although, unfortunately, the measurement of total mRNA expression in this way allows only an indirect extrapolation of protein abundance. The primary advantage of this method lies in the simultaneous comparison of the regulation of gene expression by various stimuli, between developmental stages, or others, such that one piece of evidence on protein abundance variation may be derived. Yet, in combination with other methods microarray analysis serves as a very powerful tool, as it allows for an increased validity for a conclusion on protein abundance variation.

The two most widely used methods to enrich mRNAs for microarray analysis are immunoprecipitation and polysome isolation:

- In *immunoprecipitation*, a protein of interest is precipitated from cellular extracts, or from different tissues or developmental stages, and/or upon exposure to different stimuli, using specific antibodies. The RNA is then isolated from the precipitate and used directly for microarray analysis. A comparison between arrays permits a conclusion to be drawn as to which RNA (or group of RNAs) is regulated.
- In *polysome isolation*, a sucrose gradient centrifugation is used to separate the polysomes from the lighter monosomes. The polysomal RNAs is then isolated and subjected to microarray analysis [7, 130, 131]. Again, a comparison permits a conclusion to be drawn regarding the regulation of a message, or of a group of messages. Typically, polysomal RNAs are the most likely messages in the process of being translated, and therefore represent a more direct readout of protein abundance.

6.2 Electrophoresis-Based Methods

When determining relative overall protein levels, conventional methods rely on sodium dodecylsulfate–polyacrylamide gel electrophoresis, followed by the staining of gels with silver or other reagents. These methods are of low sensitivity, with detection limits between 1 and 100 ng per band. The use of Western blotting increases the sensitivity when quantifying a protein of interest, with typical methods involving radiolabeling or chemiluminescence, followed by exposure to X-ray film or live-imaging. The main drawbacks of these methods are that they depend on the availability of specific antibodies, and can detect only one protein at a time.

A more recent development that allows comparisons of protein abundance to be made on a much larger scale is that of difference-gel electrophoresis (DiGE), which is a modification of two-dimensional (2-D) gel electrophoresis [3]. For this, up to three protein samples from different cell types are labeled with distinct fluorescent dyes, and the different samples then subjected to various stimulations and/or developmental stages. Sample separation during gel electrophoresis is based on the isoelectric points (first dimension) and the molecular weights (second dimension) of the samples. Ultimately, the gels are scanned and the samples imaged separately based on their fluorescent labels. The final 2-D profiles of the samples are then overlaid and compared, whereby differences between spots represent variations in protein abundance, or post-transcriptional and post-translational modifications. Any proteins of potential interest may be excised from the gel for identification using mass spectrometry [3].

6.3 Quantitative Proteomics

Quantitative proteomics opens the possibility for high-throughput analyses of variations in protein abundance and, to date, several such methods have been established that depend on the labeling of reagents for their efficacy [9]. One such method for profiling protein expression, termed isotope-coded affinity tagging (ICAT) [10, 132] involves all cysteine residues in the protein being labeled with a tag and either (light) ^{12}C atoms for the control protein extract, or (heavy) ^{13}C atoms for the experimental extract. The mixed samples are then subjected to trypsin digestion, followed first by chromatography to effect a size-based separation and then mass spectrometry to identify the peptide pairs. This method allows quantification of the relative ratios of individual proteins, but not the absolute protein concentrations [10]. The main benefit of ICAT is that data can be obtained for large groups of proteins [133].

Currently, an increasingly popular method for quantitative measurements of proteins is that of stable isotope labeling with amino acids in cell culture (SILAC) [4, 13, 14]. For this, essential amino acids labeled with heavy isotopes are incorporated by the cells and integrated into newly synthesized polypeptide chains. Following a certain number of cell divisions, all of the proteins will contain labeled amino acids, after which the protein populations from both the experimental (heavy) and the control (light) settings are harvested and mixed. Following trypsin digestion of the proteins, mass spectrometric analyses are carried out, with identification and quantification of the peptides [4, 13, 14].

A recent modification of the SILAC method accounts not only for changes in protein abundance but also regulation at the level of translation, by pulse labeling with heavy and medium-heavy amino acids. The so-called pulsed stable isotope labeling with amino acids in cell culture (pSILAC) technique presents a more accurate picture of protein synthesis and protein degradation. In this case, any preexisting proteins (light) can be ignored, and any differences in protein stability will not influence the analysis. Moreover, the ratio of the heavy and medium-heavy populations will reflect actual differences in protein synthesis [13, 14].

A newly developed, label-free method, which calculates Absolute Protein EXpression (APEX) levels, allows the determination of both absolute and relative protein concentrations. The method employs calculations for its three steps: (i) to estimate mass spectrometry detectability (O_i) by training data, peptide sequences, and their properties; (ii) to quantify the absolute protein abundance by using spectral counts, identification probabilities, and O_i-values; and (iii) to perform a statistical analysis for evaluating the differential expression in two distinct biological samples [2, 6].

6.4 Ribosomal Footprinting

A novel technique to quantify translation rates, and thus protein abundance, was recently developed in a yeast model system, but is expected ultimately to be applicable to any organism [134, 135]. As the ribosomes protect only a short segment of their template mRNA from nuclease digestion, this allows their exact position – their "footprint" – to be indicated [136, 137]. The optimized method of ribosomal footprinting allows a precise determination of the translating ribosomes, thus revealing the presence of upstream open reading frames, or stop codon readthrough. Together with recent advances in sequencing techniques, which allow the simultaneous analysis of a large number of footprints, a genome-wide translational profiling should be possible. The conversion of these RNA footprints into a DNA library should, in time, enable deep sequencing and the quantitative measurement of expressed proteins [135].

6.5 Single-Molecule Real-Time Imaging

Although the above-described methods are used to analyze protein expression at the time of cell lysis. it must be noted that the regulation of protein levels is a *dynamic* process. Nonetheless, recent advances in single-cell imaging and mathematical models now allow the analysis of protein abundance variations in real time [138, 139].

7 Concluding Remarks and Outlook

As computer-based predictions regarding protein abundance continue to attract attention, recent advances in data mining have opened the way to the public acquisition of data from the aforementioned high-throughput methods. One such example is the gene expression omnibus (GEO), which is a collection of genomic data obtained via microarrays and next-generation sequencing [140, 141]. In fact, based on such databases, complex dynamic networks may be established for many biological systems.

The ability to influence protein abundance by modulating mRNA stability, perhaps by employing pharmaceutical agents, may present a strategy for intervention in a variety of diseases, and in this respect large-scale screening methods represent the most promising approaches. However, high-throughput assays combined with rapidly improving algorithms for subsequent analyses offer a novel approach to the screening of therapeutics and diagnostics [142].

Acknowledgments

The authors thank Caroline Laplante for her valuable help in preparing this manuscript.

References

1 Vogel, C., Abreu, R.S., Ko, D., Le, S.Y., Shapiro, B.A., Burns, S.C., Sandhu, D., Boutz, D.R., Marcotte, E.M., Penalva, L.O. (2010) Sequence signatures and mRNA concentration can explain two-thirds of protein abundance variation in a human cell line. *Mol. Syst. Biol.*, **6**, 400.

2 Vogel, C., Marcotte, E.M. (2008) Calculating absolute and relative protein abundance

from mass spectrometry-based protein expression data. *Nat. Protoc.*, **3**, 1444–1451.

3 Tonge, R., Shaw, J., Middleton, B., Rowlinson, R., Rayner, S., Young, J., Pognan, F., Hawkins, E., Currie, I., Davison, M. (2001) Validation and development of fluorescence two-dimensional differential gel electrophoresis proteomics technology. *Proteomics*, **1**, 377–396.

4 Ong, S.E., Blagoev, B., Kratchmarova, I., Kristensen, D.B., Steen, H., Pandey, A., Mann, M. (2002) Stable isotope labeling by amino acids in cell culture, SILAC, as a simple and accurate approach to expression proteomics. *Mol. Cell. Proteomics*, **1**, 376–386.

5 Schena, M., Shalon, D., Davis, R.W., Brown, P.O. (1995) Quantitative monitoring of gene expression patterns with a complementary DNA microarray. *Science*, **270**, 467–470.

6 Lu, P., Vogel, C., Wang, R., Yao, X., Marcotte, E.M. (2007) Absolute protein expression profiling estimates the relative contributions of transcriptional and translational regulation. *Nat. Biotechnol.*, **25**, 117–124.

7 Larsson, O., Sonenberg, N., Nadon, R. (2010) Identification of differential translation in genome wide studies. *Proc. Natl Acad. Sci. USA*, **107**, 21487–21492.

8 Liu, H., Sadygov, R.G., Yates, J.R. (2004) A model for random sampling and estimation of relative protein abundance in shotgun proteomics. *Anal. Chem.*, **15**, 4193–4201.

9 Kirkpatrick, D.S., Gerber, S.A., Gygi, S.P. (2005) The absolute quantification strategy: a general procedure for the quantification of proteins and post-translational modifications. *Methods*, **35**, 265–273.

10 Greenbaum, D., Colangelo, C., Williams, K., Gerstein, M. (2003) Comparing protein abundance and mRNA expression levels on a genomic scale. *Genome Biol.*, **4**, 117.

11 Greenbaum, D., Jansen, R., Gerstein, M. (2002) Analysis of mRNA expression and protein abundance data: an approach for the comparison of the enrichment of features in the cellular population of proteins and transcripts. *Bioinformatics*, **18**, 585–596.

12 Ferguson, P.L., Smith, R.D. (2003) Proteome analysis by mass spectrometry. *Annu. Rev. Biophys. Biomol. Struct.*, **32**, 399–424.

13 Schwanhausser, B., Gossen, M., Dittmar, G., Selbach, M. (2009) Global analysis of cellular protein translation by pulsed SILAC. *Proteomics*, **9**, 205–209.

14 Schwanhausser, B., Busse, D., Li, N., Dittmar, G., Schuchhardt, J., Wolf, J., Chen, W., Selbach, M. (2011) Global quantification of mammalian gene expression control. *Nature*, **473**, 337–342.

15 Xia, Q., Hendrickson, E.L., Wang, T., Lamont, R.J., Leigh, J.A., Hackett, M. (2007) Protein abundance ratios for global studies of prokaryotes. *Proteomics*, **7**, 2904–2919.

16 Weiss, W., Görg, A. (2007) Two-dimensional electrophoresis for plant proteomics. *Methods Mol. Biol.*, **355**, 121–143.

17 VerBerkmoes, N.C., Connelly, H.M., Pan, C., Hettich, R.L. (2004) Mass spectrometric approaches for characterizing bacterial proteomes. *Expert. Rev. Proteomics*, **1**, 433–447.

18 Komili, S., Silver, P.A. (2008) Coupling and coordination in gene expression process: a systems biology view. *Nat. Rev. Genet.*, **9**, 38–48.

19 Fenselau, C. (2007) A review of quantitative methods for proteomic studies. *J. Chromatogr.*, **855**, 14–20.

20 de Sousa, R., Penalva, L.O., Marcotte, E.M., Vogel, C. (2009) Global signatures of protein and mRNA expression levels. *Mol. Biosyst.*, **5**, 1512–1526.

21 Chugh, S., Suen, C., Gramolini, A. (2010) Proteomics and mass spectrometry: what we have learned about the heart? *Curr. Cardiol. Rev.*, **6**, 124–133.

22 Carrillo, B., Yanofsky, C., Laboissiere, S., Nadon, R., Kearney, R.E. (2010) Methods for combining peptide intensities to estimate relative protein abundance. *Bioinformatics*, **26**, 98–103.

23 Brun, V., Masselon, C., Garin, J., Dupuis, A. (2009) Isotope dilution strategies for absolute quantitative proteomics. *J. Proteomics*, **72**, 740–749.

24 Bindschedler, L.B., Cramer, R. (2011) Quantitative proteomics. *Proteomics*, **11**, 756–775.

25 Bar-Even, A., Paulsson, J., Maheshri, N., Carmi, M., O'Shea, E., Pilpel, Y., Barkai, N. (2006) Noise in protein expression

scales with natural protein abundance. *Nat. Genet.*, **38**, 636–643.

26 Lacerda, C.M.R., Reardon, K.F. (2009) Environmental proteomics: applications of proteome profiling in environmental microbiology and biotechnology. *Brief. Funct. Genomic. Proteomic.*, **8**, 75–87.

27 Schneider, T., Riedel, K. (2010) Environmental proteomics: analysis of structure and function of microbial communities. *Proteomics*, **10**, 785–798.

28 Chhabra, S.R., He, Q., Huang, K.H., Gaucher, S.P., Alm, E.J., He, Z., Hadi, M.Z., Hazen, T.C., Wall, J.D., Zhou, J., Arkin, A.P., Singh, A.K. (2006) Global analysis of heat shock response in *Desulfovibrio vulgaris Hildenborough*. *J. Bacteriol.*, **188**, 1817–1828.

29 Mogk, A., Huber, D., Bukau, B. (2011) Integrating protein homeostasis strategies in prokaryotes. *Cold Spring Harbor Perspect. Biol.*, **3**, 4366.

30 Sorek, R., Cossart, P. (2010) Prokaryotic transcriptomics: a new view on regulation, physiology and pathogenicity. *Nat. Rev. Genet.*, **11**, 9–16.

31 Güell, M., Yus, E., Lluch-Senar, M., Serrano, L. (2011) Bacterial transcriptomics: what is beyond the RNA horiz-ome? *Nat. Rev. Microbiol.*, **9**, 658–669.

32 Cho, B.K., Palsson, B., Zengler, K. (2011) Deciphering the regulatory codes in bacterial genomes. *Biotechnol. J.*, **6** (9), 1052–1063.

33 Browning, D.F., Busby, S.J.W. (2004) The regulation of bacterial transcription initiation. *Nat. Rev. Microbiol.*, **2**, 57–65.

34 van Hijum, S.A.F.T., Medema, M.H., Kuipers, O.P. (2009) Mechanisms and evolution of control logic in prokaryotic transcriptional regulation. *Microbiol. Mol. Biol. Rev.*, **73**, 481–509.

35 Nie, L., Wu, G., Zhang, W. (2006) Correlation of mRNA expression and protein abundance affected by multiple sequence features related to translational efficiency in *Desulfovibrio vulgaris*: a quantitative analysis. *Genetics*, **174**, 2229–2246.

36 Lithwick, G., Margalit, H. (2003) Hierarchy of sequence-dependent features associated with prokaryotic translation. *Genome Res.*, **13**, 2665–2673.

37 Picard, F., Dressaire, C., Girbal, L., Cocaign-Bousquet, M. (2009) Examination of post-transcriptional regulations in prokaryotes by integrative biology. *C. R. Biol.*, **332**, 958–973.

38 Seshasayee, A.S., Sivaraman, K., Luscombe, N.M. (2011) An overview of prokaryotic transcription factors: a summary of function and occurrence in bacterial genomes. *Subcell. Biochem.*, **52**, 7–23.

39 Richards, J., Sundermeier, T., Svetlanov, A., Karzai, A.W. (2008) Quality control of bacterial mRNA decoding and decay. *Biochim. Biophys. Acta*, **1779**, 574–582.

40 Arraiano, C.M., Andrade, J.M., Domingues, S., Guinote, I.B., Malecki, M., Matos, R.G., Moreira, R.N., Pobre, V., Reis, F.P., Saramago, M., Silva, I.J., Viegas, S.C. (2010) The critical role of RNA processing and degradation in the control of gene expression. *FEMS Microbiol. Rev.*, **34**, 883–923.

41 Kaberdin, V.R., Singh, D., Lin-Chao, S. (2011) Composition and conservation of the mRNA-degrading machinery in bacteria. *J. Biomed. Sci.*, **18**, 23.

42 Condon, C., Bechhofer, D.H. (2011) Regulated RNA stability in the Gram positives. *Curr. Opin. Microbiol.*, **14**, 148–154.

43 Kudla, G., Murray, A.W., Tollervey, D., Plotkin, J.B. (2009) Coding-sequence determinants of gene expression in *Escherichia coli*. *Science*, **324**, 255–258.

44 Zheng, X., Hu, G.Q., She, Z.S., Zhu, H. (2011) Leaderless genes in bacteria: clue to the evolution of translation initiation mechanisms in prokaryotes. *BMC Genomics*, **12**, 361–370.

45 Nakawa, S., Niimura, Y., Miura, K.I., Gojobori, T. (2010) Dynamic evolution of translation initiation mechanisms in prokaryotes. *Proc. Natl Acad. Sci. USA*, **107**, 6382–6387.

46 Benelli, D., Londei, P. (2011) Translation initiation in Archaea: conserved and domain-specific features. *Biochem. Soc. Trans.*, **19**, 89–93.

47 Hernández, G. (2009) On the origin of the cap-dependent initiation of translation in eukaryotes. *Trends Biochem. Sci.*, **34**, 166–175.

48 Laursen, B.S., Sornsen, H.P., Mortensen, K.K. (2005) Initiation of protein synthesis in bacteria. *Microbiol. Mol. Biol. Rev.*, **69**, 101–123.

49 Scharff, L.B., Childs, L., Walther, D., Bock, R. (2011) Local absence of secondary structure permits translation of mRNAs that lack ribosome-binding sites. *PLoS Genet.*, **7**, e1002155.

50 Bastet, L., Dubé, A., Massé, E., Lafontaine, D.A. (2011) New insights into riboswitch regulation mechanisms. *Mol. Microbiol.*, **80**, 1148–1154.

51 Richards, G.R., Vanderpool, C.K. (2011) Molecular call and response: the physiology of bacterial small RNAs. *Biochim. Biophys. Acta*, **1809** (10), 525–531.

52 Allert, M., Cox, J.C., Hellinga, H.W. (2010) Multifactorial determinants of protein expression in prokaryotic open reading frames. *J. Mol. Biol.*, **402**, 905–918.

53 Plotkin, J.B., Kudla, G. (2011) Synonymous but not the same: the causes and consequences of codon bias. *Nat. Rev. Genet.*, **12**, 32–42.

54 Ermolaeva, M.D. (2001) Synonymous codon usage in bacteria. *Curr. Issues Mol. Biol.*, **3**, 91–97.

55 Hershberg, R., Petrov, D.A. (2008) Selection on codon bias. *Annu. Rev. Genet.*, **42**, 287–299.

56 Srivatsan, A., Wang, J.D. (2008) Control of bacterial transcription, translation and replication by (p)ppGpp. *Curr. Opin. Microbiol.*, **11**, 100–105.

57 Kato, T., Yoshida, H., Miyata, T., Maki, Y., Wada, A., Namba, K. (2010) Structure of the 100S ribosome in the hibernation stage revealed by electron cryomicroscopy. *Structure*, **18**, 719–724.

58 Lim, H.N., Lee, Y., Hussein, R. (2011) Fundamental relationship between operon organization and gene expression. *Proc. Natl Acad. Sci. USA*, **108**, 10626–10631.

59 Heinemeyer, W., Ramos, P.C., Dohmen, R.J. (2004) The ultimate nanoscale mincer: assembly, structure and active sites of the 20S proteasome core. *Cell. Mol. Life Sci.*, **61**, 1562–1578.

60 Groll, M., Bochtler, M. (2005) Molecular machines for protein degradation. *Chembiochem*, **6**, 222–256.

61 Macario, A.J., Conway, D., Macario, E. (2001) The molecular chaperone system and other anti-stress mechanisms in archaea. *Front. Biosci.*, **6**, D262–D283.

62 Sabate, R., Groot, N.S., Ventura, S. (2010) Protein folding and aggregation in bacteria. *Cell. Mol. Life Sci.*, **67**, 2695–26715.

63 Large, A.T., Goldberg, M.D., Lund, P.A. (2009) Chaperones and protein folding in the archaea. *Biochem. Soc. Trans.*, **37**, 46–51.

64 Ma, Q., Wood, T.K. (2011) Protein acetylation in prokaryotes increases stress resistance. *Biochem. Biophys. Res. Commun.*, **410**, 846–851.

65 Touati, D. (2000) Iron and oxidative stress in bacteria. *Arch. Biochem. Biophys.*, **373**, 1–6.

66 Murata, M., Fujimoto, H., Nishimura, K., Charoensuk, K., Nagamitsu, H., Raina, S., Kosaka, T., Oshima, T., Ogasawara, N., Yamada, M. (2011) Molecular strategy for survival at a critical high temperature in *Escherichia coli*. *PLoS ONE*, **6**, e20063.

67 Arsène, F., Tomoyasu, T., Bukau, B. (2000) The heat shock response of *Escherichia coli*. *Int. J. Food Microbiol.*, **10**, 3–9.

68 Chatterji, D., Ojha, A.K. (2001) Revisiting the stringent response, ppGpp and starvation signaling. *Curr. Opin. Microbiol.*, **4**, 160–165.

69 Sharma, U.K., Chatterji, D. (2010) Transcriptional switching in *Escherichia coli* during stress and starvation by modulation of sigma activity. *FEMS Microbiol. Rev.*, **34**, 646–657.

70 Terui, Y., Tabei, Y., Akiyama, M., Higashi, K., Tomitori, H., Yamamoto, K., Ishihama, A., Igarashi, K., Kashiwagi, K. (2010) Ribosome modulation factor, an important protein for cell viability encoded by the polyamine modulon. *J. Biol. Chem.*, **285**, 28698–28707.

71 Guisbert, E., Yura, T., Rhodius, V.A., Gross, C.A. (2008) Convergence of molecular, modeling, and systems approaches for an understanding of the *Escherichia coli* heat shock response. *Microbiol. Mol. Biol. Rev.*, **72**, 545–554.

72 Kumsta, C., Jakob, U. (2009) Redox-regulated chaperones. *Biochemistry*, **48**, 4666–4676.

73 Orioli, A., Pascali, C., Pagano, A., Teichmann, M., Dieci, G. (2011) RNA polymerase III transcription control elements: themes and variations. *Gene*, **105** (2), 263–271.

74 Pan, Y., Tsai, C.J., Ma, B., Nussinov, R. (2010) Mechanisms of transcription factor selectivity. *Trends Genet.*, **26**, 75–83.

75 Weirauch, M.T., Hughes, T.R. (2011) A catalogue of eukaryotic transcription factor types, their evolutionary origin, and species distribution. *Subcell. Biochem.*, **52**, 25–73.

76 Buratowski, S. (2005) Connections between mRNA 3′ end processing and transcription termination. *Curr. Opin. Cell Biol.*, **17**, 257–261.

77 McManus, C.J., Graveley, B.R. (2011) RNA structure and the mechanisms of alternative splicing. *Curr. Opin. Genet. Dev.*, **21**, 373–379.

78 Black, D.L. (2003) Mechanisms of alternative pre-messenger RNA splicing. *Annu. Rev. Biochem.*, **72**, 291–336.

79 Han, J., Xiong, J., Wang, D., Fu, X.D. (2011) Pre-mRNA splicing: where and when in the nucleus. *Trends Cell Biol.*, **21**, 336–343.

80 Shuman, S. (2001) Structure, mechanism, and evolution of the mRNA capping apparatus. *Prog. Nucleic Acids Mol. Biol.*, **66**, 1–40.

81 Cougot, N., van Dijk, E., Babajko, S., Séraphin, B. (2004) Cap-tabolism. *Trends Biochem. Sci.*, **29**, 436–444.

82 Gilmartin, G.M. (2005) Eukaryotic mRNA 3′ processing: a common means to different ends. *Genes Dev.*, **19**, 2517–2521.

83 Zhao, J., Hyman, L., Moore, C. (1999) Formation of mRNA 3′ ends in eukaryotes: mechanism, regulation, and interrelationships with other steps in mRNA synthesis. *Microbiol. Mol. Biol. Rev.*, **63**, 405–445.

84 Mangus, D.A., Evans, M.C., Jacobson, A. (2003) Poly(A)-binding proteins: multifunctional scaffolds for post-transcriptional control of gene expression. *Genome Biol.*, **4**, 223.

85 Floor, S.N., Jones, B.N., Gross, J.D. (2008) Control of mRNA decapping by Dcp2. *RNA Biol.*, **5**, 189–192.

86 Franks, T.M., Lykke-Andersen, J. (2008) The control of mRNA decapping and P-body formation. *Mol. Cell*, **32**, 605–615.

87 Gingold, H., Pilpel, Y. (2011) Determinants of translation efficiency and accuracy. *Mol. Syst. Biol.*, **7**, 481–494.

88 Hernández, G., Altmann, M., Lasko, P. (2010) Origins and evolution of the mechanisms regulating translation initiation in eukaryotes. *Trends Biochem. Sci.*, **35**, 63–73.

89 Jacobson, A. (1996) Poly(A) Metabolism and Translation: The Closed-Loop Model, in: Hershey, J.W.B., Mathews, M.B., Sonenberg, N. (Eds) *Translational Control*, Cold Spring Harbor Laboratory Press, pp. 451–480.

90 Hentze, M.W., Gebauer, F., Preiss, T. (2007). cis-Regulatory Sequences and Trans-Actin Factors in Translational Control, in: Mathews, M.B., Sonenberg, N., Hershey, J.W.B. (Eds) *Translational Control in Biology and Medicine*, Cold Spring Harbor Laboratory Press, pp. 269–296.

91 Mazumder, B., Seshadri, V., Fox, P.L. (2003) Translational control by the 3′-UTR: the endes specify the means. *Trends Biochem. Sci.*, **28**, 91–98.

92 Pesole, G., Grillo, G., Larizza, A., Liuni, S. (2000) The untranslated regions of eukaryotic mRNAs: structure, function, evolution and bioinformatic tools for their analysis. *Brief. Bioinformatics*, **1**, 236–249.

93 Pesole, G., Mignone, F., Gissi, C., Grillo, G., Licciulli, F., Liuni, S. (2001) Structural and functional features of eukaryotic mRNA untranslated regions. *Gene*, **276**, 73–81.

94 Barreau, C., Paillard, L., Osborne, H.B. (2006) AU-rich elements and associated factors: are there unifying principles? *Nucleic Acids Res.*, **33**, 7138–7150.

95 Martin, K.C., Ephrussi, A. (2009) mRNA localization: gene expression in the spatial dimension. *Cell*, **136**, 719–730.

96 Cerruti, H., Casas-Mollano, A. (2006) On the origin and functions of RNA-mediated silencing: from protists to man. *Curr. Genet.*, **50**, 81–99.

97 Sun, M., Hurst, L.D., Carmichael, G.G., Chen, J. (2005) Evidence for a preferential targeting of 3′-UTRs by *cis*-encoded natural antisense transcripts. *Nucleic Acids Res.*, **33**, 5533–5543.

98 Pratt, A.J., MacRae, I.J. (2009) The RNA induced silencing complex: a versatile gene-silencing machine. *J. Biol. Chem.*, **284**, 17897–17901.

99 Mukhopadhyay, D., Houchen, C.W., Kennedy, S., Dieckgraefe, B.K., Anant, S. (2003) Coupled mRNA stabilization and translational silencing of cyclooxygenase-2 by a novel RNA binding protein, CUGBP2. *Mol. Cell*, **11**, 113–126.

100 Kapp, L.D., Lorsch, J.R. (2004) The molecular mechanics of eukaryotic translation. *Annu. Rev. Biochem.*, **73**, 657–704.

101 Sonenberg, N., Hinnebusch, A. (2009) Regulation of translation initiation in eukaryotes: mechanisms and biological targets. *Cell*, **136**, 731–745.

102 Jackson, R.J., Hellen, C.U., Pestova, T.V. (2010) The mechanism of eukaryotic translation initiation and principles of its regulation. *Nat. Rev. Mol. Cell Biol.*, **11**, 113–127.

103 Sonenberg, N., Hinnebusch, A.G. (2007) New modes of translation control in development, behavior, and disease. *Mol. Cell*, **28**, 721–729.

104 Kozak, M. (1984) Compilation and analysis of sequences upstream from the translational start site in the eukaryotic mRNAs. *Nucleic Acids Res.*, **12**, 857–872.

105 Kozak, M. (1986) Point mutations define a sequence flanking the AUG initiator codon that modulates translation by eukaryotic ribosomes. *Cell*, **44**, 283–292.

106 Lukaszewicz, M., Feuermann, M., Jérouville, B., Stas, A., Boutry, M. (2000) *In vivo* evaluation of the context sequence of the translation initiation codon in plants. *Plant Sci.*, **154**, 89–98.

107 Hinnebusch, A.G. (2011) Molecular mechanism of scanning and start codon selection in eukaryotes. *Microbiol. Mol. Biol. Rev.*, **75**, 434–467.

108 Nakagawa, S., Niimura, Y., Gojobori, T., Tanaka, H., Miura, K. (2008) Diversity of preferred nucleotide sequences around the translation initiation codon in eukaryotes genomes. *Nucleic Acids Res.*, **36**, 861–871.

109 Kozak, M. (2002) Pushing the limits of the scanning mechanism for initiation of translation. *Gene*, **299**, 1–34.

110 Berthelot, K., Muldoon, M., Rajkowitsch, L., Hughes, J., McCarthy, J.E. (2004) Dynamics and processivity of 40S ribosome scanning on mRNA in yeast. *Mol. Microbiol.*, **51**, 987–1001.

111 Hochstrasser, M. (2006) Lingering mysteries of ubiquitin-chain assembly. *Cell*, **124**, 27–34.

112 Yang, Z., Klionsky, D.J. (2010) Eaten alive: a history of macroautophagy. *Nat. Cell Biol.*, **12**, 814–822.

113 Yang, Z., Klionsky, D.J. (2010) Mammalian autophagy: core molecular machinery and signaling regulation. *Curr. Opin. Cell Biol.*, **22**, 124–131.

114 Holcik, M., Sonenberg, N. (2005) Translational control in stress and apoptosis. *Nat. Rev. Mol. Cell Biol.*, **6**, 318–327.

115 Dowling, R.J., Fonseca, B.D., Sonenberg, N. (2010) Dissecting the role of mTOR: lessons from mTOR inhibitors. *Biochim. Biophys. Acta*, **1804**, 433–439.

116 Caron, E., Ghosh, S., Matsuoka, Y., Ashton-Beaucage, D., Therrien, M., Lemieux, S., Perreault, C., Roux, P.P., Kitano, H. (2010) A comprehensive map of the mTOR signaling network. *Mol. Syst. Biol.*, **6**, 453–468.

117 Tavernarakis, N. (2008) Ageing and the regulation of protein synthesis: a balancing act? *Trends Cell Biol.*, **18**, 228–235.

118 Hipkiss, A.R. (2007) On why decreasing protein synthesis can increase lifespan. *Mech. Ageing Dev.*, **128**, 412–414.

119 Zid, B.M., Rogers, A.N., Katewa, S.D., Vargas, M.A., Kolipinski, M.C., Lu, T.A., Benzer, S., Kapahi, P. (2009) 4E-BP extends lifespan upon dietary restriction by enhancing mitochondrial activity in *Drosophila*. *Cell*, **139**, 149–160.

120 Hernández, G. (2008) Was the initiation of translation in early eukaryotes IRES-driven? *Trends Biochem. Sci.*, **33**, 58–64.

121 Pacheco, A., Martinez-Salas, E. (2010) Insights into the biology of IRES elements through riboproteomic approaches. *J. Biomed. Biotechnol.*, **2010**, 458927.

122 Balvay, L., Soto Rifo, R., Ricci, E.P., Decimo, D., Ohlmann, T. (2009) Structural and functional diversity of viral IRESes. *Biochim. Biophys. Acta*, **1789**, 542–557.

123 Connor, J.H., Lyles, D.S. (2002) Vesicular stomatitis virus infection alters the eIF4F translation initiation complex and causes dephosphorylation of the eIF4E binding protein 4E-BP1. *J. Virol.*, **76**, 10177–10187.

124 Connor, J.H., Lyles, D.S. (2005) Inhibition of host and viral translation during vesicular stomatitis virus infection. eIF2 is responsible for the inhibition of viral but not host translation. *J. Biol. Chem.*, **280**, 13512–13519.

125 Walsh, D., Perez, C., Notary, J., Mohr, I. (2005) Regulation of the translation initiation factor eIF4F by multiple mechanisms in human cytomegalovirus-infected cells. *J. Virol.*, **79**, 8057–8064.

126 Lasko, P. (2009) Translational control during early development. *Progr. Mol. Biol. Transl. Sci.*, **90**, 211–254.

127 Thompson, B., Wickens, M., Kimble, J. (2007) Translational Control in Development, in: Mathews, M.B., Sonenberg, N., Hershey, J.W.B. (Eds.) *Translational Control in Biology and Medicine*, Cold Spring Harbor Laboratory Press, pp. 507–544.

128 Cho, P.F., Poulin, F., Cho-Park, Y.A., Cho-Park, I.B., Chicoine, J.D., Lasko, P., Sonenberg, N. (2005) A new paradigm for translational control: inhibition via 5′–3′ mRNA tethering by Bicoid and the eIF4E cognate 4EHP. *Cell*, **121**, 411–323.

129 Fodor, S.P., Read, J.L., Pirrung, M.C., Stryer, L., Lu, A.T., Solas, D. (1991) Light-directed, spatially addressable parallel chemical synthesis. *Science*, **251**, 767–773.

130 Arava, Y. (2003) Isolation of polysomal RNA for microarray analysis. *Methods Mol. Biol.*, **224**, 79–87.

131 Arava, Y., Wang, Y., Storey, J.D., Liu, C.L., Brown, P.O., Herschlag, D. (2003) Genome-wide analysis of mRNA translation profiles in *Saccharomyces cerevisiae*. *Proc. Natl Acad. Sci. USA*, **100**, 3889–3894.

132 Gygi, S.P., Rist, B., Gerber, S.A., Turecek, F., Gelb, M.H., Aebersold, R. (1999) Quantitative analysis of complex protein mixtures using isotope-coded affinity tags. *Nat. Biotechnol.*, **17**, 994–999.

133 Han, D.K., Eng, J., Zhou, H., Aebersold, R. (2001) Quantitative profiling of differentiation-induced microsomal proteins using isotope-coded affinity tags and mass spectrometry. *Nat. Biotechnol.*, **19**, 946–951.

134 Ingolia, N.T. (2010) Genome-wide translational profiling by ribosome footprinting. *Methods Enzymol.*, **470**, 119–142.

135 Ingolia, N.T., Ghaemmaghami, S., Newman, J.R., Weissman, J.S. (2009) Genome-wide analysis *in vivo* of translation with nucleotide resolution using ribosome profiling. *Science*, **324**, 218–223.

136 Steitz, J.A. (1969) Polypeptide chain initiation: nucleotide sequences of the three ribosomal binding sites in bacteriophage R17 RNA. *Nature*, **224**, 957–964.

137 Wolin, S.L., Walter, P. (1988) Ribosome pausing and stacking during translation of a eukaryotic mRNA. *EMBO J.*, **7**, 3559–3569.

138 Janicki, S.M., Tsukamoto, T., Salghetti, S.E., Tansey, W.P., Sachidanandam, R., Prasanth, K.V., Ried, T., Shav-Tal, Y., Bertrand, E., Singer, R.H., Spector, D.L. (2004) From silencing to gene expression: real-time analysis in single cells. *Cell*, **116**, 683–698.

139 Shav-Tal, Y., Singer, R.H., Darzacq, X. (2004) Imaging gene expression in single living cells. *Nat. Rev. Mol. Cell Biol.*, **5**, 855–861.

140 Barrett, T., Troup, D.B., Wilhite, S.E., Ledoux, P., Evangelista, C., Kim, I.F., Tomashevsky, M., Marshall, K.A., Phillippy, K.H., Sherman, P.M., Muertter, R.N., Holko, M., Ayanbule, O., Yefanov, A., Soboleva, A. (2011) NCBI GEO: archive for functional genomics data sets - 10 years on. *Nucleic Acids Res.*, **39**, D1005–D1010.

141 Edgar, R., Domrachev, M., Lash, A.E. (2002) Gene Expression Omnibus: NCBI gene expression and hybridization array data repository. *Nucleic Acids Res.*, **30**, 207–210.

142 Cheneval, D., Kastelic, T., Fuerst, P., Parker, C.N. (2010) A review of methods to monitor the modulation of mRNA stability: a novel approach to drug discovery and therapeutic intervention. *J. Biomol. Screen.*, **15**, 609–622.

Part II
Systems Biology of Evolution

Systems Biology: Advances in Molecular Biology and Medicine, First Edition. Edited by Robert A. Meyers.

6
Genetic Variation and Molecular Darwinism

Werner Arber
University of Basel, Department of Microbiology, Biozentrum, Klingelbergstrasse 70, 4056 Basel, Switzerland

Systems Biology: Advances in Molecular Biology and Medicine, First Edition. Edited by Robert A. Meyers.

Keywords

Biological evolution
A nondirected, dynamic process of diversification resulting from the steady interplay between spontaneous mutagenesis and natural selection.

DNA rearrangement
Results from mostly enzyme-mediated recombination processes, which can be either intramolecular or intermolecular.

Evolution gene
The protein product of this gene acts as a generator of genetic variations and/or as a modulator of the frequency of genetic variations.

Gene acquisition
A result of the horizontal transfer of genetic information from a donor cell to a receptor cell. With bacteria this can occur in transformation, conjugation or phage-mediated transduction.

Natural selection
A result of the capacity of living organisms to cope with the encountered physicochemical and biological environments. Largely depending on its genetic set-up and physiological phenotype, each organism may have either a selective advantage or a selective disadvantage as compared to the other organisms present in the same ecosystem.

Spontaneous mutation
Defined as any alteration of nucleotide sequences occurring to DNA, without the intended intervention of an investigator. The term *mutation* is used here as a synonym of genetic variation.

Transposition
DNA rearrangement mediated by a mobile genetic element such as a bacterial insertion sequence (IS) element or a transposon.

Variation generator
An enzyme or enzyme system, the mutagenic activity in the generation of genetic variation has been documented.

The comparison of DNA sequences of genes and entire genomes offers interesting insights into the possible evolutionary relatedness of the genetic information of living organisms. Together with a relatively rich database from experimental microbial genetics, conclusions can be drawn on the molecular mechanisms by which genetic variations are spontaneously generated. A number of different specific mechanisms contribute to the overall mutagenesis. Here, these mechanisms are grouped into three natural strategies of the spontaneous generation of genetic variations: local changes of DNA sequences; the intragenomic rearrangement of DNA segments; and the acquisition of foreign DNA by horizontal gene transfer. These three strategies have different qualities with regards to their contributions to the evolutionary process. As a general rule, none of the known mechanisms producing genetic variants is clearly directed. Rather, the resulting alterations in the inherited genomes are more random. In addition, usually only a minority of resulting variants provide a selective advantage. Interestingly, in most of the molecular mechanisms involved, the products of so-called "evolution genes" are involved as generators of genetic variation and/or as modulators of the frequencies of genetic variation. The products of evolution genes operate in tight collaboration with nongenetic factors, such as the structural flexibilities and chemical instabilities of molecules, chemical and physical mutagens, and random encounters. All of these aspects, by contributing to the spontaneous generation of genetic variations, together form the core of the theory of molecular evolution, which brings the Neo-Darwinism to molecular Darwinism. The philosophical and practical inferences of this knowledge will be discussed.

1 Introduction

Traditionally, evolutionary biology has devoted its major attention to the comparison of phenotypical traits of higher organisms, both of those actually living and of those that have become extinct (e.g., paleontological fossils). The resulting theory of descent is, together with other criteria, at the basis of the systematic classification of living organisms. Darwin's theory of natural selection brought a new element into the understanding of the long-term development of forms of life. Natural selection is the result of the coping of the organisms with the encountered living conditions which are dependent on both the environmental physico-chemical conditions and the activities of all living forms in a particular ecological niche. The Darwinian theory of evolution also postulated that intrinsic properties of life are not entirely stable and principally identical for all organisms of a given species. In the so-called modern synthesis, in which evolutionary biology and genetics became integrated, transmissible phenotypic variations representing the substrate of natural selection were explained as being due to genetic variations (or mutations). Shortly thereafter, deoxyribonucleic acid (DNA) was identified as the carrier of genetic information, and DNA is therefore also the target for mutagenesis. Within the past few decades, a rapid development of molecular genetics with novel research strategies leading to genomics, the sequencing of entire genomes and functional studies of genes and their products, has paved the way to hitherto inaccessible knowledge on the basis of life and its multiple manifestations. This also relates to the process of molecular evolution. A synoptical insight into the various molecular mechanisms contributing to the generation of genetic variations represents a molecular synthesis between the Neo-Darwinian theory and molecular genetics. This synthesis toward molecular Darwinism can confirm the Darwinian evolution at the molecular level.

2 Principles of Molecular Evolution

The principles of molecular evolution to be outlined here are founded on:

1. The Neo-Darwinian theory of evolution, with its three pillars of genetic variation, natural selection, and isolation.
2. The solidly established microbial genetics database.
3. DNA sequence comparisons with bioinformatic tools.
4. Physico-chemical knowledge of the reactivity, the conformational flexibility, and the chemical stability of biologically active molecules.

2.1 Evolutionary Roles of Genetic Variation, Natural Selection, and Isolation

The long-term maintenance of any form of life requires a relatively high stability of its genetic information. However, rare, occasional genetic variations occur in all organisms, and these give rise to mixed populations of organisms with the parental genome and organisms with one or more alterations in their genome. Such populations are steadily submitted to natural selection, yet experience shows that – in general – favorable genetic variations are considerably less frequent than unfavorable variations, which provide a selective disadvantage. Indeed, genetic variants with unfavorable genetic alterations will, sooner or later, be eliminated from propagating populations, which will become enriched with organisms carrying favorable genetic variations. It should be noted that, by far, not all alterations in the nucleotide sequences of a genome will lead to a change in the phenotype of the organism. However, such silent and neutral mutations may later become physiologically relevant in conjunction with still other, upcoming DNA sequence changes.

Natural selection is by no means a constant element; rather, it varies both in time and in space due to variations not only in the physico-chemical environmental conditions but also in the life activities of the many different organisms that are present in a particular ecological niche and which form an ecosystem. Since a genetic variation may of course also affect the influence that the organism exerts on the other organisms present in the same ecosystem (consider weeds and pathogenicity effects, as well as beneficial, synergistic effects), any novel mutation may influence not only the life of the concerned organism itself but also the lives of other cohabitants of the same ecological niche.

The third pillar – besides genetic variation and natural selection – of biological evolution is isolation, which evolutionary biologists define as two different types. The first type is geographic isolation, which may seriously reduce the number of potential habitats for an organism, while the second type is termed *reproductive isolation*. For example, two distantly related diploid organisms may not be fertile in sexual reproduction. However, reproductive isolation can also be seen in a wider definition to seriously limit the possibility of the horizontal transfer of segments of genetic information between two different types of organism.

As summarized in Fig. 1, genetic variation drives biological evolution; indeed, a complete genetic stability would render any evolutionary process impossible. A very high frequency of genetic variation would rapidly lead to the extinction of the concerned organisms because of the stated prevalence of unfavorable mutations in the spontaneous generation of genetic variations. It is natural selection, together with the available sets of genetic variants, which determines the direction of biological evolution or, in other words, the directions in which the branches of the "tree of evolution" grow. Finally, the geographic and reproductive isolations modulate the evolutionary process.

2.2 Molecular Mechanisms of the Generation of Genetic Variation

The concept to be presented here requires the reader to question some long-established textbook knowledge, such as the claim that spontaneous mutations would largely result from errors, accidents, and illegitimate processes. We defend here the alternative view that living nature actively cares for biological evolution:

1. by making use of intrinsic properties of matter such as a certain degree of chemical instability and of structural flexibility of molecules; and
2. by having developed genetically encoded systems, the products of which are involved in the generation of genetic variations and in modulating the frequencies of genetic variation.

It might be relevant to mention here that the term *mutation* is defined differently in classical genetics and in molecular genetics. In classical genetics, a mutant is any variant of a parental form showing in its phenotypic properties some alteration that becomes transmitted to the progeny. In contrast, it has become a habit in molecular genetics to call any alteration in the parental nucleotide sequence of the genome a mutation, whether it has phenotypic consequences or not. There is good reason to assume that, in most spontaneously occurring mutagenesis events, the specific mechanism involved will not pay attention to whether the sequence alteration at the involved target site will cause a phenotypic alteration or not. For studies on mechanisms and on the statistics of their occurrence, it is thus indicated to follow the molecular genetics definition of the term mutation, which we use here as a synonym of genetic variation. The term "*spontaneous mutation*" will be used to label any type of DNA sequence alteration unintended by the investigator. Yet, this definition says nothing about whether a mutation relates to a phenotypic change.

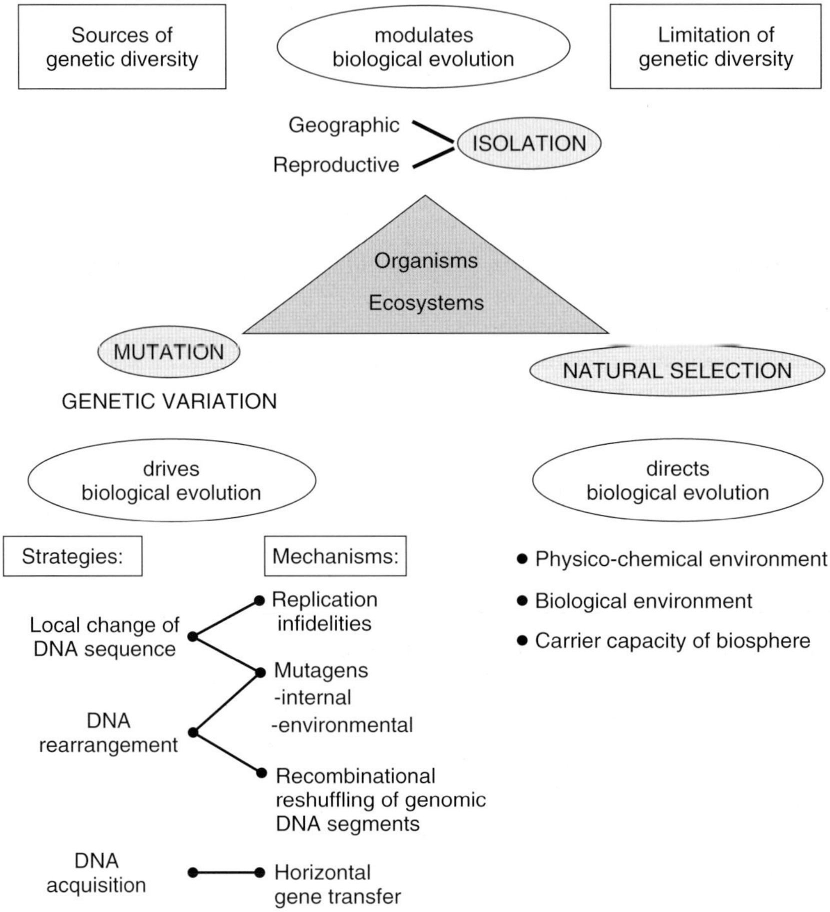

Fig. 1 Synoptical presentation of major elements of the theory of molecular evolution. A number of specific mechanisms contribute, each with its own characteristics to the four groups of mechanisms of genetic variation listed. Each of the specific mechanisms follows one (and sometimes more than one) of the three principal, qualitatively different strategies of genetic variation.

At present, research investigations into genetic variation follows two main strategies:

1. The increasing availability of entire and partial genome sequences offers excellent means to compare regulatory sequence elements, specific domains of genes, entire genes, groups of genes, and entire genomes with regard to DNA sequence homologies and genome organization. Within a given species, this can reveal a genetic polymorphism, whilst between more-or-less related species it can provide a reliable measure of the evolutionary relatedness. For example, the molecular clock – an indicator of evolutionary relatedness – is based on single nucleotide alterations. Sequence comparisons can often suggest how sequence alterations might have

occurred during the course of past evolution.

2. A more reliable insight into the generation of genetic variations can be gained by the observation of individual processes of mutagenesis. In view of the large size of genomes, and of the rare and random occurrence of spontaneous DNA sequence alterations, this approach is relatively difficult. However, a rich database is already available from microbial genetics, particularly from bacteria and their viruses and plasmids. Their relatively small genomes are haploid, so that phenotypic effects caused by genetic variation normally become rapidly manifest. With appropriate selection and screening techniques this can allow the identification of occasional, functionally relevant mutations in populations. On the other hand, investigations into structural alterations in the genomes of individual bacterial colonies – for example, by the study of restriction fragment length polymorphism (RFLP) – can reveal when and where on the genome a DNA rearrangement must have occurred.

A quite solid, general result of this type of experimental investigation reveals that, in the spontaneous generation of genetic variations, it is not only a single mechanism that is at work. Rather, a number of mechanistically different processes contribute to the overall mutagenesis, and selected examples will be discussed below. Interestingly, a critical evaluation of the situation shows that the specific mechanisms of mutagenesis often depend both on nongenetic elements and on the activities of specific enzymes – the products of so-called "evolution genes." The multitude of the thus-identified, distinct mechanisms which contribute to the formation of genetic variations can be grouped into three qualitatively different natural strategies (Fig. 1):

1. Local changes in the DNA sequences.
2. The rearrangement of DNA segments within the genome.
3. The acquisition by horizontal transfer of a DNA segment originating in another type of organism.

Selected examples for each of these strategies will be discussed in Sections 4–6.

3
Genetic Variation in Bacteria

Several seminal discoveries, based largely on investigations carried out with microorganisms between 1943 and 1953, were essential for the subsequent development of molecular genetics:

1. It was realized that bacteria and bacteriophages have genes that can mutate and recombine, and that spontaneous mutations in microorganisms normally arise independently of the presence of selective agents [1, 2]. It was also learned that the genetic information of bacteria, and of some bacteriophages, is carried in DNA molecules rather than in other biological macromolecules, such as proteins [3].
2. The newly discovered phenomena of DNA transformation, bacterial conjugation and bacteriophage-mediated transduction revealed natural means of horizontal gene transfer between different bacterial cells [4].
3. It was seen that horizontal gene transfer has natural limits, including systems of host-controlled modification,

which today are known as *DNA restriction-modification systems* [5].

4. Mobile genetic elements were identified as sources of genetic instability, and were seen to represent mediators of genetic rearrangements [6]. While such rearrangements are often caused by transposition, alterations in the genome can also result from the integration of a bacteriophage genome into the genome of its bacterial host strain, which is thereby rendered lysogenic.

It was at the end of this fruitful period of discovery, in 1953, that the structural analysis of DNA molecules led to the double-helix model [7]. The filamentous structure of DNA molecules made it clear how genetic information could be contained in the linear sequences of nucleotides. The double helical nature of the model also offered an understanding of semiconservative DNA replication at the molecular level, and thus of information transfer into progeny.

Many classical microbial genetic investigations were carried out with *Escherichia coli* K-12, the genome of which is a single circular DNA molecule (chromosome) of about 4.6×10^6 base pairs (bp). In periods of growth, the rate of spontaneous mutagenesis is about 10^{-9} per bp and per generation, which represents one new mutation in every few hundred cells in each generation. Today, *E. coli* has several well-studied bacteriophages and plasmids, and these materials continue to facilitate many investigations into life processes in these bacteria.

Under good growth conditions, the generation time of *E. coli*, measured between one cell division and the next, is very short, on the order of 30 min. Upon exponential growth, this leads to a multiplication factor of 1000 every 5 h; thus, a population of 10^9 cells, representing 30 generations, is reached from an inoculum of a single cell in only 15 h. This rapid growth rate greatly facilitates population genetic studies and, in turn, investigations into the evolutionary process.

On the filiform DNA molecules of *E. coli* and its bacteriophages and plasmids, the genetic information is relatively densely stored as linear sequences of nucleotides or base pairs. Genes depend on the presence of continuous sequences of base pairs (reading frames) that encode specific gene products, usually proteins, and of expression control signals that ensure the occurrence of gene expression at the relevant time with the required efficiency. Mutations can affect reading frames as well as control signals, both of which represent specific DNA sequences. In addition, some specific DNA sequences relate to the control of the metabolism of the DNA molecules and, in particular, of their replication.

In the following sections, selected examples will be described of mechanisms that contribute, each in its specific manner, to the formation of genetic variants. The examples will be grouped into the three above-mentioned major natural strategies (Fig. 1) although, as will be seen later, some of the specific mechanisms involve more than one of these strategies.

4
Local Changes in the DNA Sequences

The process of DNA replication is one of the important sources of genetic variation, and may depend both on the structural features of the substrate DNA and on the functional characteristics of the replication fork [8, 9].

Some of the "infidelities" of DNA replication are likely to depend on tautomeric forms of nucleotides – that is, a structural flexibility which is inherent to these organic compounds [10]. Base-pairing depends on specific structural forms, and conformational changes of nucleotides can result in a mispairing if short-lived, unstable tautomeric forms are "correctly" used in the synthesis of the new complementary strand upon DNA replication. For this reason, mutations that result from this process are considered not as errors but rather as "infidelities." DNA replication is, indeed, one of the sources of nucleotide substitution, and this plays an important role in the evolutionary development of biological functions.

Another source of mutagenesis is represented by an inherently low degree of chemical instability of the nucleotides. For example, cytosine can undergo oxidative deamination to become uracil; upon replication, this gives rise to an altered base pairing and results also in nucleotide substitution.

Other replication infidelities that may relate to slippage in the replication fork can result in either the deletion or the insertion of one or a few nucleotides in the newly synthesized DNA strand. If such mutations occur within reading frames for protein synthesis, the phenotypic effect may be drastic. Indeed this is the case when, in the protein synthesized from a gene affected by a frameshift mutation, the amino acid sequence downstream of the site of mutation differs greatly from that of the nonmutated product. In addition, the size of such a mutated protein is usually altered, depending on the chance occurrence of an appropriate stop codon in the new reading frame.

Proofreading devices and other enzymatic repair systems prevent replication infidelities from producing mutations at intolerably high rates. Generally, these act rapidly after replication by screening for imperfect base-pairing in the double helix. Successful repair thereby requires a specific means to distinguish the newly replicated DNA strand from its template – the complementary parental strand. Because of these correction activities, many primary mispairings are removed before they have the opportunity to become fixed as mutations. As a consequence, DNA repair systems are able to modulate the frequencies of mutagenesis.

The genetic information of some viruses – and also, on occasion, segments of genetic information of chromosomal origin – may pass through RNA molecules, which may later become retrotranscribed into DNA. To date, no efficient repair systems are known to act at the level of RNA; indeed, RNA replication shows a higher degree of infidelity than DNA replication. In consequence, genetic information that becomes replicated as RNA molecules generally suffers increased mutation rates.

A relatively large number of internal and environmental chemicals exert mutagenic effects by means of molecular mechanisms that, in many cases, are well understood and often cause local DNA sequence changes. Some intermediate products of the normal metabolic activities of a cell may be mutagenic, and may thus contribute to spontaneous local mutagenesis. The mutagens of the environment include a multitude of chemical compounds, as well as ultraviolet radiation and some physico-chemical constraints such as elevated temperature, by influencing the chemical stability of the nucleotides. Each of these mutagens and mutagenic conditions contributes in a specific way to the generation of genetic variations.

Again, many of the potential sequence alterations brought about by internal and environmental mutagens are efficiently repaired by enzymatic systems. Since, however, the efficiency of such repair is rarely 100%, evolutionarily relevant mutations persist.

5 Intragenomic DNA Rearrangements

Various recombination processes are well known to mediate DNA rearrangements, which often result in new nucleotide sequences.

Whilst in haploid organisms, general recombination is not essential for propagation, it does influence genetic stability at the population level in various ways as a generator of new sequence varieties. For example, it can bring about sequence duplications and deletions by acting at segments of homology that are carried at different locations in a genome. General recombination is also known to act in the reparation of damage caused to DNA by ionizing radiation. In this case, an intact genome can be assembled from the undamaged segments of sister copies of the chromosome by homologous recombination. The best known contribution of general recombination to genetic diversity is meiotic recombination, bringing about random recombinants between the paternal and the maternal chromosomes in diploid organisms [11].

Two other widely spread types of recombination system are described in greater detail in Sections 5.1 and 5.2, namely site-specific recombination and transposition, both of which are known to contribute to genetic variation. Still other recombination processes, such as that mediated by DNA gyrase, can – at least for the time being – be best grouped as "illegitimate recombinations." This group may contain several different molecular mechanisms that act with very low efficiency and have remained, at least in part, unexplained.

5.1 Site-Specific DNA Inversion at Secondary Crossover Sites

Genetic fusions represent the results of joining together segments of two genes (gene fusions) or of two operons (operon fusions) that are not normally together. An operon is a set of often functionally related genes that are copied into messenger RNA (i.e., transcribed) as a single unit. As a result of this organization, these genes are coordinately regulated – that is, they are turned on or off at the same time. Therefore, in an operon fusion one or more genes are put under a different transcription control, although the genes *per se* remain unchanged. In contrast, gene fusion results in a hybrid gene composed of sequence motifs and often functional domains originating in different genes.

In site-specific DNA inversion, a DNA segment bordered by specific DNA sequences acting as sites of crossing over becomes periodically inverted by the action of the enzyme DNA invertase [12]. Depending on the location of the crossover sites, DNA inversion can give rise to gene fusion or to operon fusion. The underlying flip-flop system can result in microbial populations composed of organisms with different phenotypic appearances: if, for example, the DNA inversion affects the specificity of phage tail fibers, as is the case with phages P1 and μ of *E. coli*, phage populations with two or even more different host ranges will result [13, 14].

Occasionally, a DNA sequence that deviates considerably from the efficiently used crossover site – a so-called "secondary crossover site" – can serve in DNA inversion, which thus involves a normal crossover site and a secondary crossover site [15, 16]. This process results in novel DNA arrangements, many of which may not be maintained because of lethal consequences or reduced fitness; however, if a few new sequences are beneficial for the life of the organism, these may be selectively favored. This DNA rearrangement activity can thus be regarded as evolutionarily important. As many different DNA sequences can serve in this process as secondary crossover sites (though at quite low frequencies), site-specific DNA inversion systems may act as variation generators in large populations of microorganisms. Thus, it has been postulated that this evolutionary role of DNA inversion systems may be more important than their much more efficient flip-flop mechanism, which can at most help a microbial population to more readily adapt to two different, frequently encountered environmental conditions. In fact, other strategies could also be used for the latter purpose.

Computer-aided comparisons of DNA sequences quite often reveal that independent genes may consist of a particular domain with high homology and of other DNA sequences which show no significant signs of relatedness. DNA inversion using secondary sites of crossing over is a potential source of such mosaic genes. DNA inversion can span over relatively large distances in DNA molecules [17], and has the advantage of not losing any DNA sequences located between the two sites of crossing over. Although deletion formation represents another source for gene fusion, it has the disadvantage that the DNA sequences between the sites of crossing over are usually eliminated.

5.2 Transposition of Mobile Genetic Elements

Nine different mobile genetic elements have been found to reside, often in several copies, in the chromosome of *E. coli* K-12 derivatives. This adds up to an occupation of about 1% of the chromosomal length by such insertion sequences (ISs; also termed *IS elements*). At rates on the order of 10^{-6} per individual IS element and per cell generation, these mobile genetic elements undergo transpositional DNA rearrangements. These include the simple transposition of an element and more complex DNA rearrangements such as DNA inversion, deletion formation, and the cointegration of two DNA molecules [6, 18]. Because of different degrees of specificity in the target selection upon transposition, the IS-mediated DNA rearrangements are neither strictly reproducible nor fully random. Transposition activities thus also act as variation generators. In addition to DNA rearrangements mediated by the enzyme transposase, which is usually encoded by the mobile DNA element itself, other DNA rearrangements simply take advantage of extended segments of DNA homologies at the sites of residence of identical IS elements, at which general recombination can act. Altogether, IS elements represent a major source of genetic plasticity for microorganisms.

Transposition occurs not only in growing populations of bacteria, but also during prolonged phases of rest. This effect is readily seen with bacterial cultures stored at room temperature in stabs (small vials containing a small volume of growth medium in agar). The stabs are inoculated with a drop of a bacterial culture taken up with a platinum loop, which is inserted ("stabbed") from the top to the bottom of the agar. After an overnight incubation, the stab is tightly sealed and stored at room temperature. Most strains of *E. coli* are viable in stabs during several decades of storage. That IS elements exert transpositional activities under these storage conditions is easily seen, as follows.

A stab can be opened at any time, a small portion of the bacterial culture removed, and the bacteria well suspended in liquid medium. After appropriate dilution, the bacteria are spread on a solid medium and grown overnight. Individual colonies are then isolated. The DNA from such subclones is extracted and fragmented with a restriction enzyme; the DNA fragments are then separated using gel electrophoresis. Subsequently, Southern hybridization with appropriate probes can be used to determine whether different subclones reveal restriction fragment length polymorphisms (RFLP), which are indicative of the occurrence of mutations during storage.

If this method is applied to subclones isolated from old stab cultures, and if DNA sequences from residential IS elements serve as hybridization probes, an extensive polymorphism is revealed. However, no or only minimal polymorphism occurs with hybridization probes from unique chromosomal genes. Good evidence is available that transposition represents a major source of this genomic plasticity observed in stabs, which at most allow for a very residual growth at the expense of dead cells. Thus, it can be concluded that the enzymes promoting transposition are steadily present in the stored stabs. Indeed, the IS-related polymorphism increases linearly with time of storage for periods of up to 30 years. In fact, in a culture of *E. coli* strain W3110 analyzed after 30 years of storage, each surviving subclone had suffered on average about a dozen RFLP changes, as identified with hybridization probes from eight different residential IS elements, of which IS5 was the most active. Lethal mutations could, of course, not be identified in this study [19–21].

Lethal mutations that affect essential genes for bacteriophage reproduction can be accumulated in the prophage state of the phage genome in its lysogenic host. Such mutants can be screened for their inability to produce infective phage particles upon induction of phage reproduction. Experiments were carried out with *E. coli* lysogenic for a phage P1 derivative grown in batch cultures at 30 °C for about 100 generations, allowing for alternative periods of growth and rest. Most of the independent lethal mutants could thereby be identified as being caused by the transposition of an IS element from the host chromosome into the P1 prophage that is maintained in its host as a plasmid. In these experiments, IS2 was the most active element, and was mainly inserted into a few preferred regions of the P1 genome, but each time at another site [22]. The insertion targets used did not show any detectable homology or similarity with each other. This is another good example of an enzymatically mediated variation generator, since the experiment as such identifies IS transposition as a major source of lethal mutagenesis [16].

Currently, there is no evidence available that bacterial mobile genetic elements would play an essential role in the bacterial life span extending from one cell division to the next. However, these elements are major players in the evolution of bacterial populations; as seen here, they clearly contribute to intragenomic DNA rearrangements. Depending on the target sequences involved, the resulting mutations may often be lethal by interrupting essential reading frames or expression control regions. Although favorable mutations may be relatively rare, they can contribute to evolutionarily advantageous developments of the genome. That mobile genetic elements also play important roles in the natural strategy of DNA acquisition will be detailed in Section 6.

6 DNA Acquisition

While the mutagenesis mechanisms belonging to the strategies of local changes in the DNA sequences (Section 4) and of intragenomic DNA rearrangements (Section 5) are exerted within the genome and can affect any part of the genome, an additional strategy of spontaneous sequence alterations depends on an external source of genetic information. In DNA or gene acquisition, genetic information indeed originates from an organism other than that undergoing mutagenesis [23]. In bacteria, DNA acquisition can occur by means of transformation, conjugation, or virus-mediated transduction. In the latter two strategies of horizontal gene transfer, a plasmid or a viral genome, respectively, acts as natural gene vector.

The association and dissociation of chromosomal genes with natural gene vectors often arises from transpositional activities, and from general recombination acting at IS elements that are at different chromosomal locations. These mechanisms have been well studied, with conjugative plasmids and bacteriophage genomes serving in specialized transduction. For example, composite transposons – which are defined as two identical IS elements flanking a segment of genomic DNA (often with more than one gene unrelated to the transposition process) – are known occasionally to transpose into a natural gene vector and, after their transfer into a receptor cell, to transpose again into the receptor chromosome. Hence, together with other mechanisms – such as site-specific and general recombination – transposition represents an important promoter of horizontal gene transfer.

Several natural factors seriously limit gene acquisition. Transformation, conjugation, and transduction depend on the surface compatibilities of the bacteria involved. Furthermore, when the donor DNA penetrates into receptor cells it is very often confronted with restriction endonucleases; these enzymes cause fragmentation of the invading foreign DNA, which is subsequently completely degraded [5]. Before fragments become degraded, however, they are recombinogenic and may find the chance to incorporate all or part of their genetic information into the host genome. Therefore, the role of restriction systems can be interpreted as follows: they keep the rate of DNA acquisition low, whilst at the same time stimulating the fixation of relatively small segments of acquired DNA to the receptor genome. This strategy of acquisition in small steps can best offer microbial populations the chance to occasionally extend their biological capacities, without any extensive risk of disturbing the functional harmony of the receptor cell by acquiring too many different functions at once. These considerations have their relevance at the level of selection exerted on the hybrids resulting from horizontal DNA transfer. In fact, this selection represents one of the last steps in the acquisition process.

DNA acquisition by horizontal gene transfer is a particularly interesting source of new genetic information for the receptor bacterium, because the chance that the acquired DNA will exert useful biological functions is quite high. Most likely, it has already assumed the same functions in the donor bacterium.

An interesting hypothesis links the universality of the genetic code with the important role played by horizontal gene transfer in the evolutionary development of the living world [24]. According to this view, those organisms which use the most common genetic language would in the long term be able to profit best from the increasing worldwide pool of genetic functions, under pressure to adapt to changes in their living conditions.

7
The Three Natural Strategies Generating Genetic Variations Contribute Differently to the Evolutionary Process

Biological evolution is a systemic process [25] and, as outlined above, many different and specific mechanisms contribute to generate genetic diversity that represents, at any time, the substrate for natural selection. The building-up of functional complexity is a stepwise process, in which many random attempts at genetic alteration become rapidly rejected, while relatively few novel sequences are approved as favorable by natural selection and are maintained and amplified. The genome can thus be seen as a "cabinet" in which key informations from favorable historical developments are stored. Stepwise, additional favorable information is added. In the context of changing selective conditions, stored information having lost its functionally beneficial relevance may be deleted or favorably altered. As has been seen, a multitude of different mechanisms are behind this dynamic process, and for a better understanding of events the identified mechanisms have been grouped into three major natural strategies of genetic variation – local changes in the DNA sequences, intragenomic DNA rearrangements, and DNA acquisition – all of which have different qualities with regard to their contributions to biological evolution.

The local DNA sequence change is probably the most frequently involved strategy of genetic variation. Indeed, its frequency – which depends primarily on intrinsic properties of matter, chemical instability, and conformational flexibility – would be intolerably high if it could not be modulated by the efficient enzymatic systems of DNA repair. Local sequence changes bring about nucleotide substitution, the deletion, and insertion of one or a few base pairs, or a local scrambling of a few base pairs. These sequence changes can contribute to a stepwise improvement of a biological function. It must be borne in mind that the functional test for such improvement is carried out by natural selection. Alternatively, a long series of stepwise local sequence changes could be expected to bring about a novel biological function. However, this type of long-term process can achieve high efficiency only when natural selection begins to be exerted on such an upcoming function.

In contrast, the reshuffling of DNA segments within the genome can be considered as "tinkering" with the existing elements, whereby favorable gene fusions and operon fusions may occasionally result. DNA rearrangement may also be the source of gene duplication and higher amplification, both of which are widely recognized contributions to the evolutionary progress.

The evolutionarily high effectiveness of horizontal gene transfer has already been highlighted in Section 6. As a matter of fact, DNA acquisition allows the recipient organism to share in the success of evolutionary developments made by others.

In drawing the evolutionary tree of bacteria, DNA acquisition should be accounted for by more or less randomly adding temporal horizontal shunts between individual branches [16]. It must be borne in mind that, usually, only small DNA segments flow through such shunts in horizontal gene transfer.

Strictly speaking, several of the specific mechanisms of genetic variation employ more than one of the three natural strategies described above. In the transposition of IS elements, for example, a chromosomal DNA segment consisting of the mobile genetic element can undergo a translocation and thereby become inserted at a new target site. As a rule, the target sequence will as a consequence be duplicated, and this usually involves a few nucleotides. Hence, this transposition event will consist of both a DNA rearrangement and a local sequence change.

To date, it appears that most of the well-studied microbial strains employ, in parallel, each of the three natural strategies in the generation of genetic variations. In addition, bacteria very often use not only one but also several different specific mechanisms for mutagenesis by each of the strategies. Dissimilar specific mechanisms often function with different efficiencies, as reflected by their contribution to the overall mutation rate. For any given strategy, it might be less relevant as to which specific mechanism is at work than that the particular strategy finds its application with an evolutionarily useful efficiency. In other words, specific mechanisms may substitute for each other within a strategy, at least to some degree. However, this rule does not apply between the strategies because of the difference in the qualities of their evolutionary contributions.

The efficiency displayed by a given specific mechanism of spontaneous mutagenesis may depend on both internal (e.g., the availability of enzymes that mediate mutagenesis) and external factors (e.g., environmental stress). It should also be noted that some mechanisms may act more or less randomly along a DNA molecule, while other mechanisms may demonstrate regional or site preferences for their activities. In view of these considerations, there is a tendency to assume that an evolutionarily fit (or well-prepared) organism should best be able to use a few specific mutagenesis mechanisms for each of the three strategies to generate genetic variations. The meaning of the term "evolutionary fitness" will be explained in Section 8.

8 Evolution Genes and Their Own Second-Order Selection

The attentive reader will have seen in the description of some specific mechanisms contributing to the spontaneous mutagenesis that, besides a number of nongenetic factors, specific products of genes are very often at work. These gene products can belong to systems for the repair of DNA damage and will, in this case, modulate the frequency of mutagenesis. Similarly, restriction enzymes seriously reduce both the chance of DNA acquisition and the size of a DNA segment that may eventually be acquired by the recipient cell. Other gene products such as transposases and other mediators of DNA recombination act as generators of genetic variations. Since variation generators and modulators of the frequencies of genetic variation are evolutionary functions, the underlying genetic information is referred to as *evolution genes*. In the microbial world, these evolution genes generally play no essential role in the physiology of individual lives moving by cell division from one generation to the next. Under laboratory conditions, neither restriction-modification systems nor enzymes for DNA rearrangements are required for the propagation of bacteria. The role of these enzyme systems is primarily evolutionary, and this becomes manifest at the level of populations.

It is assumed that evolution genes are themselves also submitted to selective pressure. However, such selection cannot follow the rules of direct selection for improvements of essential functions, such as those of housekeeping genes. Rather, the selection for the presence and improvement of a variation generator will be exerted at the level of populations. Clearly, it will also be an individual that may one day undergo a mutation improving an evolutionary function. That function will also be exerted in its progeny, in which appropriate genetic variants of genes for directly selected products will be either more or less abundant. Any genetic alteration that affects an evolution gene and proves at longer-term of higher evolutionary value will be maintained, and will provide to the carrier of the involved gene an evolutionary benefit. In the long run this will lead to a fine-tuning of the evolutionary functions of both variation generators and of modulators of the frequency of genetic variation. The underlying indirect selection based on the cells ability to provide genetic variants at a well-balanced level is termed *second-order selection* [26].

It is important to be aware that some gene products may exert their essential functions for the benefit of both the life of the individual and the evolutionary progress of the population. In these cases, it is assumed that evolutionary selection is exerted for both types of function, and will eventually bring the gene to a fine-tuned state to carry out its functions for each of the different purposes on an optional basis. However, as noted above, a number of gene products involved in genetic variation are inessential for the lives of individual bacterial cells. Similarly, the products of many housekeeping genes are inessential for biological evolution.

9
Arguments for a General Relevance of the Theory of Molecular Evolution for All Living Organisms

Largely based on evidence from microbial genetics, it has so far been postulated that the products of a number of evolution genes contribute each in their specific way to the generation of genetic variants at evolutionarily useful frequencies. Thereby, the sources of mutagenesis may relate either to the activity of the evolution gene product itself (e.g., a transposase) or to a nongenetic factor (e.g., a chemical mutagen or an intrinsic structural flexibility of a nucleotide). In many cases, nongenetic factors and the products of evolution genes cooperate in the formation of genetic variants at physiologically tolerable and evolutionarily beneficial levels. This is, for example, the case in spontaneous mutagenesis by an environmental mutagen, when some of the primary damage on the DNA is successfully repaired while other damage may lead to a fixed mutation.

The theory of evolution postulates that life on Earth started almost four billion years ago with primitive, unicellular microorganisms. It was during the first two billion years that microbes must have developed the basis for the actual set-up of evolutionary strategies and the underlying evolution genes. It can be postulated that the acquired evolutionary capacities could have allowed some microbial populations to undergo a division of labor in increasingly stable associations of cells, and this development might later have led to multicellular organisms. In this type of development, the evolutionary fitness of the involved organisms might have been an important precondition. At still later stages of further evolutionary development, the three natural strategies for generating genetic variations (see Sects 4–7) must have continued exerting their evolutionary influence, together with some other factors such as the formation of symbiotic associations [27]. As a matter of fact, considerable evolutionary relevance is attached to the endosymbiosis of higher organisms with bacteria. Such situations of cohabitation may form an ideal condition for occasional horizontal gene transfer between the close associates.

Clearly, there is still need to conduct research into the spontaneous generation of genetic variation in higher organisms, ideally at the level of the genomes. While this is already quite difficult in microorganisms, it is of increased perplexity with the much larger genomes of higher organisms. However, sequence comparisons today offer fruitful ways of searching for sequence homologies, sequence similarities and single nucleotide polymorphisms, as well as for traces of intragenomic DNA rearrangements and of the horizontal transfer of genetic information. The data available to date are in support of the principles of molecular evolution outlined in Section 2, which are likely to be valid for any type of living organism.

Some of the general evolutionary strategies developed in microorganisms must, ultimately, have also been very useful for the developmental and physiological processes at somatic levels of higher organisms. The generation of antibody diversity in the immune systems of vertebrates by genetic rearrangements, and so-called somatic mutagenesis, is a good example. Another example is the enzymatic repair of DNA damage caused in somatic cells by external mutagens such as ultraviolet (UV) irradiation.

These considerations illustrate that, whichever gene function may prove to be useful for whatever particular purpose, it may be evolutionarily retained and, during the course of time, further fine-tuned. This principle has already been encountered in the microbial world, where the presence of multifunctional enzymes (i.e., those working both for the physiology of the cells and for an evolutionary task) has been postulated to be evolutionarily improved, both by direct and second-order selection for the various functions. Indeed, this may also be the case in higher organisms.

10
Systemic Aspects of Biological and Terrestrial Evolution

As noted above, the updated knowledge on molecular Darwinism clearly indicates that biological evolution is a systemic process of enormous dimensions [25], and it is possible to trace the different types of more or less related organisms back to their common origin. Moreover, in their future evolutionary development, any organism may – at one time or another – also benefit from the acquisition of genetic information from another type of organism by horizontal gene transfer, as has already occurred during the past evolution. Hence, living organisms not only have a common past – as previously postulated by Charles Darwin – but also a common future, at least to some degree.

Natural selection is a highly complex, permanent process, and substrates for such selection are by far not limited to genetically determined phenotypic traits. Rather, it is becoming apparent that phenotypic traits can become modulated by epigenetic impacts which may, in turn, depend to a large extent on environmental conditions.

In any ecosystem, many quite different organisms cohabit and can mutually influence each other enormously. Most often, such interdependencies take the form of symbiosis, where the different types of organism involved generally help each other, so that all participants can benefit. These aspects play their role in worldwide natural selection. Only rarely does such cohabitation bring benefit to only one of the interacting partners, and in extreme cases the second partner may suffer from a close interaction. This is the case for pathogenesis, which may be regarded as a form of negative selection for the suffering partner.

Today, it is clear that in the long term, the inanimate world is in no way absolutely stable. Both, cosmic and terrestrial evolutions affect the terrestrial living conditions to some degree, and this in turn exerts an influence on habitats and also on the ecological communities in these habitats.

During the course of their cultural evolution, human beings have learned to counteract, to some extent, certain influences of negative natural selection. As is the case for any living organism, humans form an integral part of the worldwide system of evolving communities of forms of life in a developing world. A deeper scientific insight and comprehension of the process of evolution can serve as guidance for future cultural evolution if there is a tendency to avoid causing disturbances to the inherited treasure of the evolutionary pathway. It is important to be aware that the natural processes of evolution in this relatively stable world occur extremely slowly, but steadily. But, this is not a reason to ignore evolutionary constraints; rather, the rate of evolution should serve as guidance for the responsible, sustainable usage of the natural diversities of forms of life and of habitats.

11 Conceptual Aspects of the Theory of Molecular Evolution

With the information provided in Section 2.2 in mind it is fair to re-state that, for the time being, evolution genes and evolution functions represent a concept rather than being a fully proven fact. This concept is based on a particular approach to interpreting the numerous experimental data available. In the following subsections, a brief analysis will first be provided of the difficulties encountered when clarifying the situation in a scientific debate. This will be followed by highlighting the philosophical and more practical values of a deeper understanding of the molecular processes that drive biological evolution.

11.1 Pertinent Scientific Questions

In the history of scientific investigations, biologists have often sought evidence that living organisms could specifically modify, or adapt, their genetic information in order to better cope with upcoming changes in their living conditions. Most of these attempts have failed to provide the expected response. In some cases – where a certain degree of specific adaptation was observed – specific causal explanations have occasionally been found following more extensive investigations. There is at present, however, no good scientific evidence for a general rule that genetic alterations would always be directed towards a specific goal. This situation favors the view that spontaneous mutations affect DNA more or less randomly, which is in line with the general observation that only a minority of spontaneous mutants will prove to be favorable under the living conditions encountered.

The postulation that evolution genes can act as generators of genetic variations may be surprising in this context, since it is not consistent with the widely held concept of genetic information being a strict program for the fulfillment of a specific task. Whilst this definition does apply to many housekeeping genes (the products of which efficiently catalyze a reaction that reliably yields always the same product), it is not the way that a variation generator functions – inefficiently and yielding a different product from case to case (a good example of this is the transposition of mobile genetic elements). Today, not all scientists see in genetic variation the primary function of a mobile genetic element; rather, some interpret IS translocation – which often goes hand-in-hand with replication of the element – as a selfish activity. In fact, this interpretation considers mobile genetic elements as parasites, with the argument that their activity would often harm the host cell.

This discussion shows that the concept of evolution genes cannot easily be defended by referring to scientific evidence; the concept reflects rather an attitude of the observer of natural events. According to the view defended in this chapter, Nature actively cares for biological evolution. The products of evolution genes are actively involved in generating different types of genetic variants at frequencies which insure not only a genetic stability that is required to maintain the concerned form of life, but also a low frequency of genetic variations as the driving force of evolution. This interpretation recognizes biological evolution as an essential principle of self-organization of life on Earth. In other words, the living world benefits from the natural potency to evolve at any level of life-form.

Another pertinent question with no easy scientific answer relates to the evolutionary function of viruses. Some viruses are clearly identified to serve (on occasion) as gene vectors in horizontal gene transfer, and some also temporarily integrate their genome into the host genome. This situation relates to the lysogenic state of bacteria as well as to endogenous viruses such as retroviruses that reside in many higher organisms. Again, it might be questioned whether primarily, these viruses fulfill evolutionary functions for the evolutionary development of their hosts, or whether they should rather be regarded as parasites, which may carry out some evolutionary function by accident.

While prokaryotic organisms have genomes which are relatively densely packed with functional genes, many higher organisms have extended segments of intergenic DNA sequences, some of which are highly repetitive. Some of these noncoding sequences are highly homogeneous with regard to their nucleotide composition. While the biological roles played by noncoding regions are still poorly understood, it has been postulated that compositional constraints may influence natural selection [28, 29]. These aspects, which have not been detailed in this chapter, may more specifically relate to the molecular evolution of higher organisms, in addition to the principles outlined here, largely on the basis of evidence from microbial genetics.

11.2 Philosophical Values of the Knowledge on Molecular Evolution

One of the central questions of human curiosity is to know where life – and, more specifically, human life – comes from. The Darwinian Theory opposed the idea of a specific act of creation for each particular form of life, but rather proposed an alternative explanation of a steady evolutionary development which implied the descent of actual species from common ancestors. Hence, the directions of evolution are provided by natural selection acting steadily on all available forms of life, including all present variants. Until recently, there was no specific scientific explanation as to how genetic variants are generated, but with recently developed research strategies it has become possible for molecular genetics to fill this gap. The branch of science termed *molecular evolution* explains that there is not just a single source for genetic variants; rather, many different specific mechanisms contribute to the generation of genetic variants at low frequencies. These mechanisms follow one – and sometimes more than one – basic strategy of evolutionary development. These include local changes in the DNA sequences, rearrangements of the DNA segments within the genome, and the acquisition of segments of foreign DNA by horizontal gene transfer. As a general rule, spontaneous mutagenesis is not specifically directed but is – at least to some extent – a random process, so that only a minor fraction of spontaneous genetic variants will be favorable for the organism concerned and thus provide it with a selective advantage. Nevertheless, new knowledge on the precise molecular mechanisms of the generation of genetic variations provides strong evidence that, in many cases, specific enzymes are involved: these are the products of evolution genes. Such products operate in tight collaboration with nongenetic factors, which may be either intrinsic properties of matter or environmental conditions. This view of the evolutionary process represents the core of a theory of molecular evolution, and it can be seen as an extension of neo-Darwinism to the molecular level (for further detail, see Refs [30–35]).

It should be clearly stated that the theory of molecular evolution does not explain the origin of life. It can, however, explain that biological evolution exerted in all living beings is a steady, dynamic process that is actively promoted not only by intrinsic properties of matter, but also by the intervention of products of evolution genes or more generally of evolutionary functions of many different gene products. A recently published book devoted to these exciting insights is entitled *Darwin in the Genome* [36].

The high philosophical value of this extension of our world view is obvious, and merits extensive discussion and evaluation with regards to its various cultural dimensions. One interesting aspect is the implied duality of the genome. Indeed, evolution genes are located together with all the other genes on the genome, and also on accessory DNA molecules such as plasmids and viral genomes. Whilst it is most likely that a major part of the gene products carries out functions that will benefit the cell and the individual (often a multicellular organism), it is also probable that a minority of gene products will be involved in the biological evolution of the population concerned. It should be noted, however, that the generation of a novel genetic variant clearly also occurs in an individual cell. Yet, this act of creation has only a small chance of bringing to the cell concerned (and to its descendants) a selective advantage. More often, the mutation will be unfavorable and render the life of the organism more troublesome. However, while the spontaneous rate of mutagenesis remains low (mostly due to an intervention of the evolution genes that are fine-tuned for their activities), unfavorable mutations are tolerable at the level of propagating populations.

Incidentally, the situation described here offers a possible explanation to the rather difficult theodicean question: "Why does God, despite His love for the human creature, admit that physically evil events such as a mutation causing an inheritable disease can occur to individuals?" As a matter of fact, in Genesis the Creation is described as a stepwise process, and the descendants of Adam and Eve are not considered to be clones. This implies the permanent evolutionary expansion of the diversity of life-forms [37]. It is also stated in Genesis that God evaluated this system as "good," and hence from a Biblical point of view biological evolution occurs according to God's intention to amplify life diversity on planet Earth. Occasional unfavorable mutations that affect rare individuals among populations represent a sacrifice brought to the creative force residing in the natural potency for molecular evolution. Creation is, therefore, a permanent process.

In brief, the genetic information contained in each genome – of bacteria as well as of all higher organisms – represents an internal duality. It serves individuals for the fulfillment of their individual lives, and it serves populations for a slow but steady expansion of life-forms, and thus of biodiversity.

11.3
Aspects Relating to Practical Applications of Scientific Knowledge on Molecular Evolution

Today, living organisms occupy an amazing variety of ecological niches on planet Earth; these include extreme physico-chemical conditions such as elevated temperatures, high pressure, and also quite unusual compositions of chemical elements. Yet, despite the intrinsic potency of the living world to expand on an evolutionary basis, the carrying capacity of Earth for life can be estimated as in the order of 10^{30} living cells. Whilst this is indeed a very large number, it seriously limits the free expansion of life in its various forms.

The following reflections should help to illustrate this statement. An adult human being carries in the order of 10^{13} cells; consequently, today's human population occupies about 10^{23} cells of the available 10^{30}. In fact, this is close to the average available to each of the estimated 10^7 different species of organisms on Earth. By extending the reflections made in Section 3, regarding the propagation of bacteria by cell division, it can be concluded that an inoculum of a single bacterial cell would, in theory, lead to the production of 10^{30} cells within only 50 h. In reality, however, growth will be halted at a much earlier stage by a lack of nutrition. Nonetheless, this reflection illustrates very well the enormous internal forces involved in the expansion of a given form of life.

Similarly, a high potency for evolutionary expansion toward more diversity resides in the mechanisms of molecular evolution that are described in this chapter. These mechanisms can serve us as a basis to better understand both the origin and the steady – though slow – replenishment of biodiversity, as well as the internal limits set to the evolutionary expansion. This knowledge can – and should – be used increasingly as a background for any measures taken in view of the protection of biodiversity and of habitats for diverse forms of life. Not least, a better understanding of the evolutionary process may be helpful to render the development of agricultural and related practices more sustainable.

Genetic engineering offers ample new, sustainable possibilities to produce medicinal drugs, to provide foods of higher quality, and to reduce the nocive impact of the human civilization on the environment [38]. Serious reservations made by large parts of the human population impede many of the proposed biotechnological applications. A part of these concerns refers to the unpredictable long-term effects of genetically modified organisms (GMOs) that are released into the environment, such as in agricultural applications. Hence, scientific assessments of long-term and in particular evolutionary effects of such applications are required, and knowledge of the natural strategies of molecular evolution can provide a good basis for such investigations [39]. In fact, in genetic engineering DNA sequence alterations are brought about within the genome by site-directed mutagenesis in studies of the biological functions of specific genes. In addition, well-defined segments of DNA are introduced into other organisms, with a view either to amplify a particular DNA segment or to harvest a specific gene product. GMOs can also directly serve in applications as genetically modified food, and for bioremediation by microorganisms. Yet, a candid comparison of the practices involving genetic engineering with the natural strategies of generation of genetic variations reveals a high degree of similarity. Typically, the amounts of nucleotides or lengths of DNA sequences involved in these genetic modifications – both in genetic engineering and in natural genetic variation – are of the same order of magnitude. Depending on the strategy involved, they may concern one to a few base pairs, or in other instances a DNA segment containing a sequence domain or one to a few genes, both in intragenomic DNA rearrangements and in the horizontal transfer of DNA between two different organisms. Thus, it can be principally expected that the long-term evolutionary risks of GMOs compare with biohazards intrinsic to the natural process of biological evolution. Similar risks may also be inherent to classical breeding techniques.

These considerations ask for a more integral, holistic, and critical evaluation of the impact of past and present human activities on the natural process of biological evolution. Such assessments should address any human impact on genetic variation, natural selection, and isolation. Based on a long-term historical knowledge, the foundations of life and its evolutionary development on Earth are relatively robust. However, whilst this is good news for humankind, attention should not be deflected from maintaining a responsible and well-intentioned use of scientific knowledge in attempts to create a more easy and comfortable lifestyle.

References

1 Lederberg, J. (1947) Gene recombination and linked segregation in *E. coli*. *Genetics*, **32**, 505–525.

2 Luria, S.E., Delbrück, M. (1943) Mutations of bacteria from virus sensitivity to virus resistance. *Genetics*, **28**, 491–511.

3 Avery, O.T., MacLeod, C.M., McCarty, M. (1944) Studies on the chemical nature of the substance inducing transformation of pneumococcal types: induction of transformation by a desoxyribonucleic acid fraction isolated from *Pneumococcus* type III. *J. Exp. Med.*, **79**, 137–158.

4 Kucherlapati, R., Smith, G.R. (Eds) (1988) *Genetic Recombination*, American Society for Microbiology, Washington, DC.

5 Arber, W. (1965) Host-controlled modification of bacteriophage. *Annu. Rev. Microbiol.*, **19**, 365–378.

6 Shapiro, J.A. (1983) *Mobile Genetic Elements*, Academic Press, New York.

7 Watson, J.D., Crick, F.H.C. (1953) Molecular structure of nucleic acids: a structure for deoxyribose nucleic acid. *Nature*, **171**, 737–738.

8 Drake, J.W. (1991) Spontaneous mutation. *Annu. Rev. Genet.*, **25**, 125–146.

9 Echols, H., Goodman, M.F. (1991) Fidelity mechanisms in DNA replication. *Annu. Rev. Biochem.*, **60**, 477–511.

10 Watson, J.D., Crick, F.H.C. (1953) The structure of DNA. *Cold Spring Harbor Symp. Quant. Biol.*, **18**, 123–131.

11 West, S.C. (1992) Enzymes and molecular mechanisms of genetic recombination. *Annu. Rev. Biochem.*, **61**, 603–640.

12 Glasgow, A.C., Hughes, K.T., Simon, M.I. (1989) Bacterial DNA Inversion Systems, in: Berg, D.E., Howe, M.M. (Eds) *Mobile DNA*, American Society for Microbiology, Washington, DC, pp. 637–659.

13 Sandmeier, H., Iida, S., Arber, W. (1992) DNA inversion region Min of plasmid p15B and Cin of bacteriophage P1: evolution of bacteriophage tail fiber genes. *J. Bacteriol.*, **174**, 3936–3944.

14 Sandmeier, H. (1994) Acquisition and rearrangement of sequence motifs in the evolution of bacteriophage tail fibers. *Mol. Microbiol.*, **12**, 343–350.

15 Iida, S., Hiestand-Nauer, R. (1987) Role of the central dinucleotide at the crossover sites for the selection of quasi sites in DNA inversion mediated by the site-specific Cin recombinase of phage P1. *Mol. Gen. Genet.*, **208**, 464–468.

16 Arber, W. (1991) Elements in microbial evolution. *J. Mol. Evol.*, **33**, 4–12.

17 Rozsa, F.W., Viollier, P., Fussenegger, M., Hiestand-Nauer, R.M., Arber, W. (1995) Cin-mediated recombination at secondary crossover sites on the *Escherichia coli* chromosome. *J. Bacteriol.*, **177**, 1159–1168.

18 Galas, D.J., Chandler, M. (1989) Bacterial Insertion Sequences, in: Berg, D.E., Howe, M.M. (Eds) *Mobile DNA*, American Society for Microbiology, Washington, DC, pp. 109–162.

19 Naas, T., Blot, M., Fitch, W.M., Arber, W. (1994) Insertion sequence-related genetic variation in resting *Escherichia coli* K-12. *Genetics*, **136**, 721–730.

20 Arber, W., Naas, T., Blot, M. (1994) Generation of genetic diversity by DNA re-arrangements in resting bacteria. *FEMS Microbiol. Ecol.*, **15**, 5–14.

21 Naas, T., Blot, M., Fitch, W.M., Arber, W. (1995) Dynamics of IS-related genetic rearrangements in resting *Escherichia coli* K-12. *Mol. Biol. Evol.*, **12**, 198–207.

22 Sengstag, C., Arber, W. (1983) IS2 insertion is a major cause of spontaneous mutagenesis of the bacteriophage P1: non-random distribution of target sites. *EMBO J.*, **2**, 67–71.

23 Lorenz, M.G., Wackernagel, W. (1994) Bacterial gene transfer by natural genetic transformation in the environment. *Microbiol. Rev.*, **58**, 563–602.

24 Arber, W. (2006) The evolutionary strategy of DNA acquisition as a possible reason for a universal genetic code. *Hist. Philos. Life Sci.*, **28**, 525–532.

25 Arber, W. (2009) Systemic aspects of biological evolution. *J. Biotechnol.*, **144**, 242–244.

26 Weber, M. (1996) Evolutionary plasticity in prokaryotes: a panglossian view. *Biol. Philos.*, **11**, 67–88.

27 Magulis, L. (1981) *Symbiosis in Cell Evolution: Life and its Environment on the Early Earth*, Freeman, San Francisco, CA.

28 Bernardi, G. (2000) The compositional evolution of vertebrate genomes. *Gene*, **259**, 31–43.

29 Bernardi, G. (2007) The neo-selectionist theory of genome evolution. *Proc. Natl Acad. Sci. USA*, **104**, 8385–8390.

30 Arber, W. (1993) Evolution of prokaryotic genomes. *Gene*, **135**, 49–56.

31 Arber, W. (1995) The generation of variation in bacterial genomes. *J. Mol. Evol.*, **40**, 7–12.

32 Arber, W. (2000) Genetic variation: molecular mechanisms and impact on microbial evolution. *FEMS Microbiol. Rev.*, **24**, 1–7.

33 Arber, W. (2002) Evolution of prokaryotic genomes. *Curr. Top. Microbiol. Immunol.*, **264/I**, 1–14.

34 Arber, W. (2003) Elements for a theory of molecular evolution. *Gene*, **317**, 3–11.

35 Caporale, L.H. (Ed.) (1999) Molecular Strategies in Biological Evolution. *Ann. N. Y. Acad. Sci.*, **870** (Special issue).

36 Caporale, L.H. (2003) *Darwin in the Genome: Molecular Strategies in Biological Evolution*, McGraw-Hill, New York.

37 Arber, W. (2007) Sagesse Traditionelle et Savoir Récemment Acquis sur L'évolution Biologique, in: Staune, J. (Ed.) *La Science, L'homme et le Monde – les Nouveaux Enjeux*, Presses de la Renaissance, Paris, pp. 127–149.

38 Arber, W. (2009) The impact of science and technology on the civilization. *Biotechnol. Adv.*, **27**, 940–944.

39 Arber, W. (2002) Molecular evolution: comparison of natural and engineered variations. *Pontif. Acad. Sci. Scr. Varia*, **103**, 90–101.

7
Systematics and Evolution

Jeffrey H. Schwartz
University of Pittsburgh, Departments of Anthropology and History and Philosophy of Science, 3302 Posvar Hall, University of Pittsburgh, Pittsburgh, PA 15260, USA

Systems Biology: Advances in Molecular Biology and Medicine, First Edition. Edited by Robert A. Meyers.

Keywords

African apes
Chimpanzees, bonobos, and gorillas.

Anthropoidea
The so-called "higher primates"; the subordinal rank that subsumes New and Old World monkeys, plus the hominoids.

Clade
A group of related organisms; a group of organisms united by common ancestry.

Derived feature
A relative character state determined by how restrictively shared a feature under study is.

Hominids
Humans and their fossil relatives.

Hominoids
The group of anthropoid primates that includes gibbons and siamangs, chimpanzees, gorillas, orangutans, and humans.

Homology
Similarity due to common ancestry.

Large-bodied hominoids
Orangutans, humans, and African apes.

Lesser (small-bodied) hominoids
The gibbons and siamangs.

Monophyletic group
A clade.

Outgroup
A taxon that is outside of (a sister taxon of) the focal group of phylogenetic reconstruction.

Phylogeny reconstruction
The generation of theories of evolutionary relationships among taxa.

Primitive feature
A relative character state determined by how broadly shared a feature under study is.

Primitive retention
A feature that is shared by the members of a group by inheritance from a common ancestor.

Sister taxon
A taxon that is the closest relative of another.

Systematics
The study of the pattern and relationships of organisms, including the identification of species and delineation of groups of taxa.

Taxic
Relating to taxa.

Taxon (plural: taxa)
Any taxonomic rank; for example, species, genus, family, order, class.

Studies that rely on genetic and molecular information to answer evolutionary questions fall roughly into two categories: (1) the reconstruction of evolutionary relationship (including the timing of lineage divergence); and (2) the emergence of a novel structure. The former endeavor is referred to as "*molecular systematics,*" and as "*molecular anthropology*" when applied to humans and other primates. The assumption that differences accrue within lineages after they diverge from a common ancestor leads to the assumption that the degree of molecular similarity is reflected in the degree of evolutionary relatedness. Whether or not such change occurs in a clock-like manner remains a matter of debate. The reliability of molecular, rather than morphological, studies to inform about evolutionary relatedness also relies on the appeal of a "law of large numbers": it is possible to analyze thousands, or even millions, of molecules compared to a pitifully small number of morphological features.

The concern of developmental evolutionary biology ("evo-devo") is the identification of molecular elements – for example, transcription factors or other types or classes of proteins and "genes," as well as epigenetic and non-Mendelian factors – that provide insight about the processes of development of structure and form. Rather than imparting significance to a particular RNA, DNA, or amino acid sequence, the focus is on signal transduction pathways and the regulation of development, namely, the origination of structure.

In this chapter, an attempt will be made to summarize the main aspects of these fields of inquiry, including their assumptions, and to suggest avenues of future inquiry that might bring these disciplines together to provide some better information about evolution.

1 The Beginning of Molecular Systematics

The first inquiries into systematics and phylogenetic reconstruction via the investigation of the "blood relationship" of organisms can be traced to a few individuals, of whom the best known is George H. F. Nuttall [1]. In the research that preceded his monograph of 1904, *Blood Immunity and Blood Relationship: A Demonstration of Certain Blood-Relationships Amongst Animals by Means of the Precipitin Test for Blood,* Nuttall sought to demonstrate that the degree of similarity between animals in their blood serum proteins was a reflection of the closeness of their evolutionary relatedness. The idea was simple: produce antisera (antibodies or antigens) to blood serum proteins of one animal; mix the antiserum with the blood serum proteins of another animal; and measure the amount of precipitin that settled out. The more profound the precipitation, the greater the similarity (because of a greater number of antibody–antigen binding sites) and, consequently, the closer the evolutionary relationship between the two organisms being compared.

The strength of the antiserum played an important role in determining just how many species might belong to the same group. On the rationale that these species' blood proteins would be more similar, a weak antiserum was then expected to provoke reactivity in the sera of the most closely related organisms; conversely, a stronger solution would produce a precipitate when combined with the sera of a greater number of related animals. The more distantly related the groups under comparison, the less would be the expected reactivity between serum and antiserum. In addition to strength of reactivity, Nuttall thought that the reaction rate was also an indicator of closeness of relatedness. Ultimately, he decided that, "... the zoological relationships between animals are best demonstrated by means of powerful antisera." Consequently, he concluded,

> "... *if we accept* the degree of blood reaction as an index of the degree of blood-relationship *within the Anthropoidea (the so-called higher primates, New and Old World monkeys plus the hominoids, which includes gibbons, chimpanzees, gorillas, orangutans, and humans), then we find that the Old World apes are more closely allied to man than are the New World apes, and this is exactly in accordance with the opinion expressed by Darwin.*"

Although Nuttall believed that his approach accurately revealed the evolutionary relationships of animals – with the exception of a reference in 1922 by the paleontologist W. K. Gregory [2] to Nuttall having demonstrated "... the anthropoid heritage of man" – his investigations remained largely unacknowledged until the 1960s, when various reports were made which claimed to have demonstrated the evolutionary relationships of organisms through the study of elements of their biochemistry.

2 The Molecular Assumption

Perhaps the most influential of these publications was by Emile Zuckerkandl and Linus Pauling [3] on fetal and adult hemoglobin, the sequences of which – as an "evolutionary afterthought" – they compared between human, gorilla, horse, and fish. Although the comparison of gill- with lung-bearing animals might be questioned (because hemoglobin is a metabolically active protein), Zuckerkandl and Pauling were able to demonstrate that the fish was more dissimilar to humans than to the horse, while the gorilla was the most similar to humans in hemoglobin sequence. Since this pattern of taxic distance mirrored the morphologically accepted scheme of evolutionary relatedness of these animals, Zuckerkandl and Pauling suggested that their observations "... can be understood at once if it is *assumed* (emphasis added) that in the course of time the hemoglobin-chain genes duplicate, (and) that the descendants of the duplicate genes 'mutate away' from each other." From this assumption, they felt justified in proposing the additional assumption: "[O]ver-all similarity must be an expression of evolutionary history," with descendants "mutating away" and becoming "gradually more different from each other." In other words, the more time that elapses after a lineage's divergence, the greater the molecular difference its succession of descendents will accumulate. Consequently, following in the footsteps of Nuttall, Zuckerkandl and Pauling proposed that evolutionary closeness was synonymous with degree of molecular similarity. Also assumed in this nexus of assumption was the notion of constant and gradual unilinear change.

There is, however, nothing among Zuckerkandl and Pauling's studies to justify the notion that "over-all similarity" is "... an expression of evolutionary history." They merely assumed such a correspondence, as they had equally assumed that difference reflected not only change but, also and more specifically, the same type of change which, in turn, was achieved at a constant and gradual rate. Nevertheless, to inject a note of objectivity into the discussion, the discovery of 100 nucleotide differences between two taxa does not translate into the rate at which presumed substitutions occurred. Rather, additional information is required, such as a yardstick of the time at which these differences began to accumulate. However, it must then also be assumed that the rate of substitution over this period of time was gradual, and that differences between taxa did not arise during a few concentrated phases of replacement.

Nevertheless, Zuckerkandl and Pauling's effort resulted in the concretization of the "molecular assumption" – continual molecular change and its continual accumulation over evolutionary time – which would thereafter become the foundation of molecular systematics. As Caccone and Powell [4] would write in 1989: "Virtually all molecular phylogenetic studies... have a major underlying assumption: the genetic similarity or difference among taxa is an indication of phylogenetic relatedness. Lineages that diverged more recently will be genetically more similar to one another than will be lineages with more ancient splits." If such an assumption is embraced, the rest may follow logically – although not necessarily because of biological demonstration. One of the major extensions of the "molecular assumption" is the notion of a "molecular clock," the existence of which was promoted initially by Sarich and Wilson.

Beginning with their first major paper in 1966, Sarich and Wilson [5] sought to elucidate the evolutionary relationships of primates, and especially of humans and the large-bodied apes. For this, they used the technique of micro-complement fixation (MCF), which requires only minute amounts of serum and antiserum for study of immunological reactivity. Their molecule of choice was one of the larger blood serum proteins, albumin. Sarich and Wilson then translated the degree of reactivity achieved between albumin and anti-albumin into an "index of dissimilarity," which was subsequently referred to as "immunological distance" (ID). The closer an ID value was to 1.0, the greater the overall molecular similarity and thus, given an assumed equivalence between ID and evolutionary relationship, the presumption of closeness of relatedness between the organisms under study. As an apparent check on the validity of this approach, Sarich and Wilson conceived the "test of reciprocity." For example, if chimpanzee serum was crossreacted with anti-human serum (produced by injecting another animal, such as a rabbit or chicken, with human serum) the first time around, the human serum would then be crossreacted with anti-chimpanzee serum the next time to see if the resultant ID values were reasonably similar. Predictably, the test of reciprocity usually confirmed the initial immunological finding.

The supposed evolutionary arrangement of the primates that Sarich and Wilson achieved was more consistent with the general pattern of primate relationships based on comparative morphology than it was in disagreement. Because of this concordance, they concluded that "... the MCF data are in qualitative agreement with the anatomical evidence, on the basis of which the apes, Old World monkeys, New World monkeys, prosimians, and nonprimates are placed in taxa which form a series of decreasing genetic relationship to man." In 1967, in their second article on this topic, Sarich and Wilson [6] argued that molecular change must have occurred at a constant rate among all major groups of primates. In the third and last article of this series, which was also published in 1967, they concluded that the small amount of difference they detected between hominoids in their albumin reflected little molecular change and that this, in turn, implied that little time had elapsed since the separation of the hominoid lineages [6]. Since, as they claimed, molecular change ticked away at a constant rate, the ID-values represented a "molecular clock" that could reveal the times at which the various hominoids – actually any species – diverged and went off on their own evolutionary paths.

After calibrating the molecular clock on the basis of the paleontologists' interpretation of fossils as to when Old World monkey and hominoid lineages might have separated, Sarich and Wilson calculated divergence dates of about 10 million years ago for the gibbon lineage, eight million years ago for the orangutan lineage, and a mere five million years ago for the human and African ape lineages. The major implication of this calculation was that fossils older than five million years, such as *Ramapithecus* (now referred to the genus *Sivapithecus*) from about 12–14 million-year-old deposits in Indo-Pakistan could not, as paleontologists concluded from study of morphology, be hominid [7].

However, while the molecular systematists [8–13] came to be in general agreement on the premise underlying the molecular assumption – that degrees of similarity reflect degrees of evolutionary closeness – not all of them embraced the notion of a molecular clock that "ticked" at a constant rate. Among the most vocal dissenters was Goodman who, during the early 1960s, was arguably the leading proponent of Nuttall's work [14]. Although Goodman went beyond claiming that molecular similarity indicated a very close relationship between humans and the African apes to advocating that these three hominoids should also be placed in the same taxonomic family, Hominidae (a position that for years caused many comparative primate anatomists and paleontologists to be skeptical of molecular systematics in general), he nonetheless tried to accommodate the paleontologists' identification of pre-five million-year-old fossils as being hominid.

In contrast to Sarich (who in 1971 was the first molecular systematist to reject morphology as being phylogenetically informative), Goodman accepted the paleontologically derived date of at least 14 million years as the time of divergence between hominids and their potential ape relatives. As a consequence, though, Goodman then had to interpret the immunological and biochemical data so that they would be compatible with a deep time scale of hominoid diversification. He did so by constructing a selectionist argument to explain the apparent acceleration or deceleration in rates of molecular change that *de facto* must have occurred given the paleontologically established dates for the earliest appearances of each of the large-bodied hominoids. Thus, for instance, if the various hominoid lineages had indeed originated between 18 and 14 million years ago, the extreme similarity between the various extant hominoids in their albumin had to have resulted from a slowdown in the rate of molecular change of that particular blood serum protein [15–17].

Goodman [17] proposed that the inferred deceleration in rate of molecular change in large-bodied hominoid albumin was related to the fact that, although they are similar to other anthropoid primates in developing hemochorial placentation (the most intimate of placental modes in the approximation of maternal and fetal blood systems), large-bodied hominoids differ from other anthropoids in having very long gestation periods. He argued that, in animals with hemochorial placentation but with fairly short gestation periods, there would be enough slack in the system to allow unimpeded molecular change to occur and accrue. The fetus would be born before the mother's immune system could produce antibodies to it. A large-bodied hominoid's gestation period, however, would be sufficiently long not only for maternal–fetal immunological incompatibility to build up, but for maternally produced antibodies to diffuse through the placenta, with deleterious consequences to the fetus.

The dilemma which then had to be confronted was: How could an animal possess both hemochorial placentation and a long gestation period? Goodman's answer was that natural selection must have acted to reduce the possibility of immune responses from the mother toward her fetus's proteins. Consequently, there had to have been a slowdown of molecular change that was commensurate with a prolongation of the gestation period. Although albumin had been the basis of Sarich and Wilson's constant-rate molecular clock, Goodman used the same molecule as the exemplar of how the molecular clock could run at different rates.

Concurrent with the various approaches – immunological reactivity as well as gel-column electrophoretic separation – that were brought to bear during the 1960s and 1970s on the determination of evolutionary relationships was an increased interest in protein sequences. The first major study was Zuckerkandl and Pauling's 1962 analysis of hemoglobin, and for some years thereafter hemoglobin was the most intensely studied molecule. Beginning in 1973, however, a research group headed by A. E. Romero-Herrera analyzed myoglobin sequences across a number of primarily mammalian taxa. As more taxa were added to the study, the resultant, most "parsimonious" arrangement of the hominoids placed the gibbon rather than the orangutan closer to a human-African clade [18]. Only by arguing for a more complicated scheme of myoglobin "evolution" in hominoids – including, for instance, the condition that there had to have been various "back-mutations" to a state similar to the presumed unchanged state – could the more commonly accepted arrangement of divergences be arrived at: gibbon first, then the orangutan, and then a human-African ape group.

While protein sequencing was seen by some (especially by Zuckerkandl as late as 1987) as *the* only molecular level available for purposes of determining evolutionary relationships (primarily because amino-acid sequences were taken as being less subject to the pitfalls that affect DNA and which thus arise from analysis of DNA, such as insertions, deletions, and back-mutations) [19], there was a growing consensus that the best systematic information still lay at an even deeper "genetic" level – DNA. Since, as evidenced in the original organisms of DNA scrutiny – bacteria – base differences at the third position of a codon (a nucleotide triplet) did not yield different resultant amino acids, the concern of those who questioned the efforts of DNA-sequence and who focused on multicellular organisms was that a demonstration of similarity in DNA sequences could produce "false" phylogenies, because different nucleotide sequences could yield similar protein sequences. Nevertheless, because of an incorrect inference from what was known during the 1950s and 1960s of the bacterial genome – and especially because of Jacob and Monod's [20] demonstration of the adaptive role of the *lac*-operon in bacterial gene function and regulation – the groundwork appeared laid to translate directly to multicellular organisms what was then understood of the bacterial genome.

The primary facts derived from study of bacteria were that:

1. A codon coded for a specific amino acid.
2. A particular sequence of codons produced a particular sequence of amino acids and, thus, a particular protein.
3. A change in a nucleotide in the third position did not alter the amino acid produced by the codon while a nucleotide change at the first or second positions of a codon did, which thus altered the resultant protein.
4. There were "controller" molecules that inhibited or allowed gene expression depending on the experimental environment (e.g., such as the *lac*-operon inhibiting or allowing gene expression depending on the bacteria being in lactose rich or poor environments).

A change in DNA function and gene product – whether directly with nucleotide substitution or inhibition – allowed the affected bacteria to adapt instantaneously to alterations in their immediate environment. Moreover, because in a neo-Darwinian model, adaptation is part of an ongoing continuum that eventually leads to "evolution" or to new species, and every step can be conceived as change, the extrapolation from bacteria to higher organisms seemed perfectly reasonable.

Since by the 1950s the American-based, neo-Darwinian-cum-fruit fly population genetics of the "modern evolutionary synthesis" [21] had risen above those emergent in Germany [22] and Britain [23] (for primarily polemic rather than scientific reasons [24]), the language that dominated evolution used the words "change," "adaptation," and "evolution" interchangeably. Of course, this synonymy was consistent with the adopted Darwinian model in which change that would result in what could be called *evolution* was the long-term, cumulative effect of change at generational levels that constituted adaptation. In other words, the process of evolution was actually the process of adaptation protracted beyond generations to geological time [25, 26]. If the model worked for bacteria, in which the sequence of DNA more often than not had a direct effect through the generation or dis/inhibition of gene sequences that generate metabolically active proteins or enzymes, then the model should also work for multicellular organisms – that is, for plants and metazoans. Consequently, the phrase "DNA is the blueprint of life" became the molecular mantra of the evolutionary and biological sciences and was applied to all life forms. In turn, this further validated using molecular data for the purpose of determining evolutionary relationships through the also seemingly bolstered "molecular assumption."

Also at this time, the discovery that nucleotide change at the third base position did not alter amino acid, and thus protein sequence, led Kimura [27] to propose what became known as the "*neutral mutation*" theory, in which the "molecular assumption" yielded a theory of phylogenetic relatedness when applied to the comparison of third-position nucleotides among taxa [28]. The additional benefit of the "neutral mutation model," it was claimed, was that because changes at the third base position did not produce phenotypic change (as extrapolated from bacterial production of metabolically active proteins and enzymes), such changes were "neutral" or "invisible" with regard to natural selection. Moreover, since the Darwinian and then neo-Darwinian conception of "evolution" was predicated on organisms being constantly pushed one way or another by selection to adapt to their presumably every-changing surroundings, what better source could there be for inferring the phylogenetic pattern of life than molecules to which selection was blind?

Unfortunately, it would not be known for more than two decades that the bacterial genome was not a surrogate for the genomes of multicellular organisms: the former is approximately 98% coding and 2% noncoding, while the metazoan genome is virtually the reverse [29]. Even in plants, the noncoding regions dominate the genome (e.g., Ref. [30]). Consequently, while nucleotide sequence is important in bacteria – with approximately 98% of their genome allowing rapid adaptation to surrounding circumstances through myriad potential changes in metabolically active proteins and enzymes that only populations of millions can provide, and only approximately 2% regulating gene function – it is not important in multicellular organisms, especially metazoans. In the latter case, only about 2–3% of the genome is coding as in bacteria, and thus plays a role in organismal adaptation (e.g., changes in hemoglobin or albumin); the vast majority of the metazoan genome engages in the regulation of developmental genes, which is fundamental to the emergence of a functionally integrated, viable organism. Consequently, while continual change in bacteria yields a palette of potential adaptation, in multicellular organisms such instability in the predominant developmentally integrated molecular domain would likely produce death, and only on the rare occasion a novel, functional feature or system – that is, the basis of new species. Indeed, while the conception of continual change embedded in the "molecular assumption" derives from bacteria (and viruses), the observed, molecular changes in these organisms are only adaptational (the first known 2+ billion-year-old bacterium resembles a present-day bacterium). However, it is not a surrogate for a model of evolution or the emergence of organismal novelty. Clearly, if this significant difference had been known, the extrapolation from bacteria to plants – and especially to metazoans – would never have happened.

Nevertheless, without the understanding that hindsight provides, it seemed during the 1970s that if the metazoan genome was generally similar to the bacterial genome in presenting a strict relationship between a gene (a sequence of nucleotides) and a protein, or between a gene/s and a structure – this being the dominant element of the genome – then "getting" to the DNA itself would reveal the mysteries of the innumerable blueprints of life embodied in the vast array of known taxa. In addition, because DNA sequences were composed of thousands upon thousands of bases, the apparent massive scale of the comparison had the added appeal of being able to provide an overwhelming amount of phylogenetically relevant information [31].

3 DNA Hybridization

Although DNA sequencing was possible by the late 1970s, it was an expensive and laborious procedure, which made the endeavor itself – much less the comparison of DNA sequences for phylogenetic purposes – somewhat prohibitive. There was, however, another way in which DNA sequence information could be achieved, namely DNA–DNA hybridization.

The theory behind using DNA–DNA hybridization (or just DNA hybridization) as an approximation of overall similarity in DNA sequences among taxa lay in the fact that, if the two strands that form the helical structure of nuclear DNA were dissociated (which could be accomplished by subjecting it to heat), then their complementary bases would reassociate (re-anneal) and the two strands would recreate their original helical organization. However, it is not only two cleaved, but originally helically arranged, DNA strands from the same individual that will re-anneal. Analogous to the geneticist Dobzhansky's discovery earlier in the twentieth century that chromosomes of different lengths from different species of fruit fly will attempt to align [32], any two strands of DNA – whether from different individuals of even from different species – will also attempt to anneal.

As in Dobzhansky's fruit-fly experiments, in which the larger chromosome would loop and fold so that whichever of its loci were also present in the shorter one would match up, single strands of DNA (derived from heat-splitting or the melting of double helices) from different organisms would attempt to recombine or hybridize at complementary base positions. The more complementarity between hybridized DNA strands, the greater the intensity of heat needed to break the bonds holding them together. Consequently, it seemed logical to conclude that the heat (ΔT) it took to disassociate a DNA–DNA hybrid (i.e., the more thermally stable the hybrid was), the more complementary the annealed strands were in their nucleotide sequences. From this apparent demonstration, the molecular assumption could then be invoked to explain how higher

ΔT-values, which were interpreted as representing greater molecular similarity, were also a reflection of closer evolutionary relatedness. However, as logical as this thought experiment might have seemed, it relies on an assessment of overall similarity in which the identities of the bases or sequence positions that underlie differences in ΔT-values remain unknown. Conversely, similar ΔTs in different taxa might not be the result of the same regions hybridizing.

One of the first applications of DNA hybridization to evolutionary questions was reported in 1968 by Britten and Kohne [33], who were primarily concerned with learning more about genomes. They discovered that the genomes of higher organisms – but not of bacteria or viruses – contained "... hundreds of thousands of copies of DNA sequences." However, not only did repeated DNA sequences represent a considerable portion of a genome, but Britten and Kohne also believed that they were "... trivial and permanently inert." Only a small fraction of a genome was composed of unrepeated, unique, single-copy DNA sequences, which apparently constituted the active or functional elements of that genome.

Britten and Kohne also speculated on pathways or mechanisms that might produce genomic change which, they argued, must be considered on two different levels: change in nucleotide sequence; and the origin of new families of nucleotide sequences. As Zuckerkandl and Pauling had assumed for differences in hemoglobin, Britten and Kohne proposed that changes in nucleotide sequences – which would be identified as point mutations – occur slowly over time. This model, however, would explain only the "divergence of pre-existing families" of nucleotide sequences, not the introduction of new families of sequences, as was generally assumed to be the case by molecular systematists. The introduction of new families of sequences, Britten and Kohne alternatively suggested, must "... result from relatively sudden events," that is, "saltatory replications." Accordingly, "... saltatory replications of genes or gene fragments occurring at infrequent intervals during geologic history might have profound and perhaps delayed results on the course of evolution." Although Britten and Kohne's distinctions between types of molecular change were not embraced by other molecular systematists, a version of their suggestion of "saltatory replication" appears to have been borne out with the subsequent identification of vertebrate regulatory genes involved in segmentation – *Hox-a, Hox-b, Hox-c*, and *Hox-d* – as replicates of the orthologous *Antennapedia* gene identified in fruit flies [34].

One of the early attempts at using DNA hybridization to reconstruct relationships among the primates was described in 1976 by Benveniste and Todaro [35], whose primary focus was the distribution among mammals of the type C viral gene, the presence of which was found to be particularly characteristic of animals found in Asia, including the orangutan and the gibbon but which, unexpectedly, was also present in humans. After reviewing these data, Benveniste and Todaro turned to DNA hybridization to resolve the relationships of various primates. Relying on the results of previous studies, they estimated "... the effect of mismatched base-pairs on thermal stability" as being "... between 0.7 and 1.7 °C per 1% altered pairs." Since, depending on the animal, nuclear DNA consists of anywhere from 10 to 100 million nucleotide pairs, every 1% of difference in nucleotide sequence between

DNA hybrids would result in a drop of about 1.0 °C of the heat necessary to melt the hybrid molecule.

Although in their text Benveniste and Todaro emphasized a 1.1% sequence difference between humans and both African apes, and a 2.4% difference between humans and orangutans, their published ranges of ΔT-values demonstrated overlap between all large-bodied hominoids (although the actual identity of nucleotide difference and similarity that are measured in percentages are unknown). Further, since presumed homologous DNA sequences are not of similar length (i.e., they differ in the number of nucleotides) in all animals, including large-bodied hominoids (see below for humans and orangutans), it is not known when the DNA strands hybridize how a long sequence from one species aligns with a shorter sequence of another species.

Perhaps because DNA hybridization experiments seemed to provide a means of getting closer to the "most informative" genetic level, various questions regarding what, exactly, similarity in ΔT-values meant were not explored. Other questions were, however.

During the 1980s, Sibley and Ahlquist suggested that a major problem with all earlier DNA hybridization experiments was that only a limited number of cross-hybridizations had been tested [31, 36]. Sibley and Ahlquist asserted that this was why, in previous studies, various evolutionary relationships could not be resolved, particularly between humans and the African apes. Although morphological support for this latter theory of relationship was slim at best, molecular systematists often grouped these three hominoids together. But, within this presumed relationship the majority of the molecular studies failed to determine which two of the three hominoids were most similar and thus, presumably, most closely related. Sibley and Ahlquist also felt that another source of error in the earlier studies had resulted from the use of all of an animal's DNA, repeated as well as single-copy, to form hybrids. They argued that in order to achieve a greater accuracy in results, all repeated DNA had to be removed (since it was redundant), and this would then leave only single-copy DNA to work with.

Before tackling the evolutionary relationships of the large-bodied hominoids, Sibley and Ahlquist had spent years applying their DNA hybridization technique to the phylogeny of birds, during the course of which they developed the notion of a uniform average rate (UAR) of genomic change. UARs, it was believed, characterized the nature of molecular change – and thus also evolutionary change – in all taxa. The device of a UAR also negated debates over constant versus irregularly running molecular clocks, which derived from focusing on individual molecules. As Sibley and Ahlquist saw it, the entire genome could change at a UAR, even though individual molecules may change at very different rates.

Sibley and Ahlquist argued that the power of DNA hybridization in "discovering" the phylogenetic relationships of organisms came from sampling the entire genome of taxa being compared. Further, since an organism's genome is composed of millions of nucleotides, the "law of large numbers" provides the checks and balances necessary to rule out "false" similarities or parallelisms. Accordingly, since DNA strands will only anneal and hybridize along complementary stretches of nucleotide sequences, there should be no question about the degree of homology between the DNA strands. Conversely, sequences that did not line up were simply not homologous. At least, that was the argument.

The melting temperatures of DNA hybrids produced numbers, which Sibley and Ahlquist converted into phylogenetic distances using a procedure they termed "*average linkage*." Average linkage begins "... by clustering the closest pair or pairs of taxa," after which "... one links the taxa which have the smallest average distance to any existing cluster," and so on, until "... all taxa are linked." The underlying assumption, which permits the clustering of the most similar pairs, is that the DNA hybridization technique "... measures the net divergence between the homologous nucleotide sequences of the species being compared."

Whilst many aspects of Sibley and Ahlquist's phylogenetic reconstructions of birds were consistent with theories of relatedness derived from study of morphology [31], there were some significant differences regarding not only the broader phylogenetic relationships of, but also the details of relatedness among, the flycatchers. By the 1980s, however, it had become commonplace when molecular and morphological phylogenies were in discord to give priority to the former. As Caccone and Powell [5] later pointed out, once the molecular assumption is accepted, overall similarity becomes *de facto* synonymous with closeness of relatedness. Indeed, whether is it a presumed albumin clock or a UAR, the assumption of ongoing molecular change validates the molecular assumption which, in turn, demands that a molecular phylogeny is correct in its entirety, even if contradicted by other information.

Sibley and Ahlquist's "law of large numbers" and UAR appealed to most molecular systematists. And this appeal was most likely the primary reason that paleontologists finally succumbed to the notion that molecular phylogenies had ultimate authority when morphologically based theories conflicted with them. Not all molecular systematists were convinced, however. In particular, Templeton [37] objected to the use of DNA hybridization as a window on phylogenetic relationships, because it did not allow the polarity of the similarity to be determined, which is necessary to address the question: Is similarity due to distant or recent common ancestry?" Indeed, only by identifying the actual sequences of nucleotides can an attempt be made to infer the significance of demonstrated similarity and dissimilarity. Sibley and Ahlquist, however, countered that there was "... no reason to expect data derived from base sequences to improve on those from amino acid sequences, which have produced contradictory results." In their support, Sibley and Ahlquist also quoted Friday [38] (who had collaborated with Romero-Herrera on myoglobin sequencing): "Phylogenetic conclusions derived from a study of nucleotide sequences will be subject to the same suspicions as those derived from amino acid sequences."

The irony here, though – as continues to be the case in attempts to sequence the human, chimpanzee, or any other metazoan genome (e.g., Refs [39–42]) – is that the only way in which a check can be provided on an inferred DNA sequence is by sequencing the protein(s) it produces. That is, enzymatically active proteins that can be identified precisely because their physiological function can be identified. Nevertheless, during the early 1980s the bacterial genome remained the model for interpreting metazoan genomes although, in recognition of Jacob and

Monod's *lac*-operon experiments, the language used to discuss metazoan genomes increasingly distinguished between structural and regulatory genes. What was lacking, of course, was serious attention to a fundamental difference between bacteria and metazoans. Namely, metazoans (and plants) develop into adult organisms as well as adapt physiologically throughout their lives, whereas bacteria only adapt to prevailing conditions.

4
Mitochondrial DNA

As the field of molecular systematics expanded its sphere of investigation during the 1970s, DNA located outside the nucleus, in the cell's mitochondria, began to attract attention [11, 13, 43, 44]. In 1979, Brown, George, and Wilson [44] noted that there was greater difference between humans and a taxically very small sampling of Old World monkeys in terms of their mitochondrial DNA (mtDNA) than in their nuclear DNA. By assuming that human and Old World monkey lineages diverged over 20 million years ago (as estimated by paleontologists from studies of fossil primates), Brown *et al.* concluded that differences between humans and Old World monkeys could be explained if it was *assumed* that nucleotide substitutions occurred at a faster pace – five- to tenfold more rapidly – in mtDNA than nuclear DNA (i.e., if mtDNA "evolved" five- to tenfold times faster than nuclear DNA). From this assumption, Brown and colleagues calculated that mtDNA data are more accurate than nuclear DNA for studying evolutionary events that occurred "... within the past 3–10 million years."

In subsequent reports (e.g., Ref. [43]), Brown *et al.* analyzed the mtDNA of the large-bodied hominoids which, primarily because of Sarich and Wilson's albumin-based molecular clock, were assumed to have diverged within this 3- to 10-million-year period and thus, because of this assumption, were amenable to mtDNA analysis. Although the reported data demonstrated the fewest differences (interpreted as substitutions) between humans and chimpanzees, the authors concluded that humans were related to the sister-group of African apes. In light of the molecular assumption, however, the data should have yielded a human–chimpanzee sister group, as Ruvolo and her collaborators (including Brown) later concluded in their analysis of mtDNA [45].

Nevertheless, the allure of using mtDNA to resolve evolutionary relationships within the window of $\leq$10 Mya was, and continues to be, based on two additional assumptions [46]. First, unlike nuclear DNA, mtDNA is supposedly inherited only through the maternal line and thus is not subject to the complexities that occur through the recombination of maternal and paternal DNA. A second assumption is the existence of a "hotspot" in the D-looped configuration of mtDNA. In addition, because the D loop was promoted as the primary site of molecular activity (with its identification as the "hypervariable zone" adding to this mystique), molecular systematics afforded it priority in providing evidence of evolutionary change. Although crucial to the interpretation of mtDNA in phylogenetic reconstruction, these assumptions are likely incorrect.

The introduction of paternal mtDNA into fertilized eggs has, however, occasionally been reported. For example, during the early 1990s, Wilson and coworkers [47] showed this to occur commonly in mice as part of the sperm's tail, the cells of which contain mtDNA, penetrates the ovum's membrane. More recently, Maynard-Smith and coworkers calculated that there had to have been recombination of maternal and paternal mtDNA during human evolution [48]. Indeed, this should cause systematists to reconsider the seeming unassailable "fact" that humans and chimpanzees are closely related (see also Hagelberg's [46] review of mtDNA, paternal inheritance, and recombination).

With regard to the existence of hypervariable sites representing regions of preferentially high mutation rates that therefore lend themselves to phylogenetic analysis, as Hagelberg points out, "... there is no direct evidence of hypervariability," although "... most researchers believe that anomalous patterns of DNA substitution are best explained by mutation." But, she writes, "... because the notion of hypervariability fits with the received view of mtDNA clonality (maternal inheritance only), anomalies are seldom questioned." Furthermore, there is not an "average" or "typical" mtDNA sequence for any taxon (humans included), because individuals vary from each other, and this should focus attention on the fact that the mitochondria of a cell may contain mtDNA with different sequences (see also Refs [49, 50]). With regard to phylogenetic interpretation, since the intracellular role of mtDNA is essentially metabolic (as is the bacterial genome), it does not necessarily follow that the demonstration of sequence similarity between different individuals of different taxa will reflect only their common descent rather than simply independently acquired and seemingly similar metabolic solutions to environment conditions – that is, similar metabolic adaptations.

Hagelberg provides an example of the potential conundrum. Depending on which subject's mtDNA is used, it is possible to reconstruct "... our most-recent female common ancestor... [as having] lived just 6000 years ago, a date more consistent with Biblical Eve than Mitochondrial Eve." As Hagelberg cautions: "The picture is far from simple, and it is clear that extreme care must be taken in the interpretation of mtDNA phylogenetic trees in the face of possible recombination.... There are enough unexplained patterns in mtDNA data to warrant reassessment of the conclusions of many mtDNA studies."

5 DNA Sequences

In spite of Sibley and Ahlquist's and Friday's denying greater reliability to nucleotide than to protein sequence data for reconstructing phylogenetic relationships, the increasing ease with which DNA could be sequenced was not lost on molecular systematists. Both nuclear and mtDNA were sequenced and the interpretation of similarity or difference in whatever sequences were compared still predicated on the molecular assumption. The weight of DNA sequence data as being key to deciphering evolutionary relationships assumed particular significance among many molecular systematists, because of the supposed information content of nuclear DNA in general. As Bruce and Ayala wrote in 1979, in a study on blood serum proteins [9]: "Information macromolecules – that is, nucleic acids and proteins – document evolutionary history... [Thus] degrees of similarity in such macromolecules reflect, on the whole, degrees of phylogenetic propinquity."

Almost coincident with an increasing popularity of comparing DNA sequences for the purposes of inferring evolutionary relationships came the cautionary notes. Of particular interest in this regard is the crucial question of alignment of sequences from different taxa that differed in length which, in 1991, Lake [51] was among the first to address. As mentioned above, not all DNA sequences chosen for comparison, if actually homologous (e.g., representing the same "gene" or segment), are of equal length. Thus, a priori decisions must be imposed as to how to subdivide the shorter sequence in order to try to align its nucleotides with those of the longer sequence. Typically, the alignment of compared sequences is presented in the literature without justification of the assumptions underlying the alignment, upon which definitive conclusions of phylogenetic relatedness were then based. Nevertheless, as Lake warned, "The order of sequence alignment can bias the selection of tree topology." In addition to assumptions underlying how a short sequence is "chopped up" so that its bases align with those of a longer sequence is the issue of whether, having performed this surgery on the shorter sequence, the differences must then be explained between shorter and longer sequences as the result of nucleotide base deletion or insertion.

In his book, *What it Means to be 98% Chimpanzee*, Marks [52] provides an example of these problems with nuclear DNA sequences from a human and from an orangutan that illustrates three different ways in which their bases (C, T, G, A) can be aligned. Clearly, each solution of the problem presents different problems of interpretation, but from the outset the assumption is that the 40 bases in the human sequence actually have homologous counterparts in the 54 bases in the orangutan sequence (Fig. 1).

As Marks comments:

"Tabulate the differences. The top one invokes five gaps and six base substitutions; the middle has only two gaps but nine base substitutions. And the bottom one has five gaps and only three base substitutions. The three pairs of sequences differ in the assumptions about which base in one species corresponds to which base in the other. While we might, by Occam's Razor, choose the alignment that invokes the fewest inferred hypothetical evolutionary events, we still have

```
Human       CCTCCGCCGCGCCG    CTCCGC   GCCGCCGGGCA         CGGCC          CCGC
Orangutan   CC            GTCGCCTCCGCCACGCCGCGCCACCGGGCCGGGCCGGCCCGGCCCGCCCCGC

Human              CCTCCGCCGCGCCGCT         CCGCGCCGCCGGGCACGGCCCCGC
Orangutan   CCGTCGCCTCCGCCACGCCGCGCCACCGGGCCGGGCCGGCCCGGCCCGCCCCGC

Human       CCTCCGCCGCGCCG    CTCCGCGCCGCCGGG CAC  GGCC                   CCGC
Orangutan               CCGTCGCCTCCGCCACGCCGCGCCACCGGGCCGGGCCGGCCCGGCCCGCCCCGC
```

Fig. 1 Different ways in which to align the 40 bases in the human DNA sequence with the 54 bases in the orangutan DNA sequence.

to decide whether a gap 'equals' a substitution. Does the bottom one win because it has a total of only eight differences? Or might the middle one win because a gap should be considered rare and thereby 'worth,' say, five base substitutions?"

"The problem is that we cannot know which is 'right,' and the one we choose will contain implicit information about what evolutionary events have occurred, which will in turn affect the amount of similarity we tally. How similar is this stretch of DNA between human and orangutan? There may be seven differences or there may be eleven differences, depending on how we decide the bases correspond to each other across the species – and that is, of course, assuming that a one-base gap is also equivalent to a five-base gap and to a base substitution."

"In a more general sense, however, the problem of taking quantitative *estimates of difference between entities that differ in* quality *is prevalent throughout the genetic comparison of human and ape. The comparison of DNA sequences presupposes that there are corresponding, homologous sequences in both species, which of course there must be if such a comparison is actually being undertaken. But other measurements have shown that a chimpanzee cell has 10% more DNA than a human cell. (This doesn't mean anything functionally, since most DNA is functionless.) But how do you work that information into the comparison, or into the 99.44% similarity [between human and chimp]?" (comment added).*

These concerns have not, however, been widely appreciated by molecular systematists, and especially by molecular anthropologists, who not only portray the analysis of DNA sequences as neutral and objective but also use an a priori assumption of relatedness to inform the way in which they interpret similarity in sequences they have aligned according to other a priori assumptions. Exemplary in this regard is the multiple DNA sequence analysis that Ruvolo and her collaborators reported in 1997, in which they sought to resolve the presumably real dilemma of humans being more closely related not to both but to one African ape rather than the other [53]. In pursuing this question, these authors assumed first that the orangutan was the sister taxon of a clade or evolutionary group consisting of humans, chimpanzees, and gorillas. Consequently, in whatever nucleotides orangutans differed from the rest by default had to be interpreted as primitive relative to similarities between humans and one of the African apes.

With the ever-growing popularity of the parsimony-based phylogenetic computer program phylogenetic analysis using parsimony (PAUP), it became common practice to first "root" a phylogenetic analysis in a taxon that was first chosen as the primitive outgroup, which of course is identified a priori as the taxon that diverged before the others [12, 13, 45, 53–62]. Rooting parsimony or any other clustering analysis (for example, nearest-neighbor joining or maximum likelihood, which are essentially similar to Sibley and Ahlquist's linking technique) in a particular taxon is a requisite of making the algorithm "work." Nevertheless, this procedure artificially determines character polarity since, by definition, the outgroup (the taxon in which the tree is rooted) is defined from the outset as primitive in its entirety relative to the presumed ingroup (i.e., the taxa of focal interest). In turn, the taxa of phylogenetic interest are predetermined as being derived in whatever ways they differ from their predetermined primitive sister taxon [63].

The widespread use of this algorithm-based interpretation of nuclear and mtDNA, as well as of protein sequences, presents its own set of problems and assumptions. Consider the molecular assumption: Since molecules are supposedly continually changing, and this change is supposedly continually being accumulated as a lineage proceeds through time, the degree of molecular similarity is then taken as reflecting the relative recency of lineage divergence. Accordingly, each lineage accumulates its own unique array of molecular change, which should result in a lineage becoming increasingly distinctive (i.e., increasingly different) the longer it is in existence. Although tautological, this assumption explains why more recently diverged taxa are more similar than more anciently divergent lineages. On the other hand, in order to root an algorithm for purposes of generating presumed phylogenetic relationships, it must be assumed that the taxon chosen as the earlier-divergent outgroup is totally primitive relative to the taxa to which it is a priori supposed to be the sister taxon. Yet, it is the molecular assumption that validates the use of overall similarity as the key to resolving phylogenetic relationships by contrasting it with the unique differences that earlier divergent lineages accumulated along their own, unique evolutionary trajectories. Clearly, both assumptions cannot be correct simultaneously. Either the earlier divergent taxa or lineages did not change, but remained primitive (which is the logical extension of identifying a taxon as the outgroup in which to root a computer analysis), or they did change by accumulating their own suites of molecular difference (the basis of the molecular assumption). In the latter case they are derived in some aspects of their sequence (and uniquely so, for that matter, because of their individually distinct molecular histories) and not primitive *in toto* relative to the taxa to which they are being compared.

In the realm of morphological systematics, according to Hennigian or cladistic principles, overall similarity is not *de facto* a clue to evolutionary relatedness. Rather, similarity must be sorted into features that reflect a hierarchy of inheritance: primitive features from ancient ancestors; and derived features from recent ancestors [64].

Since the pattern of life is a hierarchy of nested sets of smaller and smaller clades (groups of taxa related by dint of sharing a common ancestor = a monophyletic group), that which is considered primitive versus that which is considered derived depends on the level in the hierarchy of nested clades under investigation. Primitive features (i.e., features retained in descendants) do not elucidate the relationships of these taxa; only derived features can do this. It is also important conceptually to recognize that a derived feature at one level of the hierarchy is a primitive retention at another. There is no theoretical reason why this approach to systematics cannot be applied to molecular data. The major difficulty is that molecular systematists would have to sample and compare a wide range of taxa. This is the only way in which relative primitiveness and derivedness can be determined. It cannot be justified from a priori assumptions of directionality, as underlies the molecular assumption, or by choosing an outgroup on the basis of its presumed evolutionary relationship to other taxa. However, even from the outset it is also crucial to realize that shared similarity does not translate directly into a demonstration of relatedness. Taxa may be similar not because they inherited changes that distinguished a recent common ancestor, but rather because they share primitive retentions – that is, they share features that have not changed through a succession of ancestors (e.g., Ref. [65]).

Nevertheless, it has become popular for the authors of molecular studies on the relatedness of taxa to classify their work as "cladistic" (e.g., Refs [53–56]). One argument is that the nucleotide bases – C, G, T, A – represent alternative character states [53]. To the systematically uninformed this may appear to be a logical conclusion, but it is misleading for the simple reason that a nucleotide does not represent a character state; each nucleotide – C, G, T, A – is an identifiable molecule that is not phylogenetically a variant of another [65]. Phylogenetically relevant alternative molecular character states would be better represented by comparison, for example, of arrangements of "gene" sequences with regard to *cis* and *trans* elements, patterns of introns and exons and of methylation of transposons or other elements, and pathways of molecular communication [66].

Another argument for molecular studies being cladistic derives from the claim that molecular similarity is equivalent to synapomorphy; that is, shared similarities represents shared derived character states (e.g., Refs [45, 53, 55, 56]). Yet, in reality this conclusion is merely a restatement of the circularity imbedded in the molecular assumption: the most recently diverged taxa share more recently accumulated (= derived) molecular states. Thus, the identification of shared derived molecular states occurs a posteriori, after the algorithm of choice has clustered taxa on the basis of inferred greater or lesser degrees of similarity, and also typically after one has rooted his/her tree in a particular taxon (which defines the latter taxon as primitive in all characters relative to the taxa of interest, which then are *de facto* united on the basis of their "shared derived" features). Although this "procedure" is widely used by molecular systematists, this is not how cladistic analysis proceeds. Rather, the hypothesizing of primitive versus derived character stats occurs prior to generating theories of relationship – which is the only way in which such a methodology can actually be employed.

The assumption of continual molecular change – whether through point mutations affecting nuclear or mtDNA, or altering amino acids in protein sequences – is also of interest. Recall that Zuckerkandl and Pauling framed the "molecular assumption" as a way of *explaining* the pattern of their data: "... over-all similarity *must* (emphasis added) be an expression of evolutionary history," with descendants "mutating away" from each other, becoming "... gradually more different from each other." Consequently, this assumption demands that earlier divergent taxa accumulate their own molecular differences while the most recently divergent taxa are similar because of their longer shared history of accumulated molecular changes and, after divergence, their shorter time of independent molecular change. The assumptions of molecular clocks and UAR are predicated on this notion. Nevertheless, it must be recognized that any of these devices – for example, molecular assumption, molecular clocks, UARs – are merely attempts to explain how something might have come to be as observed in its current static state. Further, molecular clocks especially were based on analyzing similarity and/or difference, first in metabolically active proteins and then in the DNA sequences from which they are generated. None of these assumptions has been demonstrated as being relevant to the noncoding regions of multicellular organisms.

Two other points need to be mentioned. The first is the unspoken, but clearly real, assumption that the major locus of molecular change is gametogenesis, and thus mutation is always and immediately available to offspring (e.g., Ref. [67]). Underappreciated, however, is the fact that the only source of constant molecular change is ultraviolet (UV) radiation, which induces an effective mutation rate of 10^{-8} to 10^{-9} [68]. Furthermore, UV-derived mutation is *random* with regard to somatic cells, sex cells, and their molecular constituents. Thus, while there might appear to be concordance between the reality of the physical world in which UV-provoked mutation is relatively constant and the conception of a constantly "evolving" molecular world, this is an illusion. For one thing, since most multicellular organisms contain at any one time more somatic than sex cells, the likely bias is that a mutation will affect somatic rather than sex cells.

The notion of *constant and accumulative* mutation affecting sex cells is of further interest because it also contradicts the basic tendency of cells to remain in homeostasis. As seen, for example, in the roles of heat-shock (stress) proteins – which serve to maintain cell membrane physical states through lipid transport, eliminating reading errors that occur during transcription and translation, DNA repair, chaperoning other proteins, and ensuring proper folding of proteins as they emerge from the ribosomes – the basic properties of a cell are dedicated to resisting and eliminating the potential for change. Intuitively this makes sense, for unabated molecular change would undermine the integrity of cell function and more probably lead to organismal death rather than to adaptively advantageous organismal change [68].

6 Repeated (Retro)Transposons

Transposons, or retrotransposons, represent the latest category of molecular discovery to be promoted as providing *the* ultimate window on the evolutionary relationships of multicellular taxa. Reminiscent of the earlier conception of "jumping genes" – genes that could be inserted anywhere in the genome – retrotransposons represent nucleotide sequences of differing lengths that appear to have been inserted into what have been deemed the evolutionarily more conservative nucleotide sequences. Shortly after their discovery, it was noted that there seem to be shorter interspersed nucleotide elements (SINEs) and longer interspersed nucleotide elements (LINEs). Since the apparent phenomenon of "jumping genes" and the insertion of nucleotide sequences into a sequence far removed were so startling and unpredicted, a version of the "molecular assumption" was articulated to accommodate them [69–71]. Namely, any such insertion could only happen once, and thus the identification of apparently similar insertions among an array of taxa had to reflect their evolutionary relatedness. An unrecognized consequence of this assumption, however, is that the molecular assumption (i.e., continual molecular change via nucleotide substitution) is invalidated.

The application of SINEs and LINEs to questions of phylogenetic relatedness was first employed in 1999 by Nikaido and colleagues [69], who analyzed a small sample of (cet)artiodactyls. In this study, attention was focused on hippopotami, which they united with cetaceans on the basis of the assumptions they made about transposon behavior: (1) Independent insertion events must occur between exactly the same two base pairs; (2) Parallel changes involving SINEs must be extremely rare; (3) Deletions of inserted elements will hardly ever occur; and (4) Reversals involving SINEs must be extremely rare. Although these assumptions were applied to a new source of molecular data, they clearly parallel those imposed by Zuckerkandl and Pauling's molecular assumption on the interpretation of overall molecular similarity. However, one question that should be addressed prior to accepting the transposon–insertion assumption is the following: Since retroviruses mirror the "stochastic" insertion behavior of SINEs and are subject to deletion and homoplasy, why should SINEs be exempt from these otherwise commonplace phenomena?

Nikaido *et al.*'s analysis was based solely on phenetic molecular similarity between camel, pig, peccary, chevrotain, deer, giraffe, sheep, cow, hippo, humpback whale, and beaked whale, without any potential insight that could be provided by outgroup comparison. Their inferred relationship was this: Camel is related to a pig–peccary group; this group is related to a chevrotain–peccoran group; and this entire group is related to a hippopotamus–whale group. Nevertheless, the consideration of unique morphological structures – for example, the number of stomach chambers and a multitude of dental, cranial, and postcranial features – strongly corroborates the hypothesis that hippopotami (whale-related or not) are the sister relatives of a group in which sister taxa pig–camel are the sister of a chevrotain–peccoran sister group. Considering the developmental underpinnings of the morphologies that yield this phylogenetic pattern of relationship, it is difficult to reject, even in the face of the assumption that insertions can only occur once and only once. The molecular development of structure continuum is difficult to ignore, especially as it is not based on a priori assertions, as is the transposon–insertion assumption.

Hillis' [72] commentary on Nikaido *et al.*'s article is significant: "Almost every new molecular approach to phylogenetic inference has been ballyhooed as capable of 'revolutionizing' the field... [but] no one technique is a perfect solution for all phylogenetic problems"; "... even if convergence and reversal are extremely rare for SINE/LINE insertion events, the characters are not immune from problems of lineage sorting of ancestral polymorphisms"; "... examples of... convergence have been demonstrated in experimental... studies"; "... although SINE/LINE insertion studies provide a welcome... source of phylogenetic data, they are not magic bullets."

I agree with Hillis' criticism, and would apply it more broadly. There is no logical reason to believe in the existence of a "phylogenetic magic bullet," molecular, or morphological, yet the desire to find one persists. For example, while Ruvolo rejected much of her data as unreliable for purposes of phylogenetic reconstruction of hominoid relationships because they did not yield an "accepted" phylogeny, she [53] nonetheless proclaimed in the face of tautology that the data she did embrace as producing the "right" phylogeny clearly demonstrated a human–chimpanzee relationship. Yet, Salem *et al.* [70] maintained that until they entered the fray with their SINE (Alu repeat) analysis, and proved conclusively a human–chimpanzee sister relationship, a (presumed) human–chimpanzee–gorilla trichotomy had never been resolved. Since, as with the molecular assumption, the internal consistency of SINE-focused assumptions is resistant to testing with other molecular data, the undeniable molecular–structural continuum makes morphologically and developmentally based phylogenies not only powerful alternative hypotheses, but also sources for testing assumptions of molecular behavior that were rapidly elevated to the status of fact [66, 73].

7
"Evo-Devo"

In 1975, King and Wilson [74] surveyed studies on blood serum proteins and DNA hybridization in the context of a potentially close evolutionary relationship between humans and chimpanzees. Although they are cited as having demonstrated this relationship, this was not their intention nor conclusion; rather, it was a matter of practicality. As they stated: "... the only two species which have been compared by all of these methods are chimpanzees... and humans," and thus "... a good opportunity is... presented for finding out whether the molecular and organismal estimates of distance agree." And, indeed, humans and chimpanzees apparently differed by about 1% in the small amount of their total genetic material that had been analyzed. King and Wilson concluded that "... all the biochemical methods agree in showing that the genetic distance between humans and the chimpanzee is probably too small to account for their substantial organismal differences."

To explain how humans and chimpanzees could be virtually identical in their genes, but markedly different in their anatomy, King and Wilson suggested that these two morphologically disparate hominoids must differ in genes that regulate development. (Recollect that the then-known bacterial genome was seen as genes or gene sequences that were primarily engaged in coding for metabolically active protein and enzymes but, Jacob and Monod's [20] *lac*-operon experiments notwithstanding, much less so in the regulation of gene function.) To their credit, especially during the mid-1970s, King and Wilson were aware of the difference between gene expression and the *regulation* of gene expression. Unfortunately, because the metazoan genomic role model was bacterial, the extrapolation was from adaptive molecular change in bacteria to anticipated structural adaptive change in metazoans, with adaptive gene expression guided by a minimal intrusion of gene regulation.

Subsequently, studies on the regulation of development expanded exponentially, not only in terms of distinguishing between regulatory and structural genes, but also and especially with regard to the array of molecules that induce gene transcription and communicate via signal transduction pathways to produce structure. Interestingly, those animals that have been studied in depth – such as the fruit fly, zebrafish, frog, chick, mouse, human – demonstrated a commonality of "homeotic genes" (which contribute to segmental patterning and segment identify) [75–81]. In turn, through their protein products (transcription factors), homeotic genes control or at least affect gene expression. Time and time again during development, the same proteins (e.g., various growth factors, *trans*-inducing, and bone-modifying proteins) and regulatory genes (and their products) are co-opted to produce what in adult organisms are different morphologies [75, 82, 83].

In 1994, Wolpert [84] summarized the situation: "During development, differences are generated between cells in the embryo that then lead to spatial organization (pattern formation), changes in form, and the generation of different cell types. Genes control development by controlling cell behavior." But one should not be too gene-centric in envisioning the emergence of form from genes and gene products alone. For, while there might be genetic regulation of some cells' activities, the results (e.g., cellular asymmetry, cell membrane elasticity or rigidity, compressive forces) might produce physical or mechanical responses which may not themselves be genetically based, but which, nonetheless, greatly affect cell geometry and ultimately organismal form [85]. "Cell behavior," to return to Wolpert, "has many different levels of meaning."

As early as the 1970s, Løvtrup [86] had argued the importance of epigenesis in development, especially the fact that changes in properties of the fertilized egg can alter the chronology and spatial organization of patterns of cellular diversification. Since the larger clades of multicellular organisms possess the same types of cell, and the same genomes within cells, as well as the same chemical substances that form the immediate environments of cells, variation in the spatial and chronological organization of cellular differentiation must be one of the keys to the emergence of evolutionary novelty. For instance, whether a cell divides symmetrically or asymmetrically (this can be affected simply by the positions of the chromosomes relative to the center or periphery of the cell) can greatly impact the spatial relationships of cells and, therefore, eventually have an effect on organismal shape (e.g., Ref. [87]). As Alberch [88] and subsequently Newman [85] have emphasized, the development of organismal form and structure is ultimately a function of the physical and mechanical properties of cells' sizes, shapes, spatial relationships, and their consequent physical constraints on neighboring cell, and thus on organismal form. One of the "hot" topics these days is "breaking cell symmetry" (e.g., Ref. [89]). For example, imagine how cells that divide equally (as, for example, in spermatogenesis) will "pack" versus those that divide asymmetrically (as, for example, in oogenesis).

The application of an understanding of the regulation of, and physical constraints on, development to evolutionary questions has given rise to the field of "evo-devo" (evolution and development), in which the interrelationship between the "genetics" and "epigenetics" of development is a primary focus. Indeed, Løvtrup's concern with the influence on metazoan development of differences in cellular differentiation during gastrulation appears to be even more germane to understanding the emergence of form and its persistence across taxa as well as of the emergence of differences in form, whether their expression constitutes variation (individual differences) or diversity (species differences).

8
Positional Information and Shape

One of the ongoing questions in developmental biology is how cells acquire positional information (PI), in terms not only of entire structures (e.g., where limbs will grow), but also of how cells acquire and express information necessary and critical to the emergence of regional shapes of structure (e.g., the different segments of a limb) [84, 90–96].

In invertebrates, the wing (e.g., as in a fruit fly) and limb (e.g., as in the brine shrimp-like crustacean, *Artemia*) positioning and segmentation involves activation of the regulatory genes *nubbin* and *apterous* [97–99]. In vertebrates, various regulatory genes, especially *Hox* genes as well as *dlx* (distal-less), are recruited in the determination of limb positioning and segmentation [90, 100, 101]. In fish and tetrapods, the *Hoxd11-13* genes are expressed along the posterior margin of the enlarging limb bud, although only in tetrapods does this homeodomain expand anteriorly across the distal end of the limb bud [80]. Additionally, in vertebrates, *Hox* genes are not only recruited in the formation of a segmented trunk but through regional activation, in the segmentation of the hindbrain [91]. Activation elsewhere in the developing organism of one of these regulatory genes, *Sonic hedgehog*, also in part underlies forebrain segmentation [102].

Eye development is also of interest in this regard. For whilst what had originally been conceived as one of a number of "master-control genes," such as *Pax-6* [76], were subsequently found to participate in signal transduction pathways leading to eye formation in invertebrates (typically multilensed, rigid) as well as in vertebrates (single lensed, deformable) [103], vertebrates have at least one element in common. Even though the vertebrates studied to day differ as to when in ontogeny and how often, and in how many, different somatic regions the *Rx* gene (which is also recruited in fruit fly eye development) is activated, it is always expressed in the vertebrate forebrain [104]. Thus, at one level, it can be hypothesized that the last common ancestor of vertebrates was characterized by early activation of the *Rx* gene in the presumptive forebrain, and that differences between taxa are the result of differences in other aspects of *Rx* gene expression. For example, the different proteins in the lenses of amphibians and mammals may be due in part to downstream or upstream effects of the *Rx* gene being expressed later on in development in the amphibian (frog) retina, whereas in mammals (mouse) the *Rx* gene is expressed early on in the presumptive optic organ itself.

In addition to considering morphological differences in light of differences in regional as well as in overlapping domains of regulatory gene expression, it is becoming increasingly clear that differences in fields of molecular gradients (morphogenetic fields) also play a role. As Lovejoy *et al.* [105] hypothesized in 1999 in their discussion of the evolution of human pelvic form, "... if a particular PI gradient were to span *n* cell diameters, and those cells defined the ultimate anteroposterior dimension of the presumptive ilium (superoinferior in the adult human), then a slight increase in the steepness of its slope would cause that signal to span fewer cells, 'distorting' the presumptive anlagen and substantially altering downstream adult morphology." In other words, although a process that seemingly required myriad steps, "... the transformation of the common ancestral pelvis (in its entirety) into that of early hominids may have been as 'simple' as a slight modification of a gradient" [comment added]. Thus, in addition to differences in gene expression and pathways of molecular communication, as well as in the physical and mechanical consequences of cellular organization, morphological novelty in metazoans (and also presumably plants) may also be significantly affected by altering the effective domains of morphogenetic fields.

Yet, what is the source of differences in gene or molecular gradient expression? Lovejoy *et al.* have suggested that there is no need to seek the answer in mutation, which is a position that Carroll later also argued [106].

9 "Mutation"

The concept of "mutation" is as slippery as that of a "gene," as the term means different things to different research groups. Moreover, incredibly, the different concepts give the appearance of "working" in their disparate intellectual contexts. Mutations were basically constrained intellectually by conceptions that for decades reiterated "textbook" lists of point mutation, gene duplication, and chromosomal rearrangement. The latter was basic to the earlier experimental studies and theoretical considerations of the fruit-fly geneticist, Dobzhansky [21, 25, 32], whose emphasis on chromosomal rearrangement as a potential source of evolutionary novelty was subsequently adopted and modified by the developmental biologist, Goldschmidt, in his theory of systemic mutation. Goldschmidt argued that this would lead to the abrupt appearance of novel forms through altering differential thresholds of gene expression [22].

One of Goldschmidt's major theoretical thrusts, however, was to distinguish between "micromutation" (leading to variation, adaptation, and the survival of species = microevolution) and "macromutation" (leading to the origin of novel features and potentially new species = evolution). The small mutations that fruit-fly population geneticists, such as Morgan (e.g., Refs [107, 108]), had argued underlay small phenotypic changes that Goldschmidt identified as micromutations and which, he argued, led only to the survival of species, not to their origin. The latter phenomenon required a more profound source of genetic disturbance – sufficient to reorganize development – and for that Goldschmidt turned in part to chromosomal rearrangement, but even more so to altering developmental thresholds. The logic here is understandable: If the chromosome theory was correct (that, indeed, units of heredity or genes were contained on chromosomes – like Morgan's diagrams of beads in a necklace), then manipulating them on a grand scale (producing a systemic mutation) might yield evolutionarily significant novelty – that is, a new species. A major problem with Gold-Schmidt's theory, however, was that even though he posited that more one than individual would bear the novelty, he did not provide a mechanism by which more than one individual would be the bearer of the novelty and, thus, of the systemic mutation underlying it. Unfortunately, this left Goldschmidt vulnerable to the slurs and ad hominem attacks of the "modern evolutionary synthesizers" in their unusually influential volumes, *Genetics and the Origin of Species* [21], *Systematics and the Origin of Species* [26], and *Tempo and Mode* [109].

Point mutations, which commonly are the result of UV radiation, are random with regard to affecting either somatic or sex cells. In addition, if they do not interfere with cell function, point mutations are not a reliable source of potential genetic and subsequently morphological novelty. Indeed, it appears that point mutations do not often cause any noticeable effect, either genetically or phenotypically (see discussion in Ref. [68]). Gene duplication – as seen, for instance in the emergence of *Hoxa-d* – does sometimes occur, but knockout experiments have demonstrated that duplication typically reflects redundancy, and thus is not a source of phenotypic novelty.

In may be true that the manipulation of thyroxin or retinoic acid levels during ontogeny can affect the size of an organism, or the shapes of some of its features [110, 111]. However, in contrast to maternal influences on methylation in embryonic cells [112] or other epigenetic and/or non-Mendelian factors [113], these perturbations impact an individual only during its lifetime, and should not be expected to be repeated across generations and under the influence of fluctuating environmental stimuli (e.g., diet, temperature, amount of daylight), unless the source and effect of the epigenetic influence is itself transgenerational and reiterated *in toto*. Consequently, the questions remain: How does a genetic or cellular change remain "fixed" or constant?, and How do many individuals come to bear it?

It is important to realize that there is a difference between change at the molecular and/or genetic level and what is perceived as phenotypic change. Common in the literature on the genetics of evolution is the mistake of conflating the two as both constituting macromutation – a misconception that derives from the confusion that Dobzhansky [21] introduced when using the terms micromutation and macromutation in an assault on Goldschmidt. Nevertheless, especially with the increasing awareness from molecular biology that there are not "genes for" features [114, 115], a vigilant approach must be undertaken in distinguishing between what appears morphologically to have been the result of a macromutation (e.g., developing feathers instead of scales) and the underlying genetic–epigenetic interactive pathway.

10
Toward a Theory of Evolutionary Change

As mentioned earlier, in contrast to bacteria, in which approximately 98% of the genome codes for metabolically active enzymes and proteins and there is a direct correlation between DNA sequence and gene products, in metazoans only 2–3% of the genome is coding, while the rest is noncoding, being represented primarily by introns and exons and control and promoter regions [29]. It is the alternative splicing of introns, and whether transcription occurs in "sense," "anti-sense" (= reverse transcription), or combined "sense–anti-sense" directions [116], together with the coordinated interaction of regulatory molecules as well as epigenetic and other non-Mendelian factors [75, 112–115, 117], that underlies much of an organism's development.

In short, the unbridgeable divide between *the* molecular and *the* morphological that molecular systematists proclaim when assigning to the former priority in deciphering evolutionary relationships (e.g., Refs [8, 45, 53, 62, 118, 119]) is nonexistent [66, 120]. It is a myth – unless the decision is made to ignore biology and equate the predominantly coding genomes of bacteria (which are constantly changing, sometimes with adaptive significance) with the predominantly noncoding genomes of metazoans (which are intimately tied to the regulation of organismal development) [68, 121]. Rather, in metazoans there is a developmentally integrated hierarchy of instructional information that spans basic gene regulatory networks (GRNs) fundamental to establishing basic body plans to differentiation gene batteries (DGBs) involved in the terminal differentiation of tissues and structures [122]. Moreover, because of this developmental continuum from "the molecular" to "the morphological," the false dichotomy of "molecules versus morphology" dissolves, and morphology is irrevocably resituated in considerations of evolutionary relationships [66, 73].

The emergence of taxic diversity, from the establishment of major clades to the subclades and species of which they are comprised, would require profound alterations of deeply entrenched GRN signaling pathways [66]. In contrast, interindividual variation, which typifies the members of any species, would result from different intensities of expression (hyper- or hypo-) of transcription factors and extant signaling pathways. Furthermore, as the vast majority of the metazoan genome is involved in developmental processes, were molecular change ("mutation") to continually occur (the assumption underlying the interpretation of DNA or other molecular sequence similarity for purposes of phylogenetic reconstruction), then the integrity of the myriad, finely tuned signaling pathways and mechanisms that maintain DNA homeostasis would be continually disrupted [68]. The result would therefore more likely be death (gametic, zygotic, or otherwise) than the persistence and gradual change of organisms [68, 123]. The task at hand, then, is to try to articulate a biologically viable mechanism that first can provide the potential for genetic novelty.

One of the most important, yet long-forgotten, elements that must be incorporated into any model of evolutionary change was first recognized in the early twentieth century by Bateson [124]: namely, since mutations arising in the dominant (immediately expressed) state tend to be lethal, most mutations must arise in the recessive (unexpressed) state. This simple realization formed the common foundation of the otherwise very different evolutionary formulations of Fisher [125], Haldane [126], and Wright [127–129]. These three mathematical population geneticists also recognized that, if most mutations arise in the recessive state, they will spread "silently" from one generation to the next, increasing in frequency, until the number of individuals heterozygous for the mutation is sufficiently dense that these individuals will produce homozygotes for the mutation that, in turn, will produce more like themselves as heterozygotes continue to produce homozygous offspring. Since, as Haldane [126], and especially Wright [127–129] emphasized, individuals in wild populations will most likely mate with individuals in their proximity, this natural tendency toward some amount of inbreeding will have the effect of speeding up the process of heterozygosis, and thus the production of homozygotes for the mutation.

Unlike Wright and Haldane, who would let this process proceed unfettered, Fisher [125] was invested in the notion that heterozygosity represented the state of greater fitness, and imagined a point at which a modifier would convert the recessive to the dominant state, well prior to the population producing homozygotes for the mutation. In his review of Fisher's "Dominance Theory," Wright [130, 131] raised the question of how such a modifier could target only one recessive allele rather than affecting all recessive alleles.

Curiously, while Wright's models of genetics in small versus large populations – with the emphasis being less on mutation than on novel allelic combinations and with provisos for rapid as well as slow transformation – formed the core of Dobzhansky's [21, 25] contributions to the "modern evolutionary synthesis," Fisher's [125] model of speciation eventually dominated evolutionary biology, probably for the following reasons. The second edition of Fisher's work [132] was likely known to a larger audience and seen as being much more compatible with that of Mayr's [26, 133–136] "biological species definition" than Wright's favored model of novel allelic combinations increasing rapidly in frequency in small, partially isolated peripheral populations that would "inject" their novel genetic combinations into the major population.

Fisher's model, on the other hand, emphasized the "fissioning" of a large population into two that would slowly diverge genetically, and consequently phenotypically, under different selection conditions. No doubt another reason for Fisher's model becoming the more popular was that Wright's mathematical creation – path analysis – was extraordinarily difficult for the "average" evolutionary biologist to comprehend (see Ref. [137]). For present purposes, though, what is crucial is that the shared starting point of Fisher's, Wright's, and Haldane's models (Haldane's relying on inbreeding in small, totally isolated peripheral populations) disappeared from discussion, being replaced by the popular post-synthesis Darwinian model: that a mutation occurs (presumably in the dominant, expressed state) and selection acts on it (presumably on its immediately available phenotypic expression) (e.g., Ref. [67]).

At the very least, this commonly accepted model promulgates two false notions: (1) that there are "genes for" structure; and (2) that mutations typically arise in the dominant state without lethal consequence. Yet, if Bateson's original hypothesis and its subsequently expanded connotations are resuscitated, the following scenario becomes available [123, 138]. One or some number of mutations unrelated in time and space will arise in the recessive state. If any of these occur during gametogenesis, they can be passed on to the next generation and have the opportunity to spread "silently" from one generation to the next, and throughout the population. With heterozygote saturation, homozygotes for the "mutation" can be produced, and from here the process continues as more homozygotes are produced. At some point, as clearly must happen, the recessive converts to the dominant state.

Of evolutionary consequence, though, is the realization that while it will take some number of generations for the "mutation" to spread throughout the population (the rapidity of the spread being also dependent on degree of inbreeding), the organismal novelty expressed in the homozygotes will appear suddenly in one generation and in some number of individuals (i.e., more than one). If the resultant novelty is not lethal, it will be perpetuated as more individuals with it will be produced. Interestingly, this model of the sudden origin of organismal novelty is consistent with the ever-present "gaps" in the fossil record, which are a nuisance only if it is expected a priori that species change through a gradual process of transformation. If, however, the "sudden origins" model presented here is embraced, then neontological and paleontological observations are consistent [123, 138]. Indeed, during the period when this model of evolutionary change was being formulated, was unexpectedly demonstrated in a laboratory population of mice in which pups with a mutation of *Hoxd13*-expressed synpolydactyly [139]. However, if as mentioned earlier, cells possess myriad mechanisms – largely through "housekeeping" molecules such as heat-shock proteins (HSPs) – to maintain DNA homeostasis, the question is: How is the potential for change introduced at the molecular level?

The molecular biologist Maresca and I have proposed that the opportunity for genetic novelty may lie in overstressing cells to the extent that their HSPs can no longer maintain genetic homeostasis: that is, they cannot fulfill their roles as chaperones, respond to the needs of the cell membrane and, perhaps most importantly for this discussion, properly fold other proteins and assist in DNA repair [68]. Although first identified in heat-shock experiments, HSPs can be affected by a variety of stresses, including diet (saturated versus unsaturated fatty acids), wind, aridity, and cold. Because most multicellular plants and animals possess HSPs, the theory is more widely applicable than metazoan-centric Darwinian and neo-Darwinian models of evolutionary change, the latter of which rely on unjustifiable extrapolations from fruit-fly population genetics.

Since most multicellular organisms (Antarctic fish being a well-known exception) possess a window of heat-shock response, they can "adapt" to normal fluctuations in their environmental circumstances by shifting the window of response (see Ref. [68]). Moreover, in this regard, it is possible to refer to organisms as adapting to their surrounding circumstances – but only within the constraints of adaptational readjustment their biology allows. If environmental change exceeds this window of adaptation or heat-shock response (as when seasons change), most organisms will "reset" this window, often in less than two months. However, until this window of response is reset, the stress induces an increase in HSP production. If, on the other hand, there is a spike in environmental stress that exceeds an organism's ability to reset its heat-shock response, then HSP function will fail and opportunities for introducing genetic novelty will emerge – especially as a result of incorrect protein folding and inefficient DNA repair. In the former situation, incorrectly folded proteins may, for example, no longer recognize (or be recognized by) promoter or enhancer regions to which they would normally bind, although they may now be capable of binding to different sites. An obvious result of this could be the activation or inhibition of a "gene" or "genes" and, thus, the creation of one or more potentially novel developmentally significant signal transduction pathways. With regard to inefficient DNA repair, genetic novelty of a different type can be introduced, with obvious potential consequences. Since in both cases the environmental stress likely be geographically restricted rather than global, the circumstances exist for more (perhaps many more) than one individual of a species to be affected, albeit not necessarily in the same way.

A note of interpretive caution is, however, necessary. First, the effects of extreme environmental spikes on HSPs must be actualized during gametogenesis since, if they are not, offspring will not be affected. Second, if these inherited, genetically based novelties do not kill the individuals that inherit them, they will probably not be expressed; that is, these genetic novelties will be in the "recessive" state. Consequently, there will not be an immediate phenotypic expression of these potential genetic changes. In addition, it must be borne in mind that, during periods of "silently spreading" genetic novelty, there could be other environmental spikes that would contribute to the pool of potential for genetic novelty and also then phenotypic novelty (however defined). Superficially, this process – or at least the sudden appearance of phenotypic novelty – may seem macromutational, but clearly it is not – at least in the original sense of Goldschmidt, or even that of Dobzhansky. Indeed, something as simple as slight changes in protein folding could have major cascading morphological effects.

11 Molecules and Systematics: Looking Toward the Future

It may be widely believed (and even be true at some level), as Sean Carroll [106] has reiterated, that "... genomes diverge as a function of time." However, the observation that genomes may be different (in whatever ways difference, and similarity, may be identified and defined) does not in and of itself provide a window on how observed differences were achieved. No doubt, some difference is due to the infrequent and random effects of UV radiation. In addition, genomic difference may be due to failures in DNA repair. There may be something intuitively appealing about Sibley and Ahlquist's [31, 36] "law of large numbers" – the idea that organisms are closely related because they share "lots" of their genome. Yet, as Marks [52] points out, humans share about 25% of their genome with bananas. Essentially, then, there is nothing inherent in a demonstration of genomic difference or similarity that translates directly into the "molecular assumption," and thus into a theory of evolutionary relatedness.

Can, then, molecular information be useful in systematics and phylogenetic reconstruction? The answer is "yes,", but it will have to be investigated and subsequently comprehended at the level of cell biology and pathways of molecular communication. As King and Wilson [74] came close to predicting decades ago, it is not through the study of molecular or genomic similarity of organisms, but rather through the investigation of those elements that underlie organismal development, that an understanding of their biology will finally be recognized. This makes sense, for if something as simple as the inactivation or deletion of a transcriptional enhancer can result in a caudal repositioning of the sacrum, or if the expansion of a morphogenetic gradient can transform in its entirety a narrow pelvic girdle with tall, thin ilial blades into a broad, deep, and squat structure, then it is by seeking to identify similarities or differences in these molecular events that the molecular basis of morphology can be more profitably explored and, consequently, the evolutionary relationships of complex organisms identified.

Take, for example, recent experiments in which bird beak size and shape were manipulated [140, 141]. When the expression of the underlying regulatory molecules and transcription factors was altered, not only did the beak size change but so also did all of the cranial structures attendant to it (soft and hard tissue, including vascularization and innervation), simultaneously and in a coordinated, biologically integrated pattern. Thus, even though the surface of the details of signaling pathways underlying development of structure may not yet have been scratched, it is suggested that an air of confidence be retained when identifying morphological differences between taxa. In the context of the broader comparison indicating that a particular structure is unique to a specific taxon, the end result of a specific change at the molecular level of development is also being observed. Morphological similarity simply reflects similar signaling pathways.

Clearly, there must be a continuation in questioning and expanding the current perceptions of what is, or will be, evolutionarily revealing at the molecular level. Witness, for instance, the steady increase in the number of studies that demonstrate virtual molecular identity between taxa that are morphologically very different, and yet also express the authors' continual amazement at this apparent contradiction. As Carroll [106] pointed out regarding the importance placed on the human- and chimpanzee-genome projects (especially as so much money was poured into them in the hope that forthcoming comparisons would instantaneously provide answers to any and all biomedical questions), to demonstrate molecular similarity does not translate into a deciphering of the pathways that make these organisms so different in hard- and soft-tissue anatomy, physiology, reproductive biology, cognitive abilities, and behavior. Here, the "law of large numbers" fails to enlighten because, in the metazoan world only about 2–3% of the genome maintains a one-to-one correspondence between a "gene" (a strict linear sequence of nucleotides bound by start and stop codons) and a "gene product" (a metabolically active protein or enzyme). Instead, the vast majority of the metazoan genome is dedicated to the orchestration of development – which obviates the relevance of models of unfettered and random molecular change. Furthermore, in multicellular animals, RNA essentially directs the "show" – for example, in reading selected bases and in splicing specific introns – as it composes different proteins from the same stretches of DNA.

When The International Chimpanzee Chromosome 22 Consortium [39–41] compared human chromosome 21 with its apparent ortholog in the chimpanzee, chromosome 22, they expressed surprise in finding that these hominoids differ by 83% in their amino acid sequences, and also that this large difference is produced from very similar DNA sequences. This should serve as a lesson. Furthermore, while the Consortium's claim of comparing human and chimpanzee genomes may have been correct, in reality the only parts that could be compared with any accuracy (because inferred DNA sequences are verified from the amino acid and protein sequences they generate) were from the coding region; that is, a portion of only 2–3% of the genome. For the small fractions of noncoding regions that were compared, the sequences could only be inferred. Furthermore, strict sequence comparisons do not reflect the fact that it is the specific *three-dimensional* structure of a noncoding region that gives it functional significance [142].

There may be a certain psychological appeal to embracing a "law of large numbers" that a comparison of chromosomes, and especially of entire genomes, purportedly provides. However, ultimately this molecular level may not be the evolutionarily informative Holy Grail that some systematists have been seeking.

References

1 Nuttall, G.H.F. (1904) *Blood Immunity and Blood Relationship*, Cambridge University Press, Cambridge.
2 Gregory, W.K. (1922) *The Origin and Evolution of the Human Dentition*, Williams and Wilkins, Baltimore.
3 Zuckerkandl, E., Pauling, L. (1962) Molecular Disease, Evolution, and Genic Heterogeneity, in: Kasha, M., Pullman, B. (Eds) *Horizons in Biochemistry*, Academic Press, New York, pp. 189–225.
4 Caccone, A., Powell, J.R. (1989) DNA divergence among hominoids. *Evolution*, **43**, 925–942.
5 Sarich, V., Wilson, A.C. (1966) Quantitative immunochemistry and the evolution of primate albumins: micro-complement fixation. *Science*, **154**, 1563–1566.
6 Sarich, V.M., Wilson, A.C. (1967) Rates of albumin evolution in primates. *Proc. Natl Acad. Sci. USA*, **58**, 142–148.
7 Sarich, V. (1971) A Molecular Approach to the Question of Human Origins, in: Dolhinow, P., Sarich, V. (Eds) *Background for Man*, Little, Brown, Boston, pp. 60–81.
8 Bruce, E.J., Ayala, F.J. (1979) Phylogenetic relationships between man and the apes: electrophoretic evidence. *Evolution*, **33**, 1040–1056.
9 Felsenstein, J. (1988) Phylogenies from molecular sequences: inference and reliability. *Annu. Rev. Genet.*, **22**, 521–565.
10 Ferris, S.D., Wilson, A.C., Brown, W.M. (1981) Evolutionary tree for apes and humans based on cleavage maps of mitochondrial DNA. *Proc. Natl Acad. Sci. USA*, **78**, 2431–2436.
11 Hasegawa, M., Yano, T. (1984) Maximum likelihood method of phylogenetic inference form DNA sequence data. *Bull. Biometr. Soc. Jpn*, **5**, 1–7.
12 Hasegawa, M., Kishino, H., Yano, T. (1985) Dating of the human-ape splitting by a molecular clock of mitochondrial DNA. *J. Mol. Evol.*, **22**, 160–174.
13 Lowenstein, J.M., Molleson, T., Washburn, S.L. (1982) Piltdown jaw confirmed as orang. *Nature*, **299**, 294.
14 Goodman, M. (1962) Immunochemistry of the primates and primate evolution. *Ann. N. Y. Acad. Sci.*, **102**, 219–234.
15 Goodman, M., Tashian, R.E. (1976) *Molecular Anthropology*, Plenum, New York.
16 Goodman, M. (1981) Decoding the pattern of protein evolution. *Prog. Biophys. Mol. Biol. (Oxford)*, **37**, 105–164.
17 Goodman, M., Braunitzer, G., Stangl, A., Schrank, B. (1983) Evidence on human origins from haemoglobins of African apes. *Nature*, **303**, 546–548.
18 Romero-Herrera, A.E., Lehmann, H., Castillo, O., Joysey, K.A., Friday, A.E. (1976) Myoglobin of the orangutan as a phylogenetic enigma. *Nature*, **261**, 162–164.
19 Zuckerkandl, E. (1987) On the molecular evolutionary clock. *J. Mol. Evol.*, **26**, 34–46.
20 Jacob, F., Monod, J. (1959) Genes of structure and genes of regulation in the biosynthesis of proteins. *C R Hebd. Seances Acad. Sci.*, **249**, 1282–1284.
21 Dobzhansky, T.H. (1941) *Genetics and the Origin of Species*, Columbia University Press, New York.
22 Goldschmidt, R.B. (1940) *The Material Basis of Evolution*, Yale University Press, New Haven.
23 Huxley, J. (1940) *The New Systematics*, Clarendon Press, Oxford.
24 Schwartz, J.H. Systematics and Evolutionary Biology: uneasy bedfellows? *Revista Mexicana de Biodiversidad*, **82**, 1–18.
25 Dobzhansky, T. (1937) *Genetics and the Origin of Species*, Columbia University Press, New York.
26 Mayr, E. (1942) *Systematics and the Origin of Species*, Columbia University Press, New York.
27 Kimura, M. (1968) Evolutionary rate at the molecular level. *Nature*, **217**, 624–626.
28 Kimura, M. (1985) *Neutral Theory of Molecular Evolution*, Cambridge University Press, Cambridge.
29 Eisen, J.A. (2000) Assessing evolutionary relationships among microbes from whole-genome analysis. *Curr. Opin. Microbiol.*, **3**, 475–480.
30 Goff, S.A., Ricke, D., Lan, T.-H., Presting, G., Wang, R., Dunn, M., Glazebrook, J., Sessions, A., Oeller, P., Varma, H., Hadley, D., Hutchison, D., Martin, C., Katagiri, F., Lange, B.M., Moughamer, T., Xia, Y., Budworth, P., Zhong, J., Miguel, T., Paszkowski, U., Zhang, S., Colbert, M., Sun, Wei-lin, Chen, L., Cooper, B., Park, S.,

Wood, T.C., Mao, L., Quail, P., Wing, R., Dean, R., Yu, Y., Zharkikh, A., Shen, R., Sahasrabudhe, S., Thomas, A., Cannings, R., Gutin, A., Pruss, D., Reid, J., Tavtigian, S., Mitchell, J., Eldredge, G., Scholl, T., Miller, R.M., Bhatnagar, S., Adey, N., Rubano, T., Tusneem, N., Robinson, R., Feldhaus, J., Macalma, T. (2002) A draft sequence of the rice genome (*Oryza sativa* L. ssp. *japonica*). *Science,* **296**, 92–100.

31 Sibley, C.G., Ahlquist, J.E. (1983) Phylogeny and classification of birds based on the data of DNA-DNA hybridization, in: Johnston, R.F. (Ed.) *Current Ornithology,* Vol. 1, Plenum Press, New York, pp. 245–292.

32 Dobzhansky, T. (1935) *Drosophila miranda,* a new species. *Genetics,* **20**, 377–391.

33 Britten, R.J., Kohne, D.E. (1968) Repeated sequences in DNA. *Science,* **161**, 529–540.

34 Duboule, D. (2007) The rise and fall of *Hox* gene clusters. *Development,* **134**, 2549–2560.

35 Benviste, R.E., Todaro, G.J. (1976) Evolution of type C viral genes: evidence for an Asian origin of man. *Nature,* **261**, 101–108.

36 Sibley, C.G., Ahlquist, J.E. (1984) The phylogeny of the hominoid primates, as indicated by DNA-DNA hybridization. *J. Mol. Evol.,* **20**, 2–15.

37 Templeton, A.R. (1981) Phylogenetic inference from restriction endonuclease site maps with particular reference to the evolution of human and apes. *Evolution,* **37**, 221–244.

38 Friday, A.E. (1981) Hominoid evolution: the nature of biochemical evidence, in: Stringer, C.B. (Ed.) *Aspects of Human Evolution: Society for the Study of Human Biology Symposium Series.* Taylor & Francis, London, pp. 1–25.

39 Chimpanzee Sequencing and Analysis Consortium (2005) Initial sequence of the chimpanzsee genome and comparison with the human genome. *Nature,* **437**, 69–87.

40 International Human Genome Sequencing Consortium (2004) Finishing the euchromatic sequence of the human genomes. *Nature,* **431**, 931–945.

41 International Human Genome Sequencing Consortium (2001) Initial sequencing and analysis of the human genome. *Nature,* **409**, 860–925.

42 Greene, R.E., Krause, J., Briggs, A.W., Maricic, T., Stenzel, U., Kircher, M., Patterson, N., Li, H., Zhai, W., Fritz, M.H.-Y., Hansen, N.F., Durand, E.Y., Malaspinas, A.-S., Jensen, J.D., Marques-Bonet, T., Alkan, C., Prüfer, K., Meyer, M., Burbano, H.A., Good, J.M., Schultz, R., Aximu-Petri, A., Butthof, A., Höber, B., Höffner, B., Siegemund, M., Weihmann, A., Nusbaum, C., Lander, E.S., Russ, C., Novod, N., Affourtit, J., Egholm, M., Verna, C., Rudan, P., Brajkovic, D., Kucan, Z., Gušic, I., Doronichev, V.B., Golovanova, L.V., Lalueza-Fox, C., de la Rasilla, M., Fortea, J., Rosas, A., Schmitz, R.W., Johnson, P.L.F., Eichler, E.E., Falush, D., Birney, E., Mullikin, J.C., Slatkin, M., Nielsen, R., Kelso, J., Lachmann, M., Reich, D., Pääbo, S. (2010) A draft sequence of the Neanderthal genome. *Science,* **238**, 710–722.

43 Brown, W.M., Prager, E.M., Wang, A., Wilson, A.C. (1982) Mitochondrial DNA sequences of primates: tempo and mode of evolution. *J. Mol. Evol.,* **18**, 225–239.

44 Brown, W.M., George, M.J., Wilson, A.C. (1979). Rapid evolution of mitochondrial DNA. *Proc. Natl Acad. Sci. USA,* **76**, 1967–1971.

45 Ruvolo, M., Disotell, T.R., Allard, M.W., Brown, W.M., Honeycutt, R.L. (1991) Resolution of the African hominoid trichotomy by use of a mitochondrial gene sequence. *Proc. Natl Acad. Sci. USA,* **88**, 1570–1574.

46 Hagelberg, E. (2003) Recombination or mutation rate heterogeneity? Implications for mitochondrial Eve. *Trends Genet.,* **19**, 84–90.

47 Gyllensten, U., Wharton, D., Josefsson, A., Wilson, A.C. (1991) Paternal inheritance of mitochondrial DNA in mice. *Nature,* **352**, 255–257.

48 Awadella, P., Eyre-Walker, A., Maynard Smith, J. (1999) Linkage disequilibrium and recombination in hominid mitochondrial DNA. *Science,* **286**, 2545–2525.

49 Scheffler, I.E.A. (2000) A century of mitochondrial research: achievements and perspectives. *Mitochondrion,* **1**, 3.

50 Lane, N. (2005) *Power, Sex, Suicide,* Oxford University Press, Oxford.

51 Lake, J.A. (1991) The order of sequence alignment can bias the selection of tree topology. *Mol. Biol. Evol.*, **8**, 378–385.

52 Marks, J. (2003) *What it Means to be 98% Chimpanzee*, University of California Press, Berkeley.

53 Ruvolo, M. (1997) Molecular phylogeny of the hominoids: inferences from multiple independent DNA sequence data sets. *Mol. Biol. Evol.*, **14**, 248–265.

54 Czelusniak, J., Goodman, M., Moncrief, N.D., Kehoe, S.M. (1990) Maximum parsimony approach to construction of evolutionary trees from aligned homologous sequences. *Methods Enzymol.*, **183**, 601–615.

55 Goldberg, A., Wildman, D.E., Schmidt, T.R., Hüttemann, M., Goodman, M., Weiss, M.L., Grossman, L.I. (2003) Sister grouping of chimpanzees and humans as revealed by genome-wide phylogenetic analysis of brain gene expression profiles. *Proc. Natl Acad. Sci. USA*, **100**, 5873–5878.

56 Goodman, M., Porter, C.A., Czelusniak, J., Page, S.L., Schneider, H., Shoshani, J., Gunnell, G., Groves, C.P. (1998) Toward a phylogenetic classification of primates based on DNA evidence complemented by fossil evidence. *Mol. Phylogenet. Evol.*, **9**, 585–598.

57 Hedges, S.B., Kumar, N.S. (2003) Genomic clocks and evolutionary timescales. *Trends Genet.*, **19**, 200–226.

58 Hedges, S.B., Kumar, S., Tamura, K., Stoneking, M. (1992) Human origins and analysis of mitochondrial DNA sequences. *Science*, **255**, 737–739.

59 Horai, S., Hayasaka, K., Kondo, R., Tsugane, K., Takahata, N. (1995) Recent African origin of modern humans revealed by complete sequence of hominoid mitochondrial DNAs. *Proc. Natl Acad. Sci. USA*, **92**, 532–536.

60 Marshall, C.R. (1991) Statistical tests and bootstrapping: assessing the reliability of phylogenies based on distance data. *Mol. Biol. Evol.*, **8**, 386–391.

61 O'hUigin, C.O., Satta, Y., Takahata, N., Klein, J. (2002) Contribution of homoplasy and of ancestral polymorphism to the evolution of genes in anthropoid primates. *Mol. Biol. Evol.*, **19**, 1501.

62 Stauffer, R.L., Walker, A.C., Ryder, O.A., Lyons-Weiler, M. (2001) Human and ape molecular clocks and constraints on paleontological hypotheses. *J. Hered.*, **92**, 469–474.

63 Schwartz, J.H. (2005) Molecular systematics and evolution, in: Meyer, R.A. (Ed.) *Encyclopedia of Molecular Cell Biology and Molecular Medicine (EMCBMM)*, Wiley-VCH Verlag GmbH, Weinheim, pp. 515–540.

64 Hennig, W. (1966) *Phylogenetic Systematics*, University of Chicago Press, Chicago.

65 Schwartz, J.H. (2008) Cladistics, in: Regal, B. (Ed.) *Icons of Evolution*, Greenwood Press, Westport, pp. 517–544.

66 Schwartz, J.H. (2009) Organismal innovation, in: O'Brien, M.J., Shennan, S.J. (Eds.) *Innovations in Cultural Systems: Contributions from Evolutionary Anthropology*, MIT Press, Cambridge, pp. 53–67.

67 Williams, G.C. (1966) *Adaptation and Natural Selection: A Critique of Some Current Evolutionary Thought*, Princeton University Press, Princeton.

68 Maresca, B., Schwartz, J.H. (2006) Sudden origins: a general mechanism of evolution based on stress protein concentration and rapid environmental change. *Anat. Rec. B New Anat.*, **289**, 38–46.

69 Nikaido, M., Roonery, A.P., Okada, N. (1999) Phylogenetic relationships among cetartiodactyls based on insertions of short and long interspersed elements: hippopotamuses are the closest extant relatives of whales. *Proc. Natl Acad. Sci. USA*, **96**, 10261–10266.

70 Salem, A.-H., Ray, D.A., Jinchuan, X., Callinan, R.A., Myers, J.S., Hedges, D.J., Garber, R.K., Witherspoon, D.J., Jorde, L.B., Batzer, M.A. (2003) Alu elements and hominid phylogenetics. *Proc. Natl Acad. Sci. USA*, **100**, 12787–12791.

71 Schmitz, J., Roos, C., Zischler, H. (2005) Primate phylogeny: molecular evidence from retroposons. *Comp. Genomics Mol. Evol.*, **108**, 26–37.

72 Hillis, D.M. (1999) SINEs of the perfect character. *Proc. Natl Acad. Sci. USA*, **96**, 9979–9981.

73 Schwartz, J.H. (2009) Reflections on systematics and phylogenetic reconstruction. *Acta Biotheor.*, **57**, 295–305.

74 King, M.-C., Wilson, A.C. (1975) Evolution at two levels in humans and chimpanzees. *Science*, **188**, 107–116.

75 Gerhart, J., Kirschner, M. (1997) *Cells, Embryos, and Evolution: Toward a Cellular and Developmental Understanding of Phenotypic Variation and Evolutionary Adaptability*, Blackwell, Malden.

76 Halder, G., Callaerts, P., Gehring, W. (1995) Induction of ectopic eyes by targeted expression of the *eyeless* gene in *Drosophila*. *Science*, **267**, 1788–1792.

77 Quiring, R., Walldorf, U., Kloter, U., Gehring, W.J. (1994) Homology of the eyeless gene in *Drosophila* to the Small eye gene in mice and *Aniridia* in humans. *Science*, **265**, 785–789.

78 Zákány, J., Duboule, D. (1996) Synpolydactyly in mice with a targeted deficiency in the HoxD complex. *Nature*, **384**, 69–71.

79 Sordino, P., van der Hoeven, F., Duboule, D. (1995) Hox gene expression in teleost fins and the origin of vertebrate digits. *Nature*, **375**, 678–681.

80 Duboule, D. (1994) How to make a limb? *Science*, **266**, 575–576.

81 Duboule, D., Dollé, P. (1989) The structural and functional organization of the murine HOX gene family resembles that of *Drosophila* homeotic genes. *EMBO J.*, **8**, 1497–1505.

82 Gilbert, S. (2006) *Developmental Biology*, Sinauer Associates Inc., Sunderland.

83 Kirschner, M.W., Gerhart, J.C. (2005) *The Plausibility of Life: Resolving Darwin's Dilemma*, Yale University Press, New Haven.

84 Wolpert, L. (1994) Do we understand development? *Science*, **266**, 571–572.

85 Newman, S.A., Forgacs, G., Müller, G. (2006) Before programs: the physical origination of multicellular forms. *Int. J. Dev. Biol.*, **50**, 289–299.

86 Løvtrup, S. (1974) *Epigenetics: A Treatise on Theoretical Biology*, John Wiley & Sons, Inc., New York.

87 Lowe, C.J., Wray, G.A. (1997) Radical alterations in the roles of homeobox genes during echinoderm evolution. *Nature*, **389**, 718–721.

88 Oster, G., Alberch, P. (1982) Evolution and bifurcation of developmental programs. *Evolution*, **36**, 444–459.

89 Schifmann, Y. (2006) Symmetry breaking and convergent extension in early chordate development. *Biophys. Mol. Biol.*, **92**, 209–231.

90 Cohn, M.J., Patel, K., Krumlauf, R., Wilkinson, D.G., Clarke, J.D.W., Tickle, C. (1997) Hox9 genes and vertebrate limb specification. *Nature*, **387**, 97–101.

91 Figdor, M., Stern, C. (1993) Segmental organization of embryonic diencephalon. *Nature*, **363**, 630–634.

92 Hunt, P., Gulisano, M., Cook, M., Sham, M.H., Faiella, A., Wilkinson, D., Boncinelli, E., Krumlauf, R. (1991) A distinct Hox code for the branchial region of the vertebrate head. *Nature*, **353**, 861–864.

93 Lumsden, A., Krumlauf, R. (1996) Patterning the vertebrate neuraxis. *Science*, **274**, 1109–1115.

94 Pourquié, O. (2003) The segmentation clock: converting embryonic time into spatial pattern. *Science*, **301**, 328–330.

95 Summerbell, D. (1981) Evidence for regulation of growth, size and pattern in the developing chick limb bud. *J. Embryol. Exp. Morphol.*, **65**, 129–150.

96 Wolpert, L. (1980) Positional information and pattern formation in limb development, in: Pratt, R.M., Christiansen, R.L. (Eds) *Current Research Trends in Prenatal Craniofacial Development*, Elsevier, North Holland, pp. 89–101.

97 Averoff, M., Cohen, S.M. (1997) Evolutionary origin of insect wings from ancestral gills. *Nature*, **385**, 627–630.

98 Patel, N.H. (1994) Developmental evolution: insights from studies of insect segmentation. *Science*, **266**, 581–590.

99 Ronshaugen, M., McGinnis, N., McGinnis, W. (2002) Hox protein mutation and macroevolution of the insect body plan. *Nature*, **415**, 914–917.

100 Davidson, D.R., Crawley, A., Hill, R.E., Tickle, C. (1991) Position-dependent expression of two related homeobox genes in developing vertebrate limbs. *Nature*, **352**, 429–431.

101 Tarchini, B., Duboule, D., Kmita, M. (2006) Regulatory constraints in the evolution of the tetrapod limb anterior–posterior polarity. *Nature*, **443**, 985–988.

102 Rubenstein, J.L.R., Martinez, S., Shimamura, K., Puelles, L. (1994) The

embryonic vertebrate forebrain: the prosomeric model. *Science*, **266**, 578–580.
103 Gehring, W.J. (2002) The genetic control of eye development and its implications for the evolution of the various eye-types. *Int. J. Dev. Biol.*, **46**, 65–73.
104 Mathers, P., Grinberg, A., Mahon, K., Jamrich, M. (1997) The Rx homeobox gene is essential for vertebrate eye development. *Nature*, **387**, 604–607.
105 Lovejoy, C.O., Cohn, M.J., White, T.D. (1999) Morphological analysis of the mammalian postcranium: a developmental perspective. *Proc. Natl Acad. Sci. USA*, **96**, 13247–13252.
106 Carroll, S.R. (2003) Genetics and the making of *Homo sapiens*. *Nature*, **422**, 849–857.
107 Morgan, T.H. (1935) *The Scientific Basis of Evolution*, W. W. Norton & Company, New York.
108 Morgan, T.H., Sturtevant, A.H., Muller, H.J., Bridges, C.B. (1926) *The Mechanism of Mendelian Heredity*, Henry Holt, New York.
109 Simpson, G.G. (1944) *Tempo and Mode in Evolution*, Columbia University Press, New York.
110 Crockford, S.J. (2003) Thyroid rhythm phenotypes and hominid evolution: a new paradigm implicates pulsatile hormone selection in speciation and adaptive changes. *Comp. Biochem. Physiol. Part A*, **135**, 105–129.
111 Huxley, J.S. (1923) Time relations in amphibian metamorphosis, with some general considerations. *Sci. Prog.*, **17**, 606–618.
112 Rassoulzadegan, M., Grandjean, V., Gounon, P., Vincent, S., Gillot, I., Cuzin, F. (2006) RNA-mediated non-Mendelian inheritance of an epigenetic change in the mouse. *Nature.*, **441**, 469–474.
113 Lolle, S.J., Victor, J.L., Young, J.M., Pruitt, R.E. (2005) Genome-wide non-mendelian inheritance of extra-genomic information in *Arabidopsis*. *Nature*, **434**, 505–509.
114 Pearson, H. (2006) What is a gene? *Nature*, **441**, 399–401.
115 Stotz, K. (2006) With 'Genes' like that, who needs an environment? Postgenomics's argument for the 'Ontogeny of Information'. *Philos. Sci.*, **73**, 905–917.
116 Ast, G. (2005) The alternative genome. *Sci. Am.*, **292**, 59–65.
117 González-Reyes, A., Elliott, H., St Johnston, D. (1995) Polarization of both major body axes in *Drosophila* by gurken-torpedo signalling. *Nature*, **375**, 654–658.
118 Pilbeam, D. (2000) Hominoid systematics: the soft evidence. *Proc. Natl Acad. Sci. USA*, **97**, 10684–10686.
119 Pilbeam, D.R. (1996) Genetic and morphological records of the Hominoidea and hominid origins: a synthesis. *Mol. Phylogenet. Evol.*, **5**, 155–168.
120 Schwartz, J.H. Organismal biology, molecular systematics, and phylogenetic reconstruction, in: Masters, J., Gamba, M., Génin, F. (Eds) *Leaping Ahead: Advances in Prosimian Biology*, Springer Science, New York, in press.
121 Schwartz, J.H., Maresca, B. (2006) Do molecular clocks run at all? A critique of molecular systematics. *Biol. Theory*, **1**, 1–15.
122 Davidson, E.H., Erwin, D.H. (2006) Gene regulatory networks and the evolution of animal body plans. *Science*, **311**, 796–800.
123 Schwartz, J.H. (1999) *Sudden Origins: Fossils, Genes, and the Emergence of Species*, John Wiley & Sons, Inc., New York.
124 Bateson, W. (1913) *Problems of Genetics*, Yale University Press, New Haven.
125 Fisher, R.A. (1930) *The Genetical Theory of Natural Selection*, Oxford University Press, Oxford.
126 Haldane, J.B.S. (1932) *The Causes of Evolution*, Harper & Brothers, New York.
127 Wright, S. (1929) Evolution in a Mendelian population. *Anat. Rec.*, **44**, 287.
128 Wright, S. (1931) Evolution in mendelian populations. *Genetics*, **16**, 97–159.
129 Wright, S. (1932) The roles of mutation, inbreeding, crossbreeding and selection in evolution, in: Jones, D.F. (Ed), *Proceedings of the Sixth International Congress of Genetics*, vol. 1, Ithaca, New York, pp. 356–366.
130 Wright, S. (1929) Fisher's dominance theory. *Am. Nat.*, **63**, 274–279.
131 Wright, S. (1930) Review of the genetical theory of natural selection by R.A. Fisher. *J. Hered.*, **21**, 349–356.
132 Fisher, R.A. (1958) *The Genetical Theory of Natural Selection*, Dover Publications, New York.
133 Mayr, E. (1949) Speciation and Systematics, in: Jepsen, G.L., Simpson, G.G., Mayr, E. (Eds) *Genetics, Paleontology, and Evolution*, Atheneum, New York, pp. 281–298.
134 Mayr, E. (1953) Comments on evolutionary literature. *Evolution*, **7**, 273–281.

135 Mayr, E. (1963) *Animal Species and Evolution*, The Belknap Press of Harvard University Press, Cambridge.

136 Mayr, E. (1969) *Principles of Systematic Zoology*, McGraw-Hill, New York.

137 Provine, W.B. (1971) *The Origins of Theoretical Population Genetics*, The University of Chicago Press, Chicago.

138 Schwartz, J.H. (1999) Homeobox genes, fossils, and the origin of species. *Anat. Rec. B New Anat.*, **257**, 15–31.

139 Johnson, K.R., Hope, O.S., Donahue, L.R., Ward-Bailey, P., Bronson, R.T., Davisson, M.T. (1998) A new spontaneous mouse mutation of *Hoxd13* with a polyalanine expansion and phenotype similar to human synpolydactyly. *Hum. Mol. Genet.*, **7**, 1033–1038.

140 Abzhanov, A., Protas, M., Grant, B.R., Grant, P.R., Tabin, C.T. (2004) *Bmp4* and morphological variation in beaks in Darwin's finches. *Science*, **305**, 1462–1465.

141 Wu, P., Jiang, T.-X., Suksaweang, S., Widelitz, R.B., Chuong, C.-M. (2004) Molecular shaping of the beak. *Science*, **305**, 1465–1466.

142 Parker, S.C.J., Hansen, L., Abaan, H.O., Tullius, T.D., Margulies, E.H. (2009) Local DNA topography correlates with functional noncoding regions of the human genome. *Science*, **324**, 389–392.

8
Evolution of the Protein Repertoire

Aaron David Goldman[1,2], Jeremy A. Horst[3,4], Ling-Hong Hung[4], and Ram Samudrala[4]
[1]Princeton University, Department of Ecology and Evolutionary Biology, 106A Guyot Hall, Princeton, NJ 08542, USA
[2]NASA Astrobiology Institute, Center for Ribosomal Origins and Evolution at Georgia Institute of Technology, NASA Postdoctoral Program, 225 North Ave. NW, Atlanta, GA 30332, USA
[3]University of California San Francisco School of Pediatric Dentistry, Department of Orofacial Sciences, 707 Parnassus Ave, San Francisco, CA 94143, USA
[4]University of Washington, Department of Microbiology, 960 Republican Street, Seattle, WA 98109, USA

Systems Biology: Advances in Molecular Biology and Medicine, First Edition. Edited by Robert A. Meyers.

Keywords

Allosteric regulation
The modification of enzyme structure and activity by actions on a region of the enzyme other than the active site

Convergent evolution
The evolution of unrelated lineages toward a common structure and/or function

Cooperativity
The synergistic property of some proteins or multiprotein complexes wherein the activity of one site increases or decreases the activity of the others

Directed evolution
The synthetic production of a desired protein or RNA function through multiple rounds of artificial selection and replication

Divergent evolution
The evolution of related lineages away from a common ancestral structure and/or function

Gene cluster
A group of multiple genes with related function in close genomic proximity

Gene duplication
A process of molecular evolution in which genes or parts of genes are copied into a new location on the genome

Homolog
One of a pair or group of features related by a common ancestor

Homologous recombination
The exchange of DNA between either separate molecules or separate regions of the same molecule facilitated by regions of similar or identical nucleotide sequences

Horizontal gene transfer
The acquisition and incorporation of new genetic information by a living organism through processes other than inheritance

Metabolic network
A combination of metabolic pathways that achieve a general physiological process

Metabolic pathway
A series of enzymatic reactions involved in the stepwise conversion of initial substrates into final products

Metabolome
The complete set of metabolites found in a single organism or community of organisms

Multiprotein complex
A group of either stably or transiently bound proteins that form a single structural and/or functional unit

Neofunctionalization
The evolution of a new protein function that is wholly different from that of its progenitor

Nonsynonymous substitution
A genetic mutation in the protein-coding region of a gene in which one nucleotide is replaced for another, and the identity of the encoded amino acid is changed

Operon
A group of genes, proximally situated on the genome, and transcribed together on a single mRNA molecule

Ortholog
One of a pair or group of homologous genes or gene products found in separate species

Paralog
One of a pair or group of homologous genes or gene products found within the same genome

Phylogenomics
The study of the evolutionary history of paralogous genes within a single species

Phylogeny
An evolutionary history of genes or organisms inferred by comparing and clustering gene sequences, protein sequences, or other features

Point mutation
A genetic mutation in which a single nucleotide is either inserted, deleted, or substituted for another

Protein domain
One of multiple modules within a protein polymer that forms structure and imparts function somewhat independently, and has a distinct evolutionary history from the rest of the protein

Protein primary structure
A description of protein structure referring to the amino acid sequence of a protein

Protein quaternary structure
A description of protein structure referring to the combination of individual protein polymers into multiprotein complexes

Protein repertoire
The complete set of proteins produced by an organism (similar to 'Protein repertoire')

Protein secondary structure
A description of protein structure referring to common three-dimensional substructures of amino acids, often interacting through hydrogen bonds

Protein tertiary structure
A description of protein structure referring to the arrangement of secondary structure elements and the loops that connect them into low-potential energy conformations

Protein–protein interaction network
A graphical representation that forms networks out of physical associations between proteins

Proteome
The complete set of proteins found in a single organism

Pseudogene
An erstwhile coding sequence that is no longer functional and not under evolutionary selection

Regulatory network
The combination of regulatory events to form larger patterns of gene expression

Subfunctionalization
The retention of only part of an ancestral function over the course of divergent evolution

Synonymous substitution
A genetic mutation in the protein-coding region of a gene in which one nucleotide is replaced for another, but the amino acid sequence is not altered

Trans-splicing
The ligation of multiple RNA transcripts from different chromosomal loci to form a single mature mRNA

Transcriptome
The complete set of mRNA transcripts found in a single organism

A protein is at once the translation product of a gene, a polymer of amino acids folded in three-dimensional space, a molecular machine imparting a function, and a small – but often critical – component of the cell's physiology. Each of these facets of a protein is inherently linked to the others; yet, each is impacted by evolutionary processes differently. The evolution of the protein repertoire involves a broad variety of phenomena, many of which are understood in considerable detail. The aim of this chapter is to provide a panoramic survey of the subject, and to relate the molecular mechanisms of evolution to their effects on proteins, the genes that encode them, and the physiological networks that they populate.

1
The First Proteins

The early Earth was likely to have had a number of prebiotic environments in which amino acids were available [1, 2]. The famous Miller–Urey experiments [3], which simulated lightning on the primordial Earth, were able to produce standard amino acids such as glycine, alanine, serine, aspartate, valine, glutamine, and phenylalanine, as well as several nonstandard amino acids [4, 5]. Alanine, glycine, and some nonstandard amino acids have also been found in carbonaceous chondrite asteroids, which implies that amino acids were being formed even as the solar system took shape [6]; subsequently, geochemical mechanisms, rather than biosynthesis, may have led to the natural assembly of peptides from these amino acids [7]. However, given that such peptides were produced without a genetic system, it is difficult to see how they could have evolved by Darwinian mechanisms, or how they could be direct ancestors of modern biological proteins. Even so, naturally occurring peptides may have played an important role in prebiotic chemistry and perhaps drove the evolution of the early genetic system towards the development of protein biosynthesis as it is known today.

The first genetic system was most likely based on RNA or an RNA-like precursor that produced RNA-based enzymes, or ribozymes, from the replication of genetic RNA molecules [8]. The enzymatic and structural capabilities of a nucleotide-based metabolism would have been limited by the narrow range of chemistry provided by nucleosides and the mutual repulsion of the charged phosphate backbone. Certainly, the metabolisms of such life forms would have been much more reliant on surrounding geochemistry than most cellular life is today. The establishment of a system to translate RNA molecules into more functionally capable proteins would also have given primitive organisms a greater control over their own physiology.

Several arguments have been put forth to support the emergence of protein translation preceding the establishment of the DNA genome. The core components of the modern translation system are composed of RNA, including messenger RNA (mRNA), transfer RNA (tRNA), and the RNA components of the ribosome (ribosomal RNA; rRNA) that catalyze peptide bond formation. In contrast, the core catalysis of DNA transcription and replication is performed solely by proteins. Moreover, it seems unlikely that deoxyribonucleotides – which are the monomers of DNA – were available until the development of ribonucleotide reductase protein enzymes [9].

In addition to its core RNA components, the modern translation system relies on several important classes of proteins that charge the tRNAs with amino acids, incorporate these within the ribosome structure, modify the tRNA and rRNA molecules, and regulate the translation process. These proteins also appear to have evolved rapidly during the RNA–protein stage of life's development [10]. Seemingly, all modern life forms share a remarkably similar translation system, which indicates that the latter had already reached a high level of sophistication by the time of the divergence of the last universal common ancestor [10, 11].

2 Organization of the Modern Protein Repertoire

Proteins are encoded on DNA molecules as sequences of nucleotides, with each amino acid represented by a codon of three nucleotides (Fig. 1). This genetic code is either identical or only marginally different between all known organisms [12, 13]. The products of protein-coding genes are linear polymers of, usually, between 200 and 500 amino acids, that fold into three-dimensional (3-D) structures. The amino acids, themselves, possess different functional groups (also termed "*R groups*" or "*side chains*") that vary in their size, their hydrophobicity, and their potential to form ionic or covalent bonds (Fig. 2). It is this diversity of chemical moieties which provides proteins with their functional and structural capabilities.

Individual protein polymers are often composed of more than one module, each of which fold, function, and evolve independently from one another [17, 18]. These modules are generally termed "*domains*," although in structural biology they may be referred to as "*folds*." It is difficult to establish a direct correlation between the current databases of structural folds and

1st \ 2nd	U	C	A	G	3rd
U	Phe	Ser	Tyr	Cys	U
U	Phe	Ser	Tyr	Cys	C
U	Leu	Ser	Ter	Ter	A
U	Leu	Ser	Ter	Trp	G
C	Leu	Pro	His	Arg	U
C	Leu	Pro	His	Arg	C
C	Leu	Pro	Gln	Arg	A
C	Leu	Pro	Gln	Arg	G
A	Ile	Thr	Asn	Ser	U
A	Ile	Thr	Asn	Ser	C
A	Ile	Thr	Lys	Arg	A
A	Met	Thr	Lys	Arg	G
G	Val	Ala	Asp	Gly	U
G	Val	Ala	Asp	Gly	C
G	Val	Ala	Glu	Gly	A
G	Val	Ala	Glu	Gly	G

Fig. 1 The genetic code and its evolution during and after the origin of life. Ikehara *et al.* [14], Higgs [15], and Trifonov [16] each employed independent methods to identify a three-stage development of the genetic code that took place during the origin of life. Codons colored in red are unanimously identified by all three methods to have been in the original codon set. Codons colored in orange are predicted by at least two methods to have been developed in the second phase of code evolution. While the genetic code has remained nearly stagnant following the origin of life, some codon reassignments are either observed in mitochondrial codes (blue) or in both mitochondrial and nuclear codes (green) [13].

the databases of functional domains [19]. The term "domain" implies an evolutionary relationship between modules with similar amino acid sequences, whereas the term, "fold" does not imply such a relationship between modules of similar structure. In spite of these differences, both concepts reveal an important pattern of modularity present within proteins.

This modular organization is also observed at larger-sized scales. Many proteins are heterogeneous complexes that are composed of multiple protein polymers and are formed through noncovalent surface interactions between two or more peptide chains. Enzymatic protein complexes may function either as a single catalytic unit or as a group of catalytic units, the reaction kinetics of which are cooperatively linked to one another. In fact, structural protein complexes may be so large that they form major infrastructures, such as cytoskeletal elements.

Functional modularity is also an important characteristic of the protein repertoire. Physiological features such as metabolic pathways, signaling cascades, and gene expression pathways act as modules within larger functional networks [20]. The cell may sense and react to an external stimulus through a signaling cascade, thereby activating one or more gene-expression pathways that produce a metabolic response to the stimulus. The metabolic

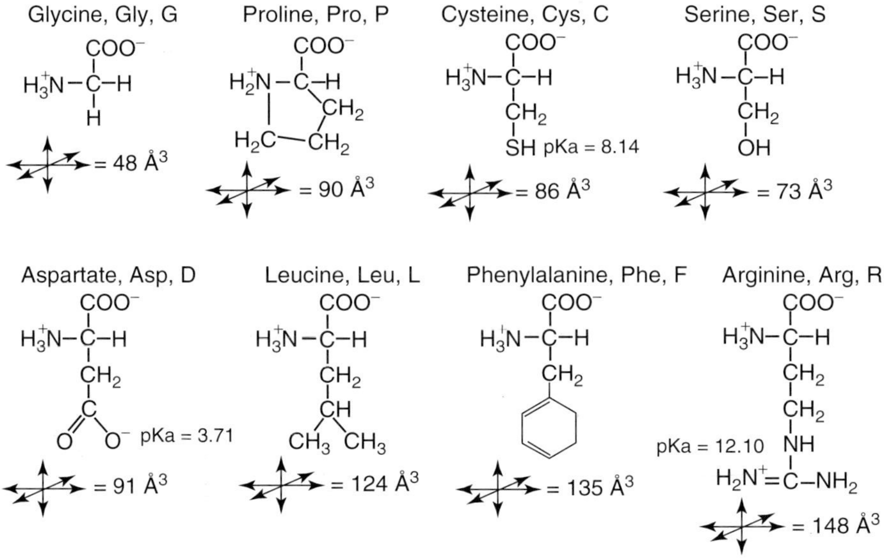

Fig. 2 An abridged list of amino acids highlighting some of the major differences in functional groups. Amino acids may be identified by their full name, a three-letter code, or a one-letter code. The diversity of volumes and chemical moieties found in amino acid side chains gives proteins an incredible range of functional capabilities.

response may, itself, be composed of multiple metabolic pathways, and from this perspective cellular physiology might be viewed as the result of multiple functional modules acting in concert. The relationship between levels of modularity in the protein repertoire is illustrated schematically in Fig. 3.

Taken together, these structural protein complexes and functional modules of multiple proteins account for the majority of cellular phenomena. The sequence, structure, and function of proteins respond to different selection pressures at each level of modular organization. The function of a protein may be evolutionarily plastic while its functional network is highly conserved. The development of a new protein fold architecture may be evolutionarily rare, yet the reorganization of pre-existing folds within a protein is evolutionarily common. In every case, the evolution of the protein repertoire is united, in that any conserved change is the result of evolutionary processes responding to selection pressures.

3 Protein Sequence and Its Evolution

3.1 Evolution of the Genetic Code

The canonical genetic code is composed of 64 possible codons encoding the 20 L-amino acids and several termination signals. Almost all of the 20 amino acids are encoded by more than one codon; for example, leucine, arginine, and serine are encoded by six codons each, while glycine, alanine, proline, valine, and threonine are encoded by four codons each. Only tryptophan and histidine are encoded by one codon. Some codons and their associated tRNAs will confer a greater efficiency or accuracy than others [23],

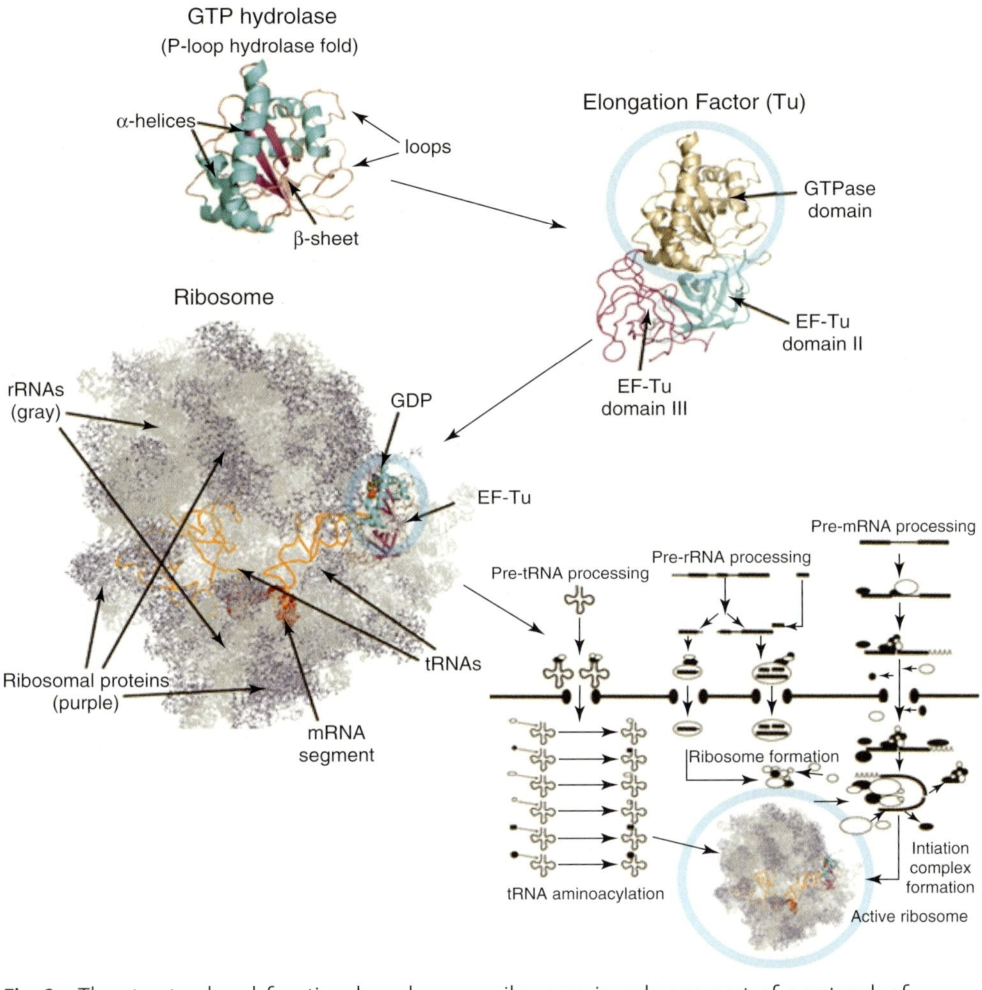

Fig. 3 The structural and functional modularity of the protein repertoire, illustrated by the GTP-hydrolyzing protein, EF-Tu, and its role in translation. The GTPase domain is one of three domains in the EF-Tu protein. Active EF-Tu interacts with the ribosome, itself, a major complex of proteins and RNA molecules, to promote elongation of the growing peptide chain. The EF-Tu-bound ribosome is only one part of a network of enzymes and components constituting translation. Ribosome and EF-Tu structure were produced by Schmeing *et al.* [21] (pdb IDs 2y0u and 2y0v). Pathway data and images are based on pathway maps from the KEGG database (pathway IDs ko03013, map00970) [22].

and such physicochemical and physiological constraints lead to the use of some codons over other synonymous, but suboptimal, codons. These biases vary from one taxonomic group to the next and, as a result, different species will have different nucleotide frequencies within their genomes. This is commonly expressed as the percentages of guanines and cytosines, or "GC%."

Changes in the canonical genetic code, itself, are rare evolutionary events that can be placed on phylogenetic trees [13]. The reassignment of a codon has the capability to alter the amino acid identity of every protein in the organism. Two models have been proposed that permit changes in the genetic code. In one model, such changes involve the reallocation of a termination codon or an amino acid codon that has gone out of use [24]. If such a codon already appears infrequently in protein-coding sequences, its reassignment may have few or no deleterious effects. The second model proposes that an intermediate state of codon ambiguity can exist during codon reassignment [25]. These two models are not mutually exclusive, however, and genetic code evolution may occur by either process.

In some cases, codon reassignments can lead to the use of a nonstandard amino acid. Selenocysteine, for example, is found in some members of all three domains of life [26] and is encoded by a shared UGA codon that can also encode a termination. An additional sequence element in the mRNA is required to direct the addition of a selenocysteine instead of the termination of translation. Another nonstandard amino acid, pyrrolysine, is employed by the methanogenic archaeal family, Methanosarcinaceae [27], and also appears to have been acquired by the bacterium, *Desulfitobacterium hafniense* through horizontal gene transfer [28, 29]. Pyrrolysine is encoded by the codon, UAG, which encodes a termination in the canonical genetic code. While these examples are few in number, they demonstrate that even features of cellular life as fundamental as the genetic code, and the complement of amino acids encoded by it, are subject to evolutionary change.

3.2
Inferring Protein Evolution from Sequence

Compared to other features of the protein repertoire, the evolution of protein sequences is relatively straightforward to analyze. Two steps are required to reconstruct the evolutionary history of a group of protein sequences. First, a multiple sequence alignment must be performed to obtain a measurement of the evolutionary distance between the proteins. Then, a phylogenetic tree can be inferred based upon these distances (this process is illustrated in Fig. 4). When performed over large databases, sequence alignment and subsequent phylogenetic inference may reveal the evolutionary history of protein families, their prevalence in Nature, and the presence and organization of conserved domains within them.

The similarity between a pair of protein sequences can be assessed by sequence alignment methods that insert gaps into each sequence, so that identical or similar amino acids line up correctly between them. Sequence alignments can be solved computationally by *dynamic programming*, an algorithmic strategy that is also used to correct spelling and recognize speech. Examples include the Needleman–Wunsch alignment algorithm, which finds the best global alignment between protein sequences [30], and the Smith–Waterman alignment algorithm which finds similar regions of both proteins and aligns them locally [31]. Unfortunately, both of these algorithms are computationally expensive. Heuristic methods, such as Basic Local Alignment Search Tool (BLAST) are much faster and only marginally less accurate [32]. Multiple sequence alignments may also be produced for a family of related sequences [33], whereby the protein sequences are

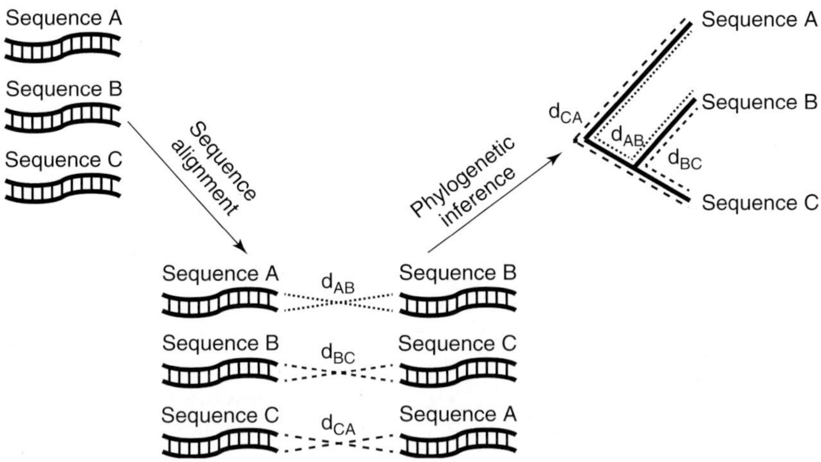

Fig. 4 The process of phylogenetic inference from sequence alignment. First, sequences are compared in a multiple sequence alignment and evolutionary distance scores are calculated between each pair of sequences. Phylogenies are then inferred based on these distance scores. The length of branches between one protein sequence and another is proportional to the distance between them.

aligned to statistical models (most notably hidden Markov models) that represent groups of protein sequences [34]. A comparison between two groups of proteins can then be achieved by aligning the hidden Markov models representing each group [35].

All alignment algorithms require a scoring matrix that defines a score for every possible amino acid substitution and the introduction of a gap. An unlikely match between two dissimilar amino acids may receive a low score, while a match between similar amino acids may receive a high score. The insertion of a gap usually corresponds to a score somewhere in between these two extremes. Although many scoring matrices exist [36], each one is, in itself, a hypothesis about which amino acids are more likely to be substituted for others over evolutionary time. One common scoring matrix is BLOSUM (BLOcks of amino acid SUbstitution Matrix), the scores of which are based on observed amino acid variations in related families compared to what would be expected if substitutions were to occur at random [37]. These profile-based scoring systems are more sensitive to distant relationships between proteins.

Several models exist for the construction of phylogenetic trees from matrices of alignment distance scores. One of the first methods, known as *"neighbor-joining,"* produces a tree by joining lowest-distance taxa into a single taxon and repeating this operation until the tree is resolved [38]. Currently, one of the most popular approaches is to estimate trees by a maximum likelihood method that assesses and optimizes the tree topology and branch length over many iterations [39]. An assessment of the confidence of branches within the tree can be produced by bootstrapping, in which the multiple sequence alignment is resampled and the tree is recalculated hundreds or thousands of times. A high confidence is given to branch points that appear in all or most of the trees from resampled data.

3.3
The Organization of Protein Sequences

The evolution of protein sequences can be described from the perspective of genes or the domains within them. Proteins that share a common evolutionary history may only share some common domains, the others having been lost during the course of evolution. Proteins that are otherwise unrelated may share a single common domain acquired, for example, through homologous recombination. For example, the evolutionarily related nucleotide and nucleoside kinase domains SCOP (Structural Classification of Proteins) ID c.37.1 can be found in a range of GTPase, ATPase, and CoA kinase enzymes, as well as metabolic and regulatory proteins, which may be composed of vastly different combinations of other domains. The evolution of protein domains is, therefore, partly independent of the remainder of the protein sequence.

Based on a survey of complete genomes from 22 species, Zhang [40] found the mean lengths of proteins to be 270 ± 9, 330 ± 5, and 449 ± 25 amino acids, in Archaea, Bacteria, and Eukarya, respectively. A later study by Brocchieri and Karlin [41] surveyed the proteomes of 81 organisms and found the median protein lengths to be 247, 267, and 361 amino acids in Archaea, Bacteria, and Eukarya, respectively. The results of studies of evolutionarily related proteins between Archaea, Bacteria, and Eukarya have suggested that these length differences are mostly localized to the termini of the proteins [42] and between conserved structural units within the protein [43].

The continuity of the protein-coding sequence within a gene can be broken in a number of ways. In Eukarya, and to a much lesser extent in Archaea, the protein-coding regions of the gene are interrupted by intron sequences, which are both present in the genome and transcribed into pre-mRNA. Introns are spliced out of the pre-mRNA to produce the mature mRNA, which is then translated by the ribosome. In some cases, the same protein-coding sequence may be processed differently to produce different products. This phenomenon of differential intron splicing was first discovered in adenovirus [44, 45], but has since been observed in eukaryotic organisms [46]. The ability of gene sequences to be discontinuous is further demonstrated by the phenomenon of *trans*-splicing, in which introns from different coding regions are spliced together into a single mature mRNA. Two distinct forms of *trans*-splicing exist that both bear resemblance to normal splicing pathways [47].

An even more striking example of gene discontinuity is found in ciliates, a phylum of single-celled Eukarya. In the ciliate order, Stichotrichida, the germline genome is scrambled such that gene segments are found fragmented, inverted, and translocated throughout the genome. The genome is rearranged during development to form the active somatic genome [48, 49]. Both genomes are retained during the lifetime of the organism, but only the unscrambled genome is active. Both, *trans*-splicing and genome rearrangement decidedly demonstrate that a protein-coding gene is not always a single continuous sequence encoded on a single DNA molecule.

Reflecting the functional modularity of their products, some protein-coding genes are organized into functional units within the genome. *Operons* are a genetic architecture common in Bacteria and Archaea and present in Eukarya. Genes within operons are cotranscribed as a single molecule

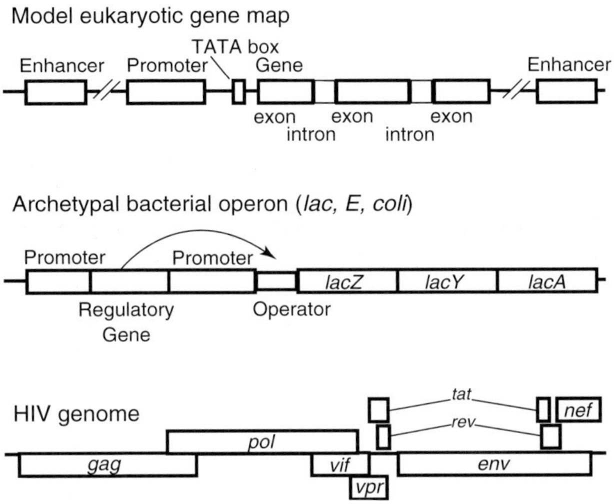

Fig. 5 Several examples of gene organization. Eukaryotic genes may be partitioned into exons, with intron sequences interrupting the coding sequence. A promoter attenuates the expression of the gene locally, while enhancers attenuate the expression of the gene from distant regions of the genome that can act proximally in 3-D space. Operons found in Bacteria and Archaea, with some examples in Eukarya, are groups of genes that are cotranscribed as a single mRNA. A regulatory region is usually located upstream of the genes. In some viruses, gene segments are observed to overlap or to be embedded within other genes, as demonstrated here by the HIV genome.

of mRNA, and this organization ensures an even stoichiometry of gene products, no matter what the level of gene expression. The operon usually contains a region of promoters and operators that modulate translation of multiple genes downstream. The gene products of an operon may impart different functions within a physiological pathway, or they may encode subunits of the same multiprotein complex. Some examples of gene organization within the DNA genome are illustrated in Fig. 5.

3.4 Genetic Mechanisms of Protein Evolution

For the most part, evolution acts on the functions of proteins. An anthropomorphic perspective would be that natural selection is primarily concerned with the molecular function of the protein, as it pertains to the whole fitness of the organism. Evolution must also act on the structural features that impart a protein's function. Enzymes, for example, require an active site structure that can seize the substrate and discharge the product. In cytoskeletal proteins, membrane channels, viral coats and many other examples, the structure of the protein complex is, itself, the primary function. Generally, the sequences of protein-coding genes are only under genome-wide constraints such as tRNA minimization and GC%. However, while the primary targets of natural selection are protein structure and function, the actual changes must take place on the protein-coding sequence.

Small-scale changes within the protein-coding sequence occur as point mutations that substitute one nucleotide for another, as well as insertions or deletions of one or more nucleotides. Point mutations may change the protein sequence in a number of ways. If the new codon created by a point mutation encodes the same or a similar amino acid, then the structure and function of the protein should not change at all. However, if the new codon encodes an amino acid of drastically different size or charge, then the structure and function of the protein may be lost or dramatically altered. If the new codon encodes a termination of translation, then the protein product will be truncated and may lose function altogether.

Insertions and deletions of a small number of nucleotides may also alter the structure and function of the protein product by varying degrees. Any insertion or deletion not in multiples of three will change the reading frame of the sequence so that all of the following codons are altered. A small insertion or deletion that changes the reading frame will drastically change the structure and function of the protein product, unless the mutation occurs sufficiently late in the coding sequence and only a small number of subsequent codons are altered. Any non-neutral mutation will most often confer a loss of function and only rarely will it produce a gain of function that is beneficial to the organism.

3.5
Genomic Mechanisms of Protein Evolution

Within the genome, proteins or parts of proteins may be duplicated through homologous recombination or transposon excision and reintegration. Gene duplication is a major source of genome growth, and accounts for an estimated 50% of new genes in prokaryotes and 90% of new genes in eukaryotes [50]. Duplicated proteins functionally diverge from the original protein, but often play a role in the same metabolic pathway or multiprotein complex [50, 51].

In a similar process with a very different outcome, homologous recombination and transposon translocation may add segments of one gene into another, or remove segments from a larger gene [52]. Rather than duplicating a gene, this process generates alternate domain organizations within existing genes, with selection pressure seeming to favor the promiscuity of some domains over others [53]. Whole-genome duplications are perhaps the most drastic evolutionary mechanism affecting genome content, and are thought to be ephemeral evolutionary events that are often lost quickly due to strong negative selection [54, 55]. However, in some cases they generate polyploid genomes in which the same gene may have multiple divergent alleles within a single individual.

For the most part, an individual organism acquires the contents of its genome through inheritance from a parent organism or, in the case of sexually dimorphic organisms, from two parent genomes. New genes may also be acquired through *horizontal gene transfer*, wherein extrachromosomal DNA is passed from one organism to another. Horizontal gene transfer can occur through the exchange of a plasmid, temperate virus, or transposable element [56, 57]. In an extreme case, endosymbionts – such as the chloroplast and the mitochondrion – bring with them an entire bacterial genome such that, following the assimilation of an endosymbiont, the nuclear genome begins to accrete genes from the original bacterial genome [58]. The process of nuclear accretion is complete in some hydrogenosomes (an organelle of some protists that is related to the mitochondrion) [58].

The genome of an organism can be altered through processes that may be as slight as a point mutation or as significant as a whole-genome duplication. It is important to note that, while evolution acts by way of these changes to the genome, the selection pressures that drive that action can be viewed from a variety of perspectives. Charles Darwin famously described evolution as selection for or against individual organisms, while his contemporary, Alfred Russell Wallace, emphasized selection on traits within populations. More recently, the Gaia hypothesis proposed by James Lovelock described a common evolution of the entire biosphere and its geological context, while Richard Dawkins' "selfish gene" concept characterized evolution based on the self-promotion of single genes. Given the complexity of the biosphere and its evolution, each of these perspectives is in one way or another correct; however, none of them is complete.

4 Protein Structure and Its Evolution

4.1 Levels of Protein Structure

Protein structure is classified at four distinct levels of increasing size and complexity. The sequence of amino acids that constitutes the protein polymer strand is called the *primary structure*, and this can easily be identified from the nucleotide sequence by employing the genetic code. The primary structure of a protein polymer folds into discrete and regular units of 3-D structure that are mostly facilitated by hydrogen bonds; these structural elements are referred to as the *secondary structure*. The two most common types of secondary structure are alpha-helices [59] and beta-sheets [60], both of which are formed through hydrogen bonding between N-H and C=O groups along the peptide backbone. An alpha-helix is a spiraling helix that turns once every four amino acids, with hydrogen bonding occurring vertically along the helix. A beta-sheet is composed of multiple strands that run either parallel or antiparallel to one another, with the hydrogen-bonding groups facing perpendicular to the beta-strand. Other forms of secondary structure, including several variations of helices, are found less regularly in protein structures. Regions of the protein polymer that do not form a secondary structure element are called *loops*.

The protein polymer's *tertiary structure* is a combination of multiple secondary structures and loops, formed into stable structural modules. The individual domains often fold into their tertiary structures independently, and may fold in the same way even if they are artificially separated from the rest of the protein. This is not always the case, however, as some tertiary structure modules may be structurally influenced by neighboring modules [61]. Cytoplasmic protein folds often contain a core of mostly hydrophobic amino acids that is rigid enough to maintain structure but flexible enough to perform their function [62]. The outer layer of the fold is usually composed of hydrophilic or charged amino acids. If the fold is not cytoplasmic, but rather traverses a membrane, then the outer layer will typically be composed of hydrophobic amino acids that interact with the hydrophobic membrane [63, 64]. Functional sites are often formed within inward extensions of the outer layer that permit solvent and ligands to occupy the resulting concavity. The tertiary structure is also held together through a range of chemical bonds of varying strengths; hydrogen bonds may occur between pairs of amino acid side chains, as well as between the side chains and the peptide backbone. *Ionic bonds* between charged atoms may form within the hydrophobic core of a protein fold, while *covalent bonds* can form between the sulfur atoms of cysteine side chains.

Transient and constitutive multiprotein complexes combine multiple proteins into a single structural unit. The structural combination of more than one protein polymer is referred to as the *quaternary structure*. Although the nature of physical interactions between protein polymers is not completely understood, alanine scanning experiments, in which the natural amino acids are serially substituted for alanines, have been used to identify "hot spots" which are essential to protein–protein interactions but comprise only a fraction of the interface [65]. These hot spots are enriched for certain amino acids, especially tryptophan, arginine, and lysine [66], and are somewhat enriched for loops over alpha or beta secondary structures [65]. Unfortunately, this scanning method may be prone to false-positive results if the alanine substitution either disrupts the local structure or completely prevents the native fold from forming; in this case the protein–protein interaction would be disturbed as a consequence.

4.2
Protein Structure *In Vivo*

Protein folding can also be thought of as the consequence of potential energy minimization. Protein structures exhibit a range of conformational stabilities determined by the energy landscape of the protein and its environment. As functional molecules, proteins must be rigid enough to hold structure yet remain flexible enough to perform their function. Proteins are often described using the analogy of a machine where, at every scale, moving parts are essential to ensure a correct function.

The flexibility and stability of proteins are determined mostly by the strength of the hydrophobic core, and in part by hydrogen-bond interactions [67, 68]. Protein flexibility and stability can be determined experimentally by using the same methods employed to resolve their structures. Debye–Waller factors are a measure of electron density spread in X-ray diffraction (XRD)-based structures, and can be used to estimate motion throughout the protein. However, these electron densities may also be explained as artifacts of the unnatural crystallization of proteins required by X-ray crystallographic methods [69]. Structure determination using nuclear magnetic resonance (NMR) provides lower-resolution results than does X-ray crystallography, but is able to capture the protein structure in solution at multiple time points. Today, emerging NMR techniques allow research groups to probe protein motion at the milli- to pico-second timescales [70, 71].

The evolution of thermophile proteins provides another window into the nature of protein stability. Proteins of hyperthermophilic organisms are exposed to high temperatures and, usually, also to high pressures – both of which may dramatically destabilize a protein's structure. The structures of hyperthermophile proteins are generally superimposable with their nonthermophilic counterparts, and their sequences are 40–85% similar [67]. Moreover, the core amino acids appear to exhibit a high degree of conservation, while solvent-accessible areas and loops show much more variation. This observation indicates that the free energy change which drives the adoption of each fold is principally produced by core interactions, while the outer layer has secondary structural features that adapt to the particular environment.

Large-scale conformational changes in protein structure are often a natural part of protein function and functional regulation. Rather than refolding the protein, conformational changes generally involve a small number of hinge regions that move secondary or tertiary structural elements to new locations with respect to one another [72]. In some cases, these conformational changes are required to bind a substrate or some other binding target and, in doing so, the bound conformation becomes the lowest energy conformation [73]. In allosteric regulation, dynamic changes between different structural conformations are produced by the binding of an effector ligand [74] in order to alter the activity of a nonadjacent site. Thus, both function and functional repression can be achieved through the alteration of protein structure.

4.3 Evolution of Protein Structure

Protein structure exhibits a complex evolution that is not fully understood. The convergence upon similar local structural motifs through different general fold architectures [75] suggests that local structure is evolutionarily plastic, while global fold architecture is generally conserved. The triosephosphate isomerase (TIM) barrel fold is an extreme example in which a single fold architecture is observed to catalyze reactions classified under five of the six main enzyme categories in the Enzyme Commission classification system [76]. It is not known whether members of the TIM barrel fold, and other functionally eclectic folds like it, have evolved by divergence from a common ancestor or from the convergence of multiple unrelated lineages upon a common structure.

Databases that organize and classify protein structures, such as the SCOP database [77], have shown that many fold architectures can be created by multiple, seemingly unrelated, sequences. The relationship of the number of sequences to the number of folds that they encode in each SCOP release is presented in Fig. 6. In this case, the protein structure seems to be resilient to drastic sequence changes over the course of evolution. In contrast, synthetic biology has shown that fold transformation can occur given only a few amino acid substitutions [78, 79].

One explanation for this observed size disparity between sequence and structure spaces is that the selection of two different protein families towards a common function results in their convergence on a common fold architecture. A recent survey of nonhomologous proteins with common function identified many cases in which two different folds converged upon similar active site geometries [75]. Similarly, a study of structural mimicry between pathogen and host proteins has shown that the structural similarity between the original protein and its mimic is either local or superficial [80]. In other words, structural mimics do not actually converge on the same global fold architecture. Thus, functional and structural convergence is not observed to affect the fold architecture, but rather local motif architectures or surface geometries.

The trend may also be explained by the evolution of a common lineage in which certain groups have diverged apart to such a degree that they have lost any noticeable sequence identity. In this

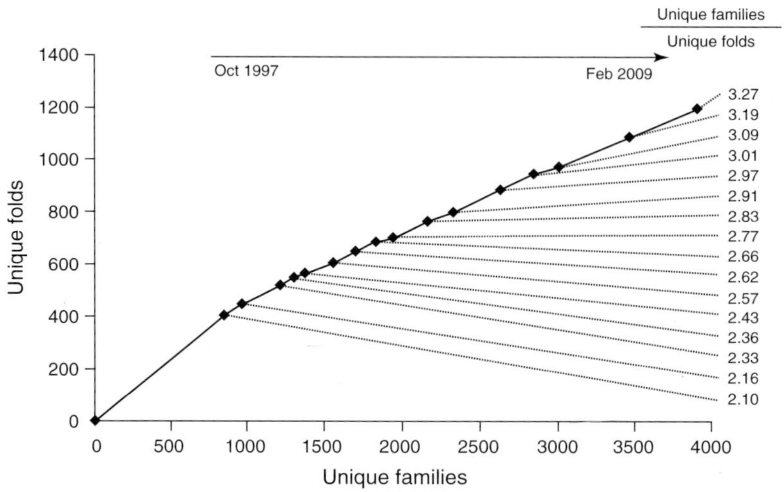

Fig. 6 A plot of the number of fold structures and sequence families within each release of the SCOP database [77]. In the SCOP system, proteins of the same fold exhibit a common arrangement of secondary structure elements to create a certain 3-D architecture. Proteins of the same SCOP family share a readily detectable sequence similarity with one another. With each release, the ratio of sequence space to fold space increases. This trend implies that, in Nature, a larger number of disparate sequences encodes a smaller number of folds.

model, the fold architectures of homologous proteins remain conserved through the course of evolution, while their sequences and functions diverge. More sensitive sequence-clustering techniques have shown that many sequence families are distantly related to one another, even though their homology cannot be detected through normal sequence comparison methods [81]. It is possible that other homologous groups are too divergent to be detected by even this clustering method.

The evolution of new folds is thought to be rare. Synthetic protein design has shown, however, that it is possible to artificially create a stable fold that is not found in Nature [82]. Perhaps arriving at a novel fold by natural selection is an unlikely event [83], and thus fold space is undersampled by evolution; instead, evolution may more easily acquire novel functions through the repurposing of pre-existing folds. However, while this model is intriguing it is so far, unproven. Because members of the same fold may not appear homologous by sequence similarity, the SCOP system makes no evolutionary claims regarding members of a common fold. It could be the case that some fold structures represent a single common origin and others – perhaps the lower-complexity folds – do not. Either way, fold architectures appear to be the most evolutionarily conserved feature of the protein repertoire.

5
Protein Function and Its Evolution

5.1
Types of Protein Function

Proteins are the primary functional macromolecules of the cell. Typically, nonproteinaceous macromolecules such as DNA, RNA, sugars, and lipids, are synthesized or scavenged by protein enzymes and transporters. Likewise, the metabolism of every cell results from catalysis by protein enzymes, controlled through networks of sensory and regulatory proteins. In addition, the structure of cells is supported by protein cytoskeletal elements that also coordinate the movement of the cell's constituents from one region of the cell to another. Signaling molecules used for intercellular communication are produced by protein enzymes in one cell and sensed by ligand-binding proteins of a neighboring cell, setting off a signaling cascade that is orchestrated by proteins.

This range of functional capabilities arises from the variety of structures that proteins can take, and the molecules that they can bind. The secondary, tertiary, and quaternary structures, as well as the dynamisms between these structural elements, may be adjusted over the course of evolution to complement specific molecules, to perform certain functions, and to respond to specific regulatory stimuli. The chemical diversity provided by the 20 canonical amino acids allows proteins to bind molecules with a range of functional groups.

Protein binding to small-molecule ligands plays a role in many important cellular functions; small-molecule transport, either across membranes or through the cytoplasm, is often facilitated by ligand-binding proteins, some of which are also involved in chemosensation. The resulting sensory information is often employed in the regulation of metabolic pathways related to the ligands, or in chemotactic responses towards or away from the ligands. Intercellular communication is achieved by a special subset of chemosensory proteins that bind to signaling molecules produced by neighboring cells.

Enzymatic catalysis is a special example of binding in which one or more bound molecules are altered in their composition or isomeric state (Fig. 7). An enzyme lowers the activation energy of the chemical reaction that it catalyzes, by coordinating components of the highest energy reaction intermediate with amino acid side chains in the enzyme active site. In many cases, the active site will also bind a *cofactor* that is involved in the reaction. Such cofactors may be inorganic, as in the case of metalloenzymes [84], where it is the metal ions that often perform the catalysis while the protein serves as a structural scaffold supporting any interactions between the metal ion and the substrate. Many enzymes also use organic cofactors that are produced metabolically and are derived largely from amino acids and nucleotides; these cofactors may exhibit a range of chemical moieties and functions beyond their progenitor molecules [85].

Proteins may also function by way of their interactions with other proteins or nucleic acids. Many structural elements of the cell, such as those of the cytoskeleton, are in fact very large, dynamic multiprotein complexes; indeed, the binding of a such a protein to an enzyme may serve to promote or inhibit that enzyme's activity [91]. Nucleic acid binding is exhibited by many of the proteins involved in DNA replication and transcription, as well as in protein translation. Many of these proteins may also have regulatory roles, either promoting or inhibiting one of these essential processes. Nucleic acid binding may also serve a structural role, as demonstrated by histones that bind DNA to form chromatin, and ribosomal proteins

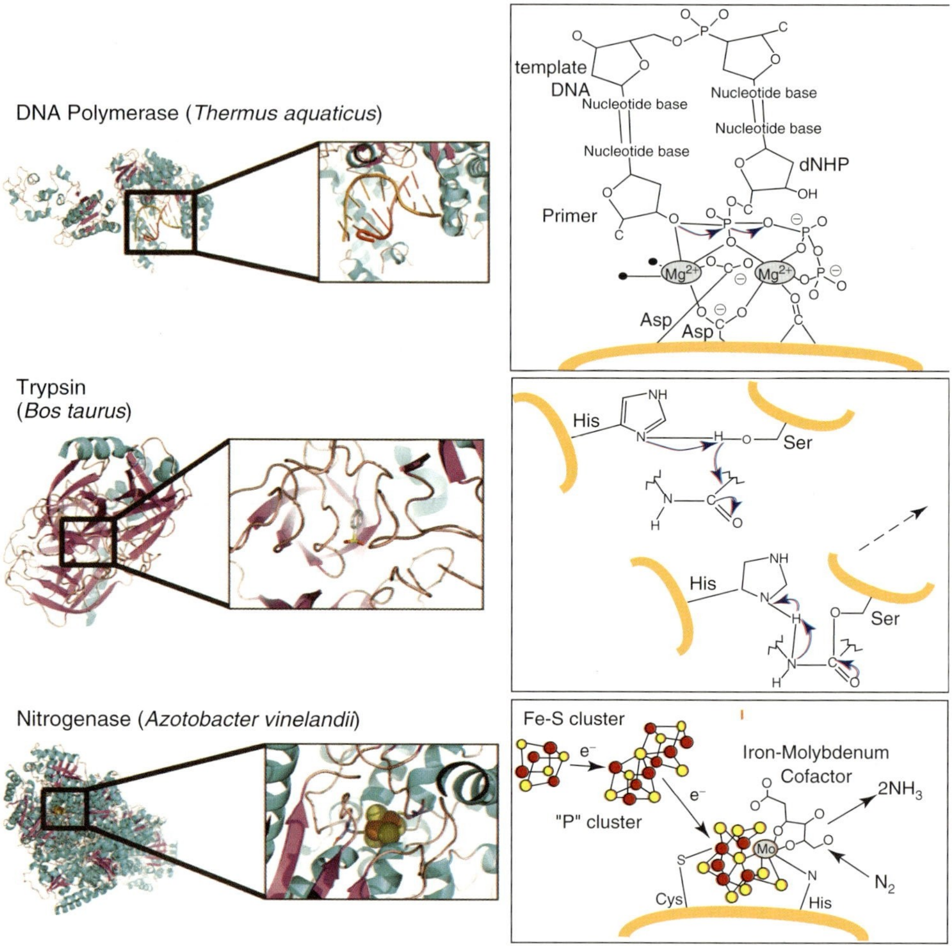

Fig. 7 Examples of three different enzymes, their active sites, and their catalytic mechanisms. DNA polymerase is a phosphotransferase that polymerizes deoxyribonucleotides along a DNA template [86, 87]. Trypsin is a peptide hydrolase that uses a serine side chain to break peptide bonds and a histidine side chain to stabilize the intermediate [88]. Nitrogenase is an oxidoreductase that converts molecular nitrogen into ammonia through the redox chemistry of a metal–sulfur cluster [89, 90]. In all three cases, kinetically unfavored reactions are catalyzed by stabilizing reaction intermediates.

that bind rRNA to enhance ribosome structure and function.

The function of a multidomain protein is a combination of the subfunctions of its domains. For example, both classes of aminoacyl tRNA-synthetases (enzymes that attach the correct amino acid to tRNAs) generally have a catalytic domain, an anticodon-binding domain, and often other domains related to RNA binding [92]. Many multiprotein complexes exhibit a similar functional modularity. For example, hemoglobin is composed of four subunits (two copies each of two homologous peptides); the binding of one oxygen to one of the subunits in the complex then alters the subsequent binding kinetics for the remaining subunits [93]. This type of association is said to be "cooperative," and shows that modularity in protein sequence and structure may also have very clear functional consequences.

5.2 Functional Networks in Physiology

The modern cell is dependent not only on the functions of proteins, but also on the precise timing and location of these functions. Physiological networks are the emergent property of this precise execution of protein functions in the cell. The ability to identify networks with high-throughput experimental methods and to analyze them computationally has led to a systems approach to molecular biology becoming possible [94]. Thus far, three distinct categories of physiological networks have been investigated in detail, namely metabolic networks, protein–protein interaction networks, and regulatory networks:

- *Metabolic networks* are composed of smaller metabolic pathways. In the cell, a metabolic pathway is generated by a series of enzymes that convert one metabolite into another in a progression of stepwise intermediate reactions. As data, metabolic pathways are usually represented graphically as a series of metabolites (nodes) connected by the enzymes that consume or produce them (edges) [94, 95]. While it is difficult, and also somewhat arbitrary, to define where one pathway ends and another begins [95], the pathway model of metabolism remains a useful tool for organizing the modularity of metabolic networks.
- *Protein–protein interactions* form a second type of functional network. These interactions include proteins in the same multiprotein complex, as well as transient regulatory or signaling interactions [96]. The study of protein–protein interaction networks is attractive because, unlike metabolic reactions, such interactions can be deduced by using high-throughput methods [94, 96] such as two-hybrid systems [97] and protein microarray analysis [98]. Specific function, like that required to produce metabolic networks, may be deduced from protein–protein interactions algorithmically [99].
- *Regulatory networks* involve the coordination of gene expression in response to intracellular and extracellular signals. A single regulatory protein may either induce or repress the expression of multiple genes that are functionally related [100]. These regulatory proteins often also induce other regulatory proteins that control the expression of a subset of genes related in function to the others [101]. In this respect, regulation also has a modular organization. Regulatory networks may be inferred from high-throughput methods that identify differential gene expression under varying conditions or stages in the life cycle. These methods most notably include DNA microarray analysis [102] and high-coverage sequencing of mRNA or cDNA [103].

Of course, these three categories of networks do not exist independently from one another. For example, metabolic networks come into existence by way of expression networks that control the production of the requisite enzymes. If only half of the enzymes in a metabolic pathway are expressed at a given time, then the products of the partial pathway will clearly be different from that of the whole. A full understanding of cellular and organismal physiology requires the integration of these network data [104]; hence, as data for each of these network categories become more abundant and encompass more species, their integration into a single framework will become increasingly important.

5.3 Evolution of Protein Function

Protein sequence duplication is a major driver of protein function evolution, with the sequences of newly duplicated genes being observed to evolve at a rapid pace [105]. While most of these duplicates become inactive pseudogenes, some may be expressed and remain under selection because they either amplify the dosage of the gene product or have some altered function [106]. The same mechanisms of duplication may produce novel functions by introducing new domains into existing proteins. The process of duplication has evident effects on functional networks; for example, a new gene function will alter its preexisting network, while a redundant gene function will affect the expression of genes within its network.

As noted above, functionally analogous enzymes may often have both unrelated sequence and unrelated structure [75]. In a recent study conducted by Almonacid *et al.* [107], the reaction mechanisms for 95 pairs of functionally analogous enzymes that have no detectable sequence similarity were compared. That mechanistic similarity was observed in only 44% of the pairs indicated that functional convergence is a regular occurrence in evolution, and does not require the unrelated proteins to share a common structure or mechanism.

The evolutionary conservation of molecular networks has only recently been open to investigation, due to the relatively recent growth of experimentally verified network data and the relatively low number of taxa for which complete networks have been determined experimentally [94]. Computational simulations of network evolution have shown that, while the outputs of a metabolic network (metabolites, biosynthetic products, energy, etc.) can be conserved during evolution, the components of the network may be very evolutionarily plastic [108]. The ancestry of common networks across diverse taxa may be better understood in the near future through the development of tools for predicted metabolic networks from annotated genomes [109] and the creation of network phylogenies [110].

During the course of evolution, functional networks may be rewired by a number of phenomena, including gene loss, gene duplication, horizontal gene transfer, changes in environmental inputs, and the exaptation of gene products from one network to another [94]. The convergent evolution of analogous networks towards the same phenotypic output is observed in essential metabolic requirements such as carbon fixation, for which six distinct metabolic pathways are known [111]. In the case of carbon fixation, the pathways themselves can be remarkably different. While this is just one example, it demonstrates that evolutionary convergence towards a similar or identical output may be much more common than convergence upon the same pathway.

6 Protein Evolution in Human Hands

Proteins are, in essence, nanoscale machines that are naturally produced by evolutionary processes. Protein design methods that simulate these processes have major potential as nanoscale engineering platforms, and recent technologies have indeed put this concept to work. Today, cutting-edge laboratory techniques can mimic evolution by combining methods of mutation, selection, and replication [112], while computational approaches may imitate evolution by evaluating Monte Carlo simulations with sophisticated scoring functions based on biological constraints [113].

6.1 *In Vitro* Protein Evolution

The first step in the development of *in vitro*, directed evolution was the invention of the phage display sampling method. Phage display uses complete viruses to express gene-bearing plasmids and to select those that bind a particular substrate. One of the first uses of phage display included the sampling of a library of randomly produced antibodies, during which one of the selected proteins was shown to be capable of binding a predetermined target with a higher affinity than any of the natural antibodies [114]. Whilst this type of experiment does not reproduce the complete evolutionary process, it does provide the selection component of the evolutionary process.

The generation of large libraries of mutated protein sequences is another essential component of *in vitro* evolution. Originally, site-directed mutagenesis was developed to investigate the mutations of individual amino acid residues [115], and in one heroic study a total of 4044 such mutations was generated and examined [116]. In this case, some enhanced functional capabilities were achieved by a few mutants, while a modest majority exhibited a significant loss of function. Homologous recombination is a greater generator of genetic diversity than point mutations, both in Nature and *in vitro* [50]. Automated homologous recombination and the selection of its products were shown to produce significantly improved function compared to mutant proteins generated through point mutations [117]. When only a refinement of enzyme function is required, the polymerase chain reaction (PCR) using an error-prone DNA polymerase is also an effective method for generating mutations [118].

In order to truly simulate evolution, the selection of mutants must be iterative so that the selected proteins are replicated, mutated, and run through the selection regime again. This can be achieved by tethering mRNA to the protein that it encodes, so that the encoding gene of the selected proteins may be saved for the next round of mutation and selection [119]. Typically, this method is performed on immobilized substrates and is only able to select ligand-binding functions. An adaptation of this method is capable of selecting ligase enzymes by immobilizing one substrate and attaching the reverse transcription primer to the other, thus ensuring that only those enzymes which perform the ligation are copied [112]. The current trajectory of *in vitro* evolution methodology and its expansion to more protein functions is sure to have a profound effect on medicine, nanotechnology, and general industry in the near future.

6.2 Computational Protein Evolution

The computational design of proteins also simulates evolutionary processes. The computational protein design process is generally the reverse of protein structure and function prediction [113], both of which are based on evolutionary constraints. In comparative protein structure prediction, templates from homologs are used to model a structure from a sequence, assuming the evolutionary conservation of a common structure between the sequence and its structural template [120]. Scoring functions are then used to measure the model's correctness [121, 122].

Computational approaches to protein design begin with a target structure or function, and employ these methods to search for sequences that are likely to achieve such design. In the past, because sequence space is too large to sample fully, design experiments have targeted either a single fold [82] or the placement of specific functional groups [123]. The creation of a predesigned fold architecture not observed in Nature was achieved by Kuhlman *et al.* [82] through Monte Carlo simulations and fold prediction, which represent a computational mimicry of the mutation/selection process. A similar design method was used to create novel protein functions with some success, selecting structures that formed around a target substrate. Enzymes produced by this method have only achieved a 10^6-fold increase in reaction rate, whereas natural enzymes typically increase reaction rates up to 10^{14}-fold [124].

An alternative method of computational protein design begins with an *in vitro* step [125], in which phage display is first used to identify the affinity of synthetic peptides for specific ligands. A scoring function is then trained on this dataset to separate the strongly and weakly binding peptides. Monte Carlo simulations using this scoring function in the selection process will generate peptides that display a higher binding affinity to the ligand than any peptides obtained from the initial phage display experiments.

At present, although computational methods are less effective than *in vitro*-directed evolution, they will always have the benefit of speed. Computational design methods may be most useful in tandem with *in vitro* evolution to generate starting sequences and to refine the final products. Alternatively, a breakthrough in methodology will lead to computational design that is solely capable of the task. Either way, evolution has shown itself to be a prodigious inventor, and protein design platforms capable of mimicking the evolutionary process should have enormous potential for a variety of medical, industrial, and scientific applications.

7
Lessons from the Evolution of the Protein Repertoire

The protein repertoire spans the entire physiology of the cell, from genetic sequence to metabolic function. Moreover, its evolution repeatedly demonstrates that a single selection pressure may manifest itself differently in each facet of the protein repertoire, and at different scales of size. Yet, even within this complexity, a pattern of modularity is beginning to emerge in gene sequence, protein structure, and molecular function. In addition, a similar pattern of modularity is found at multiple size scales, within a single protein, as a combination of proteins, and within major physiological networks.

Under scrutiny, the common conceptual framework of a protein repertoire, as defined by individual genes and proteins, does not always work. Genes that encode proteins are often discontinuous, overlapping, or are formed from sequences from multiple DNA molecules. Discrete units of structure and function may be defined not only by individual proteins, but also by domains within proteins or complexes of multiple proteins. Instead of individual proteins, the organizational principle of the protein repertoire appears to be a modularity that manifests itself in a multitude of phenomena. Not surprisingly, the evolution of the protein repertoire is not determined by a conceptual framework, but rather by the pressure to evolve populations to satisfy selection by any means.

References

1 Brack, A. (2007) From interstellar amino acids to prebiotic catalytic peptides: a review. *Chem. Biodivers.*, **4**, 665–679.
2 Zaia, D.A.M., Zaia, C.T.B.V., de Santana, H. (2008) Which amino acids should be used in prebiotic chemistry studies? *Orig. Life Evol. Biosph.*, **38**, 469–488.
3 Miller, S.L. (1953) Production of amino acids under possible primitive earth conditions. *Science*, **117**, 528.
4 Johnson, A.P., Cleaves, H.J., Dworkin, J.P., Glavin, D.P., Lazcano, A., Bada, J.L. (2008) The Miller volcanic spark discharge experiment. *Science*, **322**, 404.
5 Parker, E.T., Cleaves, H.J., Dworkin, J.P., Glavin, D.P., Callahan, M., Aubrey, A., Lazcano, A., Bada, J.L. (2011) Primordial synthesis of amines and amino acids in a 1958 Miller H2S-rich spark discharge experiment. *Proc. Natl Acad. Sci. USA*, **108**, 5526–5531.
6 Cronin, J.R., Moore, C.B. (1971) Amino Acid analyses of the Murchison, Murray, and Allende carbonaceous chondrites. *Science*, **172**, 1327–1329.
7 Huber, C., Eisenreich, W., Hecht, S., Wachterhauser, G. (2003) A possible primordial peptide cycle. *Science*, **301**, 938–940.
8 Gilbert, W. (1986) Origin of life: the RNA world. *Nature*, **319**, 618.
9 Freeland, S.J., Knight, R.D., Landweber, L.F. (1999) Molecular evolution: do proteins predate DNA? *Science*, **286**, 690–692.
10 Goldman, A.D., Samudrala, R., Baross, J.A. (2010) The evolution and functional repertoire of translation proteins following the origin of life. *BMC Biol. Dir.*, **5**, 15.
11 Fox, G.E. (2010) Origin and evolution of the ribosome. *Cold Spring Harbor Perspect. Biol.*, **2**, a003483.
12 Crick, F.H.C. (1968) The origin of the genetic code. *J. Mol. Biol.*, **38**, 367–379.
13 Knight, R.D., Freeland, S.J., Landweber, L.F. (2001) Rewiring the keyboard: evolvability of the genetic code. *Nat. Rev. Genet.*, **2**, 49–58.
14 Ikehara, K., Omori, Y., Arai, R., Hirose, A. (2002) A novel theory on the origin of the genetic code: a GNC-SNS hypothesis. *J. Mol. Evol.*, **54**, 530–538.
15 Higgs, P.G. (2009) A four-column theory for the origin of the genetic code: tracing the evolutionary pathways that gave rise to an optimized code. *BMC Biol. Direct.*, **4**, 16.
16 Trifonov, E.N. (2009) The origin of the genetic code and of the earliest oligopeptides. *Res. Microbiol.*, **160**, 481–486.
17 Sonnhammer, E.L.L., Kahn, D. (1994) Modular arrangement of proteins as inferred from analysis of homology. *Protein Sci.*, **3**, 482–492.
18 Riley, M., Labedan, B. (1997) Protein evolution viewed through *Escherichia coli* protein sequences: introducing the notion of a structural segment of homology, the module. *J. Mol. Biol.*, **268**, 857–868.
19 Majumdar, I., Kinch, L.N., Grishin, N.V. (2009) A database of domain definitions for proteins with complex interdomain geometry. *PLoS ONE*, **4**, e5084.
20 Pereira-Leal, J.B., Levy, E.D., Teichmann, S.A. (2005) The origins and evolution of functional modules: lessons from protein complexes. *Philos. Trans. R. Soc. Lond. B: Biol. Sci.*, **361**, 507–517.
21 Schmeing, T.M., Voorhees, R.M., Kelley, A.C., Ramakrishnan, V. (2011) How mutations distant from the anticodon affect the fidelity of decoding. *Nat. Struct. Mol. Biol.*, **18**, 432–436.
22 Kanehisa, M., Goto, S. (2000) KEGG: Kyoto encyclopedia of genes and genomes. *Nucleic Acids Res.*, **28**, 27–30.
23 Hershberg, R., Petrov, D.A. (2008) Selection on codon bias. *Annu. Rev. Genet.*, **42**, 287–299.
24 Osawa, S., Jukes, T.H. (1989) Codon reassignment (Codon Capture) in evolution. *J. Mol. Evol.*, **28**, 271–278.

25 Schultz, D.W., Yarus, M. (1996) Exposition of the 'ambiguous intermediate' hypothesis, which suggests that the genetic code changes through a state in which some codons have more than one meaning. *J. Mol. Evol.*, **42**, 597–601.

26 Stadtman, T.C. (1996) Selenocysteine. *Annu. Rev. Biochem.*, **65**, 83–100.

27 Hao, B., Gong, W., Ferguson, T.K., James, C.M., Krzycki, J.A., Chan, M.K. (2002) A new UAG-encoded residue in the structure of a methanogen methyltransferase. *Science*, **296**, 1462–1466.

28 Srinivasan, G., James, C.M., Krzycki, J.A. (2002) Pyrrolysine encoded by UAG in Archaea: charging of a UAG-decoding specialized tRNA. *Science*, **296**, 1459–1462.

29 Herring, S., Ambrogelly, A., Carla, R., Polycarpo, C.R., Söll, D. (2007) Recognition of pyrrolysine tRNA by the *Desulfitobacterium hafniense* pyrrolysyl-tRNA synthetase. *Nucleic Acids Res.*, **35**, 1270–1278.

30 Needleman, S.B., Wunsch, C.D. (1970) A general method applicable to the search for similarities in the amino acid sequence of two proteins. *J. Mol. Biol.*, **48**, 443–453.

31 Smith, T.F., Waterman, M.S. (1981) Identification of common molecular subsequences. *J. Mol. Biol.*, **147**, 195–197.

32 Altschul, S.F., Madden, T.L., Schffer, A.A., Zhang, J., Zhang, Z., Miller, W., Lipman, D.J. (1997) Gapped BLAST and PSI-BLAST: a new generation of protein database search programs. *Nucleic Acids Res.*, **25**, 3389–3402.

33 Carrillo, H., Lipman, D.J. (1988) The multiple sequence alignment problem in biology. *J. SIAM Appl. Math.*, **48**, 1073–1082.

34 Eddy, S.R. (1998) Profile hidden Markov models. *Bioinformatics*, **14**, 755–763.

35 Soding, J. (2005) Protein homology detection by HMM-HMM comparison. *Bioinformatics*, **21**, 951–960.

36 Kawashima, S., Pokarowski, P., Pokarowska, M., Kolinski, A., Katayama, T., Kanehisa, M. (2008) AAindex: amino acid index database, progress report. *Nucleic Acids Res.*, **36**, D202–D205.

37 Eddy, S.R. (2004) Where did the BLOSUM62 alignment score matrix come from? *Nat. Biotechnol.*, **22**, 1035–1036.

38 Saitou, N., Nei, M. (1987) The neighbor-joining method: a new method for reconstructing phylogenetic trees. *Mol. Biol. Evol.*, **4**, 406–425.

39 Guindon, S., Lethiec, F., Duroux, P., Gascuel, O. (2005) PHYML online – a web server for fast maximum likelihood-based phylogenetic inference. *Nucleic Acids Res.*, **33**, W557–W559.

40 Zhang, J. (2000) Protein-length distributions for the three domains of life. *Trends Genet.*, **16**, 107–109.

41 Brocchieri, L., Karlin, S. (2005) Protein length in eukaryotic and prokaryotic proteomes. *Nucleic Acids Res.*, **33**, 3390–3400.

42 Kurland, C.G., Canbäck, B., Berg, O.G. (2007) The origins of modern proteomes. *Biochimie*, **89**, 1454–1463.

43 Wang, M., Kurland, C.G., Caetano-Anollés, G. (2011) Reductive evolution of proteomes and protein structures. *Proc. Natl Acad. Sci. USA*, **108**, 11954–11958.

44 Chow, L.T., Gelinas, R.E., Broker, T.R., Roberts, R.J. (1977) An amazing sequence arrangement at the 5' ends of adenovirus 2 messenger RNA. *Cell*, **12**, 1–8.

45 Berget, S.M., Moore, C., Sharp, P.A. (1977) Spliced segments at the 5' terminus of adenovirus 2 late mRNA. *Proc. Natl Acad. Sci. USA*, **74**, 3171–3131–5.

46 Black, D.L. (2003) Mechanisms of alternative pre-messenger RNA splicing. *Annu. Rev. Biochem.*, **72**, 291–336.

47 Blumenthal, T., Thomas, J. (1998) Cis and trans mRNA splicing in *C. elegans*. *Trends Genet.*, **4**, 305–308.

48 Prescott, D.M. (2000) Genome gymnastics: unique modes of DNA evolution and processing in ciliates. *Nat. Rev. Genet.*, **1**, 191–198.

49 Landweber, L.F. (2007) Genetics. Why genomes in pieces? *Science*, **318**, 405–407.

50 Teichmann, S.A., Babu, M.M. (2004) Gene regulatory network growth by duplication. *Nat. Genet.*, **6**, 492–496.

51 Pereira-Leal, J.B., Levy, E.D., Kamp, C., Teichmann, S.A. (2007) Evolution of protein complexes by duplication of homomeric interactions. *Genome Biol.*, **8**, R51.

52 Babushok, D.V., Ostertag, E.M., Kazazian, H.H. Jr (2007) Current topics in genome evolution: molecular mechanisms of new gene formation. *Cell. Mol. Life Sci.*, **64**, 542–554.

53 Basu, M.K., Poliakov, E., Rogozin, I.B. (2009) Domain mobility in proteins: functional and evolutionary implications. *Brief Bioinform.*, **10**, 205–216.

54 Sémon, M., Wolfe, K.H. (2007) Consequences of genome duplication. *Curr. Opin. Genet. Dev.*, **17**, 505–512.

55 Van de Peer, Y., Maere, S., Meyer, A. (2009) The evolutionary significance of ancient genome duplications. *Nat. Rev. Genet.*, **10**, 725–732.

56 Krishnapillai, V. (1996) Horizontal gene transfer. *J. Genet.*, **75**, 219–232.

57 Koonin, E.V., Makarova, K.S., Aravind, L. (2001) Horizontal gene transfer in prokaryotes: quantification and classification. *Annu. Rev. Microbiol.*, **55**, 709–742.

58 Martin, W., Herrmann, R.G. (1998) Gene transfer from organelles to the nucleus: how much, what happens, and why? *Plant Physiol.*, **118**, 9–17.

59 Pauling, L., Corey, R.B., Branson, H.R. (1951) The structure of proteins: two hydrogen-bonded helical configurations of the polypeptide chain. *Proc. Natl Acad. Sci. USA*, **37**, 205–211.

60 Richardson, J.S. (1977) Beta-sheet topology and the relatedness of proteins. *Nature*, **268**, 495–500.

61 Grishin, N.V. (2001) Fold change in evolution of protein structures. *J. Struct. Biol.*, **134**, 167–185.

62 Cheng, G., Qian, B., Samudrala, R., Baker, D. (2005) Improvement in protein functional site prediction by distinguishing structural and functional constraints on protein family evolution using computational design. *Nucleic Acids Res.*, **33**, 5861–5867.

63 Pautsch, A., Schulz, G.E. (1998) Structure of the outer membrane protein A transmembrane domain. *Nat. Struct. Biol.*, **5**, 1013–1017.

64 de Planque, M.R., Killian, J.A. (2003) Protein-lipid interactions studied with designed transmembrane peptides: role of hydrophobic matching and interfacial anchoring. *Mol. Membr. Biol.*, **20**, 271–284.

65 Ofran, Y., Rost, B. (2007) Protein–protein interaction hotspots carved into sequences. *PLoS Comput. Biol.*, **3**, e119.

66 Bogan, A.A., Thorn, K.S. (1998) Anatomy of hot spots in protein interfaces. *J. Mol. Biol.*, **280**, 1–9.

67 Vieille, C., Zeikus, G.J. (2001) Hyperthermophilic enzymes: sources, uses, and molecular mechanisms for thermostability. *Microbiol. Mol. Biol. Rev.*, **65**, 1–43.

68 Gromihaa, M.M., Selvarajb, S. (2004) Inter-residue interactions in protein folding and stability. *Prog. Biophys. Mol. Biol.*, **86**, 235–277.

69 Hinsen, K. (2008) Structural flexibility in proteins: impact of the crystal environment. *Bioinformatics*, **24**, 521–528.

70 Ishima, R., Torchia, D.A. (2000) Protein dynamics from NMR. *Nat. Struct. Biol.*, **7**, 740–743.

71 Mittermaier, A., Kay, L.E. (2006) New tools provide new insights in NMR studies of protein dynamics. *Science*, **312**, 224–228.

72 Hayward, S. (1999) Structural principles governing domain motions in proteins. *Proteins: Struct. Funct. Genet.*, **36**, 425–435.

73 Kumar, S., Ma, B., Tsai, C.-J., Wolfson, H., Nussinov, R. (1999) Folding funnels and conformational transitions via hinge-bending motions. *Cell Biochem. Biophys.*, **31**, 141–164.

74 Kern, D., Zuiderweg, E.R.P. (2003) The role of dynamics in allosteric regulation. *Curr. Opin. Struct. Biol.*, **13**, 748–757.

75 Petrey, D., Fischer, M., Honig, B. (2009) Structural relationships among proteins with different global topologies and their implications for function annotation strategies. *Proc. Natl Acad. Sci. USA*, **106**, 17377–17382.

76 Nagano, N., Orengo, C.A., Thornton, J.M. (2002) One fold with many functions: the evolutionary relationships between TIM barrel families based on their sequences, structures and functions. *J. Mol. Biol.*, **321**, 741–765.

77 Andreeva, A., Howorth, D., Chandonia, J., Brenner, S.E., Hubbard, T.J.P., Chothia, C., Murzin, A.G. (2007) Data growth and its impact on the SCOP database: New developments. *Nucleic Acids Res.*, **36**, D419–D425.

78 Horst, J., Samudrala, R. (2009) Diversity of protein structures and difficulties in fold recognition: the curious case of protein G. *F1000 Biol. Rep.*, **1**, 69.

79 Yadid, I., Kirshenbaum, N., Sharon, M., Dym, O., Tawfik, D.S. (2010) Metamorphic proteins mediate evolutionary transitions

of structure. *Proc. Natl Acad. Sci. USA*, **107**, 7287–7292.

80 Elde, N.S., Malik, H.S. (2009) The evolutionary conundrum of pathogen mimicry. *Nat. Rev. Microbiol.*, **7**, 787–797.

81 Alva, V., Remmert, M., Biegert, A., Lupas, A.N., Söding, J. (2010) A galaxy of folds. *Protein Sci.*, **19**, 124–130.

82 Kuhlman, B., Dantas, G., Ireton, G.C., Varani, G., Stoddard, B.L., Baker, D. (2003) Design of a novel globular protein fold with atomic-level accuracy. *Science*, **302**, 1364–1368.

83 Caetano-Anollés, G., Wang, M., Caetano-Anollés, D., Mittenthal, J.E. (2009) The origin, evolution and structure of the protein world. *Biochem. J.*, **417**, 621–637.

84 Karlin, K.D. (1993) Metalloenzymes, structural motifs, and inorganic models. *Science*, **261**, 701–708.

85 Fischer, J.D., Holliday, G.L., Rahman, S.A., Thornton, J.M. (2010) The structures and physicochemical properties of organic cofactors in biocatalysis. *J. Mol. Biol.*, **403**, 803–824.

86 Eom, S.H., Wang, J., Steitz, T.A. (1996) Structure of Taq polymerase with DNA at the polymerase active site. *Nature*, **382**, 278–281.

87 Steitz, T.A. (1999) DNA polymerases: structural diversity and common mechanisms. *J. Biol. Chem.*, **274**, 17395–17398.

88 Birktoft, J.J., Blow, D.M. (1972) Structure of crystalline α-chymotrypsin. V. The atomic structure of tosyl-α-chymotrypsin at 2 Å resolution. *J. Mol. Biol.*, **68**, 187–240.

89 Schindelin, H., Kisker, C., Schlessman, J.L., Howard, J.B., Rees, D.C. (1997) Structure of ADP:AIF-stabilized nitrogenase complex and its implications for signal transduction. *Nature*, **387**, 370–376.

90 Seefeldt, L.C., Hoffman, B.M., Dean, D.R. (2009) Mechanism of Mo-dependent nitrogenase. *Annu. Rev. Biochem.*, **78**, 701–722.

91 Laskowski, M. Jr, Kato, I. (1980) Protein inhibitors of proteinases. *Annu. Rev. Biochem.*, **49**, S9J–626.

92 Cusack, S. (1997) Aminoacyl-tRNA synthetases. *Curr. Opin. Struct. Biol.*, **7**, 881–889.

93 Lee, A.W., Karplus, M. (1983) Structure-specific model of hemoglobin cooperativity. *Proc. Natl Acad. Sci. USA*, **80**, 7055–7059.

94 Yamada, T., Bork, P. (2009) Evolution of biomolecular networks: lessons from metabolic and protein interactions. *Nat. Rev. Mol. Cell Biol.*, **10**, 791–803.

95 Hatzimanikatis, V., Li, C., Ionita, J.A., Broadbelt, L.J. (2004) Metabolic networks: enzyme function and metabolite structure. *Curr. Opin. Struct. Biol.*, **14**, 300–306.

96 Cagney, G. (2009) Interaction networks: lessons from large-scale studies in yeast. *Proteomics*, **9**, 4799–4811.

97 Fields, S., Song, O. (1989) A novel genetic system to detect protein-protein interactions. *Nature*, **340**, 245–246.

98 Hall, D.A., Ptacek, J., Snyder, M. (2007) Protein microarray technology. *Mech. Ageing Dev.*, **128**, 161–167.

99 McDermott, J., Bumgarner, R.E., Samudrala, R. (2005) Functional annotation from predicted protein interaction networks. *Bioinformatics*, **21**, 3217–3226.

100 Zhou, D., Yang, R. (2006) Global analysis of gene transcription regulation in prokaryotes. *Cell. Mol. Life Sci.*, **63**, 2260–2290.

101 Lee, T.I., Rinaldi, N.J., Robert, F., Odom, D.T., Bar-Joseph, Z., Gerber, G.K., Hannett, N.M., Harbison, C.T., Thompson, C.M., Simon, I., Zeitlinger, J., Jennings, E.G., Murray, H.L., Gordon, D.B., Ren, B., Wyrick, J.J., Tagne, J.B., Volkert, T.L., Fraenkel, E., Gifford, D.K., Young, R.A. (2002) Transcriptional regulatory networks in *Saccharomyces cerevisiae*. *Science*, **298**, 799–804.

102 Schena, M. (1996) Genome analysis with gene expression microarrays. *BioEssays*, **18**, 427–431.

103 Sultan, M., Schulz, M.H., Richard, H., Magen, A., Klingenhoff, A., Scherf, M., Seifert, M., Borodina, T., Soldatov, A., Parkhomchuk, D., Schmidt, D., O'Keeffe, S., Haas, S., Vingron, M., Lehrach, H., Yaspo, M.L. (2008) A global view of gene activity and alternative splicing by deep sequencing of the human transcriptome. *Science*, **321**, 956–960.

104 McDermott, J., Samudrala, R. (2003) BIOVERSE: functional, structural, and contextual annotation of proteins and proteomes. *Nucleic Acids Res.*, **31**, 3736–3737.

105 Long, M., Betrán, E., Thornton, K., Wang, W. (2003) The origin of new genes: glimpses from the young and old. *Nat. Rev. Genet.*, **4**, 865–875.

106 Conant, G.C., Wolfe, K.H. (2008) Turning a hobby into a job: how duplicated genes find new functions. *Nat. Rev. Genet.*, **9**, 938–950.

107 Almonacid, D.E., Year, E.R., Mitchell, J.B.O., Babbitt, P.C. (2010) Quantitative comparison of catalytic mechanisms and overall reactions in convergently evolved enzymes: implications for classification of enzyme function. *PLoS Comput. Biol.*, **6**, e1000700.

108 Rodrigues, J., Wagner, A. (2009) Evolutionary plasticity and innovations in complex metabolic reaction networks. *PloS Comput. Biol.*, **5**, e1000613.

109 Karp, P.D., Paley, S.M., Krummenacker, M., Latendresse, M., Dale, J.M., Lee, T.J., Kaipa, P., Gilham, F., Spaulding, A., Popescu, L., Altman, T., Paulsen, I., Keseler, I.M., Caspi, R. (2010) Pathway tools version 13.0: integrated software for pathway/genome informatics and systems biology. *Brief Bioinform.*, **11**, 40–79.

110 Mano, A., Tuller, T.T., Béjà, O., Pinter, R.Y. (2010) Comparative classification of species and the study of pathway evolution based on the alignment of metabolic pathways. *BMC Bioinformatics*, **11**, S38.

111 Atomi, H. (2002) Microbial enzymes involved in carbon dioxide fixation. *J. Biosci. Bioeng.*, **94**, 497–505.

112 Seelig, B., Szostak, J.W. (2007) Selection and evolution of enzymes from a partially randomized non-catalytic scaffold. *Nature*, **448**, 828–831.

113 Baker, D. (2006) Prediction and design of macromolecular structures and interactions. *Philos. Trans. R. Soc. Lond. B: Biol. Sci.*, **361**, 459–463.

114 Clackson, T., Hoogenboom, H.R., Griffiths, A.D., Winter, G. (1991) Making antibody fragments using phage display libraries. *Nature*, **352**, 624–628.

115 Flavell, R.A., Sabo, D.L., Bandle, E.F., Weissmann, C. (1974) Site-directed mutagenesis: generation of an extracistronic mutation in bacteriophage Q beta RNA. *J. Mol. Biol.*, **89**, 255–272.

116 Markiewicz, P., Kleina, L.G., Cruz, C., Ehret, S., Miller, J.H. (1994) Genetic studies of the lac repressor. XIV. Analysis of 4000 altered *Escherichia coli* lac repressors reveals essential and non-essential residues, as well as "spacers" which do not require a specific sequence. *J. Mol. Biol.*, **240**, 421–433.

117 Stemmer, W.P. (1994) Rapid evolution of a protein in vitro by DNA shuffling. *Nature*, **370**, 389–391.

118 Camps, M., Naukkarinen, J., Johnson, B.P., Loeb, L.A. (2003) Targeted gene evolution in *Escherichia coli* using a highly error-prone DNA polymerase I. *Proc. Natl Acad. Sci. USA*, **100**, 9727–9732.

119 Roberts, R.W., Szostak, J.W. (1997) RNA–peptide fusions for the *in vitro* selection of peptides and proteins. *Proc. Natl Acad. Sci. USA*, **94**, 12297–12302.

120 Moult, J. (2005) A decade of CASP: progress, bottlenecks and prognosis in protein structure prediction. *Curr. Opin. Struct. Biol.*, **15**, 285–289.

121 Samudrala, R., Moult, J. (1998) An all-atom distance-dependent conditional probability discriminatory function for protein structure prediction. *J. Mol. Biol.*, **275**, 895–916.

122 Zhang, Y., Skolnick, J. (2004) Scoring function for automated assessment of protein structure template quality. *Proteins*, **57**, 702–710.

123 Siegel, J.B., Zanghellini, A., Lovick, H.M., Kiss, G., Lambert, A.R., St Clair, J.L., Gallaher, J.L., Hilvert, D., Gelb, M.H., Stoddard, B.L., Houk, K.N., Michael, F.E., Baker, D. (2010) Computational design of an enzyme catalyst for a stereoselective bimolecular Diels–Alder reaction. *Science*, **329**, 309–313.

124 Baker, D. (2010) An exciting but challenging road ahead for computational enzyme design. *Protein Sci.*, **19**, 1817–1819.

125 Oren, E.E., Tamerler, C., Sahin, D., Hnilova, M., Seker, U.O.S., Sarikaya, M., Samudrala, R. (2007) A novel knowledge-based approach for designing inorganic binding peptides. *Bioinformatics*, **23**, 2816–2822.

Part III
Modeling of Biological Systems

9
Chaos in Biochemistry and Physiology

Miguel Antonio Aon[1], *Sonia Cortassa*[1], *and David Lloyd*[2]
[1]*Johns Hopkins University, Department of Medicine, 720 Rutland Avenue, Baltimore, MD 21205, USA*
[2]*Cardiff University, Microbiology (BIOSI 1), Cardiff, Wales, CF10 3AT, UK*

Systems Biology: Advances in Molecular Biology and Medicine, First Edition. Edited by Robert A. Meyers.

Keywords

Scaling
A quantity scales if it changes according to a power law, the exponent of which is insensitive to the details of the processes involved, a feature usually referred to as *universality*.

Scale-free
Scale-free refers to the geometry or dynamic of structures or processes with no single characteristic scale (space) or frequency (temporal).

Critical
Applies to states or phenomena characterized by their extreme sensitivity to small changes or perturbations.

Fractal
An object of geometric, statistical, or dynamical nature that obeys power laws of the form $M(L) \propto L^D$, with D as the non-integer fractal dimension. Mandelbrot introduced "fractal geometry" and defined a fractal as "... a shape made of parts similar to the whole in some way." This last property is called *self-similarity*, and represents a hallmark of fractals.

Emergent behavior
Novel and sometimes surprising macroscopic properties arising from the self-organizing capacity of the internal structure and dynamics of a biological system. Self-organization appears in nonlinear open systems (cells, organisms, ecosystems) that perform out of thermodynamic equilibrium; that is, they are constantly being driven by an exchange of matter, energy, and information with the environment.

Chaotic
Chaotic is a dynamic, non-periodic behavior that describes "turbulent" motions in time, and that exhibits sensitive dependence on initial conditions.

Deterministic chaos
Deterministic chaos is a mechanism which can generate random-looking time series from simple deterministic systems. This complicated behavior can be a consequence of very simple equations that describe the dynamics. Such a system is highly dependent on its initial conditions and is unpredictable in the longer term (i.e.) for more than a few cycles) as small divergence becomes progressively amplified.

Noise
Noise refers to random time series originating from stochastic fluctuations of events of a molecular or physical nature. In certain contexts noise may represent a signal.

Mitochondria
Mitochondria are subcellular organelles, the major energy-producers of the cell as well as the source of important signaling functions, including *apoptosis* (programmed cell death).

Network
The collective organization of an ensemble of entities, or groups of these ensembles, in space (structural or topological) and time (dynamic).

Dynamic fractals
Self-similar statistical objects described in time series; that is, the time-dependent behavior of an observable (e.g. mitochondrial membrane potential NADH concentration). These statistical fractals exhibit scale-free dynamics; that is, they span a wide range of temporal scales simultaneously (e.g. from milliseconds to several minutes), in such a way that the short-term fluctuations are related to longer-term trends in the dynamic behavior. As a result, what affects them on one time scale affects them on all time scales: this is a fundamental property of dynamic fractals. The long-term correlation in mitochondrial dynamic behavior can be interpreted as "memory", since their status in the present is dependent on their condition in the past, a fact underscored by their nonrandom statistical behavior.

Definition of the Subject, and Its Importance
Considered in the abstract since Poincaré's formulation of the three-body problem in 1892, the first practical demonstration of chaos was by Lorenz in 1963, while investigating a crude model of atmospheric convection. That an irregular dynamic behavior can be produced from purely deterministic equations is a basic characteristic of nonlinear systems. Thus, apparently noise-like behavior can arise from the intrinsic dynamics of a system, and not be the mere product of random environmental effects. The existence of chaotic behavior with its extreme sensitivity to initial conditions limits the long-term predictability that is pervasive throughout physical, chemical, and biological systems in the real world.

Chaos as an emergent phenomenon in complex, self-organizing systems, belongs with fractals, critical phenomena, and network topology and dynamics, to the group of concepts that, when integrated with Systems Biology, constitute the core of the Complex Systems Approach developed in this chapter.

1 Introduction

Poincaré (1892) realized that the motion of a third small planet orbiting in the gravitational field of two massive planets would be highly complicated and not simply solved (the intractable "three-body problem") using Newton's laws of motion [1]. Lorenz first discovered the practical importance of chaos in 1963 [2]. The Lorenz equations, a system of three coupled nonlinear differential equations, also provide an approximate description of the behavior of a horizontal fluid layer heated from below. This system creates convection currents such that, if the heating is sufficiently intense, and take place in a turbulent manner, they will mimic some of the behavioral features of the Earth's atmosphere [3]. The sensitive dependence to initial conditions in the Lorenz equations, and the ability of their solutions to generate a "strange" attractor (the Lorenz attractor), constitute the hallmarks of chaos [2, 4, 5].

That an enzyme-catalyzed reaction may exhibit chaotic behavior was first demonstrated by Olsen and Degn in 1977 [6], and the theoretical bases for these observations were eventually reviewed by the same investigators [7]. A three-variable model comprising two allosteric enzymes coupled in series and activated by their respective products can exhibit chaotic behavior through a plethora of dynamic regimes (e.g., oscillations, birhythmicity) [8]. A three-variable system with feedback inhibition exhibits limit cycle behavior that can be extended into a chaotic regime if a second-order reaction with one of the variables is added [9].

The first experimental demonstration of chaos in a biochemical system was observed during the reduction of oxygen with NADH in the presence of peroxidase from horseradish in a stirred-tank reactor with continuous supply of oxygen [6]; later, in the same enzymatic system, a period-doubling cascade leading to chaos was demonstrated [10]. Other early experimental demonstrations of chaos are represented by a glycolytic oscillator in cell-free extracts of yeast, forced with sinusoidal substrate input [11, 12], and in a continuously stirred-tank reactor with simultaneous inflows of extract and glucose [13] (for a review, see Ref. [14]).

Chaos was assigned two apparently opposed roles in physiology – either that it signified pathological conditions, "dynamical diseases" [15], or that health is necessarily characterized by highly irregular and complex dynamics [16]. In the case of "dynamical diseases," chaos is seen as a manifestation of a loss of control over physiology, leading to the abnormal types of dynamics associated with pathology. In the other view, the reverse is postulated – that is, disease and even aging are associated with less complexity and more regularity. In this latter context, chaos is considered to indicate robust health because it provides the organism with an "information-rich (broadband) state" and "spectral reserve"; that is, the possibility for performing variable outputs [17]. Evidence for disparate roles of chaos in both healthy physiology and pathophysiology has been reported (for reviews, see Refs [18] and [19]).

Here, we review chaos in biochemistry and physiology in the context of "Systems Biology" and complex systems. Thus, we highlight the awareness of conceptual links between chaos and fractals, on the one hand, and networks, and nonlinear dynamics of self-organizing complex systems, on the other hand. An overall view of the nature of specific complex systems helps to highlight the potential and limitations of conjecture on the underlying laws of motion of functional biological networks from the nonlinear analysis of time series.

2
Systems Biology and the Complex Systems Approach: Chaos in Context

The re-emerging field of physiology, now widely referred to as "*Systems Biology*," reflects the major undergoing shift between *analytical* and *integrative* philosophies in biology [20]. In this major transition, which has taken place over the past decade, the importance of the concept of complex systems and whole-system approaches has become dominant [20, 21]. Cells, organisms, social-, economic-, and eco-systems are complex because they consist of a large number of usually nonlinearly interacting parts; they also operate in multiple spatial and temporal scales. By exchanging energy, matter, and information with their environment, complex systems exhibit emergent properties. Chaos is one of the possible emergent outcomes from the intrinsic dynamics of a deterministic system.

Fundamentally different from noise, the detection of chaos demands the use of refined mathematical techniques. Noise and chaos are almost always coexistent in the real world, and the rigorous differentiation of these fundamentally distinct phenomena remains a challenge. Randomness of chaotic origin is only apparent because it arises in relatively simple, deterministic, systems described by at least three differential equations [4, 5], or by a single difference equation, the logistic equation utilized in description of population dynamics [22]. The deterministic nature of chaos allows for prediction, even if only for short time intervals, and also for control. Noise cannot be predicted and can be controlled in a very limited and restricted sense [23].

Emergent properties – those single most distinguishing features of complex behavior in a system – manifest themselves as novel, and unexpected, macroscopic spatiotemporal patterns. Most importantly, they cannot be anticipated in any way from the behavior of the isolated components of that system. There is nothing inexplicable about emergence in complex systems: it arises from self-organizing principles resulting from nonlinear mechanisms and the continuous exchange of energy, matter and information with the environment [21, 24–30]. Self-organization is one of the main mechanisms that generate organized complexity, and this is expressed as disparate types of nonlinear outputs (e.g., oscillations, chaos) in a wide range of systems [31], including biological ones [24, 26, 32].

Currently developing methods (the so-called "-omics" of high-throughput technologies), combined with the growing facility for constructing mathematical models of complicated systems, constitute the core of "Systems Biology" [33, 34]. For the sake of an approximation to the study of complex systems, a *Complex Systems Approach*, emphasizing the analysis of interconnections and relationships rather than component parts, has been proposed [21] (Fig. 1). The Complex Systems Approach integrates Systems Biology to nonlinear dynamic systems analysis, and uses the modern concepts of chaos, fractals, critical phenomena, and networks [5, 35–39]. The Complex Systems Approach is needed since the focus in biology and medicine – and physiology in particular – is shifting toward studying the properties of complex networks of reactions and processes of different nature, and how these control the behavior of cells and organisms in health and disease [40–42].

A basic concept in the Complex Systems Approach is self-organization of thermodynamically nonequilibrium and kinetically nonlinear open systems – that is, those that exchange energy, matter and information with the environment. Such systems include cells, organs and organisms, as well as social, economic and ecological systems [24, 26, 27, 30, 43]. The analysis of time series represents a

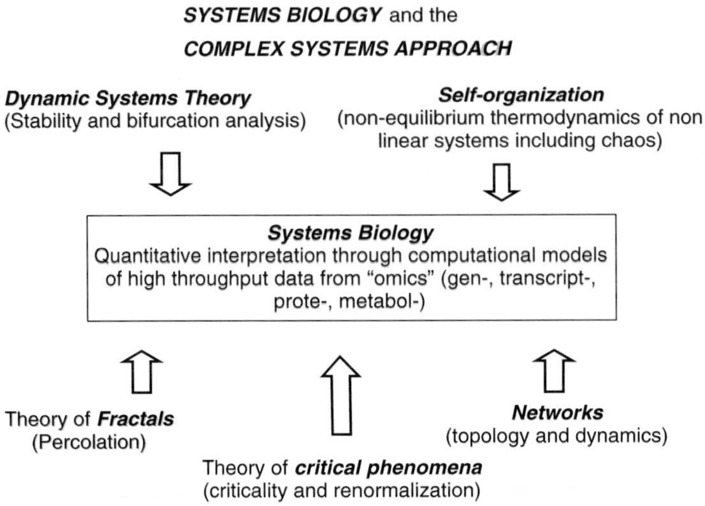

Fig. 1 The Complex Systems Approach. This approach, which emphasizes the analysis of interconnections and relationships rather than of component parts, integrates Systems Biology to nonlinear dynamic systems analysis, comprising the modern concepts of chaos, fractals, critical phenomena, and networks (see Sect. 2). Reproduced with permission from Ref. [21]; © 2009, Springer, New York.

major challenge for conjectures on the underlying dynamics of systems as complex as the heart or the brain. With a plethora of methods, nonlinear time series analysis has become the predominant tool for reconstruction of the systems dynamics in state space; this is especially so for chaotic systems [19, 23, 44, 45]. Nonlinear dynamic stability and bifurcation analyses of computational models, represented by coupled differential equations, allow the exploration of the qualitative behavior of complex systems [46, 47]. This ability is essential because complex systems exhibit a large number of parameters defining a broad parametric space; these control the behavior of numerous interacting variables. Progress in nonlinear dynamics and nonlinear time series analysis has reached a stage where fruitful applications to heart and brain dynamics have become a reality.

Another key concept in the Complex Systems Approach arises from the topologic and dynamic properties of networks. A wide range of systems, including the Internet, power grids, ecological and economic, and social webs of collaboration as well as metabolic maps and clustered cardiac mitochondria, can also be described as "networks." Despite obvious differences between these systems, they all share common features in terms of network properties. For instance, networks exhibit scale-free topologies; that is, most of the nodes in a network will have only a few links, and these will be held together by a small number of nodes exhibiting high connectivity [35, 48]. Scale-free dynamics in evolutionarily widely divergent cellular types such as yeast and heart cells is also possible. Such scale-free dynamics, characterized as *dynamic fractals*, manifest as multi-oscillatory events with frequency outputs over a broad range of temporal scales spanning at least three orders of magnitude, and described by inverse power laws [49–51].

Scale invariance is another important concept. As stated above, a complex system contains a large number of nonlinearly interrelated elements that operate in multiple spatial and temporal scales [24, 52]. The scale-free property (also termed scale-invariance or scale-independence), which implies that no one scale in space or time is able to characterize the evolution of the system, is present in a wide class of complex systems [23, 51, 53, 54]. Mathematically, the fractal, scale-free nature of complex dynamic behavior results in inverse power laws, in the form of $1/f$ noise. Long-ranged spatiotemporal correlations may develop when a system exhibits scale-free behavior, and so is able to support fluctuations over scales ranging from those of intermolecular forces to those billions of times longer [24, 31]. As a system approaches its critical point – that is, a point or time where an abrupt change occurs – the variables that describe its behavior, such as the correlation length, begin to obey power laws, and no longer depend on detailed mechanisms. Everything is determined by the universality class of the phase transition: those in the same class have the same critical exponents [31, 38, 55, 56]. Chaos and critical behavior display universality, since universal numbers [57, 58] or a few critical exponents [37, 56] characterize them. In principle, the $1/f$-types of noise and chaos respectively do not have much in common. Thus chaotic dynamics has a broadband power spectrum [59] but shows antipersistence (Hurst coefficient, $H < 0.5$); that is, an increasing trend in the past makes probable a decreasing trend in the future, and vice versa. In contrast to this is the persistence of $1/f$ noise, with $H > 0.5$ ($H = 0.5$ corresponds to uncorrelated random signal) [53, 60, 61]. However, at "the edge of chaos" – that is, the border between predictable periodic behavior and unpredictable chaos – it is possible to generate power laws [62], and $1/f$ noise from simple iterative maps of the logistic equation by fine-tuning the parameters of the system [53]. Under these conditions, chaos, colored noise and inverse power laws meet each other.

3 Reconstructing the Underlying Dynamics of Complex Systems

> *"It is one thing to know the laws of Nature, but quite another to know the outcomes of those laws. The outcomes of laws of Nature are far more complicated than the laws themselves . . . " (Barrow, 1999) [63].*

One of the most important expectations of nonlinear dynamic systems analysis is to be able to infer the underlying dynamics (e.g., laws of motion in the form of equations) of a complex system (such as the brain or heart) from the time series of an output variable (e.g., voltage). One interpretation of Barrow's quotation above suggests that the dynamics of our observable (e.g., electroencephalogram, EEG, or electrocardiogram, ECG or EKG) corresponds to one of the many solutions of the equations (laws) that determine the dynamics of such systems. Another consequence is that a mathematical model cannot unambiguously define an underlying mechanism.

Three possible strategies to circumvent this difficult problem, addressed by nonlinear dynamic systems analysis, have been proposed:

1. To reconstruct an equivalent attractor from the time series that has the

same dynamical properties as the true attractor.

2. To postulate a deterministic mathematical model that enables simulation of the observed dynamics, and the prediction of nonobserved dynamics followed by experimental verification.
3. Based on the property that a variable from a strongly interacting nonlinear deterministic system has information about every other variable, it may be conjectured that the entire system is encoded in a single variable. Alternatively, it may be conjectured that a properly processed observable (variable) will reveal changes in all the variables in the system, thus waiving the necessity for measuring them all.

As applied to the elucidation of chaotic dynamics, the three strategies are not mutually exclusive, but can be combined. In fact, the strongest evidence for chaotic behavior comes from situations in which there is a mathematical model for the dynamics that shows both periodic and chaotic dynamics as the parameters are varied [15]. The existence of chaotic dynamics can be ascertained if this mathematical model fulfils the following criteria: (i) it is deterministic (i.e., does not possess stochastic terms); (ii) its dynamics exhibit bifurcations consistent with the transition to chaos; (iii) embedding of the time series produced by the model shows evidence of a strange attractor; and (iv) it is able to predict experimentally observed chaos [64]. This process is top-down if it starts from strategy (1), or bottom-up if from strategy (2); both processes are possible and should be used. The appeal of a third strategy (3) [23] is that having experimental access to a significant variable in the system can provide crucial information about the system (e.g., in the heart, the easily accessible PQRST complex in an ECG; see immediately below).

Two of the three above-described strategies (1 and 3) have the obvious limitation (implicit in Barrow's quotation) that the time series of the variable observed is but one of the solutions of the system. This observation is reinforced by one main advantage of strategy (2); for example, that periodic solutions apart from chaotic solutions are also possible when the parameters are varied. With regards to strategy (2), even if it is possible to simulate and predict chaotic behavior with a deterministic system of differential equations, this does not mean that they constitute the true laws of motion of the system. This limitation is particularly crucial in unimaginably complex systems, such as those of the brain or heart.

The main limitation underlying all these approaches is given by the true heterarchical (as opposed to hierarchical) nature of complex systems [54]. Heterarchical control and regulation in complex networks of reactions is bidirectional – that is, it is bottom-up as well as top-down. In a heterarchy, as applied to metabolic and transport networks (Fig. 2), every reaction, metabolite, ion and process contributes, albeit to different extents, to the overall control and regulation of the network, thus undermining the existing concepts of "central controllers" and "rate-limiting" processes as well as cause–effect relationships [33, 34, 41]. Indeed, there are crucial output variables in systems as complex as the human heart. An example of the latter is the action potential (AP), where the first derivative, dV/dt, of the membrane potential, V (in millivolts), gives the usual QRS complex flanked by P and T waves in a normal ECG (Fig. 3). Another example is the force of contraction exerted by the heart during systole of

Fig. 2 Heterarchical control and regulation in the network of energetic and electromechanical processes of the heart cell. Heterarchical, as opposed to hierarchical, is the manner in which control and regulation is exerted in complex networks. In heterarchical control, all edges (e.g., enzymes, channels) and nodes (e.g., metabolites, ions) control, and are controlled by, all other edges and nodes involved in energetic, mechanical, electric, and transport processes. The desired degree of detail in the description can be applied by zooming in on the network of integrated cardiomyocyte function (Excitation-Contraction coupling and Mitochondrial Energetic; ECME model; see Refs [33, 65]), according to a generalized matrix method utilized for calculating its overall structure of control and regulation. More precisely, the top panel of the figure shows a scheme of the overall network of energetic and electromechanical processes from which it is possible to zoom in on the mitochondrion, with its own reaction network (lower left panel) or yet deeper into the tricarboxylic acid (TCA) cycle (lower right panel), with its specific biochemical circuitry. Of note is that the TCA cycle was considered as an aggregated step in the two other schemes. The ability to zoom in and out the network also reflects that heterarchical control and regulation is bidirectional; that is, bottom-up as well as top-down. In the top panel, the rectangular (ion or metabolites) or circular (myofibril conformations) boxes indicate state variables of the ECME network. The boxes depict a light-blue background when the state variables participate in conservation relationships [ATP/ADP; creatine/creatine-P (Cr/CrP); NAD^+/NADH; TRPN/TRPN-Ca, and the various conformations of myofibrils]. Ionic species are indicated on a dark-blue background. Hexagonal boxes denote inputs (ions or carbon substrate) that correspond to parameters in the model. Arrowheads point to the products of the numbered processes, whereas lines without arrowheads indicate inputs to those processes (see Ref. [33] for further details). In the mitochondrial energetics scheme, the TCA cycle was considered as a single step in the stoichiometric matrix. In the scheme of the TCA cycle (lower right panel), the solid lines represent mass–energy transformation reactions, whereas regulatory interactions (with negative signs indicating inhibition, and positive signs activation) are denoted by dashed lines. Processes accounted for by the ECME and ME models are numbered according to the following key:

Number	*Abbreviation*	*Name*	*Number*	*Abbreviation*	*Name*
1	TCA	Tricarboxylic acid cycle	15	I_{NCX}	Current through the sarcolemmal Na^+ Ca^{2+} exchanger
2	V_{RC}	Respiratory electron transport	16	I_{Cab}	Background Ca^{2+} current
3	HNe	Respiratory chain proton pumping	17	$ATPase_c$	Constitutive cytosolic ATPase
4	HFe	Succinate-driven proton pumping	18	I_{Na}	Na^+ inward currents
5	ATPsy	Mitochondrial ATP synthase	19	LCC	L-type Ca^{2+} current
6	Hu	Proton pumping through ATP synthase	20	J_{Rel}	Ca^{2+} release from ryanodine receptor (RyR)
7	Leak	Proton leak	21	J_{xfer}	Ca^{2+} transport from subspace into cytoplasm
8	ANT	Adenine nucleotide translocator	22	I_K	Outward potassium currents
9	Ca_{uni}	Mitochondrial Ca^{2+} uniporter	23	–	Ca^{2+} association–dissociation to troponin (TRPN)
10	V_{NCE}	Mitochondrial Na^+ Ca^{2+} exchanger	24	CK_i	Mitochondrial creatine kinase
11	AM-ATP	Myofibrillar ATPase	25	–	Creatine species transport
12	PMCA	Sarcolemmal Ca^{2+} ATPase	26	CrKc	Cytosolic creatine kinase
13	SERCA	Sarcoplasmic reticulum Ca^{2+} ATPase	27–35	–	Transitions between tropomyosin conformations
14	I_{NaK}	Current through the Na^+ K^+ ATPase			

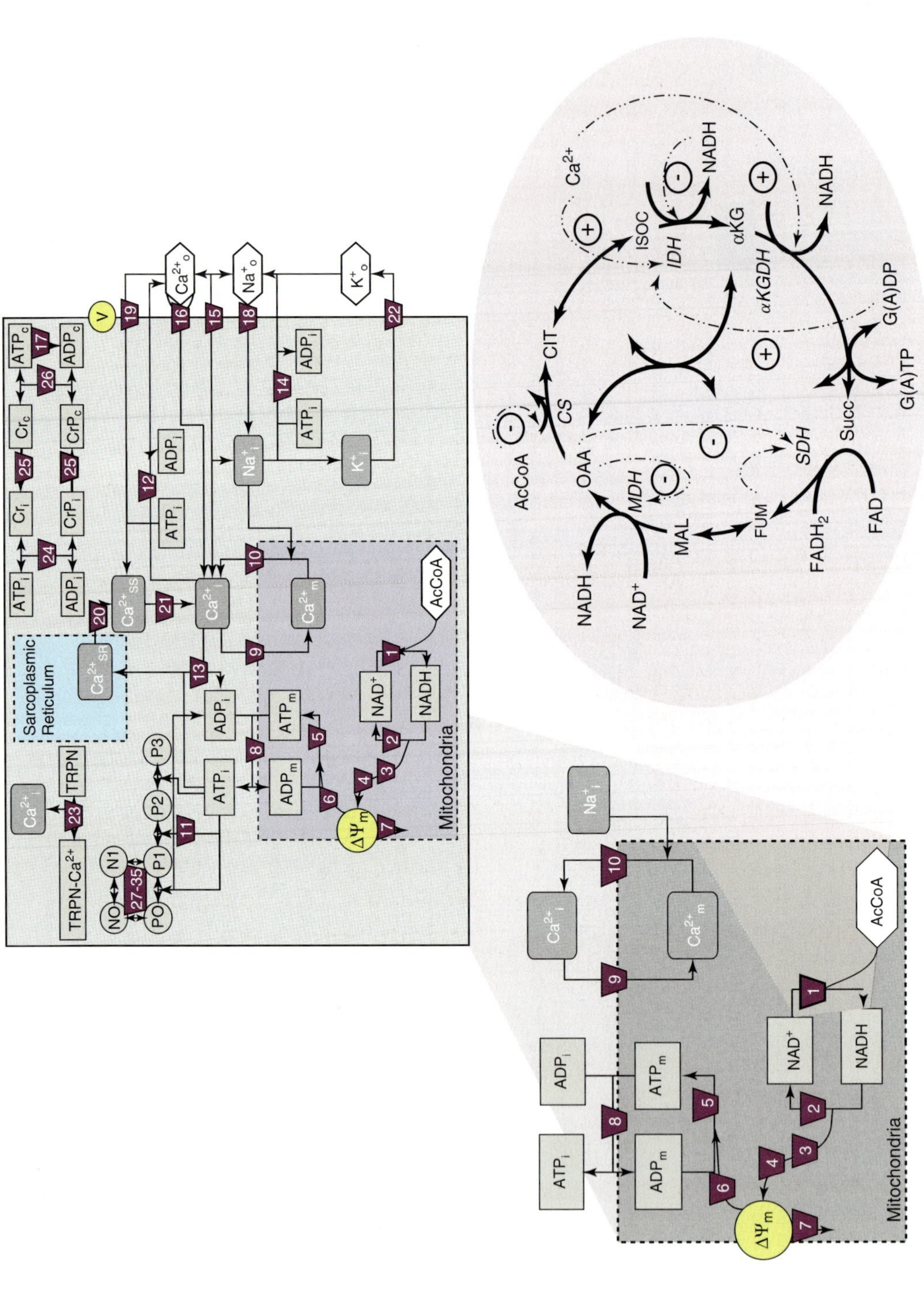
Sarcoplasmic Reticulum
Mitochondria
TRPN
TRPN-Ca2+
ATPc
ADPc
Crc
CrPc
Cri
CrPi
ATPi
ADPi
Ca2+SS
Ca2+i
Ca2+SR
Ca2+m
Ca2+o
Na+o
Na+i
K+o
K+i
ATPm
ADPm
NAD+
NADH
AcCoA
V
NO
N1
PO
P1
P2
P3
27-35
ΔΨm
AcCoA
CS
CIT
ISOC
IDH
αKG
αKGDH
Succ
SDH
FUM
MAL
MDH
OAA
NADH
NAD+
FADH2
FAD
G(A)DP
G(A)TP
Ca2+

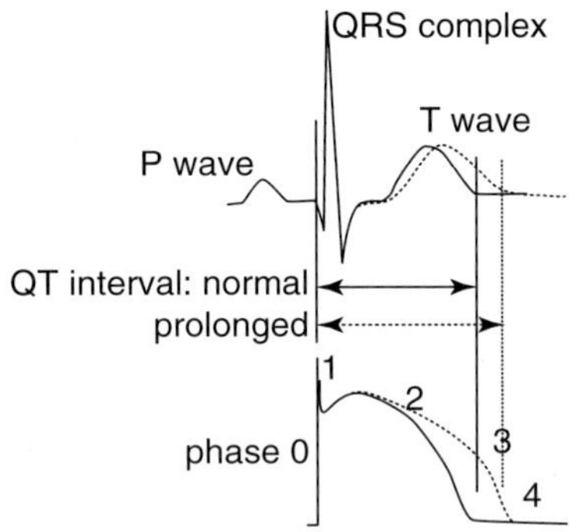

Fig. 3 Relationship between the electrocardiogram (ECG; top) and the action potential in the ventricle (lower). In the ECG, the P-wave corresponds to the depolarization of the atrium, the QRS complex to the depolarization of the larger ventricles, and the T-wave to the repolarization phase. A brief description of the different phases shown in the action potential is: 0: Na^+ channels open; 1: Na^+ channels closed; 2: Ca^{2+} channels open and fast K^+ channels closed; 3: Ca^{2+} channels closed and slow K^+ channels open; 4: resting potential. For more details, see Fig. 4 and its legend.) Redrawn from Ref. [66].

the contraction–relaxation (diastole) cycle. However, a complex array of channels underlie the AP, and their dynamics results from a sequential as well as a simultaneous interplay of Na^+, Ca^{2+}, and K^+ currents through the sarcolemmal channels (Fig. 4). The mitochondrial energetic processes fueling the electrical and contractile machineries of the heart cell are manifest on a slower time scale (few seconds) compared to the electrical processes (milliseconds) (Fig. 4). Figures 2 and 4 represent an example of a complex system (a heart cell) which reveals an essential feature: the multiple temporal and spatial scales involved; these appear as nested networks (networks within networks) of interacting dynamic systems at all levels (Fig. 2), but with different relaxation times (Fig. 4). These crucial traits are reflected in the dynamics (time series) of key integrative variables, from which some inferences about the underlying network function can be made. Also, they are at the base of the self-similar, fractal, dynamics exhibited by a wide class of complex systems (see Sects. 2 and 7).

Not losing sight of these important considerations, the task for the Complex Systems Approach (Fig. 1) [21] is not hopeless because we now have:

1. The ability and the computational power to mathematically model very complicated systems, and to analyze their control and regulation (Fig. 2), as well as predict changes in qualitative behavior.
2. An arsenal of theoretical tools (each with its own plethora of methods).
3. High-throughput technologies that allow the simultaneous monitoring of an enormous number of variables.
4. Powerful imaging methods and online monitoring systems that provide the means to study living systems at high spatial and temporal resolution for several variables simultaneously.
5. The possibility of employing sufficiently detailed bottom-up mathematical models that may help rationalize the use of key integrative variables, such as the membrane potential of cardiomyocytes or neurons, in top-down,

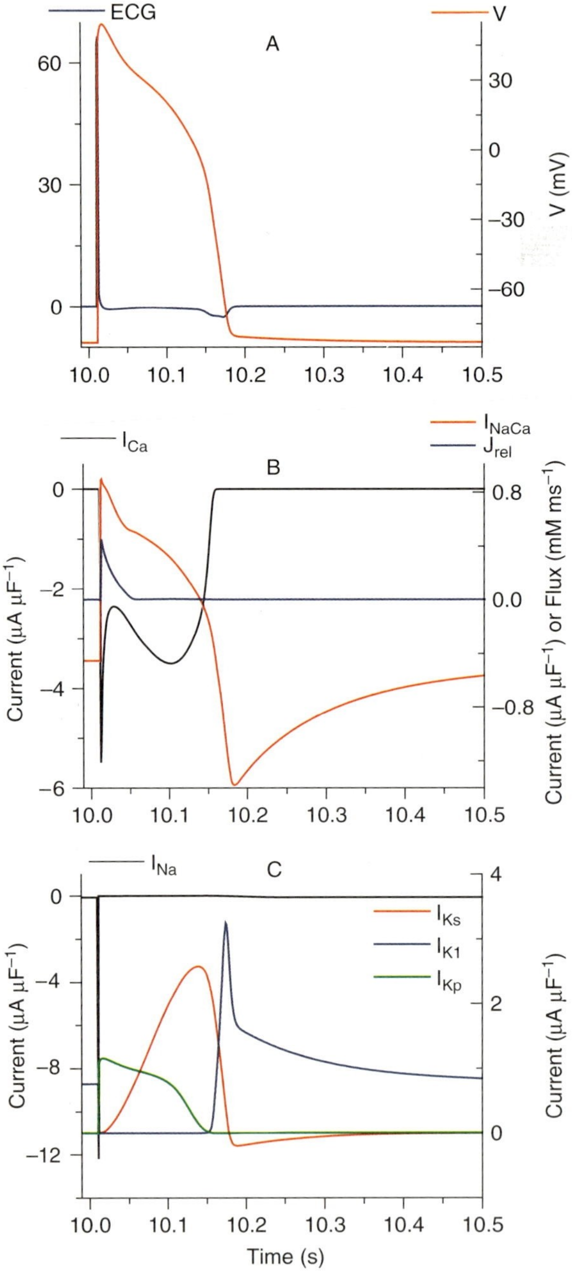

ECG
V
A
60
30
0
30
0
−30
−60
V (mV)
10.0
10.1
10.2
10.3
10.4
10.5
I_Ca
I_NaCa
J_rel
B
0
−2
−4
−6
0.8
0.0
−0.8
Current (μA μF−1)
Current (μA μF−1) or Flux (mM ms−1)
I_Na
C
I_Ks
I_K1
I_Kp
0
−4
−8
−12
4
2
0
Time (s)

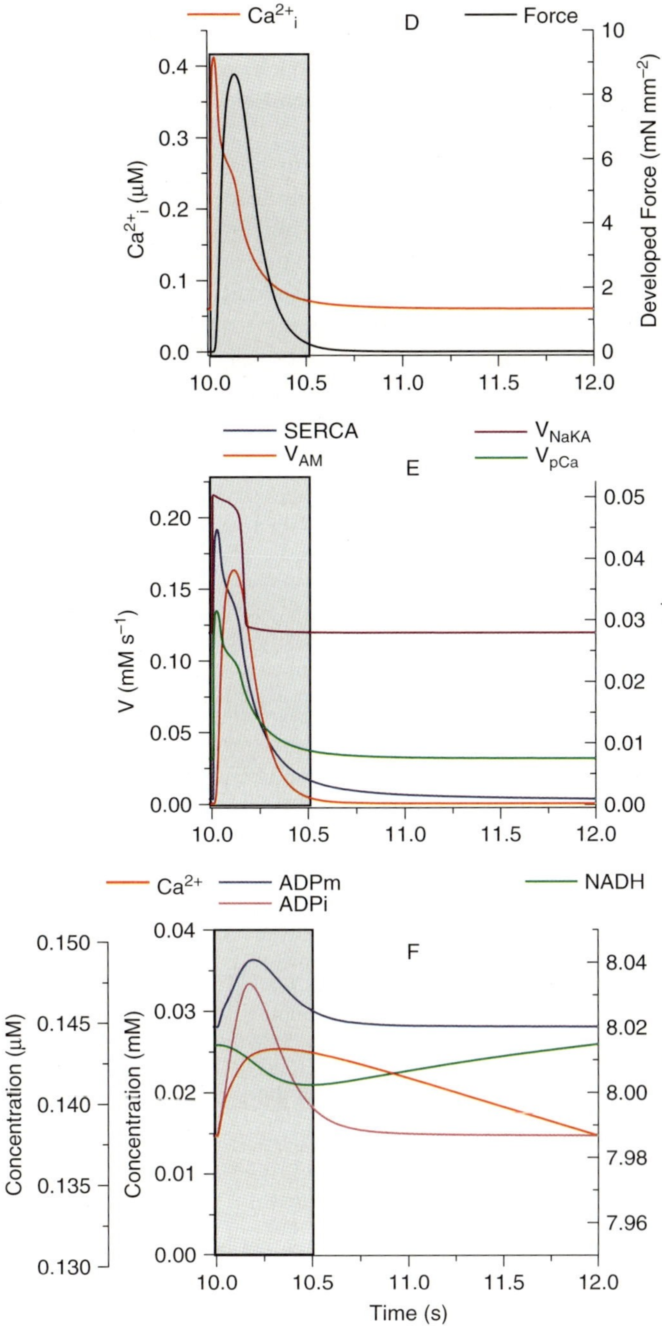

Ca2+i
D
Force
Ca2+i (μM)
Developed Force (mN mm−2)
SERCA
VAM
VNaKA
VpCa
E
V (mM s−1)
Ca2+
ADPm
ADPi
NADH
F
Concentration (μM)
Concentration (mM)
Time (s)

conceptual, models with a few state variables.[1)]

4 Chaos, Randomness, and (Colored) Noise

The intrinsic dynamics of a system governed by linear equations can only lead to exponentially decaying (or growing) or (damped) periodic solutions. In such a system where small causes lead to small effects, all irregular dynamic behavior must be attributed to some random, uncontrolled, environmental input in the form of noise. In nonlinear systems, chaotic or irregular dynamic behavior is a possible output that results despite the intrinsic deterministic dynamics of a system. Therefore, nonlinear chaotic systems can produce very irregular time series with purely deterministic equations of motion in an autonomous way – that is, without exogenous time-dependent inputs [44].

Chaos may thus be defined as "... non-periodic dynamics in deterministic systems with bounded dynamics and sensitive dependence to initial conditions" [4, 5, 64]. In order to be bounded and unstable at the same time, the trajectory of a dissipative dynamical system has been described as a "strange attractor" (Fig. 5). An *attractor* is a restricted region

1) A conspicuous example of this is the Hodgkin–Huxley equations, first used in an integrated experimental–theoretical approach to elucidate the AP dynamics of the squid giant axon) [67, 68].

Fig. 4 Time course of the main variables reflecting the state of electrical, mechanical, and energetic processes during a single cardiomyocyte beat, as computed by an integrated model of cardiomyocyte function (ECME model). When the ECME model is stimulated every 2 s, the action potential (AP) takes place during the first 200 ms [panel (a), right axis]. From the first derivative of the AP, the ECG trace can be obtained [panel (a), left axis]. During the AP, Na^+ channels are activated at the depolarization phase [panel (c), left axis], which enables the gating of L-type Ca^{2+} channels [panel (b), left axis] followed by Ca^{2+} release from internal stores in the sarcoplasmic reticulum (SER) via ryanodine receptors, J_{rel} [panel (b), right axis]. Ca^{2+} is then recaptured into the SER via SERCA activity [panel (e), left axis] or transported out of the myocyte through the $Na^+ - Ca^{2+}$ exchanger, I_{NaCa} [panel (b), right axis]. The characteristic plateau phase of the AP reflects a balance between the inward current, largely through L-type Ca^{2+} channels, and the outward current, largely through delayed rectifier K^+ channels. The potassium currents I_{Ks}, I_{K1}, and I_{KP} [panel (c), right axis] participate in the repolarization of the membrane potential to its resting, 80 mV, (diastolic) level. The Ca^{2+} transient in the cytoplasm [panel (d), left axis] triggers myofibril contraction, developing isometric force [panel (d), right axis], which is reflected by the ATP consumption of the ATPase activity associated with the formation of acto-myosin cross-bridges, V_{AM} [panel (e), left axis]. In addition to V_{AM} and SERCA activity, the other processes that contribute to the demand of ATP are the activity of the Na^+ K^+ ATPase (V_{NaKA}) and the sarcolemmal Ca^{2+} pump (I_{pCa}) [panel (e), right axis]. ADP in both compartments – the cytoplasm, ADP_i [panel (f), left axis] and the mitochondria, ADP_m [panel (f) left, axis] – exhibit a transient increase which reflects the cytoplasmic increase in ATP consumption during systole and uptake by the mitochondrion, activating respiration and decreasing NADH levels [panel (f), right axis]. Mitochondrial Ca^{2+} [panel (f), second left axis] allows a recovery from this transient mismatch between energy demand and supply, visualized as a restoration of NADH [panel (f), right axis] through the TCA cycle activity. The shaded area in panels (d–f) corresponds to the same time range zoomed in panels (a–c). Parameters and initial conditions of the simulation were as described elsewhere [65].

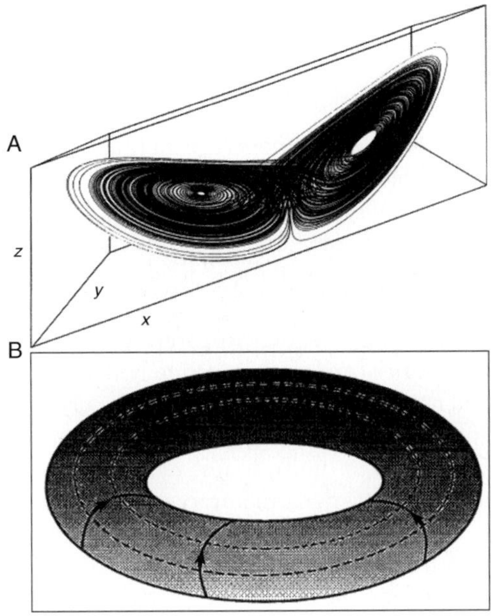

Fig. 5 Strange and toroidal attractors. (a) The strange attractor shown corresponds to the so-called Lorenz attractor, and resembles a butterfly (hence the "butterfly effect") [2, 4]. It has three dimensions (*x*, *y*, and *z*), and consists of two lobes. The trajectory makes many circuits around one lobe, and then switches to the other lobe after exceeding some critical distance from the center; (b) A toroidal attractor is described by two frequencies that may be either in phase or out of phase with each other, thus defining a periodic or quasiperiodic torus in phase space, respectively. The quasiperiodic route to chaos can pass at least three critical thresholds (through Hopf bifurcations), with increase in control parameter: low value, the system is in a steady state (fixed point, frequency zero); larger value, limit cycle (one frequency, first threshold); even larger value, quasiperiodicity (two frequencies, second threshold); much larger value, chaos (with or without intermediate quasiperiodic three frequency regime, third threshold). Reproduced from Williams, G.P. (2003) *Chaos Theory Tamed*, Joseph Henry Press, Washington D.C. [45].

of the possible phase space (see Sect. 5) to which all the trajectories settle or approach over time. *Strange attractors* characterize chaotic dynamics when represented in phase space. It is well known that a strange attractor can be quantitatively described as a fractal, where the fractal dimension is a non-integer that is less than the dimension, n, of the phase space (e.g., $n = 3$ for three-dimensional, that is, three-variable phase space) [2, 5, 15, 44, 45, 57].

As pointed out by Glass [64], if the above definition of chaos is applied to real systems, the deterministic aspect poses a potentially controversial issue because real systems are not strictly deterministic, the main reason being the existence of *noise* which is both *intrinsic* and *extrinsic* to the system. Noise can represent the extrinsic, unpredictable, influence of the environment, somehow coupled to the system, thus representing an unknown forcing. However, real systems also exhibit intrinsic noise as a result of their inherent functional behavior. Further, "noise" is inherent in the uncertainties of the measuring methods. Thus, many natural and social phenomena are neither completely ordered and predictable, nor completely random and unpredictable [23]; this mixture of order and randomness characterize behavioral traits in all complex systems.

Therefore, irregular, apparently random time series can be attributed to both chaos and noise, and this fact demands the use of specific tests to distinguish between them. Unlike chaos, noise is not related to any deterministic dynamic process, and while it is true that a low-dimensional chaotic attractor will generate a fractal time series, so too will processes other than white noise exhibiting power-law spectra, such as colored noise [59, 69]. Colored noises are clearly not necessarily related to any deterministic dynamic process.

The intrinsic type of noise, generated by the autonomous dynamics of biochemical processes through stochastic fluctuations of molecular events, may be biochemically and physiologically meaningful [70, 71]. Channel gating and synaptic transmission are noisy, nonstationary and nonlinear processes produced by fluctuations in, for example, currents, ion concentrations, and postsynaptic receptors [72]. Under certain circumstances, noise and Brownian motion can facilitate the transmission of information via a mechanism known as stochastic resonance (SR). This describes processes whereby the addition of a random function – often called the "*noise*" – to a weak information-carrying signal can enhance the detectability of the signal (e.g., trains of APs or signals generated by a neuronal assembly) by some nonlinear system. It can also intensify the information content of the system's output [73, 74] (Fig. 6). SR may help systems

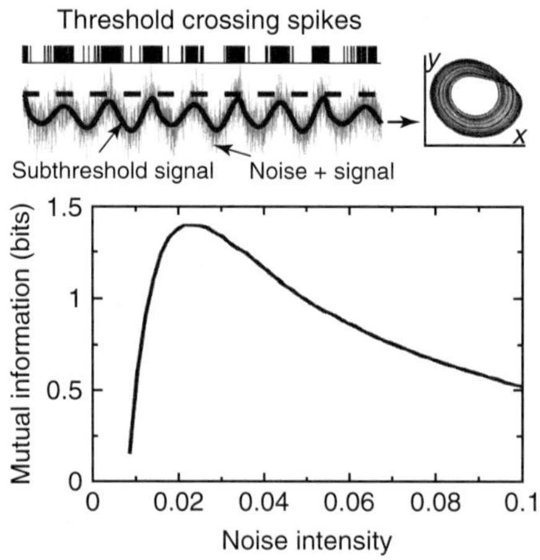

Fig. 6 Noise-enhanced information in a threshold system. The threshold is the distance between the mean of the subthreshold stimulus (thick solid line) and the threshold (dashed line). The stimulus, $x(t)$, was generated from the Rossler system, the attractor $x(t)$ versus $y(t)$ of which is shown at the right. The irregular light line represents Gaussian noise. The lower panel is the mutual information between spike trains generated by noise alone and noise plus stimulus. The optimal noise intensity is $D_{opt} \sim 0.022$, for which the information is a maximum. Reproduced from Moss, F., Ward, L.M., Sannita, W.G. (2004) *Clin. Neurophysiol.*, **115**, 267–281 [74].

use chemical energy and nonequilibrium fluctuations to drive directed motion via fluctuation-driven transport, for example in biomolecular motors [73]. The best system performance is achieved with optimum noise intensity; noise intensities at values below a threshold cause the signal to be barely or not at all detectable, whereas extreme randomness may overwhelm the information content [74, 75]. In its simplest manifestation – referred to as "*threshold SR*" or "*nondynamical SR*" – SR results from the concurrence of a threshold, a subthreshold stimulus, and noise (Fig. 6); all of these ingredients are ubiquitous in Nature [74]. Thus, in a photosensitive version of the Belousov–Zhabotinsky reaction, noise-sustained waves support the propagation of large-scale spatiotemporal patterns, which resemble calcium waves in cultured hippocampal astrocytes [76]. Experimental and theoretical data have shown that SR enhances phase locking and coherence, and also promotes synchronization in neuronal systems [72].

The functional effects of noise may represent the normal behavior of physiological systems [77, 78], and their dynamics generate time series that are characterized as colored noise. For this class of processes there is a one-to-one correspondence between the power-law dependence of the spectrum and fractal behavior; that is, the power-law index is related to the fractal dimension [23, 59, 69]. These fractal processes are scale free – that is, they do not exhibit a characteristic time scale and, unlike white or brown noise, they exhibit long-term memory. The latter property is fundamental for differentiating inherent physiological system's dynamics from "extrinsic noise" or short-term random fluctuations produced by the interaction with the environment [23]. Although, in principle, colored noises are not related to any deterministic dynamic process, as in the case of "deterministic chaos," colored noise-like behavior may arise from deterministic oscillatory dynamics [49].

5 Nonlinear Time Series Analysis: Conceptual Theoretical and Analytic Tools for Chaos Detection and Characterization

One means of progressing from the observations of a system with unknown properties to a better understanding of the dynamics of the underlying system is through nonlinear time series analysis [19, 44]. This is a more or less "top-down" approach, starting with the output of the system, and working back to the state space, attractors, and their properties.

The concept of *phase space* representation is the hallmark of nonlinear dynamical time series analysis, and most of chaos theory deals with phase space [44, 45]. Phase space or state space is an abstract mathematical space in which x- y- (two-dimensional) or x- y- z- (three-dimensional) coordinates represent the variables needed to specify the phase (or state) of a dynamical system (Fig. 7). The phase space includes all of the instantaneous states that the system can have, and as such is a complement to the time series plot with the additional power to reveal hidden patterns or structure [45]. The succession of instantaneous states of a dynamic system describe a *trajectory* (time path) in phase space, and a family of different trajectories that can evolve for the same system constitutes the *phase space portrait* or phase portrait.

Trajectories will converge to a subspace of the phase space, if a system dynamics is observed for a sufficiently long period (i.e., after the transients die

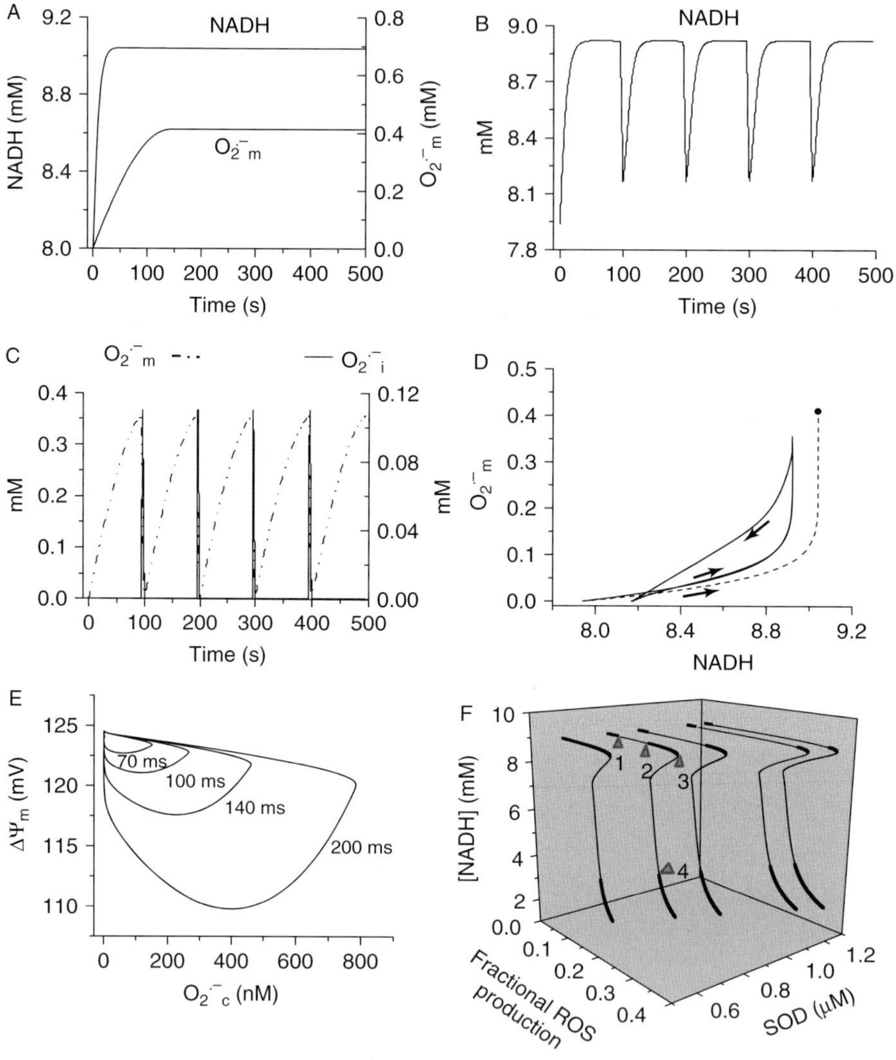

A
NADH
NADH (mM)
$O_2^{\cdot-}{}_m$ (mM)
$O_2^{\cdot-}{}_m$
Time (s)
B
NADH
mM
Time (s)
C
$O_2^{\cdot-}{}_m$
$O_2^{\cdot-}{}_i$
mM
Time (s)
D
$O_2^{\cdot-}{}_m$
NADH
E
$\Delta\Psi_m$ (mV)
70 ms
100 ms
140 ms
200 ms
$O_2^{\cdot-}{}_c$ (nM)
F
[NADH] (mM)
Fractional ROS production
SOD (μM)

out); this subspace is referred to as the *attractor* of the system (intuitively, because it "attracts" trajectories from all possible initial conditions). Unlike linear systems, which can only exhibit *point* attractors, nonlinear systems display a wide repertoire of them in phase space, including *limit cycles* or closed loops (from oscillations), *torus* of "donut" shape (quasiperiodic dynamics from superposition of different periods), and *strange* with fractal geometry (deterministic chaos) [19] (Figs 5 and 7).

Embedding a time series is one of the most important tools in chaos theory, and the first and most crucial step in nonlinear analysis is to reconstruct an attractor in the phase space of the underlying system. The procedure of *embedding* allows the reconstruction of an equivalent attractor of the underlying dynamical system. Two different embedding procedures exist: (1) Time-delay embedding; and (2) Spatial embedding [19, 45] (for technical details, see Refs [44, 45]).

If the system from which the measurements were taken has an attractor, and if the embedding dimension m is sufficiently high (more than twice the dimension of the systems attractor), then the

Fig. 7 Time series, phase space, phase portrait, and bifurcation diagram. In order to illustrate some concepts and analytic tools introduced in Sect. 5, the simulation results obtained with an integrated model of mitochondrial energetics, calcium dynamics, reactive oxygen species (ROS) production, and ROS scavenging are presented [46, 47]. This model (either isolated or integrated to other cellular processes) has been extensively validated against reported experimental evidence, and allowed predictions to be made that were later confirmed under different experimental conditions [47, 49, 65, 79–81]. Shown are the evolution of NADH and mitochondrial superoxide, $O_2^{\bullet -}{}_m$, toward (a) steady (i.e., a fixed-point attractor) or (b,c) oscillatory (i.e., limit-cycle) states, respectively; (d) The phase space plot of NADH and $O_2^{\bullet -}{}_m$ for the steady (dashed) and oscillatory (continuous) solutions. The change from a fixed-point attractor to limit-cycle behavior was achieved simply by increasing the concentration of respiratory chain carriers, while keeping constant all other parameters [47]. (e) Phase portrait plot showing the relationship between $\Delta\Psi_m$ and the superoxide anion released to the periplasmic mitochondrial space, $O_2^{\bullet -}{}_c$, during the depolarization phase of the oscillation. Plotted are the trajectories followed by several limit-cycles corresponding to oscillatory periods ranging from 70 to 200 ms after sequentially changing only one parameter; that is, the rate of ROS scavenging (see also panel f); (f) A bifurcation diagram of the state-variable NADH as a function of ROS production (fractional ROS production) and scavenging [superoxide dismutase (SOD) concentration]. The stability and type of steady states exhibited by the model were computed by continuously varying the parameters indicated. The results obtained were represented in a bifurcation diagram, which in this case consisted of an upper branch, in which NADH was predominantly reduced, and a lower branch, in which NADH was mainly oxidized. Thick lines indicate domains of stable steady-state behavior, whereas thin lines denote either unstable or oscillatory states. A stable oscillatory domain, embedded within the upper branch, emerged as the SOD concentration increased. The eigenvalues obtained from the stability analysis can be further analyzed to obtain a detailed description of the transitions at the borders of the steady states (marked by arrowheads and numbers) [47]. In the upper branch, 1 and 2 indicate Hopf bifurcations delimiting the oscillatory region (thin line), and were characterized by eigenvalues with two pairs of complex conjugates, one pair showing positive real part. In the stable regions of the diagram (thick lines), all real negative eigenvalues were found, and one or two pairs of complex conjugates, but with negative real parts; 3 and 4 denote limit points. Modified from Cortassa, S., Aon, M.A., Winslow, R.L., O'Rourke, B. (2004) *Biophys. J.*, **87**, 2060–2073 [47].

series of reconstructed vectors constitutes an "equivalent attractor." This equivalent attractor has the same dynamical properties (dimension, Lyapunov spectrum, entropy) as the true attractor [82]. This means that valuable information can be obtained about the dynamics of the system, even if direct access to all the systems variables is not available (see Sect. 3).

Once the attractor has been reconstructed it can be characterized, the simplest way to do this being to visualize it with a phase portrait (as defined above) or a Poincaré section. A Poincaré section is a cross-section transverse to the "flow" or bundle of trajectories in the phase portrait. For example, depending on the type of attractor, the Poincaré section shows one point for a limit cycle (two points for a two-period limit cycle) or a distribution of many points with some type of pattern, but complex geometry as in chaotic (strange) attractors. Thus, one main advantage of a Poincaré section is that it can indicate the type of attractor and its internal structure. It also simplifies the study of the attractor because it has at least one fewer Euclidean dimension than the attractor's phase space (e.g., an attractor in a two-dimensional x-y-phase space can be cross-sectioned with a line of Euclidean dimension 1; thus, the phase space dimension is 2 and the Euclidean dimension of the Poincaré section is 1).

The three most basic measures of attractors are the correlation dimension, the Lyapunov exponents, and the entropy. For the *correlation dimension* [83], which is the most commonly used, the key point is that the double logarithmic plot of the correlation sum (or correlation integral), C_ε, for a particular radius, ε, versus different (increasing) ε, follows a power law in the form of a straight line according to which $C \propto \varepsilon^{\nu}$, where the exponent ν is the slope and corresponds to the correlation dimension.

Lyapunov exponents quantitate the average rate of convergence or divergence of two neighboring trajectories in phase space. Their values can be negative, zero, or positive; a negative Lyapunov exponent means that two trajectories run close to one another, whereas a positive exponent results from trajectory divergence. As positive Lyapunov exponents appear only in chaotic dynamics, their presence is one of the most important indicators of chaos (see e.g., Ref. [45]). A Lyapunov exponent averages local divergences and/or convergences from many points over the entire attractor, and as such is not a local but rather a global quantity. Many of the problems involved in the computation of the correlation dimension, such as the correct choice of embedding parameters, length, and stationality of the time series and noise, are also relevant to the computation of Lyapunov exponents.

The *entropy* of an attractor is the rate of information loss of its dynamics. Information entropy is always positive and finite, but to identify chaos an entropy rate is needed (the average entropy per unit time). At limiting conditions of time and of box size (i.e., sampled phase space) the entropy is called Kolmogorov–Sinai (K-S), $H_{\text{K-S}}$, which quantifies how chaotic a dynamical system is; $H_{\text{K-S}}$ is zero for a deterministic system that is not chaotic, a positive constant for a chaotic system, and infinite for a random process [45].

When the classic measures described above are computed from filtered, noisy, nonstationary time series of limited duration and precision, spurious results can occur, such as a false suggestion of low-dimensional structure or chaos [84]. In order to overcome these problems, Stam (2005) suggested computing the measures but refraining from any interpretation of deterministic chaos, to utilize surrogate data for validation, and to use novel nonlinear measures that do not make strong assumptions about the nature of the underlying dynamics [19].

6 Periodic and Non-Periodic Dynamics

As expressed above, chaotic solutions are but one of the possible dynamic outcomes in nonlinear systems of differential equations describing a particular phenomenon. A nonchaotic regime includes fixed points, limit cycles and quasiperiodic motion (see Fig. 7). Changes among these different types of attractor occur through bifurcations; these are abrupt changes in the stability of a system dynamics as reflected by a radically different qualitative behavior (e.g., steady, oscillatory, chaotic) at critical values of the control (bifurcation) parameter [24] (Fig. 7).

Oscillatory systems – the attractor of which is represented in phase space by a limit cycle – constitute a much more typical dynamic outcome of nonlinear systems, that is often robust against perturbations. Thereby, one substantial difference between oscillatory and chaotic dynamics is that the former do not exhibit sensitive-dependence on initial conditions. Following two different initial conditions, the trajectories in nonchaotic regimes either move closer together describing limit cycles (periodic attractors), or they remain equidistant describing toroidal attractors (these combine two or more limit cycles and can be either periodic or quasiperiodic) (Figs 5 and 7). Oscillatory systems will evolve to the same attractor if changes in the initial conditions belong to the basin of attraction of the limit cycle. Unlike oscillatory systems, chaotic systems will exhibit an extreme sensitivity to initial conditions, since even a slight change will make the trajectories diverge widely (exponentially), although bounded to a strange attractor in phase space.

A trajectory within a chaotic regime is non-periodic – that is, it never repeats itself. The ensemble of trajectories within the attractor is multilayer, they never cross each other, and geometrically they can be frequently described as a fractal – that is, a self-similar structure regardless of any change in scale. In this sense, any chaotic dynamical system can be thought of as the superposition of a large number of unstable periodic orbits. Within the great number of unstable limit cycles (with different periods) embedded in a strange attractor some may be selected (at least theoretically) to a stable periodic output by small perturbations of their parameters [85, 86]. The *saddle-node* bifurcation is the typical mechanism by which, inside a chaotic regime, a stable periodic solution can occur after a small change of the control parameter [44]. This creates a pair of periodic orbits – one stable and one unstable – thus generating a type of intermittency, switching between regular and irregular behavior. Another type of bifurcation that can occur within a chaotic attractor is a *super-critical Hopf bifurcation*, where at a certain parameter value a stable fixed point becomes unstable, generating a limit cycle (topologically speaking, a circle). This one-dimensional curve can itself bifurcate into a two-frequency torus by another Hopf bifurcation (see Fig. 5). This is the route to chaos by quasiperiodicity [44].

The identification of unstable orbits is also an important step in controlling chaotic systems [87, 88]. Mechanistically, chaos can be controlled by a combination of periodic perturbations in a feedback control circuit with delay [87, 89]. A controlled chaotic attractor can, potentially, be a multioscillator that can be tuned into different time domains (see Sect. 7.2). The synchronization of multioscillatory states, implying controlled chaotic behavior of selected orbits [90–93], appears to be an essential property governing cell division [93], the circadian clock [90], and the coordination of metabolism and transcription across a population of single-celled organisms, as shown for continuous cultures of the yeast, *Saccharomyces cerevisiae* [32, 94, 95].

7 Biochemical and Physiological Chaos

7.1 Emergent Phenomena in Networks at (Sub) Cellular, Tissue, and Organ Levels

The existence of self-organization, scaling, and critical phenomena, in defined biological systems under normal and pathological conditions such as in the heart [96–98] and in the brain [19, 99], suggests that there exists a clear universality in physical concepts as applied to biology.

Self-organized, emergent phenomena, in the form of waves of mitochondrial membrane potential, $\Delta\Psi_m$, depolarization followed by sustained collective oscillatory dynamics, have been extensively documented in the heart at (sub) cellular [49, 79–81, 98, 100–104], tissue, and organ levels [96, 97, 105] (Fig. 8). Under metabolically stressful conditions (i.e., lack of substrate, oxidative stress), cardiomyocytes exhibit sustained oscillations in key energetic and redox mitochondrial variables – $\Delta\Psi_m$, NADH, and reactive oxygen species (ROS). These collective oscillations – which involve at least 60% of the mitochondrial network – are synchronized by ROS through a ROS-induced ROS release [79, 106], diffusion-based mechanism [80].

The mitochondrial network oscillates under physiological conditions, although in a high-frequency low-amplitude mode [49], a surprising finding predicted by a mitochondrial oscillator computational model [47]. Unlike in the pathophysiological situation, physiological mitochondrial oscillations in $\Delta\Psi_m$, NADH, and ROS exhibit a scale-free character; that is, they do not show a characteristic frequency but rather a family of them span a wide range of frequencies of at least three orders of magnitude (from milliseconds to several minutes in the heart) [49, 50, 107] (Fig. 9). Correlations in time series with self-similar scaling behavior suggest that events in different time scales (e.g., milliseconds to hours) are tied together in an orderly statistical sequence [23]. The power spectrum of such data usually follows an inverse power law proportional to $1/f^{\beta}$ with β as the spectral exponent ($\beta = 0$, 1, or 2 for white, pink, or brown noise, respectively). The spectral analysis of the time series of $\Delta\Psi_m$ (a crucial cellular variable) is described by an inverse power law of the form $1/f^{\beta}$ with $\beta = 1.79$, and thus intermediate between brown and pink noises [108] (Fig. 9). This fractal organization and long-term memory in mitochondrial network energetics has been characterized as *dynamic fractals*; that is, statistical fractals that exhibit scale-free dynamics, spanning a wide range of temporal scales simultaneously, in such a way that the short-term fluctuations are related to longer-term trends in the dynamic behavior. As a result, what affects one time scale affects them all: this is a fundamental property of dynamic fractals [23, 50, 51, 100, 107].

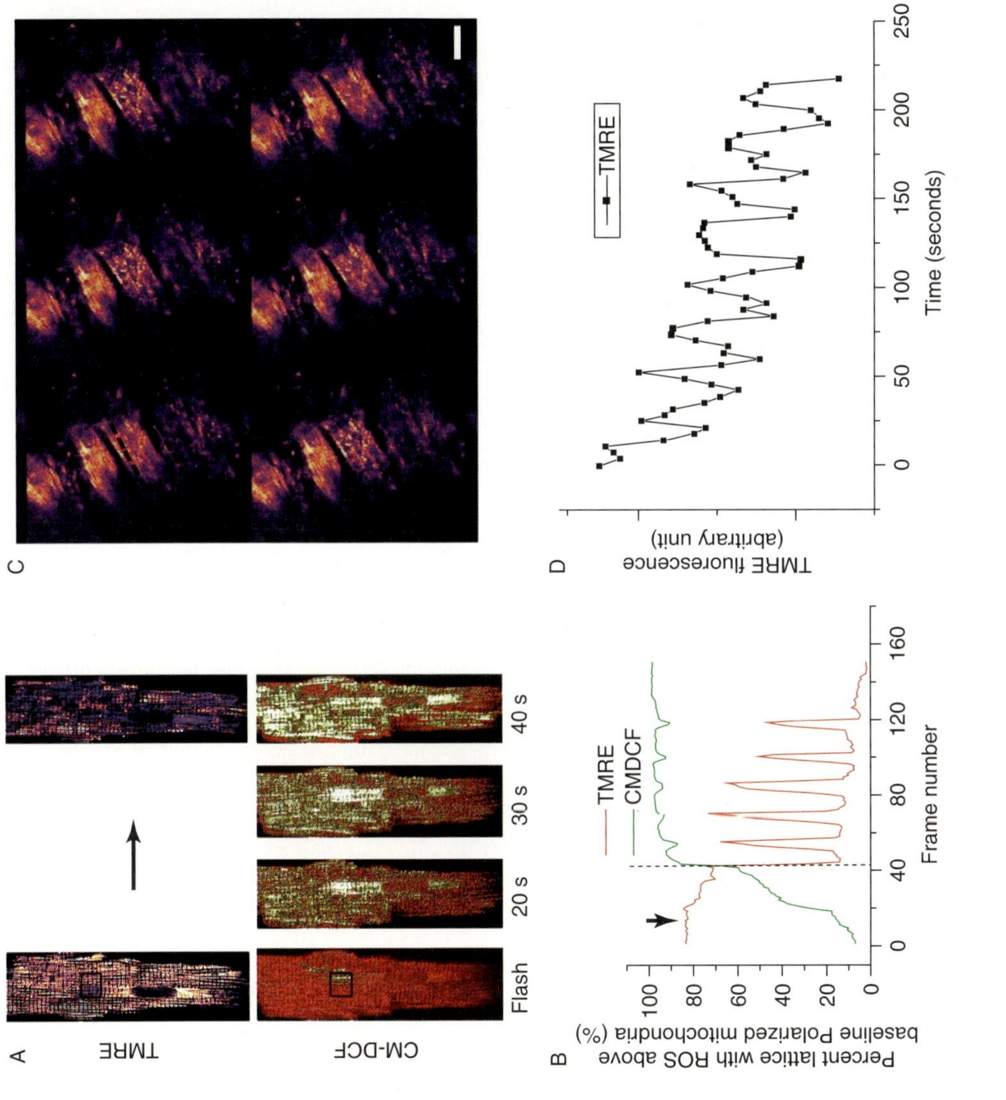
A
TMRE
CM-DCF
Flash
20 s
30 s
40 s
B
Percent lattice with ROS above baseline Polarized mitochondria (%)
TMRE
CMDCF
Frame number
C
D
TMRE fluorescence (abritrary unit)
TMRE
Time (seconds)

The cardiac mitochondrial network exhibits the transition from physiological to pathophysiological behavior with all the characteristics of systems at critical state – that is, a mitochondrial percolation cluster attains criticality at a certain threshold level of ROS [98] (Fig. 8). The transition that occurs at the percolation threshold is self-organized, and occurs with all the traits of universality – that is, with similar critical exponents as predicted by percolation theory [37, 38, 56], and the mitochondrial cluster exhibits fractal organization [55, 98, 108].

Recent data obtained in working hearts further ascertain the results found in cells, tissue, and organ levels. In intact perfused rat heart, a novel technique of high-resolution $\Delta\Psi_m$ imaging revealed complex spatial patterns and dynamically fluctuating changes in $\Delta\Psi_m$ during global ischemia (Fig. 10). These $\Delta\Psi_m$ fluctuations developed into actively propagating waves of mitochondrial depolarization/ collapse (velocity $\sim$20 $\mu m\,s^{-1}$, similar to those described in cells) [98, 102] across the epicardial surface (4 mm^2 area with cellular spatial resolution) [109]. Complex spatiotemporal gradients of $\Delta\Psi_m$ during metabolic stress produced by ischemic injury could be also visualized in hearts exhibiting left ventricular hypertrophy (LVH) [110]. These authors also showed that arrhythmia propensity increases in response to altered metabolic properties in LVH, resulting in heterogeneous levels of $\Delta\Psi_m$. These heterogeneities may ultimately modulate myocardial excitability and contribute to the formation of zones of conduction block, consistent with the recently suggested proposal of metabolic sink/block [96, 97]. This proposal states that $\Delta\Psi_m$ loss occurring on reperfusion when coupled with activation of sarcolemmal K_{ATP} channels may create spatial and temporal AP heterogeneity that

Fig. 8 Mitochondrial criticality and oscillations in energetics at the percolation threshold in isolated cardiomyocytes and intact whole heart. (a) The mitochondrial lattice of cardiomyocytes, as imaged by two-photon laser scanning fluorescence microscopy, was labeled with a membrane potential marker (top panels; tetramethyl rhodamine ethyl ester (TMRE)-labeled cell, intensity scaled from blue to white) and a reactive oxygen species (ROS)-sensitive fluorescent probe [lower panels; CM-H_2DCFDA signal, scaled from red (reduced) to yellow (oxidized)] [79, 98]; (b) The time course of mitochondrial membrane potential, $\Delta\Psi_m$, depolarization (indicated by the arrow) and development of the mitochondrial spanning cluster as mitochondrial ROS accumulate [lower panels of (a)] after a local flash (indicated by the white square). Criticality is reached when about 60% ($p_c = 0.56 \pm 0.07$) of the mitochondria have CM-DCF [5-(-6)-chloromethyl-2,7-dichlorohydrofluorescein] fluorescence levels 20% above baseline [98]. This coincides with the expected theoretical percolation threshold, p_c, of 0.596 [38, 56]. The fractal dimension, D_f, of the mitochondrial cluster ($D_f = 1.85$) as revealed by the ROS probe [panel (a), bottom row, right) was also consistent with percolation theory ($D_f = 1.89$) [24, 55, 98]; (c,d) $\Delta\Psi_m$ oscillations in myocytes of the intact heart perfused with Ca^{2+}-free buffer and imaged with two-photon laser scanning fluorescence microscopy [105]. The $\Delta\Psi_m$ signal (TMRE fluorescence) started to oscillate spontaneously after 2–3 min of normoxic perfusion with Ca^{2+}-free Tyrode solution. Sustained oscillations in $\Delta\Psi_m$ were observed for several minutes in the cell [indicated by the yellow dashed line in panel (c)]. Both, intracellular and intercellular heterogeneity of $\Delta\Psi_m$ is evident in the epicardial optical sections. The scale bar in panel (c) = 20 μm. Adapted from Aon, M.A., Cortassa, S., O'Rourke, B. (2004) *Proc. Natl Acad. Sci. USA*, **101**, 4447–4452 [98], and Slodzinski, M., Aon, M.A., O'Rourke, B. (2008) *J. Mol. Cell. Cardiol.*, **45**, 650–660 [105].

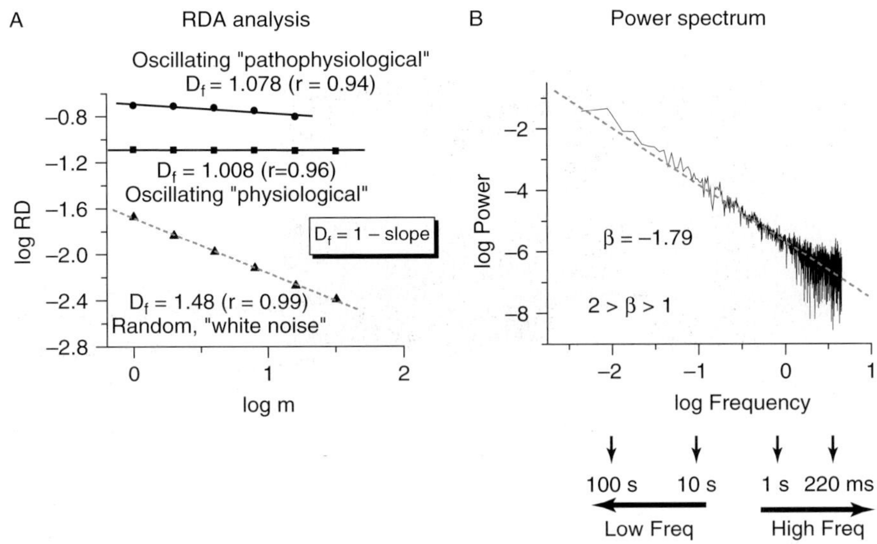

Fig. 9 Statistical analysis of $\Delta\Psi_m$ time series of the mitochondrial network of living cardiomyocytes loaded with tetramethylrhodamine methyl ester (TMRM), showing that it functions as a highly correlated network of oscillators. (a) Relative dispersional analysis (RDA) as a function of the aggregation parameter, *m*, gives a fractal dimension, D_f, close to 1.0 for myocytes showing large oscillations in $\Delta\Psi_m$ (TMRM, fluorescence) or those under physiological conditions, indicating a deterministic control process, while D_f is 1.5 for a completely random process; (b) Power spectral analysis (PSA) of $\Delta\Psi_m$ time series after fast Fourier transform also reveals a broad spectrum of oscillation in normally polarized mitochondria with a spectral exponent, $\beta = 1.79$, while a random process gives a $\beta = 0$. This means that there is no relationship between the amplitude and the frequency in a random signal. Reproduced from Aon, M.A., Cortassa, S., O'Rourke, B. (2007) On the network properties of mitochondria, in: V. Saks (ed.) *Molecular System Bioenergetics: Energy for Life*, pp. 111–135. Wiley-VCH Verlag GmbH & Co. KGaA, Weinheim, Germany [42].

can be a substrate of ventricular re-entry. Within this rationale, it was postulated that the failure of mitochondrial energetics can create "metabolic sinks" in the reperfused myocardium that may constitute sites of functional conduction block [96, 97, 111].

This is reminiscent of heart rate regulation in a normal subject, and suggests the intriguing possibility that a change in mitochondrial network dynamics could entrain pathological consequences for the cell and the organism. Evidence has been provided that, in the heart, multifrequency oscillations in rhythm and amplitude create a highly unstable prefibrillatory condition that presages fibrillation [112–114].

7.2 Chaos, Multi-oscillatory Systems, and Inverse Power Laws

In recent years, circahoralian (~40 min) respiratory oscillations in yeast have been shown to be linked to an ultradian (less than 24 h) clock [52, 115]. Microarray analysis showed a genome-wide oscillation in transcription, with expression maxima at three temporally separated

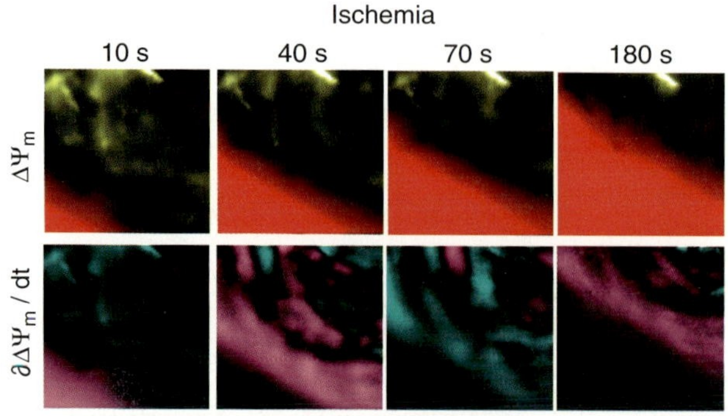

Fig. 10 Spatiotemporal fluctuations of mitochondrial membrane potential during global ischemia in the intact perfused rat heart. Successive contour maps of the mitochondrial membrane potential, $\Delta\Psi_m$ (above) and its first derivative (below) acquired at 10 s intervals across a 4×4 mm^2 mapping field during global no-flow ischemia in a representative rat heart. Detailed analysis of the spatial distribution of $\Delta\Psi_m$ and its temporal first derivative ($\partial\Delta\Psi_m/dt$) revealed the existence of neighboring regions with alternating positive and negative deflections in their $\Delta\Psi_m$ slope (i.e., ripples). These ripples exhibited peak-to-peak distances ranging from 500 to 1000 μm, therefore spanning several adjacent cardiomyocytes (mean size $\sim 150 \times 35$ μm). These data illustrate the presence of spatially and temporally discordant kinetics of $\Delta\Psi_m$, reflecting out-of-phase changes in the kinetics of $\Delta\Psi_m$ (frequency of ~0.1 Hz) between adjacent regions of ischemic tissue. Color scale in $\Delta\Psi_m$ contour maps (top row): baseline (black); depolarization (red); hyperpolarization (yellow). Color scale in $\partial\Delta\Psi_m/dt$ contour maps (bottom row): baseline (black); positive slopes (turquoise); negative slopes (purple). Figure courtesy of Dr F.G. Akar. Modified from Ref. [109].

clusters on the 40 min time-base. During the first temporal cluster (reductive phase, low respiration), approximately 4700 transcripts (of ca. 5300) were expressed, whereas the remaining ca. 600 were detected in the second temporal cluster (oxidative phase, high respiration). The detection of synchronous bursts of DNA replication, coincident with the phase of decreasing respiration rates, suggests temporally separated windows for biochemical reactions otherwise incompatible because of the sensitivity to ROS and oxidative stress exhibited by DNA synthesis [52]. This evident timekeeping function of the ultradian clock uncovers its powerful biological role, which appears to have been evolutionarily conserved across a wide range of phyla [94].

The synchronized yeast cultures exhibited other frequencies in the ultradian domain, beyond the circahoralian. Multi-oscillatory states of O_2, CO_2 levels in the culture exhibiting widely different periods (13 h, 40 min, and 4 min) were observed [95]. Interestingly, the phase space representation of three measured gases (O_2, CO_2, H_2S) from a long-term *S. cerevisiae* culture was compatible with a chaotic attractor (Fig. 11), as characterized by the existence of a positive Lyapunov exponent [95], and a broadband range power spectrum [50]. The 36 000 discrete time points obtained by mass spectrometric measurement represent the best available evidence for chaos in an intact living system.

An analysis of O_2 and CO_2 time series showed perfect correlation over the

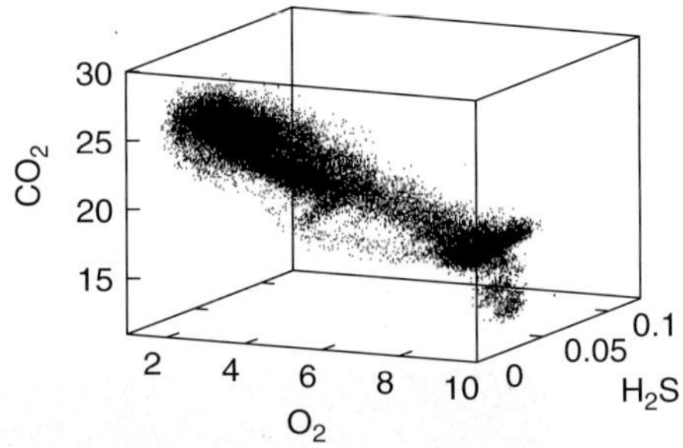

Fig. 11 A strange attractor observed experimentally in the dynamics of gas exchange in a self-synchronized continuous culture of the yeast *Saccharomyces cerevisiae*. The strange (metabolic) attractor was obtained by membrane inlet mass spectrometric (MIMS), monitoring m/z values at 32, 34, and 44 (dissolved oxygen, O_2, carbon dioxide, CO_2, and hydrogen sulfide, H_2S, respectively) relative to m/z 40 (argon), and collecting data points every 12 s for each gas (36 000 in total; see Ref. [95]). The attractor represents the set of biochemical states, as reflected in the concentrations of dissolved O_2, CO_2, and H_2S, through which the yeast culture system wanders during the complex oscillations. The strange attractor from yeast metabolism was reconstructed using time-delayed embedding (delay of 12 points and embedding dimension of 4) for the time series of O_2. The reconstructed attractor appeared topologically similar to that obtained experimentally, exhibiting a capacity dimension of 1.9 ± 0.08, in reasonable agreement with 2.09 ± 0.07 obtained with the full three-dimensional data set shown in this figure. A positive leading Lyapunov exponent of 0.752 ± 0.004 h^{-1}, well away from zero and therefore distinct from quasiperiodic dynamics, completed the characterization of chaotic dynamics in the O_2 time series.

broad range of temporal scales among multi-oscillatory states, according to relative dispersional analysis (RDA). As expected for a time series exhibiting self-similar scaling by RDA, an inverse power law proportional to $1/f^{\beta}$ was obtained after power spectral analysis (PSA) ($\beta = 1.95$ or 1.45 for O_2 and CO_2 signals, respectively) [50]. The inverse power law behavior shown by yeast physiology in aerobic continuous culture is a manifestation of self-similar, fractal behavior in its dynamics, consistent with long-term temporal correlation that can be interpreted as "memory," as opposed to random behavior [23, 116]. The broad frequency distribution, and the long-term temporal correlations exhibited by the physiological behavior of yeast, allows a change in one time scale to be felt across the frequency range from the intracellular to the intercellular; in the latter case, this occurs most likely by synchronization among single-cell oscillators [117] through chemical communication, for example, diffusible metabolites such as acetaldehyde [118], H_2S [119], and ROS [50].

The existence of a chaotic attractor including multiple oscillators correlated over a wide range of frequencies is suggestive of controlled chaos (Fig. 12). A controlled chaotic attractor implying the stabilization of selected orbits would be compatible with the synchronization of multi-oscillatory states [90, 91], which appears to be essential for the coordination of metabolism and transcription across a population of single-celled organisms [52, 90–92, 94, 95].

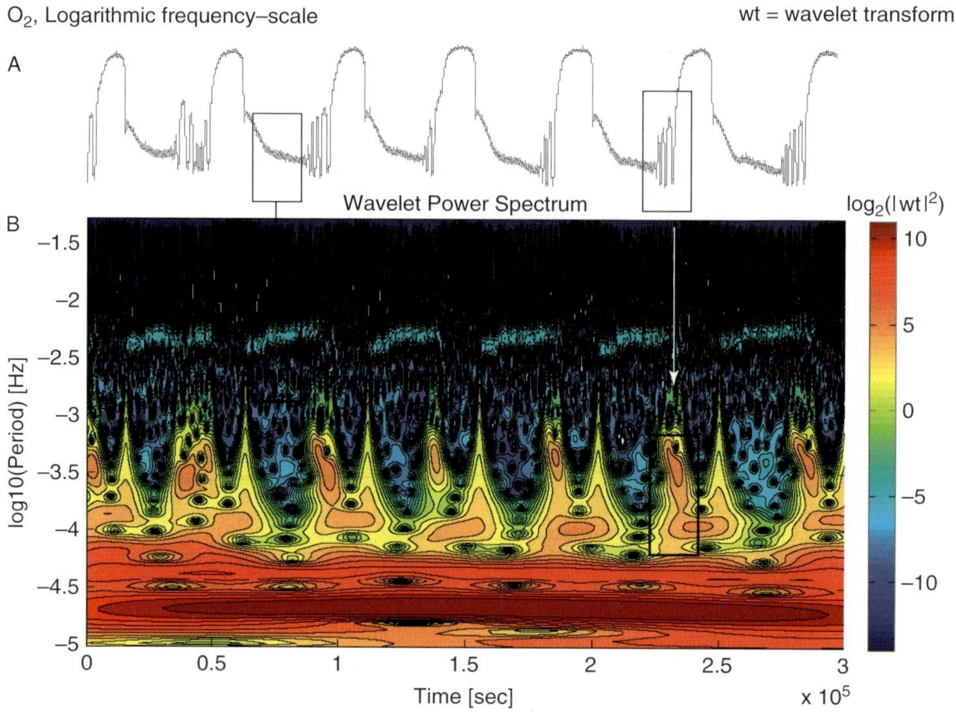

Fig. 12 Wavelet analysis of the O_2 signal time series obtained from the self-organized multi-oscillatory continuous culture of *S. cerevisiae*. (a) Relative MIMS signal of the $m/z = 32$ component, which corresponds to O_2, versus time. The time scale in panel (a) is not shown, but corresponds to hours after the start of fermentor continuous operation (see Refs [50, 94] for more detail). The large-amplitude oscillation showed substantial cycle-to-cycle variability, with a mean ($\pm$SD) period of 13.6 ± 1.3 h ($n = 8$). The other two evident oscillation periods are $\sim$40 min (circahoralian bursts) and $\sim$4 min; (b) Depicted is the corresponding logarithmic absolute squared wavelet transform over logarithmic frequency and time. As a form of time–frequency representation, the wavelet transform expands signals in terms of wavelets by breaking the signal down into different scale components [102]. At any time, the wavelet transform uncovers the predominant frequencies: there are periodically recurring frequency contents at about 3.1–10 mHz (−2.5 to − 2.0 on the logarithmic frequency scale, light blue: corresponding to the few min period range) and at about 0.1–1 mHz (−4 to −3 on the logarithmic frequency scale, yellow: corresponding to the circahoralian range). There is also an overall low-frequency component (10–30 μHz, −5 to −4.5 on the logarithmic frequency scale, dark red: corresponding to the several hours period range). Of note is the complex and fine dynamic structure revealed by wavelet analysis. The boxes and arrows show the correspondence between the time series and the wavelet plot. Wavelet analysis was carried out using the Morlet wavelet with wavelet software in Matlab v7.1.0.246 (R14), as described previously [102]. The figure and wavelet analysis were provided courtesy of Dr F. T. Kurz, affiliated with Charité Universitatsmedizin, Berlin, Germany, and Massachusetts General Hospital, Charlestown, USA.

8
Chaos in Dynamics of Heart and Brain?

The application of these ideas to heart and brain function in (patho) physiology remains an active area of research [19, 64, 99]. Progress in nonlinear dynamics and time series analysis has reached a stage, where fruitful applications to electroencephalography and magnetoencephalography have become a reality.

The overall brain as well as heart dynamics appears to be governed by nested networks of interacting dynamical systems [19, 41, 65, 81, 120]. Under normal conditions, the degree of interaction and extent of synchronization appears weak, although finite, and this bestows a certain structure to spontaneous organ dynamics. However, under pathological conditions, the coherent interaction becomes strong and the overall dynamics less complex [97].

Electrocardiographic data display a scale-free, broadband frequency spectrum of the QT interval [16] (see Fig. 3), unlike in certain cardiac disease states where there is "loss of spectral reserve." Similarly, brain activity in the awake state is not random noise but rather shows scale-free dynamics that reveals fluctuations of synchronization among the dynamics of functional networks at local as well as global levels [19]. This scale-free dynamics, which in the case of brain might even be preserved under pathological conditions, reflects critical dynamics near a phase transition. The latter is characterized by multiple outputs over a broad range of spatial and temporal scales, as shown for other systems (see Sects. 5 and 7).

In the case of the heart, the dynamic complexity underlying heart rate variability, as revealed by nonlinear time series analysis, is given by different physiological mechanisms. These include processes at the cellular level, the influence of respiration on the heart rate, coronary vasomotor adjustment to cardiac metabolism and perfusion pressure, and the interactions of the multiple feedback loops regulating the cardiovascular system [64, 121]. By using quantitative methods from nonlinear dynamic analysis, it was concluded that coronary circulation exhibits chaotic nonlinear behavior in flow motion under basal conditions. The nature of the chaotic attractor was responsive to the blockade of nitric oxide synthase or the antagonism of purinergic receptors, suggesting that a lack of certain cellular regulators (e.g., nitric oxide, AMP, ADP, ATP, and adenosine) makes the coronary flow less predictable [121].

In a recent "focused" issue of the journal *Chaos* (vol. 19; see Ref. [64]), the question of whether the normal heart rate is chaotic was addressed in several articles. The application of old and new methods of nonlinear time series analysis suggests that the topic remains controversial (see e.g., Ref. [122]). This issue of *Chaos* describes the application of new methods of time series analysis to help elucidate the complex dynamical features of heart rate variability (HRV).

The recently introduced "noise titration" method was able to identify signatures of chaotic dynamics in HRV in health and in congestive heart failure (CHF) [123]. In healthy hearts, chaos appears to depend on circadian-sleep/wake cycles, but in CHF chaos it is transient, even though it is currently not possible to differentiate reliably between deterministic or stochastic chaos (i.e., that arising from techniques of monitoring).

Chaos has been demonstrated in recent modeling studies of early cardiac "afterdepolarizations" (EADs). These are voltage oscillations that occur during the repolarizing phase of the AP, and they have been implicated as the cause of a certain type of lethal cardiac arrhythmia [124]. This model exhibits EAD behavior when driven periodically within a certain range of pacing cycle length (PCL), and chaotic dynamics was found at intermediate PCLs; on the other hand, behavior is periodic at both slow and fast pacing [124]. Over a limited range of tissue sample, smaller than a critical value, chaotic EADs may synchronize globally. However, beyond the critical size EADs no longer synchronize, and this results in regions of partial synchronization that shift in time and space [125]. The experimental verification of model predictions obtained in rabbit hearts exposed to oxidative stress induced with H_2O_2, exhibited multiple shifting foci causing polymorphic tachycardia and fibrillation [125].

Studies conducted in the brain over several years have shown that the initial hypothesis of a low dimensional chaotic attractor underlying brain dynamics is too simple; the only type of brain state that comes close to this is the brain dynamics of an epileptic seizure [19]. The emerging trend from nonlinear analysis is to identify to what extent nonlinear phenomena – such as the level of synchronization between different network elements – contribute to a particular brain state.

9 Concluding Remarks: The Status and a Prospective for Chaos

One of the main contributions of chaos theory has been to show that irregular dynamic behavior can be produced as a consequence of simple and tractable rate equations in an autonomous way – that is, without exogenous time-dependent inputs. This conceptual breakthrough allowed an establishment of the principle that aperiodic or noise-like behavior can originate from the intrinsic dynamics of a system, and not be the product of uncontrolled environmental influences.

Chaos has also established limits to our predictive capabilities, ranging from weather forecasting to sudden death from heart or brain injury. Where there is no chaos, can we make perfect predictions? Well, no; the next limit being complexity. A remarkable difference between some simple physical systems and biological systems is the large number of variables involved along with numerous feed-back and feed-forward loops, which define nonlinear interactions (Figs 2 and 4). The latter, in conjunction with the continuous exchange of energy, matter, and information with the environment, are the sources of the observed complex dynamic outputs. Deterministic chaos is one possible consequence. The mathematical tractability and solution of these systems can only be approached by computer simulation based on numerical integration of, sometimes large, systems of equations. Some of the most satisfactory demonstrations of chaos have been achieved in the simplest difference (logistic) equation [22], employed as a simplistic model of population dynamics of organisms.

The recent discovery of scaling, criticality and colored noise behavior in mitochondrial networks, at a fundamental level of cellular organization, has introduced new approaches to the understanding of complexity in biology. As it relates to energetics, life and death [21], these experiments and model constructs demonstrate that physicists were already anticipating fundamental behavior in biology.[2] However, there are limits of similarity between physics and biology with respect to the use of these principles. Biologists need to know mechanistic details, as these will determine whether we would be able to deal with, for example, heart arrest or injury, or prevent diseases. Biologists aim to search for and characterize the principles of universal physical behavior, but must take into account many biologically relevant details [21, 126]. Furthermore, diversity is an essence of biological systems that have evolved over billions of years in a stochastic environment.

Already well into the twenty-first century, the search for and characterization of chaotic dynamics has evolved into the study of complex dynamic behavior in heart and the brain, both in health and disease. Chaos is not alone, because *la renaissance* of network theory [35, 127], and its widespread application to networks of different nature – that is, metabolic, social, economic, communication, and ecosystems – constitute the foundation of Systems Biology [128, 129]. Yet, neither the refurbished network theory nor the rather simplistic version of Systems Biology – which originally was defined as the integration of high-throughput (-omic) technologies and computer modeling – are able to fully address complex biological systems without integrating the concepts of self-organization in thermodynamically nonequilibrium nonlinear systems [26, 54], fractals [36], and critical phenomena [37, 38, 56, 61]. This integration has been proposed through the Complex Systems Approach [21], of which the basic concepts are self-organization, the topologies and dynamics of networks, and scaling, in a wide class of complex systems.

The pervasive and fundamental importance of the concept of scaling for biology and biochemistry is such because it shows the intrinsic interdependence among the multiple spatiotemporal functional scales exhibited by living systems. Geometric and dynamic fractals through scaling and self-similarity [23, 116] capture the essential feature of function correlation among the different scales of space and time exhibited by living systems, from cells to ecosystems [24, 43, 51, 54, 55, 115, 130, 131].

The scale-free dynamics, with multiple outputs across several time scales (expressed as oscillatory frequencies), exhibited both by the heart and yeast systems, suggest that intracellular network *dynamic organization* manifests in the form of *dynamic fractals*. This adds to the already well-established scale-free architecture demonstrated in network connectivity [35, 48]. In the cellular realm, dynamic organization encompasses both the architectural and the topological views of network analysis, obeying two basic and distinctive characteristics of living systems: "operational closure" and self-organization [42]. The concept of *operational closure* was introduced more than 35 years ago by Varela, Maturana and Uribe [132] to describe the capacity of living systems for the autodynamic construction of their maintenance and reproduction

2) In fact, when studying critical phenomena in first-order (melting–freezing of water) or second-order (spontaneous magnetization of iron at the Curie point) phase transitions, physicists have already found that at critical points, fluctuations occur on all length scales, the system becoming infinitely sensitive to fluctuations – the slightest imbalance suffices to tip it one way or the other [31]. Power laws whose critical exponents become independent of systems details, and thus universal, and spatial fractal organization, describe systems at the critical state, no matter how complex [50, 97].

(autopoietic networks). This key concept, together with that of *dissipative structures* through self-organization [27], define living networks as continually self-producing in an iterative construction; that is, they produce those components and are themselves produced by their components. The continuous action of intrinsic biological dynamics, with the potential for shifting among attractors at bifurcation points, define the homeodynamic condition [26]. This is opposed to homeostasis, which as rigorously defined is in fact only a particular case of the more general homeodynamic condition [51, 54]. Dynamic fractals arising from nonlinear time series, as orderly statistical sequences tying events across different time scales (from milliseconds to hours), express the fundamental fractal organization of biological timekeeping; that is, what affects one time scale affects them all [23, 56]. Moreover, the influence of past to present events in a biological system can be understood through dynamic fractals, since short-term fluctuations are somehow related to longer-term trends in the dynamic behavior of a system. In this sense, the unpredictability of chaos as a source of novelty, and the long-term memory feature of dynamic fractals, may represent essential ingredients of biological timekeeping, necessarily highly conserved over evolutionary time.

Acknowledgments

These studies were supported by NIH grants **R01-HL091923-01** and **P01-HL081427**. The authors gratefully acknowledge Dr F.G. Akar (Mount Sinai Hospital, NY, USA) and Dr F.T. Kurz (Charité Universitätsmedizin, Berlin, Germany, and Massachusetts General Hospital, Charlestown, USA) for their kind contribution of Figs 10 and 12.

References

1 Poincare, J.H. (1892) Les Methodes Nouvelles de la Mechanique Celeste, English translation (1967), NASA, Washington, DC.

2 Lorenz, E.N. (2001) *The Essence of Chaos*, University of Washington Press, Seattle.

3 Ruelle, D. (1980) Strange attractors. *Math. Intell.*, **2**, 126–137.

4 Gleick, J. (1988) *Chaos: Making a New Science*, Penguin Books, New York.

5 Lorenz, E.N. (1963) Deterministic nonperiodic flow. *J. Atmos. Sci.*, **20**, 130–141.

6 Olsen, L.F., Degn, H. (1977) Chaos in an enzyme reaction. *Nature*, **267**, 177–178.

7 Olsen, L.F., Degn, H. (1985) Chaos in biological systems. *Q. Rev. Biophys.*, **18**, 165–225.

8 Decroly, O., Goldbeter, A. (1982) Birhythmicity, chaos, and other patterns of temporal self-organization in a multiply regulated biochemical system. *Proc. Natl Acad. Sci. USA*, **79**, 6917–6921.

9 Baier, G., Sahle, S. (1998) Homogeneous and spatio-temporal chaos in biochemical reactions with feedback inhibition. *J. Theor. Biol.*, **193**, 233–242.

10 Geest, T., Steinmetz, C.G., Larter, R., Olsen, L.F. (1992) Period doubling bifurcations and chaos in an enzyme reaction. *J. Phys. Chem.*, **96**, 5678–5680.

11 Markus, M., Kuschmitz, D., Hess, B. (1984) Chaotic dynamics in yeast glycolysis under periodic substrate input flux. *FEBS Lett.*, **172**, 235–238.

12 Markus, M., Kuschmitz, D., Hess, B. (1985) Properties of strange attractors in yeast glycolysis. *Biophys. Chem.*, **22**, 95–105.

13 Nielsen, K., Sorensen, P.G., Hynne, F., Busse, H.G. (1998) Sustained oscillations in glycolysis: an experimental and theoretical study of chaotic and complex periodic behavior and of quenching of simple oscillations. *Biophys. Chem.*, **72**, 49–62.

14 Aon, M.A., Cortassa, S., Lloyd, D. (2000) Chaotic dynamics and fractal space in biochemistry: simplicity underlies complexity. *Cell Biol. Int.*, **24**, 581–587.

15 Glass, L., Mackey, M.C. (1988) *From Clocks to Chaos. The Rhythms of Life*, Princeton University Press, Princeton.

16 Goldberger, A.L., West, B.J. (1987) Applications of nonlinear dynamics to clinical

cardiology. *Ann. N.Y. Acad. Sci.*, **504**, 195–213.

17 Goldberger, A.L., Rigney, D.R., West, B.J. (1990) Chaos and fractals in human physiology. *Sci. Am.*, **262**, 42–49.

18 Elbert, T., Ray, W.J., Kowalik, Z.J., Skinner, J.E., Graf, K.E., Birbaumer, N. (1994) Chaos and physiology: deterministic chaos in excitable cell assemblies. *Physiol. Rev.*, **74**, 1–47.

19 Stam, C.J. (2005) Nonlinear dynamical analysis of EEG and MEG: review of an emerging field. *Clin. Neurophys.*, **116**, 2266–2301.

20 Aon, M.A., Cortassa, S. (2005) Metabolic dynamics in cells viewed as multilayered, distributed, mass-energy-information networks, in: *Encyclopedia of Genetics, Genomics, Proteomics and Bioinformatics*, Wiley Interscience, New York. Available at: http://onlinelibrary.wiley.com/doi/10.1002/047001153X.g308206/abstract.

21 Aon, M.A., Cortassa, S. (2009) Chaotic Dynamics, Noise and Fractal Space in Biochemistry, in: Meyers, R. (Ed.) *Encyclopedia of Complexity and Systems Science*, Springer, New York, pp. 476–489.

22 May, R.M. (1976) Simple mathematical models with very complicated dynamics. *Nature*, **261**, 459–467.

23 West, B.J. (1999) *Physiology, Promiscuity and Prophecy at the Millennium: A Tale of Tails*, World Scientific, Singapore.

24 Aon, M.A., Cortassa, S. (1997) *Dynamic Biological Organization. Fundamentals as Applied to Cellular Systems*, Chapman & Hall, London.

25 Haken, H. (1983) *Synergetics. An Introduction*, Springer-Verlag, Berlin, Heidelberg.

26 Lloyd, D., Aon, M.A., Cortassa, S. (2001) Why homeodynamics, not homeostasis? *ScientificWorldJournal*, **1**, 133–145.

27 Nicolis, G., Prigogine, I. (1977) *Self-Organization in Nonequilibrium Systems: From Dissipative Structures to Order Through Fluctuations*, John Wiley & Sons, Inc., New York.

28 Yang, J.H., Yang, L., Qu, Z., Weiss, J.N. (2008) Glycolytic oscillations in isolated rabbit ventricular myocytes. *J. Biol. Chem.*, **283**, 36321–36327.

29 Yang, L., Korge, P., Weiss, J.N., Qu, Z. (2010) Mitochondrial oscillations and waves in cardiac myocytes: insights from computational models. *Biophys. J.*, **98**, 1428–1438.

30 Yates, E.F. (1987) *Self-Organizing Systems. The Emergence of Order*, Plenum Press, New York.

31 Ball, P. (2004) *The Self-Made Tapestry: Pattern Formation in Nature*, Oxford University Press, Oxford.

32 Lloyd, D. (2009) Oscillations, synchrony and deterministic chaos. *Prog. Bot.*, **70**, 70–91.

33 Cortassa, S., O'Rourke, B., Winslow, R.L., Aon, M.A. (2009) Control and regulation of mitochondrial energetics in an integrated model of cardiomyocyte function. *Biophys. J.*, **96**, 2466–2478.

34 Cortassa, S., O'Rourke, B., Winslow, R.L., Aon, M.A. (2009) Control and regulation of integrated mitochondrial function in metabolic and transport networks. *Int. J. Mol. Sci.*, **10**, 1500–1513.

35 Barabasi, A.L. (2003) *Linked*, Plume, New York.

36 Mandelbrot, B.B. (1983) *The Fractal Geometry of Nature*, W.H. Freeman, New York.

37 Sornette, D. (2000) *Critical Phenomena in Natural Sciences: Chaos, Fractals, Self Organization and Disorder: Concepts and Tools*, Springer, Berlin, Heidelberg.

38 Stauffer, D., Aharony, A. (1994) *Introduction to Percolation Theory*, Taylor & Francis, London.

39 Strogatz, S.H. (2000) From Kuramoto to Crawford: exploring the onset of synchronization in populations of coupled oscillators. *Physica D*, **143**, 1–20.

40 Aon, M.A. (2010) From isolated to networked: A paradigmatic shift in mitochondrial physiology. *Front. Physiol.*, **1**, 1–3.

41 Aon, M.A., Cortassa, S. (2011) Mitochondrial network energetics, control, regulation and emergence in heart physiology and pathophysiology. *Adv. Exp. Med. Biol.* (in press).

42 Aon, M.A., Cortassa, S., O'Rourke, B. (2007) On the Network Properties of Mitochondria, in: Saks, V. (Ed.) *Molecular System Bioenergetics: Energy for Life*, Wiley-VCH Verlag GmbH, Weinheim, pp. 111–135.

43 Lloyd, D., Rossi, E.L. (2008) Epilogue: A New Vision of Life, in: Lloyd, D., Rossi, E.L. (Eds) *Ultradian Rhythms from Molecules to*

Mind, Springer Science+Business Media B.V., New York, pp. 431–439.

44 Kantz, H., Schreiber, T. (2005) *Nonlinear Time Series Analysis*, Cambridge University Press, New York.

45 Williams, G.P. (2003) *Chaos Theory Tamed*, Joseph Henry Press, Washington, DC.

46 Cortassa, S., Aon, M.A., Marban, E., Winslow, R.L., O'Rourke, B. (2003) An integrated model of cardiac mitochondrial energy metabolism and calcium dynamics. *Biophys. J.*, **84**, 2734–2755.

47 Cortassa, S., Aon, M.A., Winslow, R.L., O'Rourke, B. (2004) A mitochondrial oscillator dependent on reactive oxygen species. *Biophys. J.*, **87**, 2060–2073.

48 Barabasi, A.L., Oltvai, Z.N. (2004) Network biology: understanding the cell's functional organization. *Nat. Rev. Genet.*, **5**, 101–113.

49 Aon, M.A., Cortassa, S., O'Rourke, B. (2006) The fundamental organization of cardiac mitochondria as a network of coupled oscillators. *Biophys. J.*, **91**, 4317–4327.

50 Aon, M.A., Roussel, M.R., Cortassa, S., O'Rourke, B., Murray, D.B., Beckmann, M., Lloyd, D. (2008) The scale-free dynamics of eukaryotic cells. *PLoS ONE*, **3**, e3624.

51 Yates, F.E. (1992) Fractal applications in biology: scaling time in biochemical networks. *Methods Enzymol.*, **210**, 636–675.

52 Lloyd, D., Murray, D.B., Klevecz, R.R., Wolf, J., Kuriyama, H. (2008) The Ultradian Clock (40 min) in Yeast, in: Lloyd, D., Rossi, E.L. (Eds) *Ultradian Rhythms From Molecules to Mind*, Springer Science+Business Media B.V., New York, pp. 11–42.

53 Gisiger, T. (2001) Scale invariance in biology: coincidence or footprint of a universal mechanism? *Biol. Rev. Camb. Philos. Soc.*, **76**, 161–209.

54 Yates, F.E. (1993) Self-Organizing Systems, in: Boyd, C.A.R., Noble, D. (Eds) *The Logic of Life: The Challenge of Integrative Physiology*, Oxford University Press, New York, pp. 189–218.

55 Aon, M.A., O'Rourke, B., Cortassa, S. (2004) The fractal architecture of cytoplasmic organization: scaling, kinetics and emergence in metabolic networks. *Mol. Cell. Biochem.*, **256-257**, 169–184.

56 Schroeder, M. (1991) *Fractals, Chaos, Power Laws: Minutes From an Infinite Paradise*, W.H. Freeman and Company, New York.

57 Cvitanovic, P. (1989) *Universality in Chaos*, Adam Hilger, Bristol, New York.

58 Feigenbaum, M.J. (1980) Universal behavior in nonlinear systems. *Los Alamos Sci.*, **1**, 4–27.

59 West, B.J. (1990) *Fractal Physiology and Chaos in Medicine*, World Scientific, Singapore.

60 Bak, P. (1996) *How Nature Works: The Science of Self-Organized Criticality*, Copernicus Publications, New York.

61 Feder, J. (1988) *Fractals*, Plenum Press, New York.

62 Huberman, B.A., Rudnick, J. (1980) Scaling behavior of chaotic flows. *Phys. Rev. Lett.*, **45**, 154–156.

63 Barrow, J.D. (1999) *Impossibility: The Limits of Science and the Science of Limits*, Vintage, London.

64 Glass, L. (2009) Introduction to controversial topics in nonlinear science: is the normal heart rate chaotic? *Chaos*, **19**, 028501.

65 Cortassa, S., Aon, M.A., O'Rourke, B., Jacques, R., Tseng, H.J., Marban, E., Winslow, R.L. (2006) A computational model integrating electrophysiology, contraction, and mitochondrial bioenergetics in the ventricular myocyte. *Biophys. J.*, **91**, 1564–1589.

66 Roden, D.M., Balser, J.R., George, A.L., Anderson, M.E. (2002) Cardiac ion channels. *Annu. Rev. Physiol.*, **64**, 431–475.

67 Hodgkin, A.L., Huxley, A.F. (1952) A quantitative description of membrane current and its application to conduction and excitation in nerve. *J. Physiol.*, **117**, 500–544.

68 Weiss, J.N., Qu, Z., Garfinkel, A. (2003) Understanding biological complexity: lessons from the past. *FASEB J.*, **17**, 1–6.

69 Osborne, A.R., Provenzale, A. (1989) Finite correlation dimension for stochastic systems with power-law spectra. *Phys. D.*, **35**, 357–381.

70 Levens, D., Gupta, A. (2010) Molecular biology. Reliable noise. *Science*, **327**, 1088–1089.

71 Rao, C.V., Wolf, D.M., Arkin, A.P. (2002) Control, exploitation and tolerance of intracellular noise. *Nature*, **420**, 231–237.

72 Traynelis, S.F., Jaramillo, F. (1998) Getting the most out of noise in the central nervous system. *Trends Neurosci.*, **21**, 137–145.

73 Astumian, R.D., Moss, F. (1998) Overview: the constructive role of noise in fluctuation driven transport and stochastic resonance. *Chaos*, **8**, 533–538.

74 Moss, F., Ward, L.M., Sannita, W.G. (2004) Stochastic resonance and sensory information processing: a tutorial and review of application. *Clin. Neurophysiol.*, **115**, 267–281.

75 Kosko, B. (2006) *Noise*, Viking, New York.

76 Jung, P., Cornell-Bell, A., Moss, F., Kadar, S., Wang, J., Showalter, K. (1998) Noise-sustained waves in subexcitable media: from chemical waves to brain waves. *Chaos*, **8**, 567–575.

77 Glass, L. (2001) Synchronization and rhythmic processes in physiology. *Nature*, **410**, 277–284.

78 West, B.J. (2006) *Where Medicine Went Wrong: Rediscovering the Path to Complexity*, World Scientific Publishing Co. Pte Ltd., Singapore.

79 Aon, M.A., Cortassa, S., Marban, E., O'Rourke, B. (2003) Synchronized whole cell oscillations in mitochondrial metabolism triggered by a local release of reactive oxygen species in cardiac myocytes. *J. Biol. Chem.*, **278**, 44735–44744.

80 Zhou, L., Aon, M.A., Almas, T., Cortassa, S., Winslow, R.L., O'Rourke, B. (2006) A reaction-diffusion model of ROS-induced ROS release in a mitochondrial network. *PLoS Comput. Biol.*, **6**, e1000657.

81 Zhou, L., Cortassa, S., Wei, A.C., Aon, M.A., Winslow, R.L., O'Rourke, B. (2009) Modeling cardiac action potential shortening driven by oxidative stress-induced mitochondrial oscillations in guinea pig cardiomyocytes. *Biophys. J.*, **97**, 1843–1852.

82 Takens, F. (1981) Detecting Strange Attractors in Turbulence, in: Rand, D.A., Young, L.-S. (Eds) *Dynamical Systems and Turbulence*, Springer-Verlag, Berlin, pp. 366–381.

83 Grassberger, P., Procaccia, I. (1983) Measuring the strangeness of strange attractors. *Physica D*, **9**, 189–208.

84 Rapp, P.E., Albano, A.M., Schmah, T.I., Farwell, L.A. (1993) Filtered noise can mimic low-dimensional chaotic attractors. *Phys. Rev. E.*, **47**, 2289–2297.

85 Garfinkel, A., Spano, M.L., Ditto, W.L., Weiss, J.N. (1992) Controlling cardiac chaos. *Science*, **257**, 1230–1235.

86 Shinbrot, T., Ditto, W., Grebogi, C., Ott, E., Spano, M., Yorke, J.A. (1992) Using the sensitive dependence of chaos (the "butterfly effect") to direct trajectories in an experimental chaotic system. *Phys. Rev. Lett.*, **68**, 2863–2866.

87 Moss, F. (1994) Neurobiology. Chaos under control. *Nature*, **370**, 596–597.

88 Ott, E., Grebogi, C., Yorke, J.A. (1990) Controlling chaos. *Phys. Rev. Lett.*, **64**, 1196–1199.

89 Meucci, R., Labate, A., Ciofini, C. (1998) Experimental techniques for controlling chaos in lasers. *Int. J. Bifurcation Chaos*, **8**, 1759–1768.

90 Lloyd, A.L., Lloyd, D. (1993) Hypothesis: the central oscillator of the circadian clock is a controlled chaotic attractor. *Biosystems*, **29**, 77–85.

91 Lloyd, A.L., Lloyd, D. (1995) Chaos: its significance and detection in biology. *Biol. Rhythm Res.*, **26**, 233–252.

92 Lloyd, D., Lloyd, A.L. (1994) Hypothesis: a controlled chaotic attractor constitutes the central oscillator of the circadian clock. *Biochem. Soc. Trans.*, **22**, 322.

93 Lloyd, D., Lloyd, A.L., Olsen, L.F. (1992) The cell division cycle: a physiologically plausible dynamic model can exhibit chaotic solutions. *Biosystems*, **27**, 17–24.

94 Lloyd, D., Murray, D.B. (2007) Redox rhythmicity: clocks at the core of temporal coherence. *BioEssays*, **29**, 465–473.

95 Roussel, M.R., Lloyd, D. (2007) Observation of a chaotic multioscillatory metabolic attractor by real-time monitoring of a yeast continuous culture. *FEBS J.*, **274**, 1011–1018.

96 Akar, F.G., Aon, M.A., Tomaselli, G.F., O'Rourke, B. (2005) The mitochondrial origin of postischemic arrhythmias. *J. Clin. Invest.*, **115**, 3527–3535.

97 Aon, M.A., Cortassa, S., Akar, F.G., Brown, D.A., Zhou, L., O'Rourke, B. (2009) From mitochondrial dynamics to arrhythmias. *Int. J. Biochem. Cell Biol.*, **41**, 1940–1948.

98 Aon, M.A., Cortassa, S., O'Rourke, B. (2004) Percolation and criticality in a

mitochondrial network. *Proc. Natl Acad. Sci. USA*, **101**, 4447–4452.

99 Stam, C.J. (2010) Use of magnetoencephalography (MEG) to study functional brain networks in neurodegenerative disorders. *J. Neurol. Sci.*, **289**, 128–134.

100 Aon, M.A., Cortassa, S., O'Rourke, B. (2008) Mitochondrial oscillations in physiology and pathophysiology. *Adv. Exp. Med. Biol.*, **641**, 98–117.

101 Brady, N.R., Elmore, S.P., van Beek, J.J., Krab, K., Courtoy, P.J., Hue, L., Westerhoff, H.V. (2004) Coordinated behavior of mitochondria in both space and time: a reactive oxygen species-activated wave of mitochondrial depolarization. *Biophys. J.*, **87**, 2022–2034.

102 Kurz, F., Aon, M.A., O'Rourke, B., Armoundas, A. (2010) Spatio-temporal oscillations of individual mitochondria in cardiac myocytes reveal modulation of synchronized mitochondrial clusters. *Proc. Natl Acad. Sci. USA*, **107**, 14315–14320.

103 O'Rourke, B., Ramza, B.M., Marban, E. (1994) Oscillations of membrane current and excitability driven by metabolic oscillations in heart cells. *Science*, **265**, 962–966.

104 Romashko, D.N., Marban, E., O'Rourke, B. (1998) Subcellular metabolic transients and mitochondrial redox waves in heart cells. *Proc. Natl Acad. Sci. USA*, **95**, 1618–1623.

105 Slodzinski, M.K., Aon, M.A., O'Rourke, B. (2008) Glutathione oxidation as a trigger of mitochondrial depolarization and oscillation in intact hearts. *J. Mol. Cell. Cardiol.*, **45**, 650–660.

106 Zorov, D.B., Filburn, C.R., Klotz, L.O., Zweier, J.L., Sollott, S.J. (2000) Reactive oxygen species (ROS)-induced ROS release: a new phenomenon accompanying induction of the mitochondrial permeability transition in cardiac myocytes. *J. Exp. Med.*, **192**, 1001–1014.

107 Aon, M.A., Cortassa, S., O'Rourke, B. (2008) Is there a Mitochondrial Clock? in: Lloyd, D., Rossi, E.L. (Eds) *Ultradian Rhythms from Molecules to Mind: A New Vision of Life*, Springer-Verlag, New York, pp. 129–144.

108 Aon, M.A., Cortassa, S., Akar, F.G., O'Rourke, B. (2006) Mitochondrial criticality: a new concept at the turning point of life or death. *Biochim. Biophys. Acta*, **1762**, 232–240.

109 Lyon, A., Joudrey, P.J., Jin, D., Nass, R.D., Aon, M.A., O'Rourke, B., Akar, F.G. (2010) Optical imaging of mitochondrial function uncovers actively propagating waves of mitochondrial membrane potential collapse across intact heart. *J. Mol. Cell. Cardiol.*, **49**, 565–575.

110 Jin, H., Nass, R.D., Joudrey, P.J., Lyon, A.R., Chemaly, E.R., Rapti, K., Akar, F.G. (2006) Altered spatiotemporal dynamics of the mitochondrial membrane potential in the hypertrophied heart. *Biophys. J.* **98**, 2063–2071.

111 O'Rourke, B., Cortassa, S., Aon, M.A. (2005) Mitochondrial ion channels: gatekeepers of life and death. *Physiology (Bethesda, MD)*, **20**, 303–315.

112 Garfinkel, A., Chen, P.S., Walter, D.O., Karagueuzian, H.S., Kogan, B., Evans, S.J., Karpoukhin, M., Hwang, C., Uchida, T., Gotoh, M., Nwasokwa, O., Sager, P., Weiss, J.N. (1997) Quasiperiodicity and chaos in cardiac fibrillation. *J. Clin. Invest.*, **99**, 305–314.

113 Lab, M.J. (1997) Fibrillation, chaos and clinical control. *Nat. Med.*, **3**, 385–386.

114 Verrier, R.L., Nearing, B.D., Lovett, E.G. (1997) Complex oscillatory heart rhythm: a dance macabre. *J. Clin. Invest.*, **99**, 156–157.

115 Lloyd, D., Murray, D.B. (2005) Ultradian metronome: timekeeper for orchestration of cellular coherence. *Trends Biochem. Sci.*, **30**, 373–377.

116 Bassingthwaighte, J.B., Liebovitch, L.S., West, B.J. (1994) *Fractal Physiology*, Oxford University Press, New York.

117 Aon, M.A., Cortassa, S., Lemar, K.M., Hayes, A.J., Lloyd, D. (2007) Single and cell population respiratory oscillations in yeast: a 2-photon scanning laser microscopy study. *FEBS Lett.*, **581**, 8–14.

118 Murray, D.B., Klevecz, R.R., Lloyd, D. (2003) Generation and maintenance of synchrony in Saccharomyces cerevisiae continuous culture. *Exp. Cell Res.*, **287**, 10–15.

119 Lloyd, D. (2006) Hydrogen sulfide: clandestine microbial messenger? *Trends Microbiol.*, **14**, 456–462.

120 Lusis, A.J., Weiss, J.N. (2003) Cardiovascular networks: systems-based approaches

to cardiovascular disease. *Circulation*, **121**, 157–170.

121 Trzeciakowski, J., Chilian, W.M. (2008) Chaotic behavior of the coronary circulation. *Med. Biol. Eng. Comput.*, **46**, 433–442.

122 Freitas, U., Roulin, E., Muir, J.F., Letellier, C. (2009) Identifying chaos from heart rate: the right task? *Chaos*, **19**, 028505.

123 Wu, G.Q., Arzeno, N.M., Shen, L.L., Tang, D.K., Zheng, D.A., Zhao, N.Q., Eckberg, D.L., Poon, C.S. (2009) Chaotic signatures of heart rate variability and its power spectrum in health, aging and heart failure. *PLoS ONE*, **4**, e4323.

124 Tran, D.X., Sato, D., Yochelis, A., Weiss, J.N., Garfinkel, A., Qu, Z. (2009) Bifurcation and chaos in a model of cardiac early afterdepolarizations. *Phys. Rev. Lett.*, **102**, 258103.

125 Sato, D., Xie, L.H., Sovari, A.A., Tran, D.X., Morita, N., Xie, F., Karagueuzian, H., Garfinkel, A., Weiss, J.N., Qu, Z. (2009) Synchronization of chaotic early afterdepolarizations in the genesis of cardiac arrhythmias. *Proc. Natl Acad. Sci. USA*, **106**, 2983–2988.

126 Vicsek, T. (2001) *Fluctuations and Scaling in Biology*, Oxford University Press, New York.

127 Strogatz, S.H. (2001) Exploring complex networks. *Nature*, **410**, 268–276.

128 Butts, C.T. (2009) Revisiting the foundations of network analysis. *Science*, **325**, 414–416.

129 Ehrenberg, M., Elf, J., Hohmann, S. (2009) Systems Biology: Nobel Symposium 146. *FEBS Lett.*, **583**, 3881.

130 Aon, M.A., Cortassa, S. (1993) An allometric interpretation of the spatio-temporal organization of molecular and cellular processes. *Mol. Cell. Biochem.*, **120**, 1–13.

131 Brown, J.H., West, G.B., Enquist, B.J. (2000) *Scaling in Biology*, Oxford University Press, New York.

132 Varela, F., Maturana, H., Uribe, R. (1974) Autopoiesis: the organization of the living systems, its characterization and a model. *Biosystems*, **5**, 187–196.

10
Computational Biology

Thomas Lengauer, Mario Albrecht, and Francisco S. Domingues
Max Planck Institute for Informatics, Department of Computational Biology and Applied Algorithmics, Campus E1 4, 66123, Saarbrücken, Germany

Systems Biology: Advances in Molecular Biology and Medicine, First Edition. Edited by Robert A. Meyers.

Keywords

Sequence alignment
A computational procedure, by which two molecular sequences (DNA, RNA, or protein) are compared and letters in the two sequences are matched in pairs, one letter in each sequence, indicating a certain type of correspondence (evolutionary, structural, or other). Alignment can break the sequences into pieces inserting gaps into either sequence to afford the optimal matching.

Phylogeny reconstruction
A computational procedure, by which different biological items, for example, species, genes, or proteins are arranged in a (mathematical) tree or tree-like form indicating similarity, evolutionary relationship, or ancestry.

Molecular docking
A computational procedure which infers the structure of a complex between (mostly two) molecules.

Protein design
Modification of a protein sequence in order to optimize certain biotechnologically interesting properties of the protein, such as thermostability.

Protein structure prediction
A computational procedure which infers the three-dimensional structure of a protein from the protein sequence.

Protein function prediction
A computational procedure which infers aspects of the function of a protein from the sequence or structure of that protein.

Molecular network
Representation of often very large number of pairwise interactions of molecules. The interactions can be of different types, such as molecular binding or functional association.

Computational immunology
A subarea of computational biology which concerns the analysis of molecular aspects of the immune system, including the identification of epitopes and analysis of molecular interactions of the immune system.

Bioinformatical resistance analysis
A computational procedure which infers the resistance of a viral strain against an antiviral drug, and the effectiveness of antiviral therapy from the viral genotype.

During recent years, biological research has become increasingly based on large-scale experimentation such that data may be collected on an organismic scale. These data are voluminous, they are often very noisy, and their interpretation – and the configuration of the experiments involved – necessitates complex computer analysis. The respective computer methods are themselves an object of intensive research in a scientific discipline known as "*computational biology*" or "*bioinformatics.*" Computational biology has a wide variety of facets that range from experiment configuration and low-level data analysis to computer-generated hypotheses.

1 Introduction

During recent decades, biology has experienced a dramatic shift, moving from a science that was mostly phenomenological and taxonomic in character to a subject that, today, is based mainly on quantitative analysis. The molecular basis of living systems has moved increasingly into the focus of biological research while, during recent years, a type of experimentation has become significant that is targeted at the large-scale collection of data. Rather than analyze very local systems, such as a single gene or protein and its effects, the aim of this type of experimentation – which is also referred to as "*screening*" – is to cover all (or at least a substantial part) of a certain type of data that can be acquired for an organism. The first screening projects to address genome sequencing were started during the late 1980s, when a decision was taken by the US Congress to sequence the complete human genome – a task which took approximately 15 years to complete. Along the way, it was essential that the necessary experimental technology was developed, and "tested" on a series of increasingly complex organisms. Those genomes which, when sequencing was completed, accentuated the amazing progress that had been made in genome technology are listed in Table 1. Today, at the time of writing of this chapter, the finished drafts of more than 1600 complete genome sequences of bacteria, archea and eukaryota are available.[1)]

The availability of genome sequences from human and many other organisms was expected to provide a deep biological understanding of the molecular workings of cells, tissues, organs, and organisms, as well as to provide insights into different medical conditions. Indeed, it was generally believed that genome sequencing would bring about major breakthroughs in medicine, pharmacology, and biotechnology. Yet today, whilst this perspective still holds, it would appear that the speed with which biological insights can be gained has been greatly overestimated, and medical breakthroughs are taking much longer than was expected [1]. Genome sequencing has, nevertheless, resulted in many success stories, including the development of revolutionary technologies for biological research [2], the provision of novel insights into genome organization [3, 4] and into evolutionary history [5, 6], and the identification of genetic influences on complex traits by genome-wide association studies [7]. It is also fair to say that "genomic medicine" has finally arrived, with the first cases having been reported of whole-genome/exome sequencing of individual patients, thereby providing a clear benefit for diagnosis and therapy [8]. Indeed, it is hoped that the clinical use of whole-genome sequencing will become routine within the next 10 to 20 years.

The sequencing of small genomes does not involve major bioinformatics efforts (although the interpretation of the genome sequence does). However, as the genomes become larger and more complex, the assembly of the sequence data has involved substantial bioinformatics, with major efforts being needed to assemble the genomes of *Drosphila melanogaster*, *Mus musculus*, and *Homo sapiens*. The bioinformatics needs of genome-sequencing projects are summarized in Sect. 2.

The availability of a complete genome sequence immediately raises a multitude of research questions that can be addressed with the help of bioinformatics methods. In Fig. 1, these questions

1) See http://img.jgi.doe.gov/.

Tab. 1 Landmark genome-sequencing projects.

Organism	*Colloquial name*	*Genome length (kb)*	*Year sequenced*
H. influenzae [9]	Bacterium (ear infections)	1 830	1995
S. cerevisiae [10]	Baker's yeast	12 069	1996
E. coli (K12) [11]	Bacterium in the intestine	4 639	1997
C. elegans [12]	Nematode, worm	97 000	1998
D. melanogaster [13]	Fruitfly	137 000	2000
A. thaliana [14]	Mouse ear cress	115 428	2000
M. musculus [15]	Mouse	2 497 000	2002
H. sapiens [16]	Human	2 866 000	2003

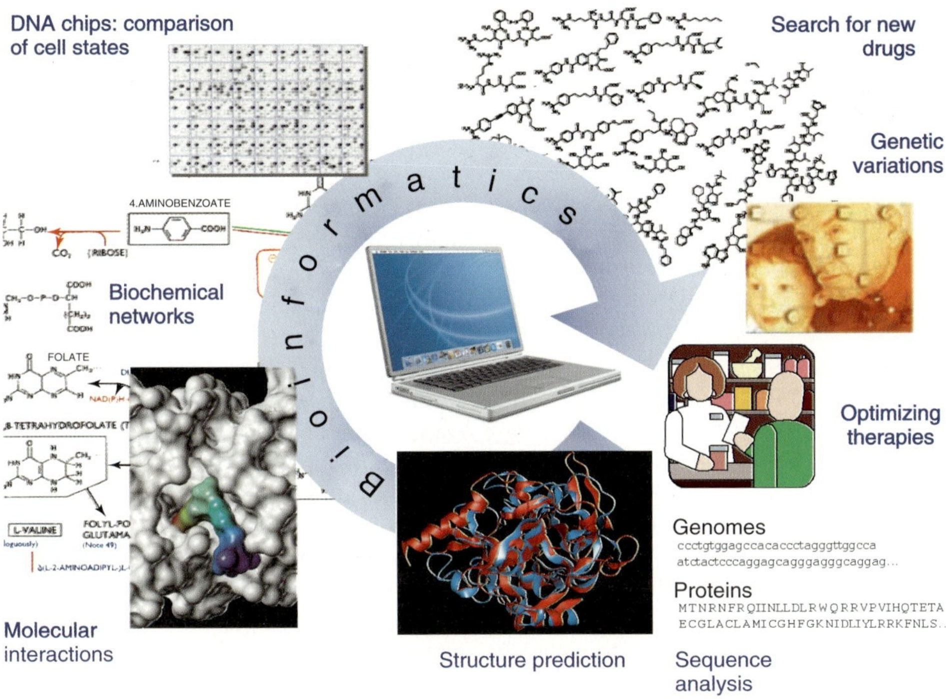

Fig. 1 Bioinformatics – from the genotype to the phenotype.

are arrayed in order from the genotype (5 o'clock direction) to the phenotype (counterclockwise up to the 4 o'clock direction). Whilst some of these questions can be addressed without resorting to additional experimental data, for other questions additional experimental data are required. These points will be briefly reviewed, and more detail provided, in the following sections.

Among the main points are the identification of genes in the genome sequence and the analysis of noncoding regions (Sect. 3), while the prediction of protein structure from protein sequence is detailed in Sect. 4. The analysis of molecular interactions, on the basis of the structure of the involved molecules, forms the foundation for understanding binding properties, and also serves as a starting point for drug design (see Sect. 5). The processes of life involve complex biochemical networks composed of pathways and cycles of intermolecular interactions; the bioinformatics analysis of these networks is detailed in Sect. 6.

As all cells of an organism have the same genome, additional data are required in order to differentiate between different tissues and cell states. As mRNA expression data represent the present-day approach to this type of information, the bioinformatics methods for analyzing these data are discussed in Sect. 7. The determination of protein function is a major aim of modern biology, in general, and of bioinformatics in particular. Protein function signals are contained in many diverse biological data; thus, protein function prediction is an integrative task, and will be discussed in Sect. 8. Computer-based approaches to drug design are detailed in Sect. 9, and bioinformatics approaches to analyzing genetic variations in organisms, both in humans and in microbial and viral parasites and infectious agents, are summarized in Sect. 10. The chapter concludes with a general appraisal of the potential and limitations of bioinformatics (see Sect. 11).

2
Sequencing Genomes

The basis of the bioinformatics support for sequencing genomes requires only a brief explanation. As the current experimental methods for genome sequencing can handle a sequence of only a few hundred base-pairs length at a time, the genomic or chromosomal sequence must be cut into fragments of approximately that length before sequencing can be started (the fragments may be longer, but only their ends will then be sequenced). When the fragments have been sequenced the original genome sequence must be pieced together from the fragments, and it is this assembly process that requires critical bioinformatics support.

The basic fragment assembly problem is quite simple. The sequencing process generates a multiple coverage of the complete genome sequence (or of a well-defined part of it that has been identified in the previous process of constructing a physical map of the genome) with small fragments of several hundred base-pairs in length. The coverage of the genome sequence must be sufficient to ensure (with high probability) that the segments overlap along the complete genome sequence (for the human genome, a sevenfold coverage was required [17]). The fragment assembler then aligns the fragments by comparing their sequences to form the complete genome. If adjacent fragments are unique and overlap along a stretch of sufficient length, then the alignment is also unique. The only variations between different fragments covering the same part of the genome sequence can originate from reading errors or polymorphisms (in a heterozygous individual), and there are only few of those. Therefore, the alignment problem is much simpler than the alignment of different (especially of evolutionarily only distantly related) sequences. Fragment assembly experiences its major complication by the occurrence of long repeats in the genome sequence. It is evident that, in the presence of repeats, the above-described fragment assembly process is corrupted. A more detailed overview of fragment assembly is provided in Ref. [18].

Fragment assembly has experienced its culmination and has achieved the most dynamic progress during the process of sequencing the human genome. Such sequencing was achieved by two different groups concurrently, using different experimental procedures, with both groups presenting a draft of the human genome sequence in February 2001.

The method used by the public consortium is occasionally referred to as hierarchical shotgun (HS) sequencing [19]. When using this method, the genomic DNA is cut into fragments by partial digestion with restriction enzymes, and the fragments are then used to produce libraries of bacterial artificial chromosome (BAC) clones that are between 150 and 200 kb in length. A physical map of these BACs is constructed in order to identify those clones that minimally overlap and are therefore suitable for sequencing. The selected BACs are then subcloned into shotgun libraries (libraries of short randomly cut pieces of the clone) and sequenced. Finally, the genome is reconstructed using data from the overlapping sequences, guided by the map. The HS approach provides an accurate finished genome, but requires substantial additional preliminary laboratory work to create the BAC libraries. The draft assembly of the public consortium was computed by Jim Kent at the University of California, Santa Cruz [20].

An alternative method – whole-genome shotgun (WGS) sequencing – was used by Craig Venter's group at Celera Genomics to provide a draft sequence of the human genome [17]. In this approach, shotgun sequencing is applied to the whole genome, by directly sequencing clones that are 2, 10, and 50 kb in length. Additional information is derived from known markers that are used for anchoring the assembled contigs (continuous stretches of sequences arising from overlapping clones) to the genome. The genome assembler for Celera was written by Gene Myers and his bioinformatics team [21, 22]. (Other algorithms for fragment assembly in shotgun sequencing have been presented in Refs [23, 24].)

In general, the choice of sequencing method depends on the complexity of the genome under study, and the desired level of completeness of the genome sequence. The WGS approach is well established for shorter genomes without many repeats, especially for prokaryotic genomes. For long and repetitive genomes, the WGS approach is effective in obtaining a draft covering perhaps 80% of the genome sequence. Unfortunately, finishing a genome is difficult when using this approach.

The WGS approach was since applied to drafting the dog genome [25], and also probes from the Sargasso Sea containing many organisms [26]. A hybrid approach, which utilizes aspects of both methods, was used for finishing the human and the mouse genome [15]. The public consortium finished the human genome sequence in April 2003 [16].

Whilst the availability of new-generation sequencing technologies brings new challenges and applications, these new technologies generate large amounts of data that require new approaches to data storage, management, and processing [27]. New sequencing technology is also becoming more accessible to greater numbers of investigators for many application areas, including the analysis of genetic variation, transcriptome analysis, the analysis of DNA–protein interactions, and metagenomics [28–30]. Computational tools are also being developed to process the large amounts of sequence data generated with these new sequencing technologies [31].

3
Molecular Sequence Analysis

Whilst a wide variety of questions can be asked regarding a genome sequence, the analyses employed are typically either low-level in nature (e.g., analyzing the information content of a sequence, or finding repetitive patterns) or high-level (e.g., finding genes and analyzing the upstream regulatory regions). Various other widely used packages are also available to perform both low- and high-level analyses, an example being the EMBOSS package [32]. Only the most important and widespread sequence analysis problems will be addressed in the following subsections.

3.1
Sequence Alignment

The most basic sequence analysis method is that of sequence alignment, whereby two (pairwise alignment) or more (multiple alignment) similar nucleic acid or protein sequences are arrayed such that corresponding nucleotides and amino acid residues, respectively, are matched to each other. In this context, *correspondence* may have different meanings. Often, *evolutionary* correspondence is referred to, in the sense that the two monomers matched have the same evolutionary origin, although on occasion *structural* correspondence may be mentioned, in the sense that the two matched monomers play the same role inside the molecular structure. The sequence alignment method can be applied to DNA, RNA, and protein sequences alike. An alignment of the human hemoglobin α_1- and β-chains is shown in Fig. 2a, while Fig. 2b depicts a multiple alignment of myoglobin and hemoglobin sequences from different organisms. The additional gap character (-) serves to introduce insertions and deletions during the alignment, such that corresponding positions in the different sequences match each other. There are exponentially many alignments in n and k between k sequences of length n. In order to find the desired alignment, a cost is attributed to each alignment; the alignment with maximum or minimum cost is then sought, depending whether the scoring function represents a similarity of the sequences or the distance between them. When discussing the cost maximization of a pairwise alignment of protein sequences for concreteness, the cost function is a sum of terms, there being a term for each alignment column that does not contain gap characters (match/mismatch terms), and also terms penalizing for gaps (gap terms). The cost terms are usually derived statistically. In the evolutionary interpretation of the alignment score, the match/mismatch terms are collected in a so-called "substitution matrix," and reflect the expectation to see the mutations witnessed by an alignment column [33]. A basic simplifying assumption is that the mutations witnessed by different alignment columns are independent. This results in quite efficient alignment algorithms that are based on the dynamic programming paradigm [34]; the gap terms are then usually set heuristically or by fitting them to a dataset of accepted alignments [35]. Pairwise alignments can align complete sequences (global alignment), and can be used to identify shorter sequence fragments in longer sequences (local alignment). The runtime of the

```
Hemoglobin alpha-1   1 MVLSPADKTNVKAAWGKVGAHAGEYGAEALERMFLSFPTTKTYFPHF-D  48
                       :.|:|.:|:.|.|.||||  :.:|.|:|||.|:::.:|.|:.:|..| |
Hemoglobin beta      1 MVHLTPEEKSAVTALWGKV--NVDEVGGEALGRLLVVYPWTQRFFESFGD 48

Hemoglobin alpha-1  49 LSH-----GSAQVKGHGKKVADALTNAVAHVDDMPNALSALSDLHAHKLR 93
                       ||.     |:::||:|||||.:|:::::||:|::..::::||:||.:||:
Hemoglobin beta     49 LSTPDAVMGNPKVKAHGKKVLGAFSDGLAHLDNLKGTFATLSELHCDKLH 98

Hemoglobin alpha-1  94 VDPVNFKLLSHCLLVTLAAHLPAEFTPAVHASLDKFLASVSTVLTSKYR 142
                       |||.||:||::.|:..||.|:..||||:|:|:.:|.:|:|:..|:.||:
Hemoglobin beta     99 VDPENFRLLGNVLVCVLAHHFGKEFTPPVQAAYQKVVAGVANALAHKYH 147
```

(a)

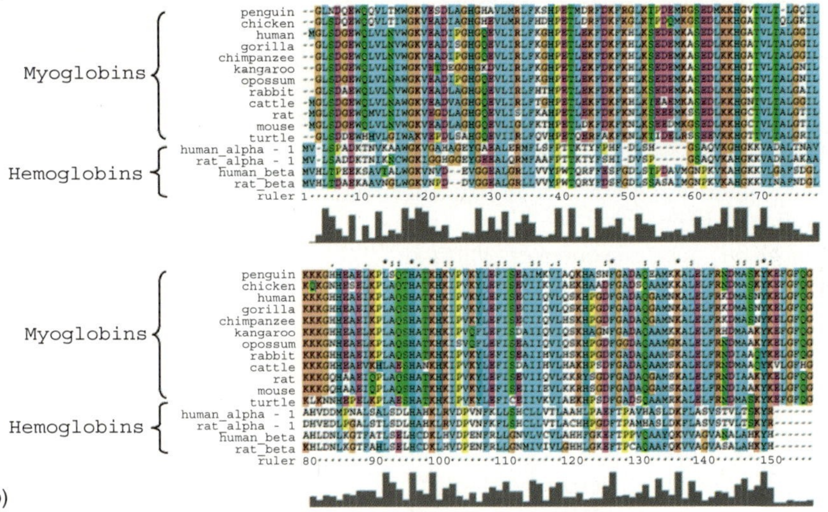

(b)

Fig. 2 (a) Global alignment of the human hemoglobin alpha-1 and hemoglobin beta sequences. The numbers at the left and right sides of the alignment represent sequence positions at the end of each row. The signs in the middle row signify similarity between the matched amino acid residues: | identical, : similar,. Not similar; (b) Multiple alignment of several myoglobin and hemoglobin sequences. The alignment was generated with ClustalW. The colors represent chemical groups of amino acid residues. The bar graph at the bottom of the figure represents the degree of conservation of the alignment columns.

alignment algorithm grows linearly as the product of the length of the two sequences to be aligned. If the aim is to screen against a large database of sequences, however, this may be too much and faster heuristics based on a gapless alignment of locally similar stretches must be resorted to. The most widely used alignment program – BLAST – is based on this approach [36], and owes its popularity not only to its efficiency but also to the fact that it returns a significance value (*p*-value) with the alignment that quantifies how improbable it is to observe the sequence similarity witnessed by the alignment. The more improbable an alignment, the smaller is its *p*-value and the more significant the finding, which indicates that the proteins are evolutionarily related (homologous). BLAST was the first major bioinformatics tool to return a significance value that is based on solid statistical theory [37].

Multiple alignments are more difficult to optimize than pairwise alignments, mainly because the straightforward extension of a pairwise gapped alignment algorithm leads to an exponential procedure and raises the problem of having to derive many cost terms based on scarce data. These problems can be circumvented, however, by using heuristics. For example, PSI-BLAST [38] employs a heuristic based on BLAST in order to identify sequences that are related distantly to a query sequence and align them in multiple fashion. Typically, PSI-BLAST takes a query sequence and computes a multiple alignment of sequences taken from a sequence database. A statistically well-founded approach to multiply aligning n given query sequences is then provided by the so-called Hidden Markov models (HMMs), a stochastic procedure by which a certain Markov chain is trained to a family of related protein sequences. The result is a probabilistic weighting of the state transitions and output behavior of the Markov chain that causes it to generate new sequences of the same protein family with a significantly higher probability than other sequences. This chain can then be used to multiply align protein sequences of that family and to determine whether the new chains belong to the family, or not [39]. With HMMs, an alignment of k sequences of length n can be obtained in time $O(kn^2)$. Other popular multiple alignment programs include ClustalW [40, 41], T-COFFEE [42], MAFFT [43], MUSCLE [44], and PROBCONS [45] (see also Sect. 3.2). Databases of manually curated multiple alignments can be used to benchmark these alignment methods [46, 47].

Sequence alignments form the basis of analyzing relationships between DNA, RNA, and protein sequences, respectively. On the DNA level, alignment algorithms are used to identify genes [48] and to analyze regulatory regions [49] whereas, on the protein level, the relationships to be searched for may be either evolutionary [50–52] or structural [53, 54]. The description of the type of relationship being sought is contained in the scoring matrix. Whereas, early approaches used the same amino acid substitution matrix for all alignment columns, newer approaches employ a so-called position-specific scoring matrix (PSSM); that is, they adapt the scoring matrix to the specific needs of an alignment column. For example, in structure alignment each column stands for a local structural neighborhood in the two proteins to be aligned. This neighborhood has specific features that can be reflected by a scoring matrix tailored to this neighborhood. The same phenomenon can be taken into account, when aligning DNA sequences of upstream regulatory regions of genes, in order to identify binding sites for transcription factors (TFs). Consequently, PSSMs are also used in this context [55, 56]. The HMMs form the basis of a number of protein family classifications [57–59], and have also been applied to gene finding and to the analysis of promoter regions [60].

3.2
Phylogeny Construction

In phylogeny construction, the main interest is to learn about the evolutionary relationship of a set of sequences. The basic concept is to generate a tree, the leaves of which are the sequences to be related, and the edges of which have appropriate lengths such that the unique path between two sequences in the tree approximates the respective evolutionary distance as accurately as possible. A phylogenetic tree of the myoglobin sequences used in Fig. 2b

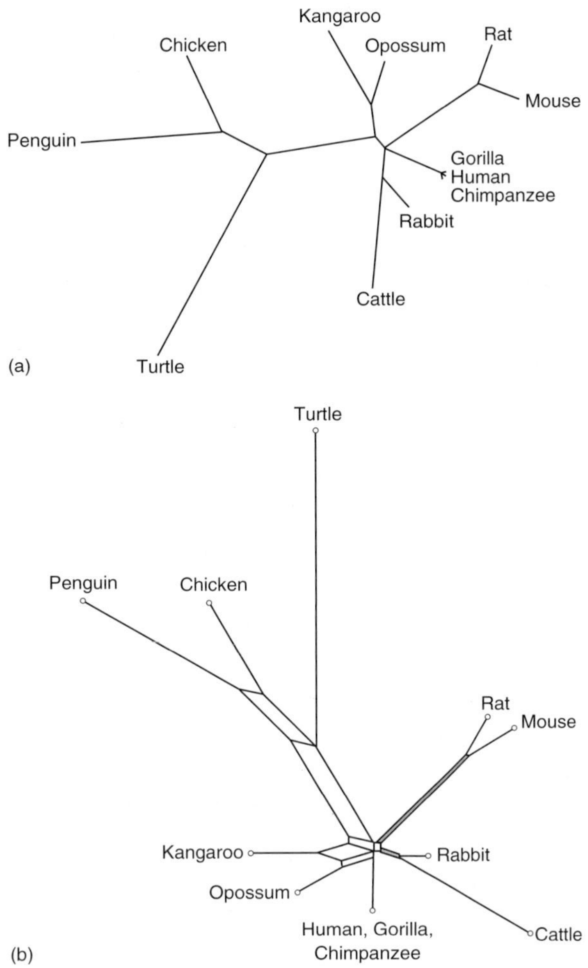

Fig. 3 (a) A phylogenetic tree of the myoglobin sequences aligned in Fig. 2b; (b) A split diagram of the myoglobin sequences aligned in Fig. 2b.

is shown in Fig. 3a. In this case, it is desirable to annotate the interior vertices of the tree with "ancestral" sequences, such that the sequences which annotate the vertices adjacent to an edge align with a score that equals the length of the edge. The tree will then correspond to a multiple alignment of the given sequences and the inferred ancestral sequences.

In general, there are two types of phylogenetic tree – directed and undirected. *Directed* trees have a root representing the ancestor of all leaves in the tree, with edges leading from the ancestor to descendant species, where the edge length represents the time taken to produce the descendants. In general, however, most trees are *undirected,* because it is difficult to estimate edge directions and lengths; in this case, the trees have no root.

Several methods have been identified for the construction of a phylogenetic tree:

- The first approach is *distance-based,* where the evolutionary distance between two sequences can be represented by the score of their pairwise alignment, or by some other measure of distance (e.g., the distance can be estimated with a maximum-likelihood method). A tree is searched for that best approximates the resulting distance matrix of the set of sequences involved. The optimization problem of finding a tree that best fits this distance measure given by the pairwise alignments of the sequences to be related is difficult (NP-hard), and so can be solved only approximately [34]. The problem of finding such a tree is intimately related to the problem of finding a high-quality multiple alignment of the set of sequences. Thus, methods are available that combine these two tasks. For instance, ClustalW (see also Sect. 3.1) constructs a multiple alignment and a phylogenetic tree bottom-up by successively clustering sequences that are most closely related. The interior nodes of the tree are annotated with sequence profiles that give the preferences for amino acids (or nucleotides) at each position in the respective cluster. These profiles can be turned into "ancestral" sequences by using consensus methods.

- The *maximum parsimony* approach aims at inferring a tree that explains the evolutionary relationship between a set of sequences by introducing as few mutations as possible. This is a strong assumption that is not realized by Nature in general, but serves to simplify the problem of generating phylogenies.
- The *maximum likelihood* approach is based on a probabilistic model of evolution, and aims at generating the phylogeny that is most likely, given the observed data. It is generally much more time-consuming than the other approaches, but has the advantage of being founded on a statistically more sound and generalizable theoretical basis [61]. PHYLIP[2] and PAUP*[3] are two widely used software libraries for constructing phylogenies.[4]

Phylogenies are difficult to validate, because the biological process that generates them is difficult to model and observe. This process comprises two parts: the stochastic process of *sequence variation*; and the highly complex and unstructured process of *selection*. Because of the complexity of this situation, each accepted phylogeny is the result of a consensus among biologists rather than an imperative consequence of the given sequence data. Specifically: (i) the alignment score reflects evolutionary distance only incompletely; and (ii) the hypothesis that the phylogeny is tree-shaped is not valid in many biological settings, for example, because recombination events can take place. The split-decomposition method [62] takes account of the fact that the distance measure derived from the sequence alignments need not be tree-like. The method generates so-called "split diagrams" that are more tree-like if the underlying distance measure allows for this, but which show deviation from the tree shape where they occur. A split diagram of the myoglobin sequences from Fig. 2b is shown in Fig. 3b. Recombination results in a new genotype with sequence combinations that differ from those present in the ancestor organisms. An iterative recombination then shuffles the sequences and masks the signals of the evolutionary relationship. Ancestral recombination graphs (ARGs) are the counterpart of evolutionary trees in the presence of recombination,

2) http://evolution.genetics.washington.edu/phylip.html.
3) http://paup.csit.fsu.edu/.
4) For a list of phylogeny programs, see http://evolution.genetics.washington.edu/phylip/software.html.

but they are difficult to compute and the mathematical theory of ARG has not been developed to a stage where it can be applied to large sequence sets [63].

A comprehensive overview of the methods available for constructing phylogenies is provided in Ref. [64], while a more recent overview is available in Ref. [65]

3.3 "Identifying" Genes

The art of "identifying" genes represents one of the grand challenges of computational biology [66, 67]. The structure of a eukaryotic gene is depicted in Fig. 4. The identification of genes entails the elucidation of all parts of the gene, and not only the coding regions; exons must be distinguished from introns, transcription and translation initiation, and termination sites must be identified. The prediction of genes is more challenging in eukaryotes than in prokaryotes because of splicing and alternative splicing. In human, for example, there are at least two splice variants for most protein-coding genes [68].

There are basically two different approaches to identifying genes in genomic sequences, namely content sensors and signal sensors [69]:

- *Content sensors* aim at classifying DNA segments into types such as coding and noncoding. The so-called *extrinsic* (or *homology-based*) content sensors do this by comparison of the DNA sequence with biologically confirmed known gene sequences. Sequence alignment methods are used for this purpose. With the growing number of genomes, this approach may be used to render the gene-finding procedure more accurate [70], or even to capture aspects of the evolutionary development of genes [71]. If sequences similar to the query sequence are not available, then the gene regions must be distinguished in their own right; that is, by the *intrinsic* properties of short sequence fragments such as compositional bias or codon usage (*ab initio* approaches). Statistical classifiers use these properties as input, and are trained on biologically characterized sequences (see Refs [72, 73]). HMMs are also used to model the intrinsic properties of gene sequences [74].
- The *signal sensor* approach tries to locate functional sites in genes such as splice sites and binding sites for TFs. One popular approach is to describe the neighborhood of functional sites with a position-specific weight matrix that is trained on a multiple alignment of the

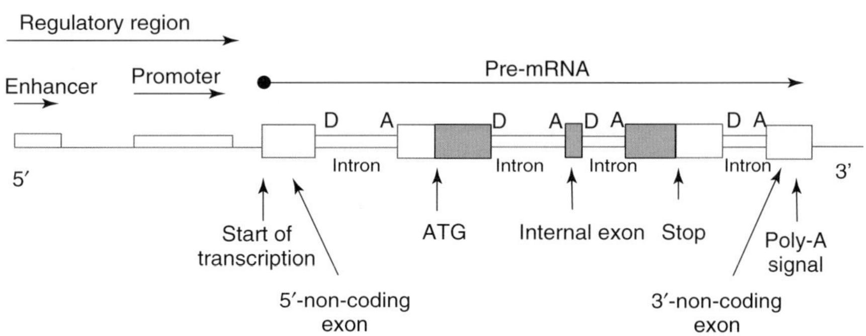

Fig. 4 The structure of a eukaryotic gene. Only the shaded parts represent coding regions.

neighborhoods of confirmed functional sites [75, 76]. The alignments used for deriving such matrices are gapless. By using HMMs, however, it is possible to introduce gaps to the respective alignment [76]. In fact, most present-day programs used for finding functional sites employ this technology.

The accuracy of gene prediction programs can be measured on the nucleotide and on the exon level [77]. For mammalian sequences, the accuracy of the best gene prediction programs is above 90% for correctly predicting coding nucleotides, but on the exon level the accuracy is markedly lower, with less than 80% of the exons having all of their exon boundaries predicted correctly. The correct prediction of complete genes is even lower: less than 50% of predicted genes correspond exactly to the actual genes [78, 79]. The problem here lies with predicting the exact limits of the coding regions; moreover, gene prediction accuracy drops dramatically as the density of genes decreases and the introns become longer. Overall, exact gene prediction in eukaryotes remains an unsolved problem.

The identification of intron–exon borders is necessary, but not sufficient, to determine the sequence of the coded proteins, as multiple protein forms can be produced by alternative splicing. Nonetheless, an initial and very encouraging success has been reported in deciphering alternative splicing [80].

3.4 Analyzing Regulatory Regions

Regulatory regions are the regions in DNA that lie in the (mostly upstream) neighborhood of the genes, and which contain binding site for the transcription machinery that is responsible for expressing the gene; therefore, these regions harbor crucial information on the regulation of gene expression. Polymorphisms in these regions are responsible for many of the small and large differences between organisms of the same species, and also for many predispositions and diseases; consequently, an understanding of their properties is extremely important. The most intensively studied regulatory regions are the promoters which harbor the binding sites of TFs. These exhibit a modular architecture; that is, they are composed of several TF-binding sites, each of which is a short segment of DNA (length 10–15 nucleotides), with the different binding sites being located at varying distances from one another. The characteristics of each of the blocks can be captured with a weight matrix (see also Sect. 3.3), while the modular architecture must be captured with specifically tailored organizational models which, today, tend to be tailored to specific classes of promoters and are not generally applicable. As a consequence, statistical methods that are capable of analyzing intrinsic sequence properties are also employed [49], where the methods used are not unlike those used for gene identification. As the TF-binding sites are short and highly conserved, a special approach to finding them is based on the analysis of almost identical short sequence segments in a set of DNA sequences [81, 82]. The methods used in practice today have been reported to find up to between 80% and 90% of the existing promoters, but to output a significant number of false positives [83]. Recently, however, advances have been made on finding TF binding sites based on the analysis of molecular structures [84].

Enhancers, silencers, and matrix-attached regions represent other types of regulatory region for which predictions have been attempted using a similar technology. Compared to protein-based bioinformatics, the analysis of regulatory regions remains at an early stage of development, and is especially difficult because the processes of molecular recognition based on proteins binding to DNA are very intricate and cannot yet be analyzed on a structural level. Thus, all information must be derived via the rather indirect methods of sequence analysis and comparison.

3.5 Finding Repetitive Elements

As noted in Sect. 2, repeats are a characteristic and complicating feature of many complex eukaryotic genomes, in that they present a problem with genome assembly but they also require specific methods of sequence analysis. However, a few programs have been developed specifically to identify different types of repeats in genomes [85].

Two general categories of repeats have been identified:

- *Tandem repeats* consist of repeating segments adjacent to one another; in satellite DNA the repeats are larger than 100 bp and they are continuously arrayed in stretches that can span millions of base pairs. The repeat unit is smaller in microsatellites or simple sequence repeats (SSRs), usually one to six nucleotides. Microsatellites have been associated with several genetic disorders. Minisatellites are tandem repeats with repeat unit lengths intermediate between those of SSRs and satellites.
- *Dispersed repeats* generally consist of transposable elements that can move or copy themselves to different genome locations, and account for most of the repeats in the larger eukaryotic genomes.

3.5.1 Repeat-Finding Programs

A wide variety of repeat-finding programs has been identified, the most obvious type being based on the observation that repeats tend to throw off sequence analysis programs; thus, they are located and masked out in a first step. Repeat masking programs such as RepeatMasker[5] use a dictionary of known repeats and find all exact or approximate matches to the dictionary entries in the genome sequence under investigation:

- Repeat masking programs can only be used to rediscover repeat patterns; a more challenging task is to find repeats, given only the genome sequence.
- Tandem repeats represent a very important subtype of repeats, where two exactly matching short sequences immediately follow each other. Some programs have been created that specialize in finding tandem repeats [86].
- The greatest challenge is to find general repeats without prior knowledge, by using only the genome sequence as a basis. These programs usually find exact repeats in a first step, after which the repeats are used as seeds for finding so-called "degenerate repeats" that are similar, but not exactly alike. The programs REPuter [87] and MUMmer [88] store the genome sequence in a data structure called a *suffix tree* that enables the rapid identification of exact

5) http://ftp.genome.washington.edu/RM/RepeatMasker.html.

repeats; the FORRepeat program uses so-called *factor oracles* for this purpose. One important feature of repeat finding programs is an effective visualization of the output.

Repeat-finding programs can be exercised for a multitude of tasks:

- **Checking genome assemblies:** Assembly programs (see Sect. 2) are imperfect, and thus the sequence assemblies they produce may contain errors. Repeat finding programs have detected palindromic repeats in the human genome sequence that were due to an incorrect assembly.
- **Low copy repeats in human diseases:** Several human diseases are associated with deletions or duplications of specific genomic regions. Such repeat patterns can be identified by a repeat finding program.
- **Checking the uniqueness of hybridization probes:** This application is technological. For a microarray, the aim is to deposit sequence probes that are unique in the genome under consideration. This uniqueness can be ensured with a repeat finder.
- **Comparative genomics:** The concatenation of two different genomes and then seeking repeats with varying rates of divergence allows for a comparison of the two genomes with different grades of similarity.

3.6 Analyzing Genome Rearrangements

With the availability of a growing number of genomes, the comparison of different genomes is becoming an increasingly valuable source of insight into the evolutionary development of species. Indeed, it is emerging that Nature rearranges genomes in intricate ways to form new genomes, but that the indivisible operations that Nature uses to rearrange genomes involve *translocation* (i.e., the integration of a sequence fragment at another location in the same or a different chromosome), *inversion* (i.e., the translocation and integration of a fragment in the opposite orientation), as well as *fusion* and *fission* of chromosomes. The main concern here is with rearrangement events on the species level – not those on the level of the individual.

The comparison of two genomes from different species aims at exhibiting which parts in genome 1 correspond to which parts in genome 2, and thereby to reveal the respective sequence of rearrangement operations. Of course, it is essential to deal with mutations that distinguish the two sequences at the same time although, in general, this is a sequence analysis problem similar to sequence alignment (in spirit), but with the rearrangement operations as basic operations instead of mutation and gap insertion. Thus, a cost can be attributed to each operation (as occurred in sequence alignment), and the cost of the sequence of operations transforming one genome sequence into another can be optimized. Algorithmically, the problem is more complex than that of sequence alignment (see Chapter 10 in Ref. [89] and Ref. [90]). One basic ingredient of the algorithms is the *breakpoint graph*, which represents the available options for a rearrangement.

The problem becomes even more difficult if several genomes are to be compared, and the existence of a phylogeny is being sought [91]. Typically, these algorithmic developments make several biologically unjustifiable assumptions. Recent comparative investigations of the genomes of human and mouse have revealed that the rearrangements of genomes occur at different scales. For example, intra-chromosomal micro-rearrangements may affect sequence segments of less than 1 Mb in length, whereas macro-rearrangements affect much longer sequence segments, and can occur within a chromosome as well as between chromosomes. The algorithmic technology has been adapted to this finding [92].

Currently, the computational analysis of genome rearrangements is on the verge of transcending a basic algorithmic investigation to generating useful insights into the evolution of species genomes [93, 94], and the relationship between genome rearrangement and disease [95].

4 Molecular Structure Prediction

In many cases, the three-dimensional (3-D) structure of a molecule can provide essential clues to its molecular function. This is the case especially for proteins and RNA, which exercise their functional role by binding to other molecules in a unique and predictable fashion; this occurs especially for enzymes and molecules involved in regulation and information transfer within the cell, or between cells. While DNA has a very uniform double-helical structure, RNA and proteins can fold to create a large variety of structures. These molecules are stable in the sense that they retain their overall fold, but they can vary their structure in subtle ways. Examples of this phenomenon include the domain movements of allosteric proteins or the phenomenon of induced fit, by which the surface of the binding site of a protein is changed subtly in order to accommodate its binding partner.

It is the close relationship between molecular structure and function that has caused structure prediction problems to have acquired such great relevance. Although, in general, the structures of proteins and/or RNA cannot be reliably predicted, the field of protein structure prediction is currently at a more advanced state than is that of RNA structure prediction.

4.1 Protein Structure Prediction

In computational biology, the problem of protein structure prediction has often been called a *Grand Challenge*, and has received much attention from both the scientific community and the media. Indeed, over the years the computer manufacturers have regarded this problem as a main motivator for the development of high-performance computers. The probable reason for such popularity is that, although the problem is quite easily communicated, a solution has still not been identified. At this point, three versions of the problem – all of which have received wide scientific attention – will be discussed. Whilst each of these versions takes the protein sequence as the primary input, some of them employ additional inputs, such as the evolutionary neighborhood of the query protein or a database of known protein structures.

Today, a large number of structure prediction methods are available on the internet, together with sophisticated means for their constant evaluation. In addition, on a biennial basis, the Critical Assessment of Structure Prediction (CASP) methods contest [96]) stages an international experiment of predicting protein structures that are unknown at the time of prediction, but which will have been resolved by the time that the prediction is evaluated.

4.1.1 Secondary Structure Prediction of Proteins

The secondary structure prediction problem aims at deciding, for each residue in the protein sequence, whether it is located in an α-helix (H), a β-strand (E), or in neither (C); the latter regions are referred to as *coil regions*. Variants of the problem exist which distinguish among different helices (α-helix, 3_{10}-helix, etc.) and subdivide the coil regions further, for example, by specifying different types of turns (such variants will not be discussed at this point, however).

In modern methods, the first step is to scan through a protein sequence database, to retrieve homologs of the query protein, and to compute a multiple alignment of all retrieved proteins. This provides a critical additional input to the structure prediction procedure such that, for each sequence position there is not only a single residue but a profile of residues that the related proteins provide. The introduction of this critical information was the key ingredient for boosting structure prediction accuracies above 70% during the early 1990s [97]. Given this information, a variety of classification methods can be used to predict the residues states H, E, C, among which neural nets [97, 98] are the most popular variant. The best of these methods can achieve an accuracy of about 80%, though the ends of secondary structure elements tend to be predicted more unreliably than their central regions. In order for the profile to be unbiased, the sequence database from which the profile is drawn must be unbiased, in the sense that there are no protein families that are either over- or under-represented in the database.

The accuracy of secondary structure prediction methods is assessed by comparing their prediction to secondary structure assignments that are automatically derived from resolved 3-D protein structures. Different methods for this purpose, such as DSSP [99] and STRIDE [100], differ in their assignments; it is due to these differences that the prediction accuracy cannot be expected to increase much above 90%.

4.1.2 Similarity-Based Protein Structure Prediction

The most successful approach to the prediction of 3-D protein structure utilizes a database of resolved protein structures – the so-called "template structures" – as the secondary input. Rather than assembling the protein structure from scratch from the sequence, the question is repeatedly asked as to whether the protein structure attains any of the template structures. These similarity-based approaches to protein structure modeling [101] exploit the tendency of evolution to conserve protein structure more than sequence. In particular, it has been observed that two proteins are most likely to share a common overall structure if at least one-fourth of the residues in one protein is matched in a pairwise alignment with residues of the same type in the other protein (25% or greater sequence identity) [102].

The question then, is how likely will the sequence of the query protein attain a given template structure? To answer this point, we perform an alignment of the query sequence with the sequence of the template structure and rely on a suitable cost function to select the best alignment. In general, only reasonably different structures will be collected in the template database. Alignments can be generated using a position-specific amino-acid substitution matrix (PSSM) that represents the structural preferences for a residue on the query sequence to be located at the location of the matched residue in the template structure [53, 54]. As in secondary structure prediction, this approach can be enhanced by introducing information from evolutionarily related proteins in the form of a sequence profile on the side of the query sequence, as well as the template sequence plus additional information, for example, stemming from secondary structure predictions [103].

The alignment of the query sequence with the sequence of a template structure is known as *protein threading* [104]. Aligning the query sequence to a template protein is only the first step of producing a full-atom protein structure model for the query protein. While this is a critical step when a similarity range between the query and template sequence is less than 40%, it is a trivial exercise in high similarity ranges, above perhaps 70%. Invariably, a faulty alignment will lead to an incorrect protein model.

The alignment provides only the structure of part of the protein backbone. Gaps in the alignment represent parts of the query sequence that cannot be mapped onto the template structure (if gaps occur in the template sequence), or tears and rips in the backbone model of the query sequence (if gaps occurs in the query sequence). The former gaps mostly stem from loops in the query protein that have no counterpart in the template structure; these loops must be modeled in a separate loop modeling step. The modeling of loops can be achieved by taking loop templates from a database of structures of protein fragments, or by using energy-optimization methods. Rips in the backbone of the query protein must be mended, and finally the side chains of the query protein must be attached to the backbone. The side-chain conformation can be predicted using a database of side-chain rotamers, or by energy optimization. Similarity-based protein modeling tools combine these different steps in different ways, using different algorithmic procedures such as discrete optimization, nonlinear optimization, and constraint programming [105–107].

The score of the alignment indicates the level of preference for the query sequence to attain the template structure. Ideally, the template that can be aligned to the query sequence with the highest alignment score should provide the structural model for the query protein. However, as the alignment scores reflect structural preference only inadequately, this model selection procedure is fallible, and the score must be accompanied by a confidence value that rates how well the prediction can be believed. Often, confidence scores are based on statistical significance that rates the unlikeliness that an alignment can be obtained by chance. However, other – more heuristic – choices of confidence values have also proven effective [108].

Improvements in the quality of protein structure predictions rest mainly on improvements of the scoring function, and this can be achieved in two ways. The first method is to identify ever-more suitable functional forms for scoring. In scoring, a trade-off must be made between the number of terms to include into the function and the computational efforts for scoring. Although interaction terms between a growing number of partners within the protein afford a more accurate scoring (if the interaction terms are correct), more time will be required to perform the computations. The second method, given a fixed functional form, is to improve the scoring parameters that typically are fitted to existing protein structures. Thus, with the growing number of protein structures available, these parameters can be refitted regularly and should automatically lead to better scoring functions, given that the functional form for scoring is suitable.

The performance of protein-threading methods is typically assessed by *fold recognition* benchmarks, where the aim is to retrieve the template structure that is most similar to the structure of the query protein. The performance can be quantified in terms of the number of correctly assigned folds or, in a more detailed fashion, by rating the quality of the alignment on which the fold assignment is based. The accuracy of protein-threading methods depends highly on the similarity between the query sequence and the template protein; accuracy will be poor in the case of low sequence similarity, but much better in the case of high sequence similarity. If the sequence identity is above 70%, then this approach is likely to afford a high-quality structure model of the target protein. Full-atom models become more unreliable with lower sequence similarity, and in many application scenarios it is helpful simply to predict the backbone structure so as to identify the structure family of the query protein. This can be achieved using current methods, even if the sequence identity between the template and target is significantly below 25% [109]. Current protein-threading methods can provide a significance estimate that indicates if the structural model is to be trusted.

While today, similarity-based modeling is quite successful, it should be noted that this approach cannot be used to discover as-yet unseen protein structures. Rather, it can be used only to rediscover structures that have been seen previously, as attained by different protein sequences. Unfortunately, this approach has considerable limitations, as all of the types of protein structure found in Nature have not yet been uncovered and there is, as yet, no consensus regarding the extent of "structure space" that has been already explored [110].

4.1.3 *Ab Initio* Protein Structure Prediction

Ab initio structure prediction does not use structural templates but rather attempts to construct the protein structure from the sequence, from scratch. This affords not only the rediscovery of known template folds attained by new protein sequences, but also the actual discovery of new (not previously seen) folds. Currently, there are basically two approaches to this problem:

- The use of energy minimization, which is of quite limited effect, the only reports having been made on an ability to model short stretches (a few dozen residues) in small proteins.
- During the late 1990s, David Baker and colleagues proposed the modeling of protein structures from structural templates of small protein fragments (3-mers and 9-mers) that were taken from resolved protein structures [111, 112]. This assembly procedure is equipped with an effective statistical scoring function [113] while, in a second phase, the predicted conformations are refined, by placement of side-chain atoms and using a high-resolution scoring function [114]. This so-called ROSETTA method has also been applied to improving secondary structure predictions [115] and in designing new protein structures (see Sect. 5.5).

4.2
RNA Secondary Structure

Although, in principle, the aspects of RNA structure prediction are similar to those encountered with proteins, the procedure is less mature than that of protein structure prediction. RNA structure prediction differs from protein structure prediction in two main ways: (i) that the secondary structure patterns of RNA are different from those of proteins (the RNA secondary structure is composed of double-helical base-pairing patterns within the RNA sequence); and (ii) that the 3-D structure prediction of RNA is hampered by a lack of data for resolved RNA structures, as compared to proteins. Today, over 73 000 protein structures are available, compared to fewer than 3500 RNA structures (see http://www.pdb.org/).

One intensively studied problem encountered on the way to determining RNA 3-D structures was that of the RNA secondary structure prediction [116]. RNA exhibits a secondary structure due to the base-pairing that occurs within the molecule; typically, the usual Watson–Crick base pairs (A–U, C–G) occur preferentially, though other pairings (U··G, A··G) and even (U··U) may also occur. In a typical RNA secondary structure (Fig. 5) the RNA secondary structure can be represented as a planar (more exactly, an outerplanar) graph, if the usually few base pairings violating this shape are disregarded. (These base pairings are referred to as *pseudo-knots* [117, 118]; see Fig. 5.)

There are basically two approaches to RNA secondary structure prediction:

- *Comparative sequence analysis*, in which the likely locations of base-pairings are deduced from correlated mutations in the respective positions along multiple alignments of homologous RNA sequences. Although, traditionally, this process was mostly performed by hand, more recent algorithmic versions of this scheme have been presented [119–121].
- An *energy-based approach*, in which an additive energy model is assumed where each local structural feature (e.g., a stacked base-pair and each of the various bulges and loops in the secondary structure; see Fig. 5) is attributed a certain energetic contribution. These energetic values are deduced from experiments and simple model assumptions. A combinatorial optimization algorithm then searches for the RNA secondary structure with the lowest energy. This optimization can be achieved efficiently by using dynamic programming, if the RNA structure is assumed to be tree-like (as described above). This method was pioneered by Zuker and Stiegler [122] and it is quite popular as it lends itself to many generalizations. In fact, different trade-offs can be selected between the complexity of the energy model and the runtime of the minimization algorithm [123]. Consequently, it is possible to compute not only the optimal but also a set of near-optimal foldings, and to display them together in a so-called *energy dot plot* in order to identify the putatively most stable regions of the structure [124]. Finally, as the energy contributions from local structure elements can be chosen to be temperature-dependent, a "movie" can be assembled of the energetically most favorable RNA structures as the temperature rises (or falls). Thermodynamic quantities such as the heat capacity can also be computed [125], using the same efficient dynamic programming approach. Each of these methods disregards pseudoknots, however. These

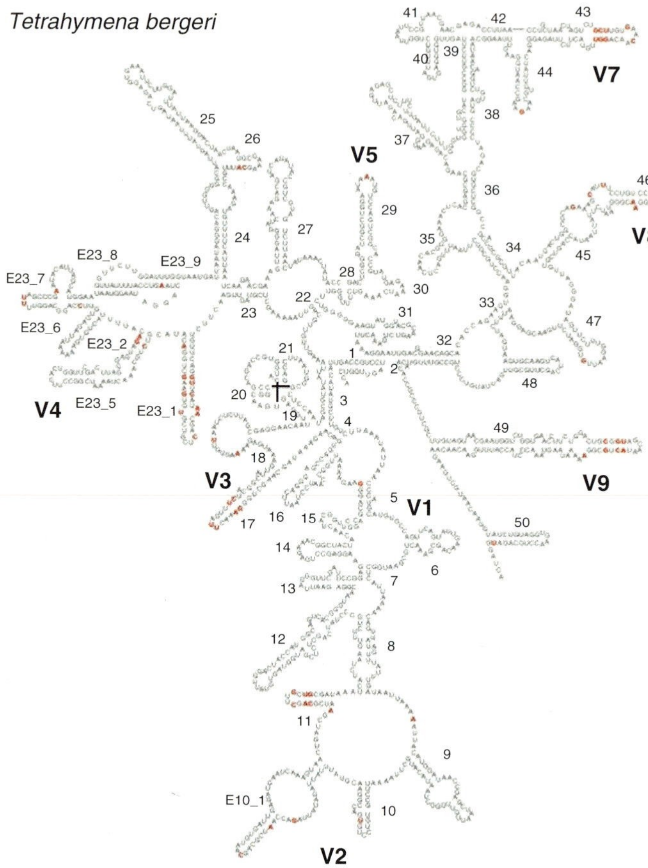

Fig. 5 Secondary structure of the RNA from the small ribosomal subunit of *Tetrahymena bergeri*. The thick lines (close to the numbers 19, 20, and 21) denote pseudoknots.

must be factored in by special algorithms, making the computation more expensive [126, 127].

Until now, RNA tertiary structure predictions have received much less attention than protein structure predictions [128]. Nevertheless, interest in the field of RNA is steadily increasing, with considerable success having been recently achieved using specialized energy-scoring functions [129].

5 Analysis of Molecular Interactions

The analysis of molecular interactions is a major component of understanding protein function, as well as a basic technology for the identification of new drugs (in this case, the term *interaction* means the noncovalent binding of two or more molecules). At this point, interactions between only two molecules will be discussed, with distinction being made between large molecules (e.g., macromolecules, such as proteins, DNA, or RNA) and small molecules (e.g., metabolites or drugs). The main case under discussion will relate to the macromolecule being a protein; accordingly, protein–protein docking problems will be considered with regards to the formation of a complex between two proteins, and protein–ligand docking problems to analyze the situation of a (small-molecule) ligand binding to a protein. These two problems differ both in terms of their relevance and of their bioinformatics solution.

5.1 Protein Ligand Docking and Drug Screening

Docking problems may be viewed as vehicles for understanding molecular interactions from a basic research point of view. The protein–ligand docking problem is also of central interest because of its application in drug design [130], and this is why the subject continues to attract much more attention from the pharmaceutical industry than do protein–protein docking problems. This application also places important requirements on protein–ligand docking software. Specifically, the demand is on high-throughput systems since, in order to screen large databases of putative drug molecules, each instance of the protein–ligand docking problem must be resolved in as short a time as possible. Today, the relevant computing times for drug screening by docking are in the second-to-minute range for docking programs that analyze full ligand and restricted protein flexibility. The ligand is generally assumed to be drug-like, with molecular weights up to about 500 Da and up to a dozen rotatable bonds. Peptides, which may be larger and have a greater number of rotatable bonds, exercise the limits of general protein–ligand docking software and thus require special approaches.

Two main issues must be accounted for in a docking program. The first issue is to find the geometric fit between the two interacting molecules. This entails identifying the correct conformation of the (generally highly flexible) ligand and the protein; this will generally involve the analysis of induced-fit conformational changes in the protein if the software allows. In general, a docking program will generate several putative protein–ligand complexes that are feasible geometrically. Consequently, a second selector is required in order to distinguish the correct binding mode(s) from the incorrect; such a selector provides an estimate of the differential free energy of binding between the complexed and the uncomplexed molecules. According to the laws of thermodynamics, the target is the conformational state that will minimize this energy score. The energy score can also be used to distinguish good from bad lead compounds; typically, the lower the energy score the stronger will be the binding affinity of the ligand to the protein, and the more interesting the ligand will be as a putative lead compound. Although energy scoring is at the heart of docking, it is also the least well understood part of docking. Likewise, in order to abide by the strict time constraints, simplified energy models must generally be used, which results in inaccurate energy estimates. The introduction of protein flexibility and improving the scoring function remain the two main challenges in the field, although progress has been made in these areas. The result of docking a small-molecule ligand into a protein is shown in Fig. 6.

A number of widely used docking programs (e.g., DOCK [131], GOLD [132],

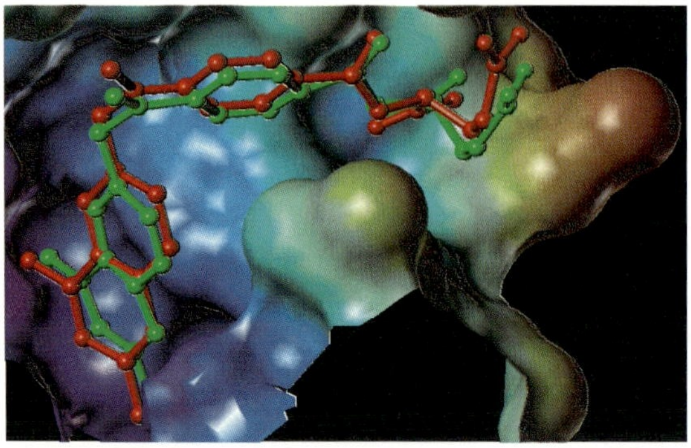

Fig. 6 Docking of the drug methotrexate into the active site of the protein dihydrofolate reductase. Green: observed crystal structure. Red: docking result computed by the docking program FlexX.

FlexX [133], Fred (OpenEye Software), AutoDock [134], LigandFit (Accelrys), ICM [135]) employ a few successful approaches to the subproblems of protein–ligand docking in different combinations (for reviews of the present state of the art in this field, see Refs [136–138]). For handling ligand flexibility, some programs precompute a limited set of rigid conformers that are docked into the binding site of the protein, one after another, whereas other programs build up the ligand incrementally inside the binding pocket of the protein, from more or less rigid fragments. Yet other programs employ more generic conformational sampling schemes based on genetic algorithms, Monte Carlo techniques, or molecular dynamics. Those docking programs that build up the ligand inside the binding pocket of the protein lend themselves to efficient screening procedures on combinatorial ligand libraries. In such a case, the alternatives of the so-called (functional) R-groups in the ligand scaffold can be screened during the build-up procedure, which may in turn produce a significant increase in the screening throughput [139, 140].

Approaches for estimating binding affinity [141] can be grouped into Class 1 methods, which are relatively accurate but slow, and Class 2 methods, which are fast but only approximate. The molecular dynamics (MD) simulations (as used in Class 1 methods) compute atomic forces with a force field, and represent protein, ligand, and solvent atoms explicitly. Although these more accurate methods are able to compute the binding free energies, they are generally too slow for practical use when screening large ligand databases. In contrast, Class 2 methods such as FlexX operate much faster since only the ligand conformations are sampled, mainly because they use a rigid protein representation and an implicit description of solvation instead of explicit solvent atoms. Consequently, large compound databases can be screened using these methods within a reasonable time. In general, the Class 2 methods are unable to accurately predict ligand binding affinity, although they have been very successful in predicting the correct structure of the bound ligand [137, 142]. Today, additional methods are available which are more accurate than Class 2 methods, but faster than Class 1 methods. These incorporate MD simulations and protein flexibility [143]. These intermediate "end point" approaches are faster than Class 1 methods, because the intermediate stages between bound and unbound are not taken into consideration.

As an aside, it should be noted that a substantial spectrum of methods is available for drug screening in the case that the structure of the protein is not known. In this case, the query for the search through ligand databases is not the protein, but rather a compound that is known to bind to the binding pocket of the protein; compounds will then be sought that are similar to the query compound. The notion of molecular similarity is critical here; this is usually based on one or a set of molecular descriptors – that is, abstractions of the molecule that entail information about relevant molecular properties. Currently, different types of descriptor are in use, with the most popular consisting of binary vectors that first list (sometimes up to one hundred thousand) molecular properties and then code, for each property, whether the molecule has that property. In this case, molecular similarity analysis is reduced to a comparison of binary vectors, which can be achieved very quickly, such that the resulting screening procedures will have a throughput of several tens of thousands of molecules per second [144]. More complex descriptors involve a topological analysis of the molecule; these incur more complex similarity analysis algorithms that have a throughput of only about 100 molecules per second although, with such descriptors, it is possible to handle the conformational flexibility of the ligands [145]. Some overviews of approaches to drug screening by molecular similarity are available in Refs [146–148].

These initial drug screening procedures seek only *lead compounds* that bind (more or less) tightly to the protein and. In order to convert such a lead compound into a drug, its so-called "ADMET" properties must first be optimized. In the term ADMET, the "A" represents absorption (via the intestinal barrier, if given orally), "D" represents distribution (throughout the body, involving delivery to the site where the drug is thought to act), "M" represents metabolism (the tendency of the drug to be metabolized before it reaches the site of action), "E" represents excretion (of the drug from the body), and "T" represents the drug's toxicity. Computational procedures for assessing these properties are summarized in Refs [149, 150].

5.2
Protein–Protein Docking

As protein–protein docking is central to understanding the regulatory events in living systems, the protein–protein docking problem represents a vehicle for understanding the inner workings of cells, and the molecular basis of diseases. As proteins are not preferred drugs, the protein–protein docking problem cannot (at least, at present) be used to identify new drugs, although it can form an aspect of identifying target proteins. An example of this is that the binding of two proteins is supposedly inhibited by a drug. The screening aspect is less important for the protein–protein-docking problem than for the protein–ligand-docking problem. One could think of docking a protein against all structurally resolved or modeled proteins of an organism, but the number of resulting instances of the docking problem is still orders of magnitude smaller for protein–protein docking (below 10^5) than for drug screening (many millions, and more).

Typically, protein–protein binding sites are larger than protein–ligand binding sites, since the energetic differences between the bound and unbound complex are usually higher for drugs (which are supposed to "stick" to the protein) than for transient protein complexes (which re-dissociate). Protein flexibility is quite essential for reliable protein docking, and desolvation phenomena are likely to be more relevant here than in protein–ligand docking. Taken together, these points lead to protein–protein docking being an exceptionally difficult variant of the docking problem.

In general, approaches to protein–protein docking are built on a procedure for rigid docking, and incorporate an induced fit on top of this algorithmic basis. Methods based on Fourier analysis or Fourier transform [151], geometric matching [152] and genetic algorithms [153] each serve as approaches for the rigid-docking part, while the induced fit can be incorporated rudimentarily by "softening" the protein surface by using appropriate energy potentials, or explicitly by sampling different side-chain conformations. In general, the energy-scoring problem is the same for protein–protein docking as it is for protein–ligand docking. If a rudimentary scoring function is used, then the ranking of the different complexes generated by the docking program by energy will be inaccurate, and a rescoring of the high-ranking complexes using a more accurate scoring function may be necessary. The explicit sampling of side-chain conformations can be accomplished using a combinatorial search procedure [154], or with genetic algorithms [155]. The dead-end elimination algorithm has been identified as an effective method for reducing the number of conformers to be searched [156, 157].

Current state-of-the-art protein–protein-docking procedures include ICM-DISCO [158, 159], 3D-DOCK [160, 161], and ZDOCK/RDOCK [162–164]. Each of these programs (and others) took part in the CAPRI (Critical Assessment of Prediction of Interactions[6)]) protein–protein docking contest that was first held in 2001 [165, 166]. A recent review of protein–protein docking is available in Ref. [167].

5.3 Protein Interactions Involving DNA

Knowledge of protein-DNA interactions is important if the regulatory events that occur within the cell are to be understood. Because of the high regularity of the DNA double-helix structure, the usually large structural changes in the protein, and the importance of the solvent in protein–DNA binding, this version of the docking problem is an especially difficult variant of the protein–protein-docking problem. Nonetheless, attempts have been made either to adapt existing protein–protein-docking software to protein–DNA docking [168], or to develop methods that are specially tailored to this problem [169, 170].

One special case of protein–DNA interaction is the binding of transcription factors (TFs, which regulate gene expression) to their DNA-binding sites. The TFs bind to conserved DNA segments that can be represented as sequence motifs, and methods akin to protein motif finding have been used to identify binding sites for specific TFs [83]. Recently, advances have been made in the identification of TF-binding sites based on the analysis of molecular structure [84]. A further problem that has also attracted attention is the binding between DNA and the nucleosomes, which form the basic packing units of DNA in eukaryotes. The patterns in which the nucleosomes are positioned along the

6) http://capri.ebi.ac.uk/.

DNA are functionally relevant, as they can influence transcription and therefore determine gene expression [171–173]. In, general, computational methods for modeling TF and nucleosome binding rely on additional data obtained from chromatin immunoprecipitation (ChIP) experiments that have allowed an experimental identification of the binding sites for TFs. Today, genome-wide high-throughput versions of this technology are available (ChIP-chip [174], ChIP-seq [175]).

5.4 Protein Design

Several computational approaches have been proposed for designing proteins with specific folds, binding sites, or enzymatic activities. These rational design approaches provide opportunities for further advancing and testing the current understanding of protein structure and function. Studies of both protein design and engineering have also been driven by practical applications in biotechnology and medicine.

Previously, the identification of protein sequences that adopt a specified protein structure has been a major challenge in protein design. To a certain extent, this problem is the reverse of structure prediction (see Sect. 4), and the computational approaches developed to address these two problems are closely related. A first landmark success in this field was the computational design of a new sequence that adopts a template fold corresponding to a known small structural motif [176]. An automated method was used for screening possible amino acid sequences and corresponding amino acid side-chain conformations for a given protein backbone structure. The successful computational design of a full-size protein with a human defined novel fold was a second breakthrough [177]. The method used iterates between sequence design [178] and structure prediction steps. The structure prediction step (see Sect. 4.1.3) consists of a search through the conformational space in the neighborhood of the design model (this step is required in order to avoid a specified fold that is not designable). The resulting designed protein proved to be stable, with a native structure that was very close to the design model, as determined using X-ray crystallography. Additional achievements in this field have been reviewed [179–181].

Another major challenge is the design of enzymes capable of catalyzing a specified chemical reaction; this field continues to attract considerable interest, given the potential applications in industrial organic synthesis. Indeed, considerable success has been reported in the design of a specific enzyme function with catalytic antibodies and with directed evolution [182]. More recently, computational design methods have also been successfully applied to the design of enzymes with new catalytic activities [183].

If the desired new type of activity already occurs in an homologous template enzyme, then an alignment between the enzyme to be modified and the homologous template can assist in the design process. Such an alignment supports the identification of candidate substitutions that alter enzyme activity in a desired way. Other computational methods have been proposed for enzyme design in the absence of an enzyme homologous to the template, and applied to the design of new types of catalytic activity not yet observed in naturally occurring enzymes. In general, these methods are used first to determine the structure of a local active site that stabilizes the transition state of the chemical reaction to be catalyzed, after which the modeled active site can be placed into a suitable protein scaffold. Hashing techniques have been proposed for searching through large sets of protein scaffolds that are appropriate for placement of the defined catalytic site [184]. Recently, this approach has also been used to design a new type of enzyme that can catalyze a type of chemical reaction that is of great interest to the synthetic organic chemist [185], where this reaction cannot be catalyzed by any known naturally occurring enzyme. Computational methods have also been employed successfully in assisting experimental approaches to enzyme design. An example of this is structure prediction and ligand docking, which have been applied to provide initial structural models for use in enzyme optimization procedures by directed evolution [186].

6 Molecular Networks

While pairwise interactions of molecules are essential components of cellular function, the inner workings of cells and organisms become apparent only at the level of sophisticated interaction networks. In recent years, the amount of experimental data for molecular interactions has increased substantially due to new and improved high-throughput measurement techniques [174, 187]. Examples are yeast two-hybrid screens for detecting protein–protein interactions, affinity purification coupled with mass spectrometry for identifying complexes of proteins, and ChIP for measuring DNA binding by proteins. Today, comprehensive datasets of molecular interactions are available for most model organisms, including microbes, yeast, flies, and worms, while human interaction data accumulate at an accelerating pace.

This increase of available data has been paralleled by important methodological advances in network bioinformatics, which are aimed primarily at acquiring an understanding of large-scale, genome-wide networks and intertwined, dynamic cellular processes in the context of systems biology and medicine [188]. Therefore, computational methods for the analysis and modeling of biochemical pathways and interaction networks play a major role towards the visionary goal of simulating a cell (as in the E-Cell Project [189]), or even an organ or organism in the healthy or disease state (as in the Virtual Physiological Human initiative [190]). The biochemical networks in cells are very complex, and the intracellular and intercellular processes that exist either within or between organisms – such as pathogens and the human host – necessitate extensive investigation of not only these networks, but also their dynamic connections.

One problem of obtaining reliable answers to biological questions from network models is that the experimental interaction data contain a significant numbers of errors, both false negatives and false positives [191]. This generally occurs because the laboratory procedures do not fully emulate the conditions under which the proteins bind in living cells and, as a consequence, the measurement techniques cannot clearly distinguish, for instance, between transient binding partners and those that bind to each other over longer time periods. Binding events that occur only under certain physiological states are also difficult to measure. Nevertheless, as detailed below, great successes have been achieved when utilizing the resultant networks

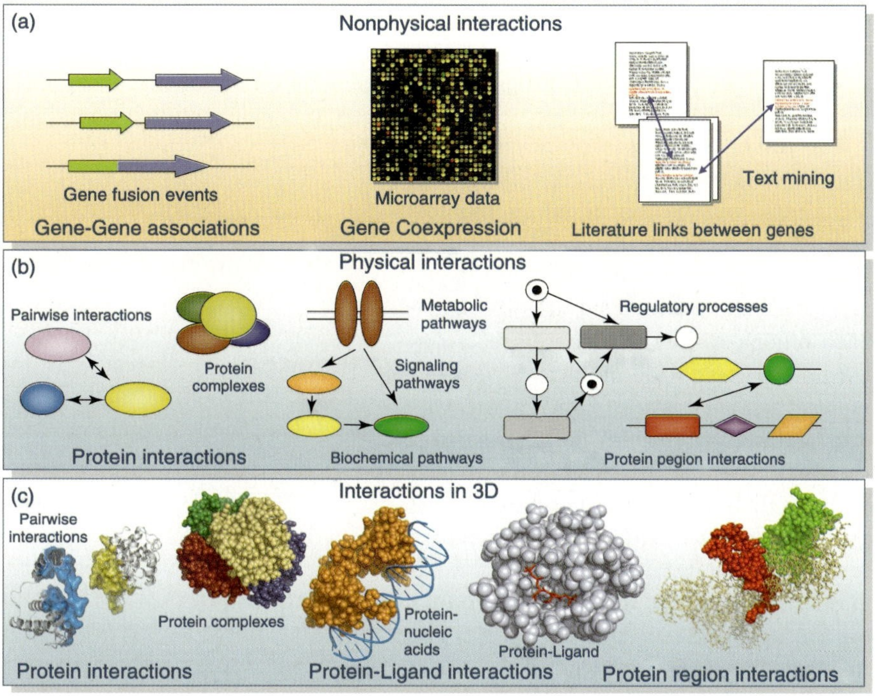

Fig. 7 Different types of networks are formed by physical and nonphysical interactions. (a) Merely functional associations result in nonphysical interaction networks and functional gene–gene relationships; (b) Direct contacts between molecules form physical networks and biochemical pathways of signaling, regulatory, and metabolic nature; (c) The consideration of 3-D structure data adds to the observed complexity of molecular interactions at the atomic level. Figure adapted from Ref. [193].

in gaining insight into the molecular systems.

6.1 Different Types of Network

Biological networks are commonly modeled and illustrated by nodes representing molecular entities (depicted as dots) and interaction edges (visualized as lines) between the nodes. The full set of molecular interactions in a cell defines the *interactome*. Depending on the connections between the biological entities such as genes, proteins, and small molecules, different types of network can be distinguished (see Fig. 7), and also have their specially targeted databases and analysis methods [192–194].

While gene networks are normally based on functional relationships between genes and the encoded products [195], protein networks often result from physical contacts of interacting molecules. Pairwise protein interactions with other proteins and molecules such as nucleic acids and small ligands are differentiated from protein complexes, which are assembled sets of proteins that bind to each other [194]. Frequently, proteins form specific biomolecular pathways to perform metabolic or signaling functions, or they function in regulatory processes as TFs by binding to the promoter regions of genes [196]. The additional inclusion of structural information provides even more molecular details of interacting protein residues and atoms [197].

The three main types of physical protein network – metabolic, regulatory, and signaling – will be discussed in the following subsections. It is important to note that networks of all three types are closely interconnected and influence each other considerably; therefore, a discernible, system-wide biological meaning is best achieved by integrating the various datasets of experimental measurements [198]. In particular, in order to deal with the distinct structural and functional properties of the networks, bioinformatics research has developed a variety of modeling and simulation methods, depending on the network size and type [199]. For instance, network reconstruction methods depend much on the respective network entities, and make use of the experimentally measured statistical dependencies to reverse-engineer specific small pathways [200, 201]. Other bioinformatics methods utilize sequence-structure features of proteins such as known binding sites, and genomic as well as evolutionary relationships between interaction partners within or between organisms in order to predict genome-wide protein interaction networks [194, 202].

6.2 Metabolic Networks

The metabolic network of a cell summarizes the set of reactions inside the cell that metabolize small molecules (the *metabolites*) with the help of catalytic proteins (the *enzymes*). Coherent sequences of successive reactions form *pathways*. If a pathway closes in on itself, it is termed a *metabolic cycle*. The objective of a pathway or a cycle can be to synthesize or degrade a macromolecule or a metabolite, or to convert or release usable energy. Metabolic networks are very uniform and composed of elements that represent chemical reactions (see Fig. 8). Each element consists of two vertices termed the *educt* and the *product*, an edge signifying the reaction labeled with the catalyzing enzyme, and possibly attached cofactors.

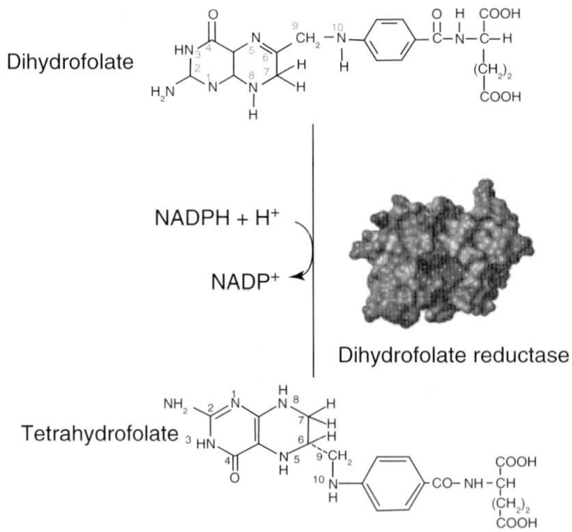

Fig. 8 Illustration of a metabolic reaction. The educt is dihydrofolate, and the product tetrahydrofolate. The reaction-catalyzing enzyme is dihydrofolate reductase. This reaction is blocked by the inhibitor bound to the reductase, as in Fig. 6.

The reactions can be reversible or irreversible, with the latter case being denoted by a pointed arrow in the unique direction of the reaction, resulting in a directed interaction network. The composition of such elements at the metabolite nodes yields complex networks such as shown in Fig. 9. The regular nature of metabolic networks affords the opportunity to build large uniform databases, and also to devise systematic analytical procedures for these networks. In order to completely describe a single reaction (such as that shown in Fig. 6a), it is first necessary to

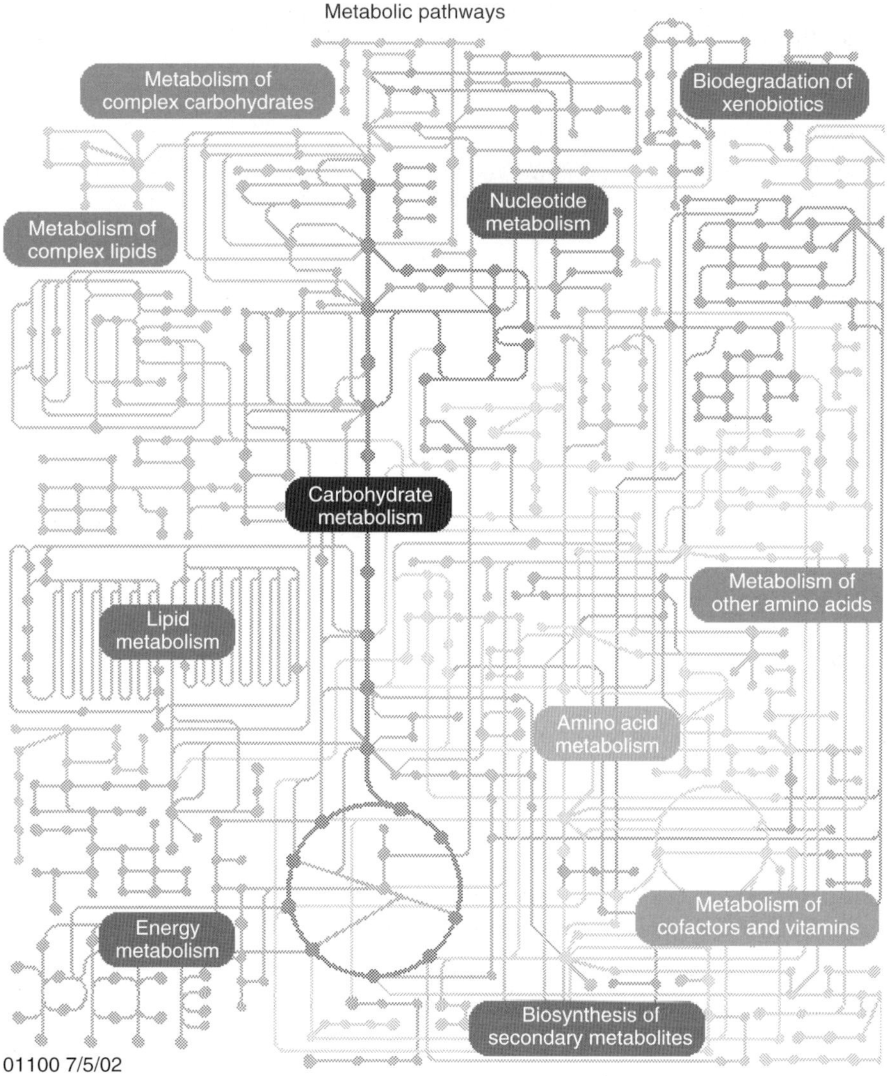

Fig. 9 Schematic diagram of part of the metabolic network of *Escherichia coli*. http://www.genome.ad.jp/kegg/pathway/map/map01100.html.

supply the kinetics of the reaction – that is, the concentration gradients of the involved molecular partners in terms of their starting concentrations. While the form of many reactions is known today, the reaction kinetics are a sorely missing piece of data for many reactions. The large-scale modeling of dynamic metabolic networks remains a challenge for the future [203].

A wealth of metabolic network databases is available for different organisms, and these are listed at BioCyc/MetaCyc [204], EuPathDB [205], GeneDB [206], MicroScope [207], and other web sites. Two well-known databases particularly targeted to human are KEGG [208] and Reactome [209]. Typically, the databases provide data records and images describing the pathways, although databases are also available (e.g., BRENDA [210] and IntEnz [211]) that concentrate on the individual reaction. All metabolic databases are equipped with their own set of tools for querying the databases, for navigation, and for visualization of the search results.

These databases are sufficiently comprehensive to enable comparisons of the factual or putative topology of metabolic pathways in different species with fully sequenced genomes. Such comparisons have been undertaken for several metabolic pathways [212], partly involving genomic context methods for protein function prediction (see Sect. 8.3). Furthermore, mathematical methods have been developed to analyze the topological structure of pathways. Flux balance analysis uses linear optimization to assign flux values to the edges of a metabolic network that maximize some objective function under steady-state conditions, for example, growth per substrate uptake [213]. In contrast, metabolic flux analysis decomposes pathways into elementary flux modes [214, 215], which are the minimal components of a pathway that can act coherently and independently under steady-state conditions. Both types of analysis aim at yielding insight into the biological structure of a metabolic network from its topology. Software exists to perform this analysis on medium-sized pathways [216], but if the steady-state as a special case does not apply, more general analysis methods have to be used based on ordinary differential equations that model the reaction kinetics in the pathway under consideration. Extensive software has been developed for this purpose [217–220]. As the required data are missing in many situations, metabolomics and large-scale metabolite screening projects are aimed at removing this deficit [221, 222].

6.3
Regulatory and Signaling Networks

The regulatory and signaling networks are generally thought of as being that part of the biochemical networks inside the cell that facilitates information flow and communication, rather than converting matter. Chromatin remodeling and the binding partners of TFs affecting gene expression play a major role in regulatory processes. In signal transduction pathways, post-translational protein modifications control the different signaling branches of protein interaction cascades. Both types of networks are much more difficult to approach than the metabolic network, and much less is known of the molecular relationships, one reason for this being that the networks are more inhomogeneous. One molecule can affect another in many ways, including agonistic, antagonistic, activating, deactivating, and inhibitory effects.

TRANSPATH [223] and TRANSFAC [224] are prominent databases of signal transduction pathways and TF binding targets. The web interface to the databases includes tools for querying and visualization, as well as an integrated tool for the analysis of microarray data in the context of the regulatory and signaling networks (see Sect. 7). Basically, there are two ways to study regulatory and signaling networks. The first approach is to simulate and analyze established networks, for which a straightforward method is to model molecular relationships with the same framework of differential equations that is used for the nonstationary analysis of metabolic networks [225]. Whilst, given sufficient data, this approach is general and can be accurate, it may incur performance problems and lacks the model hierarchy that biologists usually attribute to biochemical networks. In contrast, qualitative simulation and permutation methods simplify the general differential equations model to simpler function forms, or even to discrete models [226, 227]. The other, much more challenging, question is to uncover as-yet unknown molecular interactions and pathways from experimental data, and to model their spatiotemporal network dynamics [228]. In disease, accurate and complete network models of cellular processes bear the promise of explaining drug effects and clinical outcomes [229].

6.4 Approaches to Analyzing Interaction Networks

A multitude of computational methods can be applied to analyze molecular interaction networks [194, 230], some of which are described briefly here:

- *Topological analysis* identifies important players in the network, for instance, molecules that have a large number of interaction partners and thus are essential for network function [231]. For this purpose, many efficient algorithms are available to compute a variety of topological network measures [232].
- *Evolutionary analysis* studies the conservation of networks between organisms or tissues, and the presence or lack of nodes and interaction edges in the networks [212]. This research often necessitates the construction of network alignments that matches, between two related networks, conserved entities represented by the network nodes as well as their interactions. In this regard, network alignments are methodologically related to sequence alignments [233].
- The *functional analysis* of networks refers to understanding the actual molecular function of interactions, for instance, by determining the interacting protein regions [234] or by finding interaction modules consisting of sets of collaborating molecules. Examples are biomolecular pathways that transduce cellular signals or protein complexes that function as molecular machines [235, 236]. In order to dissect networks into smaller building blocks of interacting molecules, many clustering techniques are available that can create subnetworks based on statistical properties as well as evolutionary, functional, and topological similarity measures [237].
- *Structural analysis* adds experimental or predicted structure data of interacting molecules to networks, yielding a very detailed view of interactions [197].

- *Dynamic analysis* concentrates on interaction changes in networks under different conditions and spatiotemporal states, for instance, caused by post-translational modifications of signaling proteins or by the reception of external stimuli [201, 238].
- *Quantitative analysis* includes continuous values such as gene expression levels, protein abundance measurements, and kinetic parameters into network models. Mathematical modeling approaches with these data include differential equations and stochastic approaches [201, 239].
- *Qualitative analysis* calculates statistical confidence scores on individual interactions. The resulting quality scores often take the experimental technique used to detect the interaction or the number of replicates performed into account. Additionally, there are other quality measures based on further biological sources of evidence such as structural or functional similarity, evolutionary conservation, network context, and curated data sets serving as the "gold standard" [191, 193, 240].
- Finally, *visual analysis* makes use of visualization tools to mine network data [241]. To this end, different colors and shapes of network nodes and edges can be applied, for example, to map gene expression data on protein interaction networks [242].

7 Analysis of Expression Data

During the early 1990s, the first developments were undertaken towards measuring the profile of genes transcribed in a cell in a certain cell state. This information goes way beyond the information contained in the genome sequence, as it can help to distinguish between different tissues and different cell states within the same tissue (e.g., healthy/diseased) and, consequently, it was suggested to be the basis for an entirely new approach towards the diagnosis and therapy of diseases. Technically, the most convenient way to measure gene expression profiles is on the basis of the transcribed mRNA, rather than protein. The reason for this is the much more uniform molecular form of RNA and DNA, which lends itself more conveniently to laboratory procedures than do proteins. There is, however, a trade-off in choosing mRNA as the level on which gene expression is to be measured. On the one hand, mRNA expression levels do not always correlate well with protein expression levels, which are governed by the regulation of protein translation and degradation, in addition to transcriptional regulation. In addition, post-translational modifications of proteins which have an essential effect on protein function cannot be measured on the mRNA level. Nonetheless, because of the technical accessibility mRNA expression measurement became very popular, first on the basis of expressed sequence tags (ESTs) and then on the level of complementary DNA (cDNA) microarrays [243, 244].

The advent of cDNA microarrays during the late 1990s created a large surge of statistics and bioinformatics activities targeted both at configuring the arrays and at interpreting the resultant data. Soon, not only the potential but also the limits of microarray measurements became apparent. Today, the original microarray technology is continually replaced by deep-sequencing methods (see Sect. 7.5). Whilst this affords a more accurate measurement of the abundances of specific transcripts, it does not solve the above-mentioned problems regarding the limited correlation of mRNA expression and protein expression [245].

Today, many of the high hopes of making protein function accessible directly through gene expression measurements have not materialized. Nonetheless, it is becoming clear that diagnostic procedures progress significantly through expression measurements and, in concert with other biological data, such measurements can help to acquire some insight into the biological roles of genes and proteins.

7.1
Configuration of Experiments and Low-Level Analysis

The configuration of microarray experiments is a nontrivial technical issue that has been taken too lightly during the early days of the technology. Today, the importance and difficulty of the consistent preparation of a microarray experiment has gained widespread attention [246]. This includes rigorous quality control of the purified RNA populations, appropriate configuration of the probes on the chip to reliably select the desired cDNAs and avoid inadequate or nonspecific hybridization [247], to conduct a sufficient number of replications of the experiment [248], and to carry out sufficient logging of the experiment. For the latter purpose, the MIAME standard (minimum information about a microarray experiment) has been developed [249]. The public repository of expression data, ArrayExpress, which follows this standard is in existence and growing in volume [250]. The National Center for Computational Biology (NCBI) is providing another database of expression data, the Gene Expression Omnibus (GEO) [251].

The configuration of a microarray experiment is closely related to the low-level analysis of the data generated by the experiment [252, 253]; by this is meant the procedure which turns the raw readouts of the experiment into a number signifying expression level, in terms of RNA copies per cell, or of fold changes based on a reference profile. The low-level analysis must incorporate and appropriately model the variances in expression levels stemming from technological and biological sources of error. The respective statistical procedure is termed the *normalization* of expression data, and several methods for normalizing expression data have been described, with the choice of appropriate method being a central aspect of low-level analysis. Some overviews of basic normalization methods are provided in Refs [254, 255], while more modern methods are detailed in Refs [256, 257].

The result of low-level analysis is an *expression data matrix* of normalized expression levels. This matrix has the dimensions #probes $\times$ #samples, where the number of probes (genes) is usually very large (several tens of thousands) while the number of samples (experiments) is quite low (dozens to a few hundred). Thus, the expression data matrix is quite unbalanced, and this may have strong ramifications for the analysis of expression data.

The final step in low-level analysis is to decide exactly what is a differentially expressed gene. To this end, a threshold value of change in expression level must be fixed, above which a gene is regarded as being "conspicuous." This is a statistical question which must be resolved in terms of a model of background variation specific to both the gene and to the experiment.

7.2
Classification of Samples

The considerably easier problem is to classify the (few) samples based on the expression levels of the (many) probes. The related application problem is a diagnostic one: to distinguish different types or stages of a disease based on the related expression data. This problem is of high clinical relevance, and was investigated as soon as the microarray technology became available. An especially widely studied type of disease in this context has been cancer, with one of the first studies to classify microarray samples targeting two types of leukemia – acute myeloid leukemia (AML) and acute lymphoblastic leukemia (ALL) – which are difficult to distinguish histologically [258]. Subsequently, many investigations of different types of cancer were conducted using this approach.

When classifying samples, a variety of statistical learning techniques can be used. *Supervised learning* techniques employ labeled data – in this case samples that are known to originate from patients who have contracted a certain type of disease. A statistical classifier is then generated that classifies future samples. Learning methods include decision trees, neural nets, support vector machines, and others [259, 260]. Depending on the types of disease to be distinguished, a high classification accuracy (sometimes >90%) can be achieved, which in turn means that the type of disease can be determined quite reliably on the basis of mRNA expression data. This situation has been substantiated quite impressively for a wide variety of cancers [261]. One important part of supervised learning is that of *feature selection,* which amounts to the identification of small sets of genes the expression levels of which are indicative of the type of disease. Usually, it is possible to identify a few dozen genes whose expression levels can determine the disease type with almost the same accuracy as the classifier operating on the expression levels of all genes. In lucky cases, genes selected in this way may also be candidates for primary actors in the disease process.

Feature selection is a more general way of determining genes that are relevant for a disease than detecting differentially expressed genes, as it takes sets of genes into account rather than just single genes.

In contrast to supervised learning techniques, unsupervised learning techniques cluster the samples into groups that may reproduce known disease types or uncover new types [262] (for a review, see Ref. [263]).

7.3
Classification of Probes

Elucidating the role of genes in the disease process using expression data is much more difficult, and in many cases may even be hopeless. The reason for this is, again, the shape of the expression data matrix. Although attempts to investigate the differences between a large number of genes, based on very few experiments, may be considered ill-informed, various approaches have nevertheless been made in this direction.

The first approach involved clustering, the rationale being that coexpressed genes may be in functional association, although it soon became apparent that such a procedure would be quite inaccurate [264]. Consequently, a proposal was made to analyze expression data together with the upstream regulatory regions of the genes since, after all, it was these regions that determined transcriptional regulation. The rationale was that coexpressed genes which share functional sites in their upstream regulatory regions should be under the same or a similar regulatory regime, and therefore share a functional relationship [265–269]. This idea has now been extended to include an explicit model of the involved regulators [270].

Upstream regulatory regions are not the only type of information that can be used to enhance the analysis of expression data; rather, functional context – as exhibited by membership in a common metabolic or regulatory pathway – could also be exploited. By analyzing the expression levels of genes in the same pathway in concert, it is possible to obtain a statistical model of this pathway. Moreover, if this model is significant in the statistical sense, this might be an indication that the pathway plays a role in the disease process. Furthermore, genes with unknown functions can be tested against the pathway model, with a significant test response indicating a functional relationship of the gene with the pathway [271, 272]. This approach has been extended by including not only data from biochemical networks but also functional relationships witnessed by the scientific literature (see Sect. 8.6). Protein interaction data represent another supplementary source of information (see Sect. 8.3), such data have been analyzed in concert with expression data to elucidate molecular pathways [273, 274]. Here, the working hypothesis is that the protein products of genes exhibiting the same expression profile tend to interact. In all cases, the analytical methods are, again, statistical learning schemes, and especially if several sources of information needed to be integrated then the quite general machinery of Bayesian networks proves very effective (although Boolean networks and other graphical models are also used). However, as the relevant statistical learning procedures can produce regulatory networks that may be quite abstract and not closely linked to the molecular regulation process [275], their interpretation presents certain problems (for a review. see Ref. [276]).

7.4 Analyzing Transcriptomes with RNA-Seq

Although microarrays have been applied successfully in the analysis of the transcriptome, these hybridization-based approaches have many limitations, including high background levels, limited dynamic range, and problematic comparisons of expression levels.

Sequencing-based approaches provide an alternative to microarrays for characterizing the transcripts and quantifying gene expression levels. Although the traditional Sanger cDNA sequencing was used to characterize gene transcripts, this approach in general is not quantitative, is expensive, and has a low throughput [277]. Whilst tag-based methods are high-throughput and provide precise gene expression levels, they are quite expensive and provide only restricted coverage of the transcriptome [278]. The availability of new sequencing technologies led to the development of RNA-Seq, which provides a better high-throughput approach for mapping and quantifying transcriptomes and has been used to investigate transcription in human and in different models organisms [279–282], offering many advantages over previous methods [283]. RNA-Seq is capable of discovering new transcripts, and can be applied to analyze the transcriptome of organisms for which the genome sequences have not yet been determined. RNA-Seq is a powerful tool for determining the transcriptional structure of genes, in particular the start sites, the ends, and splicing patterns. In comparison to microarrays, RNA-Seq has very low background signal and a much larger dynamic range over which transcripts can be detected. RNA-Seq is also quite accurate for quantifying expression levels with high reproducibility. Considerable efforts have also been made to address the particular challenges associated with RNA-Seq, such as handling read mapping uncertainty or obtaining reliable estimates of expression levels for novel splicing isoforms [284–286].

In a typical RNA-Seq experiment [283, 287] a RNA sample is converted to a library of cDNA fragments that are sequenced with a new-generation sequencing method. The resulting sequencing "reads" are then aligned to a reference genome and the transcripts reconstructed (an align-then-assemble strategy). Alternatively, the transcripts are first assembled *de novo*, and only after they are aligned to the genome to delineate intron and exon structures and any variations between alternatively spliced transcripts (*de novo* assembly strategy). In contrast to the first align-then-assemble approach, the *de novo* alternative is still applicable in cases where a reference genome is missing or incomplete. However, it requires high read coverage and it might fail to correctly assemble less-abundant transcripts.

Today, a considerable range of bioinformatics tools is available for processing RNA-Seq data [288]. Typically, tools such as ABySS [289] and Velvet [290] are used for the *de novo* assembly of short sequence reads, while other methods such as Scripture [291] and Cufflinks [292] have been proposed which follow an align-then-assemble approach. Different strategies are used for mapping splice junctions; in this case, the ungapped sequence reads can be mapped easily across splice junctions if the splice junctions are known from an annotated reference genome, or from predicted splice junctions. Some align-then-assemble implementations are capable of building candidate exon–exon splice junctions, and to predict whether a sequence read is spliced [293, 294].

Gene expression levels can be measured by summing the number of sequence reads obtained for that gene with RNA-Seq. In practice, these read counts are normalized by the mRNA length and the number of million mappable reads, which results in the number of reads per kilobase per million (RPKM) [280]. Such RPKM values are comparable within a sample although, in order to compare different samples, they must be further post-processed in a similar way to microarray intensities [295].

7.5 Beyond RNA

As noted above, mRNA represents an intermediate step from the genotype (DNA) to the phenotype (post-translationally modified protein), and has been chosen for expression profiling mainly for technical reasons. Whilst the measurement of protein abundances is more difficult, due to the greater molecular diversity of proteins, it can be afforded by applying proteomics technologies that, today, are mostly based on mass spectrometry (MS) [296]. The MS technology is associated with its own set of computational low-level analysis methods that pertain to the identification of protein fragments in a mixture created by the fragmentation process associated with most MS procedures, including the assembly of protein fragments to form entire proteins and the estimation of protein abundances. The downstream bioinformatics procedures employed to interpret such abundance data share many characteristics with those used for gene expression analysis [297].

On the other hand, epigenetic modifications of the DNA, and of the proteins associating with DNA to form and regulate chromatin, have attracted increasing attention, with research investigations focused on cellular regulation. This area also has a bioinformatical aspect (for reviews, see Refs [298, 299]).

8
Protein Function Prediction

The elucidation of protein is function a central goal in molecular biology. Relatively little is known at present of the function of many proteins, with data relating to functional annotation currently available for only 70% of human proteins.[7] Many of these annotations are based on quite spurious evidence, and probably some of them lead in the wrong direction. Although today many genomes have been sequenced, an understanding of the biology of these genomes remains far off. The details of various efforts to determine protein function, employing a mixture of experimental and bioinformatics methods, are summarized in the following subsections.

8.1
What Is Protein Function?

Protein function is a colorful term that can mean quite different things in different contexts. Occasionally, the localization of a protein (cytosolic, membrane-standing, extracellular, etc.) may be taken as a qualification of its function, whereas in another context a protein's function may be understood in terms of its binding partners and the reaction that it may catalyze. In yet other situations the role of the protein in a complex biological process (cell cycle, apoptosis, etc.) is taken as its function. Clearly, the development of an appropriate ontology – that is, a structured vocabulary when referring to protein function – represents the first step towards a systematic analysis of this issue [300].

At this point, mention should be made of three attempts to provide ontologies for protein function:

- The oldest system is the hierarchical classification of enzymes by the Enzyme Commission (EC)[8] [301], which is based on the reactions that the enzymes catalyze and applies only to enzymes.
- A wider variety of protein function was covered by the MIPS Functional Catalog[9] in the context of annotation of the yeast genome [302], although this catalog has since been extended to other species. Similar to the classification by the EC, the MIPS system exhibits a hierarchical tree-like structure (with up to six levels, compared to four for the EC code).
- The most extensive effort to create an ontology for protein function was made by the international Gene Ontology (GO) Consortium[10] [303]. Again, the ontology is hierarchical, but this time there are three hierarchies in order to distinguish between the three fundamentally

7) http://www.ebi.ac.uk/GOA/proteomes.html.

8) http://www.chem.qmul.ac.uk/iubmb/enzyme/.

9) http://mips.helmholtz-muenchen.de/proj/funcatDB/search_main_frame.html.

10) http://www.geneontology.org/.

different approaches to protein function discussed above, namely cellular component, molecular function, and biological process. In addition, instead of a tree hierarchy, a directed acyclic graph is used, such that a child node could have different parent nodes.

The ontologies are expanded continuously, and the function annotations of genes and proteins in the databases and in the literature are linked to the ontologies. This whole body of knowledge continues to provide an increasingly voluminous "gold standard" by which computational methods for protein function prediction can be evaluated. Nevertheless, the structure of current ontologies provides only a structured vocabulary, and not a guideline for the functional annotation of proteins. Although the GO consortium provides an annotation guide with its ontology,[11] further investigations will need to be made to create a highly expressive and computer-digestible framework for representing protein function. Consequently, as the first larger regulatory networks are just beginning to be understood [304], initial efforts are also being undertaken to establish appropriate standards [305] and to develop approaches to generate models of the respective biological processes [306, 307]. An example of an application within a supra-cellular scenario is provided in Ref. [308].

8.2 Function from Sequence

Several methods have been developed to predict protein function from the protein sequence. Among these methods, some explore the evolutionary relationships between genes and proteins, while others rely on the identification of conserved motifs or domains. In addition, other methods employ different sequence features to predict function.

In principle, it would be expected that two proteins with sequences that are very similar globally would also have a similar or even identical function; however, this is not always true. Although a significant sequence similarity indicates that the proteins are evolutionarily related (homologous), and that they might share a similar function, the relationship between homology and function conservation is intricate and subject to debate [309–311]. In particular, protein function is not always shared between homologous proteins. It is important to distinguish between orthologous and paralogous proteins in this context:

- *Orthologous proteins* occur in different species where the respective genes diverged following a speciation event (where a species evolves into two distinct species), therefore, orthologous proteins are likely to have the same function.
- *Paralogous proteins* can occur in the same species and arise by a gene duplication event, which also allows for function divergence.

There are two ways in which to deal with the problem of identifying orthologous proteins. The first approach is to compute an all-to-all pairwise sequence similarity of the entire homologous family, and then to assume that two homologous proteins in two different organisms are likely to be orthologs if they are mutually most similar to each other within the homologous family. This is the principle behind the COG database of orthologs [312, 313].

11) http://www.geneontology.org/GO.annotation.html.

This approach provides only a crude approximation to orthology, however, and therefore the resulting COGs database must be manually curated.

The second approach is less heuristic but mathematically more involved, and is based on the principle that an accurate distinction between orthology and paralogy must incorporate a phylogenetic analysis [314, 315]. Here, the strategy is to reconcile the evolutionary tree of protein sequences of the homologous family with the respective evolutionary species tree. The sequence tree is then mapped into the species tree and the putative gene duplications are located. This amounts to a discrete optimization problem where the required number of gene duplications is minimized [316].

An alternative approach for inferring function from sequence is to identify usually very short sequence motifs that are highly indicative of function, and then to use them as functional signatures. Such motifs often represent the evolutionary constraints on a protein functional site within a set of homologous proteins. For example, the well-known ATP-binding site motif, the P-loop, is characterized by a sequence fragment of length 8 respecting the motif [AG]- $\times$ (4)-G-K-[ST]. Here, the characters in brackets denote the choices of amino acid residues according to the one-letter code at a single sequence position. Single letters represent unique amino-acid residues (no choice), and $\times$(4) denotes a stretch of four adjacent arbitrary amino acid residues. Motifs carrying information on protein function can be found by multiply aligning sequences of homologous proteins that are known to have the same function, and then selecting highly conserved sequence regions in the alignment. A motif can then be derived as some type of consensus of these regions.

The PROSITE database was the first collection of this type [317, 318] where initially the motifs where mainly manually generated. Subsequently methods have been introduced to derive the motifs automatically [319]. Here, the issue of sensitivity versus specificity is central, in that a motif that covers many of the sequences with the respective function necessarily also generates *false positives* (protein sequences that contain the motif but do not share the function). To exclude such false positives means to reduce coverage – that is, losing *true positives* (proteins which have a function that is to be described by the motif). However, several specific motifs (with few false positives) can together cover many true positives, in the sense that each true positive contains at least one of the motifs. A database of motifs based on this idea is described in Ref. [320]. The PSSMs or HMMs (see also Sects 3.1, 3.3, 3.4, and 4.1.2) are more general representations of conserved regions in protein sequences. With these, longer sequence fragments (e.g., whole protein domains) are typically described, and they are also easy to derive from multiple alignments. Many databases have been developed around this paradigm [321], with most of the sequence signatures having been integrated into the InterPro database [322] and linked to the functional terms defined in GO. A review of the different types of sequence signature for predicting function is available in Ref. [323].

Finally, supervised learning methods – most of which are based on neural nets – have been developed for predicting several aspects of protein function, including cellular localization (through the analysis of signal peptides and the prediction of transmembrane helices) and post-translational modification features (glycosylation and phosphorylation sites).

One server offering many such methods is available from the Center for Biological Sequence Analysis at the Technical University of Denmark, Lyngby.[12] One of the methods, known as *ProtFun*, makes a comprehensive classification of proteins with respect to their function (e.g., enzyme class or participation in a biological process such as amino acid biosynthesis or energy metabolism) based on general sequence features that can be calculated (e.g., sequence length, charge, amino acid composition) and predictions through other methods on the server. Some of the resulting classifiers achieve a high (90%) level of accuracy [324, 325].

8.3 Genomic Context Methods

Even in the absence of mRNA expression or protein interaction data, knowledge of protein function based on sequence data alone can be acquired in several ways. The main prerequisite is that the complete genome is present, in order that both the presence and absence of a protein can be reasoned. Currently, with many completely sequenced genomes being available (especially in the prokaryotic domain) these so-called "genomic context methods" are quite effective [326]. As a rule of thumb, approximately 30 genomes are needed to infer functional associations in proteins on the basis of genome sequence. Since, in the case of eukaryotes, this number has not yet quite been reached, genomic context methods are currently less effective in the eukaryotic domain.

In general, genomic context methods uncover functional associations between proteins that are more general than if the proteins were simply to be binding partners. The proteins may also be functionally associated on the basis that they take part in the same biological process (cell cycle, apoptosis, etc.), or that they cooperate to produce a phenotype. The latter situation includes genetic association, such as the concept of synthetic lethality where a simultaneous knockout of two genes produces lethal offspring, while the respective two single knockouts lead to viable offspring. The following subsections describe some of the currently available genomic context methods.

8.3.1 Gene Neighborhood and Gene Order

Two proteins in two different species are more likely to be functionally associated if they occur in close proximity and in the same order along the genome in many species. This is more valid in prokaryotes, where groups of genes are often found in operons where they are jointly regulated. Two variants of this method have been developed; the first method respects the order in which the genes occur along the genome [327], while the second method does not [328]. A generalization of this method to more general conservation patterns that present cycles of association that alternate between homology and neighborhood has been presented by Kolesov *et al.* [329, 330].

8.3.2 Domain Fusion

Two proteins are likely to be functionally associated if their genes are located next to each other in some organisms, but are fused into a multidomain protein in others. The rationale here is that the two domains in the fused protein are most likely involved in the same biological process and bind to each other, and that this association is likely to be maintained

12) http://www.cbs.dtu.dk/services/.

in organisms in which the genes are not fused [331, 332].

8.3.3 Phylogenetic Profiles

This method depends on the observation that two proteins which are either present in a species together or absent together, have a higher likelihood of being functionally related [333]. The cross-species comparison is based on an orthology test that has originally been reduced to a minimum value for the evolutionary distance of the two proteins that is supposed to represent orthology. More recent and effective versions employ gradual levels of evolutionary distance.

These methods are available in several servers, most notably the STRING web resource. During recent years, STRING has added additional sources of information on protein function, such molecular pathways, gene expression data, protein interactions, and even scientific publications. The result is a comprehensive resource for integrating diverse post-genome era information on protein function [334].

8.4 Function from Structure

Protein structure harbors detailed information on molecular function, mainly because it is the structure of a protein that causes it to act in the way it does. Nonetheless, the automatic deduction of information on protein function from protein structure is difficult, and not yet extensively developed. Protein structure by itself is not indicative of protein function, since folds are often reused for different functions through gene duplications or convergent evolution. On the other hand, the same function (e.g., enzymatic class) can be realized with several protein folds [335]. A detailed orthology analysis can help to resolve this situation (see Sect. 8.2). Beyond this, remote homologies can hint at functional relationships, if not identity in function. For instance, two remotely homologous enzymes can share the same reaction mechanism, while substrate specificity is not conserved [310]. In general, two types of problem are addressed in structure-based function prediction; the localization of functional sites; and the characterization of molecular function (as discussed below) (for reviews on this topic, see Refs [336, 337]).

8.4.1 Localization of Functional Sites

Functional sites comprise residues that are mostly located at the protein surface. These functional residues bind to small molecules (ligands) or to other proteins or macromolecules, and are subject to strong evolutionary constraints imposed by molecular function requirements. Commonly, the functional sites are less tolerant to mutations than are other surface regions, and thus they tend to be more conserved in sequence. Several computational methods have been proposed for predicting functional site localization, based on the identification of regions at the surface with high sequence conservation [338, 339]. In addition, it has been shown that small ligands tend to bind to proteins at the largest surface cleft [340]; consequently, some methods have been proposed that measure protein surface geometry in order to identify putative binding sites at the largest clefts [341–343]. Additional methods have been suggested that explore different functional site properties [344–349]. Proteins also tend to interact with other proteins, with several types of protein–protein interactions with unique properties having been described [350]. These properties have been used in various ways as the basis for identifying protein-binding sites [351].

8.4.2 Characterization of Molecular Function

Just as sequence alignment serves as the methodical basis for analyzing the similarities and evolutionary relationships between protein sequences, structure comparison serves this purpose for protein structures. Many methods for comparing protein backbone structure have been developed over the past 20 years [352, 353], and most of these have aligned the two proteins structurally. In other words, the amino acid residue pairs are matched, one in each protein, to correspond to each other spatially (i.e., to take each other's place in the protein structure). Whilst this alignment problem is related to sequence alignment, the scoring function must now represent structural similarity rather than sequence evolution, and it is therefore different. It is also more difficult to assume an independence between alignment columns (see also Sect. 3.1), since a match between two residues will be determined by the structural similarity between their respective surroundings. These structure comparison methods have been particularly successful in detecting remote homology between two proteins when the sequence similarity is not significant, because in evolution the protein structure will be more conserved than the protein sequence. A common approach for predicting the function of uncharacterized proteins with a known structure is by identifying remote homology using these methods, combined with an analysis of the structural alignment of the putative functional sites [354]. Automated methods have also been proposed for inferring the molecular function of uncharacterized proteins, based on the extent of backbone structural similarity to annotated proteins [355, 356].

When proteins share similar molecular functions, their functional sites tend to have similar physico-chemical environments, and a similar geometry. Based on this principle, several comparative methods have been developed for inferring molecular function based on local structure similarity at functional sites [357–361]. Integrated pipelines for function prediction that combine several structure-based methods have also been implemented and applied successfully to the functional annotation of uncharacterized proteins for which the structures were determined within structural genomics initiatives [362]. A promising alternative strategy to this relies on ligand docking for predicting the substrate of an enzyme of unknown function [363].

8.5 Text Mining

The most voluminous data source for protein function information is the literature, but this is also the most difficult to "digest" by computer. Text mining aims to overcome this problem and can be subdivided into two major disciplines, namely information retrieval and information extraction. Whereas, information retrieval is concerned with retrieving documents according to user-defined criteria, the more ambitious task of information extraction attempts to ascertain facts about pre-specified types of events, entities, or relationships described in the literature. The efficient retrieval of relevant literature according to a given search criterion is essential to biomedical research, as well as to clinical practice. In this regard, PubMed[13)]

13) http://www.ncbi.nlm.nih.gov/pubmed/.

is a powerful document-retrieval system that provides access to the most comprehensive biomedical literature database [364]. Text mining methods have been established predominantly in the newswire domain. Whereas articles in that domain aim at a general audience, biomedical articles usually target a small community of domain experts and, thus, are more difficult to interpret. The adaptation of methods to the characteristics of the biomedical literature has been described in several available reviews [365–368].

In the biomedical context, one of the ultimate goals is the correct extraction of relationships among biological and medical entities, such as proteins, genes, functional annotations, and diseases. Current text mining methods for biomedical literature face a few challenges, the first of which relates to general natural language processing problems, such as tracking references to one object throughout the text, or correctly identifying negations or hypotheses. In addition, there are domain-specific issues, such as the highly variable gene and protein nomenclature, where new terms and abbreviations are constantly being created and where multiple meanings for a word may result in ambiguities. These challenges have been addressed and several text mining methods have been proposed for predicting subcellular localization [369, 370], for detecting protein–protein interactions [371–374], for the functional annotation of genes/proteins [375], and for the reconstruction of gene-regulatory events [376]. Other implementations address the extraction of specific biological attributes, such as phosphorylation sites [377], mutations [378], or even genetic markers and their association with disease [379].

Nevertheless, text mining remains an exploratory field that requires improvements to be made in its methods, and an evaluation standard to be developed [380]. In general, information retrieval methods provide sufficient performance to be of immediate use [372]. The results of information extraction, however, should be assessed in the context of additional data, or simply taken as starting points for database entries maintained by domain experts. In order to foster an exchange of ideas and to determine the most suitable algorithms for retrieval and extraction tasks, critical assessment contests must be implemented. Towards this goal, the BioCreAtIvE (Critical Assessment for Information extraction Systems in Biology)[14] competition was launched in 2003 to enable the comparison of text mining methods in the biomedical domain [381], with new events being held biennially [382].

9 Computational Biology of Diseases

Thus far, computational biology has been presented almost exclusively from a methodical perspective (a notable exception was in Sect. 5.5, on protein design). Yet, computational biology also has a very "applied" side, with manifold purposes ranging from *biotechnological*, such as supporting the modification of microorganisms to most efficiently perform certain biotechnological tasks, such as producing desired substances or degrading undesired substances, to *fundamental*, such as unveiling the molecular bases of intra- or inter-cellular processes, to *medical*, such as facilitating the diagnosis, prognosis, and therapy of diseases.

14) http://www.biocreative.org/.

9.1
Assessing Disease Risk

Genotypic information on the individual genome, such as disease-associated sequence variants and copy number variations, may be valuable for assessing disease risk and patient susceptibility to especially complex diseases involving multiple genes. Examples of diseases that continue to be studied intensively in this regard include cancer, neurodegenerative diseases (e.g., Alzheimer's and Parkinson's disease), and inflammatory bowel diseases, notably Crohn's disease and ulcerative colitis. Currently, bioinformatics prediction methods are valuable for interpreting the functional effects of sequence variations and suggesting follow-up experiments for further studies of the molecular disease process [383]. For example, disease-associated amino acid variants can be located in a 3-D structural model of the affected protein interaction domain, such that possible dysfunctions and impaired protein interactions can be hypothesized [384]. Another application of computational methods is the functional evaluation of resistance mutations that evolve in viruses to combat antiviral drugs. Residue interaction networks derived from 3-D structures can also be analyzed to provide an understanding of mutational effects on the drug-binding sites of viral proteins [385].

The long lists of candidate disease genes are frequently the result of large-scale studies such as gene expression profiling, RNA interference screens, or genome-wide association studies for diseases. The experiments do not allow the direct pinpointing of the responsible disease genes without follow-up experiments due to experimental noise, the sheer size of the human genome, and the lack of statistical power. Here, computational prioritization methods are required to rank dozens (or even hundreds) of candidate genes with respect to their likely disease involvement for increasing the success of candidate validation experiments, and saving experimental costs and time [383]. To this end, many methods make use of topological and functional interrelationships and appropriate similarity measures between known disease genes and candidates [386, 387]. The overriding biological assumption is that the perturbation of genes and proteins that are closely associated in an interaction network cause similar phenotypes and diseases. In this context, another objective is to construct integrated disease networks that combine genotypic and phenotypic information, allowing the assessment of disease risk and the discovery of both biomarkers and drug targets [388, 389].

9.2
Supporting the Prevention of Diseases

An example of how computational biology might support the prevention of diseases derives from the field of computational immunology and vaccinology. In computational immunology, the interactions between a pathogen and the host's immune system can be monitored. The immune system components – notably in the major histocompatibility complex (MHC; occasionally referred to as the human leukocyte antigen complex; HLA) – are responsible for presenting to the cellular surface any short peptides that have been acquired from foreign proteins. The MHC–molecule complex (and the peptide to which it is bound) then bind to a specific receptor complex at the T-cell surface. If the complex formation between the MHC–peptide complex and the B-cell or T-cell receptor elicits an immune reaction, the corresponding peptide from the pathogen is termed an *epitope* or an *antigen*. Subsequently, the immune reaction facilitates various ways in which the pathogen can be attacked, including the synthesis of immune cells targeted specifically towards the elimination of any cells carrying specific antigens.

Among humans, the MHC molecules are highly variable, with each individual having a specific ensemble of basic MHC molecules that are capable of binding to specific antigens, but not to others. In this way, humans and their pathogens have been able to coevolve with a basic genetic diversity such that neither species can extinguish the other. The evolutionarily more recent part of the immune system, termed the *adaptive immune system*, can "learn" to synthesize immune cells to attack specific antigens which it has come into contact with previously. This "learning process," which is triggered by a first infection with the pathogen, forms the basis of a *vaccination*. In this case, antigens from the pathogen against which the vaccination is targeted are applied; a first immune reaction is then elicited that kick-starts the learning process and protects the individual against a subsequent infection.

One critical step in vaccination is to select the epitopes that are to be administered with the vaccine, while another is to decide in which form the epitopes are to be administered. The epitope selection process is the most amenable to optimization by computer, the objective being to select a set of epitopes that best covers a genetically divergent pathogen population. Epitope selection involves criteria such as: (i) being present in a large part of the pathogen population; and (ii) having an ability to be bound by the MHC molecules (HLA alleles) present in a large part of the human population. An estimation of the binding affinity of the epitope to the MHC molecule is an important part of the analysis, and can be made with statistical models based on validated sets of binding and nonbinding molecules [390] or, alternatively, on the basis of detailed docking studies between the MHC molecule and the peptide. The former approach is better suited for screening large sets of epitopes, the latter is computationally more intensive [391]. Docking to MHC molecules is especially difficult, because the ligand is highly flexible and often does not have a unique conformation in the binding pocket of the MHC molecule. However, it is possible to combine both approaches to yield screening methods that are potentially more accurate, especially in the case of alleles for which few measured binding affinities are available [392].

Binding of the peptide to the MHC molecules represents only one step in a cascade that begins with the proteins of the pathogen being cleaved, and ends with the MHC–peptide complex being transported to the cellular surface, where it is bound to the immune cell receptor. Bioinformatics software solutions that target each of these steps together are available [393], while the respective methods can be configured for seeking suitable epitopes for vaccines [394].

9.3 Supporting the Diagnosis and Prognosis of Diseases

The advent of high-throughput molecular screening techniques has afforded an improved determination of disease type and stage at the molecular level; a prime example of this is cancer.

In the past, classical cancer diagnosis and tumor staging was performed on the basis of the visual inspection of histological sections. However, with the advent of molecular screening technologies, such analyses can now be made on the basis of gene or protein expression profiles [395], sequence and copy number variations along the genome [396], and epigenetic patterns such as DNA methylation profiles in the tumor tissue [397]. The respective molecular signatures (biomarkers), which may be highly informative regarding the type and stage of the tumor [398], often consist of sequence or modification patterns in particular genomic regions, and abundances of particular molecules (e.g., indicative protein levels in the blood); alternatively, they may include particular patterns in interaction subnetworks [386]. The classification of the expression profiles of patients' samples is achieved using statistical methods that incorporate models for various types of experimental error which either originate from the technology or are inherent in the variance of the biological system [263]. Here, disease classification is based on statistical models that can discriminate between different disease classes and stages [399]. The staging of a tumor can incorporate statistical models of genomic decay as the disease progresses [396]. The analysis of expression data that first became available during the late 1990s has provided statistics research worldwide with a "substantial push in the application-oriented direction."

9.4
Supporting the Therapy of Diseases

Today, personalized medicine has advanced to the area of selecting patient-specific drug therapies [400], and is now highly developed for certain types of cancer, as well as anti-HIV therapies. In cancer patients, computational methods can help to classify tumors based on molecular measurements such as genome-wide gene expression data; however, alternative methods are employed to understand the biological effects of the patient genotype on the efficacy of drug therapy [401, 402].

HIV is a rapidly evolving virus, for two main reasons: (i) the viral reverse transcriptase protein which facilitates copying of the viral RNA genome to DNA lacks a proofreading mechanism, such that copying mistakes are made; and (ii) during the course of the infection there is an especially high turnover of viral particles (up to 10^9 particles per patient per day). Taken together, these variations result in an extensive population-genetic experiment being conducted in which viral variants that are resistant to current drug treatments are generated and selected for. Combinations of antiviral drugs with different drug targets and/or modes of action are administered in order raise the evolutionary barrier for escape of the virus to acquire resistance. When a drug therapy becomes ineffective a new set of drugs must be selected. Laboratory testing for viral resistance is not applicable to the clinical routine, for reasons that involve high costs, urgency of treatment, and accessibility to therapy. Consequently, the selection of a new therapy must be based on knowledge of the sequence of the viral genome of the presently dominant strain(s). The sequences of genomic regions coding for the viral drug target proteins are sufficient for this purpose. The interpretation of the viral genome is facilitated either by a set of expert rules that are applied in special computer-based expert systems, or by a statistical software that sieves information related to viral resistance from a

Drug	RF(*)	z-score	Scored mutations(**)
ZDV	257.276	9.945	215Y 210W 41L
ddI	4.057	6.087	184V 121H 178L 215Y 177E
d4T	2.477	4.594	215Y 178L 118I 184V 210W 121H 41L
3TC	149.029	18.504	184V 41L 215Y
FTC	58.902	5.469	184V 44D
ABC	7.501	11.968	184V 215Y 210W 41L
TDF	4.613	6.684	215Y 41L 98G 184V 178L 118I 177E 135T
NVP	149.235	5.243	103N 135T 210W 98G 211K
EFV	60.034	7.450	103N 98G 135T 210W 214F 177E
SQV	3.908	4.658	63P 46I 37N 71V 72T 60E 76V 57K
IDV	52.999	11.171	46I 76V 63P 82A 61E 62V 60E 71V
RTV	46.190	12.412	46I 82A 63P 76V 60E
NFV	33.156	8.246	46I 63P 10F 3I 76V 60E 62V
APV	59.452	13.085	76V 54M 10F 46I 63P
LPV	100.108	14.521	46I 76V 10F 63P 82A
TPV	3.627	3.212	71V 46L 15V 72T 3I
DRV	10.958	7.485	54M 3I 76V 37N 71V 60E 82A
ATV	19.252	7.838	82A 46L 54M 76V 93L 71V 62V

Fig. 10 Output of the system geno2pheno (www.geno2pheno.org) analyzing HIV drug resistance. The input is a sequence of the relevant portions of the viral genome. The output has one line per drug. Column 1 holds the name of the drug. Columns 2 and 3 are the absolute and normalized resistance factor, a quantitative measure for the level of resistance of the virus to the drug. High numbers mean high resistance. The final column lists the mutations in the viral proteins affecting resistance. Mutations shown in red increase resistance, those in green reduce resistance. The depicted data are from a patient for whom a successful therapy could only be found by using this software [403].

set of clinical data [404]. The former approach is the choice for new drugs for which few clinical data are available, but the latter approach is more accurate if sufficient clinical data are available [405] (see also Fig. 10). Today, both expert systems and statistical software to predict viral resistance and therapy efficacy are used in clinical practice, and these form the basis for the administration of antiviral therapies to AIDS patients in the developed world [403]. Importantly, the systems are continually updated and extended as new drugs enter the market [406]. At present, the technology is being lifted from the classical sequencing technologies that return a consensus sequence of the collection of dominant strains (with a fraction of 20% or more) in the viral population harbored by the patient, to new, ultra-deep sequencing technologies that afford a detailed resolution of the viral population into thousands of individual genome sequences. As the resistant viral minorities may play a clinically essential role, the new technology is expected to afford an increased accuracy in the prediction of therapy efficacy.

10 Perspectives

Today, the modeling and simulation of biological systems at the molecular level remains an enormous challenge. While highly computer-intensive methods afford the analysis of local aspects of molecular structures and processes to an appreciable accuracy, the scales of space and time at which this is possible remain small (a few dozen Ångstroms in space, a few nanoseconds in time). Yet, in trying to model biological systems more globally, limits of complexity are encountered. Structure predictions based on homology of all but the most closely related proteins produce an appreciable accuracy for the backbone of the protein in no more than 70–80% of the cases. Similar problems are encountered in molecular docking for all those methods that are capable of handling an appreciable number of cases within a reasonable time. The inherent noise and redundancy, as well as the physico-chemical complexity of most biological structures and processes, is so great that we have to resort to a comparative paradigm for modeling that analyzes evolution rather than physics. This paradigm has proven very successful for analyzing the more global aspects of biological structures and processes.

The human genome sequencing project has kick-started a technological revolution that has afforded complete genomes for, by now, well over 1000 organisms (http://www.genomesonline.org/). Moreover, the resultant data have afforded a completely new approach towards the analysis of evolution – an approach based on knowledge of not only what *is* in the genome but also what is *not* in it. The absence of genes in genomes entails valuable information, for example, for the analysis of evolutionary relationships or investigations on gene function (see Sect. 8.3.3). The comprehensive data collections comprising genome sequences, genetic variation, gene expression profiles and, to a growing extent, proteomic, metabolomic, interactomic, and epigenetic data, harbor an unprecedented potential for the analysis of intra- and intercellular structures and processes. Although the comparative paradigm is the main tool, here, it is not used exclusively. Genome-scale metabolic network models, for example, can be analyzed using stoichiometric methods that model chemistry, rather than evolution.

As a tool for interpreting large sets of molecular data, computational biology has revealed deep and often surprising biological insights, including new paradigms for the architecture and evolution of biological systems. In basic research, computational biology is most effective when it is exercised in a tight loop with experimentation. In this way, experiments may be configured and optimized by using computational methods, and the data resulting from the experiments interpreted with the computer. Today, computational biology has reached well beyond basic research, having begun to generate benefits for patients, such as new biomarkers for the diagnosis and prognosis of tumors, approaches to the personalized administration of drug therapies (e.g., in cancer and AIDS), as well as new approaches to drug screening, design, and optimization.

Clearly, the future of computational biology is expected to grow, with attention focused on acquiring new types of data, including cellular imaging [407–409] and the dynamic modeling of biological processes across multiple scales, of space and time. For cellular processes that are not based on molecular components which attain a unique 3-D structure (e.g. in some proteins and in glycosylation [410]), effective modeling procedures based on statistical models of molecular ensembles are still lacking. Yet, this technical view of the field must not obscure the dominant challenges that will continue to characterize the field:

- Computational biology requires not only large and comprehensive data volumes, but data of high quality and with detailed annotation. Standards must be developed for the generation, storage, and annotation of data. At the turn of the millennium, the field of transcriptomics spearheaded such a standard termed *Minimum Information About a Microarray Experiment* [411], which has since found followers in several other areas [412]. Biological data are notoriously noisy, and part of this noise is due to the lacking maturity of often emerging measurement technologies; however, a significant part is also due to biological variation, which is not a bug but rather a feature of biological systems. To separate both sources of noise will require new developments in statistics, as will the incorporation of biological variance into the models.
- Nomenclature in biology is notoriously heterogeneous and inconsistent, and this hallmarks the need for well-developed ontologies that can provide an exchange of information on a more structured level than is possible in plain text. The Gene Ontology Consortium, which again arose around the turn of the millennium in response to the expectation of great data volumes, is the central development in this area [413]. Ontologies have also long been in use in the fields of enzymology (NC-UIBMB Enzyme Catalogue [414]), medicine (Unified Medical language System, UMLS [415]), and genetic diseases (Online Mendelian Inheritance in Man, OMIM [416]). Although, to date, many fields have created their own ontologies, the linking of these ontologies in a consistent manner represents a major challenge.
- Another urgent issue is the provision of efficient and inexpensive computer power. This would include peta-order storage space and fast internet connections for the exchange of large sets of primary data generated by new techniques such as ultra-deep sequencing and bioimaging, as well as derived data sets resulting from computational analysis and simulation. In this case, much might be learned from computational physics and astronomy, which already involve large data systems.
- In many areas, quantitative measurements of molecular processes and phenotypes are currently missing. One such example is in the field of psychiatric diseases, which is currently seeking molecular correlates for the type and intensity of the disease, the so-called endophenotypes [417]. To date, patients are being classified on the basis of qualitative information originating from interviews. In contrast, quantitative biomarkers for malignant tumors and infectious diseases exist, and have sparked progress in both diagnosis and therapy.
- Perhaps the most fundamental challenge for computational biology is to bridge the enormous span of scales in time and space at which biological processes unfold. In this respect, computational biology has been working bottom-up in this regime of scales, such that processes which occur at small scales in time and space can be handled much better than those that occur over

larger time scales. The challenge, therefore, is to identify a hierarchy of models, each of which can handle a certain scale, and to link them in a consistent manner to span all scales. Although this challenge might have a somewhat utopic feel, it is nevertheless beginning to be tackled by multinational research programs and consortia [418, 419].

The progress made to date in computational biology, and the challenges that lie ahead, are two clear reasons for the excitement generated when working in a field which lies at the center of scientific innovation and harbors the potential to obtain critical insights into previously inaccessible aspects of living systems. As in all areas of research, there is a constant confrontation with the grandeur of mother Nature's creation, and indications abound as to how far humankind is from realizing a true understanding of her schemes. New experimental techniques and large-scale measurements, in combination with subsequent computational analysis, often lead to the discovery of as-yet unknown biological facets that continually open new doors into the overwhelming complexity of animate nature. Today, the transition is being made from an age of "biological alchemy," with the first steps being taken into a quantitative understanding of life's processes. Of course, the power of humankind must not be overestimated, especially when tinkering with biological systems that have evolved over very long periods of time. Rather, an awareness of the risks that might be incurred by a lack of knowledge should prove to be an excellent preparation for making further progress in this field.

Acknowledgments

The authors are grateful to Niko Beerenwinkel, Andreas Hahn, Daniel Hanisch, Andreas Kämper, Hugo Kubinyi, Jochen Maydt, Christian von Mering, Jörg Rahnenführer, Knut Reinert, Kirsten Roomp, Oliver Sander, Tobias Sing, Ingolf Sommer, and Priti Talwar, for providing useful comments and suggestions on the text.

Note on the Second Edition on This Chapter

In this second edition, Sects 1–8 have been brought up to date at the time of printing, while Sect. 9 has been expanded due to the rapid growth of applications of computational biology in medicine. The more technical section on Bioinformatics Techniques (Sect. 10 of the first edition) has been omitted in favor of an expanded section on perspectives of the field.

References

1 Butler, D. (2010) Human genome at ten: science after the sequence. *Nature*, **465** (7301), 1000–1001.
2 Collins, F. (2010) Has the revolution arrived? *Nature*, **464** (7289), 674–675.
3 De, S., Babu, M.M. (2010) Genomic neighbourhood and the regulation of gene expression. *Curr. Opin. Cell Biol.*, **22** (3), 326–333.
4 Fedorova, E., Zink, D. (2009) Nuclear genome organization: common themes and individual patterns. *Curr. Opin. Genet. Dev.*, **19** (2), 166–171.
5 Boussau, B., Daubin, V. (2010) Genomes as documents of evolutionary history. *Trends Ecol. Evol.*, **25** (4), 224–232.
6 Hayden, E.C. (2010) Human genome at ten: life is complicated. *Nature*, **464** (7290), 664–667.
7 Manolio, T.A. (2010) Genomewide association studies and assessment of the risk of disease. *N. Engl. J. Med.*, **363** (2), 166–176.

8 Borrell, B. (2010) US clinics quietly embrace whole-genome sequencing. *Nature News*, doi: 10.1038/news.2010.465

9 Fleischmann, R.D., Adams, M.D., White, O., Clayton, R.A., Kirkness, E.F., Kerlavage, A.R., Bult, C.J., Tomb, J.-F., Dougherty, B.A., Merrick, J.M., McKenney, K., Sutton, G., FitzHugh, W., Fields, C., Gocayne, J.D., Scott, J., Shirley, R., Liu, L., Glodek, A., Kelley, J.M., Weidman, J.F., Phillips, C.A., Spriggs, T., Hedblom, E., Cotton, M.D., Utterback, T.R., Hanna, M.C., Nguyen, D.T., Saudek, D.M., Brandon, R.C., Fine, L.D., Fritchman, J.L., Fuhrmann, J.L., Geoghagen, N.S.M., Gnehm, C.L., McDonald, L.A., Small, K.V., Fraser, C.M., Smith, H.O., Venter, J.C. (1995) Whole-genome random sequencing and assembly of *Haemophilus influenzae* Rd. *Science*, **269** (5223), 496–512.

10 Goffeau, A., Barrell, B. G., Bussey, H., Davis, R.W., Dujon, B., Feldmann, H., Galibert, F., Hoheisel, J.D., Jacq, C., Johnston, M., Louis, E.J., Mewes, H.W., Murakami, Y., Philippsen, P., Tettelin, H., Oliver, S.G. (1996) Life with 6000 genes. *Science*, **274** (5287), 546, 563–567.

11 Blattner, F.R., Plunkett, G. III, Bloch, C.A., Perna, N.T., Burland, V., Riley, M., Collado-Vides, J., Glasner, J.D., Rode, C.K., Mayhew, G.F., Gregor, J., Davis, N.W., Kirkpatrick, H.A., Goeden, M.A., Rose, D.J., Mau, B., Shao, Y. (1997) The complete genome sequence of *Escherichia coli* K-12. *Science*, **277** (5331), 1453–1462.

12 The *C. elegans* Sequencing Consortium (1998) Genome sequence of the nematode *C. elegans*: a platform for investigating biology. *Science*, **282** (5396), 2012–2018.

13 Adams, M.D., Celniker, S.E., Holt, R.A., Evans, C.A., Gocayne, J.D., Amanatides, P.G., Scherer, S.E., Li, P.W., Hoskins, R.A., Galle, R.F., George, R.A., Lewis, S.E., Richards, S., Ashburner, M., Henderson, S.N., Sutton, G.G., Wortman, J.R., Yandell, M.D., Zhang, Q., Chen, L.X., Brandon, R.C., Rogers, Y.H., Blazej, R.G., Champe, M., Pfeiffer, B.D., Wan, K.H., Doyle, C., Baxter, E.G., Helt, G., Nelson, C.R., Gabor Miklos, G.L., Abril, J.F., Agbayani, A.,, An, H.J., Andrews-Pfannkoch, C., Baldwin, D., Ballew, R.M., Basu, A., Baxendale, J., Bayraktaroglu, L., Beasley, E.M., Beeson, K.Y., Benos, P.V., Berman, B.P., Bhandari, D., Bolshakov, S., Borkova, D., Botchan, M.R., Bouck, J., Brokstein, P., Brottier, P., Burtis, K.C., Busam, D.A., Butler, H., Cadieu, E., Center, A., Chandra, I., Cherry, J.M., Cawley, S., Dahlke, C., Davenport, L.B., Davies, P.,, de Pablos, B., Delcher, A., Deng, Z., Mays, A.D., Dew, I., Dietz, S.M., Dodson, K., Doup, L.E., Downes, M., Dugan-Rocha, S., Dunkov, B.C., Dunn, P., Durbin, K.J., Evangelista, C.C., Ferraz, C., Ferriera, S., Fleischmann, W., Fosler, C., Gabrielian, A.E., Garg, N.S., Gelbart, W.M., Glasser, K., Glodek, A., Gong, F., Gorrell, J.H., Gu, Z., Guan, P., Harris, M., Harris, N.L., Harvey, D., Heiman, T.J., Hernandez, J.R., Houck, J., Hostin, D., Houston, K.A., Howland, T.J., Wei, M.H., Ibegwam, C., Jalali, M., Kalush, F., Karpen, G.H., Ke, Z., Kennison, J.A., Ketchum, K.A., Kimmel, B.E., Kodira, C.D., Kraft, C., Kravitz, S., Kulp, D., Lai, Z., Lasko, P., Lei, Y., Levitsky, A.A., Li, J., Li, Z., Liang, Y., Lin, X., Liu, X., Mattei, B., McIntosh, T.C., McLeod, M.P., McPherson, D., Merkulov, G., Milshina, N.V., Mobarry, C., Morris, J., Moshrefi, A., Mount, S.M., Moy, M., Murphy, B., Murphy, L., Muzny, D.M., Nelson, D.L., Nelson, D.R., Nelson, K.A.,, Nixon, K., Nusskern, D.R., Pacleb, J.M., Palazzolo, M., Pittman, G.S., Pan, S., Pollard, J., Puri, V., Reese, M.G., Reinert, K., Remington, K., Saunders, R.D., Scheeler, F., Shen, H., Shue, B.C., Siden-Kiamos, I., Simpson, M., Skupski, M.P., Smith, T., Spier, E., Spradling, A.C., Stapleton, M., Strong, R., Sun, E., Svirskas, R., Tector, C., Turner, R., Venter, E., Wang, A.H., Wang, X., Wang, Z.Y., Wassarman, D.A., Weinstock, G.M., Weissenbach, J., Williams, S.M., Woodage, T., Worley, K.C., Wu, D., Yang, S., Yao, Q.A., Ye, J., Yeh, R.F., Zaveri, J.S., Zhan, M., Zhang, G., Zhao, Q., Zheng, L., Zheng, X.H., Zhong, F.N., Zhong, W., Zhou, X., Zhu, S., Zhu, X., Smith, H.O., Gibbs, R.A., Myers, E.W., Rubin, G.M., Venter, J.C. (2000) The genome sequence of *Drosophila melanogaster*. *Science*, **287** (5461), 2185–2195.

14 *Arabidopsis* Genome Initiative (2000) Analysis of the genome sequence of the flowering plant *Arabidopsis thaliana*. *Nature*, **408** (6814), 796–815.

15 Waterston, R.H., Lindblad-Toh, K., Birney, E., Rogers, J., Abril, J.F., Agarwal, P., Agarwala, R., Ainscough, R., Alexandersson, M., An, P. (2002) Initial sequencing and comparative analysis of the mouse genome. *Nature*, **420** (6915), 520–562.

16 Consortium, I.H.G. (2004) Finishing the euchromatic sequence of the human genome. *Nature*, **431** (7011), 931–945.

17 Venter, J.C., Adams, M.D., Myers, E.W., Li, P.W., Mural, R.J., Sutton, G.G., Smith, H.O., Yandell, M., Evans, C.A., Holt, R.A. (2001) The sequence of the human genome. *Science*, **291** (5507), 1304–1351.

18 Huang, X. (2001) Bioinformatics Support of Genome Sequencing Projects, in: Lengauer, T. (Ed.) *Bioinformatics – from Genomes to Drugs*, Wiley-VCH Verlag GmbH, Weinheim, pp. 25–48.

19 Lander, E.S., Linton, L.M., Birren, B., Nusbaum, C., Zody, M.C., Baldwin, J., Devon, K., Dewar, K., Doyle, M., FitzHugh, W. (2001) Initial sequencing and analysis of the human genome. *Nature*, **409** (6822), 860–921.

20 Kent, W.J., Haussler, D. (2001) Assembly of the working draft of the human genome with GigAssembler. *Genome Res.*, **11** (9), 1541–1548.

21 Huson, D.H., Reinert, K., Kravitz, S.A., Remington, K.A., Delcher, A.L., Dew, I.M., Flanigan, M., Halpern, A.L., Lai, Z., Mobarry, C.M. (2001) Design of a compartmentalized shotgun assembler for the human genome. *Bioinformatics*, **17** (Suppl. 1), S132–S139.

22 Huson, D.H., Reinert, K., Myers, E.W. (2002) The greedy path-merging algorithm for contig scaffolding. *J. Assoc. Comput. Machinery*, **49** (5), 603–615.

23 Pevzner, P.A., Tang, H., Waterman, M.S. (2001) An Eulerian path approach to DNA fragment assembly. *Proc. Natl Acad. Sci. USA*, **98** (17), 9748–9753.

24 Batzoglou, S., Jaffe, D.B., Stanley, K., Butler, J., Gnerre, S., Mauceli, E., Berger, B., Mesirov, J.P., Lander, E.S. (2002) ARACHNE: a whole-genome shotgun assembler. *Genome Res.*, **12** (1), 177–189.

25 Kirkness, E.F., Bafna, V., Halpern, A.L., Levy, S., Remington, K., Rusch, D.B., Delcher, A.L., Pop, M., Wang, W., Fraser, C.M., Venter, J.C. (2003) The dog genome: survey sequencing and comparative analysis. *Science*, **301** (5641), 1898–1903.

26 Venter, J.C., Remington, K., Heidelberg, J.F., Halpern, A.L., Rusch, D., Eisen, J.A., Wu, D., Paulsen, I., Nelson, K.E., Nelson, W., Fouts, D.E., Levy, S., Knap, A.H., Lomas, M.W., Nealson, K., White, O., Peterson, J., Hoffman, J., Parsons, R., Baden-Tillson, H., Pfannkoch, C., Rogers, Y.H., Smith, H.O. (2004) Environmental genome shotgun sequencing of the Sargasso Sea. *Science*, **304** (5667), 66–74.

27 Batley, J., Edwards, D. (2009) Genome sequence data: management, storage, and visualization. *Biotechniques*, **46** (5), 333–334, 336.

28 Kahvejian, A., Quackenbush, J., Thompson, J.F. (2008) What would you do if you could sequence everything? *Nat. Biotechnol.*, **26** (10), 1125–1133.

29 Metzker, M.L. (2010) Sequencing technologies-the next generation. *Nat. Rev. Genet.*, **11** (1), 31–46.

30 Horner, D.S., Pavesi, G., Castrignano, T., De Meo, P.D., Liuni, S., Sammeth, M., Picardi, E., Pesole, G. (2010) Bioinformatics approaches for genomics and post genomics applications of next-generation sequencing. *Brief. Bioinform.*, **11** (2), 181–197.

31 Flicek, P., Birney, E. (2009) Sense from sequence reads: methods for alignment and assembly. *Nat. Methods*, **6** (Suppl. 11), S6–S12.

32 Rice, P., Longden, I., Bleasby, A. (2000) EMBOSS: the European molecular biology open software suite. *Trends Genet.*, **16** (6), 276–277.

33 Vingron, M. (2007) Sequence Alignment and Sequence Database Search, in: Lengauer, T. (Ed.) *Bioinformatics – from Genomes to Therapies*, Wiley-VCH Verlag GmbH, Weinheim, pp. 57–82.

34 Gusfield, D. (1997) *Algorithms on Strings, Trees, and Sequences: Computer Science and Computational Biology*, Cambridge University Press, Cambridge, New York, xviii, 534 p.

35 Zien, A., Zimmer, R., Lengauer, T. (2000) A simple iterative approach to parameter optimization. *J. Comput. Biol.*, **7** (3), 483–501.

36 Altschul, S.F., Gish, W., Miller, W., Myers, E.W., Lipman, D.J. (1990) Basic local

alignment search tool. *J. Mol. Biol.*, **215** (3), 403–410.

37 Karlin, S., Altschul, S.F. (1990) Methods for assessing the statistical significance of molecular sequence features by using general scoring schemes. *Proc. Natl Acad. Sci. USA*, **87** (6), 2264–2268.

38 Altschul, S.F., Koonin, E.V. (1998) Iterated profile searches with PSI-BLAST – a tool for discovery in protein databases. *Trends Biochem. Sci.*, **23** (11), 444–447.

39 Eddy, S.R. (1998) Profile hidden Markov models. *Bioinformatics*, **14** (9), 755–763.

40 Chenna, R., Sugawara, H., Koike, T., Lopez, R., Gibson, T.J., Higgins, D.G., Thompson, J.D. (2003) Multiple sequence alignment with the Clustal series of programs. *Nucleic Acids Res.*, **31** (13), 3497–3500.

41 Higgins, D.G., Sharp, P.M. (1988) CLUSTAL: a package for performing multiple sequence alignment on a microcomputer. *Gene*, **73** (1), 237–244.

42 Notredame, C., Higgins, D.G., Heringa, J. (2000) T-Coffee: a novel method for fast and accurate multiple sequence alignment. *J. Mol. Biol.*, **302** (1), 205–217.

43 Katoh, K., Asimenos, G., Toh, H. (2009) Multiple alignment of DNA sequences with MAFFT. *Methods Mol. Biol.*, **537**, 39–64.

44 Edgar, R.C. (2004) MUSCLE: multiple sequence alignment with high accuracy and high throughput. *Nucleic Acids Res.*, **32** (5), 1792–1797.

45 Do, C.B., Mahabhashyam, M.S., Brudno, M., Batzoglou, S. (2005) ProbCons: probabilistic consistency-based multiple sequence alignment. *Genome Res.*, **15** (2), 330–340.

46 Thompson, J.D., Plewniak, F., Poch, O. (1999) BAliBASE: a benchmark alignment database for the evaluation of multiple alignment programs. *Bioinformatics*, **15** (1), 87–88.

47 Raghava, G., Searle, S.M., Audley, P.C., Barber, J.D., Barton, G.J. (2003) OXBench: a benchmark for evaluation of protein multiple sequence alignment accuracy. *BMC Bioinformatics*, **4** (1), 47.

48 Solovyev, V. (2001) Structure, Properties and Computer Identification of Eukaryotic Genes, in: Lengauer, T. (Ed.) *Bioinformatics – From Genomes to Drugs*, Wiley-VCH Verlag GmbH, Weinheim, pp. 59–111.

49 Werner, T. (2001) Analyzing Regulatory Regions in Genomes, in: Lengauer, T. (Ed.) *Bioinformatics – From Genomes to Drugs*, Wiley-VCH Verlag GmbH, Weinheim, pp. 113–143.

50 Gonnet, G.H., Cohen, M.A., Benner, S.A. (1992) Exhaustive matching of the entire protein sequence database [see comments]. *Science*, **256** (5062), 1443–1445.

51 Dayhoff, M.O. (1978) *Protein Segment Dictionary 78: From the Atlas of Protein Sequence and Structure*, Vol. **5**, and supplements 1, 2, and 3, National Biomedical Research Foundation, Georgetown University Medical Center, Silver Spring, MD, p. vii, 470.

52 Henikoff, S., Henikoff, J.G. (1992) Amino acid substitution matrices from protein blocks. *Proc. Natl Acad. Sci. USA*, **89** (22), 10915–10919.

53 Bowie, J.U., Luthy, R., Eisenberg, D. (1991) A method to identify protein sequences that fold into a known three-dimensional structure. *Science*, **253** (5016), 164–170.

54 Kelley, L.A., MacCallum, R.M., Sternberg, M.J. (2000) Enhanced genome annotation using structural profiles in the program 3D-PSSM. *J. Mol. Biol.*, **299** (2), 499–520.

55 Claverie, J.M., Audic, S. (1996) The statistical significance of nucleotide position-weight matrix matches. *Comput. Appl. Biosci.*, **12** (5), 431–439.

56 Kondrakhin, Y.V., Kel, A.E., Kolchanov, N.A., Romashchenko, A.G., Milanesi, L. (1995) Eukaryotic promoter recognition by binding sites for transcription factors. *Comput. Appl. Biosci.*, **11** (5), 477–488.

57 Bateman, A., Birney, E., Cerruti, L., Durbin, R., Etwiller, L., Eddy, S.R., Griffiths-Jones, S., Howe, K.L., Marshall, M., Sonnhammer, E.L. (2002) The Pfam protein families database. *Nucleic Acids Res.*, **30** (1), 276–280.

58 Letunic, I., Goodstadt, L., Dickens, N.J., Doerks, T., Schultz, J., Mott, R., Ciccarelli, F., Copley, R.R., Ponting, C.P., Bork, P. (2002) Recent improvements to the SMART domain-based sequence annotation resource. *Nucleic Acids Res.*, **30** (1), 242–244.

59 Haft, D.H., Selengut, J.D., White, O. (2003) The TIGRFAMs database of protein families. *Nucleic Acids Res.*, **31** (1), 371–373.

60 Pavlidis, P., Furey, T.S., Liberto, M., Haussler, D., Grundy, W.N. (2001) Promoter region-based classification of genes. Pacific Symposium of Biocomputing, The Orchid at Maunal Loa, Big Island, Hawaii, January 3–7, pp. 151–163.
61 Yang, Z. (1996) Phylogenetic analysis using parsimony and likelihood methods. *J. Mol. Evol.*, **42** (2), 294–307.
62 Bandelt, H.J., Dress, A.W. (1992) Split decomposition: a new and useful approach to phylogenetic analysis of distance data. *Mol. Phylogenet. Evol.*, **1** (3), 242–252.
63 Posada, D., Crandall, K.A., Holmes, E.C. (2002) Recombination in evolutionary genomics. *Annu. Rev. Genet.*, **36**, 75–97.
64 Felsenstein, J. (2003) *Inferring Phylogenies*, 1st edn, Sinauer Associates, Sunderland, MA.
65 Sawa, G., Dicks, J., Roberts, I.N. (2003) Current approaches to whole genome phylogenetic analysis. *Brief. Bioinform.*, **4** (1), 63–74.
66 Kulp, D.C. (2007) Finding Protein-Coding Genes, in: Lengauer, T. (Ed.) *Bioinformatics – From Genomes to Therapies*, Wiley-VCH Verlag GmbH, Weinheim, pp. 129–158.
67 McHardy, A.C. (2008) Finding genes in genome sequence. *Methods Mol. Biol.*, **452**, 163–177.
68 Nilsen, T.W., Graveley, B.R. (2010) Expansion of the eukaryotic proteome by alternative splicing. *Nature*, **463** (7280), 457–463.
69 Mathe, C., Sagot, M.F., Schiex, T., Rouze, P. (2002) Current methods of gene prediction, their strengths and weaknesses. *Nucleic Acids Res.*, **30** (19), 4103–4117.
70 Alexandersson, M., Cawley, S., Pachter, L. (2003) SLAM: cross-species gene finding and alignment with a generalized pair hidden Markov model. *Genome Res.*, **13** (3), 496–502.
71 Pedersen, J.S., Hein, J. (2003) Gene finding with a hidden Markov model of genome structure and evolution. *Bioinformatics*, **19** (2), 219–227.
72 Solovyev, V., Salamov, A. (1997) The Gene-Finder computer tools for analysis of human and model organisms genome sequences. *Proc. Int. Conf. Intell. Syst. Mol. Biol.*, **5**, 294–302.
73 Snyder, E.E., Stormo, G.D. (1995) Identification of protein coding regions in genomic DNA. *J. Mol. Biol.*, **248** (1), 1–18.
74 Burge, C., Karlin, S. (1997) Prediction of complete gene structures in human genomic DNA. *J. Mol. Biol.*, **268** (1), 78–94.
75 Brunak, S., Engelbrecht, J., Knudsen, S. (1991) Prediction of human mRNA donor and acceptor sites from the DNA sequence. *J. Mol. Biol.*, **220** (1), 49–65.
76 Reese, M.G., Eeckman, F.H., Kulp, D., Haussler, D. (1997) Improved splice site detection in Genie. *J. Comput. Biol.*, **4** (3), 311–323.
77 Burset, M., Guigo, R. (1996) Evaluation of gene structure prediction programs. *Genomics*, **34** (3), 353–367.
78 Rogic, S., Mackworth, A.K., Ouellette, F.B. (2001) Evaluation of gene-finding programs on mammalian sequences. *Genome Res.*, **11** (5), 817–832.
79 Rogic, S., Ouellette, B.F., Mackworth, A.K. (2002) Improving gene recognition accuracy by combining predictions from two gene-finding programs. *Bioinformatics*, **18** (8), 1034–1045.
80 Barash, Y., Calarco, J.A., Gao, W., Wang, X., Shai, O., Blencowe, B.J., Frey, B.J. (2010) Deciphering the splicing code. *Nature*, **465** (7294), 53–59.
81 Pevzner, P.A., Sze, S.H. (2000) Combinatorial approaches to finding subtle signals in DNA sequences. *Proc. Int. Conf. Intell. Syst. Mol. Biol.*, **8**, 269–278.
82 Buhler, J., Tompa, M. (2002) Finding motifs using random projections. *J. Comput. Biol.*, **9** (2), 225–242.
83 Werner, T. (2007) Analyzing Regulatory Regions in Genomes, in: Lengauer, T. (Ed.) *Bioinformatics – From Genomes to Therapies*, Wiley-VCH Verlag GmbH, Weinheim, pp. 159–196.
84 Liu, L.A., Bader, J.S. (2009) Structure-based ab initio prediction of transcription factor-binding sites. *Methods Mol. Biol.*, **541**, 23–41.
85 Haas, B.J., Salzberg, S.L. (2007) Finding Repeats in Genome Sequences, in: Lengauer, T. (Ed.) *Bioinformatics – From Genomes to Therapies*, Wiley-VCH Verlag GmbH, Weinheim, pp. 197–234.
86 Benson, G. (1999) Tandem repeats finder: a program to analyze DNA sequences. *Nucleic Acids Res.*, **27** (2), 573–580.
87 Kurtz, S., Schleiermacher, C. (2001) REPuter: the manifold applications of repeat

analysis on a genomic scale. *Nucleic Acids Res.*, **29** (22), 4633–4642.

88 Delcher, A.L., Kasif, S., Fleischmann, R.D., Peterson, J., White, O., Salzberg, S.L. (1999) Alignment of whole genomes. *Nucleic Acids Res.*, **27** (11), 2369–2376.

89 Pevzner, P.A. (2000) *Computational Molecular Biology, An Algorithmic Approach*, MIT Press, Cambridge, MA.

90 Tesler, G. (2002) Efficient algorithms for multichromosomal genome rearrangements. *J. Comput. Syst. Sci.*, **65** (3), 587–609.

91 Sankoff, D., Blanchette, M. (1999) Phylogenetic invariants for genome rearrangements. *J. Comput. Biol.*, **6** (3-4), 431–445.

92 Pevzner, P., Tesler, G. (2003) Genome rearrangements in mammalian evolution: lessons from human and mouse genomes. *Genome Res.*, **13** (1), 37–45.

93 Blaxter, M. (2003) Comparative genomics: two worms are better than one. *Nature*, **426** (6965), 395–396.

94 Eichler, E.E., Sankoff, D. (2003) Structural dynamics of eukaryotic chromosome evolution. *Science*, **301** (5634), 793–797.

95 Raphael, B.J., Volik, S., Collins, C., Pevzner, P.A. (2003) Reconstructing tumor genome architectures. *Bioinformatics*, **19** (Suppl. 2), II162–II171.

96 Moult, J., Fidelis, K., Kryshtafovych, A., Rost, B., Tramontano, A. (2009) Critical assessment of methods of protein structure prediction – Round VIII. *Proteins*, **77** (Suppl. 9), 1–4.

97 Rost, B., Sander, C. (1993) Prediction of protein secondary structure at better than 70% accuracy. *J. Mol. Biol.*, **232** (2), 584–599.

98 Jones, D.T. (1999) Protein secondary structure prediction based on position-specific scoring matrices. *J. Mol. Biol.*, **292** (2), 195–202.

99 Kabsch, W., Sander, C. (1983) Dictionary of protein secondary structure: pattern recognition of hydrogen-bonded and geometrical features. *Biopolymers*, **22** (12), 2577–2637.

100 Frishman, D., Argos, P. (1995) Knowledge-based protein secondary structure assignment. *Proteins*, **23** (4), 566–579.

101 Dunbrack, R.L. Jr (2007) Homology Modeling in Biology and Medicine, in: Lengauer, T. (Ed.) *Bioinformatics – From Genomes to Therapies*, Wiley-VCH Verlag GmbH, Weinheim, pp. 297–350.

102 Sander, C., Schneider, R. (1991) Database of homology-derived protein structures and the structural meaning of sequence alignment. *Proteins*, **9** (1), 56–68.

103 McGuffin, L.J., Jones, D.T. (2003) Improvement of the GenTHREADER method for genomic fold recognition. *Bioinformatics*, **19** (7), 874–881.

104 Jones, D.T., Miller, R.T., Thornton, J.M. (1995) Successful protein fold recognition by optimal sequence threading validated by rigorous blind testing. *Proteins*, **23** (3), 387–397.

105 Sanchez, R., Sali, A. (1997) Evaluation of comparative protein structure modeling by MODELLER-3. *Proteins*, **41** (Suppl. 1), 50–58.

106 Dunbrack, R.L. Jr (1999) Comparative modeling of CASP3 targets using PSI-BLAST and SCWRL. *Proteins*, **43** (Suppl. 3), 81–87.

107 Schwede, T., Kopp, J., Guex, N., Peitsch, M.C. (2003) SWISS-MODEL: an automated protein homology-modeling server. *Nucleic Acids Res.*, **31** (13), 3381–3385.

108 Sommer, I., Zien, A., Von Öhsen, N., Zimmer, R., Lengauer, T. (2002) Confidence measures for protein fold recognition. *Bioinformatics*, **18** (6), 802–812.

109 Sommer, I. (2007) Protein Fold Recognition Based on Distant Homologs, in: Lengauer, T. (Ed.) *Bioinformatics – From Genomes to Therapies*, Wiley-VCH Verlag GmbH, Weinheim, pp. 351–388.

110 Levitt, M. (2007) Growth of novel protein structural data. *Proc. Natl Acad. Sci. USA*, **104** (9), 3183–3188.

111 Simons, K.T., Kooperberg, C., Huang, E., Baker, D. (1997) Assembly of protein tertiary structures from fragments with similar local sequences using simulated annealing and Bayesian scoring functions. *J. Mol. Biol.*, **268** (1), 209–225.

112 Bonneau, R., Strauss, C.E., Rohl, C.A., Chivian, D., Bradley, P., Malmstrom, L., Robertson, T., Baker, D. (2002) De novo prediction of three-dimensional structures for major protein families. *J. Mol. Biol.*, **322** (1), 65–78.

113 Simons, K.T., Ruczinski, I., Kooperberg, C., Fox, B.A., Bystroff, C., Baker, D. (1999) Improved recognition of native-like protein structures using a

combination of sequence-dependent and sequence-independent features of proteins. *Proteins*, **34** (1), 82–95.

114 Bonneau, R. (2007) De Novo Structure Prediction: Methods and Applications, in: Lengauer, T. (Ed.) *Bioinformatics – From Genomes to Therapies*. Wiley-VCH Verlag GmbH, Weinheim, pp. 389–418.

115 Meiler, J., Baker, D. (2003) Coupled prediction of protein secondary and tertiary structure. *Proc. Natl Acad. Sci. USA*, **100** (21), 12105–12110.

116 Hofacker, I., Stadler, P.F. (2007) RNA Secondary Structures, in: Lengauer, T. (Ed.) *Bioinformatics – From Genomes to Therapies*. Wiley-VCH Verlag GmbH, Weinheim, pp. 439–490.

117 van Batenburg, F.H., Gultyaev, A.P., Pleij, C.W. (2001) PseudoBase: structural information on RNA pseudoknots. *Nucleic Acids Res.*, **29** (1), 194–195.

118 Han, K., Byun, Y. (2003) PSEUDOVIEWER2: visualization of RNA pseudoknots of any type. *Nucleic Acids Res.*, **31** (13), 3432–3440.

119 Parsch, J., Braverman, J.M., Stephan, W. (2000) Comparative sequence analysis and patterns of covariation in RNA secondary structures. *Genetics*, **154** (2), 909–921.

120 Hofacker, I.L., Stadler, P.F. (1999) Automatic detection of conserved base pairing patterns in RNA virus genomes. *Comput. Chem.*, **23** (3-4), 401–414.

121 Grate, L. (1995) Automatic RNA secondary structure determination with stochastic context- free grammars. *Proc. Int. Conf. Intell. Syst. Mol. Biol.*, **3**, 136–144.

122 Zuker, M., Stiegler, P. (1981) Optimal computer folding of large RNA sequences using thermodynamics and auxiliary information. *Nucleic Acids Res.*, **9** (1), 133–148.

123 Lyngso, R.B., Zuker, M., Pedersen, C.N. (1999) Fast evaluation of internal loops in RNA secondary structure prediction. *Bioinformatics*, **15** (6), 440–445.

124 Zuker, M. (1989) On finding all suboptimal foldings of an RNA molecule. *Science*, **244** (4900), 48–52.

125 McCaskill, J.S. (1990) The equilibrium partition function and base pair binding probabilities for RNA secondary structure. *Biopolymers*, **29** (6-7), 1105–1119.

126 Lyngso, R.B., Pedersen, C.N. (2000) RNA pseudoknot prediction in energy-based models. *J. Comput. Biol.*, **7** (3-4), 409–427.

127 Cai, L., Malmberg, R.L., Wu, Y. (2003) Stochastic modeling of RNA pseudoknotted structures: a grammatical approach. *Bioinformatics*, **19** (Suppl. 1), I66–I73.

128 Major, F., Thibault, P. (2007) RNA Tertiary Structure Prediction, in: Lengauer, T. (Ed.) *Bioinformatics – From Genomes to Therapies*, Wiley-VCH Verlag GmbH, Weinheim, pp. 491–539.

129 Das, R., Karanicolas, J., Baker, D. (2010) Atomic accuracy in predicting and designing noncanonical RNA structure. *Nat. Methods*, **7** (4), 291–294.

130 Rarey, M., Degen, J., Reulecke, I. (2007) Docking and Scoring for Structure-based Drug Design, in: Lengauer, T. (Ed.) *Bioinformatics – From Genomes to Therapies*, Wiley-VCH Verlag GmbH, Weinheim, pp. 541–600.

131 Ewing, T.J., Makino, S., Skillman, A.G., Kuntz, I.D. (2001) DOCK 4.0: search strategies for automated molecular docking of flexible molecule databases. *J. Comput. Aided Mol. Des.*, **15** (5), 411–428.

132 Jones, G., Willett, P., Glen, R.C. (1995) Molecular recognition of receptor sites using a genetic algorithm with a description of desolvation. *J. Mol. Biol.*, **245** (1), 43–53.

133 Rarey, M., Kramer, B., Lengauer, T., Klebe, G. (1996) A fast flexible docking method using an incremental construction algorithm. *J. Mol. Biol.*, **261** (3), 470–489.

134 Morris, G.M., Goodsell, D.S., Halliday, R.S., Huey, R., Hart, W.E., Belew, R.K., Olson, A.J. (1998) Automated docking using a lamarckian genetic algorithm and an empirical binding free energy function. *J. Comput. Chem.*, **19** (14), 1639–1662.

135 Abagyan, R., Totrov, M., Kuznetsov, D. (1994) ICM-a method for protein modeling and design: applications to docking and structure prediction from the distorted native conformation. *J. Comput. Chem.*, **15** (5), 488–506.

136 Klebe, G. (2006) Virtual ligand screening: strategies, perspectives and limitations. *Drug Discov. Today*, **11** (13-14), 580–594.

137 Warren, G.L., Andrews, C.W., Capelli, A.M., Clarke, B., LaLonde, J., Lambert, M.H., Lindvall, M., Nevins, N., Semus, S.F., Senger, S., Tedesco, G., Wall, I.D.,

Woolven, J.M., Peishoff, C.E., Head, M.S. (2006) A critical assessment of docking programs and scoring functions. *J. Med. Chem.*, **49** (20), 5912–5931.

138 Yuriev, E., Agostino, M., Ramsland, P.A. (2011) Challenges and advances in computational docking: 2009 in review. *J. Mol. Recognit.*, **24** (2), 149–164.

139 Rarey, M., Lengauer, T. (2000) A recursive algorithm for efficient combinatorial library docking. *Perspect. Drug Discov. Des.*, **20**, 63–81.

140 Lamb, M.L., Burdick, K.W., Toba, S., Young, M.M., Skillman, A.G., Zou, X., Arnold, J.R., Kuntz, I.D. (2001) Design, docking, and evaluation of multiple libraries against multiple targets. *Proteins*, **42** (3), 296–318.

141 Guvench, O., MacKerell, A.D. Jr (2009) Computational evaluation of protein-small molecule binding. *Curr. Opin. Struct. Biol.*, **19** (1), 56–61.

142 Kim, R., Skolnick, J. (2008) Assessment of programs for ligand binding affinity prediction. *J. Comput. Chem.*, **29** (8), 1316–1331.

143 Foloppe, N., Hubbard, R. (2006) Towards predictive ligand design with free-energy based computational methods? *Curr. Med. Chem.*, **13** (29), 3583–3608.

144 Matter, H. (1997) Selecting optimally diverse compounds from structure databases: a validation study of two-dimensional and three-dimensional molecular descriptors. *J. Med. Chem.*, **40** (8), 1219–1229.

145 Rarey, M., Dixon, J.S. (1998) Feature trees: a new molecular similarity measure based on tree matching. *J. Comput. Aided Mol. Des.*, **12** (5), 471–490.

146 Bajorath, J. (2001) Selected concepts and investigations in compound classification, molecular descriptor analysis, and virtual screening. *J. Chem. Inf. Comput. Sci.*, **41** (2), 233–245.

147 Chen, X., Reynolds, C.H. (2002) Performance of similarity measures in 2D fragment-based similarity searching: comparison of structural descriptors and similarity coefficients. *J. Chem. Inf. Comput. Sci.*, **42** (6), 1407–1414.

148 Lengauer, T., Lemmen, C., Rarey, M., Zimmermann, M. (2004) Novel technologies for virtual screening. *Drug Discov. Today*, **9** (1), 27–34.

149 Ekins, S., Boulanger, B., Swaan, P.W., Hupcey, M.A. (2002) Towards a new age of virtual ADME/TOX and multidimensional drug discovery. *J. Comput. Aided Mol. Des.*, **16** (5-6), 381–401.

150 Yu, H., Adedoyin, A. (2003) ADME-Tox in drug discovery: integration of experimental and computational technologies. *Drug Discov. Today*, **8** (18), 852–861.

151 Katchalski-Katzir, E., Shariv, I., Eisenstein, M., Friesem, A.A., Aflalo, C., Vakser, I.A. (1992) Molecular surface recognition: determination of geometric fit between proteins and their ligands by correlation techniques. *Proc. Natl Acad. Sci. USA*, **89** (6), 2195–2199.

152 Norel, R., Lin, S.L., Wolfson, H.J., Nussinov, R. (1995) Molecular surface complementarity at protein-protein interfaces: the critical role played by surface normals at well placed, sparse, points in docking. *J. Mol. Biol.*, **252** (2), 263–273.

153 Gardiner, E.J., Willett, P., Artymiuk, P.J. (2001) Protein docking using a genetic algorithm. *Proteins*, **44** (1), 44–56.

154 Althaus, E., Kohlbacher, O., Lenhof, H.P., Muller, P. (2002) A combinatorial approach to protein docking with flexible side chains. *J. Comput. Biol.*, **9** (4), 597–612.

155 Jones, G., Willett, P., Glen, R.C., Leach, A.R., Taylor, R. (1997) Development and validation of a genetic algorithm for flexible docking. *J. Mol. Biol.*, **267** (3), 727–748.

156 Desmet, J., De Maeyer, M., Hazes, B., Lasters, I. (1992) The dead-end elimination theorem and its use in protein side-chain positioning. *Nature*, **356**, 539–542.

157 Pierce, N.A., Spriet, J.A., Desmet, J., Mayo, S.L. (2000) Conformational splitting: a more powerful criterion for dead-end elimination. *J. Comput. Chem.*, **21** (11), 999–1009.

158 Fernandez-Recio, J., Totrov, M., Abagyan, R. (2003) ICM-DISCO docking by global energy optimization with fully flexible side-chains. *Proteins*, **52** (1), 113–117.

159 Fernandez-Recio, J., Abagyan, R., Totrov, M. (2005) Improving CAPRI predictions: optimized desolvation for rigid-body docking. *Proteins*, **60** (2), 308–313.

160 Gabb, H.A., Jackson, R.M., Sternberg, M.J. (1997) Modelling protein docking using shape complementarity, electrostatics and

biochemical information. *J. Mol. Biol.*, **272** (1), 106–120.

161 Carter, P., Lesk, V.I., Islam, S.A., Sternberg, M.J. (2005) Protein-protein docking using 3D-Dock in rounds 3, 4, and 5 of CAPRI. *Proteins*, **60** (2), 281–288.

162 Chen, R., Li, L., Weng, Z. (2003) ZDOCK: an initial-stage protein-docking algorithm. *Proteins*, **52** (1), 80–87.

163 Li, L., Chen, R., Weng, Z. (2003) RDOCK: refinement of rigid-body protein docking predictions. *Proteins*, **53** (3), 693–707.

164 Wiehe, K., Pierce, B., Mintseris, J., Tong, W.W., Anderson, R., Chen, R., Weng, Z. (2005) ZDOCK and RDOCK performance in CAPRI rounds 3, 4, and 5. *Proteins*, **60** (2), 207–213.

165 Janin, J., Wodak, S. (2007) The third CAPRI assessment meeting Toronto, Canada, April 20–21, 2007. *Structure*, **15** (7), 755–759.

166 Fernandez-Recio, J., Sternberg, M.J. (2010) The 4th meeting on the Critical Assessment of Predicted Interaction (CAPRI) held at the Mare Nostrum, Barcelona. *Proteins Struct. Funct. Bioinf.*, **78** (15), 3065–3066.

167 Hildebrandt, A., Kohlbacher, O., Lenhof, H.-P. (2007) Modeling Protein–Protein and Protein–DNA Docking, in: Lengauer, T. (Ed.) *Bioinformatics – From Genomes to Therapies*, Wiley-VCH Verlag GmbH, Weinheim, pp. 601–650.

168 Sternberg, M., Moont, G. (2002) Modeling Protein–Protein and Protein–DNA Docking, in: Lengauer, T. (Ed.) *Bioinformatics – From Genomes to Drugs*, Wiley-VCH Verlag GmbH, Weinheim, pp. 361–404.

169 Knegtel, R.M., Antoon, J., Rullmann, C., Boelens, R., Kaptein, R. (1994) MONTY: a Monte Carlo approach to protein-DNA recognition. *J. Mol. Biol.*, **235** (1), 318–324.

170 Zacharias, M., Sklenar, H. (1999) Conformational analysis of single-base bulges in A-form DNA and RNA using a hierarchical approach and energetic evaluation with a continuum solvent model. *J. Mol. Biol.*, **289** (2), 261–275.

171 Cairns, B.R. (2009) The logic of chromatin architecture and remodelling at promoters. *Nature*, **461** (7261), 193–198.

172 Stein, A., Takasuka, T.E., Collings, C.K. (2010) Are nucleosome positions in vivo primarily determined by histone-DNA sequence preferences? *Nucleic Acids Res.*, **38** (3), 709–719.

173 Kaplan, N., Moore, I.K., Fondufe- Mittendorf, Y., Gossett, A.J., Tillo, D., Field, Y., LeProust, E.M., Hughes, T.R., Lieb, J.D., Widom, J., Segal, E. (2009) The DNA-encoded nucleosome organization of a eukaryotic genome. *Nature*, **458** (7236), 362–366.

174 Horak, C.E., Snyder, M. (2002) ChIP-chip: a genomic approach for identifying transcription factor binding sites. *Methods Enzymol.*, **350**, 469–483.

175 Schmid, C.D., Bucher, P. (2007) ChIP-Seq data reveal nucleosome architecture of human promoters. *Cell*, **131** (5), 831–832; author reply 832–833.

176 Dahiyat, B.I., Mayo, S.L. (1997) De novo protein design: fully automated sequence selection. *Science*, **278** (5335), 82–87.

177 Kuhlman, B., Dantas, G., Ireton, G.C., Varani, G., Stoddard, B.L., Baker, D. (2003) Design of a novel globular protein fold with atomic-level accuracy. *Science*, **302** (5649), 1364–1368.

178 Kuhlman, B., Baker, D. (2000) Native protein sequences are close to optimal for their structures. *Proc. Natl Acad. Sci. USA*, **97** (19), 10383–10388.

179 Kuhlman, B., Baker, D. (2004) Exploring folding free energy landscapes using computational protein design. *Curr. Opin. Struct. Biol.*, **14** (1), 89–95.

180 Butterfoss, G.L., Kuhlman, B. (2006) Computer-based design of novel protein structures. *Annu. Rev. Biophys. Biomol. Struct.*, **35**, 49–65.

181 Lippow, S.M., Tidor, B. (2007) Progress in computational protein design. *Curr. Opin. Biotechnol.*, **18** (4), 305–311.

182 Golynskiy, M.V., Seelig, B. (2010) De novo enzymes: from computational design to mRNA display. *Trends Biotechnol.*, **28** (7), 340–345.

183 Gerlt, J.A., Babbitt, P.C. (2009) Enzyme (re)design: lessons from natural evolution and computation. *Curr. Opin. Chem. Biol.*, **13** (1), 10–18.

184 Zanghellini, A., Jiang, L., Wollacott, A.M., Cheng, G., Meiler, J., Althoff, E.A., Rothlisberger, D., Baker, D. (2006) New algorithms and an in silico benchmark for computational enzyme design. *Protein Sci.*, **15** (12), 2785–2794.

185 Siegel, J.B., Zanghellini, A., Lovick, H.M., Kiss, G., Lambert, A.R., St Clair, J.L., Gallaher, J.L., Hilvert, D., Gelb, M.H., Stoddard, B.L., Houk, K.N., Michael, F.E., Baker, D. (2010) Computational design of an enzyme catalyst for a stereoselective bimolecular Diels–Alder reaction. *Science*, **329** (5989), 309–313.

186 Savile, C.K., Janey, J.M., Mundorff, E.C., Moore, J.C., Tam, S., Jarvis, W., Colbeck, J.C., Krebber, A., Fleitz, F.J., Brands, J., Devine, P.N., Huisman, G.W., Hughes, G.J. (2010) Biocatalytic asymmetric synthesis of chiral amines from ketones applied to sitagliptin manufacture. *Science*, **329** (5989), 305–309.

187 Cusick, M.E., Klitgord, N., Vidal, M., Hill, D.E. (2005) Interactome: gateway into systems biology. *Hum. Mol. Genet.*, **14** (2), R171–R181.

188 Ideker, T., Galitski, T., Hood, L. (2001) A new approach to decoding life: systems biology. *Annu. Rev. Genomics Hum. Genet.*, **2**, 343–372.

189 Takahashi, K., Ishikawa, N., Sadamoto, Y., Sasamoto, H., Ohta, S., Shiozawa, A., Miyoshi, F., Naito, Y., Nakayama, Y., Tomita, M. (2003) E-Cell 2: Multi-platform E-Cell simulation system. *Bioinformatics*, **19** (13), 1727–1729.

190 Kohl, P., Noble, D. (2009) Systems biology and the virtual physiological human. *Mol. Syst. Biol.*, **5**, 292.

191 Ramirez, F., Schlicker, A., Assenov, Y., Lengauer, T., Albrecht, M. (2007) Computational analysis of human protein interaction networks. *Proteomics*, **7** (15), 2541–2552.

192 Carter, G.W. (2005) Inferring network interactions within a cell. *Brief. Bioinform.*, **6** (4), 380–389.

193 Blankenburg, H., Finn, R.D., Prlic, A., Jenkinson, A.M., Ramirez, F., Emig, D., Schelhorn, S.E., Buch, J., Lengauer, T., Albrecht, M. (2009) DASMI: exchanging, annotating and assessing molecular interaction data. *Bioinformatics*, **25** (10), 1321–1328.

194 Raman, K. (2010) Construction and analysis of protein-protein interaction networks. *Autom. Exp.*, **2** (1), 2.

195 Beyer, A., Bandyopadhyay, S., Ideker, T. (2007) Integrating physical and genetic maps: from genomes to interaction networks. *Nat. Rev. Genet.*, **8** (9), 699–710.

196 Frishman, D., Albrecht, M., Blankenburg, H., Bork, P., Harrington, E.D., Hermjakob, H., Jensen, L.J., Juan, D.A., Lengauer, T., Pagel, P. (2009) Protein–Protein Interactions: Analysis and Prediction, in: Frishman, D., Valencia, A. (Eds) *Modern Genome Annotation – The Biosapiens Network*, Springer Medizin Verlag, Wien, pp. 353–412.

197 Aloy, P., Russell, R.B. (2006) Structural systems biology: modelling protein interactions. *Nat. Rev. Mol. Cell Biol.*, **7** (3), 188–197.

198 Joyce, A.R., Palsson, B.O. (2006) The model organism as a system: integrating 'omics' data sets. *Nat. Rev. Mol. Cell Biol.*, **7** (3), 198–210.

199 Ng, A., Bursteinas, B., Gao, Q., Mollison, E., Zvelebil, M. (2006) Resources for integrative systems biology: from data through databases to networks and dynamic system models. *Brief. Bioinform.*, **7** (4), 318–330.

200 Papin, J.A., Hunter, T., Palsson, B.O., Subramaniam, S. (2005) Reconstruction of cellular signalling networks and analysis of their properties. *Nat. Rev. Mol. Cell Biol.*, **6** (2), 99–111.

201 Hyduke, D.R., Palsson, B.O. (2010) Towards genome-scale signalling-network reconstructions. *Nat. Rev. Genet.*, **11** (4), 297–307.

202 Shoemaker, B.A., Panchenko, A.R. (2007) Deciphering protein-protein interactions. Part II. Computational methods to predict protein and domain interaction partners. *PLoS Comput. Biol.*, **3** (4), e43.

203 Jamshidi, N., Palsson, B.O. (2008) Formulating genome-scale kinetic models in the post-genome era. *Mol. Syst. Biol.*, **4**, 171.

204 Caspi, R., Altman, T., Dale, J.M., Dreher, K., Fulcher, C.A., Gilham, F., Kaipa, P., Karthikeyan, A.S., Kothari, A., Krummenacker, M., Latendresse, M., Mueller, L.A., Paley, S., Popescu, L., Pujar, A., Shearer, A.G., Zhang, P., Karp, P.D. (2010) The MetaCyc database of metabolic pathways and enzymes and the BioCyc collection of pathway/genome databases. *Nucleic Acids Res.*, **38** (Database issue), D473–D479.

205 Aurrecoechea, C., Brestelli, J., Brunk, B.P., Fischer, S., Gajria, B., Gao, X.,

Gingle, A., Grant, G., Harb, O.S., Heiges, M., Innamorato, F., Iodice, J., Kissinger, J.C., Kraemer, E.T., Li, W., Miller, J.A., Nayak, V., Pennington, C., Pinney, D.F., Roos, D.S., Ross, C., Srinivasamoorthy, G., Stoeckert, C.J. Jr, Thibodeau, R., Treatman, C., Wang, H. (2010) EuPathDB: a portal to eukaryotic pathogen databases. *Nucleic Acids Res.*, **38** (Database issue), D415–D419.

206 Hertz-Fowler, C., Peacock, C.S., Wood, V., Aslett, M., Kerhornou, A., Mooney, P., Tivey, A., Berriman, M., Hall, N., Rutherford, K., Parkhill, J., Ivens, A.C., Rajandream, M.A., Barrell, B. (2004) GeneDB: a resource for prokaryotic and eukaryotic organisms. *Nucleic Acids Res.*, **32** (Database issue), D339–D343.

207 Vallenet, D., Engelen, S., Mornico, D., Cruveiller, S., Fleury, L., Lajus, A., Rouy, Z., Roche, D., Salvignol, G., Scarpelli, C., Medigue, C. (2009) MicroScope: a platform for microbial genome annotation and comparative genomics. *Database (Oxford)*, **2009**, bap021.

208 Kanehisa, M., Goto, S., Furumichi, M., Tanabe, M., Hirakawa, M. (2010) KEGG for representation and analysis of molecular networks involving diseases and drugs. *Nucleic Acids Res.*, **38** (Database issue), D355–D360.

209 Matthews, L., Gopinath, G., Gillespie, M., Caudy, M., Croft, D., de Bono, B., Garapati, P., Hemish, J., Hermjakob, H., Jassal, B. (2009) Reactome knowledgebase of human biological pathways and processes. *Nucleic Acids Res.*, **37** (Database issue), D619–D622.

210 Chang, A., Scheer, M., Grote, A., Schomburg, I., Schomburg, D. (2009) BRENDA, AMENDA and FRENDA the enzyme information system: new content and tools in 2009. *Nucleic Acids Res.*, **37** (Database issue), D588–D592.

211 Fleischmann, A., Darsow, M., Degtyarenko, K., Fleischmann, W., Boyce, S., Axelsen, K.B., Bairoch, A., Schomburg, D., Tipton, K.F., Apweiler, R. (2004) IntEnz, the integrated relational enzyme database. *Nucleic Acids Res.*, **32** (1), D434–D437.

212 Yamada, T., Bork, P. (2009) Evolution of biomolecular networks: lessons from metabolic and protein interactions. *Nat. Rev. Mol. Cell Biol.*, **10** (11), 791–803.

213 Raman, K., Chandra, N. (2009) Flux balance analysis of biological systems: applications and challenges. *Brief. Bioinform.*, **10** (4), 435–449.

214 Ruppin, E., Papin, J.A., de Figueiredo, L.F., Schuster, S. (2010) Metabolic reconstruction, constraint-based analysis and game theory to probe genome-scale metabolic networks. *Curr. Opin. Biotechnol.*, **21** (4), 502–510.

215 Niklas, J., Schneider, K., Heinzle, E. (2010) Metabolic flux analysis in eukaryotes. *Curr. Opin. Biotechnol.*, **21** (1), 63–69.

216 Pfeiffer, T., Sanchez-Valdenebro, I., Nuno, J.C., Montero, F., Schuster, S. (1999) METATOOL: for studying metabolic networks. *Bioinformatics*, **15** (3), 251–257.

217 Mendes, P. (1997) Biochemistry by numbers: simulation of biochemical pathways with Gepasi 3. *Trends Biochem. Sci.*, **22** (9), 361–363.

218 Goryanin, I., Hodgman, T.C., Selkov, E. (1999) Mathematical simulation and analysis of cellular metabolism and regulation. *Bioinformatics*, **15** (9), 749–758.

219 Voit, E.O. (2000) *Computational Analysis of Biochemical Systems: A Practical Guide for Biochemists and Molecular Biologists*, Cambridge University Press, Cambridge.

220 Wiechert, W. (2002) Modeling and simulation: tools for metabolic engineering. *J. Biotechnol.*, **94** (1), 37–63.

221 Wishart, D.S. (2007) Current progress in computational metabolomics. *Brief. Bioinform.*, **8** (5), 279–293.

222 Saito, N., Ohashi, Y., Soga, T., Tomita, M. (2010) Unveiling cellular biochemical reactions via metabolomics-driven approaches. *Curr. Opin. Microbiol.*, **13** (3), 358–362.

223 Krull, M., Pistor, S., Voss, N., Kel, A., Reuter, I., Kronenberg, D., Michael, H., Schwarzer, K., Potapov, A., Choi, C. (2006) TRANSPATH: an information resource for storing and visualizing signaling pathways and their pathological aberrations. *Nucleic Acids Res.*, **34** (Database issue), D546–D551.

224 Matys, V., Kel-Margoulis, O.V., Fricke, E., Liebich, I., Land, S., Barre-Dirrie, A., Reuter, I., Chekmenev, D., Krull, M., Hornischer, K. (2006) TRANSFAC and its module TRANSCompel: transcriptional

gene regulation in eukaryotes. *Nucleic Acids Res.*, **34** (Database issue), D108–D110.

225 Karlebach, G., Shamir, R. (2008) Modelling and analysis of gene regulatory networks. *Nat. Rev. Mol. Cell Biol.*, **9** (10), 770–780.

226 Peleg, M., Yeh, I., Altman, R.B. (2002) Modelling biological processes using workflow and Petri Net models. *Bioinformatics*, **18** (6), 825–837.

227 Tegner, J., Bjorkegren, J. (2007) Perturbations to uncover gene networks. *Trends Genet.*, **23** (1), 34–41.

228 Kholodenko, B.N. (2006) Cell-signalling dynamics in time and space. *Nat. Rev. Mol. Cell Biol.*, **7** (3), 165–176.

229 Pujol, A., Mosca, R., Farres, J., Aloy, P. (2010) Unveiling the role of network and systems biology in drug discovery. *Trends Pharmacol. Sci.*, **31** (3), 115–123.

230 Zhang, S., Jin, G., Zhang, X.S., Chen, L. (2007) Discovering functions and revealing mechanisms at molecular level from biological networks. *Proteomics*, **7** (16), 2856–2869.

231 Barabasi, A.L., Oltvai, Z.N. (2004) Network biology: understanding the cell's functional organization. *Nat. Rev. Genet.*, **5** (2), 101–113.

232 Assenov, Y., Ramirez, F., Schelhorn, S.E., Lengauer, T., Albrecht, M. (2008) Computing topological parameters of biological networks. *Bioinformatics*, **24** (2), 282–284.

233 Sharan, R., Ideker, T. (2006) Modeling cellular machinery through biological network comparison. *Nat. Biotechnol.*, **24** (4), 427–433.

234 Schelhorn, S.E., Lengauer, T., Albrecht, M. (2008) An integrative approach for predicting interactions of protein regions. *Bioinformatics*, **24** (16), i35–i41.

235 Nikolsky, Y., Nikolskaya, T., Bugrim, A. (2005) Biological networks and analysis of experimental data in drug discovery. *Drug Discov. Today*, **10** (9), 653–662.

236 Alon, U. (2007) Network motifs: theory and experimental approaches. *Nat. Rev. Genet.*, **8** (6), 450–461.

237 Sharan, R., Ulitsky, I., Shamir, R. (2007) Network-based prediction of protein function. *Mol. Syst. Biol.*, **3**, 88.

238 Przytycka, T.M., Singh, M., Slonim, D.K. (2010) Toward the dynamic interactome: it's about time. *Brief. Bioinform.*, **11** (1), 15–29.

239 Wilkinson, D.J. (2009) Stochastic modelling for quantitative description of heterogeneous biological systems. *Nat. Rev. Genet.*, **10** (2), 122–133.

240 Schlicker, A., Huthmacher, C., Ramirez, F., Lengauer, T., Albrecht, M. (2007) Functional evaluation of domain-domain interactions and human protein interaction networks. *Bioinformatics*, **23** (7), 859–865.

241 Gehlenborg, N., O'Donoghue, S.I., Baliga, N.S., Goesmann, A., Hibbs, M.A., Kitano, H., Kohlbacher, O., Neuweger, H., Schneider, R., Tenenbaum, D. (2010) Visualization of omics data for systems biology. *Nat. Methods*, **7** (Suppl. 3), S56–S68.

242 Emig, D., Salomonis, N., Baumbach, J., Lengauer, T., Conklin, B.R., Albrecht, M. (2010) AltAnalyze and DomainGraph: analyzing and visualizing exon expression data. *Nucleic Acids Res.*, **38** (Suppl.), W755–W762.

243 Lipshutz, R.J., Fodor, S.P., Gingeras, T.R., Lockhart, D.J. (1999) High density synthetic oligonucleotide arrays. *Nat. Genet.*, **21** (Suppl. 1), 20–24.

244 Brown, P.O., Botstein, D. (1999) Exploring the new world of the genome with DNA microarrays. *Nat. Genet.*, **21** (Suppl. 1), 33–37.

245 Morozova, O., Hirst, M., Marra, M.A. (2009) Applications of new sequencing technologies for transcriptome analysis. *Annu. Rev. Genomics Hum. Genet.*, **10**, 135–151.

246 Forster, T., Roy, D., Ghazal, P. (2003) Experiments using microarray technology: limitations and standard operating procedures. *J. Endocrinol.*, **178** (2), 195–204.

247 Tobler, J.B., Molla, M.N., Nuwaysir, E.F., Green, R.D., Shavlik, J.W. (2002) Evaluating machine learning approaches for aiding probe selection for gene-expression arrays. *Bioinformatics*, **18** (Suppl. 1), S164–S171.

248 Zien, A., Fluck, J., Zimmer, R., Lengauer, T. (2003) Microarrays: how many do you need? *J. Comput. Biol.*, **10** (3-4), 653–667.

249 Brazma, A., Hingamp, P., Quackenbush, J., Sherlock, G., Spellman, P., Stoeckert, C., Aach, J., Ansorge, W., Ball, C.A., Causton, H.C. (2001) Minimum information about a microarray experiment (MIAME)-toward standards for microarray data. *Nat. Genet.*, **29** (4), 365–371.

250 Brazma, A., Parkinson, H., Sarkans, U., Shojatalab, M., Vilo, J., Abeygunawardena, N., Holloway, E., Kapushesky, M., Kemmeren, P., Lara, G.G. (2003) ArrayExpress – a public repository for microarray gene expression data at the EBI. *Nucleic Acids Res.*, **31** (1), 68–71.

251 Edgar, R., Domrachev, M., Lash, A.E. (2002) Gene Expression Omnibus: NCBI gene expression and hybridization array data repository. *Nucleic Acids Res.*, **30** (1), 207–210.

252 Krajewski, P., Bocianowski, J. (2002) Statistical methods for microarray assays. *J. Appl. Genet.*, **43** (3), 269–278.

253 Yang, Y.H., Speed, T. (2002) Design issues for cDNA microarray experiments. *Nat. Rev. Genet.*, **3** (8), 579–588.

254 Quackenbush, J. (2002) Microarray data normalization and transformation. *Nat. Genet.*, **32** (Suppl.), 496–501.

255 Park, T., Yi, S.G., Kang, S.H., Lee, S., Lee, Y.S., Simon, R. (2003) Evaluation of normalization methods for microarray data. *BMC Bioinformatics*, **4** (1), 33.

256 Zien, A., Aigner, T., Zimmer, R., Lengauer, T. (2001) Centralization: a new method for the normalization of gene expression data. *Bioinformatics*, **17** (Suppl. 1), S323–S331.

257 Huber, W., von Heydebreck, A., Sueltmann, H., Poustka, A., Vingron, M. (2003) Parameter estimation for the calibration and variance stabilization of micorarray data. *Stat. Appl. Genet. Mol. Biol.*, **2** (1), Article 3.

258 Golub, T.R., Slonim, D.K., Tamayo, P., Huard, C., Gaasenbeek, M., Mesirov, J.P., Coller, H., Loh, M.L., Downing, J.R., Caligiuri, M.A. (1999) Molecular classification of cancer: class discovery and class prediction by gene expression monitoring. *Science*, **286** (5439), 531–537.

259 Hastie, T., Tibshirani, R., Friedman, J.S. (2001) *The Elements of Statistical Learning*, Springer-Verlag, New York.

260 Friedman, N., Kaminski, N. (2002) Statistical methods for analyzing gene expression data for cancer research. *Ernst Schering Research Foundation Workshop*, pp. 109–131.

261 Russo, G., Zegar, C., Giordano, A. (2003) Advantages and limitations of microarray technology in human cancer. *Oncogene*, **22** (42), 6497–6507.

262 Sharan, R., Elkon, R., Shamir, R. (2002) Cluster analysis and its applications to gene expression data. *Ernst Schering Research Foundation Workshop*, pp. 83–108.

263 Lottaz, C., Kostka, D., Spang, R. (2007) Classification of Patients, in: Lengauer, T. (Ed.) *Bioinformatics – From Genomes to Therapies*. Wiley-VCH Verlag GmbH, Weinheim, pp. 957–992.

264 Iyer, V.R., Eisen, M.B., Ross, D.T., Schuler, G., Moore, T., Lee, J.C., Trent, J.M., Staudt, L.M., Hudson, J., Boguski, M.S. (1999) The transcriptional program in the response of human fibroblasts to serum. *Science*, **283** (5398), 83–87.

265 Lyons, T.J., Gasch, A.P., Gaither, L.A., Botstein, D., Brown, P.O., Eide, D.J. (2000) Genome-wide characterization of the Zap1p zinc-responsive regulon in yeast. *Proc. Natl Acad. Sci. USA*, **97** (14), 7957–7962.

266 Tavazoie, S., Hughes, J.D., Campbell, M.J., Cho, R.J., Church, G.M. (1999) Systematic determination of genetic network architecture. *Nat. Genet.*, **22** (3), 281–285.

267 Werner, T. (2001) Target gene identification from expression array data by promoter analysis. *Biomol. Eng.*, **17** (3), 87–94.

268 Palin, K., Ukkonen, E., Brazma, A., Vilo, J. (2002) Correlating gene promoters and expression in gene disruption experiments. *Bioinformatics*, **18** (Suppl. 2), S172–S180.

269 Yu, H., Luscombe, N.M., Qian, J., Gerstein, M. (2003) Genomic analysis of gene expression relationships in transcriptional regulatory networks. *Trends Genet.*, **19** (8), 422–427.

270 Segal, E., Yelensky, R., Koller, D. (2003) Genome-wide discovery of transcriptional modules from DNA sequence and gene expression. *Bioinformatics*, **19** (Suppl. 1), I273–I282.

271 Zien, A., Kuffner, R., Zimmer, R., Lengauer, T. (2000) Analysis of gene expression data with pathway scores. *Proc. Int. Conf. Intell. Syst. Mol. Biol.*, **8**, 407–417.

272 Hanisch, D., Zien, A., Zimmer, R., Lengauer, T. (2002) Co-clustering of biological networks and gene expression data. *Bioinformatics*, **18** (Suppl. 1), 145S–1154.

273 Ideker, T., Thorsson, V., Ranish, J.A., Christmas, R., Buhler, J., Eng, J.K.,

Bumgarner, R., Goodlett, D.R., Aebersold, R., Hood, L. (2001) Integrated genomic and proteomic analyses of a systematically perturbed metabolic network. *Science*, **292** (5518), 929–934.

274 Segal, E., Wang, H., Koller, D. (2003) Discovering molecular pathways from protein interaction and gene expression data. *Bioinformatics*, **19** (Suppl. 1), I264–I272.

275 de Jong, H. (2002) Modeling and simulation of genetic regulatory systems. *J. Comput. Biol.*, **9** (1), 67–103.

276 Rahnenführer, J., Lengauer, T. (2007) Classification of Genes, in: Lengauer, T. (Ed.) *Bioinformatics – From Genomes to Therapies*, Wiley-VCH Verlag GmbH, Weinheim, pp. 993–1022.

277 Gerhard, D.S., Wagner, L., Feingold, E.A., Shenmen, C.M., Grouse, L.H., Schuler, G., Klein, S.L., Old, S., Rasooly, R., Good, P. (2004) The status, quality, and expansion of the NIH full-length cDNA project: the Mammalian Gene Collection (MGC). *Genome Res.*, **14** (10B), 2121–2127.

278 Harbers, M., Carninci, P. (2005) Tag-based approaches for transcriptome research and genome annotation. *Nat. Methods*, **2** (7), 495–502.

279 Nagalakshmi, U., Wang, Z., Waern, K., Shou, C., Raha, D., Gerstein, M., Snyder, M. (2008) The transcriptional landscape of the yeast genome defined by RNA sequencing. *Science*, **320** (5881), 1344–1349.

280 Mortazavi, A., Williams, B.A., McCue, K., Schaeffer, L., Wold, B. (2008) Mapping and quantifying mammalian transcriptomes by RNA-Seq. *Nat. Methods*, **5** (7), 621–628.

281 Lister, R., O'Malley, R.C., Tonti-Filippini, J., Gregory, B.D., Berry, C.C., Millar, A.H., Ecker, J.R. (2008) Highly integrated single-base resolution maps of the epigenome in *Arabidopsis*. *Cell*, **133** (3), 523–536.

282 Sultan, M., Schulz, M.H., Richard, H., Magen, A., Klingenhoff, A., Scherf, M., Seifert, M., Borodina, T., Soldatov, A., Parkhomchuk, D. (2008) A global view of gene activity and alternative splicing by deep sequencing of the human transcriptome. *Science*, **321** (5891), 956–960.

283 Wang, Z., Gerstein, M., Snyder, M. (2009) RNA-Seq: a revolutionary tool for transcriptomics. *Nat. Rev. Genet.*, **10** (1), 57–63.

284 Li, B., Ruotti, V., Stewart, R.M., Thomson, J.A., Dewey, C.N. (2010) RNA-Seq gene expression estimation with read mapping uncertainty. *Bioinformatics*, **26** (4), 493–500.

285 Ozsolak, F., Milos, P.M. (2011) RNA sequencing: advances, challenges and opportunities. *Nat. Rev. Genet.*, **12** (2), 87–98.

286 Garber, M., Grabherr, M.G., Guttman, M., Trapnell, C. (2011) Computational methods for transcriptome annotation and quantification using RNA-seq. *Nat. Methods*, **8** (6), 469–477.

287 Haas, B.J., Zody, M.C. (2010) Advancing RNA-Seq analysis. *Nat. Biotechnol.*, **28** (5), 421–423.

288 Pepke, S., Wold, B., Mortazavi, A. (2009) Computation for ChIP-seq and RNA-seq studies. *Nat. Methods*, **6** (Suppl. 11), S22–S32.

289 Birol, I., Jackman, S.D., Nielsen, C.B., Qian, J.Q., Varhol, R., Stazyk, G., Morin, R.D., Zhao, Y., Hirst, M., Schein, J.E., Horsman, D.E., Connors, J.M., Gascoyne, R.D., Marra, M.A., Jones, S.J. (2009) De novo transcriptome assembly with ABySS. *Bioinformatics*, **25** (21), 2872–2877.

290 Zerbino, D.R., Birney, E. (2008) Velvet: algorithms for de novo short read assembly using de Bruijn graphs. *Genome Res.*, **18** (5), 821–829.

291 Guttman, M., Garber, M., Levin, J.Z., Donaghey, J., Robinson, J., Adiconis, X., Fan, L., Koziol, M.J., Gnirke, A., Nusbaum, C. (2010) Ab initio reconstruction of cell type-specific transcriptomes in mouse reveals the conserved multi-exonic structure of lincRNAs. *Nat. Biotechnol.*, **28** (5), 503–510.

292 Trapnell, C., Williams, B.A., Pertea, G., Mortazavi, A., Kwan, G., van Baren, M.J., Salzberg, S.L., Wold, B.J., Pachter, L. (2010) Transcript assembly and quantification by RNA-Seq reveals unannotated transcripts and isoform switching during cell differentiation. *Nat. Biotechnol.*, **28** (5), 511–515.

293 Denoeud, F., Aury, J.M., Da Silva, C., Noel, B., Rogier, O., Delledonne, M., Morgante, M., Valle, G., Wincker, P., Scarpelli, C. (2008) Annotating genomes with massive-scale RNA sequencing. *Genome Biol.*, **9** (12), R175.

294 De Bona, F., Ossowski, S., Schneeberger, K., Ratsch, G. (2008) Optimal spliced alignments of short sequence reads. *Bioinformatics*, **24** (16), i174–i180.

295 Huber, W., von Heydebreck, A., Sueltmann, H., Poustka, A., Vingron, M. (2002) Variance stabilization applied to microarray data calibration and to the quantification of differential expression. *Bioinformatics*, **18** (Suppl. 1), 96S–104.

296 Becker, C.H., Bern, M. (2010) Recent developments in quantitative proteomics. *Mutat. Res.*, **722** (2), 171–182.

297 Clarke, R., Ressom, H.W., Wang, A., Xuan, J., Liu, M.C., Gehan, E.A., Wang, Y. (2008) The properties of high-dimensional data spaces: implications for exploring gene and protein expression data. *Nat. Rev. Cancer*, **8** (1), 37–49.

298 Bock, C., Lengauer, T. (2008) Computational epigenetics. *Bioinformatics*, **24** (1), 1–10.

299 Ongenaert, M. (2010) Epigenetic databases and computational methodologies in the analysis of epigenetic datasets. *Adv. Genet.*, **71**, 259–295.

300 Lan, N., Montelione, G.T., Gerstein, M. (2003) Ontologies for proteomics: towards a systematic definition of structure and function that scales to the genome level. *Curr. Opin. Chem. Biol.*, **7** (1), 44–54.

301 Webb, E.C. (1992) *Enzyme Nomenclature 1992. Recommendations of the Nomenclature Committee of the International Union of Biochemistry and Molecular Biology*, Academic Press, New York.

302 Mewes, H.W., Frishman, D., Gruber, C., Geier, B., Haase, D., Kaps, A., Lemcke, K., Mannhaupt, G., Pfeiffer, F., Schuller, C. (2000) MIPS: a database for genomes and protein sequences. *Nucleic Acids Res.*, **28** (1), 37–40.

303 Ashburner, M., Ball, C.A., Blake, J.A., Botstein, D., Butler, H., Cherry, J.M., Davis, A.P., Dolinski, K., Dwight, S.S., Eppig, J.T. (2000) Gene ontology: tool for the unification of biology. The Gene Ontology Consortium. *Nat. Genet.*, **25** (1), 25–29.

304 Davidson, E.H., Rast, J.P., Oliveri, P., Ransick, A., Calestani, C., Yuh, C.H., Minokawa, T., Amore, G., Hinman, V., Arenas-Mena, C. (2002) A genomic regulatory network for development. *Science*, **295** (5560), 1669–1678.

305 Hucka, M., Finney, A., Sauro, H.M., Bolouri, H., Doyle, J.C., Kitano, H., Arkin, A.P., Bornstein, B.J., Bray, D., Cornish-Bowden, A. (2003) The systems biology markup language (SBML): a medium for representation and exchange of biochemical network models. *Bioinformatics*, **19** (4), 524–531.

306 Zhu, J., Zhang, B., Smith, E.N., Drees, B., Brem, R.B., Kruglyak, L., Bumgarner, R.E., Schadt, E.E. (2008) Integrating large-scale functional genomic data to dissect the complexity of yeast regulatory networks. *Nat. Genet.*, **40** (7), 854–861.

307 Fisher, J., Henzinger, T.A. (2007) Executable cell biology. *Nat. Biotechnol.*, **25** (11), 1239–1249.

308 Fisher, J., Piterman, N., Hajnal, A., Henzinger, T.A. (2007) Predictive modeling of signaling crosstalk during *C. elegans* vulval development. *PLoS Comput. Biol.*, **3** (5), e92.

309 Fitch, W.M. (2000) Homology a personal view on some of the problems. *Trends Genet.*, **16** (5), 227–231.

310 Todd, A.E., Orengo, C.A., Thornton, J.M. (2001) Evolution of function in protein superfamilies, from a structural perspective. *J. Mol. Biol.*, **307** (4), 1113–1143.

311 Wilson, C.A., Kreychman, J., Gerstein, M. (2000) Assessing annotation transfer for genomics: quantifying the relations between protein sequence, structure and function through traditional and probabilistic scores. *J. Mol. Biol.*, **297** (1), 233–249.

312 Tatusov, R.L., Koonin, E.V., Lipman, D.J. (1997) A genomic perspective on protein families. *Science*, **278** (5338), 631–637.

313 Tatusov, R.L., Fedorova, N.D., Jackson, J.D., Jacobs, A.R., Kiryutin, B., Koonin, E.V., Krylov, D.M., Mazumder, R., Mekhedov, S.L., Nikolskaya, A.N. (2003) The COG database: an updated version includes eukaryotes. *BMC Bioinformatics*, **4** (1), 41.

314 Rasmussen, M.D., Kellis, M. (2010) A Bayesian approach for fast and accurate gene tree reconstruction. *Mol. Biol. Evol.*, **28** (1), 273–290.

315 Akerborg, O., Sennblad, B., Arvestad, L., Lagergren, J. (2009) Simultaneous Bayesian gene tree reconstruction and reconciliation analysis. *Proc. Natl Acad. Sci. USA*, **106** (14), 5714–5719.

316 Gabaldon, T. (2008) Large-scale assignment of orthology: back to phylogenetics? *Genome Biol.*, **9** (10), 235.

317 Bucher, P., Bairoch, A. (1994) A generalized profile syntax for biomolecular sequence motifs and its function in automatic sequence interpretation. *Proc. Int. Conf. Intell. Syst. Mol. Biol.*, **2**, 53–61.

318 Hulo, N., Bairoch, A., Bulliard, V., Cerutti, L., Cuche, B.A., de Castro, E., Lachaize, C., Langendijk-Genevaux, P.S., Sigrist, C.J. (2008) The 20 years of PROSITE. *Nucleic Acids Res.*, **36** (Database issue), D245–D249.

319 Nevill-Manning, C.G., Wu, T.D., Brutlag, D.L. (1998) Highly specific protein sequence motifs for genome analysis. *Proc. Natl Acad. Sci. USA*, **95** (11), 5865–5871.

320 Huang, J.Y., Brutlag, D.L. (2001) The EMOTIF database. *Nucleic Acids Res.*, **29** (1), 202–204.

321 Attwood, T.K. (2000) The role of pattern databases in sequence analysis. *Brief. Bioinform.*, **1** (1), 45–59.

322 Mulder, N.J., Apweiler, R., Attwood, T.K., Bairoch, A., Barrell, D., Bateman, A., Binns, D., Biswas, M., Bradley, P., Bork, P. (2003) The InterPro Database, 2003 brings increased coverage and new features. *Nucleic Acids Res.*, **31** (1), 315–318.

323 Brutlag, D.L. (2007) Inferring Protein Function from Sequence, in: Lengauer, T. (Ed.) *Bioinformatics – From Genomes to Therapies*, Wiley-VCH Verlag GmbH, Weinheim, pp. 1087–1121.

324 Jensen, L.J., Gupta, R., Blom, N., Devos, D., Tamames, J., Kesmir, C., Nielsen, H., Staerfeldt, H.H., Rapacki, K., Workman, C. (2002) Prediction of human protein function from post-translational modifications and localization features. *J. Mol. Biol.*, **319** (5), 1257–1265.

325 Jensen, L.J., Gupta, R., Staerfeldt, H.H., Brunak, S. (2003) Prediction of human protein function according to Gene Ontology categories. *Bioinformatics*, **19** (5), 635–642.

326 von Mering, C. (2007) Inferring Protein Function from Genomic Context, in: Lengauer, T. (Ed.) *Bioinformatics – From Genomes to Therapies*, Wiley-VCH Verlag GmbH, Weinheim, pp. 1179–1211.

327 Dandekar, T., Snel, B., Huynen, M., Bork, P. (1998) Conservation of gene order: a fingerprint of proteins that physically interact. *Trends Biochem. Sci.*, **23** (9), 324–328.

328 Overbeek, R., Fonstein, M., D'Souza, M., Pusch, G.D., Maltsev, N. (1999) The use of gene clusters to infer functional coupling. *Proc. Natl Acad. Sci. USA*, **96** (6), 2896–2901.

329 Kolesov, G., Mewes, H.W., Frishman, D. (2001) SNAPping up functionally related genes based on context information: a colinearity-free approach. *J. Mol. Biol.*, **311** (4), 639–656.

330 Kolesov, G., Mewes, H.W., Frishman, D. (2002) SNAPper: gene order predicts gene function. *Bioinformatics*, **18** (7), 1017–1019.

331 Enright, A.J., Iliopoulos, I., Kyrpides, N.C., Ouzounis, C.A. (1999) Protein interaction maps for complete genomes based on gene fusion events [see comments]. *Nature*, **402** (6757), 86–90.

332 Marcotte, E.M., Pellegrini, M., Ng, H.L., Rice, D.W., Yeates, T.O., Eisenberg, D. (1999) Detecting protein function and protein-protein interactions from genome sequences. *Science*, **285** (5428), 751–753.

333 Pellegrini, M., Marcotte, E.M., Thompson, M.J., Eisenberg, D., Yeates, T.O. (1999) Assigning protein functions by comparative genome analysis: protein phylogenetic profiles. *Proc. Natl Acad. Sci. USA*, **96** (8), 4285–4288.

334 Jensen, L.J., Kuhn, M., Stark, M., Chaffron, S., Creevey, C., Muller, J., Doerks, T., Julien, P., Roth, A., Simonovic, M. (2009) STRING 8 – a global view on proteins and their functional interactions in 630 organisms. *Nucleic Acids Res.*, **37** (Database issue), D412–D416.

335 Thornton, J.M., Todd, A.E., Milburn, D., Borkakoti, N., Orengo, C.A. (2000) From structure to function: approaches and limitations. *Nat. Struct. Biol.*, **7** (Suppl.), 991–994.

336 Domingues, F.S., Lengauer, T. (2007) Inferring Protein Function from Protein Structure, in: Lengauer, T. (Ed.) *Bioinformatics – From Genomes to Therapies*, Wiley-VCH Verlag GmbH, Weinheim, pp. 1211–1252.

337 Gherardini, P.F., Helmer-Citterich, M. (2008) Structure-based function prediction: approaches and applications. *Brief. Funct. Genomic Proteomic*, **7** (4), 291–302.

338 Nimrod, G., Schushan, M., Steinberg, D.M., Ben-Tal, N. (2008) Detection of functionally important regions in "hypothetical proteins" of known structure. *Structure*, **16** (12), 1755–1763.

339 Ashkenazy, H., Erez, E., Martz, E., Pupko, T., Ben-Tal, N. (2010) ConSurf 2010: calculating evolutionary conservation in sequence and structure of proteins and nucleic acids. *Nucleic Acids Res.*, **38** (Suppl.), W529–W533.

340 Laskowski, R.A., Luscombe, N.M., Swindells, M.B., Thornton, J.M. (1996) Protein clefts in molecular recognition and function. *Protein Sci.*, **5** (12), 2438–2452.

341 Laskowski, R.A. (1995) SURFNET: a program for visualizing molecular surfaces, cavities, and intermolecular interactions. *J. Mol. Graph.*, **13** (5), 323–330.

342 Dundas, J., Ouyang, Z., Tseng, J., Binkowski, A., Turpaz, Y., Liang, J. (2006) CASTp: computed atlas of surface topography of proteins with structural and topographical mapping of functionally annotated residues. *Nucleic Acids Res.*, **34** (Web Server issue), W116–W118.

343 Xie, L., Bourne, P.E. (2007) A robust and efficient algorithm for the shape description of protein structures and its application in predicting ligand binding sites. *BMC Bioinformatics*, **8** (Suppl. 4), S9.

344 Elcock, A.H. (2001) Prediction of functionally important residues based solely on the computed energetics of protein structure. *J. Mol. Biol.*, **312** (4), 885–896.

345 Amitai, G., Shemesh, A., Sitbon, E., Shklar, M., Netanely, D., Venger, I., Pietrokovski, S. (2004) Network analysis of protein structures identifies functional residues. *J. Mol. Biol.*, **344** (4), 1135–1146.

346 An, J., Totrov, M., Abagyan, R. (2005) Pocketome via comprehensive identification and classification of ligand-binding envelopes. *Mol. Cell. Proteomics*, **4** (6), 752–761.

347 Ko, J., Murga, L.F., Wei, Y., Ondrechen, M.J. (2005) Prediction of active sites for protein structures from computed chemical properties. *Bioinformatics*, **21** (Suppl. 1), i258–i265.

348 Nayal, M., Honig, B. (2006) On the nature of cavities on protein surfaces: application to the identification of drug-binding sites. *Proteins*, **63** (4), 892–906.

349 Capra, J.A., Laskowski, R.A., Thornton, J.M., Singh, M., Funkhouser, T.A. (2009) Predicting protein ligand binding sites by combining evolutionary sequence conservation and 3D structure. *PLoS Comput. Biol.*, **5** (12), e1000585.

350 Janin, J., Bahadur, R.P., Chakrabarti, P. (2008) Protein–protein interaction and quaternary structure. *Q. Rev. Biophys.*, **41** (2), 133–180.

351 Tuncbag, N., Kar, G., Keskin, O., Gursoy, A., Nussinov, R. (2009) A survey of available tools and web servers for analysis of protein-protein interactions and interfaces. *Brief. Bioinform.*, **10** (3), 217–232.

352 Sierk, M.L., Kleywegt, G.J. (2004) Deja vu all over again: finding and analyzing protein structure similarities. *Structure*, **12** (12), 2103–2111.

353 Hasegawa, H., Holm, L. (2009) Advances and pitfalls of protein structural alignment. *Curr. Opin. Struct. Biol.*, **19** (3), 341–348.

354 Yakunin, A.F., Yee, A.A., Savchenko, A., Edwards, A.M., Arrowsmith, C.H. (2004) Structural proteomics: a tool for genome annotation. *Curr. Opin. Chem. Biol.*, **8** (1), 42–48.

355 Ponomarenko, J.V., Bourne, P.E., Shindyalov, I.N. (2005) Assigning new GO annotations to protein data bank sequences by combining structure and sequence homology. *Proteins*, **58** (4), 855–865.

356 Weinhold, N., Sander, O., Domingues, F.S., Lengauer, T., Sommer, I. (2008) Local function conservation in sequence and structure space. *PLoS Comput. Biol.*, **4** (7), e1000105.

357 Schmitt, S., Kuhn, D., Klebe, G. (2002) A new method to detect related function among proteins independent of sequence and fold homology. *J. Mol. Biol.*, **323** (2), 387–406.

358 Spriggs, R.V., Artymiuk, P.J., Willett, P. (2003) Searching for patterns of amino acids in 3D protein structures. *J. Chem. Inf. Comput. Sci.*, **43** (2), 412–421.

359 Banatao, D.R., Huang, C.C., Babbitt, P.C., Altman, R.B., Klein, T.E. (2001) ViewFeature: integrated feature analysis and visualization. Pacific Symposium of Biocomputing, The Orchid at Mauna Loa, Big Island, Hawaii, January 3–7, pp. 240–250.

360 Gold, N.D., Jackson, R.M. (2006) Fold independent structural comparisons of

protein-ligand binding sites for exploring functional relationships. *J. Mol. Biol.*, **355** (5), 1112–1124.

361 Redfern, O.C., Dessailly, B.H., Dallman, T.J., Sillitoe, I., Orengo, C.A. (2009) FLORA: a novel method to predict protein function from structure in diverse superfamilies. *PLoS Comput. Biol.*, **5** (8), e1000485.

362 Watson, J.D., Sanderson, S., Ezersky, A., Savchenko, A., Edwards, A., Orengo, C., Joachimiak, A., Laskowski, R.A., Thornton, J.M. (2007) Towards fully automated structure-based function prediction in structural genomics: a case study. *J. Mol. Biol.*, **367** (5), 1511–1522.

363 Hermann, J.C., Marti-Arbona, R., Fedorov, A.A., Fedorov, E., Almo, S.C., Shoichet, B.K., Raushel, F.M. (2007) Structure-based activity prediction for an enzyme of unknown function. *Nature*, **448** (7155), 775–779.

364 Sood, A., Erwin, P.J., Ebbert, J.O. (2004) Using advanced search tools on PubMed for citation retrieval. *Mayo Clin. Proc.*, **79** (10), 1295–1299; quiz1300.

365 Jensen, L.J., Saric, J., Bork, P. (2006) Literature mining for the biologist: from information retrieval to biological discovery. *Nat. Rev. Genet.*, **7** (2), 119–129.

366 Krallinger, M., Valencia, A. (2007) Mining Information on Protein Function from Text, in: Lengauer, T. (Ed.) *Bioinformatics – From Genomes to Therapies*, Wiley-VCH Verlag GmbH, Weinheim, pp. 1253–1296.

367 Krallinger, M., Valencia, A., Hirschman, L. (2008) Linking genes to literature: text mining, information extraction, and retrieval applications for biology. *Genome Biol.*, **9** (Suppl. 2), S8.

368 Krallinger, M., Leitner, F., Valencia, A. (2010) Analysis of biological processes and diseases using text mining approaches. *Methods Mol. Biol.*, **593**, 341–382.

369 Brady, S., Shatkay, H. (2008) EpiLoc: a (working) text-based system for predicting protein subcellular location. Pacific Symposium of Biocomputing, Fairmont Orchid, Big Island, Hawaii, January 4–8, pp. 604–615.

370 Craven, M., Kumlien, J. (1999) Constructing biological knowledge bases by extracting information from text sources. Proceedings, International Conference on Intelligence Systems in Molecular Biology, pp. 77–86.

371 Friedman, C., Kra, P., Yu, H., Krauthammer, M., Rzhetsky, A. (2001) GENIES: a natural-language processing system for the extraction of molecular pathways from journal articles. *Bioinformatics*, **17** (Suppl. 1), S74–S82.

372 Donaldson, I., Martin, J., de Bruijn, B., Wolting, C., Lay, V., Tuekam, B., Zhang, S., Baskin, B., Bader, G.D., Michalickova, K. (2003) PreBIND and Textomy – mining the biomedical literature for protein–protein interactions using a support vector machine. *BMC Bioinformatics*, **4** (1), 11.

373 Hoffmann, R., Valencia, A. (2005) Implementing the iHOP concept for navigation of biomedical literature. *Bioinformatics*, **21** (Suppl. 2), ii252–ii258.

374 Rinaldi, F., Schneider, G., Kaljurand, K., Hess, M., Andronis, C., Konstandi, O., Persidis, A. (2007) Mining of relations between proteins over biomedical scientific literature using a deep-linguistic approach. *Artif. Intell. Med.*, **39** (2), 127–136.

375 Koike, A., Niwa, Y., Takagi, T. (2005) Automatic extraction of gene/protein biological functions from biomedical text. *Bioinformatics*, **21** (7), 1227–1236.

376 Rodriguez-Penagos, C., Salgado, H., Martinez-Flores, I., Collado-Vides, J. (2007) Automatic reconstruction of a bacterial regulatory network using natural language processing. *BMC Bioinformatics*, **8**, 293.

377 Narayanaswamy, M., Ravikumar, K.E., Vijay-Shanker, K. (2005) Beyond the clause: extraction of phosphorylation information from Medline abstracts. *Bioinformatics*, **21** (Suppl. 1), i319–i327.

378 Caporaso, J.G., Baumgartner, W.A. Jr, Randolph, D.A., Cohen, K.B., Hunter, L. (2007) MutationFinder: a high-performance system for extracting point mutation mentions from text. *Bioinformatics*, **23** (14), 1862–1865.

379 Xuan, W., Wang, P., Watson, S.J., Meng, F. (2007) Medline search engine for finding genetic markers with biological significance. *Bioinformatics*, **23** (18), 2477–2484.

380 Yeh, A.S., Hirschman, L., Morgan, A.A. (2003) Evaluation of text data mining for database curation: lessons learned from the KDD challenge cup. *Bioinformatics*, **19** (Suppl. 1), I331–I339.

381 Hirschman, L., Yeh, A., Blaschke, C., Valencia, A. (2005) Overview of BioCreAtIvE: critical assessment of information extraction for biology. *BMC Bioinformatics*, **6** (Suppl. 1), S1.

382 Leitner, F., Chatr-aryamontri, A., Mardis, S.A., Ceol, A., Krallinger, M., Licata, L., Hirschman, L., Cesareni, G., Valencia, A. (2010) The FEBS Letters/BioCreative II.5 experiment: making biological information accessible. *Nat. Biotechnol.*, **28** (9), 897–899.

383 Kann, M.G. (2010) Advances in translational bioinformatics: computational approaches for the hunting of disease genes. *Brief. Bioinform.*, **11** (1), 96–110.

384 Hampe, J., Franke, A., Rosenstiel, P., Till, A., Teuber, M., Huse, K., Albrecht, M., Mayr, G., De La Vega, F.M., Briggs, J. (2007) A genome-wide association scan of nonsynonymous SNPs identifies a susceptibility variant for Crohn disease in ATG16L1. *Nat. Genet.*, **39** (2), 207–211.

385 Welsch, C., Domingues, F.S., Susser, S., Antes, I., Hartmann, C., Mayr, G., Schlicker, A., Sarrazin, C., Albrecht, M., Zeuzem, S. (2008) Molecular basis of telaprevir resistance due to V36 and T54 mutations in the NS3-4 A protease of HCV. *Genome Biol.*, **9** (1), R16.

386 Ideker, T., Sharan, R. (2008) Protein networks in disease. *Genome Res.*, **18** (4), 644–652.

387 Schlicker, A., Albrecht, M. (2010) FunSimMat update: new features for exploring functional similarity. *Nucleic Acids Res.*, **38** (Database issue), D244–D248.

388 Schadt, E.E. (2009) Molecular networks as sensors and drivers of common human diseases. *Nature*, **461** (7261), 218–223.

389 Baudot, A., Gomez-Lopez, G., Valencia, A. (2009) Translational disease interpretation with molecular networks. *Genome Biol.*, **10** (6), 221.

390 Lundegaard, C., Lamberth, K., Harndahl, M., Buus, S., Lund, O., Nielsen, M. (2008) NetMHC-3.0: accurate web accessible predictions of human, mouse and monkey MHC class I affinities for peptides of length 8–11. *Nucleic Acids Res.*, **36** (Web Server issue), W509–W512.

391 Tong, J.C., Tan, T.W., Ranganathan, S. (2004) Modeling the structure of bound peptide ligands to major histocompatibility complex. *Protein Sci.*, **13** (9), 2523–2532.

392 Antes, I., Siu, S.W., Lengauer, T. (2006) DynaPred: a structure and sequence based method for the prediction of MHC class I binding peptide sequences and conformations. *Bioinformatics*, **22** (14), e16–e24.

393 Feldhahn, M., Dönnes, P., Thiel, P., Kohlbacher, O. (2009) FRED--a framework for T-cell epitope detection. *Bioinformatics*, **25** (20), 2758–2759.

394 Toussaint, N.C., Kohlbacher, O. (2009) OptiTope – a web server for the selection of an optimal set of peptides for epitope-based vaccines. *Nucleic Acids Res.*, **37** (Web Server issue), W617–W622.

395 Keller, A., Ludwig, N., Comtesse, N., Henn, W., Steudel, W.I., Lenhof, H.P., Meese, E. (2009) Combining gene expression signatures and autoantibody profiles in human meningioma. *Gene Ther.*, **16** (2), 184–189.

396 Rahnenführer, J., Beerenwinkel, N., Schulz, W.A., Hartmann, C., von Deimling, A., Wullich, B., Lengauer, T. (2005) Estimating cancer survival and clinical outcome based on genetic tumor progression scores. *Bioinformatics*, **21** (10), 2438–2446.

397 Mikeska, T., Bock, C., El-Maarri, O., Hubner, A., Ehrentraut, D., Schramm, J., Felsberg, J., Kahl, P., Buttner, R., Pietsch, T. (2007) Optimization of quantitative MGMT promoter methylation analysis using pyrosequencing and combined bisulfite restriction analysis. *J. Mol. Diagn.*, **9** (3), 368–381.

398 Schilsky, R.L. (2010) Personalized medicine in oncology: the future is now. *Nat. Rev. Drug Discov.*, **9** (5), 363–366.

399 Sims, A.H. (2009) Bioinformatics and breast cancer: what can high-throughput genomic approaches actually tell us? *J. Clin. Pathol.*, **62** (10), 879–885.

400 Wist, A.D., Berger, S.I., Iyengar, R. (2009) Systems pharmacology and genome medicine: a future perspective. *Genome Med.*, **1** (1), 11.

401 Wemmert, S., Ketter, R., Rahnenführer, J., Beerenwinkel, N., Strowitzki, M.,

Feiden, W., Hartmann, C., Lengauer, T., Stockhammer, F., Zang, K.D. (2005) Patients with high-grade gliomas harboring deletions of chromosomes 9p and 10q benefit from temozolomide treatment. *Neoplasia*, **7** (10), 883–893.

402 Sos, M.L., Michel, K., Zander, T., Weiss, J., Frommolt, P., Peifer, M., Li, D., Ullrich, R., Koker, M., Fischer, F. (2009) Predicting drug susceptibility of non-small cell lung cancers based on genetic lesions. *J. Clin. Invest.*, **119** (6), 1727–1740.

403 Lengauer, T., Sing, T. (2006) Bioinformatics-assisted anti-HIV therapy. *Nat. Rev. Microbiol.*, **4** (10), 790–797.

404 Prosperi, M.C., Altmann, A., Rosen-Zvi, M., Aharoni, E., Borgulya, G., Bazso, F., Sonnerborg, A., Schülter, E., Struck, D., Ulivi, G. (2009) Investigation of expert rule bases, logistic regression, and non-linear machine learning techniques for predicting response to antiretroviral treatment. *Antivir. Ther.*, **14** (3), 433–442.

405 Altmann, A., Däumer, M., Beerenwinkel, N., Peres, Y., Schülter, E., Büch, J., Rhee, S.Y., Sonnerborg, A., Fessel, W.J., Shafer, R.W. (2009) Predicting the response to combination antiretroviral therapy: retrospective validation of geno2pheno-THEO on a large clinical database. *J. Infect. Dis.*, **199** (7), 999–1006.

406 Lengauer, T., Sander, O., Sierra, S., Thielen, A., Kaiser, R. (2007) Bioinformatics prediction of HIV coreceptor usage. *Nat. Biotechnol.*, **25** (12), 1407–1410.

407 Schubert, W. (2010) On the origin of cell functions encoded in the toponome. *J. Biotechnol.*, **49** (4), 252–259.

408 Muzzey, D., van Oudenaarden, A. (2009) Quantitative time-lapse fluorescence microscopy in single cells. *Annu. Rev. Cell Dev. Biol.*, **25**, 301–327.

409 MacAleese, L., Stauber, J., Heeren, R.M. (2009) Perspectives for imaging mass spectrometry in the proteomics landscape. *Proteomics*, **9** (4), 819–834.

410 Frank, M., Schloissnig, S. (2010) Bioinformatics and molecular modeling in glycobiology. *Cell. Mol. Life Sci.*, **67** (16), 2749–2772.

411 Brazma, A. (2009) Minimum information about a microarray experiment (MIAME) – successes, failures, challenges. *Sci. World J.*, **9**, 420–423.

412 Taylor, C.F., Field, D., Sansone, S.A., Aerts, J., Apweiler, R., Ashburner, M., Ball, C.A., Binz, P.A., Bogue, M., Booth, T. (2008) Promoting coherent minimum reporting guidelines for biological and biomedical investigations: the MIBBI project. *Nat. Biotechnol.*, **26** (8), 889–896.

413 Gene Onotology Consortium (2010) The gene ontology in 2010: extensions and refinements. *Nucleic Acids Res.*, **38** (Database issue), D331–D335.

414 Tipton, K., Boyce, S. (1999) Nomenclature committee of the international union of biochemistry and molecular biology (NC-IUBMB). *Eur. J. Biochem.*, **264** (2, Suppl. 5), 610–650.

415 Bodenreider, O. (2004) The Unified Medical Language System (UMLS): integrating biomedical terminology. *Nucleic Acids Res.*, **32** (Database issue), D267–D270.

416 Hamosh, A., Scott, A.F., Amberger, J., Bocchini, C., Valle, D., McKusick, V.A. (2002) Online Mendelian Inheritance in Man (OMIM), a knowledgebase of human genes and genetic disorders. *Nucleic Acids Res.*, **30** (1), 52–55.

417 Burmeister, M., McInnis, M.G., Zollner, S. (2008) Psychiatric genetics: progress amid controversy. *Nat. Rev. Genet.*, **9** (7), 527–540.

418 Viceconti, M., Clapworthy, G., Van Sint Jan, S. (2008) The virtual physiological human – a European initiative for in silico human modelling. *J. Physiol. Sci.*, **58** (7), 441–446.

419 Hunter, P., Coveney, P.V., de Bono, B., Diaz, V., Fenner, J., Frangi, A.F., Harris, P., Hose, R., Kohl, P., Lawford, P. (2010) A vision and strategy for the virtual physiological human in 2010 and beyond. *Philos. Trans. A Math. Phys. Eng. Sci.*, **368** (1920), 2595–2614.

11
Dynamics of Biomolecular Networks

Emanuele Cozzo[1], Joaquín Sanz[1], and Yamir Moreno[1,2]
[1]University of Zaragoza, Institute for Biocomputation and Physics of Complex Systems (BIFI), 50018, Zaragoza, Spain
[2]University of Zaragoza, Department of Theoretical Physics, Faculty of Sciences, Zaragoza 50009, Spain

Systems Biology: Advances in Molecular Biology and Medicine, First Edition. Edited by Robert A. Meyers.

Keywords

Boolean network
A set of nodes which have two possible states (1 or ON and 0 or OFF) and a list of Boolean functions, each of which are assigned to a node.

Biologically relevant Boolean functions
A Boolean function specifies, at each time step, the activity of the node it is assigned to, given the activity at the previous moment of some other nodes linked to it as inputs. Among all the theoretically possible Boolean functions, there are some that can be biologically implemented, while others are biologically meaningless.

Network inference
Boolean functions can be inferred from experimental expression data.

Probabilistic Boolean network (PBN)
PBN models represent a tool to take into account the uncertainty and stochasticity of real regulatory networks. Instead of assigning just one Boolean function to each node, it is possible to assign a set of functions together with a probability for each function to contribute. The dynamics of such a network is termed a Markov chain.

Rate equation
An ordinary differential equation that, in the context of continuous modeling of molecular regulatory interactions, defines the temporal rate of production of a certain regulated substrate as a function of the concentrations of the ensemble of regulators affecting it.

AND/OR logical functions
In Boolean terms AND and OR are the simplest logical functions of two Boolean input variables. In an AND logic, the output is only true when both the inputs are true in an OR logic it is sufficient for the output to be true that, at least, one of the inputs were true.

Network motif
A small biomolecular circuit of a low number of substances (typically 2–4) interacting between them, according to a certain regulatory topological scheme of regulations (typically 2–6).

During the past few years, whilst the capacity to produce large amounts of experimental data has increased dramatically, the ability to analyze these data has not progressed at the same pace. Thus, there is a risk that this sea of acquired data will become overwhelming, and that no meaningful theoretical or phenomenological insights will be extracted from it. One way to overcome such a problem would be to study the interactions among a system's constituents as a network, involving both the structure and dynamics of that system. In this chapter, the main approaches used to model the dynamical behavior of biomolecular, regulatory gene networks are revised, whereby two different approaches are discussed and analyzed: (i) the discrete case, as represented by Boolean functions; and (ii) the continuous case, which is based on differential equations. For both scenarios, the most recent results are outlined, and details provided of the many variants that can be adopted. Finally, the future lines of research in this field are proposed.

1 Introduction

This chapter deals with networks at the level of the cell, which is the basic unit of any living organism. In the cell, all types of structural and functional processes are ruled by the intricate interaction of genes, proteins, and other molecules. At this point, due to limitations of space, the topological properties of molecular networks, or the most common metrics used to characterize the system, will not be discussed. Rather, the reader is referred to specific reports of networks (e.g., Refs [1, 2]), the dynamics of which will be focused on here.

Regulatory mechanisms among genes can be translated into mathematical language in various ways. The architecture of the cell implies a set of dynamical connections among its components that must be translated into a set of mathematical equations that captures the temporal and spatial evolution of the system. The appropriate choice of the dynamical equations will depend on the level of description required. In this sense, large-scale gene regulatory networks are usually described that make use of Boolean functions. On the other hand, when the aim is to describe simple regulatory mechanisms that involve very few genes, more detailed models – such as nonlinear differential equations – are best suited. Thus, the use of each type of description will depend on the sum of complexities with regards to the structure and dynamics of the system.

The development of quantitative models to capture the coordinated behavior of the circuitry of interacting molecules allows, ultimately, the physiological properties of the cell to be determined. For instance, the use of ordinary, partial, and stochastic differential equations has allowed the tracking and prediction of how quickly each component of a biochemical network changes with time. The extrapolation of differential equations characterizing the dynamics of small subsystems (such as gene circuits) to include an increasing degree of complexity may render the model prohibitively complicated. In such situations, or when many rate constants are unknown, it may become necessary to resort to dynamical systems theory as a reliable approach to the solution of the differential equations. Steady states, limit cycles, and other dynamical configurations can provide quantitative information regarding changes in the magnitude and direction of the state variables. This type of analysis can be translated into the physiological states of the cell (e.g., stable steady states are associated with checkpoints).

Finally, the main drawback of the differential equation-based approach is the need to recognize the (very rarely available) kinetic details of the molecular and cellular interactions. There is increasing evidence, however, that the input–output curves of many regulatory relationships are strongly sigmoidal, and can be approximated by step functions. In addition to this, regulatory networks often maintain their function even when faced with fluctuations in components and reaction rates. This allows the implementation of coarse-grained methods, such as Boolean models, that are also widely used in systems biology research.

Both approaches – namely, continuous and discrete modeling – are discussed in the following sections. More precisely, within the family of discrete models, attention will be focused on the simplest case of Boolean approaches while, with regards to continuous methods, approaches based on ordinary differential equations (ODEs) will be reviewed. Whilst these are only two possibilities of a wider family of models, they may help to provide an understanding of the main features of regulatory systems dynamics. It should be noted that this review is by no means exhaustive, as limitations of space have left no option but to make a choice.

2 Boolean Dynamics Models

Boolean networks have experienced an extensive history in modeling biological systems since the pioneering studies of Kauffman and Thomas during the early 1970s [3, 4]. These early investigations focused on the generic properties of networks and their connections with the behavior of living systems, and the evolution of biological networks. The "post-genomic revolution" – which has been based on new experimental techniques and the consequent availability of large amounts of data – has led to a resurgence in dynamical Boolean network modeling, as applied to the study of the dynamics of real biological networks [5]. Despite the great level of abstraction and simplifications of these models, they are able to provide answers to different questions, and also from more theoretical questions, concerning the origin of life and its evolution [6], and also to much more practical questions related to the actual behavior of real organisms and drug target identifications [7]. As will become evident later, there are two main approaches to extracting knowledge from a Boolean network model of regulatory networks. The first approach is to construct random Boolean networks to study and interpret their generic properties and relate them to known biological features. A second approach involves directly inferring particular structures of biological networks from experimental expression data. In this way, it is possible to reveal more detailed information on the specific organisms under study, and hence to design therapeutic interventions [8].

The modeling of gene regulatory networks via Boolean networks must take into account the fact that cellular processes occur within a very noisy environment, and use very unreliable elements. The key is then to include an increasing complexity in the models, and later to question its robustness. The possibility of Boolean network modeling relies on the sigmoidal nature of the regulatory interactions, so that the step-function approximations are natural: the gene product is either absent (below a given threshold) or present, and the gene is either off or on, leading to a "logical" description of the interactions. This simplification is the elementary version of a class of idealization that relies on discrete approximations of the nonlinear regulatory interactions. Other approximations are also possible and actually utilized, for example piece-linear approximation [9]. However, the Boolean network model is the best formally studied model system, and the model for which a wide range of reverse engineering algorithms exists. Further details of this form of modeling approach are provided in the next sections.

2.1 Boolean Formalisms

A Boolean network G(V,F) is defined by a set of nodes $V = \{x_1, \ldots, x_n\}$ which have two possible states (1 or ON, and 0 or OFF) and a list of Boolean functions $F = (f_1, \ldots, f_n)$. A Boolean function f_i specifies the activity of the node x_i at each time step, taking into account the activity at the previous moment of some other nodes linked to it (the inputs). In a biological context, the value of x_i represents the state of the expression of gene i (1 when it is expressed and 0 otherwise), while the list of Boolean functions corresponds to the set of regulatory interactions among genes. The dynamics of a Boolean network can be updated either synchronously or asynchronously, depending on whether all the nodes are updated at the same time or not, respectively. As will be seen later, the description of attractors sensibly depends on the updating choice (at the moment, attention is focused on synchronous updating).

The dynamics can be represented in the form of a directed graph, so that each node represents a state of the network, while a link between two nodes represents the possible transition from one to another, according to the list of Boolean functions. As the dynamics is deterministic (recall that synchronous updating is being assumed), each node can have only one output link, and consequently a path in this graph would be analogous to a trajectory in continuous models. In addition, due to the deterministic nature of Boolean networks and the finite number of states, there are states which are repeatedly visited. Such states, which are referred to as named "attractors," can be fixed points or cycles, while the transient states that lead to an attractor are termed its *basin of attraction*. Hence, a complete description of the dynamical properties of a Boolean network implies knowledge of the attractors and their basins of attraction – that is, of the state space. An important parameter, as will be seen later, is the cycle length; this is the number of different states that will be visited before returning to the original state in an attractor. Finally, it should be noted that Boolean network models are examples of finite dynamical systems, and can be generalized to finite fields [10].

2.2 Generic Properties of (Random) Boolean Networks and Cell Behaviors: Cell Differentiations and the Cell Cycle

The intuition of Kauffman, in his pioneering studies of gene regulatory networks, was to relate the dynamical properties of (random) Boolean networks to cellular types [11]. In other words, the goal of this approach was to interpret different cellular types in terms of different attractors of the regulatory network. Another interpretation of the attractors of a Boolean network is that they may represent different functional states of a cell (differentiation, growth, quiescence, and apoptosis) [12]. Taking together these two interpretations allows the developmental processes to be thought of in terms of an exploration of the attractors in the state space. Admittedly, it is possible to mimic the developmental process of an organism, thus simulating the dynamics of a Boolean network.

The so-called "NK-model" of Kauffman is based on a random Boolean network with N fixed nodes and K fixed inputs per node. The random points here are the wiring of the network, and the Boolean rules that are assigned to each node. Initially, Kauffman began to study such networks with the aim of gaining insights into the mechanisms of regulatory networks, and identified very interesting behaviors beyond a mere biological context. As K decreases from N to 1, there is a phase transition from chaos to order. However, for $K > 2$ the dynamics is chaotic, whereas for $K < 2$ it is ordered. Complex dynamics emerges during the transition region.

In order to characterize the chaotic and the ordered phases, two properties are scrutinized: (i) the expected length of the cycles; and (ii) the sensitivity to the initial conditions. In the chaotic regime, the cycles quickly become very long ($\sim 2^{N/2}$), so that they have no biological sense and the system then requires too much time to explore its attractors (the number of which scales as N). In contrast, in the ordered regime the cycle lengths and the number of attractors are both $N^{1/2}$, such that the network has a relatively large number of attractors that can be explored within a relatively short time. This led to the interpretation by Kauffman of the attractors as cell differentiation. In the case of the second descriptor, within the chaotic regime a small perturbation produces a cascade that propagates along the network, affecting the dynamics in random fashion. However, in the ordered regime a perturbation is rarely produced that causes the network to leave the attractor on which it is seated (when this is the case, the network falls in a near attractor). Thus, a type of stability is achieved that also allows for the possibility of exploring near attractors (this is the second interpretation).

The actual mechanism at the root of the chaos–order transition is the formation of percolating frozen cores of elements fixed in one of the two possible states, 1 or 0. In the chaotic regime, there are islands of frozen cores in a connected sea of oscillating elements, which explains the chaotic dynamics. In contrast, in the ordered regime the frozen cores have percolated while the functional islands of oscillating elements have remained, but cannot influence each other; this gives rise to the observed stability. At the transition region, the frozen cores begin to percolate, allowing the emergence of complex dynamics.

2.3
Topological and Dynamical Properties: Homeostasis, Flexibility, and Evolvability

Structural stability is a central concept in dynamical systems, and has its biological counterpart in the concept of *homeostasis* – the need of an organism not to be destroyed by "small" changes in the environment it is in contact with. In other words, homeostasis represents the capacity of the organism to self-sustain its metabolism and developmental processes. Whereas homeostasis accounts for the capacity of a living system to sustain itself, *flexibility* captures whether organisms are able to develop different functional states through epigenetic processes in a varying environment. This latter property may be related to multi-stationarity [13] – that is, the presence of different stable attractors among which the system can choose, meaning that the organism would have a wide range of responses to changes in environmental conditions.

As hypothesized by Thomas [14], and successively demonstrated [15, 16], the above-described properties can be placed in strict relation to the topological properties of the underlying regulatory network, where the key role is played by feedback (FB) loops. In the words of Thomas, in a FB loop each element exerts an influence on the evolution of all elements in the loop, including itself. In a "positive loop" the elements are positively influenced, whereas in a "negative loop" the elements are negatively influenced. If a sign is assigned to these regulatory interactions – a plus sign (+) to a positive regulation and a minus sign (−) to a negative regulation – then the sign of a loop will be determined by the parity of the negative interactions. To maintain homeostasis, the presence of at least one negative loop is required, whereas to ensure multi-stationarity at least one positive loop is required. The power of those statements lies in their transversality, as they were demonstrated in many different contexts, from differential models to Boolean models. Consequently, in order to discover genes involved in switching differentiation processes it is necessary to examine genes involved in positive loops. In addition, the different dynamical behaviors associated with the two different topologies (positive and negative loops) provide a rational means of decomposing any complex regulatory networks. In other words, loops can be used as a set of building blocks to analyze a complex network, or to construct one with given properties.

On examining the transcriptional regulatory (TR) network of *Escherichia coli* metabolism, other topological properties can be found that might account for homeostasis, flexibility, and evolvability. In fact, the network can be coupled to the environment simply by adding to the network any nodes that represent the external metabolites [17]. The resultant network is an acyclic directed graph [18], organized in a hierarchical manner, with the external metabolites being the root and metabolic genes (as they have no outgoing links) being the leaves. Between these layers are located the transcription factors (TFs). On analyzing the dynamics, it becomes clear that: there are only fixed point attractors; the basin of attraction of each attractor is the entire state space; and that the attractors' configuration depend upon the environmental conditions [17]. On the other hand, the hierarchical acyclic structure accounts for homeostasis and the control is located in the root [19]. Therefore, the configuration of the root forces the configuration of the TFs which, in turn, determines the configuration of the leaves. Moreover, due to the acyclic structure, the fixed point is stable.

If the leaves are removed from the network, then the resulting graph will be strongly disconnected and organized in a modular manner. The modules are not directly connected but rather influence the common leaves, which accounts for the flexibility [17]. Finally, the same modular structures relate flexibility to evolvability. Due to the separation among modules, changes (e.g., resulting from a mutation) in a certain module will affect only the dynamical properties of this module but not the remainder of the network.

Consequently, the organism can explore new niches without affecting its homeostasis. The previous result appears to be in contrast with the assertion of Kauffman, that networks must remain close to the transition to chaos in order to evolve, whilst this network is found deep in the ordered behavior. This apparent discrepancy does not really exist, however, because in one model the external factors are directly involved, whereas the other results refer to autonomous systems. It can be said that the organization of the attractors is a function of the autonomous network, while the changes in the environmental factors unfreeze the frozen genes and move the network to a new attractor. This mechanism represent an alternative to the "edge of chaos" hypothesis proposed by Kauffman.

2.4 Biologically Relevant Boolean Rules

In real regulatory networks, not all of the Boolean rules have the same probability. Rather, it appears that there are biologically relevant rules that form a small subset of all theoretically possible functions, and that these are related to the robustness of the network. It was first noted that these relevant rules are canalizing functions [6]; the latter is a rule where an input alone can determine the output, while the others concur to determine the output only if the canalizing input is in the noncanalizing state. A further investigation [20] showed that the large majority of biologically relevant rules are not simply canalizing, but hierarchically are canalizing. In a hierarchical canalizing function, all inputs are essential in a hierarchical manner; that is, the second canalizing input is canalizing if the first one is in a noncanalizing status, and so on.

It has been shown that a Boolean network with particular subclasses of hierarchical canalizing functions demonstrates an ordered behavior; hence, these types of Boolean rules account for robustness [21]. On analyzing a measured yeast transcriptional network, Kauffman and colleagues showed that, for the ensemble of generated models, those networks with canalizing functions were remarkably stable, whereas those with arbitrary Boolean rules were only marginally stable [22]. This was an expected result, since real organisms live in a highly noisy environment, but must still preserve their state. Besides, canalizing functions are much realistically realizable than random functions. An example of this is the "exclusive or" rule, where the output is 1 if one of the inputs is 1, but is 0 if both inputs are 1. This is evidently unrealistic for a real regulatory interaction, and consequently Boolean networks with forcing rules will show an ordered behavior. This also occurs because forcing rules increase the probability of the formation of forcing structures; sub-circuits in which a canalized state will propagate to linked elements despite the initial condition [6, 22]. Thus, forcing structures favor stability and, perhaps also the emergence of fixed points.

The aforementioned properties suggest that it would be best to adopt not a completely random distribution in modeling a real network, but rather a distribution of hierarchical canalizing functions. However, some groups have provided an alternative meaning to the notion of biologically meaningful Boolean rules [23], which takes into account that not all biologically relevant functions are forcing and that not all of the canalizing functions are biologically realizable. The meaningful functions are selected taking into account the inhibitory or activatory roles of each input. Clearly, the majority of these functions are canalizing, while the dynamical effects are the same as if just canalizing functions were used. Finally, it is important to note that the rule sum is used in many dynamical simulations:

$$\sum_j J_{ij} S_j + h > 0 \rightarrow S_i(t+1) = 1$$
$$\sum_j J_{ij} S_j + h \leq 0 \rightarrow S_i(t+1) = 0 \qquad (1)$$

where J_{ij} is "+1" if the link represents a positive regulation, "−1" for inhibitory regulations, and h is a threshold. This rule accounts only for the inhibitory/activatory nature of the interactions.

2.5
Dynamical Simulation: An Example

The yeast cell cycle is one of the best known networks, and provides a good test for a Boolean model. Such a model has been proposed by Li *et al.* [24], who constructed a network of key known regulators. The network had 11 nodes representing protein states, plus an external signal and two types of link: positive and negative, where the latter represented inhibition, repression, or degradation. The protein states were updated at each time step according to the sum rule:

$$\sum_j J_{ij} S_j > 0 \rightarrow S_i(t+1) = 1$$
$$\sum_j J_{ij} S_j < 0 \rightarrow S_i(t+1) = 0$$
$$\sum_j J_{ij} S_j = 0 \rightarrow S_i(t+1) = S_i(t) \qquad (2)$$

where $J_{ij} = J_p$ for positive links, and $J_{ij} = J_n$ for negative links. The authors also added self-links that represented the degradation of nodes that were not negatively regulated by others nodes. This very simple model was based only on the wiring diagram inferred from qualitative experimental knowledge, and on the positive/negative nature of the interaction. The predictive power of this type of model does not lie in the accurate prediction of the expression dynamics, but rather on the attractors' picture – that is, the description of the network's states and their relations through dynamical transitions.

An analysis of the dynamics of the yeast cell-cycle network showed that all of the initial states flowed into one of seven fixed points. Strikingly, one of these fixed points attracted 86% of the initial configurations, and was remarkably related to the biological stationary state G_1 in which cell growth occurred. Further exciting the G_1 state by turning on the signal node induced the network to follow a path in the phase space, which ultimately returned again to G_1. This path was also related to a biological path, and also possessed a very stable trajectory, in the sense that any nonbiological state would converge to the biological path.

2.6
Boolean Networks Inference from Experimental Data: Probabilistic Boolean Networks

Today, microarrays continue to generate the huge amounts of data that are now available. The interest in inferring networks from real experimental data derives above all from a need to understand genetic regulation in specific organisms, in order perhaps to develop therapeutic interventions in diseases such as cancer or bacterial infections. Due to the great uncertainty of this type of data, however, and to a number of underlying latent factors, reverse engineering is a very difficult task.

During the past few years a vast amount of analytical and computational tools have been developed to infer Boolean networks from experimental data, with the choice among such tools depending on the type of data at hand and the goal of the model to be constructed [25, 26]. As noted above, within a Boolean network each gene state is determined by some other genes, by means of a Boolean function. Hence, the major task is to design Boolean functions (predictors in estimation theory) from the data. The general concept here is that, given a target gene Y, and a set of input genes X_1, X_2, ...,X_n, the optimal predictor of the target Y, based on the prediction variable X_1, ...,X_n is that with the minimal error according to some probabilistic error measure. In practice, however, given the expression level of Y, the optimal predictor is that which better predicts it given the expression levels of genes X_1, X_2, ...,X_n. The theoretically optimal predictor Y is unknown and must be estimated (for a detailed analysis, see Ref. [27]).

In practical cases, due to errors and external factors, is often necessary to settle just for Boolean rules that minimize the number of mis-classifications. In this case, an important role is played by the constraints that can be imposed, including a prior knowledge of the biologically relevant functions, such as the canalizing functions [26]. The main problem with this type of inference is the inherent determinism of Boolean rules when compared to the uncertainty shown by the data, such as: the stochastic nature of gene expression; the experimental noise; and possible interacting latent variables. Such a problem may lead to a Boolean function designed on the sample data, but which may be unable to make predictions when confronted with different conditions.

One natural approach to remedy this would be to include uncertainty in the model. Specifically, a number of simple functions (those that have just a few inputs) are inferred, such that each of them has a chance to contribute. This approach led to the formulation of Probabilistic Boolean Networks (PBNs) [24] whereby, starting from a given state, the network has a given probability to jump to some other states according to the Boolean rules, and their probabilities to contribute to the flow between different states. The main advantage of this approach relies on the fact that these networks are strictly related to the framework of Markov chains, so that a whole body of rigorous results can be used.

2.7 Addition of Noise

In order to mimic random perturbations in the framework of Boolean networks, a node is typically flipped arbitrarily to its opposite state. However, as this practice is quite unrealistic, an alternative approach – consisting of extending Boolean models to be time continuous and stochastic [7] – has recently been proposed. One way to achieve this would be to add an inner concentration dynamics inside each node, keeping the discrete (1 or 0) formulation for its relation with other nodes. The input signal of each node would drive the growth or decay of a concentration variable $c_i(t)$ assigned to each node. In addition, an explicit time delay t_d, which mimicked the transition time in a real extended system, could be added. Stochastic biochemical noise can be additionally incorporated by allowing the delay to fluctuate. In this way, by using the sum rule for the input signals, it is possible to write the following differential equation for the dynamics:

$$\sum_j J_{ij} S_j (t - t_d) + h_i \geq 0 \rightarrow \tau \frac{dc_i(t)}{dt} = 1 - c_i(t)$$
$$\sum_j J_{ij} S_j (t - t_d) + h_i < 0 \rightarrow \tau \frac{dc_i(t)}{dt} = 0 - c_i(t) \quad (3)$$

It should be noted that, by introducing a threshold rule, the binary output is recovered as:

$$c(t) \geq T \rightarrow S_i(t) = 1$$
$$c(t) < T \rightarrow S_i(t) = 0 \quad (4)$$

As noted above, noise can be added at the delay $t_d \rightarrow t_d + \chi_{ij}$, where χ_{ij} is a random variable assigned to each link J_{ij}. In this case, the nodes are no longer synchronously updated and, as a consequence, most of the attractors disappear, revealing their artificial nature due to synchronous updating. Finally, it should be noted that the noisy version of the yeast model presented previously is robust [28].

3 Continuous Dynamics Models

The potential of Boolean approaches in the context of genetic regulatory dynamics modeling lies in the versatility that the strong assumptions beneath the method confer to it. Not for nothing (in the Boolean formalism) can a gene be considered only in two states of ON or OFF, being any intermediate, quantitative possibility, and necessarily neglected. In the cases in which this hypothesis cannot be considered as a reasonable statement, certain important dynamical aspects of the systems are not captured by this type of Boolean method. Consequently, there is a need for more general modeling tools that could successfully describe a wider range of dynamical behaviors. Historically, the main tool used to describe chemical kinetics under genetic regulatory processes is none other than the ODE, on which attention will be focused in Sect. 3.1.

This type of formalism considers the concentrations of DNAs, RNAs, and proteins that are involved in a genetic circuitry as dynamical variables, the temporal and continuous evolutions of which can be defined by a rate equation that is essentially an ODE:

$$\dot{p} = f(\vec{p}) \quad (5)$$

In this type of equation, the variable p_i represents the concentration of the i-th product involved in the regulatory network: it is either a protein, a RNA, or any other substance. The components of the vector $\vec{p}$ represent the whole set of concentrations of products on which the temporal evolution of p_i depends. The rate function f is highly nonlinear.

Once the different mathematical possibilities that can be adopted (always using ODEs) to model genetic regulatory interactions have been characterized by essentially modifying the rate function, f, the next stage is to further analyze the dynamics of the network motifs. Motifs are small structures composed of a reduced number of genes (proteins; typically between two and four) that regulate each other. Although the global, genome-wide topological patterns of regulatory networks are highly complex [29, 30], when examining the small scale of these systems the network motifs appear as modular devices, the dynamics of which can be characterized (at least as a general exercise) as if they were isolated systems [31–33]. Complementary results derived either from dynamic modeling, experimental procedures, and/or bioinformatics techniques conducted during the past years have supported the former proposal. In this sense, a series of investigations has been undertaken to model the function of several network motifs by successfully comparing experimental results with numerical data derived from dynamical modeling [34–40]. These findings sustain the thesis that (at least in the cases studied) the dynamics of these small genetic circuits should be associated with simple information-processing tasks such as noise reduction, sequential programming, logic operations, bistable switching, and pulse generation. Based on this point of view, network motifs may consist of "functional bricks" of the biological regulatory networks, in a similar way that transistors, diodes, and amplifiers function in electronic circuits [41]. In addition to the above-mentioned findings, several studies have confirmed that the number of network motifs of each type present in biological networks is anything but random [42]. To date, all of the regulatory networks studied have been alike with regards to having similar motifs, when compared to adequately defined null models. These results suggest that the concept of motifs as elementary information-processing devices may have Darwinian consequences, in the form of a certain evolutive pressure that is responsible for the different levels of significance observed for different motifs. These types of question, some of which remain unanswered, are discussed in Sect. 3.2.

3.1 ODE Formalisms: From Biochemistry to Mathematics

In this section, the details of the main families of ODE-based formalisms to model genetic regulatory systems are revised. At this point, it may be beneficial to emphasize the conceptual difference that exists between the two typical families of models reported in the literature. As a first step (see Sect. 3.1.1),the first types of model (grouped as "biochemical background-based" models) are discussed, the main characteristic of which is their dependence on the precise biochemical mechanisms that eventually will drive the biomolecular processes. These mechanisms are translated into specific hypothesis that are contained in the equations. In turn, a rich variety of substantially different phenomenologies translates into very different dynamical behaviors. As will be discussed, the main problem regarding this type of model is that details of the underlying precise biochemistry are not always known, and consequently it is difficult to perform useful simulations. More precisely, not knowing the rate constants that appear in the equations will always create a bottleneck when comparing experimental data and numerical results derived from model predictions. It is also important to stress the value of distinguishing between the dynamical implications relating to the different types of regulatory interaction (such as TR interactions) and protein–protein (PP) interactions (such as phosphorylation and dephosphorylation).

In Sect. 3.1.2, some alternative approaches will be revised that, from a more empirical point of view, do not incorporate sufficient detailed biochemical knowledge to reproduce the qualitative dynamics of those regulatory systems that have been characterized experimentally.

3.1.1 Biochemical Background-Based Models

As noted above, the family of models described in this section treat the concentrations of substances involved as dynamical variables that evolve, according to the regulations that each substance might receive. The temporal evolution of these concentrations follows rate ODEs with the same general form of Eq. (5). Without abandoning this scheme, these models share the same spirit, namely, the idea of selecting the precise form of the rate functions f, depending on the precise biochemical mechanisms that drive the regulatory interactions. Having said that, it is also worth highlighting the differences between the two types of regulatory interaction that are better characterized in the literature, namely TR and PP interactions. As will be seen, the dynamics associated with these groups of regulatory interactions are not exactly equivalent. More importantly, these divergences are not always exhaustively taken into account [31, 32].

Dynamic Modeling of Transcriptional Regulatory Interactions In the most simple picture, a TF is a protein that is capable of recognizing precise sequences of DNA (i.e., promoters, or even better, target DNA regions within promoters) that are very close to the points where the transcription of regulated genes begins. By binding the DNA at these specific regions, the TF essentially modifies the chemical affinity of RNA polymerase (RNAp) to these transcription starting sites of the genes under regulation. Leaving aside the details of these catalytic mechanisms, there are two important points to start the analysis:

- The TF regulates the production of the substrate protein S encoded by the gene s.
- In order to do achieve this, the TF does not interact directly with protein S; hence, kinetically the regulation dynamics does not depend on the concentration of the regulated substrate S.

The simplest form to describe the influence of the TF, T, on protein S, is then:

$$\dot{S} = k_1 f(T) - k_2 S \tag{6}$$

In Eq. (6) the first term clearly refers to the regulatory effect of the TF T on variable S that, as noted previously, does not depend on S. In order to represent an activation (inhibition), $f(T)$ must be a monotonically increasing (decreasing) function in T. In turn, the second term corresponds to the usual degradation process of the protein [43, 44]. The rate constants k_1 and k_2 are always positive.

One candidate frequently found in the literature [45] is the so-called Hill function, which takes the form:

$$H^+(T,\theta,H) = \frac{T^H}{T^H + \theta^H} \quad \text{(Activation)} \tag{7a}$$

$$H^-(T,\theta,H) = \frac{\theta^H}{T^H + \theta^H} \quad \text{(Inhibition)} \tag{7b}$$

In its physical domain, both functions are always positive and bounded: $0 \leq H^+ < 1$ and $0 < H^- \leq 1$. In addition, the activation (inhibition) function is monotonically increasing (decreasing) with the concentration of the TF T. The sigmoidal shape of these curves has the mathematical properties desired to describe the regulatory effects in each case, is in good agreement with experimental results as it has been known for more than 30 years [46, 47], and also allows the difficult task of introducing the interaction TF–DNA–RNAp directly into the model to be avoided. In Eqs (7a) and (b), the parameter θ defines the half-maximal value, while the exponent H takes into account any eventual cooperative effects by defining the steepness of the sigmoid. Thus, the temporal evolution of S, as a function of the concentration T of a certain activator (inhibitor), can be expressed as:

$$\dot{S} = k_1 \frac{T^H}{T^H + \theta^H} - k_2 S \quad \text{(Activation)} \tag{8a}$$

$$\dot{S} = k_1 \frac{\theta^H}{T^H + \theta^H} - k_2 S \quad \text{(Inhibition)} \tag{8b}$$

Moreover, any delays due to transcription and translation processes can be taken into account by simply delaying the argument of Hill functions. The importance of such delaying effects has been characterized in detail [48, 49], though its relevance was found to be only relative [50].

Dynamic Modeling of Protein–Protein Regulatory Interactions With regards to protein functionality, on frequent occasions many proteins (in both prokaryotic and eukaryotic cells) must suffer post-translational modifications mediated by other enzymes, such that their configurations are modified and their functionalities altered. In this sense, the enzymes in charge of the catalysis of these modifications act, as a matter of fact, as protein regulators. Phosphorylation–dephosphorylation reactions constitute the paradigmatic example of this type of PP interaction. The regulatory role of these interactions has been characterized, occasionally, in great detail. For example, in budding yeast, the cyclin-dependent kinases (cdKs) are enzymes which, by phosphorylating on a series of substrates (usually referred to as executory proteins; EPs) either activate or inhibit them. As shown previously, the periodic profiles of activation for these EPs, ultimately regulated by cdKs, is closely related to the correct development of the cell cycle [51].

The paradigmatic description of this regulatory mechanism (by virtue of which a certain enzyme regulates the activity of a certain substrate) is a chemical sequence of two processes. First, the enzyme and the substrate must meet at a rate k_1 to form a transient complex, the concentration of which will be denoted as C. This complex must then be broken to release the intact enzyme and the modified substrate, for example at a rate of k_3. The concentrations of the enzyme, inactive substrate and active substrate are then denoted, respectively, as E,S, and S_a. In addition, a defective break of the complex can be observed that is incapable of causing the desired modification on S; this is produced at rate k_2, such that:

$$\begin{aligned} &S + E \xrightarrow{k_1} C \xrightarrow{k_3} S_a + E \\ &C \xrightarrow{k_2} S + E \end{aligned} \tag{9}$$

With regards to these chemical reactions, it is important to note that the regulatory role of the enzyme E (unlike the case discussed above, of the TF T acting on a certain DNA binding site) has two main factors: (i) it does not modify the total concentration of substrate ($S_T = S + S_a$, which remains conserved during the process); and (ii) it depends heavily on the concentration of the substrate. This scheme is none other than the Michaelis–Menten (MM) model, the corresponding rate equations of which are as follows:

$$\dot{S} = -k_1 SE + k_2 C \tag{10a}$$

$$\dot{C} = k_1 SE - (k_2 + k_3) C \tag{10b}$$

$$\dot{E} = -k_1 SE + (k_3 + k_2) C \tag{10c}$$

$$\dot{S}_a = k_3 C \tag{10d}$$

with the constraints $E + C = E_T$ (as the enzyme can be found either free or combined with the substrate) and $S + S_a + C = S_T$, as the substrate can, in turn, be found in its free form – either active or inactive – and also combined with the enzyme to form the transient complex, C. These relationships allow Eqs (10a) and (b) to be rewritten as an independent system, by substituting into them the following expressions:

$$E = E_T - C \tag{11a}$$

$$S = S_T - S_a - C \tag{11b}$$

which yields:

$$\dot{S} = -k_1 S (E_T - C) + k_2 C \tag{12a}$$

$$\dot{C} = k_1 S (E_T - C) - (k_2 + k_3) C \tag{12b}$$

The temporal evolution of the other two variables obey the expressions $\dot{E} = -\dot{C}$ and $\dot{S}_a = -\dot{C} - \dot{S}$. The standard approach to MM dynamics [52] is to assume that the complex achieves its dynamical equilibrium very quickly, after which the temporal variations on complex concentration can be neglected: $\dot{C} \approx 0$. If it is assumed that the latter situation holds, then the stationary concentration of complex C^* can be easily obtained from the homogeneous version of Eq. (12b), yielding:

$$C^* = \frac{E_T S^*}{S^* + K_M} \tag{13}$$

where $K_M = (k_2 + k_3)/k_1$is the so-called MM constant. Once this quasi-stationary state is reached for the complex, the temporal evolution of the fractions of substrates is obtained as:

$$\dot{S} = -\dot{S}_a = -\frac{k_3 E_T S^*}{S^* + K_M} \tag{14}$$

which does not vanish. That is, even at this quasi-stationary approach, both the stationary concentrations S^* and ultimately C^* are still dependent on time, even if the temporal derivative of C is small enough to be neglected in Eq. (12b).

There is, however, an important problem related to the MM approach, as explained here, when reducing the unpleasant dynamics described in system (Eq. 10) to the easier dynamics reduced ultimately to Eq. (14). If it is considered that the complex formation stage is very fast, then it is possible to set $\dot{C} \approx 0$ for almost all occasions. However, it is in fact being assumed that two times scales are involved in the problem: (i) an initial fast phase that persists until the quasi-stationary concentration of complex is reached; and (ii) a stage in which the evolution of substrate concentration can be approximated by Eq. (14), while the complex concentration behaves approximately according to Eq. (13). In this sense, to integrate Eq. (14) – which is only valid in the second phase – the initial conditions for S would be required at the start of the second phase.

The usual approach to this problem involves considering the total amount of substrate S_T as the initial condition; that is, to consider that the concentration of substrate S is not substantially modified during the transient phase. As the variation of S is due to the substrate–enzyme reaction to form the complex until quasi-stationarity is reached, it is assumed that, during complex formation, the "limiting reactive" is the enzyme which, in mathematical terms, is none other than $E_T \ll S_T$. This approximation (which is sufficient in many experimental cases) must be reconsidered if the aim is to simulate complex regulatory networks, though it may not be a reasonable hypothesis for all PP interactions in the network. In those cases where the concentrations of the enzyme (regulator) and substrate (target) are similar, the option of integrating (Eq. 14), using S_T as the initial condition, may not hold. A more detailed explanation of how to deal with this problem (when present) is presented in the classical text of Murray [52].

The two types of regulation mechanism – transcriptional and enzymatic – are clearly not unique possibilities. It is possible that the main example of protein functionality modification, which has not been directly mentioned here, is that due to ligand–protein binding. In this sense, it can be noted that certain proteins can bind with small ligands to form transient complexes, the functionality of which is different from the free protein case. Consequently, the ligand can be seen as the regulator and the protein as the substrate, the behavior of which has to be modified. The dynamics derived from this type of process can also be described using a MM approach that is very similar – and even simpler – to that described here. One very relevant case of proteins for which functionality can be regulated by the presence of small signaling ligand molecules, is that of the TFs themselves. The chemical affinity of a large quantity of TFs to their DNA consensus sequences depends on whether the TF is free, or is combined with a certain amount of ligand molecules. As noted above, the dynamics of these phenomena is very similar to MM enzymatic dynamics, and can be checked in the appendix of Alon's book [45].

3.1.2 Empirically Based Approaches

Quite often, the approaches detailed in the previous section are less useful than might be imagined, due to the fact that in many cases it is difficult to verify any of the hypotheses that underlie the models. Consequently, some comments are in order. On the one hand, all of these rate equations have been written based on the hypothesis that the proteins involved do not interact with any other substrate, which might therefore be omitted from the scheme. However, this situation is absolutely false; for instance, a single TF such as PhoP in *Mycobacterium tuberculosis* can regulate more than 100 genes [53, 54]. Hence, problems may emerge if it is considered that, either for the case of TR interactions or PP regulations, the regulators and substrates involved in regulatory systems are under the influence of a complex pattern of interactions with many more molecules. For example, cooperative [55] competition between regulators to bind the same target [56, 57], or the so-called "zero-order ultra-sensitivity" in PP interactions [58], will affect the dynamics of the regulatory systems, albeit in a nontrivial manner. Some of the considerations that can be made regarding these points, along with the dynamics of network motifs, are discussed in Sect. 3.2.

On the other hand, the essential issue regarding the validity of these models in each particular case lies in the difficulty of determining the rate constants of the equations in advance. Whilst an alternative (as in other areas) would be to assume a reversal engineering approach to determine the constants, this would mean being forced to adopt (ab initio) a certain dynamical model that would reflect a precise biochemical mechanism, and not others. This situation – besides the fact that multiple models can adequately describe the same experimental behaviors only by adequately tuning the rate constants – would force an admission that it is not sensible to seek a greater sophistication of the models according to more detailed and precise biochemical phenomenology, at least until the experimental determination of the individual rate constants and other biochemical parameters became possible.

One family of these alternative models is that of the piecewise-linear differential equations (PLDEs). In these PLDEs, the strict biochemical details of the regulatory mechanism are ignored; rather, attention is focused on determining quasi-empirically the range of regulator concentrations that drives the expression of a certain target at each of the experimentally observed levels. Mathematically, PLDEs have the following form:

$$\dot{S}_i = f(\vec{S}) - k_i S_i \tag{15}$$

where the components S_i of vector $\vec{S}$ are the concentrations of the substrates involved in the regulatory system (the set of these substrates is called Σ). In turn, while k_i is the usual degradation rate of the i-th substrate [43, 44], the function $f(\vec{S})$ is often defined as follows:

$$f_i(\vec{S}) = \sum_{j\in\Sigma} k_{ij} b_{ij}(\vec{S}) \tag{16}$$

where the functions $b_{ij}(\vec{S})$ can be defined as a combination of sums and multiplications of step functions in which the parameters θ_k define the thresholds:

$$\chi^{+}(S_j, \theta_k) = \begin{cases} 0, S_j < \theta_k \\ 1, S_j > \theta_k \end{cases} \quad \text{or}$$

$$\chi^{-}(S_j, \theta_k) = \begin{cases} 0, S_j > \theta_k \\ 1, S_j < \theta_k \end{cases} \tag{17}$$

Essentially, when the set of rate constants k_{ij} has been defined, the functions $b_{ij}(\vec{S})$ will control the regions of the phase space at which each rate constant (or, eventually, combinations of them) drives the production rate of each substrate S_i [59–61]. This method is closely related to Boolean approaches in the sense that the ultimate cause of variation in the substrates' production rates is reduced to step-like, discontinuous functions. Nevertheless, the temporal evolution of substrates concentrations is defined continuously by Eq. (15), this being an important difference between PLDE methods and Boolean (discrete, in general) approaches. Another relevant difference is that whereas in discrete models, it is necessary to determine only an adequate updating rule in order to define the possible transitions between system states, in the case of the PLDEs the parameters (more precisely the rate constants k_{ij}, and the functions $b_{ij}(\vec{S})$) must be determined in order that the model can be defined.

There are, however, other possibilities. In a recent review, Tyson and Novak [32] exploited the potential and versatility of another ODE-based approach that was not attached to any precise biochemical mechanism, and first proposed in Ref. [62]. According to this scheme, the rate equation associated with the temporal evolution of the concentration of the i-th substrate S_i may be defined as follows:

$$\dot{S}_i = k_i \left[f\left(\theta_i W_i(\vec{S})\right) - S_i \right] \tag{18}$$

where the influence of each substrate on the production rate S_i of is no longer carried by a certain combination of step functions, but by the sigmoid:

$$f\left(\theta_i W_i(\vec{S})\right) = \frac{1}{1 + e^{-\theta_i W_i(\vec{S})}} \tag{19}$$

that varies from 0 (in the limit $W_i(\vec{S}) \ll -1/\theta_i$) to 1 (in the limit $W_i(\vec{S}) \gg 1/\theta_i$). This parameter defines the value of the function at its inflection point, whereas the variable of the function $W_i(\vec{S})$ codifies the biochemical input to the rate equation as follows:

$$W_i(\vec{S}) = \omega_{io} + \sum_j \omega_{ij} S_j \quad (20)$$

In the latter expression, the coefficients ω_{ij} measure the influence of the j-th substrate on i-th concentration. The versatility of this type of approach has been demonstrated in Ref. [32], where the main dynamical features of most studied network motifs created to date were revised. All simulations in these studies were performed following such formalism; moreover, the main dynamical behaviors observed in network motifs, including noise reduction, logic processing of signals, pulse generation, oscillations, and bi-stability, were also reproduced.

The two families of empirically inspired models presented here suppose versatile tools in dynamical modeling that can be used to reproduce not always obvious motif dynamics [32], and have even been applied (after adequate refinements) to simulate complex physiological cell events such as sporulation beginning in *Bacillus subtilis* [63]. However, within the spirit of simplifying the mathematical treatment of the problem, the differences between TR and PP interactions were not taken into account. The importance of this fact, more than in any other question relative to the dynamical differences between these two types of regulation, was that, in their typical forms presented here, these models do not take into account the possibility that only a fraction of a regulated substrate could be activated (inhibited).

3.2 Summing Nodes and Links: From Math to Systems Biology

Having outlined the main biochemical and mathematical issues, and having discussed the most widely diffused theoretical approaches to single regulatory interactions modeling, the problem – rather than being solved – is in fact about to begin. The reason for this, paradigmatically, derives from the general nonlinearity of the biochemical processes under study, and the problem can be summarized as follows. Previously in this chapter, the main mathematical tools used to perform dynamical simulations of single regulatory interactions (one regulator, one target) have been reviewed. Yet, in cell biochemistry the odds of having a protein, the concentration of which depends only on one variable are null. In attempts to understand how structures that involve even small numbers of substrates and regulations (nodes and links) can give rise to nontrivial dynamical behaviors, the most widely applied technique over the past few years has been the exhaustive study of the dynamics of network motifs. As noted above, a network motif is a regulatory system composed of a low number of nodes (genes, proteins) and links (regulations), typically between two and six. In the following subsections, the dynamics of the four best-characterized motifs – FB loops, single input modules (SIMs), single output modules (SOMs), and feedforward loops (FFLs) – are reviewed.

3.2.1 Simple but Subtle Structures: SIMs and SOMs

Perhaps the two simplest structures that can be imagine in the context of small regulatory systems are SIMs and SOMs, the

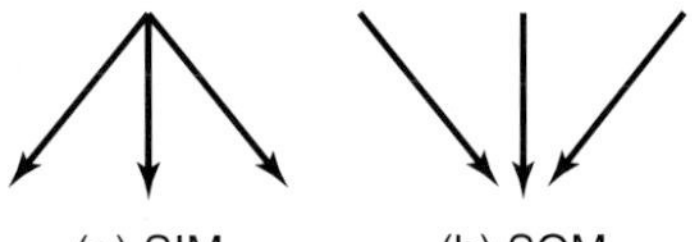

Fig. 1 (a) Single input modules (SIMs) are formed by a regulator responsible for the activity of many targets; (b) Single output modules (SOMs) consist of a single target, the expression of which depends on many regulators, either transcription factors (TFs) or regulating enzymes. The black color of all the regulations in the figure indicates generic-transcriptional or protein–protein-regulatory interactions; in subsequent figures, blue lines indicate protein–protein (PP) regulations, and red lines transcriptional regulatory (TR) interactions. The arrows represent activations, and right-angles represent inhibitions.

painless structures of which are shown in Fig 1. Despite their simplicity, these structures reproduce certain subtle dynamical properties that could pass unnoticed, but set out certain modeling issues that force the adoption of new hypotheses, in addition to the assumptions detailed in the preceding sections of the chapter. More specifically, in the case of SOM a problem arises that has not previously been encountered; namely, how can several interactions on a single substrate be modeled? Not surprisingly, multiple answers exist for this question.

Starting with TR interactions, a rich variety of behaviors can be found depending on the way in which the regulators interact with the target promoter [64, 65]. A first case would consist of two regulators that must interact with each other to effectively perform the final activation (inhibition) of target transcription. These necessary interactions can occur either before DNA-binding (the two regulators would be a factor–cofactor couple in this case) or after the independent binding of the regulators. For instance, when the DNA target regions of each single TF are close – but do not overlap – the RNAp affinity to the site can be regulated only when both TFs have been bound. This type of biochemical phenomenology drives an AND logic gate, in the sense that the presence of both regulators is required for the regulation to occur. When both regulators are activators, the response of this AND logic, three-nodes SOM, can be modeled as follows:

$$S = k_1 \frac{\left(T_1/\theta_1\right)^H}{1+\left(T_1/\theta_1\right)^H} \cdot \frac{\left(T_2/\theta_2\right)^H}{1+\left(T_2/\theta_2\right)^H} - k_2 S \tag{21}$$

where it has been assumed, for simplicity, that both regulators present a similar level of cooperation, accounted for by a common Hill exponent, H. The final term in the equation describes again a normal degradation process for the substrate.

There are, of course, many other possibilities. For example, when considering non-overlapping, independent DNA target regions for the two regulators, it is found that the activity of each regulator is independent, such that the RNAp activity on the promoter may be stimulated by any of the two factors, also independently. In this case, the device does not function as an AND gate, but as an OR gate. For this type of noncompetitive double activation – the OR logic process – the Hill terms are not multiplied but rather are summed, while the response of the device can be modeled as:

$$S = k_1 \frac{\left(T_1/\theta_1\right)^H}{1+\left(T_1/\theta_1\right)^H} + k_3 \frac{\left(T_2/\theta_2\right)^H}{1+\left(T_2/\theta_2\right)^H} - k_2 S \tag{22}$$

Finally, if the DNA target regions overlap, then a simultaneous binding of the two factors is physically impossible. This situation is referred to as "competitive binding," and although the device functions as a logical OR gate (by force, in this case), the mathematical modeling is slightly different. Again for double activation:

$$S = k_1 \frac{\left(T_1/\theta_1\right)^H}{1+\left(T_1/\theta_1\right)^H+\left(T_2/\theta_2\right)^H} + k_3 \frac{\left(T_2/\theta_2\right)^H}{1+\left(T_1/\theta_1\right)^H+\left(T_2/\theta_2\right)^H} - k_2 S \tag{23}$$

In the cases of both Eqs (22) and (23), it is sufficient to change the Hill terms in the numerator by unity to report for the eventual inhibitory activity of a regulator. It should be mentioned here that this mathematical approach offers, spontaneously, an adequate description of dual regulation. Dual TFs are regulators that act with different signs, depending on the presence of a second regulator in the cell. In this sense, by using a variant of Eq. (21) with two opposite multiplicands, this important phenomenology can be easily described. These types of modeling strategy, to reproduce different logical combinations of inputs of only one substrate, are used in Ref. [66]. Nevertheless, the virtually unbounded richness of biochemical possibilities [35, 49, 67] causes the number of possible models to be multiplied.

When dealing with PP interactions, the overview is no less simple. In the simplest case, of a three-node SOM in which there is TR interaction plus PP regulation, if both processes are activatory then the TF will enhance the total amount of substrate $S_T = S + S_a$, while a second regulator (now an enzyme) will activate the substrate ($S \rightarrow S_a$) via, for instance, a phosphorylation mechanism. On considering the dephosphorylation to be spontaneous, and to occur at a kinetic rate k_5, the only meaningful choice is that of a somehow generalized OR gate, that can be modeled as follows:

$$S_T = k_1 \frac{\left(T/\theta\right)^H}{1+\left(T/\theta\right)^H} - k_2 S \tag{24a}$$

$$\dot{S}_a = \frac{k_4\,(S_T - S_a)\,E}{K_{M4} + (S_T - S_a)} - k_5 S_a \tag{24b}$$

Several possibilities exist when considering two enzymes acting on a single substrate. In the first case, the conformational modifications may be independent, with two enzymes binding the substrate at different domains, reproducing again OR logic performances. However, the situation may also be temporally consecutive, in which case two consecutive reactions are necessary to reach a final, active, or inactive-state, the logic, which would be that of an AND gate. But the possibilities do not end here; the two enzymes might compete for a single, common binding site whereby cooperative competition introduces, again, an essential factor to be taken into account for modeling. All of these factors can be modeled by extending MM terms to each logical scheme, as performed previously for TR combinations. Definitively, the exhaustive mathematical modeling of all possible kinetic processes is beyond the objective of this chapter. Nevertheless, if the aim is to understand the typical dynamics of network motifs (even as simple as SOMs), then all of these considerations regarding the types of regulation should be carefully considered.

The dynamical modeling of the other structure depicted in Fig 1 – the SIM – is clearly easier. The interactions of a single regulator with more than one substrate are plausibly modeled like independent interactions, even when a regulator is responsible for hundreds of regulations, under the hypothesis that the total concentration of complexes (either TF–DNA complexes or PP in the case of enzymes) is negligible compared to the total amount of regulator in the cell. Even when assuming this highly simplifying approximation, experimental studies have shown that, on certain SIMs for which target genes codify substrates on a single metabolic pathway, the kinetic parameters are anything but contingent [68]. Consequently, this device operates as a type of "biochemical pipeline" on which each product appears exactly at the time it is required. The apparent simplicity of the structure hides a subtle, useful, and evolved device.

3.2.2 **Oscillators, Clocks, and Bistable Switches: FB Dynamics**

Followed the renowned investigations conducted by Elowitz and Leibler [69] in 2000 on negative FB dynamics, and by Gardner *et al.* [70] and Becksei *et al.* [71] on positive FB dynamics, discussions regarding FB loops and their rich behaviors as clocks, oscillators, and bistable switches have entered the fray on many occasions [31, 33, 42, 72]. A FB loop can be characterized topologically as a directed, closed loop of regulations. When the product of all the links involved is negative, the situation is referred to as a negative feedback (NFB) loop, whereas in the opposite case a positive feedback (PFB) loop is denoted. Starting with two-node FB loops, there exist two possible (topological) combinations that lead to PFBs: two proteins repressing each other; and two proteins activating each other. Yet, when differentiating between the different types of regulation, whether TR or PP, the number of possible combinations will be six.

With regards to their dynamical modeling, the first issue relates to the type of regulatory interaction that is being dealt with here. When considering two substances that interact with each other, this represents a TF that presents with two possible conformations – inactive and active. In addition, if an enzyme E is considered, the full activity of which arises spontaneously just after translation, no differentiation can be made between the enzyme's active and inactive forms. Initially, the transcription of enzyme E is activated by the presence of a certain TF S, acting as an external signal. The enzyme, in turn, catalyzes activation of the inactive fraction T via, for instance, a phosphorylation mechanism. To close the FB, the active form of the TF, T_a, enhances the expression of E. In that case, the active fraction of the TF, T_a, would coincide with the phosphorylated form, and the enzyme could be a kinase. Consider, for simplicity, the dephosphorylation of the active form T_a as a process that does not require any additional enzyme to occur. Finally, the total amount of TF ($T_T = T + T_a$) could be considered constant, although it certainly may depend on other signaling inputs or decay processes.

At this point, it should be noted that many other possibilities clearly exist, involving phosphates, proteins of which the active forms are not phosphorylated, and other multiple catalytic mechanisms that are not concerned with phosphate group

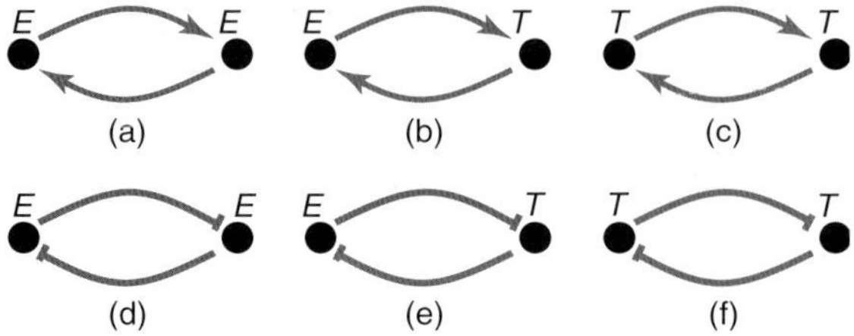

Fig. 2 (a–f) The six possible positive feedback (PFB) loops. Note the differences between the PP- and TR-motifs.

transfers. On examining Fig 2, the system previously described corresponds to Fig. 2(b). If the TR interactions are modeled by Hill terms and the PP interaction like a MM process, then the system can be described by the following system of rate equations:

$$\dot{E} = \frac{k_1 {}^{S}\!/_{\theta_1}}{1 + {}^{S}\!/_{\theta_1}} + \frac{k_2 {}^{T_a}\!/_{\theta_2}}{1 + {}^{T_a}\!/_{\theta_2}} - k_3 E \qquad (25a)$$

$$\dot{T}_a = \frac{k_4 (T_T - T_a) E}{K_{M4} + (T_T - T_a)} - k_5 T_a \qquad (25b)$$

where k_i are the mass action-derived rate constants of the processes, K_{Mi} are the Michaelis constants associated with the PP interactions, and θ_i are the parameters that define the half-maximum values of the Hill curves for TR terms. For simplicity, any cooperative effects on TR interactions have not been considered, and the MM mechanisms are standard; hence, no multisite phosphorylation processes or more complex dynamical effects have been taken into account. The binding of the TFs S and T_a on the enzyme's promoter defines a noncompetitive OR logic gate.

On the other hand, in order to describe a PFB composed by a mutual inhibition scheme, the modifications required to turn the system described in Eq. (25) (see Fig. 2(b)) into its mate (as represented in Fig. 2e) are twofold. First, T_a must inhibit the transcription of E rather than activate it. Second, if E is still to be considered as a kinase, then the active form of the TF (T_a) will no longer be the product of the reaction catalyzed by E, but rather the substrate. Hence, the effect of E consists of turning T_a on, as it is now an inhibitor. In other words, in this case the phosphorylated form is inactive, such that the noncatalyzing dephosphorylation may turn the TF active. In summarizing, the rate equations are:

$$\dot{E} = \frac{k_1 {}^{S}\!/_{\theta_1}}{1 + {}^{S}\!/_{\theta_1}} + \frac{k_2 {}^{T_a}\!/_{\theta_2}}{1 + {}^{T_a}\!/_{\theta_2}} - k_3 E \qquad (26a)$$

$$\dot{T}_a = \frac{k_4 T_a E}{K_{M4} + T_a} + k_5 (T_T - T_a) \qquad (26b)$$

For a closely similar approach to PFB modeling, see Ref. [31], whereas in Ref. [32] PFBs are also modeled using empirical models formally equivalent to Eq. (18). The essential dynamical features of these systems are discussed in these notable reviews, as well as in other many texts (see Ref. [33] for a general contextualization and Ref. [72] for a more specific treatment of bi-stability and FBs).

With regards to the other family of FB structures – the NFB loops – on examining two-node structures there is only one solution in terms of mathematical signing: the activator activates the inhibitor, which inhibits the former. By emphasizing again the ability to differentiate between possible combinations of the different regulations, only four of the six different possibilities found for PFBs are apparent, of which two are the main dynamical regimes that these types of structure offer. On the one hand, NFB loops can behave like homeostatic devices that are capable of preserving the stationary responses contained within narrow windows to wide ranges of signal concentration. This type of homeostatic regulation is most commonly employed in biosynthetic pathways.

On the other hand, NFBs can demonstrate sustained oscillations, for which mathematical requirements have been characterized in detail [73, 74]. Within the generic context of a system of two chemical species, x_1, x_2, with production rates depending on them through functions $\dot{x}_i = f(x_1, x_2)$, it has been proved that, in general, when the production rates are monotonically increasing with the concentrations (i.e., the more concentrated is a chemical compound, the higher the rate of all reactions it participates in), the trajectories are always bounded in phase space. That situation leads to limit cycle-sustained oscillations when the steady state of the system is unstable. For these two-component systems it has been proved [75] that at least three chemical reactions are necessary, one of these being autocatalytic (i.e., the production rate of one substance must depend on its own concentration) and involving at least three molecules.

Some of the biochemical conditions yielding oscillatory behaviors were enumerated by Elowitz and Leibler [69] by using a repressilator system; this was essentially a synthetic three-component NFB cycle in which each of the proteins repressed the next. The conditions identified were the presence of strong promoters, strong ribosome-binding sites, tight and cooperative repression, and similar mRNA and protein decay times. When the system adequately fulfilled these conditions, the stationary fixed point of its dynamics became unstable such that limit-cycle oscillations appeared. More recent investigations have been targeted at identifying the relationships that exist between the particular topology of the FB loops driving the processes, and the dynamical footprints associated with these oscillatory, experimental behaviors [76].

The NFB loop has been proposed as an architecture adapted to sustain oscillations in periodical series of biochemical events, such as cell cycles [37, 77, 78] and circadian rhythms [36, 79, 80]. In these studies (as reviewed by Tyson and Alon [31, 33]), the more sophisticated dynamics coupling of PFB and NFB loops was examined in greater detail. Beyond these considerations, it should be noted that certain questions regarding FB loops remain poorly studied, notably in the case of explicit dynamical implications with FBs of PP interactions rather than purely transcriptional structures. Nonetheless, the fact that purely transcriptional FBs are much more common in developmental TR networks than in sensory TR networks suggests that the dynamics of these structures may be more useful in the former situation [37].

3.2.3 FFLs: Noise Management and Pulse Generation

The final structure to be discussed is the FFL motif, which is composed of two regulators, R_1 and R_2, acting on a common substrate, S. One of these regulators, say R_1, regulates in turn the activity of the other regulator, R_2. Depending on the sign of the regulations, the FFLs have been subdivided into two groups [81], namely coherent FFLs and incoherent FFLs. In coherent FFLs, the sign of the indirect substrate regulation $R_1 \rightarrow R_2 \rightarrow S$ coincides with the sign of the direct regulation $R_1 \rightarrow S$, but for incoherent FFLs the signs of both paths are opposite. By adhering only to this sign criterion, there exist four different combinations of incoherent FFLs and four further possibilities for coherent motifs.

There is, however, another source of multiplicity, namely the possibility of dealing either with TR motifs or with PP interactions. By considering that both regulations due to R_1 are of the same type, the possibilities ascend to 32 diverse FFLs; by further considering the different logical implementations for the convergent regulations in S, the possibilities amount to 48 (a PP regulation and a TR regulation cannot define an AND gate). Finally, if self-inhibitions in the regulators are admitted, and different sensing combinations are considered (i.e., only R_1 can sense an external signal or instead both regulators can sense one signal each), then the number of combinations might reach 384.

The statistics of motifs significance has shown that FFLs are generally overrepresented [42]. In an attempt to illustrate the main cases of all possible FFLs, three noticeable examples we been selected, for different reasons: the so-called coherent type 1 FFL (Fig. 3 a); its cognate incoherent type 1 (Fig. 3 b); and an example of a non purely transcriptional FFL (Fig. 3 c).

The three interactions involved in coherent type 1 FFL are activations. So, for an AND gate implementation, and if considering both regulators as TF (denoting the respective concentrations as T_1 and T_2), its behavior can be modeled as:

$$\dot{T}_{1a} = k_1 \frac{S_o\,(T_{1T} - T_{1a})}{K_{Mo} + (T_{1T} - T_{1a})} - k_2 T_{1a} \quad (27a)$$

$$\dot{T}_2 = k_3 \frac{(T_{1a}/\theta_1)^H}{1 + (T_{1a}/\theta_1)^H} - k_4 T_2 \quad (27b)$$

$$\dot{S} = k_5 \frac{\left(T_{1a}/\theta_2\right)^H}{1 + \left(T_{1a}/\theta_2\right)^H} \cdot \frac{\left(T_2/\theta_3\right)^H}{1 + \left(T_2/\theta_3\right)^H} - k_6 S \quad (27c)$$

The dynamics driven by this scheme has, as its main features, noise-filtering abilities generated by sign-sensitive delays [81]. The scheme is simple: when the activation of T_1 via the signal S_o occurs, the transcription of T_2 starts. With regards to the substrate S, as its promoter activity is governed by an AND combination of both regulators, it is not immediately transcribed, as a certain minimum level for the concentration of T_2 must be reached. Only if the signal is persistent enough will this activation threshold in T_2 concentration be reached, and so will begin to express. If the signal suddenly shuts off, the activity of S will fall immediately as T_2 is also diluted rapidly after signal removal (see Fig 4).

For incoherent type 1 FFLs, only the regulation of S by T_2 is an inhibition. By combining again the two incoming regulations on S with an AND logic, only the rate equation for S will be changed to:

$$\dot{S} = k_5 \frac{\left(T_{1a}/\theta_2\right)^H}{1 + \left(T_{1a}/\theta_2\right)^H} \cdot \frac{1}{1 + \left(T_2/\theta_3\right)^H} - k_6 S \quad (28)$$

In this case, T_1 would act as an activator affected by the coinhibitor T_2 that, when present, would invert the

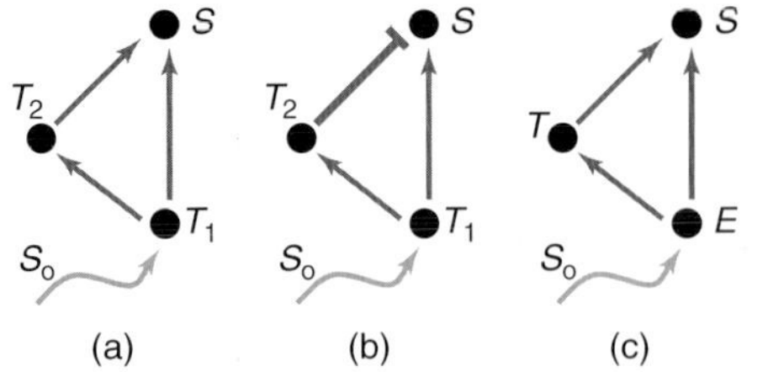

Fig. 3 (a–c) The three motifs receive an external input parametrized by the presence of a certain signaling enzymatic regulator, S_o.

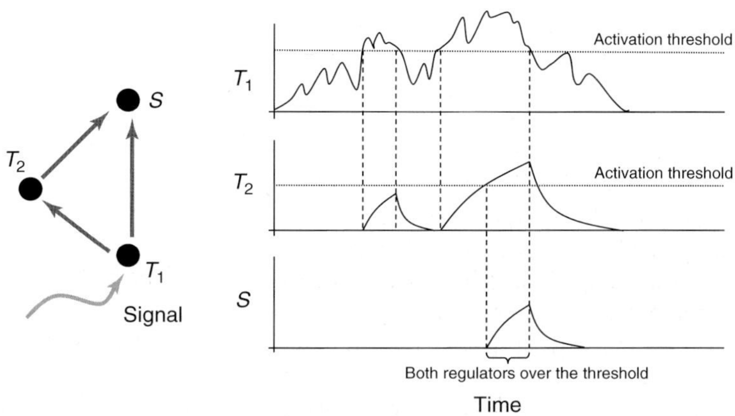

Fig. 4 The term-sensitive sign delay refers to the fact that the delay appears only as a consequence of a positive signal stimulus, but not after a negative change in signal concentration.

regulation sign from activation to inhibition. The dynamics of this system comprised two main characteristics: accelerating responses, and generating pulses. These dynamical features have been addressed experimentally in *E. coli* synthetic [39] and natural [40] regulatory systems. As highlighted in Ref. [33], the response-accelerating performance of this type of structure can be considered more important than that of negative self-inhibition, as the latter can only accelerate the production of its own regulator and genes within its operon, whereas FFLs can drive the acceleration of the activity at any operon.

Finally, mention should be made of a recent investigation conducted by Csikász-Nagy *et al.* [51], who studied cell-cycle regulation by the cyclin-dependent kinase cdK1 in budding yeast. In this case, it was established that the activity of many (more than was expected by random) of the proteins involved in cell-cycle regulation, the periodic expression of which was controlled by cdK1, were also controlled by TFs that followed a pattern that (irrespective of the signs) was precisely that of the FFL represented in Fig. 3c. Although the dynamical models proposed in Ref. [51] were simple, the studies were novel in terms of integrating TR data with PP interactions at a genome–proteome-wide level. As noted repeatedly above, the dynamics of these small modules is not always easy to anticipate, and studies of the dynamical behavior of mixed motifs may lead to more than one surprise!

3.3 Perspectives

This review of the main features of the Boolean network approach to modeling gene regulatory networks has not been exhaustive, with only selected topics having been presented. Of future interest would be an application of the Boolean network framework to the predictive modeling of real systems. For example, from a practical perspective it would be intriguing to determine whether the network could be forced to switch from one attractor to another, or the long-term behavior of the network's dynamics could be altered. In a disease such as cancer, for instance, the possibility of pushing tumor cells towards the apoptosis attractor would surely be of great therapeutic interest.

With the continuous approaches much caution must be applied, and the general proviso that what is true for purely TR motifs may not be always true for general regulatory structures should be stressed. Nonetheless, as has been emphasized above, there exist many more ways to regulate protein activities than transcriptional regulations, and their integration at a greater scale into a common framework is surely a task towards which much effort will be devoted in the near future.

In contrast, the motif approach to regulatory networks modeling relies on the fact that the true temporal profiles of protein concentrations may – in a significant proportion of cases – obey the dynamical performances of small regulatory motifs constituted only by the protein involved and a reduced number of regulators that affect it. Yet, the question remains as to whether motifs represent very general behaviors, or whether they represent scarce, cunningly chosen examples that are not susceptible to being generalized.

References

1 Boccaletti, S., Latora, V., Moreno, Y., Chavez, M., Hwang, D.-U. (2006) Complex networks: structure and dynamics. *Phys. Rep.*, **424**, 176.

2 Boccaletti, S., Latora, V., Moreno, Y. (Eds) (2009) *Handbook on Biological Networks*, Springer-Verlag, Singapore.

3 Kauffman, S.A. (1969) Metabolic stability and epigenesis in randomly constructed genetic nets. *J. Theor. Biol.*, **22**, 437–467.

4 Thomas, R. (1973) Boolean formalization of genetic control circuits. *J. Theor. Biol.*, **42**, 563–585.

5 Somogyi, R., Sniegoski, C. (1996) Modeling the complexity of genetic networks: understanding multigenic and pleiotropic regulation. *Complexity*, **1**, 45–63.

6 Kauffman, S.A. (1993) *The Origins of Order: Self-Organization and Selection in Evolution*, Oxford University Press, New York.

7 Bornholdt, S. (2008) Boolean network models of cellular regulation: Prospects and limitations. *J. R. Soc. Interface*, **5**, S85–S94.

8 Shmulevich, I., Dougherty, E.R., Zhang, W. (2002) From Boolean to probabilistic Boolean networks as model of genetic regulatory networks. *Proc. IEEE*, **90** (11), 1778–1792.

9 de Jong, H. (2002) Modeling and simulation of genetic regulatory systems: a literature review. *J. Comput. Biol.*, **9** (1), 67–103.

10 Moreno, O., Bollman, D., Aviño-Diaz, M. (2002) Finite dynamical systems, linear automata, and finite fields. Proceedings of the WSEAS International Conference on System Science, Applied Mathematics and Computer Science, and Power Engineering Systems, Copacabana, Rio de Janeiro, Brazil, pp. 1481–1483.

11 Kauffman, S.A. (1969) Homeostasis and differentiation in random genetic control network. *Nature*, **224**, 177–178.

12 Huang, S. (1999) Gene expression profiling, genetic networks, and cellular states: an integrating concept for tumorigenesis and drug discovery. *J. Mol. Med.*, **77**, 469–480.

13 Delbruck, M. (1949) Génétique du Bactériophage, in: *Unités Biologiques Dousées Continuité Génétique*, Colloques Internationaux du Centre National de la Recherche Scientifique.

14 Thomas, R. (1981) *On the Relation Between the Logical Structure of Systems and their Ability to Generate Multiple Steady States and Sustained Oscillations*, Synergetics, Springer, Berlin, Vol. **9**, pp. 180–193.

15 Plathe, E., Mestl, T., Omholt, S.W. (1995) Feedback loops, stability and multistationarity in dynamical systems. *J. Biol. Syst.*, **3**, 569–577.

16 Remy, E., Ruet, P., Thieffry, D. (2008) Graphics requirement for multistability and attractive cycles in a Boolean dynamical framework. *Adv. Appl. Math.*, **41**, 335–350.

17 Samal, A., Jain, S. (2008) The regulatory network of *E. coli* metabolism as a Boolean dynamical system exhibits both homeostasis

and flexibility of response. *BMC Syst. Biol.*, **2**, 21.

18 Shen-Orr, S., Milo, R., Mangan, S., Alon, U. (2002) Network motifs in the transcriptional regulation network of *E. coli*. *Nat. Genet.*, **31**, 64–68.

19 Balazsi, G., Barabasi, A.L., Oltvai, Z.N. (2005) Topological units of environmental signal processing in the transcriptional regulatory network of *Escherichia coli*. *Proc. Natl Acad. Sci. USA*, **102** (22), 7841–7846.

20 Harris, S., Sawhill, B., Wuensche, A., Kauffman, S. (2002) A model of transcriptional regulatory networks based on biases in the observed regulation rules. *Complexity*, **7** (4), 23–40.

21 Nikolajewa, S., Friedel, M., Wilhelm, T. (2007) Boolean networks with biologically relevant rules show ordered behavior. *BioSystems*, **90**, 40–47.

22 Kauffman, S., Peterson, C., Samuelsson, B., Troein, C. (2003) Random Boolean network models and yeast transcriptional network. *Proc. Natl Acad. Sci. USA*, **100**, 147962–114799.

23 Raeymaekers, L. (2002) Dynamics of Boolean networks controlled by biologically meaningful functions. *J. Theor. Biol.*, **218**, 331–341.

24 Li, F., Long, T., Lu, Y., Ouyang, Q., Tang, C. (2004) The yeast cell-cycle network is robustly designed. *Proc. Natl Acad. Sci. USA*, **101**, 4781–4786.

25 Takar, J., Albert, R. (2009) Boolean Networks in Inference and Dynamic Modeling of Biological Systems at the Molecular and Physiological Level, in: Boccaletti, S., Latora, V., Moreno, Y. (Eds) *Handbook on Biological Networks*, World Scientific, Singapore, Vol. **10**, pp. 59–78.

26 Shmulevich, I., Dougherty, E.R., Kim, S., Zhang, W. (2002) Probabilistic Boolean networks: a rule-based uncertainty model for gene regulatory networks. *Bioinformatics*, **18** (2), 261–274.

27 Daugherty, E.R., Bittner, M.L., Chen, Y., Sivakumar, K., Meltzer, P., Trent, J.M. (2000) General nonlinear framework for the analysis of gene interaction via multivariate expression arrays. *J. Biomed. Opt.*, **5** (4), 411–424.

28 Braunewell, S., Bornholdt, S. (2006) Superstability of the yeast cell-cycle dynamics: ensuring causality in the presence of biochemical stochasticity. *J. Theor. Biol.*, **245**, 638–643.

29 Wheng, S., Bhalla, U.S., Lyengar, R. (1999) Complexity in biological signaling systems. *Science*, **284** (5411), 92–96.

30 Babu, M.M., Luscombe, N.M., Aravind, L., Gerstein, M., Teichmann, S.A. (2004) Structure and function of transcriptional regulatory networks. *Curr. Opin. Struct. Biol.*, **14** (3), 283–291.

31 Tyson, J.J., Chen, K.C., Novak, B. (2003) Sniffers, buzzers, toggles and blinkers: dynamics of regulatory pathways in the cell. *Curr. Opin. Cell Biol.*, **15**, 221–231.

32 Tyson, J.J., Novak, B. (2010) Functional motifs in biochemical reaction networks. *Annu. Rev. Phys. Chem.*, **61**, 219–240.

33 Alon, U. (2007) Network motifs: theory and experimental approaches. *Nat. Rev. Genet.*, **8**, 450–461.

34 Ferrell, J.E., Machleder, E.M. (1998) The biochemical basis of an all-or-none cell fate switch in *Xenopus* oocytes. *Science*, **280**, 895–898.

35 Ptashne, M., Gann, A. (2002) *Genes & Signals*, Cold Spring Harbor Laboratory Press, Cold Spring Harbor.

36 Zeilinger, M.N., Farré, E.M., Taylor, S.R., Kay, S.A., Doyle, F.J. (2006) A novel computational model of the circadian clock in *Arabidopsis* that incorporates PRR7 and PRR9. *Mol. Syst. Biol.*, **2**, 58.

37 Pomerening, J.R., Sontag, E.D., Ferrell, J.E. Jr (2003) Building a cell cycle oscillator: hysteresis and bistability in the activator of Cdc2. *Nat. Cell Biol.*, **5**, 346–351.

38 Mangan, S., Zaslaver, A., Alon, U. (2003) The coherent feedforward loop serves as a sign-sensitive delay element in transcription networks. *J. Mol. Biol.*, **334**, 197–204.

39 Basu, S., Mehreja, R., Thiberge, S., Chen, M.T., Weiss, R. (2004) Spatiotemporal control of gene expression with pulse-generating networks. *Proc. Natl Acad. Sci. USA*, **101**, 6355–6360.

40 Mangan, S., Zaslaver, A., Alon, U. (2006) The incoherent feed-forward loop accelerates the response-time of the gal system of *Escherichia coli*. *J. Mol. Biol.*, **356**, 1073–1081.

41 Milo, R., Shen-Orr, S., Itzkovitz, S., Kashtan, N., Chklovskii, D., Alon, U. (2002) Network motifs: simple building blocks of complex networks. *Science*, **298**, 824–827.

42 Milo, R., Itzkovitz, S., Kashtan, N., Levitt, R., Shen-Orr, S., Ayzenshtat, V., Sheffer, M., Alon, U. (2004) Superfamilies of designed and evolved networks. *Science* **303**, 1538–1542.

43 Glass, L. (1975) Classification of biological networks by their qualitative dynamics. *J. Theor. Biol.*, **54**, 85–107.

44 Mestl, T., Plahte, E., Omholt, S.W. (1995) A mathematical framework for describing and analysing gene regulatory networks. *J. Theor. Biol.*, **176**, 291–300.

45 Alon, U. (2007) *An Introduction to Systems Biology: Design Principles of Biological Circuits*, Chapman & Hall, Boca Raton, Florida.

46 Yagil, G., Yagil, E. (1971) On the relation between effector concentration and the rate of induced enzyme synthesis. *Biophys. J.*, **11**, 11–27.

47 Yagil, G. (1975) Quantitative Aspects of Protein Induction, in: Horecker, B.L., Stadtman, E.R. (Eds) *Current Topics in Cell Regulation*, Academic Press, New York, pp. 183–237.

48 Lin, E.C.C., Lynch, A.S. (1996) *Regulation of Gene Expression in Escherichia coli*, R.G. Landes Company, Austin.

49 Wagner, R. (2000) *Transcription Regulation in Prokaryotes*, Oxford University Press, Oxford.

50 Rosenfeld, N., Elowitz, M.B., Alon, U. (2002) Negative autoregulation speeds the response times of transcription networks. *J. Mol. Biol.*, **323**, 785–793.

51 Csikász-Nagy, A., Kapuy, O., Tóth, A., Pál, C., Jensen, L.J., Uhlmann, F., Tyson, J.J. (2009) Cell cycle regulation by feed-forward loops coupling transcription and phosphorylation. *Mol. Syst. Biol.*, **5**, 236.

52 Murray, J.D. (2002) *Mathematical Biology, An Introduction*, 3rd edn, Springer, Berlin.

53 Asensio, J.G., Maia, C., Ferrer, N.L., Barilone, N., Laval, F., Soto, C.Y., Winter, N., Daffé, M., Gicquel, B., Martín, C., Jackson, M. (2006) The virulence associated two-component PhoP-PhoR system controls the biosynthesis of polyketide-derived lipids in *Mycobacterium tuberculosis*. *J. Biol. Chem.*, **281** (3), 1313–1316.

54 Walters, S.B., Dubnau, E., Kolesnikova, I., Laval, F., Daffé, M., Smith, I. (2006) The *Mycobacterium tuberculosis* PhoPR two-component system regulates genes essential for virulence and complex lipid biosynthesis. *Mol. Microbiol.*, **60** (2), 312–330.

55 Thomas, R., Kauffman, M. (2001) Multistationarity, the basis of cell differentiation and memory. *Chaos*, **11**, 170–179.

56 Segel, L. (1984) *Modeling Dynamic Phenomena in Molecular and Cellular Biology*, Cambridge University Press, Cambridge.

57 Goldbeter, A. (1997) *Biochemical Oscillations and Cellular Rhythms*, Cambridge University Press, Cambridge.

58 Iglesias, P.A., Levchenko, A. (2002) Modeling the cell's guidance system. *Sci. STKE*, **2002** (148), RE12.

59 Glass, L. (1977) Global Analysis of Nonlinear Chemical Kinetics, in: Berne, B. (Ed.) *Statistical Mechanics, Part B: Time Dependent Processes*, Plenum Press, New York, pp. 311–349.

60 Plahte, E., Mestl, T., Omholt, S.W. (1998) A methodological basis for description and analysis of systems with complex switch-like interactions. *J. Math. Biol.*, **36**, 321–348.

61 Snoussi, E.H. (1989) Qualitative dynamics of piecewise-linear differential equations: A discrete mapping approach. *Dyn. Stab. Syst.*, **4** (3-4), 189–207.

62 Mjolsness, E., Sharp, D.H., Reinitz, J. (1991) A connectionist model of development. *J. Theor. Biol.*, **152**, 429–453.

63 de Jong, H., Page, M., Hernandez, C., Geiselmann, J. (2001) Qualitative simulation of genetic regulatory networks: method and application, in: Nebel, B. (Ed.) *Proceedings of the 17th International Joint Conference on Artificial Intelligence (IJCAI-01)*, Morgan Kaufmann, San Mateo, California, pp. 67–73.

64 Yuh, C.H., Bolouri, H., Davidson, E.H. (1998) Genomic cis-regulatory logic: experimental and computational analysis of a sea urchin gene. *Science*, **279**, 1896–1902.

65 Buchler, N.E., Gerland, U., Hwa, T. (2003) On schemes of combinatorial transcription logic. *Proc. Natl Acad. Sci. USA*, **100**, 5136–5141.

66 Mangan, S., Alon, U. (2003) Structure and function of the feed-forward loop network motif. *Proc. Natl Acad. Sci. USA*, **100** (21), 11980–11985.

67 Latchman, D.S. (1990) Eukaryotic transcription factors. *Biochem. J.*, **270**, 281–289.

68 Zaslaver, A., Mayo, A.E., Rosenberg, R., Bashkin, P., Sberro, H., Tsalyuk, M.,

Surrette, M.G., Alon, U. (2004) Just-in-time transcription program in metabolic pathways. *Nat. Genet.*, **36** (5), 486–491.

69 Elowitz, M., Leibler, S. (2000) A synthetic oscillatory network of transcriptional regulators. *Nature*, **403**, 335–338.

70 Gardner, T.S., Cantor, C.R., Collins, J.J. (2000) Construction of a genetic toggle switch in *Escherichia coli*. *Nature*, **403**, 339–342.

71 Becskei, A., Seraphin, B., Serrano, L. (2001) Positive feedback in eukaryotic gene networks: cell differentiation by graded to binary response conversion. *EMBO J.*, **20**, 2528–2535.

72 Ferrell, J.E. Jr (2002) Self perpetuating states in signal transduction: positive feedback, double negative feedback and bistability. *Curr. Opin. Chem. Biol.*, **6**, 140–148.

73 Mallet-Paret, J., Smith, H.L. (1989) The Poincare–Bendixson theorem for monotone cyclic feedback systems. *J. Dyn. Differ. Equations*, **2** (4), 367–421.

74 Edelstein-Keshet, L. (2005) *Mathematical Models in Biology*, Classics in Applied Mathematics, Society for Industrial and Applied Mathematics.

75 Schnakenberg, J. (1979) Simple chemical reaction systems with limit cycle behavior. *J. Theor. Biol.*, **81**, 389–400.

76 Pigolotti, S., Krishna, S., Jensen, M.H. (2007) Oscillation patterns in negative feedback loops. *Proc. Natl Acad. Sci. USA*, **104** (16), 6533–6537.

77 Goldbeter, A. (1991) A minimal cascade model for the mitotic oscillator involving cyclin and cdc2 kinase. *Proc. Natl Acad. Sci. USA*, **88**, 9107–9111.

78 Kholodenko, B.N. (2000) Negative feedback and ultrasensitivity can bring about oscillations in the mitogen-activated protein kinase cascades. *Eur. J. Biochem.*, **267**, 1583–1588.

79 Goldbeter, A. (1995) A model for circadian oscillations in the *Drosophila* period protein (PER). *Proc. R. Soc. Lond. Ser. B.*, **261**, 319–324.

80 Leloup, J.C., Goldbeter, A. (2000) Modeling the molecular regulatory mechanism of circadian rhythms in *Drosophila*. *BioEssays*, **22**, 84–93.

81 Goentoro, L., Shoval, O., Kirschner, M.W., Alon, U. (2009) The incoherent feedforward loop can provide fold-change detection in gene regulation. *Mol. Cell*, **36**, 894–899.

12
E-Cell: Computer Simulation of the Cell

Pawan K. Dhar[1], *Kouichi Takahashi*[2], *Yoichi Nakayama*[2], *and Masaru Tomita*[2]
[1]*Symbiosis International University, School of Biomedical Sciences, Pune 411 042, India*
[2]*Keio University, Institute for Advanced Biosciences, Tsuruoka, Yamagata 997-0035, Japan*

Systems Biology: Advances in Molecular Biology and Medicine, First Edition. Edited by Robert A. Meyers.

Keywords

E-cell
Short form for electronic cell: a tool for modeling and simulating cellular pathways.

Self-supporting cell
A virtual cell with 127 genes sufficient for survival.

***In-silico* modeling**
A model building process using computers.

Mycoplasma genitalium
A microorganism with the smallest number of genes.

Virtual erythrocyte
An *in-silico* model of whole erythrocytic metabolism.

Modeling
Abstracting and connecting the elements of a system.

Simulation
A process of visualizing a model dynamically.

Cells are massively parallel and massively interactive systems. The grand challenge is to understand their structural and functional design and to use the knowledge acquired to build useful applications. In traditional settings, it was difficult to focus on more than one gene at a time, but recently developed high-throughput technologies have enabled studies to be conducted at the whole organism level. Nevertheless, data from these experiments are often noisy and require a large number of replicates to validate even a single observation. Furthermore, the statistical treatment of high-throughput data is also error-prone. To overcome the physical and conceptual limitations, there is a need to develop strategies and tools to address complex biological problems. Systems Biology studies conducted during the past decade have lent credibility to an *in-silico* approach for understanding and engineering whole-cell systems. The E-Cell platform has been specifically designed to address network-based problems. The E Cell has been used successfully to create a self-sustaining cell with 127 genes – that is, just stable enough for survival. In this chapter, some of the basic modeling concepts, their importance, the role of E-Cell, and the future challenges of the modeling community, will be discussed.

1 Introduction

A model is a representation of a system, and reflects a combination of hypotheses, evidences, and abstractions. Models are the closest replicas of actual phenomena with diagnostic and predictive abilities. Models should be easily understood, controllable, and analyzable for large and complex data. The aim of whole-cell modeling is to provide both a conceptual basis and a working methodology for studying the cell in its entirety, to replicate the known knowledge, identify unknown entities, make predictions, and to design experiments to address unanswered questions.

A cell by itself is a complete genetic and biochemical reactor holding all the information necessary to sustain life. It offers an ideal middle path between (extreme ends of) atomic interactions and whole organs. By creating a whole-cell model, it is theoretically possible to stretch out data and hypothesis in either direction. Experimental biology has now reached a stage where data analysis and interpretation are heavily dependent on *in-silico* approaches. Although the static representation of data has traditionally helped develop an overall perspective, dynamic modeling aids in better understanding of the cellular decisions (Table 1).

Broadly, whole-cell transactions may be classified into enzymatic and nonenzymatic processes. Enzymatic processes represent most of the metabolic events, while nonenzymatic processes represent gene expression and regulation, signal transduction, and diffusion. In order to create a complete virtual cell, it is important to have provisions for DNA replication and repair, transcription and its regulation, translation, energy metabolism, metabolism, cell division, signaling pathways, cell membrane dynamics (ion channels, pump, nutrients), and intracellular molecular trafficking with appropriate mathematical representations (Table 2).

Tab. 1 Basics of map construction.

Type of representation	*Description*
Linear chain	Unidirectional flow
Branched chain	Two enzymes participate in one reaction, resulting in different products
Loops	Two branches unite, forming an inherent dependency between them
Cycles	Larger loops comprising many intermediates

To accomplish the "big picture," the data should not only be of good quality but also should be treated with intuitive mathematical representations that accurately describe life *in vivo*. It is noteworthy to mention that good data is more of an exception than the rule! For modeling metabolic pathways, the data input typically consists of rate constants and concentrations. A metabolic pathway usually consists of forward and reverse reactions (uni-, bi-, ter-) of ordered/random types. The inhibitors may be intermediate compounds of the same pathway or external entities. The availability of good data makes the modeling process more or less a straightforward process, but often missing links must be identified due to incompleteness of information. Problems in doing so arise mostly due to numerical reasons – stiffness and parameter sensitivity. The main difference between data-to-model and model-to-data approaches is that, in the former case, the starting materials are substrate, enzyme, and modifier concentrations, while in

Tab. 2 Mathematical representation of cellular processes.

Process	*Dominant phenomenon*	*Typical computational scheme(s)*
Metabolism	Enzymatic reaction	DAE, S-Systems, FBA
Signal transduction	Molecular binding	DAE, stochastic, diffusion reaction
Gene expression	Molecular binding, polymerization, degradation	OOM, S-Systems, DAE, Boolean, stochastic, Bayesian, rule based
DNA replication	Molecular binding, polymerization	OOM, DAE
Cytoskeletal	Polymerization, depolymerization	DAE, particle dynamics
Cytoplasmic streaming	Streaming	Rheology, finite element method
Membrane transport	Osmotic pressure, membrane potential	DAE, electrophysiology

DAE: Differential algebraic equations; FBA: Flux balance analysis; OOM: Object-Oriented Modeling.

the latter case the kinetic constants and reaction velocities are assumed. However, the difference between these two approaches sometimes blurs, because in real-life situations modeling often involves manual data-fitting approaches to match an expected output or hypothesis. The major advantage of carrying out simulations is not only to study the system *per se*, but also to extrapolate its behavior, in the presence of a hypothetical condition – for example, a cell with many essential gene knockouts. In addition to demystifying nonintuitive phenomena, simulation allows the testing of experimentally unfeasible scenarios and reduces experimental costs. Although wet experiments are indispensable for the advancement of biological knowledge, *in-silico* modeling can help to shorten knowledge discovery. With the enormous computational power easily available today, the challenging part in modeling is more conceptual than physical.

An overview of modeling tools, online resources, and databases is provided in the next section.

2
Biological Modeling and Simulation Tools

A number of promising tools are available for studying gene expression, regulation, and metabolic pathways (Tables 3 and 4). Listed below are the partial descriptions of a few such tools.

DBsolve: URL: http://websites.ntl.com/~igor.goryanin/. DBsolve is an integrated development environment for metabolic, enzymatic, and receptor–ligand binding simulation. It is an ordinary differential equation (ODE)-based tool that also incorporates the stoichiometry of chemical reactions. The main strength of DBsolve is the calculation of steady state, fitting, and optimization options.

Gepasi: URL: http://www.gepasi.org/. Gepasi simulates the steady-state and time-course behavior of reactions over time and space, based on stoichiometry and reaction kinetics values. The program is based on ODEs. It is a very useful tool for conducting metabolic control analysis and linear kinetic stability analysis leading

Tab. 3 Cellular databases and pathways.

Name URL	*Features*
1. General Online Maps and Pathways	
IUBMB-Nicholson Minimaps http://www.tcd.ie/Biochemistry/IUBMB-Nicholson/	Comprehensive; describes regulatory and spatial features of substrates and enzymes; available in .gif, .svg, and .pdf forms
Boehringer Mannheim biochemical pathways http://www.expasy.ch/cgi-bin/search-biochem-index	Comprehensive; covers many organisms, most extensively used by researchers, available in online and paper formats
Kyoto Encyclopedia of Genes and Genomes (KEGG) http://www.genome.ad.jp/kegg/	Huge database on gene sequence, regulatory pathways, metabolism, molecular assemblies, and so on
What Is There (WIT) http://wit.mcs.anl.gov/WIT2/	Covers metabolic pathways of over 25 organisms
Enzyme and Metabolic Pathway (EMP) http://emp.mcs.anl.gov/	Includes metabolic pathways, reaction mechanisms, rate laws, and numeric data from research reports
Biopathways Consortium http://www.biopathways.org/	An open forum for developing technologies and standards for biopathways
EcoCyc http://ecocyc.org/	Houses all the *Escherichia coli* pathways with an aim of creating its functional catalog
PathDB	Plant metabolic database management system; runs on client server architecture
http://www.ncgr.org/pathdb/	
http://umbbd.ahc.umn.edu/	
METAVISTA http://www.metabolic-explorer.com	Resource for proteomic profiling, metabolic profiling, and metabolic flux analysis
2. Regulatory pathways	
KEGG regulatory pathways http://www.genome.ad.jp/kegg/regulation.html	An extension of KEGG database
BioCarta http://www.biocarta.com/	Interactive web-based resource on gene function and proteomics
Biomolecular Interaction Network Database (BIND) http://www.bind.ca/index.phtml	Describes chemical reactions, conformational changes, and protein and network interactions across various species.
Signal Pathway Database (SPAD) http://www.grt.kyushu-u.ac.jp/spad/	Signal transduction database with emphasis on protein–protein and protein–DNA interactions
Cell Signaling Networks Database (CSNDB) http://geo.nihs.go.jp/csndb/	Contains sequences, structures, functions, and reactions involved in cell signaling

(*continued overleaf*)

Tab. 3 (*continued*)

Name URL	***Features***
Munich Information Centre for Protein Sequences (MIPS) http://mips.gsf.de/proj/yeast/CYGD/db/index.html	Functional yeast genomic database
GeNet – Gene Networks Database http://www.csa.ru/Inst/gorb_dep/inbios/genet/genet.htm	Developmentally regulated gene networks
EmbryoNet http://www.csa.ru/Inst/gorb_dep/inbios/genet/embryo.htm	Developmentally regulated genetic networks
Genetic network maps http://www.csa.ru/Inst/gorb_dep/inbios/genet/access.htm	*Drosophila* embryogenesis network
Wnt signaling pathwayhttp://www.stanford.edu/~rnusse/wntwindow.html	*Drosophila* developmentally regulated Wnt signaling pathways
3. Transcription factors and expression	
TRANSPATH http://transpath.gbf.de/	Describes pathways involved in regulation of transcription factors
TRANSFAC http://transfac.gbf.de/TRANSFAC/index.html	Transcription factor database
RegulonDB http://www.cifn.unam.mx/Computational_Genomics/regulondb/	Transcription regulation and operon organization database
DBTBS http://elmo.ims.u-tokyo.ac.jp/dbtbs/	*Bacillus subtilis* promoter and transcription factor database
Saccharomyces cerevisiae Promoter Database (SCPD) http://cgsigma.cshl.org/jian/	Database on promoters and mapped regulatory regions of yeast
Axeldb http://www.dkfz-heidelberg.de/abt0135/axeldb.htm	Gene expression database of *Xenopus laevis*
NEXTDB http://nematode.lab.nig.ac.jp/	*Caenorhabditis elegans* expression database
MAGEST http://www.genome.ad.jp/magest/about.html	Ascidian expression database
4. Enzyme database	
BRENDA http://www.brenda.uni-koeln.de/	The most comprehensive database on biochemical reactions
ExPASy http://www.expasy.ch/	Database on protein sequences and structures
NC-IUBMB http://www.chem.qmw.ac.uk/iubmb/enzyme/	Enzyme nomenclature database
Ligand chemical database http://www.genome.ad.jp/dbget-bin/www_bfind?ligand	Database of chemical compounds and reactions in biological systems

Tab. 3 (*continued*)

Name URL	***Features***
NIST Thermodynamic database http://wwwbmcd.nist.gov:8080/enzyme/enzyme.html	Repository on thermodynamics of enzyme catalyzed reactions
PROCAT http://www.biochem.ucl.ac.uk/bsm/PROCAT/PROCAT.html	Specialized for 3-D active site templates of enzymes
5. Scientific literature search	
PubMed http://www.ncbi.nlm.nih.gov/entrez/query.fcgi	Huge repository of biomedical literature
Medline http://research.bmn.com/medline/search	The oldest and most comprehensive biomedical literature database
Scirus http://www.scirus.com	Extremely useful metasearch tool developed by Elsevier Science Publishers

Tab. 4 Tools for drawing pathways.

Name of the tool URL	***Target application***
Pathfinder http://bibiserv.techfak.unibielefeld.de/pathfinder/	Dynamic visualization of metabolic pathways represented as acyclic graphs
Electric arc http://home.xnet.com/~selkovjr/ElectricArc/	CAD-based, can be used to design abstract graphs to electronic circuits
Biopath http://biopath.fmi.uni-passau.de/index.html	Used for digitizing *Boehringer Biochemical Pathways*
Pathway browser http://www-pr.informatik.uni-tuebingen.de/~eiglsper/pathways/	Visualization tool, XML-based, requires Java

to determination of the steady state of a system.

Jarnac: URL: http://www.cds.caltech.edu/~hsauro/Jarnac.htm. Jarnac is a cell-modeling language for describing metabolic, signal transduction, and gene networks. It is linked to Jdesigner that the user interacts with for modeling a biochemical event.

Virtual Cell: URL: http://www.nrcam.uchc.edu/. Virtual Cell is a modeling tool that associates biochemical and electrophysiological data with microscopic image data describing subcellular locations. It is based on a strong mathematical foundation, and the results can be analyzed as images. Access to the Virtual Cell modeling software is via the Internet using a Java-based interface.

A-Cell: URL: http://www.fujixerox.co.jp/crc/cng/A-Cell/. A-Cell is a Windows-based graphical user interface (GUI) for the construction of biochemical reaction models. In addition, it has the capability of importing previously constructed models and combining them with the system.

BioQUEST: URL: http://omega.cc.umb.edu/~bwhite/ek.html. BioQUEST is a set of building blocks that run on the numerical simulation program "Extend," allowing the user to construct conduct time series analysis of biochemical reactions.

Dynafit: URL: http://www.biokin.com/. Dynafit is a software for simulating biochemical reactions. It also provides *ViraFit* for analysis of hepatitis C viral data, *BatchKi* (client server tool) for the determination of tight-binding enzyme inhibition constants, and *PlateKi* (similar to Batchki), but runs as a stand-alone application.

ModelMaker: URL: http://www.modelkinetix.com/. ModelMaker allows the modeling of continuous and discontinuous functions and stiff and stochastic systems. It also provides optimization, minimization, Monte Carlo, and sensitivity analysis.

MetaModel: URL: http://bip.cnrs-mrs.fr/bip10/modeling.htm. MetaModel 3.0 is a DOS-based program for simulating simple biochemical reactions.

DMSS: URL: http://www.bio.cam.ac.uk/~mw263/ftp/doc/ISMB99.ps. Discrete Metabolic Simulation System (DMSS) does not employ kinetic parameters, stoichiometry matrices, or flux coefficients. Instead, the rate of a reaction is modeled on the basis of competing metabolite concentrations or metabolite affinities to enzymes, including metabolite and enzyme concentrations.

E-Cell: URL: http://www.e-cell.org. E-Cell is a modeling and simulation environment. The basic concepts and applications of the E-Cell system are detailed in the following sections.

3 The E-Cell System

3.1 Introduction

The raw material for biological complexity is an immense diversity in components and rules that are employed to create and sustain life. In order to understand the underlying complexity and engineer new systems, it is necessary to create an environment that can translate biology at the level of mathematics. During the early 1990s, the concept of *in-silico* biology had just begun to appear on the horizon, but the scientific community was waiting for proof of the concept. Hence, because no system existed at that time, the decision was taken to create one and, after many trials and errors, a "virtual baby" called *E-Cell* was born.

E-Cell is a short form for "Electronic Cell" – a generic object-oriented environment for modeling and simulating molecular processes of the whole cell in user-definable models, equipped with graphical interfaces that allow observation and interaction. The E-Cell modeling approach links diverse cellular processes such as gene expression, signaling, and metabolism, to form a virtual cell framework. By using E-Cell, it is possible to create a model and also to translate this model into a simulation environment through mathematical equations. More precisely, however, it is a generic system for constructing object models of the cells that can (optionally) emulate the behavior of numeric equation solvers.

The E-Cell project was started at Keio University, Japan, in October 1996. The first working version of E-Cell was ready within three months, and the first virtual cell (*Mycoplasma genitalium*) was developed within a year. In March 2001, the beta version 1.0 of the software was publicly released under open source, and new GUI and peripheral software tools were added. Python is an ideal language for the other user-side components, where productivity and readability are demanded (Table 5). E-Cell is an open-source project: the entire documentation with source code is available from http://www.bioinformatics.org/E-Cell/. Bioinformatics.org is a nonprofit, academic-based organization committed to opening access to bioinformatics research projects. The publicly available mailing lists have been detailed in Table 6. As E-Cell is an open-source project, the expected third-party contribution includes algorithm modules, GUIs, new language bindings, and mathematical analysis modules. E-Cell (version 3.0) is a highly modularized software that can easily be extended by writing plug-ins.

Tab. 5 E-Cell versions.

Cell 1.0 (linux)	*E-Cell 2.0 (windows)*	*E-Cell 3.0 (linux)*
Entire E-Cell application including GUI written in C++	Uses a cocktail of C++, Perl, Cygwin, Java	Core portions (libecs, libemc) and simulation objects (reactors, substances, steppers) written in C++
Peripheral programs (er2eri, ss2er, rd2ch, etc.) written in Perl, Python, and yacc/lex	–	Most other components, including front end, written in Python

Tab. 6 Public E-Cell mailing lists.

Mailing lists	*Features*	*Address*
E-Cell announce	Very low traffic moderated ML for announcements regarding E-Cell projects	e-cell-announce@e-cell.org
E-Cell users	For free discussions on E-Cell	e-cell-users@e-cell.org
		http://www.e-cell.org/mailman/listinfo/e-cell-users
		http://www.e-cell.org/moin/moin.cgi
E-Cell development	For developing E-Cell 3.0	e-cell3-devel@bioinformatics.org

ML = mailing list.

3.2 Architecture of E-Cell

The description here mainly refers to E-Cell version 1.0. The E-Cell software constructs object models equivalent to a cell system or a part of a cell system employing a structured Substance-Reactor Model (SRM). In the SRM, the objects belong to one of the three fundamental object types (Primitives): Substance, Reactor, or System:

- The Substances represent amounts of a molecular species or other state variables.
- The Reactors represent cellular phenomena that result in change in the amount or value of the molecular species or the state variables.
- The Systems are used as containers for the Primitives representing functional and/or physical compartments.

In E-Cell 1.0, a cell model description is composed of two parts: the definition of subclasses of the Primitives (mainly of the Reactor); and a rule file. The rule file contains information of: (1) a list of the three Primitive objects in the model; (2) relationships among the objects (e.g., stoichiometry of reactions); and (3) parameter values for the objects (e.g., rate constants).

3.2.1 Elements of the Control Panel

The control panel includes the following elements:

- The substance window shows the quantity of a selected substance. It also allows the user to alter the quantity at will during the simulation process.
- The reactor window displays the activity of a selected reaction.
- The *activity of a reaction* is defined as the amount of product produced per second in the reaction process.
- Tracers are windows that plot the concentration of substances with time (Fig. 1).

3.2.2 Elements of the E-Cell Model

In the E-Cell system, the substance is a substrate, product, catalyst, or an ion that affects a reaction. Typically, substances include proteins, protein complexes, DNA (genes), RNA, and small molecules. The total number of molecules involved in

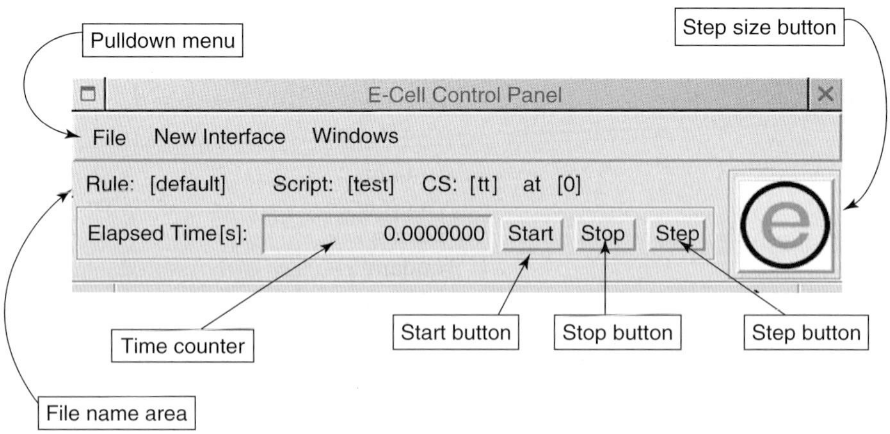

Fig. 1 Elements of the control panel.

a reaction is defined as *quantity*, while *concentration* describes the amount of substance present in a reaction space (in moles per liter). In the E-Cell, the quantity of a substance equals {Avogadro's number × concentration × volume}. The E-Cell simulation software uses the number of molecules in a sample to trace a reaction, and automatically converts concentration into quantity. The spreadsheet data file must be converted to .er text format. It is possible to use macros in the .er file to model complicated systems with ease. In E-Cell 3.0, the .er and .eri file formats are no longer used; rather, an XML-based E-Cell model description language is used. The E-Cell system extracts quantitative information from the rule file, links it up with the equations described in reactors, and plots the velocity curve on tracers.

A special characteristic of E-Cell is the *accumulator*. The Reserve Accumulator (the default feature) is used when decimal fractions are unimportant, for example, when representing an individual cell. However, the Simple Accumulator is used in situations in which the floating-point value is crucial to the interpretation of the results, for example, if the number of molecules is very large or if the simulation represents an "average cell" among a large number of cells. The Monte Carlo Accumulator is used if the simulation requires a high degree of precision in statistical analysis.

3.3 Features of E-Cell 2.0

Recently, Mitsui Knowledge Industry has released the Windows version of E-Cell (ver. 2.0). The E-Cell 2.0 is very similar to the E-Cell 1.0, except that the virtual memory function (to show the concentration in the tracer for a long time) is not implemented for now. However, the E-Cell data file contains time, mean value, maximum value, and minimum value as against time and value output of version 1.0.

To run this version, the following supporting software is required:

- *m4.exe distributed in Cygwin (http://sources.redhat.com/cygwin)
- *Java runtime environment (http://java.sun.com/products/jdk/1.2/)
- *Perl (http://www.perl.com/pub/a/ language/info/software.html)
- *C++(http://www.borland.com/ bcppbuilder/freecompiler/).

3.4 Features of E-Cell 3.0

E-Cell 3.0 is currently being developed with an aim of providing the cell simulation community with a generic and high-performance software environment. It is also Linux-based, and has a geometry information interface. It will integrate any sets of different simulation algorithms, including the Variable-Process model, differential equation-based, diffusion reaction, and particle dynamics-based approaches. One of the main highlights of the software would be integration of subsystems with different timescales. E-Cell 3.0 allows many components, driven by different simulation algorithms and different time scales, to coexist in the simulation by employing a discrete-event worldview as its fundamental formalism.

The core simulation software of E-Cell 3.0 is a set of extension modules for the Python language interpreter, written in C++/C/Python. This consists of a libecs cell modeling tool kit, an E-Cell microcore (EMC) layer, a Python language binding

(PyECS), and other peripheral Python modules. Libecs (code name: Koyurugi) is a generic object-oriented C++ class library for constructing various object-based cell models. One of the special features of Koyurugi is that the constructed cell models themselves work as simulation engines. The E-Cell Micro Core defines interfaces and implementations of the Simulator class, which provides a simple application programming interface (API) of the Koyurugi class library. PyECS is basically a Python binding of the EMC. Libecs, EMC, and a main portion of PyECS are written in C++ language.

3.5 Advantages of the E-Cell System

The E-Cell system has four main advantages:

1. The E-Cell architecture allows users easily to add components to the E-Cell simulation software in order to address individual modeling needs.
2. The E-Cell can accommodate many different types of simulation rather than follow one specific methodology; that is, it can simulate deterministic or stochastic models, either alone or together. Thus, users are able to model biological systems according to their characteristics, and incorporate diverse methods in the same model. In the first version of E-Cell, this is enabled by Reactors. Reactors are coded in C++, thereby offering users a huge bandwidth for simulating a large variety of reactions. In E-Cell 3.0, it will be possible to create new types of Reactors, Substances, and Systems (and Stepper), thus allowing more flexibility. For example, users will be able to define an integration method.
3. The E-Cell is custom-made, even for people with little or no programming knowledge.
4. The E-Cell offers efficient data management through Rule files. This is particularly useful when large amounts of data are available.

3.6 Limitations of the E-Cell System

A primary limitation of E-Cell 1.0 and 2.0 is that, at present, they do not have sophisticated concentration gradient model/simulated three-dimensional (3-D) structures and molecular dynamics. However, this constraint has been overcome in E-Cell 3.0. A second limitation is that ODEs and algebraic equations can only be calculated explicitly. However, in most cases users can make adjustments to components to incorporate other calculation methods (such as incorporating a library within the reactor that allows the implicit calculation of ODEs).

3.7 E-Cell with 127 Genes

The E-Cell with 127 genes is a hypothetical cell that contains the minimum gene set for survival (Fig. 2). For this, the genomic construction from *M. genitalium* was borrowed to build a first virtual cell to conduct what was termed "minimum cellular metabolism." This model takes up glucose from the culture medium using a phosphotransferase system, generates ATP by catabolizing glucose to lactate through glycolysis and fermentation, and then exports lactate out of the cell. The enzymes and substrates are synthesized spontaneously and degraded over time to sustain "life." The protein

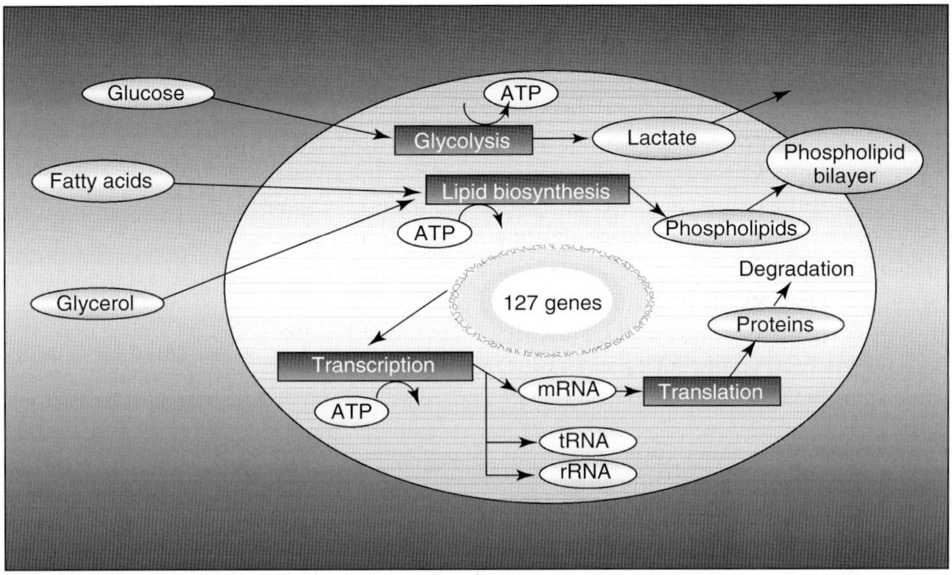

Fig. 2 E-Cell with 127 genes.

synthesis is implemented by modeling the molecules necessary for transcription and translation, namely, RNA polymerase, ribosomal subunits, rRNAs, tRNAs, and tRNA ligases. The cell also takes up glycerol and fatty acids, and produces phosphatidyl glycerol for membrane structure, using a phospholipid biosynthesis pathway. The model cell is "self-supporting," but not capable of proliferating; the cell does not have pathways for DNA replication or the cell cycle. The *Mycoplasma* ammunition used was formed of genes involved in glycolysis ($n = 9$), lactate fermentation ($n = 1$), phospholipid biosynthesis ($n = 4$), phosphotransferase system ($n = 2$), glycerol uptake ($n = 1$), RNA polymerase ($n = 6$), amino acid metabolism ($n = 2$), ribosomal L-subunit ($n = 30$), ribosomal S-subunit ($n = 19$), rRNA ($n = 2$), tRNA ($n = 20$), tRNA ligase ($n = 19$), initiation factor ($n = 4$), and elongation factor ($n = 1$). Overall, this resulted in 98 protein-coding genes and 22 RNA-coding genes. The remaining seven genes were imported from other sources. All of this was spread out into 495 reaction rules that modeled the enzymatic reactions responsible for increasing/decreasing substrate/product quantities, multisubstrate complex formation, transportations of substances, and stochastic processes, for example, transcriptional factors binding factor binding to a specific site of the chromosome.

Mycoplasma genitalium was chosen for constructing a virtual cell with the minimum number of genes for survival, because it has the smallest known genome. Its genomic sequence (580 kb) was determined in 1995. The gene set of *M. genitalium* was abstracted to accommodate only those genes required for the bare, essential cellular metabolism. At the time of developing the first version of E-Cell, 120 genes from *M. genitalium* were identified and well documented. However, in order to successfully hand-construct a self-sustaining cell, this number just fell short by seven. This shortage was made up by bringing in genes from "external sources" – four for phospholipid biosynthesis, one gene each for nucleoside phosphate kinase and nucleoside diphosphate kinase, and one for tRNA ligase. At that time, the phospholipid biosynthesis and a few other pathways in *M. genitalium* were not well characterized. The information on the kinetic properties of genes and proteins was mostly obtained from the Kyoto Encyclopedia of Genes and Genomes (KEGG) and BioCyc (previously called *EcoCyc*) databases.

3.8 Applications of the E-Cell System

The E-Cell with 127 genes predicted for the first time a sudden sharp increase in the ATP level in a glucose-starved cell, followed by the equally sharp decrease of ATP. Although the second event was expected, the first event was a major surprise and was later confirmed experimentally. This anomalous situation was explained by the fact that, although ATP production was stalled by cutting off glucose levels, it took the cell a short time to consume the intermediates for ATP production, and this resulted in a sudden increase of ATP. This example demonstrates the potential of *in-silico* modeling for generating new information. Furthermore, recent pointers have indicated a paradigm shift from experimental biology to *in-silico* biology.

Overall, the E-Cell has applications in the following areas:

Metabolic requirements: The assessment of a cell's metabolic requirements is an area that the E-Cell can successfully address. At present, *M. genitalium* is grown in a complex medium containing several chemically undefined components, including fetal bovine serum and also yeast and beef extracts. By combining knowledge of the metabolic enzymes present in a cell with information concerning protein transporters of metabolites across the cell membrane, it should be possible – by using the E-Cell model – to evaluate whether a particular defined medium can support growth.

Gene expression: E-Cell software can be used to decipher gene regulatory networks. The plan is to use *M. genitalium* to achieve this objective.

Minimal gene set: The self-sustaining E-Cell will be further extended to define the minimal set of genes required for a self-replicating cell under a specific set of laboratory conditions.

Clinical applications: Currently, investigations are being undertaken to determine the clinical applications of the E-Cell; examples include diabetes and enzyme deficiencies in erythrocytes (see Sect. 3.9).

3.9 Simulation of Erythrocyte Enzyme Deficiencies

Glucose-6-phosphate dehydrogenase (G6PD) is a key enzyme that produces NADPH in the pentose phosphate pathway (Fig. 3). Initially, G6PD converts glucose-6-phosphoric acid into 6-phosphoglucono-1,5-lactone (thus generating NADP), which is then metabolized to ribulose-5-phosphoric acid via 6-phosphogluconic acid, generating NADPH in the process. Within the erythrocyte, a major function of glutathione (GSH) is to eliminate superoxide anions and organic hydroperoxides. Peroxides are eliminated through the action of glutathione peroxidase, yielding oxidized glutathione (GSSG).

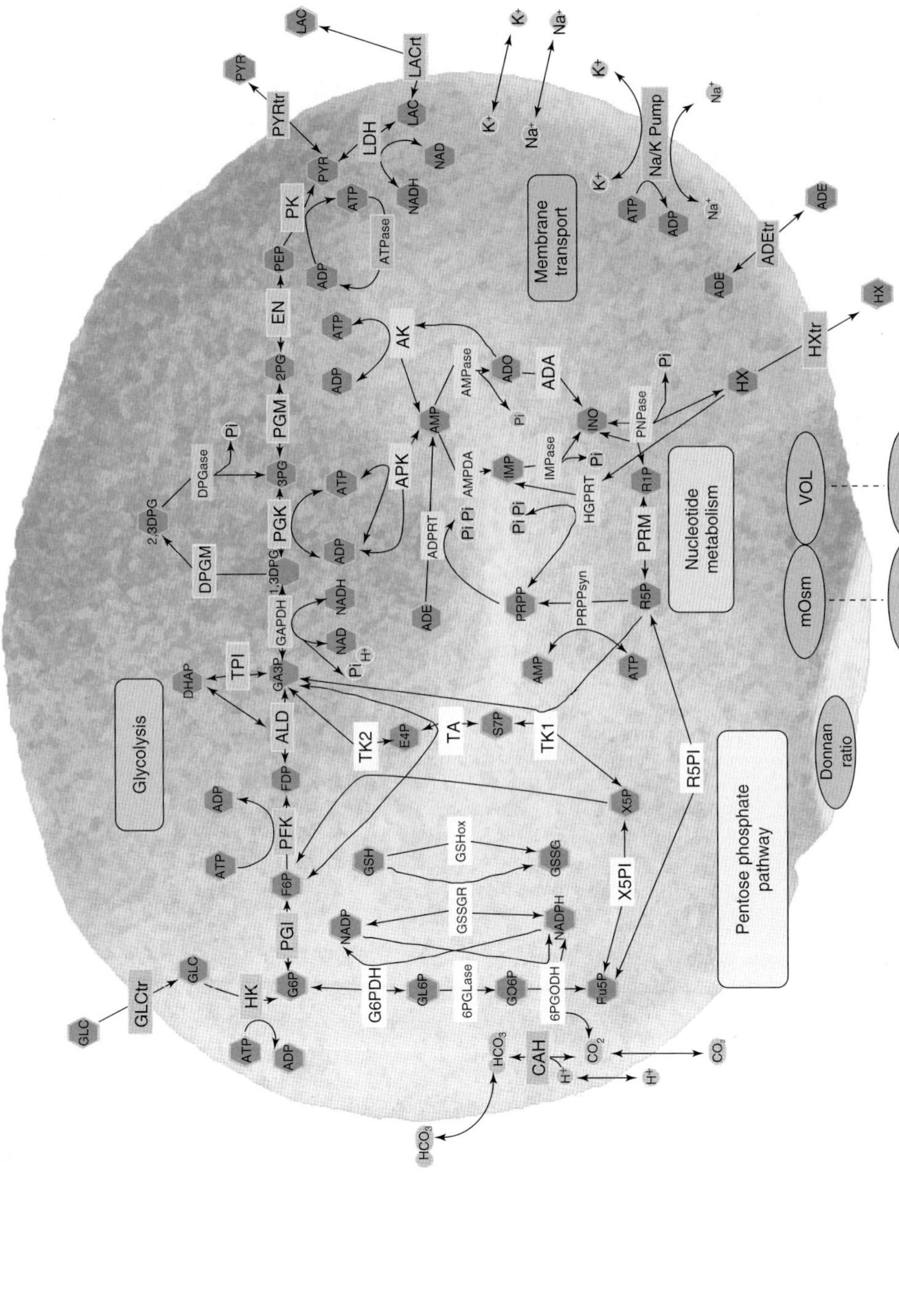

Fig. 3 The whole erythrocyte model.

A G6PD deficiency has been implemented into the E-Cell model, and the kinetic parameters have been modified in accordance with the biochemical environment of mutant cells taken from patients reporting this deficiency. The simulation experiments were carried out with steady-state concentrations corresponding to those of the normal erythrocyte. Sequential changes in the quantity of NADPH, GSH, and ATP were observed in the simulation experiments. However, the longevity of the computer model, as estimated by the concentration of ATP, was found to be much shorter than that of the "real" erythrocyte with G6PD deficiency. This difference was, presumably, due to a lack of pathways producing GSH, and of the export system for GSSG. After modification, however, the longevity of the cell and the GSH/GSSG ratio was found to have increased. These results indicate that these pathways partially compensate for the reduction of GSH and have a role in easing anemia, a condition which is related to G6PD deficiency. This result can also provide a good explanation for the fact that G6PD deficiency is the most common cause of anemia. If the deficiency with these compensation pathways had no severe disadvantage on survival, then the condition would spread through the population. When the activity of G6PD is decreased, the activity of 6-phosphogluconate dehydrogenase is increased, thereby compensating for the reduced production of NADPH. However, because either 6-phosphoglucono-1,5-lactone was not supplied, or because there was a deficiency of G6P, the 6-phosphogluconic acid supply was rapidly exhausted and the production of NADPH stopped. Consequently, the amount of NADPH began to reduce gradually and soon became exhausted. The level of GSH then began to decrease due to its conversion into GSSG. Finally, the metabolic performance of the cells worsened when the ATP became exhausted due to an inhibition of the rate-determining enzymes (due to

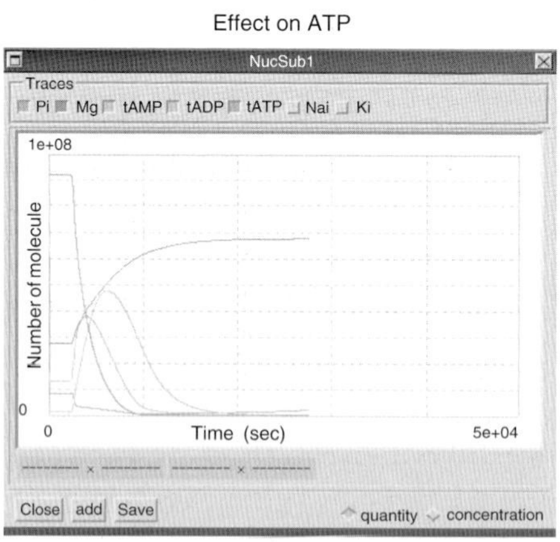

Pi: Inorganic phosphates, Mg: Free magnesium
tAMP: total AMP, tADP: total ADP, tATP: total ATP

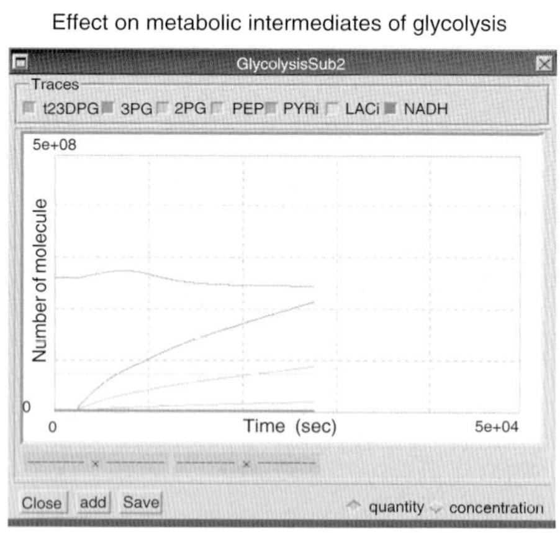

t23DPG: total 2,3-Diphosphogylcerate,
3PG: 3-Phosphoglycerate, 2PG: 2-Phosphoglycerate
PEP: Phosphoenolpyruvate, LACi: Lactate

Fig. 4 Simulation of erythrocyte deficiency.

low GSH/GSSG). This model of G6PD deficiency correlates well with the clinical situation, and may serve as a "test bed" for extending the model to other human erythrocyte metabolic disorders (Fig. 4).

Recently, an erythrocyte model with a pyruvate kinase (PK) deficiency has also been reconstructed. On creating an *in-silico* model for PK deficiency, the ATP production rate was found to decrease proportionately, leading to an eventual elimination from the system. Simultaneously, an increase in the concentration of 2-phosphogrycerate, 3-phosphoglycerate and phosphoenolpyruvate was observed, which is in agreement with the clinical presentation of the PK phenotype.

4
Practical Applications

The E-cell offers possibilities of creating new opportunities for drug target selection, based on predictive models. For example, pathway-based disease models can assist at the preclinical stage to identify any potential toxic effects of "lead compounds." If a compound targets a network hub, the possibility that such a drug would give rise to adverse side effects is quite high. However, if drug targets are found first, to be either non-hubs, terminal nodes or linkers in the network, or second, multiple weak binders which collectively bring about the effect, then such compounds will be preferred as candidate drugs. Today, many companies employ disease- and population-based drug response models to lower their R&D costs. A prior assessment of side effects/toxic effects can result in a speeding up of the drug discovery process, leading to significant savings.

By producing detailed "route maps" of the molecular circuitry of a cell, it is possible – at least in theory – to develop smarter therapeutic strategies. However, the success of this strategy depends on the completeness and accuracy of the relevant data acquired. Previously, Systems Biology has played a key role in providing an understanding of the drug gefitinib (Iressa; Astra Zeneca), for drugs to treat liver abnormalities (Pfizer), and of kinase inhibitor mechanisms (Johnson & Johnson) [1]. In general, the systems approach has resulted in better descriptions of many biological systems, leading to the possibilities of systems design and engineering and the promotion of a new discipline termed *Synthetic Biology*. In Synthetic Biology, attention is focused on the ground-up engineering of novel systems for useful applications. However, in order that Synthetic Biology becomes successful, it is first important that Systems Biology approaches continue to generate new data, and to provide new models and new descriptions of how biological components collaborate to generate distinct phenotypes.

5
Concluding Remarks

In order to understand the *whole*, it is important first to *study* the *whole*. Given the enormous complexity and data generated by the genome, proteome, transcriptome and metabolome, computer simulations are clearly indispensable for future biological research. Whether, or not, it is feasible to construct a computer model of a whole living cell remains an open question. Although attempts at whole-cell modeling were not made until the late 1990s, the importance of computer simulations of cellular metabolism has in fact been realized since the 1980s [2], with various cellular processes such as gene expression [3, 4], cell cycles [5, 6] and metabolic pathways [7–11] having been modeled and simulated independently. In order to understand the crosstalk among these seemingly "self-regulating systems," it is necessary to construct an integrated model of the cell. But, one of the major problems when constructing large-scale models is a lack of quantitative data, since most of the biological knowledge currently available is of a qualitative nature (in the form of pathway maps). Unfortunately, the quantitative data available are often noisy and not well suited to simulations [12]. Thus, a major challenge in this respect is to collect large amounts of very accurate (and preferably time series) data, to construct quantitative models, and to "train" the models with additional results acquired from the laboratory, until the simulation matches "real-life" biology [13, 14]. The problem is that, in order to achieve this objective, it is not only good data that are needed but also a novel computational and software engineering approach. This is, indeed, the "new biology of the twenty-first century."

References

1 Rubenstein, K. (2008) *Systems Biology: A Disruptive Technology. CHI Insight Pharma Reports Edition*. Cambridge Health Institute, Massachusetts, USA.

2 Tomita, M., Hashimoto, K., Takahashi, K., Shimizu, T., Matsuzaki, Y., Miyoshi, F., Saito, K., Tanida, S., Yugi, K., Venter, J.C. (1999) E-Cell software environment for whole cell simulation. *Bioinformatics*, **15**, 72–74.

3 Meyers, S., Friendland, P. (1984) Knowledge based simulation of genetic regulation in bacteriophage lambda. *Nucleic Acids Res.*, **12**, 1–9.

4 Koile, K., Overton, G.C. (1989). A qualitative model for gene expression. Proceedings of the 1989 Summer Computer Simulation Conference, Austin, Texas. Society for Computer Simulation, San Diego, California, pp. 415–421.

5 Novak, B., Tyson, J.J. (1995) Quantitative analysis of a molecular model of mitotic control in fission yeast. *J. Theor. Biol.*, **173**, 283–305.

6 Chen, K.C., Csikasz-Nagy, A., Gyorffy, B., Val, J., Novak, B., Tyson, J.J. (2000) Kinetic analysis of a molecular model of the budding yeast cell cycle. *Mol. Biol. Cell*, **11**, 369–391.

7 Palsson, B.O., Narang, A., Joshi, A. (1989) Computer model of human erythrocyte metabolism. *Prog. Clin. Biol. Res.*, **319**, 133–150.

8 Sauro, H.M. (1993) SCAMP: a general purpose simulator and metabolic control analysis program. *Comput. Appl. Biosci.*, **9**, 563–571.

9 Mendes, P. (1993) GEPASI: a software package for modeling the dynamics, steady states and control of biochemical and other systems. *Comput. Appl. Biosci.*, **9**, 563–571.

10 Mendes, P. (1997) Biochemistry by numbers: simulation of biochemical pathways with Gepasi 3. *Trends Biochem. Sci.*, **22**, 361–363.

11 Goryanin, I., Hodgman, T.C., Selkov, E. (1999) Mathematical simulation and analysis of cellular metabolism and regulation. *Bioinformatics*, **15**, 749–758.

12 Rao, C.V., Wolf, D.M., Arkin Adam, P. (2002) Control, exploitation and tolerance of intracellular noise. *Nature*, **420**, 231–237.

13 Tomita, M. (2001) Whole cell simulation: a grand challenge of the 21st century. *Trends Biotechnol.*, **19**, 205–210.

14 Takahashi, K., Yugi, K., Hashimoto, K., Yamada, Y., Pickett, C.J.F., Tomita, M. (2002) Computational challenges in cell simulation: a software engineering approach. *IEEE Intell. Syst.*, **17**, 64–71.

13
Fractals in Biology and Medicine

*Gabriele Angelo Losa**
Institute of Scientific Interdisciplinary Studies, v.F. Rusca 1, 6600 Locarno, Switzerland

* This chapter is dedicated to the memory of Professor Jean Paul Rigaut, who passed away in 2005, and who pioneered the adoption of fractal geometry in the biomedical sciences.

Systems Biology: Advances in Molecular Biology and Medicine, First Edition. Edited by Robert A. Meyers.

Keywords

Apoptosis
Apoptosis (programmed cell death) is a physiological form of death that involves a cell after complexion or exhaustion of its own function. Apoptotic cells are swiftly eliminated *in vivo* by the reticuloendothelial system, without the release of cellular agents which may damage the surrounding tissue.

Fractal
A term, coined by Mandelbrot, from the Latin adjective *fractus (fragmented, irregular)* that derives from Latin verb *frangere* meaning to break, to create irregular fragments.

Fractal (fractional) dimension
A dimension which exceeds the topological dimension, but need not be an integer.

Form invariance
The outstanding property of a fractal object: any part of it, after being scaled by an arbitrary factor appears the same as the original. In other words, it is form invariant or self-similar.

Fractalomics
A subdiscipline which includes all biological and/or natural objects that meet the criteria for a strict description of natural fractals, that is, self-similar patterns and scaling properties.

Self-similarity
A fractal invariant object is termed self-similar.

Scaling
The degree of irregularity and/or fragmentation of a fractal shape is identical at all scales of observation.

Scaling window
In contrast to mathematically generated fractals, biological structures and objects observed in Nature are self-similar within a limited range of scales. Only within this scale interval or scaling window can the scale-invariant (fractal) properties of an irregular object of finite size be observed.

Leukemia
A form of progressive malignant disease of the blood-forming organs, characterized by a distorted proliferation of leukocytes and their precursors in the blood and bone marrow. Many types of leukemia exist; these are classified according to the type of white blood cell involved.

Lymphoma
The term used for a diverse group of cancers of the lymphatic system, a connecting network of glands, organs, and vessels, the principal cell of which is the lymphocyte.

The extension of the concepts of fractal geometry towards the life sciences has led to significant progress in understanding the complex functional properties and architectural/morphological/structural features that characterize cells and tissues during ontogenesis and also in both normal and pathological development processes. It has even been argued that fractal geometry could provide a coherent description of the design principles underlying living organisms. Fractals fulfill a certain number of theoretical and methodological criteria, including a high level of organization, shape irregularity, functional, and morphological self-similarity, scale invariance, iterative pathways, and a peculiar noninteger fractal dimension (FD). Whereas, mathematical objects are deterministic invariant or self-similar over an unlimited range of scales, biological components are statistically self-similar only within a fractal domain defined by upper and lower limits – termed the *scaling window* – in which the relationship between the scale of observation and the measured size or length of the object can be established. For statistically self-similar irregular objects, the size and geometric parameters differ when inspected at increasing resolution, which reveals more details. Selected examples will contribute to depict complex biological shapes and structures as fractal entities, and also show why the application of the fractal principle is valuable for measuring the dimensional, geometrical, and functional parameters of cells, tissues, and organs occurring within the vegetal and animal realms. If the criteria for a strict description of natural fractals are met, then it follows that a "Fractal Geometry of Life" may be envisaged, and all natural objects and biological systems exhibiting self-similar patterns and scaling properties may be considered as belonging to the new subdiscipline of *fractalomics*.

1 Introduction

The Fractal Geometry of Nature [1], Benoît Mandelbrot's masterpiece evoking a new "Weltanschauung," has provided an innovative paradigm, a novel epistemological approach for interpreting the natural world and a more intelligent vision of life itself (in the etymological sense of the Latin word "intellegere"), even though it has given rise to – and is still giving rise to – controversial opinions in the scientific community while offering incentives to satisfy the curiosity of the public at large [2].

This fractal geometry, which was founded on a body of well-defined laws and coherent principles [3], including those derived from chaos theory [4], allows the recognition and quantitative description of complex shapes, images, and other figures that usually are created through unlimited iterations of a simple generator (often a mathematical motif) by means of computer-aided design (CAD).

CAD figures which were undecipherable using classical geometry were referred to as "*fractals*" because of their peculiarity, which lies in the reproducibility of their shape over a range of scales and in a noninteger topological dimension called a "*fractal dimension*", from the Latin word *fractus*. Non-Euclidean iterated figures – now including fractals – have often been considered to bear a resemblance to pathological entities or mathematical monsters despite of – or owing to – their beauty, richness and fascinating shapes [5]. Nowadays, most of these have become explicable and even familiar, since Mandelbrot's assertion that they can almost be considered as a general rule of Nature. This led Mandelbrot to conclude that, "... clouds are not spheres, mountains are not cones, coastlines are not circles, and bark is not smooth, nor does lightning travel in a straight line." Subsequently, it was noted that these virtual figures share some morphological traits and self-similar properties which could be encountered not only in elements of the inanimate world, but also (though less evident) in complex forms, functions and shapes belonging to the plant and animal realms. Living forms develop according to organized morphological patterns correlated with a complex system of functional metabolic interactions which make the accomplishment of the adaptive response possible. Iteration, self-similarity, form invariance upon scaling, nonequilibrium thermodynamics, self-organization, and energy dissipation are among the mechanisms reputed to sustain the emergence and maintenance of living forms, in contrast to those of homeostasis, linearity, smoothness, regularity, and thermodynamic reversibility pertaining to a more traditional vision based upon the concepts and rules of Euclidean geometry and adequate for an ideal world [6]. Over the past decade, a large amount of experimental evidence has been accumulated showing that biological elements do indeed express statistical self-similar patterns and fractal properties within a defined interval of scales. This is termed the "*scaling window*," in which a direct relationship between the observation scale and the measured size/length of an object or the frequency of a temporal event can be ascertained and in turn quantified by a peculiar FD [3]. In other words, the FD of a biological component remains constant within the scaling window, and serves to quantify variations in length, area or volume with changes in the dimensions of the measuring scale. However, real "fractality" exists only when the experimental scaling range covers at least two orders of magnitude, although fractality over many orders of magnitude has been observed in various natural fields [7]. Hence, defining a "scaling range" of length measurements appears to be an inescapable requisite for assessing the fractality of any biological element. Experimental evidence of a definite scale interval avoids any ambiguous assignment of objects or figures lacking that requirement, and confirms Mandelbrot's assertion that "... fractals are not a panacea; they are not everywhere" [7].

2
The Irruption of Fractal Geometry in Biology and Medicine

From the direct observations of Nature, it emerges that most cells, tissues, organs – in either the animal or vegetal worlds – are systems in which the component parts and unit fragments assemble with different levels of complexity and organization. This means that a single fragment or element may, on various scales, reproduce the whole object from which it is derived; in other words, it is self-similar, albeit in a statistical sense. Very few of these shapes can be analytically described or evaluated by using Euclidean geometry, which was developed to trace the regular and ideal geometrical forms that are practically unknown in natural and biological systems. Thus, a Fractal Geometry of Life can be envisaged and the totality of biological elements, natural objects and physiopathologic processes carrying spatial or temporal self-similar properties, can be gathered into "fractalomics." Fractalomics is proposed as a novel variety of "-omics" comprehensive of biological systems by analogy with other established subdisciplines such as genomics, where the suffix "*omics*" (derived from the Greek word "*ome*"), refers to wholeness or to completion. Although the first coherent essay on fractal geometry was published in French more than 30 years ago [8], it may be worth considering exactly how and when the "heuristic introduction" of such an innovative discipline occurred or, more pregnantly stated, as when "the irruption of fractal geometry" into the life sciences such as biology and medicine actually took place [9]. Although there is no precise date, it is generally agreed that such as introduction occurred within the "golden age" of cell biology – that is, between the 1960s and 1990s.

2.1 The Morphofunctional Complexity of Cells and Tissues

According to "the state of the art," there was a pressing need to consider the morphological complexity of cells and tissues by using a systemic approach, whilst at the same time developing instruments that could enable the accomplishment of that goal without introducing any shape approximation or smoothing – a condition which could not be satisfactorily achieved with conventional analytical methods. In fact, the latter situation – which relied on conventional disciplines such as morphometry and stereology [10] – yielded experimental data relating to the quantitative description of membranes that was usually controversial. This left many questions unresolved and thereby prevented a true consensus being reached among the investigators [11–13].

To highlight the striking debate that led to such turmoil within the biologist community, it might suffice to report the original description, proffered by an outstanding scientist in the field [14], of the first case study conducted several years earlier [15]. This related to the application of fractal geometry in cell biology; notably, the discovery that cellular membrane systems have fractal properties arose from the uncertainty of observations regarding the extent of such membranes. When the results of the first studies on the morphometry of liver cell membranes were reported, the values obtained were much higher than had been reported by others. There followed much debate as to which of these estimates was correct, and whether liver cells contained 6 or 11 m^2 of membranes per cm^3 – a quite significant difference. The question was also raised as to whether the stereological methods used were reliable, since it appeared possible that the same method might yield different results if the measurements were made under different magnifications of the electron micrographs. Ultimately, the systematic measurement of liver cell membranes

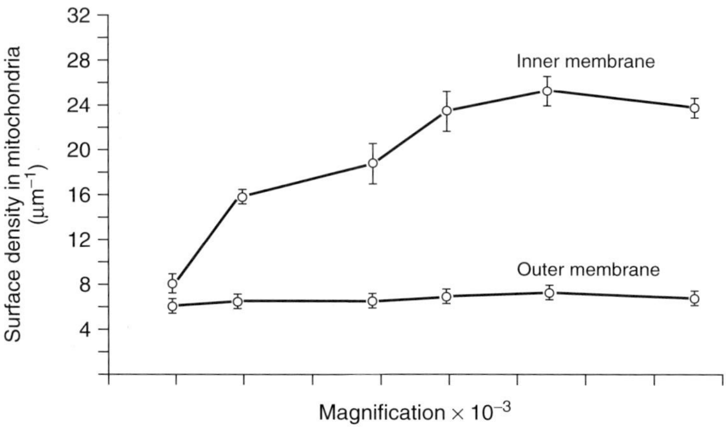

Fig. 1 Changes of surface density estimates for the outer and inner mitochondrial membranes, with increased magnification. Reproduced from Ref. [15].

revealed that the estimates of surface density increased with increased resolution [14]. Shortly after the conclusion of the experimental phase of the above-mentioned study, Mandelbrot suggested that the results should be interpreted with the likely effect of the "resolution scale," in analogy with the "Coast of Britain effect" [16]. If so, this would resolve the estimate discrepancy and explain why measurements of irregular liver cell membranes at higher magnification yielded higher values than were obtained at lower magnification [14]. It must be stressed here that the scaling effect applies mainly to cellular membranes with a folded surface or an indented profile, such as the inner mitochondrial membrane. The surface density estimate is increased with increasing magnification, yielding a FD which is consistently as high as the estimated value of 2.54. In contrast, measurement of the surface density of the outer mitochondrial membrane, which is almost smooth, was only slightly affected by the resolution effect. In fact, the estimated FD of about 2.09 was rather close to the topological dimension of 2.0, as documented in Fig. 1 [15].

The second case study, which investigated the gas exchange surface of the lung, revealed that the alveolar surface area was increased at increasing magnification, with a slope yielding an FD of 2.24 [17]. Previously, it had been reported that the inner lung surface consists of a hierarchy of successive structures: first the alveoli, then capillaries in the alveolar wall, and finally alveolar epithelial cells with wrinkled membranes, which are resolved at increasing magnification (Fig. 2). These different structures can, however, hardly be considered as self-similar structures of a unique type, since each has its own generator, is determined by its own constructive algorithm and, accordingly, it would be necessary to search for at least two or three self-similar levels> [14], which a successive systematic study could, indeed, confirm.

3
Fractal Criteria

Mandelbrot stated in his book that, "A fractal set is a set in metric space for which

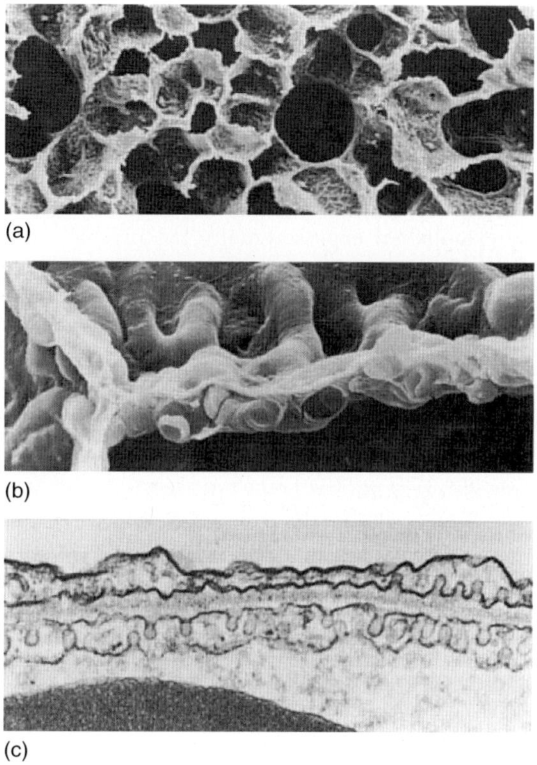

Fig. 2 Increasing microscopic resolution reveals a hierarchy of structures that form the inner lung surface. (a) Scanning electron microscopy image of alveoli around the airways. Original magnification ×75; (b) Scanning electron microscopy image of imprints of capillaries in the alveolar walls. Original magnification ×1000; (c) Transmission electron microscopy image of membrane folds of the epithelial cell forming the surface of the air–blood barrier. Original magnification ×46000. Reproduced from Ref. [17].

the Hausdorff–Besicovitch dimension D is greater than the topological dimension Dt." In Nature, a fractal object is defined by its structural properties, mainly by its lack of smoothness. Additional important properties of a fractal object are roughness or shape irregularity at every scale, high level of organization, iterative pattern, a peculiar noninteger FD, and self-similarity or scale invariance. This means that an object is referred to as being *self-similar* if any part of it, after being scaled by an arbitrary factor, appears the same as the whole object [3]. The Richardson–Mandelbrot equation provides the mathematical basis for understanding geometrical and spatial fractal structures, and for measuring and interpreting them, namely:

$$L(\varepsilon) = N(\varepsilon).(\varepsilon) \qquad (1)$$

where $L(\varepsilon)$ represents the contour (perimeter) length of the biological component under investigation, (ε) the unit length of measure, and $N(\varepsilon)$ the number of unit lengths (ε) required to cover the contour $L(\varepsilon)$. By substituting $N(\varepsilon)$ with $[lo^D\varepsilon^{-D}]$ in Eq. (1), where lo is a reference scale without influence on the determination of D, the above equation can be transformed by logarithmic procedure and rewritten as:

$$\log[L(\varepsilon)/lo] = (1 - D)\log[\varepsilon/lo] \qquad (2)$$

Equation (2) represents a dimensionless scaling power law indicating that the estimated contour, perimeter, or curve length $L(\varepsilon)$ changes as a power function of the scale unit length (ε). The dimensional exponent D is the FD which defines the nature of the curve. Mathematical fractals are invariant over an unlimited range of scales, whereas biological components are statistically self-similar only within a fractal domain or *scaling window* delimited by upper and lower scaling bounds covering at least two orders of magnitude. This also implies that such a domain must be established experimentally for any biological element investigated, and that a lowest scale of measure exists which can detect the smallest entity measurable, below which there is no physical sense to measurement. Only within region II on a log–log plot, can a straight line be drawn and its slope $(1 - D)$, as defined in the logarithmic equation $\log L(\varepsilon) = (1 - D)\log(\varepsilon)$, be used to evaluate the numerical value of the FD D (Fig. 3) [3]. The fractal scaling power law is ubiquitous in Nature [18, 19], and can be applied to the study of a wide variety of biological problems, including allometric scaling growth [20], allosteric enzyme kinetics [21, 22], metabolic rates in mammals [23], population genetics [24], tumor growth [25, 26], the modeling of drug clearance [27], the ontogenic growth of human hemopoietic stem cells (HSCs) [28], cardiac function and body size [29] and many others, as documented recently [30]. Disregarding the theoretical and experimental framework of criteria which define inexact statistical fractals (such as biological and natural criteria) could lead to a series of pitfalls in determining the FD, as noted by several groups [18, 31–39]. For a better understanding, in a certain number of reports the FD was evaluated from the slope of the straight line, but without the scaling domain being established or, occasionally, with only the lower bound scale displayed on the log–log plot (corresponding to the maximal resolution scale) being checked. Often, the straight line was drawn without the data being fitted by using an automatic procedure based on a least-squares fit algorithm or other statistical method, thus preventing an objective slope estimation. The automatic procedure enabled searches to be conducted for the widest interval within which the standard deviation of the estimated slope did not exceed a given limit, corresponding to a

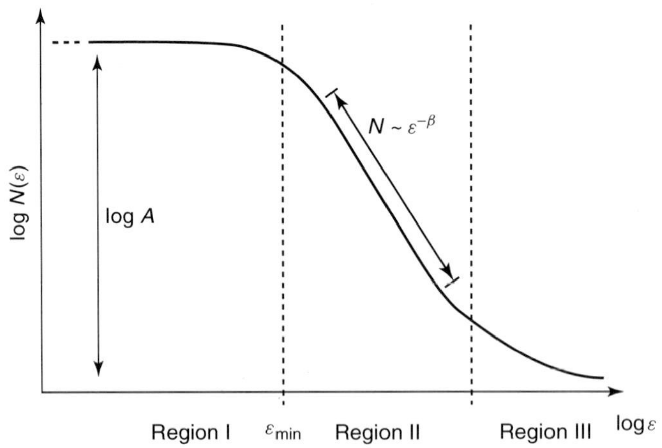

Fig. 3 The three typical regions which characterize asymptotic biological and natural fractals. The fractality (autosimilar scaling properties) is restricted to the fractal window represented by Region II. Reproduced from Ref. [41].

confidence interval ranging from 95 to 99%, whereby the FD *D* of any shape profile examined could be estimated at a fixed standard deviation [40].

Apart from the *scaling window* principle, there are two other criteria which must be fulfilled to achieve a reliable fractal analysis. The first criterion requires the adoption of a correct sampling procedure, while the second imposes that the experimental data be evaluated through an adequate statistical methodology, which rests on the type of frequency distribution with which the collected data must fit. That said, at this point neither specific criticism of specific reports (such as those cited in the references) will be made, nor comparisons made (or opinions expressed about), of data concerning similar topics, since such a task is beyond the scope of this study.

4 Fractalomics

The FD is considered as a statistical measure which correlates the morphological structural and behavioral complexity of cellular components and biological tissues [41, 42]. The FD is also a numerical descriptor which serves to measure the qualitative morphological traits and self-similar properties demonstrated by most biological elements at various levels – whether cellular, tissue, or organic – as highlighted below with several examples.

4.1 Subcellular Components and Membranes

In previous studies in which the fractal nature of cytoplasm was discussed, the fractal architecture was revealed to be a result of the iteration of an invariant simple pattern spanning several length scales, and organized into a percolation lattice with clusters emerging as fractal forms. Such a spatiotemporal cytoplasmic organization bestowed properties which amplify enzymatic activities [43, 44] and metabolic networks [45]. In immature feline oocytes with or without the *cumulus oophorus* (CO), which is known to affect oocyte developmental potential, distinct cytoplasmic components with irregular features show self-similar properties that could be described by fractal analysis [46]. Original images of oocytes collected by ovariectomy and segmented using a gray threshold procedure revealed that the highest FD of 1.91 was measured on gray-dark profiles of the cytoplasmic elements characterized by a highly connected network of lipid droplets and intracellular membranes. Unexpectedly, the FD values from the different oocyte elements were close to each other, and not influenced by the presence or absence of the CO. The fractal analysis provided an effective quantitative descriptor of the complex cytoplasm morphology which may contribute to an objective representation and reliable classification of the feline oocyte, without introducing any bias or shape approximation (Fig. 4).

In particular, at the electron microscopy level fractal analysis proved useful for objective investigations of the fine cytoplasmic structure, and also of the organization of various types of chromatin, nuclear components and other subcellular organelles, either in normal or pathological tissues and in cell cultures. For example, measurement of the FD of euchromatin and heterochromatin nuclear domains helped to discriminate lymphoid cells found in mycosis fungoides from those in chronic dermatitis [47]. The external nuclear membrane (ENM) envelope and nuclear membrane-bound

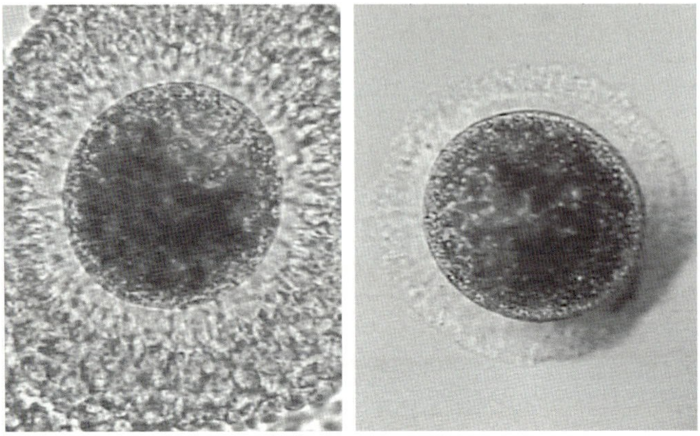

(a) (b)

Fig. 4 Feline oocytes (a) with and (b) without the cumulus oophorus. Original magnification ×40. Images reproduced by kind permission of Prof. Gionata De Vico, Federico II University of Naples.

heterochromatin (NMBHC) domains of MCF-7 human breast cancer cells, when briefly triggered by steroids such as 17β-estradiol or dexamethasone, underwent ultrastructural changes at the start of their growth which could be quantified by their peculiar FDs. In fact, after 5 min of treatment, the 17β-estradiol (1 nM) growth factor had significantly enhanced the ultrastructural irregularity or the DNA unfolding of the NMBHC domain by increasing its FD, whereas dexamethasone (1 nM), a growth antagonist, had reduced such irregularity when compared to control MCF-7 cells. Neither steroid significantly modified the ENM ultrastructure, however [48].

This fractal tool has also been employed to document the feasibility of using ultrastructural changes in the cell surface and nuclear inter(eu)chromatin to assess the early phases of apoptosis (programmed cell death) induced in human breast cancer (SKBR-3) cells by the ionophore, calcimycin. The ultrastructural changes, which involved a loss in heterochromatin irregularity (or an increased condensation of it), as quantified by a lower FD, were evident well before the detection of conventional cell markers, which were only measurable during the active phases of apoptotis [49] (Figs 5 and 6). The results of these and other reports [50–53] have indicated that the fractal analysis carried out on electron microscopic images is very efficient for the quantitative detection of cellular components and associated morphostructural changes. The results of a pioneering study, showing that rat liver cells contain intracellular membranes with irregular and self-similar traits, when observed over several scales of measurement, were also reported [15].

4.2
Leukemia and Lymphoma

The first application of fractal morphometry in nonsolid cancer came later, when human leukemia cells of lymphoid and/or myeloid origin were characterized on electron microscopic images through the quantitative measurement of membrane surface properties, which could be correlated with specific phenotype markers.

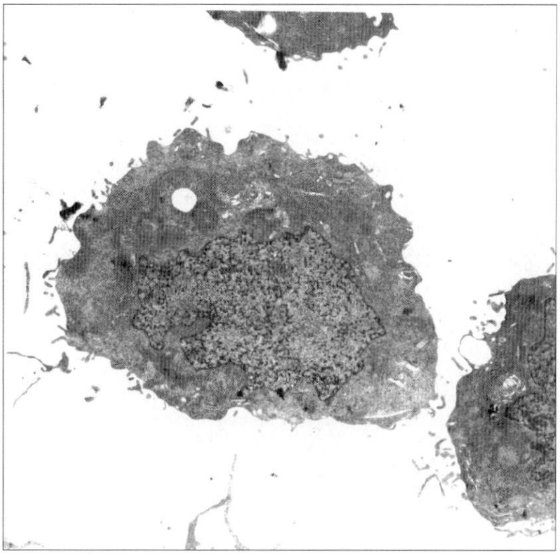

Fig. 5 Electron microscopy image of SK-BR-3 breast cancer cells (control) cultured for 24 h without 1 μM calcimycin. Original magnification ×10 580. Reproduced from Ref. [49].

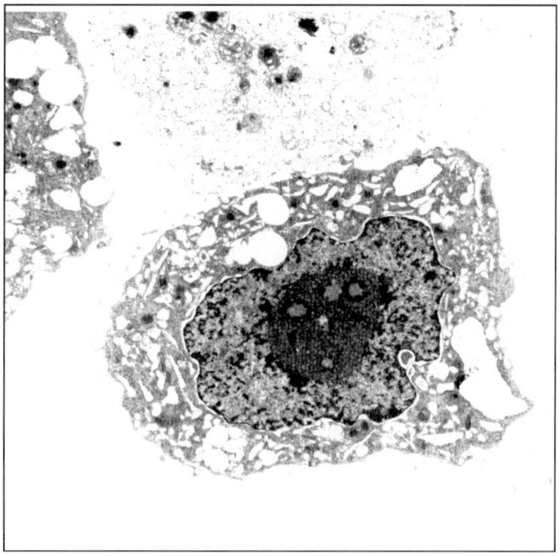

Fig. 6 Electron microscopy image of SK-BR-3 breast cancer cells cultured with 1 μM calcimycin for 24 h. Changes in plasma and nuclear membranes and other organelles occur at the start of apoptosis; these may be evaluated using fractal geometry. Original magnification ×10 580. Reproduced from Ref. [49].

Cells isolated *ex vivo* from the blood of humans with acute T-lymphoid leukemia revealed pericellular membranes with a nearly smooth outline, as documented by FD values which were significantly lower than those evaluated on the pericellular membranes of healthy blood cells (Fig. 7). Healthy lymphocytes of B-cell lineage had an FD (1.20) which was significantly different from the FD of lymphocytes of T-cell lineage – that is, CD4-T helper (1.17) and CD8-T suppressor (1.23) cells. Unexpectedly, strongly proliferating T-lymphoid leukemic cells were found to possess a plasma membrane characterized by a low FD value (1.10), close to the value measured on the plasma membrane of *in vitro*-growing lymphoblasts

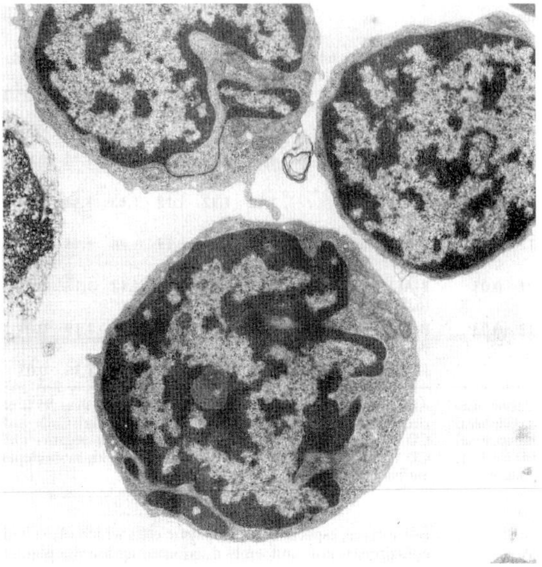

Fig. 7 Electron microscopy image of cells of acute lymphoblastic leukemia (T-ALL). Original magnification ×8000. Reproduced from Ref. [54].

derived from mature T lymphocytes triggered by phytohemagglutinin (PHA), a mitogenic lectin [54]. About 80% of acute leukemia subtypes of the B-cell lineage (c-ALL and pre-B undifferentiated phenotype) showed plasma membranes with FD-values ranging from 1.12 to 1.17, below that of the plasma membrane of differentiated B lymphocytes. The remaining cases (20%) of acute lymphoblastic B-leukemia showed a more convoluted cell surface, with FD-values of up to 1.24. Cells from hairy-cell leukemia, a chronic type of human leukemia, with a highly convoluted plasma membrane morphology and a completely different surface phenotype, displayed the highest FD-values, of between 1.32 and 1.36 [55]. The FD of scale-invariant self-similar chromatin was measured in the nuclei of blasts isolated from patients suffering from acute leukemia of the precursor B-lymphoblastic type (B-ALL). The increase in FD, together with an accentuated coarseness of the nuclear surface, reflected significant changes in the DNA methylation pattern usually localized in heterochromatin nuclear regions, and this was regarded as a poor prognostic factor for these patients [56]. The value of fractal analysis in assessing the hematological cell phenotype and defining a clinical group was confirmed some 20 years later [57]. In this case, by using conventional slide preparations, the authors analyzed the nuclei "contours" of cells belonging to the B lineage; that is, normal and reactive lymphocytes and lymphoid cells isolated from patients with chronic lymphocytic leukemia (CLL), follicular lymphoma (FL), and diffuse large B-cell lymphoma (DLBCL). The FD-values of the perinuclear membranes were found to differ significantly between the groups, and correlated with the membranes' biological properties. Notably, reactive lymphocytes had an FD-value (1.20) which was situated between that of CLL (FD = 1.25) and normal cells (FD = 1.13), while aggressive lymphoma cells had significantly higher FD-values, ranging from 1.23 (FL) to 1.31 (DLBCL). By comparing data from such reports on hematological malignancies, it transpired that cells isolated from patients with different types of leukemia and/or lymphoma have a nuclear chromatin with a roughness or complexity (high FD-values) which increases with increasing degrees of aggressiveness and malignancy, whereas pericellular membranes acted in an inverse manner and appeared smoother (low FD-value) in cells with a high degree of malignancy. From this, it could be inferred that hematological tumors do not undergo uniform neoplastic transformations, but rather demonstrate manifold metabolic and phenotype changes. This implies an either increasing or a decreasing complexity of the morphological surface, and an altered organization of cell components that is mainly dependent on the cytotype under investigation. This contrasts with an apparent type of behavior observed in several cell colonies of breast cancer origin and of experimental tumors, which were observed to obey the same dynamics of proliferation and growth and to display contours with fractal self-similar features when submitted to scaling analysis [58].

4.3 Tumor Tissues

Over the past two decades, several reports have described the heuristic importance of applying the fractal approach for the quantitative characterization of cancer tissues. The aim here is to overcome the uncertainty in current practice which involves various systems of diagnosis that mostly arise from subjective observations rather than from a quantitative procedure [3, 59–64]. In routine histology and cytology, the examination of cell nuclei and nuclear components by fractal morphometry has greatly improved the comprehension of cell behavior and the assessment of diagnosis and prognosis for various disease states [65]. The nuclear chromatin organization was effectively quantified by applying fractal morphometry in order to evaluate the degree of malignancy in human breast cytology [66], while the FD served in the discrimination of cytology smears of breast and cervical lesions [67]. More recent studies targeting cell nuclei periphery have demonstrated fractal properties which have allowed the classification of early ovarian cancers, and even provided the ability to distinguish normal from malignant liver cells [53, 68]. Relevant fields for which fractal geometry might provide an original approach for investigation, and where the FD-value represents more than an additional geometrical parameter – as claimed by followers of the "reductionistic view." Rather, it should provide information relating to cell and tissue heterogeneity, the architectural organization of organs, shape features, developmental, morphogenetic, and the growth processes of tissues and organs in healthy, pathological, and tumor conditions. Cell heterogeneity, which is known to contribute in a decisive manner to the histological grading of human breast cancer, has been examined using geostatistics and the Hurst fractal parameter [69]. As reported recently, tumor grading (as a measure of the degree of cellular differentiation) may be difficult to assess, because tumors often consist of an heterogeneous mixture of cells with varying degrees of differentiation. On inspection of breast and prostate histology specimens, the tumor structures were observed to derive from poorly differentiated cell elements that possessed a greater complexity, as characterized by a higher degree of irregularity. Subsequently, by measuring the FD-value it proved possible to quantify not only the degree of irregularity but also local variations in cellular differentiation. Such a tool could aid pathologists in grading heterogeneity, and also in determining the spatial extent of poorly differentiated regions of tumors [70, 71]. All of the above-reported examples, as well as those below, appear to indicate that the occurrence of morphogenetic dynamics, the emergence of complex patterns, and the architectural organization of active tissues and tumor masses may be driven by constructive mechanisms related to fractal principles, including the deterministic and/or random iteration of constituent units with varying degrees of self-similarity, scaling properties, and form conservation [72]. Today, the preservation of tissue architecture and the cell polarity of organs, and the eventual restoration of organized traits in tumor tissues, deconstructed and deregulated at various levels, is an emerging field of interest, it having been observed that biological entities organize with their own degrees of structural and behavioral complexity, and develop on different spatial and time scales [73–76]. For a quantitative description of all the problems mentioned above which "... take all levels of biological organization into consideration" [77], adequate tools of investigation are required based on the principles of fractal geometry. This constitutes a novel way of understanding higher-level phenomena (form generation, tissue organization and development, carcinogenesis, cell proliferation, apoptosis, or cell death) which could begin a conveyance towards the unifying frame of tissue organization field theory (TOFT) [76].

4.4 Stroma, Extracellular Matrix, and Epithelium Interaction

For many years, the interactions between the stroma, the extracellular matrix (ECM) and the epithelium have been extensively examined in various mammalian tissues, mainly because of their roles in the architectural organization of tissues [73, 74, 78–85].

Stromal tissue has major roles in the control and regulation of physiological processes and supporting the tumorigenic process [86], as has been well documented in breast cancer [87–90]. Recently, well-defined three-dimensional (3-D) models have been developed to decipher stromal–epithelial interactions which mediate mammary gland development and the formation and progression of breast cancer [91]. The outline roughness and the internal irregularity of the collagen ECM examined on biopsy specimens of livers affected by chronic diseases were evaluated using the fractal approach, which has yielded a reliable measure which is extremely useful when describing these two qualitative properties of the liver matrix [92]. It has also been shown that a quantitative evaluation of the surface FD may allow not only measurement of the complex geometrical architecture, but also modeling of the development and growth of tumor neovascular systems. It would also allow an exploration of the morphological variability of vasculatures in nature, and in particular the microvasculature of normal and adenomatous pituitary tissue [93]. Neuronal and glia cells from the brain, spinal cord neurons and retinal ganglion cells were found to show FD-values that correlated with the increase in morphological complexity, thus revealing a progressive level of morphological maturity [94–98]. Several fractal and non-fractal parameters have been considered for quantitative assessment of the vascular architecture, using a variety of test specimens and computational tools. Fractal parameters have the advantage of being scale invariant; that is, they are independent of the magnification and resolution of the images investigated, and this makes it easier to compare different set-ups and experiments [99].

In the normal human retina, the blood vessels or vascular trees were shown to exhibit a typical FD-value of 1.7 – similar to that of a diffusion-limited growth process – and this may have implications for the embryological development of the retinal vascular system [100]. A considerable amount of fluctuation was present in both artery blood flow velocity (FV) and arterial blood pressure (ABP) after subarachnoid hemorrhage (SAH). Consequently, variability and fractal analysis may each provide valuable information regarding the complexity of the human organism: fluctuations are reduced in cerebral vasospasm with a decrease in variability, suggesting a loss of complexity associated with a less favorable outcome. The decomplexification theory of illness may therefore apply to SAH [101].

More recently, FD-values have been used as characterization parameters for premalignant and malignant epithelial lesions of the floor of the mouth in humans [102, 103]. Architectural changes associated with aging of the normal oral buccal mucosa have been shown to exist between three main age ranges, by measuring the global FD of the epithelial tissue interface and the FD-values of the segmented epithelial cell borders [104]. The onset of fundamental phenomena such as growth and cell death can be adequately investigated by applying fractal geometry: a recent report revealed that micro-architectural alterations of the uninvolved colonic mucosa, shown through an increased FD-value, occurred early during an experimental colon carcinogenesis and preceded the expression of conventional biomarkers of both apoptosis and proliferation [105].

Fractal structures have also been observed in animal diseases. For example, the FD-value of the dog kidney proximal convoluted tubule, established by means of the box-counting algorithm, was used to automate its recognition in anatomy and pathology [106]. A discrimination between benign (fibroadenoma, FD $= 1.09$) and malignant (carcinoma, FD $= 1.21$) mammary tumors in dogs and cats was achieved by determining the FD-values of the inner surfaces of the mammary ducts [107].

Canine trichoblastomas constitute a class of "... benign tumor derived from or reduplicating the primitive hair germ of embryonic follicular development" [108], and represent about 25% of all epithelial skin neoplasms [109]. They are classified into ribbon types (RT), trabecular/granular types (TT/GT), and spindle types (ST). Trichoblastomas are not exclusively epithelial tumors, but rather occur as heterogeneous biological systems arising from epithelial–mesenchymal interactions (Figs 8 and 9). In these neoplasms, the epithelial component appears to be equivalent to the hair germ, while the mesenchymal component represent the dermal papilla, with both being essential for hair follicle development [110]. Despite the relevance of these aspects, the complex mesenchymal epithelial relationship and the links between molecular and morphogenetic cell signals, which may occur in trichoblastomas, remain poorly understood. Hence, fractal morphometry was applied to these canine tumors to investigate the tissue organization in relation to the distribution patterns of the epithelial/connective tissue components and the expression of β-catenin, a structural constituent between cells that anchors the

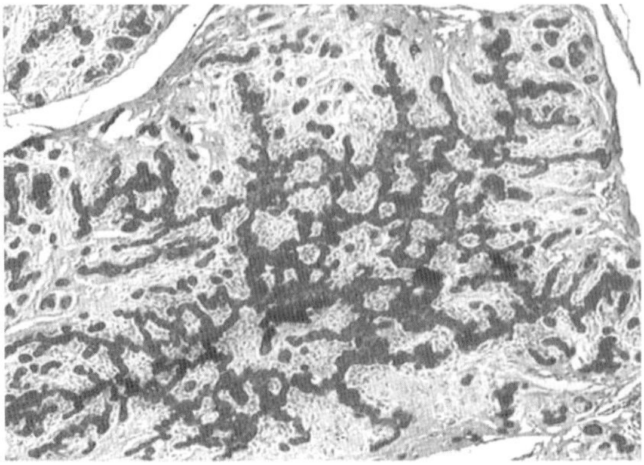

Fig. 8 Microscopic view of canine trichoblastoma of the ribbon type (RT). The epithelial component is positive (gray) for the marker cytokeratin. Original magnification ×40. Reproduced from Ref. [111].

Fig. 9 An outline of the area occupied by the neoplastic epithelium of the cell shown in Fig 8. Original magnification ×40. Reproduced from Ref. [111].

actin cytoskeleton, regulates normal cell growth and behavior, and modulates the epithelial architecture and polarity of the cells and tissues [111]. Fractal analysis was performed on the masks and outlines of epithelial tumor components segmented from gray-level threshold images taken randomly from histological trichoblastoma sections at ×10 magnification, using a program described elsewhere [42]. The morphometric results indicated that the relative amount of mesenchymal stroma (expressed as $[V_v]_m$) was higher ($p < 0.001$) in the RT cell ($46 \pm 2\%$) than in the other types ($31 \pm 5\%$ and $33 \pm 3\%$ in ST and TT/GT, respectively). All of the tumors examined showed epithelial components with irregular self-similar properties that could be distinguished by their characteristic FD-values. A subsequent tumor mask analysis revealed, unexpectedly, that the RT trichoblastomas showed low FD-values (1.75 ± 0.01) that were significantly different ($p < 0.001$) from those of the ST and TT/GC (1.78 ± 0.03 and 1.85 ± 0.02, respectively). In contrast, the trichoblastoma outlines showed absolute FD-values which were lower than those obtained from tumor masks, but these were statistically inadequate to provided any histological discrimination. β-Catenin was shown to occur in the cytoplasm and nucleus of both mesenchymal and epithelial neoplastic cells, while its expression pattern (nucleus, cytoplasm, or both) was closely related to the growth pattern morphology and the type-specific architectural organization of trichoblastoma types, as documented by the FD-value which was increased progressively on moving from RT, to ST, to TT/GT [112]. In future, trichoblastoma of the dog might serve as a reliable model to unravel the morphogenetic dynamics and organization underlying the neoplastic process and to provide, in quantitative terms, useful information with regards to the link between the molecular, cellular, and tissue changes that occur during tumor development [113].

4.5 Fractal Shapes and Image Analysis Techniques

One highly promising approach appears to be a combination of fractal analysis, to provide a quantitative description of shapes, with radiographic imaging, which has the ability to discriminate malignant from benign tumor masses, and also from normal tissue structures [114–121]. The FD of the contour of a mass may be computed either directly from the two-dimensional (2-D) contour or from a one-dimensional (1-D) signature derived from the contour. This may be useful for characterizing shape and gray-scale complexity, which may vary between benign masses and malignant tumors in mammograms [122]. Fractal theory has provided the basis for a unique software platform program, which has been developed for use in conjunction with magnetic resonance imaging (MRI), and has shown great promise in the early diagnosis and treatment of breast cancer. In a recent study where this advanced method was applied, more than 30% of the patients were shown to have additional tumors in the same breast, and in almost 10% of cases tumors were shown also to be present in the other breast. Most importantly, these tumors had not been located when using conventional mammography or ultrasound [123].

Recently, image analysis and non-Euclidean geometrical fractal analysis have been applied to describe changes in the actin cytoskeleton of neonatal cardiac fibroblasts when they respond to mechanical stress [124]. In this case, human proteins were analyzed in order to seek a link between intra-proteomic similarity and functional properties, and the relationship that emerged could be quantified as a fractal dimension [125, 126]. More interestingly, these findings highlight the modality by which information is stored in amino acid sequences [127].

5 Concluding Remarks

5.1 Understanding the Biology of Normal and Tumor Tissues

Irregularity and self-similarity under scale changes are the main attributes of the morphological complexity of cells and tissues, either normal or pathological. In other words, the shape of a self-similar object does not change when the scales of measurement change, because any part of it might be similar to the original object. The size and geometric parameters of an irregular object, however, differ when inspected at increasing resolution, revealing greater detail. Over the past three decades, significant progress has been made in understanding how to analyze irregular shapes and structures in the physical and biological sciences. Dominant influences include the important discovery by Mandelbrot of a practical geometry of nature termed fractal geometry, and the continuous improvements in computational capabilities. The application of the principles of fractal geometry – unlike conventional Euclidean geometry which was developed to describe regular and ideal geometrical shapes that are virtually unknown in Nature – enables measurement of the fractal dimension, contour length, surface area, and other dimensional parameters of almost all irregular and complex biological tissues. Over the past decade, a great deal of experimental evidence has been accumulated showing that, even in the biological world, fractal patterns could be observed within a *scaling window* – a condition that should be established experimentally for each tissue element. The fractal dimension is a quantitative descriptor that can be used alone to quantify qualitative peculiarities (such as form irregularity), to describe morphogenetic processes, and also to identify cell components, cell types and other tissues sharing different morphological traits and functional peculiarities. In contrast, mathematical statistical approaches such as the gray-level co-occurrence matrix (GLCM), polynomial equations and Fourier analysis (amongst others), require a large number of parameters to do so. Through the use of several recently reported examples, it has been possible to highlight applications of the fractal approach in measuring irregular self-similar features in normal and pathological cells and tissues, with a high degree of organized complexity and of plasticity [128]. The potential roles of the fractal approach in reassessing morphological information for a deeper insight into, and understanding of, the biology of normal tissues and tumor masses have also been determined. In particular, the fractal approach allows not only the avoidance of any approximation or simplification in analyzing real shapes and functional behaviors (and hence to describe irregular morphologic components and ultrastructural features as they are), but also – through a quantitative comparison – to show every modification that the structural features and shapes in either normal, pathological, or tumor stages may undergo over a period of time.

5.2
The Fractal Design in Systems Biology

In conclusion, two questions must be posed. First, do the numerous examples presented here concur in explaining how self-similarity or form invariance on various scales and self-organization can govern different biological processes, such as architectural organization, evolutionary developmental process, form conservation, growth, morphogenesis, and shape remodeling which emerge in all living organisms in line with the assumptions of Systems Science? [75, 129–131]. Systems Biology is the study of an organism, viewed as an integrated and interacting network of genes, proteins, biochemical reactions, and morphostructures, which give rise to life. Instead of analyzing the individual components or aspects of the organism, Systems Biology focuses on all of the components and the interactions among them, as part of one system. Systems Biology is an holistic approach [132] which aims to understand the completeness and complexity of biological systems, which means examining the structure and dynamics of cellular and organismal function, rather than identifying the characteristics of the isolated parts of a cell or organism [133]. In fact, many properties of life arise at the systems level only, as the behavior of the system as a whole cannot be explained by its constituents alone. Even in pathological situations – an example being

Alzheimer's disease (AD) – most biological complex networks (including protein interaction networks) have revealed a feature of scale-free networks with multiscale self-similarity that relates to fractals [134]. If so, then the hypothesis that the "... morphogenesis of biostructures follows fractal principles" and that "... fractal geometry is a design principle for living organisms," as recalled elsewhere [2, 14], may be envisaged for interpreting how biological phenomena and shapes come about, while being well aware that the true reality may remain undisclosed!

The second question outlines the link between fractals and biological design, which can be formulated as: "Do genes contain fractal algorithms?" [14]. This appears much more interlocutory, because genes are DNA entities that codify constructive units or templates, while fractal algorithms represent mechanisms (iteration, self-organization, environmental constraints, etc.) which Nature may eventually adopt in order to assemble self-similar dynamic units into final shapes. Whether genes do, or do not, determine biological shape directly or by following stochastic and environmental effects [135], either through a selective activation driven by transcription factors [136] or by fractal mechanisms, will likely be a matter of lively debate. It is worth noting here that genomic/functional analysis, such as the RNA-interference (RNAi) technique adapted to *Drosophila*, *Caenorhabditis elegans*, plants, and mammalian cell cultures [137], has made it possible to screen systematically for genes controlling specific cell-biological processes, including those required to influence cytoskeletal organization, to generate distinct morphologies [138] and, like *Hox* genes [139], to act as architects which control the spatial structuring of the organism through its entire development [140]. To date, however, very few genes known to control the shape of biological elements, such as in fruit and vegetables, have been cloned [141].

Acknowledgments

The authors thank Prof. Benoît Mandelbrot (University of Yale, USA), Prof. Theo Nonnenmacher (University of Ulm, Germany), and Prof. Ewald R. Weibel (University of Bern, Switzerland) for meaningful discussions and helpful comments.

References

1 Mandelbrot, B. (1983) *The Fractal Geometry of Nature*, Freeman, San Francisco.

2 Weibel, E.R. (1991) Fractal geometry: a design principle for living organisms. *Am. J. Physiol.*, **261**, 361–370.

3 Losa, G.A., Nonnenmacher, T.F. (1996) Self-similarity and fractal irregularity in pathologic tissues. *Mod. Pathol.*, **9**, 174–182.

4 Prigogine, I. (1997) *Les Lois du Chaos*, Flammarion, Paris.

5 Mandelbrot, B. (2006) Les Fractales, les Monstres et la Beauté, in: Belaubre, G. (Ed.) *L'irruption des Géométries Fractales dans les Sciences*, Editions de l'Académie Européenne Interdisciplinaire des Sciences (AEIS), Paris, pp. 317–341.

6 Losa, G.A. (2002) Fractal morphometry of cell complexity. *Biol. Forum*, **95**, 239–258.

7 Mandelbrot, B. (1998) Is Nature fractal? *Science*, **279** (5352), 783–784.

8 Mandelbrot, B.B. (1977) *Fractals: Form, Chance and Dimension*, W.H. Freeman & Company, San Francisco.

9 Belaubre, G. (Ed.) (2006) *L'irruption des Géométries Fractales dans les Sciences*, Editions de Académie Européenne Interdisciplinaire des Sciences (AEIS), Paris.

10 Weibel, E.R. (1981) Stereological methods in cell biology: where are we – where are we going? *J. Histochem. Cytochem.*, **29** (9), 1043–1052.

11 Loud, A.V. (1968) A quantitative stereological description of the ultrastructure of normal rat liver parenchymal cells. *J. Cell Biol.*, **37**, 27–37.

12 Weibel, E.R., Stäubli, W., Gnägi, H.R., Hess, F.A. (1969) Correlated morphometric and biochemical studies on the liver cell. I. Morphometric model, stereological methods and normal morphometric data for rat liver. *J. Cell Biol.*, **42**, 68–91.

13 Losa, G.A., Weibel, E.R., Bolender, R.P. (1978) Integrated stereological and biochemical studies on hepatocytic membranes. III. Relative surface of endoplasmic reticulum membranes in microsomal fractions estimated on freeze-fracture preparations. *J. Cell Biol.*, **78**, 289–308.

14 Weibel, E.R. (1994) Design of Biological Organisms and Fractal Geometry, in: Nonnenmacher, T.F., Losa, G.A., Weibel, E.R. (Eds) *Fractals in Biology and Medicine*, Vol. I, Birkhäuser Press, Basel, Berlin, Boston.

15 Paumgartner, D., Losa, G.A., Weibel, E.R. (1981) Resolution effect on the stereological estimation of surface and volume and its interpretation in terms of fractal dimensions. *J. Microsc.*, **121**, 51–63.

16 Mandelbrot, B. (1967) How long is the coast of Britain? Statistical self-similarity and fractional dimension. *Science*, **155**, 636–640.

17 Gehr, P., Bachofen, M., Weibel, E.R. (1978) The normal human lung: ultrastructure and morphometric estimation of diffusion capacity. *Respir. J.*, **32**, 121–140.

18 West, B.J., Deering, W. (1997) Fractal physiology for physicists: Levy statistics. *Phys. Rep.*, **246** (1,2), 1–100.

19 Weibel, E.R. (2002) The pitfalls of power laws. *Nature*, **417**, 131–132.

20 Dreyer, O., Puzio, R. (2001) Allometric scaling in animals and plants. *J. Math. Biol.*, **43**, 144–156.

21 Savageau, M.A. (1985) Mathematics of organizationally complex systems. *Biomed. Biochim. Acta*, **44** (6), 839–844.

22 Savageau, M.A. (1995) Michaelis-Menten mechanism reconsidered: implications of fractal kinetics. *J. Theor. Biol.*, **176**, 115–124.

23 Weibel, E.R., Bagicalupe, L.D., Schmitt, B., Hoppeler, H. (2004) Allometric scaling of maximal metabolic rate in mammals: muscle aerobic capacity as determinant factor. *Respir. J. Neurobiol.*, **140**, 115–132.

24 Vlad, M.O., Mora, F., Popa, V.T., Szedlacsek, S.E., Ross, J. (2007) Functional, fractal nonlinear response with application to rate processes with memory, allometry, and population genetics. *Proc. Natl Acad. Sci. USA*, **104** (12), 4798–4803.

25 Delsanto, P.P., Condat, C.A., Pugno, N., Gliozzi, A.S., Griffa, M. (2008) A multilevel approach to cancer growth modelling. *J. Theor. Biol.*, **250**, 16–24.

26 Guiot, C., Pugno, N., Delsanto, P.P., Deisboeck, T.S. (2008) Physical aspects of cancer invasion. *Phys. Biol.*, **4**, 1–6.

27 Pugno, N. (2008) Dimensional analysis shows that mice models could lead to drug overdosages in humans. *Cancer Ther.*, **6**, 965–968.

28 Dingly, D., Pacheco, J.M. (2007) Ontogenetic growth of the haemopoietic stem cell pool in humans. *Proc. R. Soc. Biol. Sci. Ser. B*, **274**, 2497–2501.

29 Dewey, F.E., Rosenthal, D., Murphy, D.J., Froelicher, V.F., Ashley, E.A. (2008) Does size matter? Clinical applications of scaling cardiac size and function for body size. *Circulation*, **117**, 2279–2287.

30 West, G.W., Brown, J.H. (2005) The origin of allometric scaling laws in biology from genomes to ecosystems: towards a quantitative unifying theory of biological structure and organization. *J. Exp. Biol.*, **208**, 1575–1592.

31 Rigaut, J.P. (1984) An empirical formulation relating boundary length to resolution in specimens showing "non-ideally fractal" dimensions. *J. Microsc.*, **13**, 41–54.

32 Rigaut, J.P. (1989) Fractals in Biological Image Analysis and Vision, in: Losa, G.A., Merlini, D. (Eds) *Gli Oggetti Frattali in Astrofisica, Biologia, Fisica e Matematica*, Edizioni Cerfim, Locarno, pp. 111–145.

33 Smith, T.G., Lange, G.D. Jr, Marks, W.B. (1996) Fractal methods and results in cellular morphology – dimensions, lacunarity and multifractals. *J. Neurosci. Methods*, **69**, 123–136.

34 Landini, G., Rigaut, J.P. (1997) A method for estimating the dimension of asymptotic fractal sets. *Bioimaging*, **5** (2), 65–70.

35 Jelinek, H.F., Fernandez, E. (1998) Neurons and fractals: how reliable and useful are calculations of fractal dimensions? *J. Neurosci. Methods*, **81**, 9–18.

36 Jelinek, H.F., Elston, G.N., Zietch, B. (2005) Fractal Analysis: Pitfalls and Revelations in Neuroscience, in: Losa, G.A., Merlini, D., Nonnenmacher, T.F., Weibel, E.R. (Eds) *Fractals in Biology and Medicine*, Vol. IV, Birkhäuser Press, Basel, Berlin, Boston, pp. 85–94.

37 Milosevic, N.T., Ristanowich, D. (2006) Fractality of dendritic arborization of spinal cord neurons. *Neurosci. Lett.*, **396**, 172–176.

38 Eke, A., Herman, P., Kocsis, L., Kozak, L.R. (2002) Fractal characterization of complexity in temporal physiological signals. *Physiol. Meas.*, **23**, 1–38.

39 Eke, A., Herman, P., Hajnal, M. (2006) Fractal and noisy CBV dynamics in humans: influence of age and gender. *J. Cereb. Blood Flow Metab.*, **26**, 891–898.

40 Nonnenmacher, T.F. (1994) Spatial and Temporal Fractal Patterns in Cell and Molecular Biology, in: Nonnenmacher, T., Losa, G.A., Weibel, E.R. (Eds) *Fractals in Biology and Medicine*, Vol. I, Birkhäuser Press, Basel, Berlin, Boston, pp. 22–37.

41 Nonnenmacher, T.F., Baumann, G., Barth, A., Losa, G.A. (1994) Digital image analysis of self-similar cell profiles. *J. Biomed. Comput.*, **37**, 131–138.

42 Dollinger, J.W., Metzler, R., Nonnenmacher, T.F. (1998) Bi-asymptotic fractals: fractals between lower and upper bounds. *J. Phys. A: Math. Gen.*, **31**, 3839–3847.

43 Aon, M.A., Cortassa, S. (1994) On the fractal nature of cytoplasm. *Fed. Eur. Biochem. Soc. Lett.*, **344**, 1–4.

44 Aon, M.A., O'Rourke, B., Cortassa, S. (2004) The fractal architecture of cytoplasmic organization: scaling, kinetics and emergence in metabolic networks. *Annu. Rev. Biomed. Eng.*, **6**, 427–452.

45 Goldberger, A.L., Amaral, L.A.N., Hausdorff, J.M., Ivanov, P.Ch., Peng, C.K., Stanley, H.E. (2002) Fractal dynamics in physiology: alterations with disease and aging. *Proc. Natl Acad. Sci. USA*, **99**, 2466–2472.

46 De Vico, G., Peretti, V., Losa, G.A. (2005) Fractal organization of feline oocyte cytoplasm. *Eur. J. Histochem.*, **49** (2), 151–156.

47 Bianciardi, G., Miracco, C., Santi, M.D., Perrone, A., Bruni, A., Luzi, P. (2002) Fractal dimension of lymphocytic nuclear membrane in mycosis fungoides and chronic dermatitis, in: Losa, G.A., Merlini, D., Nonnenmacher, T.F., Weibel, E.R. (Eds) *Fractals in Biology and Medicine*, Vol. II, Birkhäuser Press, Basel, Berlin, Boston, pp. 231–237.

48 Losa, G.A., Graber, R., Baumann, G., Nonnenmacher, T.F. (1998) Steroid hormones modify nuclear heterochromatin structure and plasma membrane enzyme of MCF-7 cells. A combined fractal, electron microscopic and enzymatic analysis. *Eur. J. Histochem.*, **42**, 1–9.

49 Losa, G.A., Castelli, C. (2005) Nuclear patterns of human breast cancer cells during apoptosis: characterization by fractal dimension and co-occurrence matrix statistics. *Cell Tissue Res.*, **322**, 257–267.

50 Santoro, R., Marinelli, F., Turchetti, G., Zini, N., Scotlandi, K., Falcieri, E., Cinti, C., Maraldi, N.M. (2002) Fractal analysis of chromatin during apoptosis, in: Losa, G.A., Merlini, D., Nonnenmacher, T.F., Weibel, E.R. (Eds) *Fractals in Biology and Medicine*, Vol. III, Birkhäuser Press, Basel, Berlin, Boston, pp. 77–84.

51 Marinelli, F., Santoro, R., Maraldi, N.M. (1998) Fractal analysis of heterochromatin nuclear domains in lymphocytes, in: Losa, G.A., Merlini, D., Nonnenmacher, T.F., Weibel, E.R. (Eds) *Fractals in Biology and Medicine*, Vol. II, Birkhäuser Press, Basel, Berlin, Boston, pp. 220–225.

52 When, B., Jacob, W., Van de Wouwer, G., Da Silva, F., Montironi, R., Thompson, D., Bartels, H.G., Van Daele, A., Bartels, P.H. (2002) Fractal dimension, form and shape factors for the quantification of nuclear signature profiles, in: Losa, G.A., Merlini, D., Nonnenmacher, T.F., Weibel, E.R. (Eds) *Fractals in Biology and Medicine*, Vol. III, Birkhäuser Press, Basel, Berlin, Boston, pp. 47–54.

53 Nielsen, B., Albregtsen, F., Danielsen, H.E. (2002) Fractal signature vectors and lacunarity class distance matrices to extract new adaptive texture features from cell nuclei,

in: Losa, G.A., Merlini, D., Nonnenmacher, T.F., Weibel, E.R. (Eds) *Fractals in Biology and Medicine*, Vol. III, Birkhäuser Press, Basel, Berlin, Boston, pp. 55–65.

54 Losa, G.A., Baumann, G., Nonnenmacher, T. (1992) Fractal dimension of pericellular membranes in human lymphocytes and lymphoblastic leukaemia cells. *Pathol. Res. Pract.*, **188**, 680–686.

55 Losa, G.A. (1994) Fractal properties of pericellular membrane from lymphocytes and leukaemia cells, in: Nonnenmacher, T.F., Losa, G.A., Weibel, E.R. (Eds) *Fractals in Biology and Medicine*, Vol. I, Birkhäuser Press, Basel, Berlin, Boston, pp. 190–200.

56 Adam, R.L., Silva, R.C., Pereira, F.G., Leite, N.J., Lorand-Metze, I., Metze, K. (2006) The fractal dimension of nuclear chromatin as a prognostic factor in acute precursor B lymphoblastic leukemia. *Cell. Oncol.*, **28** (1–2), 55–59.

57 Mashiah, A., Wolach, O., Sandbank, J., Uzie, I.O., Raanani, P., Lahav, M. (2008) Lymphoma and leukemia cells possess fractal dimensions that correlate with their interpretation in terms of fractal biological features. *Acta Haematol.*, **119**, 142–150.

58 Brú, A., Albertos, S., Subiza, J.L., García-Asenjo, J.L., Brú, I. (2003) The universal dynamics of tumor growth. *Biophys. J.*, **85**, 2948–2961.

59 Goldberger, A.L., West, B.J. (1987) Fractals in physiology and medicine. *Yale J. Biol. Med.*, **60** (5), 421–435.

60 Cross, S.S. (1997) The application of fractal geometric analysis to microscope images. *Micron*, **25** (1), 101–113.

61 Baish, J.W., Jain, R.K. (2000) Fractals and cancer. *Cancer Res.*, **60**, 3683–3688.

62 Landini, G., Hirayama, Y., Ti, L., Kitano, M. (2000) Increased fractal complexity of the epithelial-connective tissue interface in the tongue of 4NQO-treated rats. *Pathol. Res. Pract.*, **196** (4), 251–258.

63 Spillmann, W.B., Robertson, J.L., Huckle, W.R., Govindan, B.S., Meissner, K.E. (2004) Complexity, fractals, disease time, and cancer. *Phys. Rev. E*, **70** (06), 1911–1922.

64 Janecka, I.P. (2007) Cancer control through principles of systems science, complexity, and chaos theory: A model. *Int. J. Med. Sci.*, **4**, 164–173.

65 Muniandy, S.V., Stanlas, J. (2008) Modelling of chromatin morphologies in breast cancer cells undergoing apoptosis using generalized Cauchy field. *Comput. Med. Imaging Graph.*, **32** (7), 631–637.

66 Einstein, A.J., Wu, H.S., Sanchez, M., Gil, J. (1998) Fractal characterization of chromatin appearance for diagnosis in breast cytology. *J. Pathol.*, **185**, 366–381.

67 Ohri, S., Dey, P., Nijhawan, R. (2004) Fractal dimension in aspiration cytology smears of breast and cervical lesions. *Anal. Quant. Cytol. Histol.*, **26**, 109–112.

68 Nielsen, B., Albregtsen, F., Danielsen, H.E. (2005) Fractal analysis of monolayer cell nuclei from two different prognostic classes of early ovarian cancer, in: Losa, G.A., Merlini, D., Nonnenmacher, T.F., Weibel, E.R. (Eds) *Fractals in Biology and Medicine*, Vol. IV, Birkhäuser Press, Basel, Berlin, Boston, pp. 175–186.

69 Sharifi-Salamatian, V., Pesquet-Popescu, B., Simony-Lafontaine, J., Rigaut, J.P. (2004) Index for spatial heterogeneity in breast cancer. *J. Microsc.*, **216** (2), 110–122.

70 Tambasco, M., Magliocco, A.M. (2008) Relationship between tumor grade and computed architectural complexity in breast cancer specimens. *Hum. Pathol.*, **39**, 740–746.

71 Tambasco, M., Kouznetsov, A., Magliocco, A.M. (2008) Quantifying local variations in the architectural complexity of histology specimens. *Biol. Forum*, **101** (1), 155–158.

72 Landini, G. (2002) Pattern complexity in organogenesis and carcinogenesis, in: Losa, G.A., Merlini, D., Nonnenmacher, T.F., Weibel, E.R. (Eds) *Fractals in Biology and Medicine*, Vol. III. Birkhäuser Press, Basel, Berlin, Boston, pp. 3–13.

73 Russo, J., Linch, H., Russo, J.H. (2001) Mammary gland architecture as a determining factor in the susceptibility of the human breast to cancer. *Breast J.*, **7**, 278–291.

74 Bissell, M.J., Rizki, A., Mian, I.S. (2003) Tissue architecture: the ultimate regulator of breast epithelial function. *Curr. Opin. Cell Biol.*, **15**, 753–762.

75 Grizzi, F., Chiriva-Internati, M. (2005) The complexity of anatomical systems. *Theor. Biol. Med. Modell.*, **2**, 26–35.

76 Nelson, C.S., Jean, R.P., Tan, J.L., Liu, W.F., Sniadecki, N.J., Spector, A.A., Chen,

C.S. (2005) Emergent patterns of growth controlled by multicellular form and mechanics. *Proc. Natl Acad. Sci. USA*, **102** (33), 11594–11599.

77 Soto, A.M., Sonnenschein, C. (2005) Emergentism as a default: cancer as a problem of tissue organization. *J. Biosci.*, **30** (1), 101–116.

78 Liotta, L.A., Rao, C.N., Barsky, S.H. (1983) Tumor invasion and the extracellular matrix. *Lab. Invest.*, **49**, 636–649.

79 Bissel, M.J., Hall, H.G. (1987) Form and function in the mammary gland: the role of the extracellular matrix, in: Nevell, M., Daniel, C. (Eds) *The Mammary Gland: Development, Regulation and Function*, Plenum Publishing Corp., New York, pp. 97–146.

80 Iozzo, R.V., Cohen, I. (1993) Altered proteoglycan gene expression and the tumor stroma. *J. Cell Biol.*, **78**, 289–308.

81 Losa, G.A., Alini, M. (1993) Sulfated proteoglycans in the extracellular matrix of human breast tissues with infiltrating carcinoma. *Int. J. Cancer*, **54**, 552–557.

82 Wiseman, B.S., Werb, Z. (2002) Stromal effects on mammary gland development and breast cancer. *Science*, **296** (5570), 1046–1049.

83 Maffini, M.V., Soto, A.M., Calabro, J.M., Ucci, A.A., Sonnenschein, C. (2003) The stroma as crucial target in rat mammary gland carcinogenesis. *J. Cell Sci.*, **117**, 1495–1502.

84 Sonnenschein, C., Soto, A.M. (2004) The somatic mutation theory of cancer: growing problems with the paradigm? *BioEssays*, **26**, 1107–2004.

85 Sonnenschein, C., Soto, A.M. (2008) Theories of carcinogenesis: an emerging perspective. *Semin. Cancer Biol.*, **18**, 372–377.

86 Ingber, D.E. (2008) Can cancer be reversed by engineering the tumour microenvironment. *Semin. Cancer Biol.*, **18** (5), 356–364.

87 Kim, J.B., Stein, R., O'Hare, M.J. (2005) Tumour-stromal interactions in breast cancer: the role of stroma in tumourigenesis. *Tumor Biol.*, **26**, 173–185.

88 Provenzano, P., Kevin, P., Eliceiri, W., Campbell, J.M., Inman, D.R., White, J.G., Keely, P.J. (2006) Collagen reorganization at the tumor-stromal interface facilitates local invasion. *BMC Med.*, **4**, 38–53.

89 Beck, A.H., Espinosa, I., Gilks, C.B., van de Rijnand, M., West, R.B. (2008) The fibromatosis signature defines a robust stromal response in breast carcinoma. *Lab. Invest.*, **88**, 591–601.

90 Schnitt, S.J. (2009) The transition from ductal carcinoma *in situ* to invasive breast cancer: the other side of the coin. *Breast Cancer Res.*, **11**, 101–109.

91 Krause, S., Maffini, M.V., Soto, A.M., Sonnenschein, C. (2008) A novel 3D in vitro culture model to study stromal–epithelial interactions in the mammary gland. *Tissue Eng. C*, **14**, 261–271.

92 Grizzi, F., Ceva-Grimaldi, G., Dioguardi, N. (2001) Fractal geometry as useful tool for quantifying irregular lesions in human liver biopsy specimen. *Hum. Pathol.*, **39** (12), 1859–1860.

93 Di Ieva, A., Grizzi, F., Ceva-Grimaldi, G., Russo, C., Gaetani, P., Aimar, E., Levi, D., Pisano, P., Pancioni, F., Rodriguez y Baena, R. (2007) Fractal dimension as a quantitator of the microvasculature of normal and adenomatous pituitary tissue. *J. Anat.*, **211**, 673–680.

94 Smith, T.G. Jr, Bejar, T.N. (1994) Comparative fractal analysis of cultured glia derived from optic nerve and brain demonstrated different rates of morphological differentiation. *Brain Res.*, **634**, 181–190.

95 Bernard, F., Bossu, J.L., Gaillard, S. (2001) Identification of living oligodendrocyte developmental stages by fractal analysis of cell morphology. *J. Neurosci. Res.*, **65**, 439–445.

96 Milosevic, N.T., Ristanovic, D., Stankovic, J.B. (2005) Fractal analysis of the laminar organization of spinal cord neurons. *J. Neurosci. Methods*, **146**, 198–204.

97 Ristanovic, D., Stefanovic, B.D., Milosevic, N.T., Grgurevic, M., Stankovic, J.B. (2006) Mathematical modelling and computational analysis of neuronal cell images: application to dendritic arborization of Golgi-impregnated neurons in dorsal horns of the rat spinal cord. *Neurocomputing*, **69**, 403–423.

98 Jelinek, H.F., Milosevic, N.T., Ristanowich, D. (2008) Fractal dimension as a tool for classification of rat retinal ganglion cells. *Biol. Forum*, **101** (1), 146–150.

99 Mancardi, D., Varetto, G., Bucci, E., Maniero, F., Guiot, C. (2008) Fractal parameters and vascular networks: facts and artefacts. *Theor. Biol. Med. Modell.*, **5**, 12–20.

100 Masters, B.R. (2004) Fractal analysis of the vascular tree in the human retina. *J. Med. Eng. Technol.*, **26** (4), 303–312.

101 Soehle, M., Czosnyka, M., Chatfield, D.A., Hoeft, A., Pen, A. (2008) Variability and fractal analysis of middle cerebral artery blood flow velocity and arterial blood pressure in subarachnoid hemorrhage. *J. Cereb. Blood Flow Metab.*, **28**, 64–73.

102 Landini, G., Rippin, J.W. (1994) Fractal dimension as a characterisation parameter of premalignant and malignant epithelial lesions of the floor of the mouth, in: Nonnenmacher, T.F., Losa, G.A., Weibel, E.R. (Eds) *Fractals in Biology and Medicine*, Vol. I, Birkhäuser Press, Basel, Berlin, Boston, pp. 315–322.

103 Abu Eid, R., Landini, G. (2003) Quantification of the global and local complexity of the epithelial-connective tissue interface of normal, dysplastic, and neoplastic oral mucosae using digital imaging. *Pathol. Res. Pract.*, **199**, 475–482.

104 Abu Eid, R., Sawair, F., Saku, T., Landini, G. (2008) Architectural changes associated with ageing of the normal oral buccal mucosa. *Biol. Forum*, **1**, 131–136.

105 Roy, H.K., Iversen, P., Hart, J., Liu, Y., Koetsier, J.L., Kim, Y., Kunte, D.P. (2004) Down-regulation of SNAIL suppresses MIN mouse tumorigenesis: modulation of apoptosis, proliferation, and fractal dimension. *Mol. Cancer Ther.*, **3**, 1159–1165.

106 Gil, J., Gimeno, M., Laborda, J., Nuviala, J. (2006) Fractal dimension of dog kidney proximal convoluted tubuli sections by mean box-counting algorithm. *Int. J. Morphol.*, **24** (4), 549–554.

107 De Vico, G., Sfacteria, G., Piedimonte, G. (2002) The fractal dimension of the inner surface of neoplastic mammary ducts in mammary fibroadenomas and mammary carcinoma of dog and cat, in: Losa, G.A., Merlini, D., Nonnenmacher, T.F., Weibe, E.R. (Eds) *Fractals in Biology and Medicine*, Vol. III, Birkhäuser Press, Basel, Berlin, Boston, pp. 95–100.

108 Goldschmidt, M.H., Dunstan, R.W., Stannard, A.A., von Tscharner, C., Walder, E.J., Yager, J.A. (1998) Histological Classification of Epithelial and Melanocytic Tumours of the Skin of Domestic Animals, in: *World Health Organization. International Histological Classification of Spontaneous Animal Tumours*, Vol. III, Armed Forces Institute of Pathology, Washington, DC, pp. 22–23.

109 Abramo, F., Pratesi, F., Cantile, C., Sozzi, S., Poli, A. (1999) Survey of canine and feline follicular tumours and tumour-like lesions in central Italy. *J. Sci. Anim. Pract.*, **40** (10), 479–481.

110 Millar, S.E. (2002) Molecular mechanisms regulating hair follicle development. *J. Invest. Dermatol.*, **118**, 216–225.

111 He, H., Xi, H. (2008) Wnt/β-catenin signaling: new (and old) players and new insights. *Curr. Opin. Cell Biol.*, **20** (2), 119–125.

112 De Vico, G., Cataldi, M., Carella, F., Beltraminelli, S., Losa, G.A. (2008) The expression of β-catenin in relation to the fractal organization of canine trichoblastoma tissues. *Biol. Forum*, **101** (1), 136–138.

113 Losa, G.A., De Vico, G., Cataldi, M., Carella, F., Beltraminelli, S. (2009) Contribution of connective and epithelial tissue components to the morphologic organization of canine trichoblastomas. *Connect. Tissue Res.*, **50**, 28–29.

114 Penn, A.I., Loew, M.H. (1997) Estimating fractal dimension with fractal interpolation function models. *IEEE Trans. Med. Imaging*, **16**, 930–937.

115 Penn, A.I., Thompson, S., Brem, R., Lehman, C., Weatherall, P., Schnall, M., Newstead, G., Conant, E., Ascher, S., Morris, E., Pisano, E. (2006) Morphologic blooming in breast MRI as a characterization of margin for discriminating benign from malignant lesions. *Phys. Rep.*, **246**, 1–100.

116 Li, H., Giger, M.L., Olopade, O.I., Li, L. (2007) Fractal analysis of mammographic parenchymal patterns in breast cancer risk assessment. *Acad. Radiol.*, **14**, 513–521.

117 Bocchi, L., Coppini, G., Nori, J., Valli, G. (2004) Detection of single and clustered microcalcifications in mammograms using fractals models and neural networks. *Med. Eng. Phys.*, **26** (4), 303–312.

118 Kriege, M., Brekelmans, C.T., Boetes, C., Rutgers, E.J.T., Klijn, J.G.M. (2004) Efficacy of MRI and mammography for breast-cancer screening in women with a familial or genetic predisposition. *Lab. Invest.*, **88** (6), 591–601.

119 Chen, H.H., Prey, M.U., Babcock, D.S., Day, N.E. (1997) Breast carcinoma cell kinetics, morphology, stage, and the preclinical screen-detectable phase. *Lab. Invest.*, **2**, 9–23.

120 Megalooikonomou, V., Zhang, J., Kontos, D., Bakic, P.R. (2007) Analysis of texture patterns in medical images with an application to breast imaging. *Proc. SPIE Med. Imaging*, **6514**, 14–21.

121 Soares, P., Andruszkiewicz, P., Freire, M., Cruzl, P., Pereira, M. (2007) Self-similarity analysis applied to 2D breast cancer imaging. *Proc. Int. Conf. Systems Networks Commun.*, **1**, 1–6.

122 Rangayyan, R.M., Nguyen, T.M. (2007) Fractal analysis of contours of breast masses in mammograms. *J. Digit. Imaging*, **20** (3), 223–237.

123 Wiener, J.I., Schilling, K.J., Adami, C., Obuchowski, N.A. (2005) Assessment of suspected breast cancer by MRI: a prospective clinical trial using a combined kinetic and morphologic analysis. *Am. J. Radiol.*, **184**, 878–886.

124 Fuseler, J.W., Millette, C.F., Davis, J.M., Carver, W. (2007) Fractal and image analysis of morphological changes in the actin cytoskeleton of neonatal cardiac fibroblasts in response to mechanical stretch. *Microsc. Microanal.*, **13**, 133–143.

125 Kanduc, D., Capone, G.M. (2008) The similarity profile of the human proteome as a fractal dimension. *Biol. Forum*, **101**, 142–145.

126 Kanduc, D. (2010) Protein information content resides in rare peptide segments. *Peptides*, **31**, 983–988.

127 Kanduc, D., Capone, G., Pesce Delfino, V., Losa, G.A. (2010) The fractal dimension of protein information. *Adv. Stud. Biol.*, **2** (2), 53–62.

128 Buiatti, M., Buiatti, M. (2008) Chance versus necessity in living systems: a false antinomy. *Biol. Forum*, **101** (1), 29–66.

129 Von Bertalanffy, L. (1968) *General System Theory. Foundations, Development, Applications*, Georges Braziller Inc., New York.

130 Minati, G. (2008) *New Approaches for Modelling Emergence of Collective Phenomena*, Polimetrica, Segrate (I).

131 Konopka, A. (2007) *Systems Biology: Principles, Methods, and Concepts*, CRC Press, Boca Raton.

132 Chong, L., Ray, L.B. (2002) Whole-istic biology. *Science*, **295**, 1661.

133 Kitano, H. (2002) Systems biology: a brief overview. *Science*, **295** (5560), 1662–1664.

134 Wu, X., Huan, T., Pandey, R., Zhou, T., Chen, J.Y. (2009) Finding fractal patterns in molecular interaction networks. *Int. J. Comput. Biol. Drug Des.*, **4**, 340–352.

135 Honda, H. (1999) Genes do not determine biological shape directly. *Forma*, **14**, 287–293.

136 Nusslein-Volhard, C., Frohnhofer, H.G., Lehmann, R. (1987) Determination of anteroposterior polarity in *Drosophila*. *Science*, **238** (4834), 1675–1681.

137 Fire, A., Xu, S., Montgomery, M.K., Kostas, S.A., Driver, S.E., Mello, C.C. (1998) Potent and specific genetic interference by double-stranded RNA in *Caenorhabditis elegans*. *Nature*, **391** (6669), 806–811.

138 Kiger, A.A., Baum, B., Jones, S., Jones, M.R., Coulson, A., Echeverri, C., Perrimon, N. (2003) A functional genomics analysis of cell morphology using RNA interference. *J. Biol.*, **2** (4), 27–37.

139 Gehring, W., Qian, Y.Q., Billeter, M., Furukubo-Tokunaga, K., Schier, A.F., Resendez-Perez, D., Affolter, M., Otting, G., Wüthrich, K. (1994) Homeodomain-DNA recognition. *Cell*, **78**, 211–223.

140 Lambert, D., Rezsohasy, R. (2004) *Comment les Pattes Viennent aux Serpents. Essai Sur l'étonnante Plasticité du Vivant*, Flammarion, Paris.

141 Gonzalo, M.J., van der Knaap, E. (2008) A comparative analysis into the genetic bases of morphology in tomato varieties exhibiting elongated fruit shape. *Theor. Appl. Genet.*, **116**, 647–656.

14 Models of Cell Migration

Michael Meyer-Hermann[1,2] *and Tilo Beyer*[3]
[1]*Helmholtz Centre for Infection Research, Department of Systems Immunology, Inhoffenstr. 7, 38124 Braunschweig, Germany*
[2]*Bio Center for Life Science, University of Technology Braunschweig, Spielmannstr. 7, 38106 Braunschweig, Germany*
[3]*Institute for Molecular and Clinical Immunology, Otto-von-Guericke University, Leipziger Str. 44, 39120 Magdeburg, Germany*

Systems Biology: Advances in Molecular Biology and Medicine, First Edition. Edited by Robert A. Meyers.

Keywords

Leukocytes
White blood cells which are derived by differentiation from hematopoietic stem cells in the bone marrow. These cells form the basis of immune defense against pathogens and infections. They can be subdivided into cells associated with the innate and adaptive immune systems. Important innate immune cells comprise neutrophils, monocytes, macrophages, natural killer cells, and dendritic cells. The adaptive immune system comprises T and B cells.

Lymphocytes
Lymphocytes comprise the main cellular compartment of the adaptive immune system. While B lymphocytes (B cells) are at the heart of the antibody-mediated humoral immune response, T lymphocytes (T cells) can have different functions, ranging from killing infected cells by cytotoxic T cells or providing help in activation by T helper cells. A specific subtype of T cells are regulatory T cells, which help to control the strength of an adaptive immune response. While both, T and B cells stem from the bone marrow, B cells mature also in the bone marrow, while T cells are selected in the thymus.

Germinal center (GC)
The GC is a site of a specific B-cell-mediated immune response. In a GC reaction, the B cells are activated by a pathogen, together with T-cell help in secondary lymphoid organs (e.g., spleen and lymph nodes). Activated B cells enter a specific B-cell compartment (the B-cell follicle) in the same organ and expand while mutating the antibody coding region. After 3 days, about 10 000 B cells will have emerged that exhibit a large diversity

of antibodies. These B cells express the antibody which is subsequently tested for its affinity to the pathogen. B cells surviving this selection process either return to the state of division and mutation to enter another round of selection (a process termed recycling), or leave the GC as plasma or memory cells. Memory cells allow for a faster protection if the same or a similar pathogen is re-encountered. Plasma cells generate large amounts of antibodies to protect against an acute immune challenge. As plasma cells derived from the GC are selected for their high affinity to the pathogen, these antibodies are highly efficient compared to the first line of antibodies generated earlier. Thus, the GC gives rise to a process termed "antibody affinity maturation."

Stromal cells
The connective tissue of organs is built from stromal cells, such as fibroblasts. Stromal cells are also present in secondary lymphoid organs, and are the backbone of organ organization and cell migration.

Chemokines
Soluble or membrane-bound substances which can be detected by migrating cells. Chemokines can guide cells towards a specific organic compartment, or they may be involved in the transmigration of cells through epithelial layers. Organs such as lymph nodes exhibit a special distribution of different chemokines which allows the homing of different cell types (e.g., T and B cells) to their respective compartments.

Chemotaxis
The process of migration of cells in the direction of an increasing chemokine concentration. Chemotaxis must be distinguished from chemokinesis, where not the direction but rather the speed of the cell is controlled by chemokines.

Cell motility is the basis of life's existence; indeed, the ontogenesis of any organism relies on an active rearrangement of cells that interpret signals in their environment, and then seek specific targets with a purposeful behavior that appears to be the result of a complex decision process. In organogenesis, cell rearrangement leads to final structures of a quasi-static nature such that, when cells achieve their final positions their movement is minimal and only the maintenance of a tissue in response to stress or cell death will require local rearrangements. The situation is different for cells of the immune system, however. Leukocytes must move constantly through the entire organism to detect danger signals, to transport pathogens to sites such as secondary lymphoid organs (e.g., lymph nodes and spleen) where they are presented, and to induce adequate immune responses. In addition to this surveillance function, which is based on leukocyte motility, the secondary lymphoid organs differ from other organs in their cellular content. Leukocytes (as part of the adaptive immunity system), and the T and B cells (lymphocytes) move constantly in and out of the organs, where lymphocytes are exchanged on a daily basis. Despite such motility, lymph nodes exhibit a stable cellular organization that is achieved by a flow equilibrium of cells entering and leaving the secondary lymphoid tissues. In order to understand the emergence of tissue patterns, it is important to recognize how cells migrate, and how such migration is regulated by the key molecules involved in organogenesis. In this chapter, the various modes of cell movement, the signals inducing motility, and the resultant cell behavior are discussed. Although targeted essentially at graduates in the life sciences, physics, and mathematics, the text may also benefit groups conducting research in this field, with specific points being described in detail and open questions identified.

1 Overview

The process of cell migration was most likely first witnessed over three centuries ago, by Antonie van Leeuwenhoek [1]. Among the many single cellular organisms that he frequently observed was the amoeba, which today remains a model for cell migration, notably in the form of its representative, *Discoideum dictyostelium.* The term "amoeboid shape changes" became popular and was also applied to the morphologies of white blood cells, described in great detail by Paul Ehrlich [1]. In this chapter, attention will be focused on leukocyte motility, mainly because these cells exhibit the greatest number of motility features, although the migration of stromal cells (which may serve as a guiding backbone for leukocyte migration) will also be considered. Within the chapter, the molecular basis and phenomenology of cell migration will first be introduced, and the cytoskeleton of cells and mechanisms involved in cell migration discussed, using a three-step model as a reference for fibroblast migration. The mode of fibroblast migration is then compared to the migration of leukocytes, followed by details of how cells exert forces on a substrate. The concept of constriction rings will be introduced as an alternative to focal adhesion points.

Having described the cellular components and mechanisms required for cell migration, attention will be turned to the control of both the speed and direction of cell migration, of how cells can sense chemokines, and how the sensitivity for a specific chemokine may be regulated and integrated. Multi-photon imaging allows the intravital observation of cell migration and chemotaxis. with the acquired data suggesting a concept of persistent random walk; this provides the possibility of distinguishing chemotaxis from chemokinesis, and assumes that chemokines control cell speed rather than direction. Finally, guidance by extracellular fibers rather than by chemokines will be considered. Multi-photon imaging highlights the difficulties of interpreting these complex data sets, and the need for mathematical modeling. Among the different mathematical modeling approaches, partial differential equations (PDEs), agent-based whole-cell models, and subcellular models will be discussed. These different concepts will be compared and their suitability for specific biological problems investigated. Finally, two examples are provided of applying mathematical models to lymphoid tissue organization, the aim being to show how such modeling may produce surprising results, adding great value to the data acquired by comprehensive experimentation.

2
Phenomenology and Mechanisms of Cell Motility

2.1
Cytoskeleton and Cell Shape

In order for a cell to exhibit a mechanical response and to undergo active motion, the presence of three major components is required, namely the cell membrane, the cytoplasm, and the cytoskeleton [2]. The *cell membrane* can be considered as a flexible two-dimensional (2-D) object with mechanical properties that are, in the majority of cases, related to the tightly connected cortical shell that consists of actin and intermediate filaments. The *cytoplasm* is basically a viscous fluid that is held under pressure and bound by the *cell membrane*. In contrast to plant cells, the cytoplasm in animal cells is generally under low pressure. The mechanical response of animal cells is determined mainly by the cytoskeleton, which comprises three distinct components (Fig. 1) [2]:

- Microtubules: These are hollow cylinders which are about 25 nm in diameter and several micrometers in length. The centralized organization of microtubules radiating outwards from an area close to the nucleus predisposes them for the transport of organelles and proteins through the cell. Microtubules serve also as a structural support and as force generators, due to their polymerization dynamics. Motor proteins can shift microtubules along each other; such forces lead to chromosome separation during cell division.
- Microfilaments or actin filaments: These are formed by polymerization of the protein actin, and can assemble

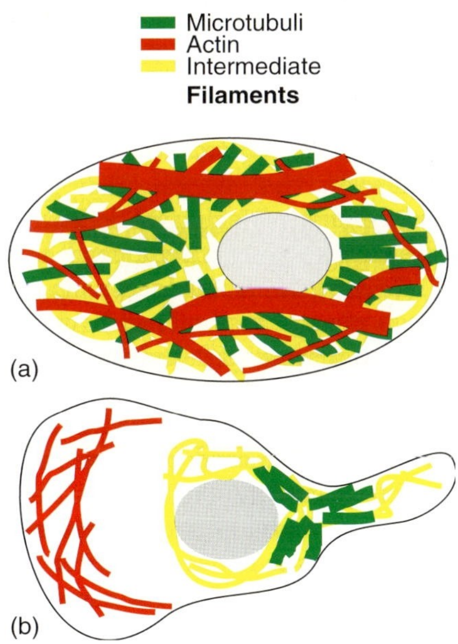

Fig. 1 The cell cytoskeleton. (a) A schematic representation of the cytoskeletal components of a stromal cell. The microtubules (green) radiate outwards from an area close to the nucleus (small gray ellipse). The microfilaments (red) form long stress fibers of various thickness that sometimes cross the whole cell. The intermediate filaments (yellow) form an extensive network which is more dense around the nucleus; (b) The cytoskeleton of a migrating lymphocyte. The most prominent difference from stromal cells is that the microtubules and intermediate filaments are retracted in a narrow region which forms the uropod (the narrow extension to the right of the figure), while the microfilaments concentrate at the opposite end of the cell, which forms the leading edge of the migrating lymphocyte.

to create filaments of considerable thickness (even thicker than microtubules); they are also able to resist stretching and compression. These filaments exhibit a dynamic redistribution by polymerization and de-polymerization; when forces are exerted on certain cell types, actin can form stress fibers within minutes. Microfilaments are thought to be the most important structure-inducing force during cell migration. Actin filaments are responsible for active deformations of the cell membrane by forming microvilli, protrusions, and pseudopods. The mechanical properties of actin filaments have been extensively studied [3].

- Intermediate filaments: The typical thickness of these filaments (9–11 nm) is between that of the actin filaments and microtubules. They are highly resistive to stretching when a large strain is applied [4, 5]. Intermediate filaments form a tight network throughout the cell body, and connect the nucleus to the other parts of the cytoskeleton, and to the cell membrane. The stiff nucleus, in combination with such connectivity, has a major influence on the mechanical properties of the cell [6], and studies show that intermediate filaments are involved in cell migration [7, 8]. Intermediate filaments may also connect cells to other cells via desmosomes, and to the extracellular matrix (ECM) via hemidesmosomes.

The morphology of migrating lymphocytes is characterized by these three elements [9–17]. At the front of the cell is the *lamellipodium*, the leading edge of which consists of one or more membrane protrusions that are generated by the polymerization dynamics of the actin network. Fingers which extend to the front are termed *filopodia*, and these give rise to *pseudopods* which are filled with cytoplasm. The lamellipodium is followed by the round main cell body with the nucleus. The rear is formed by the trailing edge which consists of a narrow, cytoplasmic, backward-directed extension called the *uropod*. The leading edge is rich in chemokine receptors, adhesion molecules, actin, and actin-related proteins, while towards the rear – between the uropod and the main cell body – the microtubule organizing center (MTOC) is located. From the MTOC, the microtubule network radiates outwards, to reduce the mechanical stiffness of the cell and allow for larger cell deformations. The results of several studies have indicated that the intermediate filament network is also retracted into the uropod [7, 15], most likely for the same reason. Intercellular adhesion molecules and myosin II are concentrated in the uropod [2].

2.2 Mechanisms of Cell Migration

Cell deformation and migration are closely correlated properties that have been well described in experiments. As cell migration appears to be a tightly orchestrated dynamics of several intracellular elements, the question might be asked as to which intracellular processes actually drives active cell migration? Among proposed processes, the three most relevant are deemed to be actin polymerization, myosin sliding, and cortical expansion [9, 18] (these mechanisms are discussed from a biophysical aspect in Sect. 3.1):

- *Actin polymerization*: This is considered to be the dominant mechanism of force generation in migrating cells.

Actin exists in two forms: soluble actin monomers; and polymerized as microfilaments. The polymerization of actin to microfilaments is a polarized process that occurs at only one end of the polymer, while the other end is depolymerized. The growth of the filaments generates an actin network that mediates the stiffness required to allow for the development of membrane protrusions and pseudopod growth. Actin polymerization extents these structures by an elongation of the microfilaments immediately beneath the membrane.

- *Myosin sliding*: This is a mechanism that shifts actin filaments along each other. The filaments of myosin-II are crosslinked with actin, which in turn stabilizes the microfilament structure and generates contractile forces. Myosin sliding is considered to be the reason for the retraction of the rear of a cell during migration, and suggests that migration forces are generated in the uropod. Cells can migrate in the absence of functional myosin-II, but at a significantly lower speed.
- *Cortical expansion*: This model explains pseudopod formation by a locally increased pressure of the cytoplasm which is induced by ion influx at the leading edge of the cell. Additionally, hydrostatic pressure induced by cell contraction at the uropod may squeeze the cell forward. Although the actin cytoskeleton is reorganized during these processes, it plays a different role than in the actin polymerization model, as it is not assumed to generate the forces that drive the cell forward.

2.3 Three Step Model of Fibroblast Migration

The reference model for mammalian cell migration are fibroblasts on 2-D substrates. The fibroblasts, which form part of the connective tissue and produce collagen, move at a rate of less than 1 μm min^{-1} on 2-D substrates, and connect each other to a network of cells and collagen fibers. The results of fibroblast experiments led to the creation of a three-step model of *mesenchymal migration* [9–12, 16, 18, 19].

The first step of migration involves the firm adhesion of one or more protrusions to the substrate (Fig. 2). Pseudopod formation at the leading edge of a cell is dependent on actin polymerization, which is also the cause of the ruffling of filopodia

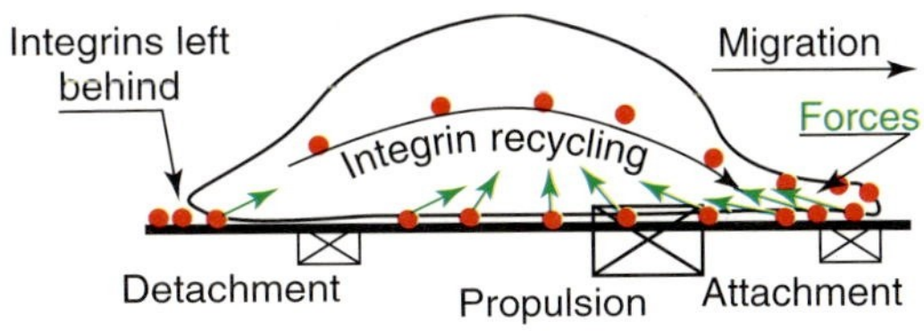

Fig. 2 The three-step model of cell migration (in this scheme, the cell moves from left to right). First, protrusions are formed at the leading edge and attach the cell to the substrate via integrins (red dots). By using these integrins as anchors, the dynamics of the cytoskeleton induces forces (green arrows) onto the contacts. The rear of the cell detaches from the substrate, either by cleaving the integrins and leaving them behind, or by modulating their affinity and keeping them in the cell membrane. The integrins at the rear are transported along the membrane or, with the help of vesicles, to the leading edge of the cell.

that precedes pseudopod formation. The firm adhesion is mediated by adhesion molecules from the integrin family, and this leads to the formation of focal adhesion points. This premature form of focal contacts develops within 1–2 min, while the mature focal adhesion site requires 10–20 min for its complete formation [20, 21]. Upon attachment, the integrins provide signals to the interior of the cell, finally leading to actin polymerization and the formation of stress fibers.

The second step in fibroblast migration is the formation of a force gradient acting on the focal adhesions at the leading edge and the adhesive bonds at the trailing edge. The stress applied to the focal adhesions induces signals that strengthen the actin polymerization and the adhesion.

The final step is detachment of the rear of the cell from the substrate. Two mechanisms have been proposed for this: (i) the cleavage of either integrin molecules or the anchoring structures in the cell; and (ii) an affinity modulation of integrins. In the latter case, the integrins become unbound from the substrate and are transported to the leading edge by integrin-coated membrane vesicles [22, 23]. This process can transport the equivalent of the cell surface in 4–8 min along the cell length.

The question then arises as to how the migration of mesenchymal cells would depend on the strength of adhesion. In agreement with intuition, migration is suppressed at both low and high adhesion levels, with cells moving slowly either because of a lack of force transmission to the substrate at the leading edge, or because of an increased sticking to the substrate at the rear [18, 22]. Typically, the cells exhibit maximum motility in a range of medium adhesion.

A second question would relate to whether this situation might change in three dimensions, or in "real" tissues. Fibroblast migration in a three-dimensional (3-D) ECM exhibits similar properties as in the 2-D case, although the leading edge is no longer flat but rather exhibits several cylindrical protrusions [9]. In addition, the cell morphology become is more bipolar or tripolar compared to the spread triangular shape in the 2-D case. In the 3-D case, contact guidance gains much greater importance, as cell migration in the ECM follows the fibers and pass through larger gaps, thus minimizing the resistance induced by the ECM. The ECM can be degraded by the action of matrix metalloproteinase (MMP) enzymes, which are produced by the migrating cell itself to provide sufficient space into which the cells can migrate [9, 17]. Consequently, it would be beneficial to investigate the stromal background of cell migration in greater detail.

2.4
Stromal Network Dynamics in Lymphoid Organs

The ECM not only serves as the mechanical backbone of connective tissue, but also plays an important role in the differential migration of leukocytes in lymphoid tissues. As such, the ECM is distributed inhomogeneously and composed of different environments. Typically, lymphoid tissue generally is organized into T zones and follicles which are dominated by two subtypes of lymphocyte, namely the T and B cells, respectively. The ECM fibers differ between the T zone and the B-cell follicles.

Stromal cells set up a dense reticular network of comparatively thick fibers in the T zone [24, 25]; these fibers have a typical spacing of 5–20 μm, such that only a small part of the lymphocyte surface is in contact with the reticular network [26, 27]. This causes the lymphocytes to adopt a more spherical shape compared to other situations [9, 18]. The T zone contains ECM fibers such as collagen types I and III [17, 28, 29], fibronectin [28, 30], tenascin [28, 30], as well as lesser amounts of collagen IV [28, 31] and laminin [28, 29, 31].

The B-cell follicles contain only a sparse ECM network, including collagen I and III, and small amounts of fibronectin and vitronectin [28, 30]. These few proteins are correlated with vessels passing through the B-cell follicle [27, 32].

Upon activation, the B cells expand in the follicles and induce the development of specific structures termed *germinal centers*, in which B cells are somatically mutated to generate high-affinity antibodies. This area appears almost completely devoid of ECM proteins [24, 25, 28, 30, 31, 33–35]. More specifically, the follicle divides into a mantle zone, which is virtually free of all ECM fibers, and the germinal center, which contains some vitronectin in the apical light zone [28] and forms that part of the germinal center distant to the T zone. A similar pattern can be observed for the proteoglycans, another component of the ECM fibers [27].

In view of the ECM properties being specific for the state of cell activation, the ECM must be considered as a dynamic structure, the local composition and density of which are changed upon encountering immune stimuli.

It is important to note that the ECM components are not distributed homogeneously within the zones; rather, most of the T-zone fibers and vitronectin accumulate at the boundary between the T zone and the B-cell follicles [28, 32]. This inhomogeneity may be a result either of mechanical displacement by the B cells in follicles, or of a locally increased ECM production [36].

Tab. 1 Integrin subunits expressed in compartments with corresponding ECM components [28].

ECM component	***Integrin***
Collagen	$\alpha_1\beta_1, \alpha_2\beta_1, \alpha_3\beta_1$
Laminin	$\alpha_1\beta_1, \alpha_2\beta_1, \alpha_3\beta_1, \alpha_6\beta_1$
Fibronectin	$\alpha_3\beta_1, \alpha_4\beta_1, \alpha_5\beta_1, \alpha_V\beta_1, \alpha_{IIb}\beta_3, \alpha_V\beta_3$
Vitronectin	$\alpha_V\beta_3, \alpha_V\beta_5$
Thrombospondin	$\alpha_4\beta_1, \alpha_5\beta_1, \alpha_V\beta_3$
Tenascin	$\alpha_V\beta_3$

The integrins guide leukocyte migration through the lymphoid tissue, with the integrin distribution, in general, reflecting the presence of the corresponding ECM components to which the integrins can bind (Table 1). The mantle zone in activated B cell follicles is mostly free of ECM, and does not contain any integrin subunit of the types $\beta_1, \beta_3, \beta_4, \alpha_1, \alpha_2, \alpha_3, \alpha_4, \alpha_5, \alpha_6, \alpha_{I/b}$, and α_v [28]. The T zone contains lymphocytes expressing the β_1 and α_4 subunits binding to vascular cell adhesion protein 1 (VCAM-1) and fibronectin. At the border of follicle and T zone, only β_1 of the above-listed integrin subunits has been identified. However, there exist differences between species; for example, monkeys show β_1 integrin subunit expression in the germinal center, whereas humans do not [28]. As yet, it remains unclear as to whether lymphocytes adapt their expression profile to the presence of the ECM components in a compartment, or the integrin expression pattern of the lymphocytes guides them to a compatible compartment: this truly is a "chicken or egg" situation.

2.5 Migration of Leukocytes in Lymphoid Organs

Leukocytes recirculate between the blood and lymphoid organs, their motility being critical for the successful detection of immune stimuli and appropriate immune responses. In addition, they perform an amoeboid migration which differs from the mesenchymal migration mode (see Sect. 2.3) in several aspects [9, 18, 19].

The results of various experiments have suggested that leukocytes move on 2-D substrates much like fibroblasts, but that migration in three dimensions can occur independent of adhesion [9]. Whereas, in mesenchymal migration the integrins are capped and leave a track of remaining adhesion receptors behind the migrating cell, the leukocytes keep their adhesion receptors with them. In this way, the formation of focal contacts and stress fibers is suppressed, and the cortical actin is diffusely distributed. Adhesive contacts are then replaced by biomechanical contacts in order to provide the anchorage needed to transmit the traction forces [37]. The contact is mediated by bleb-like structures that expand in between the ECM fibers, while the anchors may appear as lateral protrusions that hook into the ECM [38].

Microtubules play a minor role in active migration. Although microtubules are likely involved in controlling the actin–myosin interaction [39], in T cells the MTOC is formed as a rod-like shaft that bears basically all microtubules [40] (Fig. 1b). The position of the MTOC defines the polarity of the cell; typically, the MTOCs are moved "out of the way" to increase the deformability of the cell. The reversible nature of microtubule retraction is relevant when the T cells make cognate contact with the antigen-presenting cells (APCs) since, at this point, the microtubules are required to support the transport of proteins and organelles to the site of signal exchange with the adjacent cell [39, 41].

2.6 Switching between Migration Modes

The characteristics of lymphocyte migration in follicles suggest that the mesenchymal migration mode does not apply to lymphocytes; this is related to the less-organized cytoskeleton of lymphocytes, which lacks stress fibers in order that the potential of the transmigrating epithelial layers can be increased. Interestingly, fibroblasts employ an amoeboid-type of migration in the presence of the chemokine interleukin 8 (IL-8) [9], without the formation of focal adhesions. The same situation applies to tumor cells, which can switch between mesenchymal and amoeboid migration modes when the activity of the MMPs is blocked [42–44]. The switch to amoeboid migration is correlated with a change in morphology, from a constitutive elongated spindle-shape to a less-polarized, more ellipsoidal, flexible morphology. At this point, instead of dissolving the ECM the tumor cells will be deformed and squeeze through gaps in the ECM [44]. This potential transition between migration modes raises the question of whether leukocytes might also switch to a mesenchymal migration mode in a 3-D situation [14]. Notably, their migration in a 2-D situation is reminiscent of fibroblast migration [9].

3
Mechanics of Migration

The phenomenology of cell migration (as described in Sect. 2) may now be set into a biomechanical context, whereby the main point in question is exactly how the cell exerts forces onto either a substrate or a reticular network. This information is required in order to identify the possible points of control of cell motility, and is of particular interest for the development of mathematical migration theories aimed at representing the mechanics of cell migration.

3.1
Force Generation by Fibroblasts and Leukocytes

Previous experimental studies have concentrated on the actin polymerization dynamics and force generation during cell migration on 2-D substrates [45–48], with the forces of migrating fibroblasts having been determined by measuring the deformation of elastic substrates during cell migration. By knowing the mechanical properties of the substrate, it is possible to calculate the forces; typically, the stress applied at the cell–substrate contacts is in the range of 5.5 $\mathrm{nN\mu m^{-2}}$ [46] to 80–130 $\mathrm{nN\mu m^{-2}}$ [45]. The forces are exerted in all directions, with a center located directly in front of the nucleus [45, 49] (see Fig. 3). These traction forces may reach 200 nN in the case of fibroblasts, while the net force is in the order of 10–30 nN. The strongest forces are generated perpendicular to the direction of motion [45].

Recent advances in microscopy technology have enabled the determination of traction forces of mesenchymal migration in 3-D environments [51, 52]. The technique relies on the same principle as the deformation of elastic 2-D substrates, but in the 3-D case the density of fibers of a collagen gel matrix is influenced by a migrating cell. Ultimately, the size of the traction forces is found to be comparable to that in the 2-D situation. In contrast,

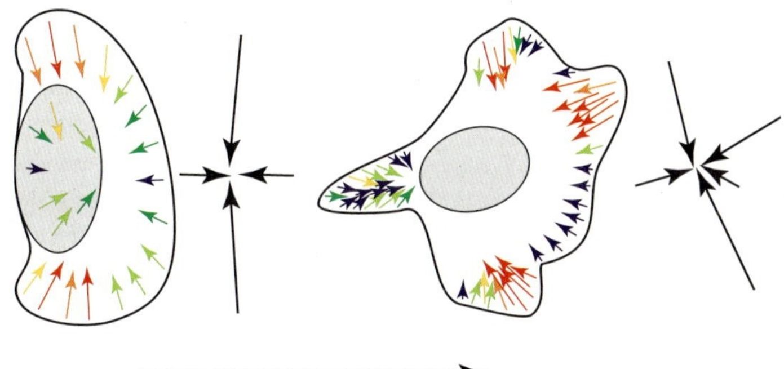

Fig. 3 The traction force pattern of a migrating keratocyte (left) and a migrating fibroblast (right). The traction forces are predominantly acting perpendicular to the direction of motion with a symmetrical pattern in keratocytes, and a tendency to a multipolar pattern in fibroblasts. This is illustrated by the resulting force vectors to the right of each panel. The center of all force vectors is located slightly in front of the nucleus. Colors from blue to red indicate increasing strength of local forces. The figures are inspired from the results provided in Refs. [45, 50].

the traction force pattern is much more complex and appears to be characterized by a superposition of a rolling motion with the planar contraction known from the 2-D situation [52].

The traction forces generated by leukocytes are about one order of magnitude lower compared to fibroblasts, and have been estimated to be in the range of 30 nN for neutrophils [10, 18], thus, in the range of the net force exerted by fibroblasts [45]. At the same time, neutrophils migrate approximately one order of magnitude faster than fibroblasts, reaching average speeds of 10 $\mu m\ min^{-1}$ and peak speeds of 30 $\mu m\ min^{-1}$ [18, 53].

Unfortunately, data acquired with neutrophils cannot be transferred directly to lymphocytes, despite both cells belonging to the class of leukocytes [54]. In contrast to T cells, in which microtubules are recollected in the MTOC at the rear of the cell [39], microtubules in neutrophils are located close to the leading edge and are less organized [54]. It has been hypothesized that this difference is related to a lack of proliferation potential in neutrophils, whereas the T cells are highly proliferative; consequently, in contrast to T cells an organized system of microtubules might not be required in neutrophils [54]. Instead, it may be essential for neutrophils to optimize their shape flexibility, which would allow them to pass narrow gaps very efficiently and to reach infected tissues. Clearly, as neutrophils represent the first line of defense against invading pathogens, agility is most essential for their function.

3.2 Lymphocyte Migration with Constriction Rings

Historically, cinematographic records ("time-lapse video microscopy" in modern terms) were generated by Lewis for both lymphocytes and neutrophils [55, 56], in order to produce a one-to-one correlation between the changes in a cell's shape and its migration behavior. Lewis was probably the first to have identified the phenomenon of the *constriction ring* in cell migration, which provides an important insight into the mechanics of leukocyte migration. This mode of migration is independent of focal adhesion points (see Sect. 2.5), and also of ECM degradation by the MMPs [17]. Instead, the cells exert forces by using the constriction ring [17, 39, 55, 57, 58] (Fig. 4), which is pressed

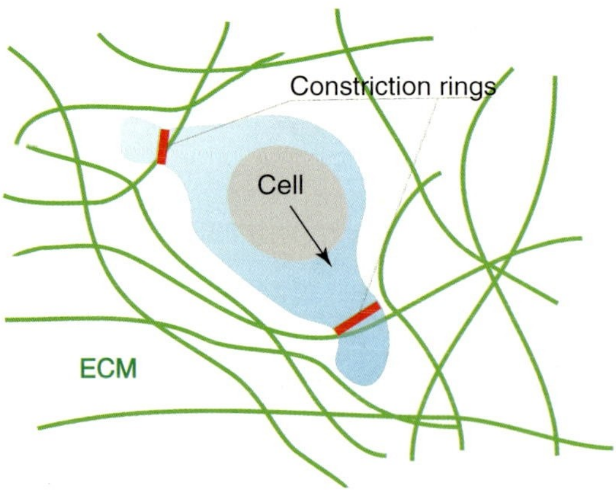

Fig. 4 Schematic representation of migrating lymphocytes. Lymphocytes form constriction rings (red) along their cell body that allow the cell to squeeze through gaps in the ECM (green lines).

along the boundary of gaps in the ECM and enables a flow of cytoplasm through the gap when the rear of the cell contracts [59] – much in the way that a person would push them self through the frame of a narrow door. When the constriction ring forms in the absence of the ECM, it has a diameter that is approximately four-fifths of the lymphocyte size [39], but it can adapt to ECM gaps as small as about 1–2 μm [44].

The dynamics of intracellular processes during lymphocyte motion can be disentangled by performing inhibition or overexpression experiments. For example, when microtubules are disrupted by drug treatment the constriction ring is shrunk to about one-fifth of the cell diameter. Subsequently, a back-and-forth oscillation of the constricting ring is observed between the two ends of the cell within a period of 2.5–4.5 min, which results in the ring having an average speed of about 0.1 $\mu m\ s^{-1}$. At the same time, membrane activity is increased on the cell pole opposite to the ring. Such oscillatory behavior has been described by a model which is based exclusively on actin polymerization dynamics [60]; however, it was also suggested that constriction ring formation would depend on the microtubules, and that factors derived from the microtubules would determine the position of the ring in the cell.

Microfilaments are present asymmetrically around the constriction ring, with an accumulation towards one cell pole, whereas the free actin is distributed homogeneously. When the microfilaments are disrupted the cell loses its overall spherical shape and resembles like the Greek letter phi (ϕ), appearing as a spherical cell body with two opposing, long and thin extensions of unequal length. When both the microtubules and microfilaments are disrupted, the cell will remain spherical and exhibit blebbing of the cell membrane. The enhancement of microtubule formation by drug treatment may also cause an elongated cell shape similar to that observed in the absence of microfilaments.

These shape characteristics have been investigated (both experimentally and theoretically) in greater detail for fibroblasts [61], which exhibit the same elongated ϕ shape as lymphocytes when the microfilaments are disrupted. The theoretical concept applied to the ϕ shape was that the cortical shell is under stress and trying to minimize the elastic energy and apparent cell surface when spanned along the microtubular rod, which in turn generates the long arms of the cell. In line with this, disrupting the microtubules also induced the same spherical structure with a blebbing surface, as in lymphocytes. Although no constriction ring was observed in these fibroblasts, the results suggested that the microtubules were responsible for the ϕ shape in lymphocytes with disrupted microfilaments.

The involvement of myosin in these shape dynamics has also been reported [39, 58]. In untreated cells, myosin is distributed almost uniformly, though with a slightly higher concentration in the cortical shell. Any subsequent disruption of the microtubules causes a relocation of the myosin to the plasma membrane, thus forming a ring or a tight cap. When the microfilaments are destroyed, the myosin is redistributed into the newly formed long extensions of the cell; this suggests that the microtubules control the distribution of myosin, which has the potential to contract microfilaments on only one side of the cell.

By taking together the observed organization of the cytoskeletal elements under

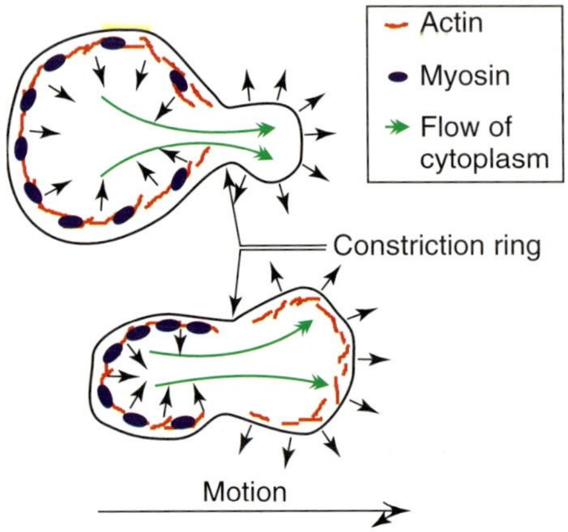

Fig. 5 Migration forces associated with constriction rings. The rear of the cell contains an active actin–myosin machinery that compresses the cytoplasm and causes its flow through the constriction ring. The resulting hydrostatic pressure induces outward-directed forces at the front of the cell. In essence, the cell is squeezing itself through the constriction ring by contracting its rear and expanding its front. The process is initialized by an asymmetry in the actin distribution, with no or little actin at the front. Upon completion of the compression/expansion cycle, the cortical shell is restored at the front of the cell, ready to form a new constriction ring.

the influence of inhibition and overexpression experiments, it was concluded that the constriction ring was, indeed, the site of force generation leading to forward protrusion and cytoplasmic flow [39, 58] according to the following model (Fig. 5). First, the constriction ring on the moving lymphocytes separates the cytoplasm into organelle-free (containing only the centrosome) and organelle-containing parts; during this process the constriction ring exhibits a wave-like motion that passes down the length of the cell [37], similar to that observed when disrupting microtubules [39]. On this occasion, however, the ring is fixed to the collagen matrix by bleb-like structures [37]. As a consequence, the microtubules enforce an actin–myosin system with activity in the direction of cell polarity. In turn, the induced distribution of myosin provides a control over the cytoplasmic flow towards the active membrane in the organelle-free part of the cytoplasm, where the new protrusions are formed. As the cytoplasm in the blebs does not join the cytoplasm squeezing through the ring, this allows the cell to move through the ring within a range of 16 to 80 s (mean 30 s) [37], as was first described by Lewis [55].

The ring oscillation involves an assembly–disassembly cycle whereby, when a ring approaches one end of the cell it leaves a cap of myosin to be used for the ring contraction. In the same way, a new ring forms at the same end and travels along the cell in the opposite direction; while on its way, the latter ring recruits myosin from the cap, and this is deposited in another cap at the opposite end of the cell, that the ring then approaches.

The natural hypothesis derived from this assembly–disassembly cycle is an association of the persistence time of lymphocytes; that is, the duration of cell movement in one direction before changing direction (see Sect. 5) with the duration of the present cycle. However, the way in which the association between myosin and chemokine receptors (e.g., chemokine receptor type 4; CXCR4) fits into this picture has yet to be resolved [62].

In summary, lymphocytes exhibit a migration mode that corresponds to a flow of cytoplasm generated by the contraction of cytoskeletal elements and biomechanical anchorage to the ECM. The complex dynamics of the aforementioned three-step model, which was derived from fibroblasts that migrate on 2-D substrates, does not apply to lymphocytes. Notably, like lymphocytes can also fibroblasts employ a migration mode that involves the use of constriction rings [58, 63, 64].

4 Chemotaxis

The existence of an *aerotaxis–chemotaxis* reaction by bacteria towards an oxygen gradient as first reported by Theodor Wilhelm Engelmann [65]. Shortly afterwards, the term "chemotaxis" was assigned to leukocytes, which appeared to produce an as-yet unknown chemical substance which caused inflammation [66]. Subsequently, this description was extended to include the definition of a living white blood cell as one that did respond to a chemotactic stimulus [67].

Today, however, *chemotaxis* is defined as "the directed migration of a cell upwards or downwards in a gradient of a substance termed a chemoattractant or a chemorepellent, respectively." Chemotaxis is regarded as one of the fundamental mechanisms that orchestrate the distribution of many cell types, and is particularly central to the motility of leukocytes. Processes such as the penetration of tissues by other cells, the homing of cells to specific tissues, and the organization of whole organs, are driven and controlled by chemotaxis. As an example, whilst the above-mentioned lymphoid tissue are known to exhibit T zones and B-cell follicles, the respective cells seek and locate their compartments in response to differentially expressed chemokines [68, 69] (i.e., chemoattractant cytokines).

At this point, attention will be focused on the chemokines CXCL (chemokine C-X-C-motif ligand) 12, CXCL13, CCL (chemokine C-C-motif ligand) 5, CCL19, CCL21, and IL-8, all of which are essential for leukocyte motility. In addition, lipids may act as chemoattractants; sphingosine 1-phosphate (S1P), in particular, is relevant to lymphocyte migration in lymphoid tissues [70–72] (chemorepellents such as semaphorins will not be discussed at this point).

Chemokines are small proteins that are about 100 amino acid residues in size, and which constitute their function *in vivo* when bound to glycosaminoglycans (GAGs) that are connected to the ECM; in contrast, soluble chemokines are barely detectable [68, 73–76]. Cells have the ability to detect chemokines in either their GAG-bound or unbound forms, by using their respective chemokine receptors. Typically, there is not a one-to-one relationship between chemokines and their receptors; rather, each chemokine can bind to several receptors, and each receptor can bind various chemokines [68].

The mechanisms used by cells to sense a chemokine gradient, how chemotaxis is regulated, and how signals are integrated are discussed in the following subsections.

4.1 Sensing Chemokines

Chemokine gradients can be sensed basically using two different strategies, namely temporal or spatial sensing:

- In *temporal sensing*, the chemokine concentrations are compared at two subsequent time points. Temporal sensing is employed by bacteria, because of their small size; typically, the difference in chemokine concentration at the front and back of the bacterial cell will be too small to be detected with chemokine receptors. Neutrophils have also been shown to employ temporal sensing [77]. The molecular basis of temporal sensing is an oscillation of the signal transduction system with a period of 10 s, such that the concentration of the chemoattractant is monitored within this period. If the concentration becomes higher between two measurements, then the direction of locomotion will remain unaltered; however, if the concentration becomes lower in two subsequent measurements, then the cell will reverse its direction. In this way, the cells will be guided up the concentration gradient. The change in direction is not an actual rotation of the cell, but rather is initiated by the generation of lamellipodia in the new direction of migration.
- In *spatial sensing*, the chemokine concentrations are compared simultaneously at the front and back of the cell. Eukaryotic cells have been shown to use spatial sensing [78], and have inspired several models for the detection of chemokine gradients [78, 79]. The steepness of the gradient was found to play an important role; for example, at a threshold concentration difference between the front and back of the cell of only 2%, the cells were able to sense the gradient. Given that chemokine receptors are located preferentially at the cell leading edge at a 10 : 1 ratio or higher, independent of the presence of chemokines [75, 80], it might be speculated that the organization of the receptor system would be more suited to the temporal rather than the spatial sensing mode. Spatial sensing would be facilitated by a more homogeneous distribution of chemokine receptors.

The absolute concentration of the chemokine is of minor importance for the directionality of the cell; thus, the mechanism is designed to identify the source of a chemokine rather than a high concentration. Nonetheless, the concentration of chemokines can control cell motility (i.e., chemokinesis) as it is the speed of cell displacement, not the direction, that is controlled by the chemokine concentration, with the response curve being typically bell-shaped. For chemokines relevant to lymphoid tissues, the maximum concentrations range between 10 and 200 nM [81–88], but below a threshold concentration of 1 nM the cells will not respond to a chemoattractant [83, 85, 89], whereas threshold values of 30 nM have also been reported [84]. The concentration of half-maximum response (EC_{50}) is approximately 1 nM for neutrophil chemoattractants, and close to their dissociation constant K_d [89]. The K_d values for other chemokines such as CCL19, CXCL12, CCL5, or IL-8 are within a similar range [75, 85, 90, 91], as are the EC_{50} threshold concentrations for leukocyte responses to these chemokines [83].

More specifically, B cells are guided by a combination of the chemokine CXCL13 and the receptor CXCR5, to home to the B-cell follicles in lymphoid tissue. The threshold concentration at which B cells respond to CXCL13 is 10 nM, and the response is suppressed at concentrations above 1 μ [92]. In the T zone of lymphoid tissue, CCL21 attracts lymphocytes carrying the receptor CCR7, with a comparatively high concentration of 100 nM being required for an optimal response [84]. Typical concentrations required for CCL21 are in the range of 1 μM [93], while the alternative CCR7 ligand CCL19 has been identified at a 100-fold lower concentration in lymphoid tissues.

The mode of chemokine sensing of lymphocytes *in vivo* remains a subject of debate, since there is no known specific molecule associated with temporal versus spatial sensing that might be either inhibited or knocked-out. The two modes of chemokine sensing might also be substituted for each other in response to external stimuli; initially, gradient sensing requires the first contact to the chemoattractant to be above a threshold concentration, but when the cells have polarized and begun to migrate they may switch to a temporal sensing mode.

4.2 Chemotaxis Regulation

The chemotactic responses of cells may be regulated on many levels, with the expression of corresponding receptors even being regulated genetically. (Note: effects at gene-expression level are not considered here; rather, attention is focused on cell migration over shorter time scales.) The characteristic period for the genetic regulation of chemokine receptors is in the range of days [70] while, on a shorter time scale, the modification of receptor activity and also receptor internalization may be more relevant. However, it should be borne in mind that early genes may respond in less than an hour [94, 95]. Neither can it be ignored that – as with chemokines [96] – chemokine receptors may also be regulated by such rapid genetic mechanisms.

The internalization of chemokine receptors may be considered as a mechanism of desensitization of cells for the corresponding ligand. In neutrophils, desensitization for a chemokine has been shown to be associated with high concentrations of that chemokine [89], with the migratory response of cells not depending critically on chemokine receptor internalization at either low or medium chemokine concentrations [68, 87, 97, 98]. These data support the view that internalization is required to redistribute the receptors on the cell surface in order to sense the direction of the gradients [68], and that at high chemokine concentrations internalization may be relevant to desensitization.

Interestingly, CCL19 – but not CCL21 – induces internalization of the chemokine receptor CCR7, which raises the question of which mechanism causes the cross-desensitization between CCL19 and CCL21, as reported previously [84]. More generally, different ligands of the same receptor induce different levels of internalization [98]. Consequently, it may be hypothesized that the ligand has an effect on the recycling of the receptor back to the plasma membrane, rather than on the internalization rate.

Internalization is also regulated by factors other than chemokines. For instance, interferon-α (IFNα) enhances the chemotactic response of B cells by decreasing chemokine receptor internalization [99]. However, IFNα does not interfere with the spontaneous internalization – that is, the background level of receptor internalization in the absence of the ligand. This further supports the view that it is not internalization itself but rather the recycling of internalized receptors which is controlled.

Naive B cells freshly isolated from tonsillar tissue failed to respond to CXCL12 and CXCL13 [100]. This implied that the presence of chemokine receptors CXCR4 and CXCR5 is necessary, but not sufficient, to induce the chemotaxis of naive B cells. The suppression of chemotaxis may be mediated by desensitization (i.e., internalization or by cross-desensitization; see Sect. 4.3), or by an intracellular regulation of the responsiveness of the cell to chemotactic signals.

Similarly, it has been shown that CCR7 – the receptor for CCL19 and CCL21 – is undetectable on the surface of freshly isolated B cells [101]. However, the fact that the protein is identified in the cytoplasm and brought to the surface in CCL21-free culture suggests receptor internalization upon exposure to CCL19 in the tissue.

As a lesson from this, it is important to realize that these *in-vitro* experiments may not be conclusive of the *in vivo* situation, because receptor internalization or other means of internalization occur on a rather short time scale. Even freshly isolated cells cannot be considered to behave as in living tissues, and there is also evidence that the chemokine flora surrounding the cells impact directly on the activation of different signaling pathways. Thus, each cell must be considered as being in a specific state of activation, as expressing a specific receptor profile, and interacting with a specific local and dynamic cytokine distribution.

4.3 Complex Responses to Multiple Chemokines

With crosstalk between different chemokines and chemokine receptors having been observed in several systems [68], it emerged that the relationship between the chemokine receptors and their ligands is much more complex than one-to-one. Some examples of inter-chemokine crosstalk are detailed below.

A cross-desensitization between chemokines that use the same receptor, such as CCL19 and CCL21 [84], has been discussed previously (see Sect. 4.2). However, cross-desensitization also occurs in the case of chemokines employing different chemokine receptors. For example, CD4 T cells perform chemotaxis in response to IL-16 [81], which can block chemotaxis in response to CXCL12 without downregulation of the corresponding receptor CXCR4 [102]. The block develops in 10–15 min and reaches maximum effect after 1 h. Conversely, CXCL12 cannot block chemotaxis in response to IL-16 but can block the response to CCL19 in lymphocytes, whereas CCL19 cannot block the response to CXCL12 [83].

The presence of multiple chemoattractants does not always lead to a block of all but one of the possible responses. The cell may also compute a vector sum of incoming signals, and in this way determine an average direction from multiple chemoattractants [103]. The integrated response will determine the direction of migration and leave the cell speed unaltered; thus, multiple chemokines induce chemotaxis, but not chemokinesis.

If multiple chemokines are provided in a given sequence, the direction is no longer the vector sum of the signals; rather, the direction of migration will be dominated by the latest chemoattractant, even if it is presented at lower concentrations or in lower gradients [103]. A natural explanation for the relevance of the sequence of chemokines is again desensitization of receptors, for example by internalization (see Sect. 4.2 and Ref. [87]).

An alternative view has been provided from results obtained with neutrophils [104], where two signaling pathways for two different receptors have been identified, suggesting a hierarchy of chemoattractants, the levels of which were named "target" and "intermediary." When multiple signals of members from one group only are provided, the cell responds by computing the vector sum; otherwise the *target* chemoattractants are preferred. Yet, the story is even more involved: the blocking of *intermediary* chemoattractants by a target chemoattractant may be incomplete [89]. Moreover, the presence of an intermediary chemoattractant can enhance chemotaxis in response to a target chemoattractant, such that the cell will respond to lower concentrations of the chemoattractant than with the target chemoattractant alone.

Which type of response results from multiple signals may depend on the signaling cascade involved – that is, on which receptors share or use a concurring cascade to induce directed cell migration. Currently, the data available for chemokines CXCL12, CCL19, CCL21, CXCL13, and S1P are not sufficiently conclusive to allow the determination of a signaling hierarchy.

Some evidence of a hierarchy has been provided, however, from studies of the behavior of newly antigen-activated naïve B cells, which upregulate their CCR7-levels by a factor of two to three and, consequently, enhance their response to CCL19 and CCL21, which are dominant in the T zone. The response to CXCL13, which normally leads to B cells being retained in the lymphoid follicles, is then balanced such that the cells migrate to the boundary of the follicle [71, 100, 101, 105–108]. This type of migration is essential for the B cells to acquire T cell help, and in doing so to initiate an antibody affinity maturation process in germinal centers. When either CXCR5 is overexpressed or CCR7 is lacking, the B cells fail to relocate to the follicle border upon antigen-stimulation and so remain in the follicle [106]. In contrast, if CCR7 is overexpressed by genetic manipulations, the B cells will locate at the border, even without antigen, and move further into the T zone upon antigen stimulation. This strongly suggests that chemotaxis is gradually increased with an increased expression of the specific receptor in relation to other receptors.

Similarly, activated T cells in the T zone upregulate their response to CXCL13 and, in doing so, relocate to the follicle border [71, 105, 109, 110] in order to meet B cells.

The fact that different chemokine receptor levels can affect the homing of lymphocytes relies on a distribution of chemoattractants away from the respective chemoattractant sources by diffusion and into other zones. Only then can the B cells in the follicle sense T-zone chemokines, and T cells in the T zone find the follicular border. The observed chemokine gradients penetrate other zones to a depth of 80–140 μm [108].

5
Multi-Photon Imaging of Lymphocyte Migration

Following the description by Rudolf Virchow of the association of leukocytes and inflammation a connection between the shape changes of leukocytes and their recruitment to sites of inflammation was recognized soon. The concept was not clear, however, as noted in a debate reported in 1887 [111] where, in essence, the demand was made for careful, high-quality microscopic techniques that could address the question where and when leukocyte migration and shape changes occur. This notion may sound familiar to those presently discussing the benefits and caveats of today's technology, and interpreting their results!

The intravital imaging of lymphocyte movement in lymph nodes was first employed at the start of the twenty-first century to analyze lymphocyte migration. The technique involves rendering a subset of lymphocytes in the crowded lymph node fluorescent, such that the cells could be tracked in the living organ, in 3-D fashion. This allowed, for the first time, the visualization of living tissues, and led from what previously was simply a hypothesis to the generation of data that would provide an excellent basis for mathematical models (see Sect. 6). The rate and mean speed of distribution, the average persistence time before changing directionality, the turning angle distribution after each repolarization, the distance that cells could reach from their starting point, and even the shape of the cells, could be measured to provide an extremely rich data set. Some results derived from multi-photon imaging related to cell motility are reviewed below, while an in-depth analysis of the experimental details is available in Ref. [112].

5.1
Persistent Random Walk and Transient Chemotaxis Model

The discovery of Brownian motion, first by Jan Ingenhousz in 1785 and again by Robert Brown in 1827 [113], suggested that cell migration might occur as the result of thermal fluctuations. From the late nineteenth century onwards, however, physicists learned much about atoms and molecules based on Brownian motion or *random walk*. In 1920, the idea of a *persistent random walk* was introduced into physics [114], while fifty years later the observation of single migrating cells led to the concept of a persistent random walk of cells [115, 116]. In contrast to Brownian motion, cells follow a certain direction for a specified time before they reorient. Today, various random walk models are applied to diverse migration processes in biology [117].

Multi-photon live imaging data have revealed some fundamental properties of the persistent random walk of lymphocytes in lymphoid tissue. Typically, T cells have a 2 min cycle between two alternating phases of migration [112, 118–121] that include: (i) an elongated phase, with a compact triangular morphology; and (ii) a resting phase, during which the cells are balling up. The distance traveled by a cell in one cycle is about 20 μm, with peak velocities of up to 30 $\mu m\ min^{-1}$ and a mean velocity of about 20 $\mu m\ min^{-1}$. During the elongated phase the migration pattern is amoeboid, with a leading edge followed by the cell body and the trailing edge, the *uropod* (see Fig. 1b).

If migrating cells collide with other cells, this causes them to move a few micrometers and induces a Brownian-like motion of lymphocytes; however, lymphocyte migration is essentially dominated by active motility [120]. The direction of motion is fairly constant over one cycle (and sometimes even over several cycles), with the direction of motion changing after each balling-up of the T cells. In this case, rather than the cell turning, new lamellipodia are formed. In the long term, the migration of T cells appears to be a random walk, with a motility coefficient of 67 $\mu m^2\ min^{-1}$ [41, 118]. Thus, the newly acquired multi-photon data were consistent with the old concept of *persistent random walk*.

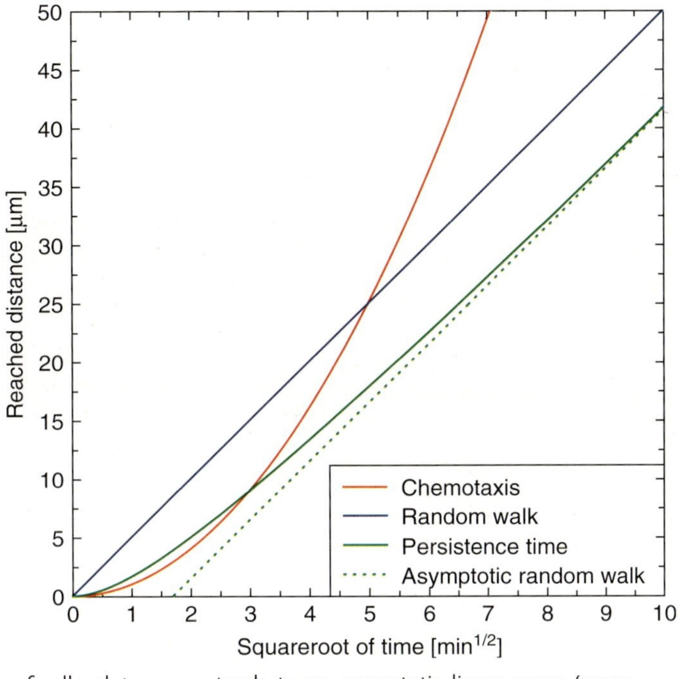

Fig. 6 The reached distance of cells plotted against the square root of time, indicates the mode of migration. Chemotaxis induces a quadratic relationship (red), while random walk leads to a linear curve (blue). If the cells perform a random walk with persistence time, the curve starts as quadratic (green) and then tends to an asymptotic linear curve (green dotted). The asymptotic linear curve has the same slope as the simple linear curve (blue versus green dotted), representing an identical motility coefficient for both hypothetical cell types.

The B cells in follicles demonstrate essentially the same behavior, but they move more slowly (10–12 µm min^{-1}) and they change direction slightly more frequently, probing their environment by using amoeboid-like cell extensions. In the long term, the migration is again a random walk with a motility coefficient of 12 µm^2 min^{-1}, which provides an order of magnitude for quantities characterizing lymphocyte migration. The exact values vary to some extent between experimental groups, and depend on the quantification methods employed [112]. Thus, the modeler must pay attention to using consistent data sets rather than combining data from different set-ups.

The finding that lymphocytes perform a persistent random walk was surprising, since the organization of cells in lymph nodes is basically controlled by chemokines that drive the T and B cells to their respective compartments [106, 107, 122]. A more expected result would have been that the distance over which cells migrate would increase linearly with time; that is, it would reflect chemotaxis (see Fig. 6). Instead, the distance was shown to be proportional to the square root of time, which is typical for diffusion-like migration patterns.

Mathematical modeling has revealed that, even when the response of cells to chemokine signals is rather powerful, the curve measured will remain compatible with a persistent random walk [123]. Rather, it is the dimensions of the area to which the moving cells are restricted that limits the measurement of an expected quadratic distance-reached curve. Typically, the cells rapidly reach the boundary of this area, and the quadratic curve is converted to a saturating form. The resulting curve is a mixture of these, and remains linear to a good approximation. However, in addition to this simple explanation, it is also important to understand the data relating to the specific dynamics of sensitization for the chemokine (this point is discussed in detail in Sect. 7.2).

5.2 Chemokinesis in Lymphoid Tissue

Normally, freshly isolated T cells will demonstrate a minimal attraction to CCL19 [85], which suggests the existence of a desensitization mechanism also for T cells. The internalization of chemokine receptors may explain why, despite the presence of CCL21 in the T zone and the receptor CCR7 on lymphocytes, chemotaxis is not the dominant mode of migration in the T zone of lymphoid tissues [120].

An alternative model is based on the distinction of chemokinesis and chemotaxis. In this scenario, chemotaxis would be the dominant pathway by which lymphocytes entered their respective compartments. In the case of T cells, CCR7 expression is indeed required for their entry into lymph nodes although, once the T cells have reached the T zone, CCL21 would induce T-cell chemokinesis rather than chemotaxis. The migratory speed of the T cells is determined by the overall level of the chemokine. The chemokine gradient will remain without effect on chemotaxis, because the T cell will have switched to a different migration mode; thus, the T cells would perform a random walk within the T zone. However, once the T cells have reached the T zone border they reactivate chemotaxis, such that they are retained in the T zone unless they were to target lymphoid tissue exit points. It remains unclear, however, why the multi-photon data pointed to a *persistent* random walk, or how often T cells in the T zone would change directions; this point is discussed in next section.

5.3 ECM Fibers Guiding Lymphocytes

Two-photon electron microscopy experiments have shown lymphocytes and the natural ECM together [108, 118–120, 124–126], where the sparse network of the reticular fibers forms the substrate of cell migration into the lymph nodes. The exact role of the substrate in cell migration is more difficult to determine, however.

It has been hypothesized that the motility of lymphocytes, and their homing to compartments, is mediated by temporal sensing (see Sect. 4.1); that is, lymphocytes reverse their direction if they enter the wrong compartments [120] and perform a random walk within those compartments. It has been shown that rather sharp gradients exist at the border of the compartment [108]: CCL21 levels are high in the T zone but virtually zero inside the follicles, whereas CXCL13 exhibits the opposite pattern. Such steep gradients can only be maintained if the chemokines are mainly presented on surfaces rather than in soluble form, and this points to a haptotaxis mechanism. However, the persistence in the direction of cell movement noted above suggests that once the cells have polarized and begun to migrate, their direction of movement is kept independent of the chemokine gradient [75]. Thus, the chemokine gradient must be detected only during short time intervals, and would then be ignored. This concept is in conflict with that of temporal sensing, in which concentrations are compared along the path of migration, at least for neutrophils [77].

This conceptual difficulty can be resolved with a more attractive scenario. In this theory, persistent random walk is derived from the backbone of the lymph node, the fibroblast reticular network (see Sect. 2.4). The T cells then move along the dendrites of this network, and have the opportunity to decide on a new direction at every crossing point of the network fibers [126]. This pattern of motility is very appealing, and it has indeed been shown (using intravital microscopy) that T cells in the T-cell-dominated zone in lymph nodes may walk along the fibers and turn at the crossing points of the network.

It is difficult to prove that the measured turning angle distribution of T cells reflects the angles of the reticular network. In principle, the measured turning angles should reflect these, but unfortunately the migrating T cells do not walk straight on the fibers but rather make turns on the fibers, or are pushed by other migrating cells. As a consequence, the measured turning angles are a superposition of the random fluctuations and fiber angles which cannot be easily disentangled. In fact, the angles of fluctuations strongly overlap with the angles from the reticular network. Interestingly, cells never invert their direction, even though the fiber on which the cell arrives at a node of the network is always available.

6 Mathematical Models of Cell Migration

Mathematical models of cell migration and chemotaxis differ in their basic approach architectures, in the type of experimental data entering the model, and in the biological questions that may be addressed with the model. The first step is to select the scale of description. If the main interest is in the populations of cells, then a PDE approach might be sufficient (see Sect. 6.1). This class of models may also include molecular signaling mechanisms, and their translation into modulations of cell behavior.

The next more-detailed level of description is the whole-cell model (see Sect. 6.2). These are mathematical descriptions of the cell path, based on representations of individual cells, such that the dynamics of deformation, of protrusions, and the uropod are not considered in detail. In some models the mechanic properties of the cells are incorporated on a phenomenological level, although it depends on the scientific question under consideration whether this degree of detail is appropriate, or not. Subcellular models are needed if interest is directed towards the relationship between cell mechanics and cell shape with cell migration (see Sect. 6.3).

6.1 Partial Differential Equation (PDE) Models

The most abstract level of modeling are ordinary differential equations (ODEs), which deal with population dynamics. These models have no spatial resolution, and treat cells as cell numbers n_i with cell interactions defined by a function f:

$$\dot{n}_i(t) = f\left(t, n_i(t), \{n_j\}\right) \tag{1}$$

where $\{n_j\}$ denotes a set of different cell populations n_j.

Adding spatial information leads to PDEs where effects such as cell migration enter via diffusive transport $(\nabla(D(\mathbf{x})\nabla c_i(t,\mathbf{x})))$ with diffusion constant D or advective transport $(\mathbf{v} \cdot \nabla c_i(t,\mathbf{x}))$ with the cell velocity $\mathbf{v}$:

$$\dot{c}_i(t,\mathbf{x}) = \nabla\left(D(\mathbf{x})\nabla c_i(t,\mathbf{x})\right) + \mathbf{v}\cdot\nabla c_i(t,\mathbf{x}) + f\left(t,\mathbf{x},c_i(t,\mathbf{x}),\{c_j(t,\mathbf{x})\}\right). \quad (2)$$

The space-coordinate $\mathbf{x}$ may be either one-, two-, or three-dimensional. This model type requires large cell numbers to make the cell concentrations $c_i(t,\mathbf{x})$ meaningful. $\{c_j(t,\mathbf{x})\}$ denote all cell types other than c_i.

Cell population models such as the Keller–Segel model [127], and its various modifications, rely on PDEs. This approach absorbs several microscopic effects in macroscopic variables. For instance, volume exclusion is considered as a phenomenological effect and not as an actual restriction in space. Read-out variables such as the speed of cells, by construction, are averaged quantities. For example, the speed distribution of cells or the distribution of cell polarization are not accessible. Thus, the comparison with the corresponding complex experimental data [108, 118, 120, 125] is restricted to the comparison of average quantities.

Several models have been developed which describe cell migration starting from a molecular basis but without attempting to derive exerted forces onto the environment. For instance, an involved PDE model coupled the cytomechanics of deformation and migration in order to describe experimental results on fibroblasts, including chemotaxis [128]. This approach was able to demonstrate how external stimuli can induce morphological instabilities, leading to the migration of cells. Another PDE model, based on the viscoelastic description of the cytoplasm and actin polymerization dynamics, has been used to compute the dependency of cell velocity on the adhesion to a substrate [129]. These studies set the adhesion-dependence of the three-step migration model (see Sect. 2.3) on a solid mathematical basis, while the nonlinear dynamics of actin polymerization led to several models of cell migration based on symmetry breaking in the system [130, 131]. However, there exist even more sophisticated models that combine the basic concepts to explain observations on cell mechanics [132].

One series of theoretical studies has considered the indirect interaction of cells via the elastic properties of the ECM [133–135]. The forces exerted by the cells onto the matrix induce a prestress which stiffens the ECM; the cells then respond to this stiffer matrix by redefining the direction of their migration, such that cell interaction and axis alignment can be described by force dipoles. Experimental evidence has been provided for this mechanism, with it having been shown that tensile stress on cells induces the growth of microtubules, which may then act as mechanosensors for the cells [136].

Previously, models for chemotaxis have mainly focused on the internal signaling requirement to sense spatial gradients [79]. In this case, one of the basic assumptions has been that global inhibition and the local excitation of the signaling cascade is an underlying mechanism for chemokine gradient sensing in eukaryotic cells. An alternative model has been developed that investigates how a cell responds to multiple chemokines, based on chemokine concentrations and gradients [137]. This model couples the microscopic behavior, which is characterized phenomenologically by a function, to the population dynamics. This approach has the potential to extract information on the chemotactic response of a single cell to multiple chemokines from the patterns observed in cell populations.

6.2
Whole-Cell Models

6.2.1 Deterministic and Stochastic Event Generators on a Lattice

A natural and computationally cheap approach to cell migration is based on a representation of space as a discrete lattice. The cells occupy single nodes on this lattice, and move according to local or global stimuli, or to cell internal programs. Each cell is represented by a finite set of node states $\mathcal{S}$, and empty space is treated as a special cell type. The simplest applications provide a set of rules to compute the state $\mathcal{S}_i^{(n+1)}$ at node i and at discrete time $n+1$ from the state $\mathcal{S}_i^n$, and the states $\{\mathcal{S}_j^n\}$ of the neighbor lattice nodes j at time n. This is a generalization of Conway's famous "game of life," which is a deterministic cellular automaton.

Several choices are available to define the neighborhood topology, depending on the lattice type and dimension. The most common is the Moore-neighborhood on 2-D square lattices, where the four adjacent horizontal and vertical neighbor nodes are considered as interacting neighbors. Other realizations of cellular automata represent cell movement more directly, where a cell at some point is moved to a neighbor site, based on certain rules; however, this is not very different from the usual cellular automata. The latter model can be mapped on the state representation by changing two lattice sites at once.

Extensions of these models use probabilities instead of deterministic rules, or replace regular lattices with irregular lattices, such as static Voronoi tessellations [138]. Diffusive substances may also be coupled to the model in so-called "hybrid cellular automata," to model the effect of soluble signals such as chemokines. For example, the crawling slug stage of *D. discoideum* has been studied many times using various methods, including hybrid cellular automata combined with PDE models [139]. An application to lymphoid tissue organization will be presented in Sect. 7.2.

In general, 2-D event generators are very appealing because they are fast and efficient. In particular, when using stochastic event generators the computational load is a limiting factor because of the need to repeat the simulations many times, and it often also excludes 3-D simulations. While this practical issue is relevant, 2-D simulations may lead to artifacts, depending on the topic under consideration. Although growth processes (e.g., tumor growth) may be well captured in 2-D fashion, in the case of highly motile cells such as lymphocytes the number of possible paths is a critical property of the system, and is different in the 2-D and 3-D cases. As a consequence, cell migration might be suppressed in 2-D fashion because of a lack of possible paths. In such situations, additional algorithms are needed in order to avoid an unrealistic inhibition of motility. For example, it has been proposed that if two cells in a 2-D simulation have vectors for the direction of migration with a negative scalar product, then the two cells will exchange position on the lattice. Such an exchange algorithm might even be important in 3-D cases [140] when the density of cells is high and the artifact of suppressed motility is inferred by the discretization of space.

6.2.2 Off-Lattice Whole-Cell Models

Instead of a lattice-based representation of cells, where each cell is associated with a node on the lattice, cells can also be represented in continuous space. In the simplest representation, a cell is shown as a point in space, with each point describing a trajectory [72, 140–142]. When the mechanical properties of the migrating cells are not particularly relevant, this approach is most suitable. However, in lymphoid tissue the situation is less clear: on the one hand the density of cells is comparably high, whereas on the other hand the lymphocytes can be easily deformed, which might compensate for the high density and justify the use of point-like cells.

Other models include certain mechanical properties, such as the compressibility and elasticity of cells, starting from a specific shape of the cell. Normally, this is a spherical or ellipsoidal shape for the equilibrium state of the cell. These objects acquire mechanical properties by using effective viscoelastic models such as the Voight or Kelvin models [143–146]. Other approaches use potentials to describe the interaction between cells, taking into account any deformations [147, 148].

An alternative approach would be to determine the cell shape from the interaction with the neighboring cells, as in the Voronoi method; in this case, the cell shape is not imposed but rather is a result of the cell's immediate environment. This procedure has been verified with experimental data acquired from plants, in skin, or in cell cultures, which show densely packed cells to have acquired very precise Voronoi-like shapes. The Voronoi-approach has been applied to tissues in both 2-D [149] and 3-D [150] situations.

Both approaches – that is, one imposing a cell shape and one deriving the cell shape from the environment – were combined [150–153] and termed Delaunay-Object-Dynamics [154], in association with the molecular dynamics (MD) methodology that follows a similar philosophy. This resulted in a model of deformable space-filling polyhedra, where the cells' shapes depended on the cell neighborhood and they exhibited internal viscoelastic dynamics. This approach is most suitable for the modeling of interacting and motile cellular compounds in tissues [155] (see Sect. 7).

6.3 Subcellular Models

Subcellular models include the cell shape and the dynamics of cell deformations in static situation, under external forces, or during the course of active cell motility, but omitting the actual intracellular reorganization of the elements of the cytoskeleton. Cytoskeleton dynamics enter the model only in the form of their resulting cell mechanical properties, or with approximations of the cytoskeleton structure. The different model architectures described below place differing emphases on the cell mechanics.

6.3.1 Cellular Potts Model

The extended Potts model [156] is a well-known and often-used approach that was derived from the Ising model [157] (which itself is well proven in physics) and was originally given the objective of simulating the sorting of a mixture of cells of two types. In this case, instead of using two spins (up and down in the Ising model), N spins equal to the number of cells are used, with a spin being attributed to each object, including the medium surrounding the cell objects. One spin state s corresponds to one single cell of type $\tau(s)$, where all lattice nodes with spin s together represent this cell and define the cell volume and shape.

The interaction of the cells is described in analogy to the Ising model with a model energy (Hamiltonian):

$$H = \sum_{i,j;i<j} J_{\tau(s_i)\tau(s_j)}(1 - \delta_{s_i,s_j}) + \lambda \sum_{s} \left[V(s,t) - V_0(\tau(s),t)\right]^2. \quad (3)$$

where $J_{\tau(s_i)\tau(s_j)}$ represents the surface energy of two neighboring sites i and j occupied by the cells s_i and s_j of type $\tau(s_i)$ and $\tau(s_j)$, respectively. The Kronecker-symbol δ_{s_i,s_j} is 1 if $s_i = s_j$, and 0 otherwise; this is used to constrain the surface energy sum to the interface between two cells. The second term is a volume constraint to keep a cell s close to a target volume $V_0(\tau(s),t)$ for the corresponding cell type $\tau(s)$. $V(s,t)$ is the volume of the cell calculated by counting all lattice sites i which belong to the cell s, that is, are in the spin state s. The minimization of the surface energy and the application of the volume constraint induce two concurring driving forces which results in a tendency to obtain a spherical cell with a volume that slightly deviates from the target volume $V_0(\tau(s),t)$. The relative weight of the surface energy and the volume conservation term is controlled by the free parameter, λ.

Changes in shape and, therefore, the movement of cells, are governed by thermodynamic interactions. The following Monte Carlo algorithm is used, which is also denoted as a Metropolis algorithm. At each time step in the simulation, a lattice site is randomly selected; the spin value of the selected site is then converted to the spin value of a (also randomly chosen) neighboring site with a probability dependent on the variation on the energy of the system. The spin would be always changed if the system was driven to a configuration with smaller energy. However, if the configuration were to end up in a state of higher energy, the exchange would be made according to the probability $\exp(-\Delta H/kT)$, where ΔH is the virtual energy shift, k is the Boltzmann constant, and T is the temperature of the system. The temperature was referred to as the *biological temperature*, and is another free parameter of this model. Its choice determines how far the system may deviate from its minimum energy state.

This algorithm of cell dynamics has an interesting property, in that cell movement is only made possible by changing the cell volume, as a subunit on the lattice must change its spin state to that of one of its neighbors. These volume fluctuations are averaged out on longer time scales, and by the volume conserving potential in the Hamiltonian equation (Eq. (3)). Therefore, these unphysical changes in cell volume associated with cell movement are negligible in the limit of many subcellular nodes per cell. However, in the limit of small numbers of subunits per cell, additional rules are required for the Potts model.

The Potts model was initially applied to describe the cell sorting of two types of cell [156]. For this purpose, the interactions in Eq. (3) have been defined as follows: bonds between lattice nodes with the same spin (i.e., inside the same cell) have energy zero, whereas at the cell boundaries there is a cell type-dependent surface energy that controls the cell sorting.

Several extensions of this model have been developed to incorporate chemotaxis [158], proliferation [159], or more complex cell shapes [160] by adding further terms to the Hamiltonian (Eq. (3)). Similarly to the stochastic event generator (see Sect. 6.2.1), the model may be coupled to a PDE system to describe soluble substances such as chemokines. In this way, it became possible to describe the life cycle of *D. discoideum* [161].

As a rule of thumb, the Potts model is most suitable to describe the asymptotic state of a system. If the dynamics of the system are interesting, or if the system remains far from the energy minimum, then the free parameters of the model λ and T will prohibit a quantitative prediction [144]. For instance, cell migration is inherently an active process of cells consuming energy, active deformation, active changes of the cytoskeleton, and active force exertion onto substrates. Such an active process is hardly associated with a modeling architecture that relies on the minimization of an Hamiltonian, and thus on passive mechanisms. The cell is not minimizing energy but rather consuming energy to be active; consequently, the Potts model will provide a good approximation to the equilibrium state of a system of motile cells, but it may fail to capture the dynamics of the system on the way to the equilibrium on a quantitative level. It should be noted that, when cellular contacts are ignored, the macroscopic limit of the Potts model is the Keller–Segel model [162].

6.3.2 Hyphasma

An alternative lattice model architecture which intrinsically includes the one- and the multi-subunit limit is the hyphasma model [41]. This model is based on forces instead of energy minimization, and was originally developed to describe lymphocyte migration in secondary lymphoid organs, in order to interpret the results found with the experimental method of intravital multi-photon imaging [118] (see Sect. 5). The intrinsic inclusion of the one- and multi-subunit limit on the model allows the lattice constant to be determined from the size of the smallest cell in the studied system, which then is described as a single volume element. This provides a faster computation when interest is focused only on differences in volume (and not on the substructure of the cells).

The hyphasma method uses a mechanism of cell movement that conserves the total volume of the cell when no growth or shrinking is intended. Cell objects are represented by the cell volume, the cell polarity, a list of cell subunits, and internal velocity states. The physical movement of the cells is formulated in terms of forces acting on subunits of the cell (in contrast to the idea of energy minimization in the Potts model; see Sect. 6.3.1). All reaction kinetics are formulated as reaction rates, and actions are taken according to probabilistic decisions. An overdamped approximation was used in order to calculate the cell displacement. A Reynolds number (the ratio of inertia and viscous forces) of below 10^{-5} justifies this approximation.

The kinetics of the cell subunits is based on two types of velocity: undirected active movement with persistence of orientation; and cell reshaping. Cell reshaping forces tends to make the cell spherical, thus, acting in the spirit of the surface energy term in the Hamiltonian of the Potts model (see Eq. (3)). All forces that reshape the cell toward a sphere (i.e., hydrostatic pressure, reduced actin filament assembly, actomyosin contraction, or membrane surface tension) are included in this single reshaping force. This force is implemented as an overall elastic force that drives the subunits of an elongated cell back to the vicinity of the current barycenter and promotes a spherical shape. This involves two unknown phenomenological parameters that effectively represent underlying mechanical properties of the cell, namely

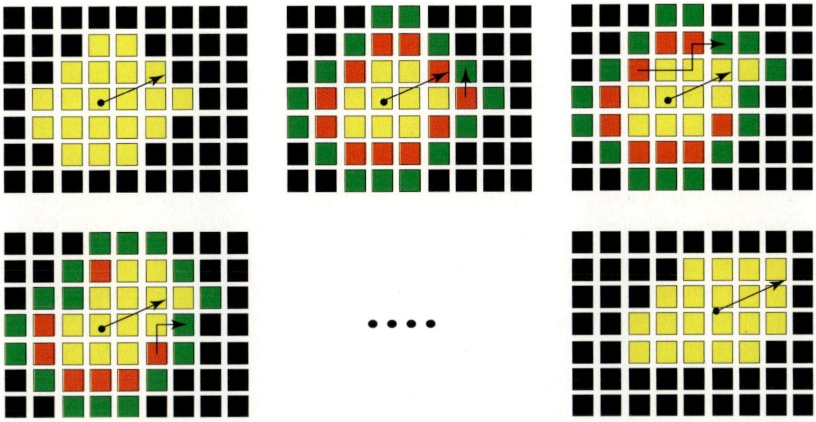

Fig. 7 A schematic description of the algorithm of cell movement in the hyphasma model. The current barycenter is denoted by the black dot, and the cell polarity by the arrow attached to it. This arrow also points the position of the *virtual barycenter* at the border of the cell. The initial cell position (upper left panel) is shown by yellow nodes, while black nodes belong to the cell's surrounding. For cell movement the nodes are further distinguished (upper central panel) into red nodes belonging to the cell border, and green nodes representing free target nodes surrounding the cell. Only the red border nodes of the cell may be moved. Red nodes are randomly chosen and moved towards the green target nodes nearest to the virtual barycenter (arrows in upper central, upper right, and lower left panels). This is repeated until all red nodes have been moved or are no longer allowed to move. A possible final configuration is shown in the lower right panel. For details, see Ref. [41].

the strength of the elastic forces, and the elongation at which the reshaping forces become half-maximal.

In contrast, active movement tends to elongate the cell as a result of the used algorithm. The direction of the active movement is determined by the polarity of the cell. The barycenter of the cell is virtually shifted in the direction of the polarity to the border of the cell. Subsequently, every subunit representing a border point of the cell is randomly moved towards free lattice points near the new barycenter. The movement is suppressed in case the movement of a border subunit would cause the subunits of the cell to disconnect from each other. The displacement of subunits is stopped when either no border subunit remains to be moved, or the barycenter of the cell has been displaced by one lattice constant (see Fig. 7). In the spirit of persistent random walk models, the polarity remains unchanged for a persistence time and then changes randomly.

By using the intravital imaging data described in Ref. [118], all parameters of cell motility such as average cell speed and the persistence time of orientation, are available. With only two unknown parameters associated with the reshaping force (see above), all measured data were reproduced [41]. This included the measured distribution of cell speed, the reached distance curve (plotted against the square root of time), the typical time course of cell speed, and the range of variation of cell speeds. Even the distribution of measured cell elongations was found without any additional fitting. One excellent feature of

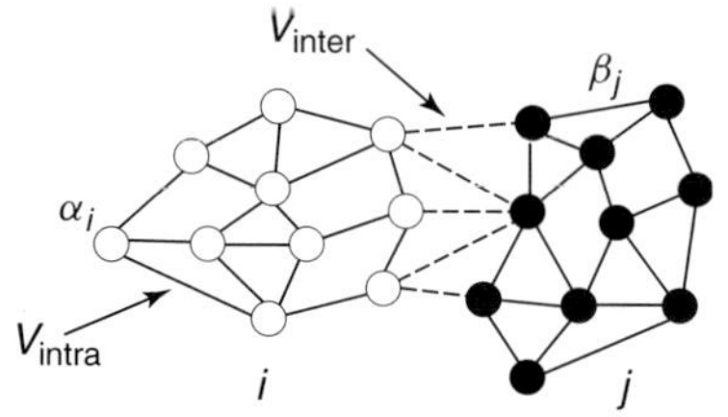

Fig. 8 Two cells (i and j) interacting according to the subcellular element model. White elements belong to cell i, and black to cell j. Intracellular interactions are represented by solid lines, while intercellular interactions are represented by dashed lines. α_i is an element of cell i, β_j an element of cell j, V_{intra} the intracellular potential and V_{inter} the intercellular potential. Figure reproduced with permission from Ref. [164]; © 2005, American Institute of Mathematical Sciences (AIMS).

this model to emerge was that it generically couples cell deformation and cell displacement in a way that is quantitatively consistent with two-photon imaging data of lymphocyte migration. However, the internal structure of the cell is amorphic, and the effect of cytoskeleton dynamics is treated only on a phenomenological level.

6.3.3 Subelement Model: An Off-Lattice Approach

Off-lattice models are widely used in the simulation of multicellular systems [149, 153, 163] (see Sect. 7), and may also be used to describe subcellular dynamics. The *subcellular element model* [164] was designed to allow the simulation of large numbers of cells, while still taking into consideration cell-shape dynamics. Subcellular particles are not confined to a lattice but rather are allowed to occupy any position in space. In the model, each cell is composed of a specific number of subcellular interacting particles (see Fig. 8). The integrity of the cell is guaranteed by attractive interactions between the subelements of the cell. Long-range chemical signaling is not included in the system, and the cells are thought to interact with each other via local biochemical interactions. The position of each subcellular element changes according to three processes: (i) a weak stochastic component; (ii) an elastic response to intracellular biomechanical forces; and (iii) an elastic response to intercellular biomechanical forces. Additionally, the movement is considered to be overdamped. The dynamics of the elements is described by Langevin equations [164].

The intracellular and intercellular interactions are characterized by two phenomenological potentials into which all relevant biological details must be encoded. Long-range interactions that may be mediated by microtubules are not covered by the approach due to the local nature of all interactions. The method provides efficient procedures to update neighborhood relations. An interesting application of this model to cell rheology was recently reported [165].

6.3.4 Tensegrity Model

An alternative approach of including the internal structure of cells is provided by the so-called "tensegrity model" [166, 167], in which a cell is approximated by a tensegrity structure. (*Tensegrity structures* are defined as systems that stabilize their

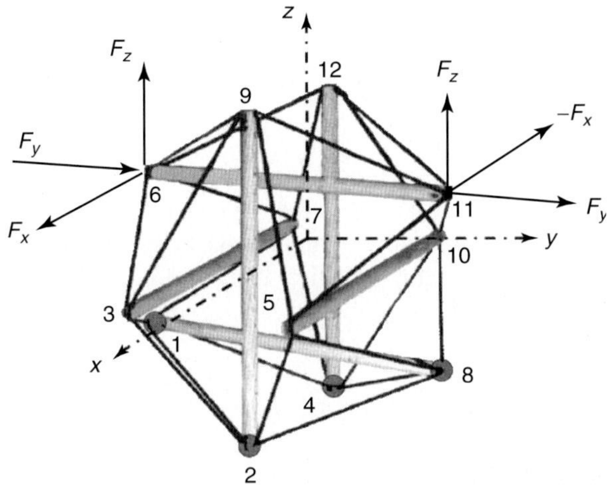

Fig. 9 Thirty-element tensegrity structure (six bars and 24 viscoelastic cables) with the external forces applied to nodes 6 and 11 are analyzed. Extension and compression forces F_z are applied along the z-axis. Shear forces F_y are applied along the y-axis. A twisting torque with opposite forces applied at node $\pm F_x$. Figure reproduced with permission from Ref. [168]; © 2002, Academic Press, Elsevier.

shape by continuous tension – that is, tensional integrity – rather than by continuous compression.) The idea was to create a cell model that relates mechanics to chemistry at the molecular level, and that can be translated into mathematical terms. Tensegrity includes two broad structural classes: prestressed, and geodesic. Both of these fail to maintain shape stability upon mechanical stress without the continuous transmission of tension forces that are thought to be sustained by cytoskeletal microfilaments and intermediate filaments, and balanced internally by microtubules and externally by adhesion to the ECM.

An experimental validation of the cellular tensegrity model requires the demonstration of three major behaviors of living cells:

- Cells, in terms of mechanics, behave as discrete networks composed of different interconnected cytoskeletal filaments.
- Cytoskeletal prestress is a major determinant of cell deformability.
- Microtubules function as compression struts and act in a complementary manner with ECM anchors to resist cytoskeletal tensional forces.

To date, however, a comprehensive and quantitative demonstration of these properties is lacking.

A tensegrity-based model was used for the theoretical analysis of the structural viscoelasticity of the cytoskeleton [168]. For the simulation, a 30-element tensegrity structure was modified in order to consider the viscoelastic properties of cables (see Fig. 9). Cables were assumed to behave like viscoelastic Voight bodies (an elastic element in parallel with a viscous dashpot). In addition, the Young modulus and cross-sectional area that characterize the cables and bars were kept constant, while the viscosity modulus of cables and the length of bars were allowed to vary.

The viscoelastic response of the system was studied over a large range of initial states of overall deformation by solving a system of differential equations, where the external-forces vector was related to both the nodal-displacement vector associated with a global-rigidity matrix, while the rate of nodal-displacement vector was associated with the global-damping matrix. Overall, the results supported the idea of a structural origin to cellular viscoelasticity, and confirmed the biological relevance of the employed concept – at least, for this specific problem.

7 Mathematical Models of Tissue Organization

There exist several model types for simulating tissue with increasing microscopic complexity [148, 163]. At this point, the aim is to show that tissue organization is inherently a multiscale problem. This can be achieved by connecting the molecular level of chemokine receptor internalization dynamics to macroscopic effects on the level of tissue patterns on the one hand (Sect. 7.1), and by setting lymphoid tissue organization into a functional context on the other hand (Sect. 7.2). Examples will be provided which demonstrate two important aspects: (i) that cell migration must be understood in the context of the features of the environment and of molecular interactions; and (ii) that mathematical modeling has the potential to bridge the scales and to provide unexpected insights.

7.1 Chemokine Receptor Internalization Can Induce Tissue Instability

As described above (see Sect. 4.2) and shown clearly in *in vitro* experiments, chemokine receptors can be internalized [98]. Yet, the impact of chemokine receptor internalization dynamics on the sensitivity for chemokine gradients, on cell motility and, ultimately, on the formation of cell aggregates remains unclear. Nonetheless, these points were subsequently investigated in an *in silico* model [169] using the off-lattice method of Delaunay-Object-Dynamics (see Sect. 6.2.2). This approach is best applied to fast lymphocytes [120, 155] in a chemokine distribution generated by a point source. The high speed of the cells' migration implies that the time scale of chemokine diffusion and cell movement are comparable, and that the chemokine distribution can no longer be modeled by a quasi-steady-state approximation.

In silico, cells are represented as soft elastic spheres in continuous 3-D space, where each cell can generate migration forces onto an ECM. Together with the friction in the tissue, this system is described by coupled Newtonian equations of motion in an overdamped approximation [144, 150]. The contact partners of cells are derived from a Delaunay triangulation [150, 155] which is locally updated in response to cell movement [170, 171]. The mechanical parameters of cell migration are chosen such that the speed distributions of motile lymphocytes in their natural environment [155] are reproduced.

The cell orientation, which determines the direction of migration and net force generation, is maintained for a persistence time T_p (see Sect. 5), derived experimentally [77, 120]. When the persistence time has passed, the cell reorients according to the local chemokine gradient; however, if the concentration difference between the front and back of the cell is below the threshold value of 2% [78] (see Sect. 4.1), the cell will switch to the mode of persistent random walk.

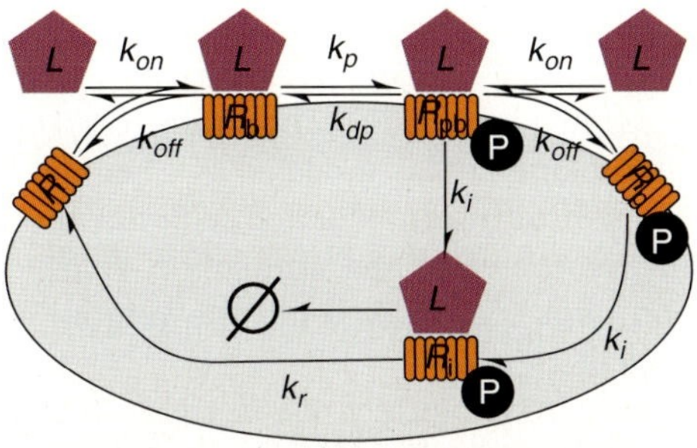

Fig. 10 Scheme of chemokine receptor dynamics. The receptor may exist in five different states: free (R_f); bound to the chemokine L (R_b); bound to the ligand L and phosphorylated (R_{pb}); ligand free but still phosphorylated (R_p); and internalized (R_i). Only the phosphorylated forms (R_p, R_{pb}) are internalized and are further on recycled back to the free receptor state R_f. The ligand that is internalized with the receptor is degraded.

The impact of cell-surface receptor dynamics on cell motility is based on the following assumptions (see Fig. 10):

- The motility signal for the cell is proportional to the ligand-induced phosphorylation of receptors.
- Phosphorylated receptors can be internalized.
- Internalized receptors are dynamically replaced at the surface by recycled or newly synthesized proteins (see Sect. 4.2).

These assumptions are translated into a set of differential equations:

$$\begin{aligned}
\dot{c} &= D\Delta c + Q - \kappa c - k_{\text{on}}(R_f + R_p)c \\
&\quad + k_{\text{off}}(R_b + R_{pb}) \\
\dot{R}_f &= -\beta R_f - k_{\text{on}} R_f c + k_{\text{off}} R_b + k_{\text{dp}} R_p + \alpha \\
&\quad + k_r R_i \\
\dot{R}_b &= -\beta R_b + k_{\text{on}} R_f c - k_{\text{off}} R_b - k_p R_b \\
&\quad + k_{\text{dp}} R_{pb} \\
\dot{R}_{pb} &= -\beta R_{pb} + k_{\text{on}} R_p c - k_{\text{off}} R_{pb} \\
&\quad + k_p R_b - (k_{\text{dp}} + k_i) R_{pb} \\
\dot{R}_p &= -\beta R_p - k_{\text{on}} R_p c + k_{\text{off}} R_{pb} \\
&\quad - (k_{\text{dp}} + k_i) R_p \\
\dot{R}_i &= k_i(R_p + R_{pb}) - k_r R_i
\end{aligned} \tag{4}$$

In this case, chemokine c is secreted with rate Q, degraded with the rate constant κ, and diffuses with diffusion constant D. Chemokine receptors can adopt five different states (see Fig. 10): free (R_f); ligand-bound (R_b); ligand-bound and phosphorylated (R_{pb}); phosphorylated (R_p); and internalized (R_i). Internalized receptors cannot bind the chemokine and are recycled back to the compartment of free surface receptors.

The dynamics of the receptors feeds back onto the concentration profile of the chemokine c by sequestration. In addition, the internalization of receptor–chemokine complexes (R_{pb}) effectively leads to degradation of the chemokines. The chemokine receptor exhibits a constitutive turnover, which is represented by the degradation rate β and the production α of free receptors. In this case, α and β are chosen such that, in the absence of the ligand, the amount of free receptor equals the receptor expression $R_{tot} = \alpha/\beta$. Some reaction rates were assumed independent of the receptor state: k_{on} and k_{off} are independent of receptor phosphorylation and k_{dp} and k_i are the same for bound and unbound chemokines.

The simulation is initialized with 5000 cells in a dynamic chemokine field generated by a static point source (Q). The chemokine field is modeled by a standard reaction diffusion equation with a source in the center of the simulation volume and with the sinks being the migrating cells themselves.

For physiologically relevant parameters (see the discussion of parameters in Ref. [169]), the following unexpected behavior is found. Initially, the cells aggregate toward the chemokine source, but they spread in the simulation volume and form smaller aggregates distant to the chemokine source. Further on, a pulsatile behavior of local aggregations and spreading cells is observed. No stable asymptotic state is found.

This instability is a result of the comparably high speed of cell migration in relation to the diffusion process of the chemokine. Local chemokine internalization leads to a depletion of the chemokine, and inverts its gradient where the cell density is sufficiently high. As the cells respond faster to the inverted gradient than the chemokine profile is restored by diffusion, the cells form local aggregates away from the central chemokine source.

The receptor dynamics of single cells in these unstable aggregates can be monitored *in silico*. It has been found that cells exhibit stable or quasi-oscillatory receptor dynamics, with the type of receptor dynamics depending on the individual fate of a cell in the dynamically changing chemokine field, and being the result of a complex, nonlinear interaction.

The instability of aggregation and the local misplaced small and dynamic aggregates are a robust feature of the simulation. However, in specific parameter regimes the expected simple and central aggregate can be restored in either of two ways. First, the parameters must be set such that the local depletion of chemokines is prevented; this is achieved by an increased chemokine source Q or by increased k_{dp} to reduce the internalization rate. Second, the diffusion constant D may be increased into a nonphysiological regime so as to quickly equilibrate the locally inverted chemokine gradients.

The impact of receptor internalization on tissue organization was investigated in a seemingly simple system of a fixed number of cells that would respond to a point source of chemokines. While the intuitive result of such a simulation would be an aggregation of cells around the point source, a variety of patterns was seen to emerge from this simple set-up. The details of the receptor dynamics of single cells have a significant impact on the cell behavior, and even on the tissue organization. This shows that multiscale systems in which the scales are not clearly separated can exhibit quite surprising effects.

7.2
Migration Patterns in Germinal Centers

Upon an immune response, the lymphoid B-cell follicles develop germinal centers containing various active B-cell subtypes. The germinal centers exhibit a specific morphology in which two zones – the "dark" and "light" zones – can be distinguished. The dark zone contains mostly dividing and mutating B cells, while the light zone contains mostly nondividing and apoptotic B cells. Selected B cells arising from this environment will differentiate to memory B cells or plasma cells, which in turn generate large amounts of antibodies. The main task of germinal centers is affinity maturation; that is, the somatic hypermutation of the antibody-encoding region of B cells and the selection of those clones with high-affinity antibodies which allow for an increased efficiency in the fight against specific pathogens [172]. It was hypothesized that affinity maturation is only possible when the B cells undergo iterative cycles of mutation and selection, the so-called "recycling hypothesis." Two issues of germinal centers – whether chemotaxis induces the two zones of germinal centers, and whether the recycling hypothesis is true – will be discussed below, on the basis of multi-photon imaging experiments (Sect. 5).

The unresponsiveness of freshly isolated cells (see Sect. 4.2) is of great importance for the interpretation of experimental data. The presence of a chemokine receptor does not imply chemotactic activity, and this is also true for germinal center B cells, which express CXCR4 or CXCR5. Indeed, this expression was shown to be required for the cell sorting into the two zones of the germinal center [173]. However, the majority of the germinal center B cells must be inactive in terms of chemotaxis, because multi-photon imaging has proved that germinal center B cells predominantly perform a random walk [174–176]. In view that B-cell chemotaxis is required to maintain the zonal structure of germinal centers, this result was rather surprising and motivated the development of mathematical models.

A stochastic hybrid event generator (Sect. 6.2.1) was used to investigate B-cell chemotaxis, affinity maturation, and the recycling hypothesis [140, 142, 177–179]. A regular lattice model for the cells was simultaneously used as a reaction-diffusion grid for the chemokines CXCL12 and CXCL13, which established stable gradients in the germinal center environment. The cell migration mode was deduced from intravital multi-photon microscopy data [174–176]. In particular, the average cell speed, the average persistence time of the direction of motion, and the distribution of turning angles were used as inputs for the simulation. The resulting speed–distribution and the relationship between the distance and square root of time were deduced from the model and compared to the experimental results in order to prove its validity.

The *in silico* result, that the persistent random walk observed in multi-photon experiments might be a superposition of finite size effects and chemokine desensitization dynamics (Sect. 5.1), led to the development of the *transient chemotaxis* concept for activated B cells in follicles [140], which was later supported by an independent investigation [142]. According to this model, a stable chemokine gradient guides the lymphocytes to their respective compartments. The sensitivity of the lymphocytes is regulated depending on the concentration of the chemokine: at above-threshold concentrations, the lymphocytes are desensitized for the chemokine (see Sect. 4.2 for the discussion of possible desensitization mechanisms) and perform a persistent random walk. However, when the cells reach a below-threshold concentration of the chemokine at the border of the compartment, they are resensitized for the chemokine. Ultimately, this model was shown to have captured the three main properties observed in lymphoid tissue:

- Chemotaxis drives lymphocytes into to their target compartment.
- Within the compartment, the lymphocytes desensitize for the chemokine and perform a persistent random walk that is consistent with the multi-photon measurements.
- Lymphocytes are retained in this compartment by chemokine resensitization.

The long-lasting discussions [172, 176, 180–182] regarding B-cell recycling between both compartments – that is, the dark and light zones – could be clarified with the help of mathematical models [140, 142, 183]. At first glance, the transmigration of B cells between zones appeared to be a rare event in 1 h tracking experiments, and the expected asymmetry between both directions of migration (due to selection in the light zone) could not be found [174–176]. Yet, mathematical models showed that the frequency of transmigration events was exactly as expected from the recycling hypothesis, and that the asymmetry would only appear in experiments in which the cells were tracked for several hours [123, 140]. This situation was subsequently confirmed in a setting with photoactivatable B cells – that is, B cells rendered fluorescent in one of the zones and tracked for 6 h [183].

The resulting picture of the germinal center assumed that B cells would divide and mutate in the dark zone, sensitize for CXCL13, and then follow the chemokine gradient to the light zone, where the chemokine is secreted by stromal cells. Subsequently, the cells would desensitize for CXCL13 and undergo a competitive selection process for high-affinity antibodies. The surviving cells would sensitize for CXCL12, which has a low concentration in the light zone, and then follow the chemokine gradient to its source in the dark zone, in order to continue dividing and mutating the antibody. This gives rise to an iterative Darwinian evolution of high-affinity clones.

It should be noted, that the cellular organization and affinity maturation are similarly described by 2-D and 3-D event generators. However, the motility data could only be appropriately modeled in the 3-D settings (see Sect. 6.2.1).

There is gathering evidence that T cells use the ECM in order to perform a persistent random walk (Sect. 5.3). But, based on this perspective, the question would then arise as to how the persistent random walk of B cells in follicles might be explained, where the ECM is only weakly expressed [24, 25, 28, 30, 31, 33–35], and whether migration would be normal in the absence of integrin expression [38]. A persistent random walk in the absence of ECM points to alternative mechanisms that induces a reorientation of the cell after a persistence time, irrespective of any fiber crossing.

8
Conclusions

Cell migration is at the heart of organ development and protection against foreign invaders. In this chapter, the most important cellular components involved in active cell motility have been discussed, and the way in which the different parts of the cellular cytoskeleton act together to define a cell's shape and its potential to migrate, have been outlined; as a consequence, different modes of motility have been identified. In order to exert forces on a substrate, either focal adhesion points or biomechanical anchors must be used. Yet, although these two mechanisms have been attributed to fibroblasts and leukocytes, respectively, evidence indicates that cells may switch between different modes of migration in order to adapt to their respective environments.

It has been shown how leukocytes are guided to their compartments, and how the different guidance modes, including chemotaxis, chemokinesis, and guidance by ECM fibers, have been identified. Strikingly, the migration patterns of T and

B cells are comparable, yet their extracellular backgrounds are quite different in terms of the composition of T- and B-cell zones. Whilst this again points to a flexibility of the guidance model, depending on functionality and environment, the migration mode of B cells in ECM-reduced follicles remains to be determined.

The tremendous power of multi-photon imaging has been demonstrated in intravital experiments that allow cells to be tracked in lymphoid tissues. At the same time, the limitations in the interpretation of the data produced have been outlined, and how the mathematical modeling of cell migration can assist in understanding these data. The different modeling approaches have been described and weighted on the basis of their strengths and limitations. Although subcellular modeling techniques continue to be developed, in most cases whole-cell models have been used to describe cell migration. Typically, these models exhibit great power in bridging the molecular to the tissue scale. It has also emerged that the functional environment is essential in order to develop predictive mathematical models.

To conclude, the functional environment determines cell migration and differentiation rather than cell intrinsic programs, which are more likely to provide a "toolbox" of possible behaviors from which cells can adaptively select.

References

1 Serafini, A. (1993) *The Epic History of Biology*, 1st edn, Perseus Publishing, New York.

2 Zhu, C., Bao, G., Wang, N. (2000) Cell mechanics: mechanical response, cell adhesion, and molecular deformation. *Annu. Rev. Biomed. Eng.*, **2**, 189–226.

3 Stricker, J., Falzone, T., Gardel, M.L. (2010) Mechanics of the F-actin cytoskeleton. *J. Biomech.*, **43**, 9–14.

4 Stamenovic, D., Ingber, D.E. (2002) Models of cytoskeletal mechanics of adherent cells. *Biomech. Model Mechanobiol.*, **1** (1), 95–108.

5 Wang, N., Tolic-Norrelykke, I.M., Chen, J., Mijailovich, S.M., Butler, J.P., Fredberg, J.J., Stamenovic, D. (2002) Cell prestress. I. Stiffness and prestress are closely associated in adherent contractile cells. *Am. J. Physiol.*, **282** (3), C606–C616.

6 Maniotis, A.J., Chen, C.S., Ingber, D.E. (1997) Demonstration of mechanical connections between integrins, cytoskeletal filaments, and nucleoplasm that stabilize nuclear structure. *Proc. Natl Acad. Sci. USA*, **94** (3), 849–854.

7 Brown, M.J., Hallam, J.A., Liu, Y., Yamada, K.M., Shaw, S. (2001) Cutting edge: integration of human T lymphocyte cytoskeleton by the cytolinker plectin. *J. Immunol.*, **167** (2), 641–645.

8 Helfand, B.T., Chang, L., Goldman, R.D. (2003) The dynamic and motile properties of intermediate filaments. *Annu. Rev. Cell Dev. Biol.*, **19**, 445–467.

9 Friedl, P., Brocker, E.B. (2000) The biology of cell locomotion within three-dimensional extracellular matrix. *Cell. Mol. Life Sci.*, **57** (1), 41–64.

10 Lauffenburger, D.A., Horwitz, A.F. (1996) Cell migration: a physically integrated molecular process. *Cell*, **84** (3), 359–369.

11 Mitchison, T.J., Cramer, L.P. (1996) Actin-based cell motility and cell locomotion. *Cell*, **84** (3), 371–379.

12 Galbraith, C.G., Sheetz, M.P. (1998) Forces on adhesive contacts affect cell function. *Curr. Opin. Cell Biol.*, **10** (5), 566–571.

13 Serrador, J.M., Nieto, M., Sanchez-Madrid, F. (1999) Cytoskeletal rearrangement during migration and activation of T lymphocytes. *Trends Cell Biol.*, **9** (6), 228–233.

14 Samstag, Y., Eibert, S.M., Klemke, M., Wabnitz, G.H. (2003) Actin cytoskeletal dynamics in T lymphocyte activation and migration. *J. Leukoc. Biol.*, **73** (1), 30–48.

15 Vicente-Manzanares, M., Sanchez-Madrid, F. (2004) Role of the cytoskeleton during leukocyte responses. *Nat. Rev. Immunol.*, **4** (2), 110–122.

16 Webb, D.J., Parsons, J.T., Horwitz, A.F. (2002) Adhesion assembly, disassembly

and turnover in migrating cells – over and over and over again. *Nat. Cell Biol.*, **4** (4), E97–100.

17 Wolf, K., Muller, R., Borgmann, S., Brocker, E.B., Friedl, P. (2003) Amoeboid shape change and contact guidance: T-lymphocyte crawling through fibrillar collagen is independent of matrix remodeling by MMPs and other proteases. *Blood*, **102** (9), 3262–3269.

18 Friedl, P., Borgmann, S., Brocker, E.B. (2001) Amoeboid leukocyte crawling through extracellular matrix: lessons from the *Dictyostelium* paradigm of cell movement. *J. Leukoc. Biol.*, **70** (4), 491–509.

19 Entschladen, F., Gunzer, M., Scheuffele, C.M., Niggemann, B., Zanker, K.S. (2000) T lymphocytes and neutrophil granulocytes differ in regulatory signaling and migratory dynamics with regard to spontaneous locomotion and chemotaxis. *Cell Immunol.*, **199** (2), 104–114.

20 Galbraith, C.G., Yamada, K.M., Sheetz, M.P. (2002) The relationship between force and focal complex development. *J. Cell Biol.*, **159** (4), 695–705.

21 Zamir, E., Katz, M., Posen, Y., Erez, N., Yamada, K.M., Katz, B.Z., Lin, S., Lin, D.C., Bershadsky, A., Kam, Z., Geiger, B. (2000) Dynamics and segregation of cell–matrix adhesions in cultured fibroblasts. *Nat. Cell Biol.*, **2**, 191–196.

22 Palecek, S.P., Huttenlocher, A., Horwitz, A.F., Lauffenburger, D.A. (1998) Physical and biochemical regulation of integrin release during rear detachment of migrating cells. *J. Cell Sci.*, **111** (Pt. 7), 929–940.

23 Pierini, L.M., Lawson, M.A., Eddy, R.J., Hendey, B., Maxfield, F.R. (2000) Oriented endocytic recycling of alpha5beta1 in motile neutrophils. *Blood*, **95** (8), 2471–2480.

24 Ohtani, O., Kikuta, A., Ohtsuka, A., Murakami, T. (1991) Organization of the reticular network of rabbit Peyer's patches. *Anat. Rec.*, **229** (2), 251–258.

25 Ushiki, T., Ohtani, O., Abe, K. (1995) Scanning electron microscopic studies of reticular framework in the rat mesenteric lymph node. *Anat. Rec.*, **241** (1), 113–122.

26 Gretz, J.E., Kaldjian, E.P., Anderson, A.O., Shaw, S. (1996) Sophisticated strategies for information encounter in the lymph node: the reticular network as a conduit of soluble information and a highway for cell traffic. *J. Immunol.*, **157** (2), 495–499.

27 Kaldjian, E.P., Gretz, J.E., Anderson, A.O., Shi, Y., Shaw, S. (2001) Spatial and molecular organization of lymph node T cell cortex: a labyrinthine cavity bounded by an epithelium-like monolayer of fibroblastic reticular cells anchored to basement membrane-like extracellular matrix. *Int. Immunol.*, **13** (10), 1243–1253.

28 Castanos-Velez, E., Biberfeld, P., Patarroyo, M. (1995) Extracellular matrix proteins and integrin receptors in reactive and non-reactive lymph nodes. *Immunology*, **86** (2), 270–278.

29 Sixt, M., Kanazawa, N., Selg, M., Samson, T., Roos, G., Reinhardt, D.P., Pabst, R., Lutz, M.B., Sorokin, L. (2005) The conduit system transports soluble antigens from the afferent lymph to resident dendritic cells in the T cell area of the lymph node. *Immunity*, **22** (1), 19–29.

30 Liakka, K.A., Autio-Harmainen, H.I. (1992) Distribution of the extracellular matrix proteins tenascin, fibronectin, and vitronectin in fetal, infant, and adult human spleens. *J. Histochem. Cytochem.*, **40** (8), 1203–1210.

31 Maatta, M., Salo, S., Tasanen, K., Soini, Y., Liakka, A., Bruckner-Tuderman, L., Autio-Harmainen, H. (2004) Distribution of basement membrane anchoring molecules in normal and transformed endometrium: altered expression of laminin gamma2 chain and collagen type XVII in endometrial adenocarcinomas. *J. Mol. Histol.*, **35** (8-9), 715–722.

32 Katakai, T., Hara, T., Lee, J.H., Gonda, H., Sugai, M., Shimizu, A. (2004) A novel reticular stromal structure in lymph node cortex: an immuno-platform for interactions among dendritic cells, T cells and B cells. *Int. Immunol.*, **16** (8), 1133–1142.

33 Ohtsuka, A., Piazza, A.J., Ermak, T.H., Owen, R.L. (1992) Correlation of extracellular matrix components with the cytoarchitecture of mouse Peyer's patches. *Cell Tissue Res.*, **269** (3), 403–410.

34 Katakai, T., Hara, T., Sugai, M., Gonda, H., Shimizu, A. (2004) Lymph node fibroblastic reticular cells construct the stromal reticulum via contact with lymphocytes. *J. Exp. Med.*, **200** (6), 783–795.

35 Satoh, T., Takeda, R., Oikawa, H., Satodate, R. (1997) Immunohistochemical and structural characteristics of the reticular framework of the white pulp and marginal zone in the human spleen. *Anat. Rec.*, **249** (4), 486–494.

36 Thomazy, V.A., Vega, F., Medeiros, L.J., Davies, P.J., Jones, D. (2003) Phenotypic modulation of the stromal reticular network in normal and neoplastic lymph nodes: tissue transglutaminase reveals coordinate regulation of multiple cell types. *Am. J. Pathol.*, **163** (1), 165–174.

37 Haston, W.S., Shields, J.M. (1984) Contraction waves in lymphocyte locomotion. *J. Cell Sci.*, **68**, 227–241.

38 Lämmermann, T., Bader, B.L., Monkley, S.J., Worbs, T., Wedlich-Söldner, R., Hirsch, K., Keller, M., Förster, R., Critchley, D.R., Fässler, R., Sixt, M. (2008) Rapid leukocyte migration by integrin-independent flowing and squeezing. *Nature*, **453** (7191), 51–55.

39 Bornens, M., Paintrand, M., Celati, C. (1989) The cortical microfilament system of lymphoblasts displays a periodic oscillatory activity in the absence of microtubules: implications for cell polarity. *J. Cell Biol.*, **109** (3), 1071–1083.

40 Ratner, S., Sherrod, W.S., Lichlyter, D. (1997) Microtubule retraction into the uropod and its role in T cell polarization and motility. *J. Immunol.*, **159** (3), 1063–1067.

41 Meyer-Hermann, M.E., Maini, P.K. (2005) Interpreting two-photon imaging data of lymphocyte motility. *Phys. Rev. E*, **71** (6, Pt 1), 061912.

42 Niggemann, B., Joseph, J., Weidt, C., Lang, K., Zaenker, K.S., Entschladen, F. (2004) Tumor cell locomotion: differential dynamics of spontaneous and induced migration in a 3D collagen matrix. *Exp. Cell Res.*, **298** (1), 178–187.

43 Sahai, E., Marshall, C.J. (2003) Differing modes of tumour cell invasion have distinct requirements for Rho/ROCK signalling and extracellular proteolysis. *Nat. Cell Biol.*, **5** (8), 711–719.

44 Wolf, K., Mazo, I., Leung, H., Engelke, K., von Andrian, U.H., Deryugina, E.I., Strongin, A.Y., Brocker, E.B., Friedl, P. (2003) Compensation mechanism in tumor cell migration: mesenchymal-amoeboid transition after blocking of pericellular proteolysis. *J. Cell Biol.*, **160** (2), 267–277.

45 Burton, K., Park, J.H., Taylor, D.L. (1999) Keratocytes generate traction forces in two phases. *Mol. Biol. Cell*, **10** (11), 3745–3769.

46 Balaban, N.Q., Schwarz, U.S., Riveline, D., Goichberg, P., Tzur, G., Sabanay, I., Mahalu, D., Safran, S., Bershadsky, A., Addadi, L., Geiger, B. (2001) Force and focal adhesion assembly: a close relationship studied using elastic micropatterned substrates. *Nat. Cell Biol.*, **3**, 466–472.

47 Galbraith, C.G., Sheetz, M.P. (1999) Keratocytes pull with similar forces on their dorsal and ventral surfaces. *J. Cell Biol.*, **147** (6), 1313–1324.

48 Oliver, T., Dembo, M., Jacobson, K. (1999) Separation of propulsive and adhesive traction stresses in locomoting keratocytes. *J. Cell Biol.*, **145** (3), 589–604.

49 Schwarz, U.S., Balaban, N.Q., Riveline, D., Bershadsky, A., Geiger, B., Safran, S.A. (2002) Calculation of forces at focal adhesions from elastic substrate data: the effect of localized force and the need for regularization. *Biophys. J.*, **83** (3), 1380–1394.

50 Munevar, S., Wang, Y., Dembo, M. (2001) Traction force microscopy of migrating normal and H-ras transformed 3T3 fibroblasts. *Biophys. J.*, **80**, 1744–1757.

51 Kraning-Rush, C.M., Carey, S.P., Califano, J.P., Smith, B.N., Reinhart-King, C.A. (2011) The role of the cytoskeleton in cellular force generation in 2D and 3D environments. *Phys. Biol.*, **8**, 015009.

52 Franck, C., Maskarinec, S.A., Tirrell, D.A., Ravichandran, G. (2011) Three-dimensional traction force microscopy: a new tool for quantifying cell-matrix interactions. *PLoS ONE*, **6**, e17833.

53 Ehrengruber, M.U., Deranleau, D.A., Coates, T.D. (1996) Shape oscillations of human neutrophil leukocytes: characterization and relationship to cell motility. *J. Exp. Biol.*, **199** (4), 741–747.

54 Entschladen, F., Zanker, K.S. (2000) Locomotion of tumor cells: a molecular comparison to migrating pre- and postmitotic leukocytes. *J. Cancer Res. Clin. Oncol.*, **126** (12), 671–681.

55 Lewis, W.H. (1931) Locomotion of lymphocytes. *Bull. Johns Hopkins Hosp.*, **49**, 29–36.

56 Lewis, W.H. (1934) On the locomotion of polymorphonuclear neutrophils of the rat in autoplasma cultures. *Bull. Johns Hopkins Hosp.*, **55**, 273–279.
57 Murray, J., Vawter-Hugart, H., Voss, E., Soll, D.R. (1992) Three-dimensional motility cycle in leukocytes. *Cell Motil. Cytoskeleton*, **22** (3), 211–223.
58 Renkawitz, J., Sixt, M. (2010) Mechanisms of force generation and force transmission during interstitial leukocyte migration. *EMBO Rep.*, **11**, 744–750.
59 de Bruyn, P.P.H. (1946) The amoeboid movement of the mammalian leukocyte in tissue culture. *Anat. Rec.*, **95** (2), 177–191.
60 Guyader, H.L., Hyver, C. (1997) Periodic activity of the cortical cytoskeleton of the lymphoblast: modelling by a reaction-diffusion system. *C. R. Acad. Sci. III*, **320** (1), 59–65.
61 Iglic, A., Veranic, P., Batista, U., Kralj-Iglic, V. (2001) Theoretical analysis of shape transformation of V-79 cells after treatment with cytochalasin B. *J. Biomech.*, **34** (6), 765–772.
62 Rey, M., Vicente-Manzanares, M., Viedma, F., Yanez-Mo, M., Urzainqui, A., Barreiro, O., Vazquez, J., Sanchez-Madrid, F. (2002) Cutting edge: association of the motor protein nonmuscle myosin heavy chain-IIA with the C terminus of the chemokine receptor CXCR4 in T lymphocytes. *J. Immunol.*, **169** (10), 5410–5414.
63 Paluch, E., Piel, M., Prost, J., Bornens, M., Sykes, C. (2005) Cortical actomyosin breakage triggers shape oscillations in cells and cell fragments. *Biophys. J.*, **89** (1), 724–733.
64 Charras, G., Paluch, E. (2008) Blebs lead the way: how to migrate without lamellipodia. *Nat. Rev. Mol. Cell Biol.*, **9**, 730–736.
65 Engelmann, T.W.I. (1881) Neue Methode zur Untersuchung der Sauerstoffausscheidung pflanzlicher und tierischer Organismen. *Botanische Ztg.*, **39**, 441–448.
66 Leber, T. (1888) Über die Entstehung der entzündung und die Wirkung der Entzündungerregenden Schädlichkeiten. *Fortschr. Med.*, **4**, 460.
67 Ross, H.C. (1909) On a combination of substances which excites amoeboid movement in leucocytes, by which living can be differentiated from dead cells. *Lancet*, **173** (4455), 152–154.
68 Moser, B., Wolf, M., Walz, A., Loetscher, P. (2004) Chemokines: multiple levels of leukocyte migration control. *Trends Immunol.*, **25** (2), 75–84.
69 Sallusto, F., Mackay, C.R. (2004) Chemoattractants and their receptors in homeostasis and inflammation. *Curr. Opin. Immunol.*, **16** (6), 724–731.
70 Lo, C.G., Xu, Y., Proia, R.L., Cyster, J.G. (2005) Cyclical modulation of sphingosine-1-phosphate receptor 1 surface expression during lymphocyte recirculation and relationship to lymphoid organ transit. *J. Exp. Med.*, **201** (2), 291–301.
71 Cyster, J.G. (2005) Chemokines, sphingosine-1-phosphate, and cell migration in secondary lymphoid organs. *Annu. Rev. Immunol.*, **23**, 127–159.
72 Grigorova, I.L., Panteleev, M., Cyster, J.G. (2010) Lymph node cortical sinus organization and relationship to lymphocyte egress dynamics and antigen exposure. *Proc. Natl Acad. Sci. USA*, **107**, 20447–20452.
73 Hoogewerf, A.J., Kuschert, G.S., Proudfoot, A.E., Borlat, F., Clark-Lewis, I., Power, C.A., Wells, T.N. (1997) Glycosaminoglycans mediate cell surface oligomerization of chemokines. *Biochemistry*, **36** (44), 13570–13578.
74 Proudfoot, A.E., Handel, T.M., Johnson, Z., Lau, E.K., LiWang, P., Clark-Lewis, I., Borlat, F., Wells, T.N., Kosco-Vilbois, M.H. (2003) Glycosaminoglycan binding and oligomerization are essential for the in vivo activity of certain chemokines. *Proc. Natl Acad. Sci. USA*, **100** (4), 1885–1890.
75 Pelletier, A.J., van der Laan, L.J., Hildbrand, P., Siani, M.A., Thompson, D.A., Dawson, P.E., Torbett, B.E., Salomon, D.R. (2000) Presentation of chemokine SDF-1 alpha by fibronectin mediates directed migration of T cells. *Blood*, **96** (8), 2682–2690.
76 Weissenbach, M., Clahsen, T., Weber, C., Spitzer, D., Wirth, D., Vestweber, D., Heinrich, P.C., Schaper, F. (2004) Interleukin-6 is a direct mediator of T cell migration. *Eur. J. Immunol.*, **34** (10), 2895–2906.
77 Albrecht, E., Petty, H.R. (1998) Cellular memory: neutrophil orientation reverses during temporally decreasing chemoattractant concentrations. *Proc. Natl Acad. Sci. USA*, **95** (9), 5039–5044.

78 Parent, C.A., Devreotes, P.N. (1999) A cell's sense of direction. *Science*, **284** (5415), 765–770.

79 Levchenko, A., Iglesias, P.A. (2002) Models of eukaryotic gradient sensing: application to chemotaxis of amoebae and neutrophils. *Biophys. J.*, **82** (1, Pt 1), 50–63.

80 Nieto, M., Frade, J.M., Sancho, D., Mellado, M., Martinez-A, C., Sanchez-Madrid, F. (1997) Polarization of chemokine receptors to the leading edge during lymphocyte chemotaxis. *J. Exp. Med.*, **186** (1), 153–158.

81 Cruikshank, W.W., Center, D.M., Nisar, N., Wu, M., Natke, B., Theodore, A.C., Kornfeld, H. (1994) Molecular and functional analysis of a lymphocyte chemoattractant factor: association of biologic function with CD4 expression. *Proc. Natl Acad. Sci. USA*, **91** (11), 5109–5113.

82 Bernardini, G., Hedrick, J., Sozzani, S., Luini, W., Spinetti, G., Weiss, M., Menon, S., Zlotnik, A., Mantovani, A., Santoni, A., Napolitano, M. (1998) Identification of the CC chemokines TARC and macrophage inflammatory protein-1 beta as novel functional ligands for the CCR8 receptor. *Eur. J. Immunol.*, **28** (2), 582–588.

83 Kim, C.H., Pelus, L.M., White, J.R., Applebaum, E., Johanson, K., Broxmeyer, H.E. (1998) CKβ-11/macrophage inflammatory protein-3β/EBI1-ligand chemokine is an efficacious chemoattractant for T and B cells. *J. Immunol.*, **160** (5), 2418–2424.

84 Willimann, K., Legler, D.F., Loetscher, M., Roos, R.S., Delgado, M.B., Clark-Lewis, I., Baggiolini, M., Moser, B. (1998) The chemokine SLC is expressed in T cell areas of lymph nodes and mucosal lymphoid tissues and attracts activated T cells via CCR7. *Eur. J. Immunol.*, **28** (6), 2025–2034.

85 Yoshida, R., Nagira, M., Kitaura, M., Imagawa, N., Imai, T., Yoshie, O. (1998) Secondary lymphoid-tissue chemokine is a functional ligand for the CC chemokine receptor CCR7. *J. Biol. Chem.*, **273** (12), 7118–7122.

86 Matloubian, M., Lo, C.G., Cinamon, G., Lesneski, M.J., Xu, Y., Brinkmann, V., Allende, M.L., Proia, R.L., Cyster, J.G. (2004) Lymphocyte egress from thymus and peripheral lymphoid organs is dependent on S1P receptor 1. *Nature*, **427** (6972), 355–360.

87 Bardi, G., Lipp, M., Baggiolini, M., Loetscher, P. (2001) The T cell chemokine receptor CCR7 is internalized on stimulation with ELC, but not with SLC. *Eur. J. Immunol.*, **31** (11), 3291–3297.

88 Yopp, A.C., Ochando, J.C., Mao, M., Ledgerwood, L., Ding, Y., Bromberg, J.S. (2005) Sphingosine 1-phosphate receptors regulate chemokine-driven transendothelial migration of lymph node but not splenic T cells. *J. Immunol.*, **175** (5), 2913–2924.

89 Tomhave, E.D., Richardson, R.M., Didsbury, J.R., Menard, L., Snyderman, R., Ali, H. (1994) Cross-desensitization of receptors for peptide chemoattractants. Characterization of a new form of leukocyte regulation. *J. Immunol.*, **153** (7), 3267–3275.

90 Lin, F., Nguyen, C.M., Wang, S.J., Saadi, W., Gross, S.P., Jeon, N.L. (2004) Effective neutrophil chemotaxis is strongly influenced by mean IL-8 concentration. *Biochem. Biophys. Res. Commun.*, **319** (2), 576–581.

91 Slimani, H., Charnaux, N., Mbemba, E., Saffar, L., Vassy, R., Vita, C., Gattegno, L. (2003) Binding of the CC-chemokine RANTES to syndecan-1 and syndecan-4 expressed on HeLa cells. *Glycobiology*, **13** (9), 623–634.

92 Gunn, M.D., Ngo, V.N., Ansel, K.M., Ekland, E.H., Cyster, J.G., Williams, L.T. (1998) A B-cell-homing chemokine made in lymphoid follicles activates Burkitt's lymphoma receptor-1. *Nature*, **391** (6669), 799–803.

93 Luther, S.A., Bidgol, A., Hargreaves, D.C., Schmidt, A., Xu, Y., Paniyadi, J., Matloubian, M., Cyster, J.G. (2002) Differing activities of homeostatic chemokines CCL19, CCL21, and CXCL12 in lymphocyte and dendritic cell recruitment and lymphoid neogenesis. *J. Immunol.*, **169** (1), 424–433.

94 Kelly, K., Underwood, B. (1986) The role of ionic signals in early gene induction during T cell activation. *Gene Amplif. Anal.*, **4**, 197–206.

95 Lau, L.F., Nathans, D. (1987) Expression of a set of growth-related immediate early genes in BALB/c 3T3 cells: coordinate regulation with c-fos or c-myc. *Proc. Natl Acad. Sci. USA*, **84**, 1182–1186.

96 Introna, M., Bast, R.C. Jr, Tannenbaum, C.S., Hamilton, T.A., Adams, D.O. (1987) The effect of LPS on expression of the early competence genes JE and KC in murine peritoneal macrophages. *J. Immunol.*, **138**, 3891–3896.

97 Arai, H., Monteclaro, F.S., Tsou, C.L., Franci, C., Charo, I.F. (1997) Dissociation of chemotaxis from agonist-induced receptor internalization in a lymphocyte cell line transfected with CCR2B. Evidence that directed migration does not require rapid modulation of signaling at the receptor level. *J. Biol. Chem.*, **272** (40), 25037–25042.

98 Neel, N.F., Schutyser, E., Sai, J., Fan, G.H., Richmond, A. (2005) Chemokine receptor internalization and intracellular trafficking. *Cytokine Growth Factor Rev.*, **16** (6), 637–658.

99 Badr, G., Borhis, G., Treton, D., Richard, Y. (2005) IFNα enhances human B-cell chemotaxis by modulating ligand-induced chemokine receptor signaling and internalization. *Int. Immunol.*, **17** (4), 459–467.

100 Roy, M.P., Kim, C.H., Butcher, E.C. (2002) Cytokine control of memory B cell homing machinery. *J. Immunol.*, **169** (4), 1676–1682.

101 Casamayor-Palleja, M., Mondiere, P., Verschelde, C., Bella, C., Defrance, T. (2002) BCR ligation reprograms B cells for migration to the T zone and B-cell follicle sequentially. *Blood*, **99** (6), 1913–1921.

102 Van Drenth, C., Jenkins, A., Ledwich, L., Ryan, T.C., Mashikian, M.V., Brazer, W., Center, D.M., Cruikshank, W.W. (2000) Desensitization of CXC chemokine receptor 4, mediated by IL-16/CD4, is independent of p56lck enzymatic activity. *J. Immunol.*, **165** (11), 6356–6363.

103 Foxman, E.F., Kunkel, E.J., Butcher, E.C. (1999) Integrating conflicting chemotactic signals. The role of memory in leukocyte navigation. *J. Cell Biol.*, **147** (3), 577–588.

104 Heit, B., Tavener, S., Raharjo, E., Kubes, P. (2002) An intracellular signaling hierarchy determines direction of migration in opposing chemotactic gradients. *J. Cell Biol.*, **159** (1), 91–102.

105 Breitfeld, D., Ohl, L., Kremmer, E., Ellwart, J., Sallusto, F., Lipp, M., Forster, R. (2000) Follicular B helper T cells express CXC chemokine receptor 5, localize to B cell follicles, and support immunoglobulin production. *J. Exp. Med.*, **192** (11), 1545–1552.

106 Reif, K., Ekland, E.H., Ohl, L., Nakano, H., Lipp, M., Forster, R., Cyster, J.G. (2002) Balanced responsiveness to chemoattractants from adjacent zones determines B-cell position. *Nature*, **416** (6876), 94–99.

107 Muller, G., Hopken, U.E., Lipp, M. (2003) The impact of CCR7 and CXCR5 on lymphoid organ development and systemic immunity. *Immunol. Rev.*, **195**, 117–135.

108 Okada, T., Miller, M.J., Parker, I., Krummel, M.F., Neighbors, M., Hartley, S.B., O'Garra, A., Cahalan, M.D., Cyster, J.G. (2005) Antigen-engaged B cells undergo chemotaxis toward the T zone and form motile conjugates with helper T cells. *PLoS Biol.*, **3** (6), e150.

109 Ansel, K.M., McHeyzer-Williams, L.J., Ngo, V.N., McHeyzer-Williams, M.G., Cyster, J.G. (1999) In vivo-activated CD4 T cells upregulate CXC chemokine receptor 5 and reprogram their response to lymphoid chemokines. *J. Exp. Med.*, **190** (8), 1123–1134.

110 Fillatreau, S., Gray, D. (2003) T cell accumulation in B cell follicles is regulated by dendritic cells and is independent of B cell activation. *J. Exp. Med.*, **197** (2), 195–206.

111 Haman, W.A. (1887) On the action and properties of the white blood-cells. Do the intra-vascular white blood-cells possess the property of amoeboid change of shape and position? is the asseverated migration of white blood-cells an observable phenomenon of inflammation? *Lancet*, **129** (3318), 671–672.

112 Cahalan, M.D., Parker, I. (2008) Choreography of cell motility and interaction dynamics imaged by two-photon microscopy in lymphoid organs. *Annu. Rev. Immunol.*, **26**, 585–626.

113 Shlesinger, M.F. (2001) Physics in the noise. *Nature*, **411** (6838), 641.

114 Fürth, R. (1920) The Brownian motion with consideration of the longevity of the direction of movement. *Z. Phys.*, **2**, 244–256.

115 Gail, M.H., Boone, C.W. (1970) The locomotion of mouse fibroblasts in tissue culture. *Biophys. J.*, **10** (10), 980–993.

116 Allan, R.B., Wilkinson, P.C. (1978) A visual analysis of chemotactic and chemokinetic

locomotion of human neutrophil leucocytes. Use of a new chemotaxis assay with *Candida albicans* as gradient source. *Exp. Cell Res.*, **111** (1), 191–203.

117 Codling, E.A., Plank, M.J., Benhamou, S. (2008) Random walk models in biology. *J. R. Soc. Interface*, **5** (25), 813–834.

118 Miller, M.J., Wei, S.H., Parker, I., Cahalan, M.D. (2002) Two-photon imaging of lymphocyte motility and antigen response in intact lymph node. *Science*, **296** (5574), 1869–1873.

119 Miller, M.J., Wei, S.H., Cahalan, M.D., Parker, I. (2003) Autonomous T cell trafficking examined in vivo with intravital two-photon microscopy. *Proc. Natl Acad. Sci. USA*, **100** (5), 2604–2609.

120 Wei, S.H., Parker, I., Miller, M.J., Cahalan, M.D. (2003) A stochastic view of lymphocyte motility and trafficking within the lymph node. *Immunol. Rev.*, **195**, 136–159.

121 Mempel, T.R., Henrickson, S.E., Von Andrian, U.H. (2004) T-cell priming by dendritic cells in lymph nodes occurs in three distinct phases. *Nature*, **427** (6970), 154–159.

122 Nishikawa, S., Honda, K., Vieira, P., Yoshida, H. (2003) Organogenesis of peripheral lymphoid organs. *Immunol. Rev.*, **195**, 72–80.

123 Meyer-Hermann, M., Figge, M.T., Toellner, K.M. (2009) Germinal centres seen through the mathematical eye: B-cell models on the catwalk. *Trends Immunol.*, **30**, 157–164.

124 Grayson, M.H., Hotchkiss, R.S., Karl, I.E., Holtzman, M.J., Chaplin, D.D. (2003) Intravital microscopy comparing T lymphocyte trafficking to the spleen and the mesenteric lymph node. *Am. J. Physiol. Heart Circ. Physiol.*, **284** (6), H2213–H2226.

125 Gunzer, M., Weishaupt, C., Hillmer, A., Basoglu, Y., Friedl, P., Dittmar, K.E., Kolanus, W., Varga, G., Grabbe, S. (2004) A spectrum of biophysical interaction modes between T cells and different antigen-presenting cells during priming in 3-D collagen and in vivo. *Blood*, **104** (9), 2801–2809.

126 Bajenoff, M., Egen, J.G., Koo, L.Y., Laugier, J.P., Brau, F., Glaichenhaus, N., Germain, R.N. (2006) Stromal cell networks regulate lymphocyte entry, migration, and territoriality in lymph nodes. *Immunity*, **25**, 989–1001.

127 Horstmann, D. (2003) From 1970 until present: the Keller-Segel model in chemotaxis and its consequences I. *Jahresber. DMV*, **105** (3), 103–165.

128 Stephanou, A., Tracqui, P. (2002) Cytomechanics of cell deformations and migration: from models to experiments. *C. R. Biol.*, **325** (4), 295–308.

129 Gracheva, M.E., Othmer, H.G. (2004) A continuum model of motility in ameboid cells. *Bull. Math. Biol.*, **66** (1), 167–193.

130 Romanovskii, Yu.M., Teplov, V.A. (1995) The physical base of cell movement. The mechanisms of self-organisation of amoeboid motility. *Physics-Uspekhi*, **38** (5), 521–543.

131 Sambeth, R., Baumgaertner, A. (2001) Autocatalytic polymerization generates persistent random walk of crawling cells. *Phys. Rev. Lett.*, **86** (22), 5196–5199.

132 Herant, M., Marganski, W.A., Dembo, M. (2003) The mechanics of neutrophils: synthetic modeling of three experiments. *Biophys. J.*, **84** (5), 3389–3413.

133 Bischofs, I.B., Schwarz, U.S. (2003) Cell organization in soft media due to active mechanosensing. *Proc. Natl Acad. Sci. USA*, **100** (16), 9274–9279.

134 Bischofs, I.B., Safran, S.A., Schwarz, U.S. (2004) Elastic interactions of active cells with soft materials. *Phys. Rev. E*, **69** (2, Pt 1), 021911.

135 Schwarz, U.S., Safran, S.A. (2002) Elastic interactions of cells. *Phys. Rev. Lett.*, **88** (4), 048102.

136 Kaverina, I., Krylyshkina, O., Beningo, K., Anderson, K., Wang, Y.L., Small, J.V. (2002) Tensile stress stimulates microtubule outgrowth in living cells. *J. Cell Sci.*, **115** (Pt 11), 2283–2291.

137 Painter, K.J., Maini, P.K., Othmer, H.G. (2000) Development and applications of a model for cellular response to multiple chemotactic cues. *J. Math. Biol.*, **41** (4), 285–314.

138 Kansal, A.R., Torquato, S., Harsh GR, I.V., Chiocca, E.A., Deisboeck, T.S. (2000) Simulated brain tumor growth dynamics using a three-dimensional cellular automaton. *J. Theor. Biol.*, **203** (4), 367–382.

139 Savill, N.J., Hogeweg, P. (1997) Modelling morphogenesis: from single cells to crawling slugs. *J. Theor. Biol.*, **184**, 229–235.

140 Figge, M.T., Garin, A., Gunzer, M., Kosco-Vilbois, M., Toellner, K.-M., Meyer-Hermann, M. (2008) Deriving a germinal center lymphocyte migration model from two-photon data. *J. Exp. Med.*, **205**, 319–329.

141 Beauchemin, C., Dixit, N.M., Perelson, A.S. (2007) Characterizing T cell movement within lymph nodes in the absence of antigen. *J. Immunol.*, **178**, 5505–5512.

142 Beltman, J.B., Allen, C.D.C., Cyster, J.G., de Boer, R.J. (2011) B cells within germinal centers migrate preferentially from dark to light zone. *Proc. Natl Acad. Sci. USA*, **108**, 8755–8760.

143 Odell, G.M., Oster, G., Alberch, P., Burnside, B. (1981) The mechanical basis of morphogenesis. I. Epithelial folding and invagination. *Dev. Biol.*, **85** (2), 446–462.

144 Palsson, E., Othmer, H.G. (2000) A model for individual and collective cell movement in *Dictyostelium discoideum*. *Proc. Natl Acad. Sci. USA*, **97** (19), 10448–10453.

145 Palsson, E. (2001) A three-dimensional model of cell movement in multicellular systems. *Future Gener. Comput. Syst.*, **17**, 835–852.

146 Dallon, J.C., Othmer, H.G. (2004) How cellular movement determines the collective force generated by the *Dictyostelium discoideum* slug. *J. Theor. Biol.*, **231** (2), 203–222.

147 Drasdo, D., Kree, R., McCaskill, J.S. (1995) Monte Carlo approach to tissue-cell populations. *Phys. Rev. E*, **52** (6), 6635–6657.

148 Drasdo, D. (2003) On Selected Individual-Based Approaches to the Dynamics in Multicellular Systems, in: Alt, W., Chaplain, M., Griebel, M., Lenz, J. (Eds) *Polymer and Cell Dynamics: Multiscale Modeling and Numerical Simulations*, Birkhaeuser, pp. 169–204.

149 Meineke, F.A., Potten, C.S., Loeffler, M. (2001) Cell migration and organization in the intestinal crypt using a lattice-free model. *Cell Prolif.*, **34** (4), 253–266.

150 Schaller, G., Meyer-Hermann, M. (2005) Multicellular tumor spheroid in an off-lattice Voronoi–Delaunay cell model. *Phys. Rev. E.*, **71** (5, Pt 1), 051910.

151 Honda, H., Tanemura, M., Nagai, T. (2004) A three-dimensional vertex dynamics cell model of space-filling polyhedra simulating cell behavior in a cell aggregate. *J. Theor. Biol.*, **226** (4), 439–453.

152 Schaller, G., Meyer-Hermann, M. (2007) A modelling approach towards an epidermal homoeostasis control. *J. Theor. Biol.*, **247**, 554–573.

153 Beyer, T., Meyer-Hermann, M. (2008) Mechanisms of organogenesis of primary lymphoid follicles. *Int. Immunol.*, **20**, 615–623.

154 Meyer-Hermann, M. (2008) Delaunay-object-dynamics: cell mechanics with a 3D kinetic and dynamic weighted Delaunay-triangulation. *Curr. Top. Dev. Biol.*, **81**, 373–399.

155 Beyer, T., Meyer-Hermann, M. (2007) Modeling emergent tissue organization involving high-speed migrating cells in a flow equilibrium. *Phys. Rev. E*, **76**, 021929.

156 Graner, F., Glazier, J.A. (1992) Simulation of biological cell sorting using a two-dimensional extended Potts model. *Phys. Rev. Lett.*, **69** (13), 2013–2016.

157 Ising, E. (1925) Beitrag zur Theorie des Ferromagnetismus. *Z. Phys.*, **31**, 253–258.

158 Jiang, Y., Levine, H., Glazier, J. (1998) Possible cooperation of differential adhesion and chemotaxis in mound formation of *Dictyostelium*. *Biophys. J.*, **75** (6), 2615–2625.

159 Stott, E.L., Britton, N.F., Glazier, J.A., Zajac, M. (1999) Stochastic simulation of benign avascular tumour growth using the Potts model. *Math. Comput. Model.*, **30**, 183–198.

160 Merks, R.M., Brodsky, S.V., Goligorksy, M.S., Newman, S.A., Glazier, J.A. (2006) Cell elongation is key to in silico replication of in vitro vasculogenesis and subsequent remodeling. *Dev. Biol.*, **289** (1), 44–54.

161 Maree, A.F., Hogeweg, P. (2001) How amoeboids self-organize into a fruiting body: multicellular coordination in Dictyostelium discoideum. *Proc. Natl Acad. Sci. USA*, **98**, 3879–3883.

162 Alber, M., Chen, N., Lushnikov, P.M., Newman, S.A. (2007) Continuous macroscopic limit of a discrete stochastic model for interaction of living cells. *Phys. Rev. Lett.*, **99** (16), 168102.

163 Galle, J., Aust, G., Schaller, G., Beyer, T., Drasdo, D. (2006) Individual cell-based

models of the spatial-temporal organization of multicellular systems – achievements and limitations. *Cytometry A*, **69**, 704–710.

164 Newman, T.J. (2005) Modeling multicellular systems using subcellular elements. *Math. Biosci. Eng.*, **2**, 611–622.

165 Sandersius, S.A., Newman, T.J. (2008) Modeling cell rheology with the subcellular element model. *Phys. Biol.*, **5**, 015002.

166 Ingber, D.E. (2003) Tensegrity I. Cell structure and hierarchical systems biology. *J. Cell Sci.*, **116** (Pt 7), 1157–1173.

167 Ingber, D.E. (2003) Tensegrity II. How structural networks influence cellular information processing networks. *J. Cell Sci.*, **116** (Pt 8), 1397–1408.

168 Canadas, P., Laurent, V.M., Oddou, C., Isabey, D., Wendling, S. (2002) A cellular tensegrity model to analyse the structural viscoelasticity of the cytoskeleton. *J. Theor. Biol.*, **218** (2), 155–173.

169 Beyer, T., Meyer-Hermann, M. (2008) Cell transmembrane receptors determine tissue pattern stability. *Phys. Rev. Lett.*, **101**, 148102.

170 Schaller, G., Meyer-Hermann, M. (2004) Kinetic and dynamic Delaunay tetrahedralizations in three dimensions. *Comput. Phys. Commun.*, **162**, 9–23.

171 Beyer, T., Schaller, G., Deutsch, A., Meyer-Hermann, M. (2005) Parallel dynamic and kinetic regular triangulation in three dimensions. *Comput. Phys. Commun.*, **172** (2), 86–108.

172 MacLennan, I.C. (1994) Germinal centers. *Annu. Rev. Immunol.*, **12**, 117–139.

173 Allen, C.D., Ansel, K.M., Low, C., Lesley, R., Tamamura, H., Fujii, N., Cyster, J.G. (2004) Germinal center dark and light zone organization is mediated by CXCR4 and CXCR5. *Nat. Immunol.*, **5** (9), 943–952.

174 Allen, C.D., Okada, T., Tang, H.L., Cyster, J.G. (2007) Imaging of germinal center selection events during affinity maturation. *Science*, **315** (5811), 528–531.

175 Schwickert, T.A., Lindquist, R.L., Shakhar, G., Livshits, G., Skokos, D., Kosco-Vilbois, M.H., Dustin, M.L., Nussenzweig, M.C. (2007) In vivo imaging of germinal centres reveals a dynamic open structure. *Nature*, **446** (7131), 83–87.

176 Hauser, A.E., Junt, T., Mempel, T.R., Sneddon, M.W., Kleinstein, S.H., Henrickson, S.E., von Andrian, U.H., Shlomchik, M.J., Haberman, A.M. (2007) Definition of germinal-center B cell migration in vivo reveals predominant intrazonal circulation patterns. *Immunity*, **26**, 655–667.

177 Meyer-Hermann, M. (2002) A mathematical model for the germinal center morphology and affinity maturation. *J. Theor. Biol.*, **216** (3), 273–300.

178 Meyer-Hermann, M.E., Maini, P.K. (2005) Back to "one-way" germinal centers. *J. Immunol.*, **174** (5), 2489–2493.

179 Meyer-Hermann, M., Maini, P.K., Iber, D. (2006) An analysis of B cell selection mechanisms in germinal centers. *Math. Med. Biol.*, **23**, 255–277.

180 Kepler, T.B., Perelson, A.S. (1993) Cyclic re-entry of germinal center B cells and the efficiency of affinity maturation. *Immunol. Today*, **14**, 412–415.

181 Oprea, M., van Nimwegen, E., Perelson, A.S. (2000) Dynamics of one-pass germinal center models: implications for affinity maturation. *Bull. Math. Biol.*, **62** (1), 121–153.

182 Meyer-Hermann, M., Deutsch, A., Or-Guil, M. (2001) Recycling probability and dynamical properties of germinal center reactions. *J. Theor. Biol.*, **210** (3), 265–285.

183 Victora, G.D., Schwickert, T.A., Fooksman, D.R., Kamphorst, A.O., Meyer-Hermann, M., Dustin, M.L., Nussenzweig, M.C. (2010) Germinal center dynamics revealed by multiphoton microscopy with a photoactivatable fluorescent reporter. *Cell*, **143**, 592–605.

15
Protein Modeling

Marian R. Zlomislic, Valentina Corradi, and D. Peter Tieleman
Department of Biological Sciences and Institute for Biocomplexity and Informatics, 2500 University Dr. NW, Calgary, AB T21N4, Canada

Systems Biology: Advances in Molecular Biology and Medicine, First Edition. Edited by Robert A. Meyers.

Keywords

Primary, secondary, tertiary, and quaternary protein structure
The amino acid sequence of a protein is its primary structure. Alpha-helices, beta-strands, and random coils are typical secondary structures that form between consecutive residues; these are defined by the hydrogen-bonding pattern observed in the protein backbone. The tertiary structure of a protein describes the relative position of all protein atoms in 3-D space. Elements of secondary structure that were far apart sequentially may now form higher-order structures, such as parallel or anti-parallel beta sheets, or helix bundles. The quaternary structure describes how multiple proteins are arranged together in larger complexes.

Ab initio modeling
Ab initio modeling is used to predict the structure of a protein, based only the amino acid sequence.

Fold recognition
Fold recognition is used to predict which fold (among approximately 1000 observed protein folds) a given sequence will adopt.

Homology modeling
Prediction of the 3-D structure of a protein, based on the known structure of a related protein.

Structural genomics
A concerted effort to determine, experimentally, the high-resolution structure of all proteins in a genome, often with a focus on proteins that are predicted to be structurally different from all structures that are already in the database. This greatly increases the likelihood that a given sequence can be modeled using homology modeling.

Molecular graphics
Modern computer graphics are used to visualize molecular structures and models. The ability to highlight different features or properties in a 3-D model offers a powerful

means of exploring molecules. Stereo hardware and immersive environments further enhance the usefulness of molecular graphics.

Poisson–Boltzmann equation
This equation describes electrostatic interactions between charges in a continuous medium (e.g., water) in the presence of a salt. It can be solved numerically, using several popular software packages, to provide insight into electrostatic interactions in proteins, between proteins and their substrates, between proteins and membranes, and in other systems. A common application is the calculation of pK_a-shifts for acidic and basic residues in proteins, due to their environment.

Docking
This term incorporates a group of methods used to study the interactions of small molecules and proteins. The process typically involves an extensive search of possible orientations of the small molecule in a binding site of the protein, and a scoring function that ranks different orientations by their likelihood of occurrence.

QM/MM simulation
A computer simulation method in which an important part of a system, such as the active site of an enzyme, is described using quantum mechanics, with the remainder of the system described using molecular dynamics methods.

Molecular dynamics simulation
A computer simulation method in which the motions of all atoms in a molecular model, including water and ions, are calculated over a period of up to one microsecond. This computationally expensive method is one of the most detailed ways of studying the dynamics of macromolecules.

Brownian dynamics simulation
This is related to molecular dynamics, but does not include the same degree of detail. Solvent, in particular, is normally represented as a continuous medium; this significantly reduces the complexity of the models and allows much longer simulations.

CASP, CAFASP, and CAPRI
Critical Assessment of Structure Prediction (CASP), Critical Assessment of Fully Automated Structure Prediction (CAFASP), and Critical Assessment of PRedicted Interactions (CAPRI) are large-scale assessments in which many research groups submit their best structure predictions on the same set of targets. This permits a critical comparison to be made between methods, and an assessment of the strengths and weaknesses of different methods and the field as a whole.

Protein modeling consists of a broad range of computational techniques to provide an understanding of the properties of proteins, and has become an integral part of structural biology and drug design. Modeling can be used to predict the secondary structure or fold of a protein based on its sequence alone, to predict the three-dimensional (3-D) structure of a protein based on knowledge of the structure of a related protein, to design new proteins, and to predict properties that depend on the experimentally determined 3-D structure of a protein. Examples of such properties include drug binding, protein–protein interactions, and interactions with elements within a protein's environment, including ions, lipids, carbohydrates, and nucleic acids. Conformational changes in proteins can be investigated using molecular dynamics simulations to provide a detailed insight into the dynamics of proteins, which is a crucial aspect of protein function. In recent developments, quantum mechanical calculations have been used frequently to study reactions in proteins. With the ever-improving power of computers, increasingly detailed aspects of protein function can now be investigated using modeling methods, at a scale and level of detail that is often very difficult or impossible to achieve experimentally. In this chapter, the main principles and techniques involved in protein modeling are introduced. Some examples are provided to highlight how protein modeling can be used to complement other methods.

1 Introduction

1.1 Primary, Secondary, Tertiary, and Quaternary Structures of Proteins

Proteins are structured at different levels [1]:

- The *primary structure* of a protein is its amino acid sequence, as encoded by DNA. All proteins are constructed from about 20 common amino acids, as well as some amino acids that are formed through chemical modification.
- The *secondary structure* of a protein is a sequence of common structural, three-dimensional (3-D) elements, or building blocks. These secondary structure elements include alpha-helices and beta-strands as the best-known elements [2], although there are many other building blocks.
- The *tertiary structure* of a protein is its 3-D structure in space, which can be thought of as placing the secondary structure elements together in space.
- The *quaternary structure* of a protein describes how multiple proteins interact through noncovalent forces to form larger protein complexes.

An example of the different levels of structure is shown in Fig. 1, based on the protein barstar [3]; it is shown in a protein complex with barnase to illustrate the quaternary structure. The minimal requirement for modeling proteins is knowledge of the primary structure and, as the primary sequence is one of the main results of genome sequencing, this is a very modest requirement. As increasing amounts of experimental structural information become available, however, modeling typically becomes more accurate by using that experimental information to identify plausible models. Any experimental information that places limitations on possible models is potentially useful [4, 5].

MKKAVINGEQIRSISDLHQTLKKELALPEYYGENLAALWDCLTGWVEYPLVLEWRQ
FEQSKQLTENGAESVLQVFREAKAEGCDITIILS
(a)

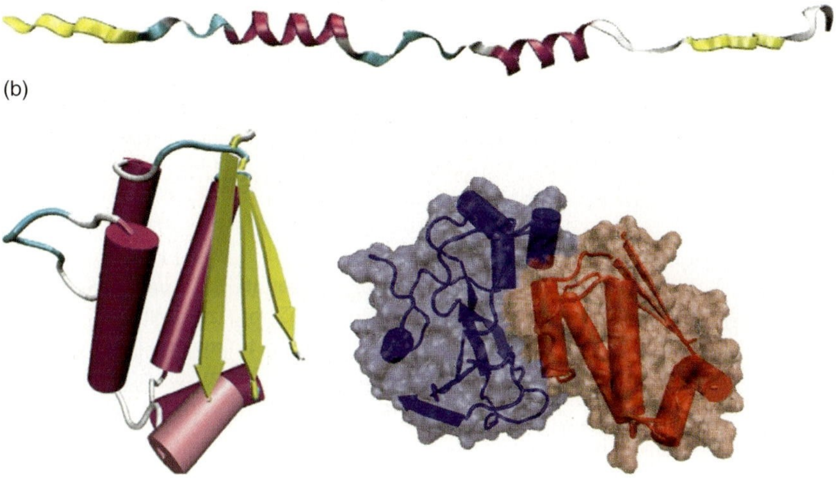

(b) (c) (d)

Fig. 1 (a) The primary sequence of the protein barstar in FASTA format; (b) The secondary structure of the first 60 residues of barstar, colored purple for helices, and yellow for beta-strands. All are shown in ribbon format; (c) The complete tertiary structure of barstar in cartoon format. The helices are now bundled together, and the beta-strands have formed a three-stranded parallel beta-sheet; (d) Example of quaternary structure: the barnase/barstar complex. Barstar (red) and barnase (blue) area shown in ribbon and spacefilling formats. Images created with VMD using the PDB entry 1B2U.

1.2 Relationship between Structure and Function

A basic assumption in protein science is that the function of a protein follows from its structure [1]. However, as proteins with very similar 3-D structures may have very different functions, the precise details of the 3-D (tertiary/quaternary) structure are very important. Nonetheless, a large amount of information is contained in the primary and secondary structures of a protein. Indeed, whilst Nature in most cases seems only to require the primary sequence, computational methods are not yet able reliably to predict the 3-D structure of proteins based on the primary sequence alone. Currently, a primary sequence is useful when comparing proteins of known function with a similar sequence, and occasionally allows the assignment of a tentative function. Knowing the secondary structure of a protein leads to a significant narrowing down of the range of possible 3-D structures. It also allows comparisons to be made with databases of the secondary sequences of known proteins, which in turn often makes it possible to recognize the 3-D fold of a protein based on the secondary structure alone. Interestingly, although – in principle – the number of possible 3-D structures is practically infinite, only about a 1000 different "folds" seem to occur in Nature. This is particularly useful for whole-genome studies:

if the primary sequence can be used, in combination with either a prediction or knowledge of the secondary structure, then the chances are reasonable that the fold of the protein can be recognized. Together, this represents a constructive step towards attaining a 3-D structure.

The 3-D structure of a protein is the most useful level for molecular modeling, as the interactions between small molecules (drugs and substrate) and an enzyme, the permeation properties of ion channels and transporters, the activation of receptors by ligand binding, and chemical reactions in enzymes and other processes that require a very detailed knowledge of atomic structure for their understanding, can all be investigated. The 3-D level is also required to study the interactions between protein subunits, or for protein–protein or protein–peptide interactions. In practice, the best means of obtaining knowledge of the 3-D structure is from experimental structure determinations, although prediction methods based primarily on homology modeling (see Sect. 2) are increasingly becoming sufficiently accurate to provide useful starting points for further computational studies.

1.3 Experimental Structure Determination Methods

Previously, several experimental methods have been developed to determine the structure of proteins. The primary structure is determined by sequencing, either of the protein itself or of the DNA that encodes the protein. The secondary structure can be measured using spectroscopic methods, as these usually determine the proportion of the various secondary structure elements. The most common techniques used for this include circular dichroism (CD) spectroscopy [6], infrared (IR) spectroscopy [7], and nuclear magnetic resonance (NMR) spectroscopy [8].

As a basis for detailed modeling problems, high-resolution 3-D structures are typically required, the two main techniques for the experimental determination of 3-D structures being X-ray crystallography [9] and NMR spectroscopy [10]. As a third method, electron microscopy has been used to resolve the structure of several membrane proteins [11].

Crystallography is a very powerful method that can be used on proteins and protein complexes of any size. Unfortunately, the main limitation of crystallography is that the protein must form regular crystals and, since not all proteins do this, the technique may occasionally force proteins into structures that may not be physiologically relevant.

NMR spectroscopy can be used to determine the structure of a protein in solution, which is usually a more realistic environment. The two main limitations of solution NMR spectroscopy are that it is difficult to apply the method in practice to proteins with a molecular weight above 40 kDa, and it is usually considerably more labor-intensive than crystallography.

These limitations have spurred the development of new methods targeted at using proteins of greater molecular weights, and at automating the process of structural calculations from a measured spectrum. *Solid-state protein NMR spectroscopy* is a relatively new method that can be used to determine the structure of proteins in ordered systems, such as membranes. Previously, electron microscopy has been used to determine 3-D structures mainly of membrane proteins, which are especially difficult to crystallize in readiness for X-ray crystallography [12]. Whilst electron microscopy is very labor-intensive, it can provide structural information on very large complexes, where the overall arrangement of large subunits can be seen at low resolution. The details of the structure can then be added from the crystal structures of the individual proteins that comprise the complex.

1.4 Role of Modeling

Modeling has many potential uses, at different levels of protein structure [13, 14]. However, as protein modeling is such a broad field, it may be beneficial first to distinguish two separate general goals.

The first set of goals involves predicting aspects of the structure of a protein, based on less-detailed information. If only the primary structure (the sequence) of a protein is known, then an attempt can be made to predict the secondary structure with reasonable accuracy, often in the 75–85% range [15]. Based on the degree of hydrophobicity of the predicted helices, it is usually possible also to predict transmembrane segments, identifying a protein as a membrane protein, with reasonable accuracy. The secondary structure may be useful to predict the general fold of a protein, which might allow a tentative function to be assigned to the protein. Predicting the tertiary structure directly from the sequence is a very difficult problem, and has a low success rate. If the high-resolution structure of a related protein is known, however, homology modeling can be used to provide a good model of the target protein. In principle, such modeling efforts make it possible to circumvent experimental structure determinations. At present, a major goal of protein structure research is to improve all of the steps in the process that leads to an accurate 3-D model, both by improving "ab initio" prediction methods and by experimentally determining sufficient protein structures that homology modeling would become feasible for most proteins in the genome. Ideally, structural models could then be obtained for all of the proteins in an organism, directly from its genome sequence. Clearly, as the number of structures solved experimentally increases, the algorithms to detect homology between remotely related proteins and built homology models will improve, such that this goal will become increasingly feasible.

The second set of goals begins with an experimentally determined high-resolution structure (or a very high-quality homology model) and uses physics-based methods to model those aspects of a protein that cannot easily be determined experimentally. This type of approach could include electrostatics calculations to determine how proteins interact with substrates or other proteins, docking calculations to identify differences in binding constants and to design new inhibitors for enzymes, or detailed molecular dynamics (MD) simulations to investigate the dynamics of proteins. This type of goal usually involves a much greater biochemical knowledge of the protein, and is often centered on specific questions: Why does mutating residue X to alanine change the binding affinity of a particular drug by a factor of 1000? Why is this ion channel selective for potassium over sodium, even though the only difference between potassium and sodium superficially is a small difference in radius?

Each set of goals is considered in greater detail, and specific examples are given, in the following sections.

2
Structure Prediction Methods

2.1
The Protein Folding Problem

One of the most challenging problems in biophysics is to understand how it is possible that proteins fold rapidly (in microseconds to seconds) into a well-defined structure when, based on their primary sequence alone, an almost infinite number of structures is possible. By conducting a systematic search of all possible conformations, Levinthal showed that folding on a realistic time scale could not occur (this is referred to as Levinthal's paradox) [16]. Subsequently, major research efforts have been devoted to understanding protein folding, with significant attention focused on the physics of simplified models [17]. Clearly, if it is possible to fold a protein *in vivo* from the sequence only, then with a correct understanding of the laws governing protein folding it should be possible to predict any protein structure computationally. Although this remains an ambitious goal, during the past few years major progress has been made in improving prediction methods at all levels of structure. Such progress has been documented strikingly in the proceedings of the semi-annual Critical Assessment of Structure Prediction (CASP) competition.

The CASP is an interesting and exciting venture involving a large scientific community, and has been held every two years since 1994 [18]. The main goal of the CASP is to obtain an in-depth and objective assessment of current abilities and inabilities in the area of protein structure prediction (http://predictioncenter.org). This involves a neutral organizing committee that collects unpublished but already-solved protein structures from experimental structure-determination groups. Specific information, such as the protein's primary sequence, is reported on a web site, and modeling groups can attempt to predict the structures. A panel of assessors judges the predictions against the real structures after which, at the CASP scientific meeting, the results are discussed, areas of progress and of problems are identified, and the progress and directions of the field as a whole are examined. This procedure has led to the CASP becoming a very fair way of establishing the merits of particular methods, and of identifying where future efforts can be most productively focused. During the past few years, the CASP has begun to introduce automated prediction software as a separate exercise (CAFASP – Critical Assessment of Fully Automated Structure Prediction) [19].

The process has also been adapted to predicting protein–protein interactions (CAPRI – Critical Assessment of PRediction of Interactions), the first edition of which emerged in 2001 [20]. More details about CAPRI can be found at http://capri.ebi.ac.uk.

These competitions provide a novice in the field of structure prediction with an excellent starting point to compare the many protein modeling programs, and also help a user to decide which program would work best with their requirements. Despite differences in the performance of these programs, many of them draw from the same fundamental approaches to protein structure prediction. The three main approaches are discussed below.

2.2
De novo/Ab initio Methods

"*De novo*" methods are a general designation for methods that predict the structure of a protein based on the primary sequence alone [21]. The ideal method to do this would be based on simple physical laws only. Although this would be a true "*de novo*" method, it is exceedingly difficult and has only been modestly successful for peptides and small proteins. The difficulty is the extremely large number of possible conformations for a polypeptide chain; predicting the structure of a protein from the sequence only indeed would be equivalent to solving the protein-folding problem. In practice, *de novo* methods incorporate a variety of information derived from the database of existing structures. This can be very direct information, such as taking the conformation of a loop with the same sequence directly from the database, or less direct, such as through heuristical rules derived from statistical analysis of all known protein structures.

2.3 Fold Recognition Methods

Fold recognition methods attempt to identify structures within a new protein (the query) by comparing its sequence to proteins with known structure (the template). This can be called the *inverse folding problem*, since the goal is to identify the probable folds that might fit the sequence, instead of trying to determine how the sequence will fold.

Several fold recognition methods are available. These are characterized by common features such as: a fold library; a scoring function, to evaluate the fit of the query sequence to the template (the fold); an algorithm to find the best alignment between the sequence and the fold; and a ranking method.

Fold libraries are derived from the Protein Data Bank (PDB; http://www.rcsb.org). A major feature here is the lack of redundancy, which provides an easier understanding of protein structure and function. Structural Classification of Proteins (SCOP), Class Architecture Topology Homologous superfamily (CATH), and Families of Structurally Similar Proteins (FSSP) are three of the most widely used databases that refer to three different methods of protein classification – manual, semi-automated, and completely automated, respectively.

In SCOP (http://scop.mrc-lmb.cam.ac.uk/scop/), the protein domains are organized hierarchically, on the basis of evolutionary criteria and similarity of their structure and sequence [22]. Human expertise is, in this case, very important to manually classify proteins at four levels:

- Family – proteins with a certain degree of similarity.
- Super family – protein families sharing similar sequences or functions/structures, because they are evolutionarily related.
- Common fold – superfamilies with structural similarities but not evolutionary related.
- Class – a class of secondary structure.

The Quantitative Structural Classification of Proteins (QSCOP) server is an extension of SCOP that allows the quantification of the structural similarity between two SCOP domains [23]. The two SCOP domains are superimposed, and the number of residue pairs that are structurally equivalent is computed and expressed as a percentage relative to the length of the larger domain. This implementation provides a better understanding of the structural diversity within SCOP.

CATH adopts a SCOP domain that is similar to SCOP, although the definition of domain here is different for these two databases [24]. Proteins are first grouped on the basis of sequence similarity, the domains are then identified, and four levels are described:

- Class (or C-level), to describe the secondary structure of each domain.
- Architecture (or A-level), to distinguish among structures of the same class with different arrangement of the secondary structure.
- Topology (or T-level), corresponding to the fold level in SCOP.
- Homologous superfamily (H-level), corresponding to SCOP superfamily level.

The A-level is assigned manually by visual inspection and taking into account the literature, while the C-level is assigned using an automated procedure described by Michie *et al.* [25]. Structure comparisons to identify related folds (T- and H-level) are performed with the program Sequential Structure Alignment Program (SSAP), a dynamic programming-based method for structural alignments [26].

FSSP, like SCOP and CATH, relates proteins using evolutionary criteria, but does not provide any classification into classes, fold families, or superfamilies [27]. The structural similarity search among proteins is performed in a completely automatic way using the distance matrix alignment (DALI) method. This is a fragment-based method that splits the input structures into hexapeptides and computes a distance matrix using the contact maps between successive fragments [28].

There are two broad approaches to fold recognition methods. The first approach is sequence-based, and relies on identifying homologous sequences; it then assumes that a strong sequence similarity equates to a strong structure and function similarity. While this may seem straightforward, current algorithms may not be sensitive enough to recognize distant homologs, and therefore the "correct" template might be missed. In addition, mutations between these sequences (including gaps and insertions) sometimes make it difficult to identify the correct sequence alignment. It can be misleading to rely on the sequence alignment alone for fold recognition. It is possible to identify proteins that share sequence identity without having any structural or functional similarity to the query, and so blind sequence alignments may not yield a successful prediction. The classification of protein sequences into families, followed by an identification of the patterns within the family such as strongly conserved residues (which might be important to function) and the pattern of mutations (for example, is a hydrophobic residue always replaced by another hydrophobic residue is a conserved mutation?), can improve the confidence of the alignment. Currently, the position-specific iterated basic local alignment search tool (PSI-BLAST) is one of the most popular homology recognition algorithms [29], although several others perform equally well.

In order to recognize the fold of a protein when the sequence homology is very low (<25%), additional structural information should be taken into account. The method of 3-D profiles was developed for this purpose by David Eisenberg in 1991 [30] whereby, in each 3-D structure, the environment of each residue is described in terms of three properties: secondary structure; solvent accessibility; and polarity of the neighboring residues. By using these parameters, the folded 3-D structure of the protein is converted into a one-dimensional (1-D) string, describing the environment class of each residue position. Statistical preferences (3-D–1-D profiles) of each amino acid can then be retrieved for each position, and the scores are then aligned with the query sequence, such that the quality of this alignment describes the structural compatibility between the sequence and the 3-D structure. When using this method, high 3-D–1-D scores can be assigned to proteins that are evolutionarily correlated and with very low sequence similarity. This is possible because the environment of each position in a protein structure is more conserved than the amino acid sequence itself.

A different method was proposed by Manfred Sippl [31]. In this case, instead of checking if the query sequence together with its structural preferences matched the structural environment of a given structure, the method relied on mean-force potentials. This is a statistical and knowledge-based method (also referred to as the *Boltzmann device*) that requires the detailed analysis of a database of experimentally resolved protein structures [32, 33]. Here, the basic assumption is that these structures represent equilibrium states, and the distribution of the distances between interacting atoms should follow the Boltzmann distribution. By monitoring this distribution, the energy of interaction between amino acid pairs as a function of their separation can be derived. By using Boltzmann's law, these frequencies can then be converted into a potential of mean force of the interaction as a function of distance. When a query sequence is given, these potentials can be used to compute the energies of all the folds of the database with respect to the query sequence itself. Only the most probable folds are then taken into account, assuming that the preferred conformations for a given sequence correspond to the conformations with the lowest energy [34]. Since their formulation, mean-force potentials have been used in several fold recognition methods, as well as in the identification of errors in solved structures [35], and in the refinement of protein structure models.

The second approach to fold recognition is structure based. The term "*threading*," as used for the first time in 1991 by David Jones, Willie Taylor, and Janet Thornton [36], is applied to this approach, which requires a detailed 3-D description of the protein structure. Here, the main objective is to fit (or "thread") the query sequence onto the backbone coordinates of known protein structures. The reliability of the final 3-D model is then evaluated on the basis of empirical potentials, derived from a dataset of proteins of known 3-D structure, and a solvation potential. Historically, in order to thread the sequence to the templates, a dynamic programming-based algorithm was used that previously had been applied to the structure comparison problem [37, 38]. In this case, a sequence alignment method is used to fit the sequence onto the templates. The best threading is then found by applying a distance criterion: the structural environment of a given residue is evaluated in terms of distances with respect to all the other residues. These distances should be constant between equivalent residues in different structures. The first program that implemented these methods was THREADER [39], an updated version of which is available at: http://bioinf.cs.ucl.ac.uk/software_downloads/threader/.

As a general rule, to identify the optimal fit of the sequence with each framework is a complex process, particularly if there are gaps and deletions among the sequences of the two proteins. It is very common to use sequence-based methods in conjunction with threading to identify the best matches. Each iteration of the threading process is scored, and this score is then used to determine the best matches. Following Anfinsen's hypothesis, that the native state of the protein is the lowest energy structure, the evaluated energy of the query's fit to the template is a strong predictor of which models are "good" within an experiment. A number of threading algorithms are available, which vary in their choice of protein model (backbone atoms versus just alpha-carbon atoms, side chains versus interaction sites) and their method of alignment, as well as their method of evaluating the energy of the proposed structure, just to name a few differences [40].

When embarking on a fold recognition study, the state of the art of current fold recognition methods can be judged from the results of the previously discussed CASP competition, and also of the Live Bench Project [41]. Unlike CASP, the targets in this exercise are recently deposited structures from the PDB. All of the methods assessed are fully automated. An expanding number of fold recognition servers (as well as meta-predictors and sequence comparison servers) participate in this weekly assessment exercise, and the most current results can be found at http://bioinfo.pl/livebench/. It is important to recognize that fold recognition methods are mostly knowledge-based, in that the predictions are based only on what is already known, and it is therefore impossible to identify any novel folds. Rather, novel fold prediction would fall into the category of an *ab initio* method, where the predictions are based primarily on physical principles.

2.4 Homology Modeling

Homology modeling (also known as *comparative modeling*) is a common method used to predict the total 3-D structure of a protein when experimental data from X-ray crystallography or NMR studies are not available for that protein [42]. It is based on the general observation that the structure of a protein during evolution is much more conserved than the sequence; hence, proteins with different sequences but still sharing a certain similarity are likely to have the same structure. "A certain similarity" can be quantified based on the database of known structures: if two proteins with more than about 50 residues have a sequence identity of approximately 20–30%, they will almost certainly have the same structure. In general terms, depending on the sequence identity between the protein target and the template, homology models can be classified into three "zones" – midnight, twilight, and safe zone – which correspond to <20%, 20–40%, and >40% of sequence identity, respectively [43, 44]. Thus, homology modeling can be used if there is a protein with a known 3-D structure (the template) that is sufficiently structurally similar to the protein whose structure is to be modeled (the target). In practice, homology modeling follows a series of steps, each of which involve choices by the modeler, or an automated computer server. Following Krieger *et al.* [45], homology modeling can be thought of as involving seven steps:

1. Template recognition and initial alignment
2. Alignment correction
3. Backbone generation
4. Loop modeling
5. Side-chain modeling

6. Model optimization
7. Model validation.

The overall strategy is first to align the sequence of the target with that of the template, to build the overall chain structure of the target, and then to consider "details." The accuracy of this alignment and the quality of the 3-D structure of the protein chosen as a template are the most important determinants of the accuracy of the final model. As a general rule, the use of a multiple alignment, where the sequences of several members of a protein family are aligned, can increase the quality of the final model. In order to identify a suitable template for the protein target, database search techniques can be applied, such as FASTA and BLAST. Multiple sequence alignment methods, such as PSI-BLAST, can accomplish the search in an iterative way, providing at the end of the process more than one protein homolog to the protein target. When several templates are identified, several factors can be used as guidelines to choose the best one, including function similarity, sequence identity/similarity percentage, or the number of missing residues in the 3-D structures.

The second step involves a manual correction of the alignment, if there is a compelling reason to do so. For example, in potassium channels there is a near-universally conserved motif TVGYG that is likely to be aligned, even if multiple-sequence alignment algorithms suggest alternative alignments. The backbone of the target is typically initially copied from the template, followed by algorithms that allow some flexible adjustments where necessary. When residues in the target and the template are identical and aligned, most programs also directly copy their side chains. This can be helpful, for example, when modeling the residues in the active site, where a specific conformation of some amino acids is required, even though this procedure is accurate only at a high sequence identity level. A major problem in homology modeling involves insertions that are not present in the template. In potassium channels, the loops between transmembrane helices can be very different between channels, even if the transmembrane segments are very similar. Similarly, the loops connecting the transmembrane helices in G-protein-coupled receptors convey much of the specificity of different receptors, but this most interesting part generally cannot be modeled by homology modeling. Loop regions are usually exposed on the surface of the proteins; their conformations may be quite flexible, and therefore difficult to resolve in crystallography, and their sequences can be quite variable. As a consequence, a 3-D template for these regions is often missing. Thus, an ability to model these loops is an important step, and a significant research focus. Several methods are available to predict the structure of loops, based on sequence information, database searches, energy-based calculations, or a combination of these approaches [42, 46]. Regions predicted via loop modeling are usually less accurate than those predicted using a 3-D structure as a template, and the level of accuracy decreases with the number of residues in the loop. When the backbone has been optimized as much as possible, side chains can be modeled and the model tested against experimental information. In many cases, these steps will have to be repeated.

It is now usually fairly easy to generate homology models, provided that a suitable template is available. Commonly used programs include Modeller [47] and

WhatIf [48], as well as automated web servers such as SwissModel [49]. The key measure of success for homology models is their experimental validation, although less than perfect models can be very useful in interpreting experiments and guiding the design of new experiments. Homology models have also been used in computer-aided drug design, providing, for example, a structural model of active sites that can be used in the development of new compounds capable of targeting them.

3 Structure-Based Modeling

3.1 Molecular Graphics-Based Methods

Over the past five to ten years, the personal computer has begun to replace the large supercomputers and dedicated workstations once needed to operate molecular visualization tools. Today, a number of molecular graphics programs are available for use on personal computers, often as freeware. Any program with basic rendering capabilities can be a valuable tool to study a particular protein. A variety of representations, illustrated in Fig. 2 for the protein barnase, can help the modeler understand the link between the protein's structure and its function. By manipulating colors and representations, it is possible to interrogate the structure to locate the positions of the conserved residues/sequences in the protein's structure, to examine the overall surface of the protein as a guide to potential protein–protein interactions, and to seek cavities that might be ligand-docking sites, just to name a few examples. The details of some commonly used programs for protein modeling are outlined in the following

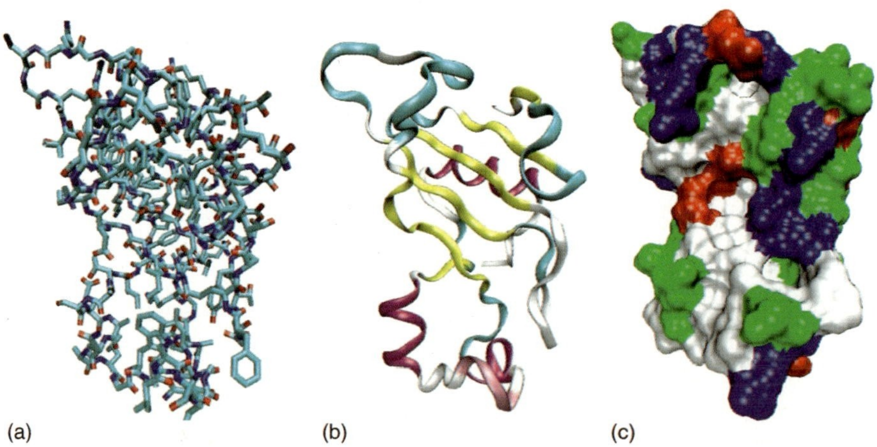

Fig. 2 Barnase rendered in three formats. (a) Colored by atom type (carbon = blue; nitrogen = navy blue; oxygen = red); (b) Colored by secondary structure features in ribbon format, where beta-sheets are colored yellow, helices purple, and turns and random coil features blue and white; (c) Rendered in surface format, colored by residue type, where charged residues are colored red or blue, polar residues are green, and nonpolar residues are white. All figures are rendered with the same view of the protein, looking into the barstar binding pocket. All figures were rendered with VMD.

section, though others are described by Tate [50], and even more are listed on the PDB web site (http://sw-tools.pdb.org/).

Rasmol is one of the most widely used programs for molecular visualization [51]. Besides the basic options in which the display of the protein can be altered from sticks to spacefilling to ribbon format, and so on, a variety of tools is available that can be used through that command line interface. Examples include the selection of specific residues of the protein, measuring distances, and highlighting parts of the protein within a certain radius. Rasmol can be used to probe all of the information contained in a pdb file, but it cannot alter that information.

With Swiss-PDB Viewer [49], protein structures can be manipulated in a number of ways. Typically, the residues can be "mutated," and dihedral angles modified. Swiss-PDB Viewer provides a graphical interface within which several protein structures can be analyzed and manipulated at the same time, and it is tightly linked to Swiss-Model. Working with these two programs greatly reduces the effort required to generate models, as it is possible to thread a protein primary sequence onto a 3-D template and obtain an immediate feedback of how well the threaded protein will be accepted by the reference structure, before submitting a request to build missing loops and refine any side-chain packing.

Visual molecular dynamics (VMD) is another very powerful visualization program which can also be used to view trajectories from MD simulations [52]. It has a broad range of representations for proteins, including molecular surfaces, electrostatic potential maps, and crystal information.

3.2 Poisson–Boltzmann/Electrostatics Calculations

Poisson–Boltzmann theory (PB) has become an important tool for studying biomolecular systems [53–55]. This type of calculation provides a view of a protein in terms of regions of positive or negative potential. PB calculations are also useful for identifying interaction sites on proteins with charged ligands or other proteins [56], and they can also be used to determine the pK_a values of the amino acids in different environments. An example of this is in the enzyme active sites, where the catalytic residues often experience large pK_a shifts due to their local environment. PB calculations also provide a basis for Brownian dynamics (BD) simulations (see Sect. 4).

PB is based on the Poisson equation, which describes electrostatic interactions in general, but also includes implicitly the presence of ions in solution. It is assumed that, at equilibrium, the distribution of mobile ions in the system can be approximated by a continuous charge density $\rho_{eq}(\mathbf{r})$, determined by the Boltzmann factor:

$$\rho_{eq}(\mathbf{r}) = \sum_i z_i e n_{0i} \exp[-z_i e\phi(\mathbf{r})/kT]$$

where i is an ionic species, z_i the charge of ion species i, n_{0i} a reference number density for species i, e is the unit charge, and $\rho(\mathbf{r})$ and $\phi(\mathbf{r})$ are the local equilibrium charge density and the average electrostatic potential, respectively. The electrostatic potential $\phi(\mathbf{r})$ is the solution of Poisson's equation:

$$\varepsilon_0 \nabla \cdot [\varepsilon(\mathbf{r}) \nabla \phi(\mathbf{r})] = -\rho_{eq}(\mathbf{r}) - \rho_{ex}(\mathbf{r})$$

Assuming a 1 : 1 electrolyte, these two equations can be combined to give:

$$\varepsilon_0 \nabla \cdot [\varepsilon(\mathbf{r}) \nabla \phi(\mathbf{r})] = 2en_0 \sinh[e\phi(\mathbf{r})/kT] - \rho_{ex}(\mathbf{r})$$

Here, the mobile charges represented by ρ_{eq} are in equilibrium, and the fixed charges on the protein or membrane are represented by ρ_{ex}. This is the full nonlinear PB equation, which can be solved numerically. It can be linearized by expanding sinh and retaining only the leading term. When there are no fixed charges, and assuming a spatially homogeneous dielectric, this simplifies to:

$$\nabla^2 \phi = \kappa^2 \phi$$

where κ^{-1} is the Debye screening length, given by:

$$\kappa^{-1} = \sqrt{\frac{\varepsilon_0 \varepsilon k T}{2e^2 n_0}}$$

The Debye screening length $\kappa^{-1} = 8$ Å for 150 mM salt at 298 K in water with a dielectric constant of 80. The screening of a central charge in bulk solution is about 80% at $\mathbf{r} = 3\kappa^{-1}$, or about 25 Å. This distance decreases to about 9 Å for 1 M salt solution.

The PB equation, in either its linear or nonlinear form, can be solved for macromolecules by a number of commonly used programs, including DelPhi [57], UHBD [58], and APBS [59].

3.3 Ligand Docking

The binding of small molecules to proteins is a key determinant of many biological processes. The ligand-docking problem is the challenge to predict where, in what orientation, and with what affinity, a small molecule will bind to a binding site in a protein [60–62]. In the ultimate situation, the aim would be to predict all three with high accuracy, for a high computational cost if necessary. However, methods are also needed that allow the rapid (but computationally inexpensive) screening of a library of millions of compounds to search for drug leads (virtual screening) that may interact favorably with potential drug targets [63, 64]. Broadly speaking, ligand-docking can be approached from two angles. In the first approach, compounds of known structure are docked into the active site of a protein of interest, after which some form of scoring function is used to measure how likely the compound will bind tightly to the protein. In the second approach, starting from the known structure of the protein, a molecule can be designed "from scratch" by fitting functional groups in the active site so as to optimize the interactions with the protein target (*de novo* drug design) [65]. The resulting "computational" molecule can then be synthesized and tested.

Several classes of scoring function exist. One class is based on a combination of interaction energies, with terms that are similar to the potential function used in MD simulations. Simpler functions based on, for example, the number of hydrogen bonds or penalties for steric clashes are also useful, and can be computationally faster. In a second class, an attempt is made to calculate directly the free energy of binding by using a parameterized function or an approximate description of the free energy of binding derived from a potential function. A third class of scoring function uses statistical information about similar molecules and general features of the binding site to rank ligand–protein complexes. The scoring of the final docked conformations is the most challenging aspect of docking-based methods [66].

In principle, a scoring function should be able to distinguish active from inactive compounds, ideally rigorously predicting the binding affinity. However, each scoring function will have specific advantages and disadvantages that can affect the accuracy of the docking results. Recently, it has been reported that consensus scoring methods can improve the quality of the final ranking [67]. In this case, a combination of multiple scoring functions is applied to produce a consensus score that is more accurate to predict the binding affinity, thus reducing the rate of false positives.

3.4 Protein–Protein Interactions

Today, protein–protein interactions are emerging as one of the most important themes in biochemistry, as they can drive and control a variety of functions, including signal transduction, transport, gene expression, and protein function regulation. Some interactions are sufficiently stable that a complete complex can be crystallized, though this is relatively rare. More often, proteins interact transiently with other proteins at some point in their lives, often in critical processes such as signaling. The PDB contains mainly single proteins, because in order for them to be crystallized they must normally be purified to a high degree, which disrupts all but the strongest noncovalent interactions.

Several computational protein–protein docking methods have been developed to predict the structure of a protein–protein complex [68, 69]. The general strategy in protein–protein docking focuses on shape complementarity, considering the remarkable degree of surface complementarity in most stable complexes [70]. This suggests that it should be possible to draw a molecular surface on both proteins (e.g., see Fig. 1d), and search for all possible relative orientations of the two proteins for the most favorable orientation. This type of search is expensive, and the challenge is to identify the most favorable orientation among many possible orientations. Shape complementarity can be computed by using several scoring functions, which provide more favorable values for orientations that optimize interactions such as hydrogen bonding, the burying of hydrophobic exposed areas, and the matching up of complementary electrostatic potentials on the surfaces of the proteins that form the complex [71].

Several protein–protein docking programs start with a rigid docking approach, and in some cases additional steps are included to take into account the flexibility of the side chains [68]. Progress in the area of protein–protein docking is monitored in the CAPRI challenge. One recent successful approach in protein–protein docking has been HADDOCK, which combines a molecular mechanics energy function, a simplified form of dynamics to enable flexibility of parts of the interface, and the incorporation of "ambiguous interaction restraints" [72]. These are "soft" restraints that, by themselves, are insufficient to guide docking but combine to yield more reliable solutions than energy functions alone, provided that experimental information is available that can be used as restraints.

4 Simulations of Protein Dynamics

The three common simulation methods used to study protein dynamics are explained in the following subsections.

4.1 Molecular Dynamics Simulations

The most common atomistic simulation technique is MD [73–75], whereby the interactions between all atoms in the system are described by empirical potentials. An example of a frequently used potential function is:

$$V\left(\mathbf{r}^N\right)=\sum_{bonds}\frac{k_i}{2}\left(l_i-l_{i,0}\right)^2+\sum_{angles}\frac{k_i}{2}\left(\theta_i-\theta_{i,0}\right)^2 + \sum_{torsions}\frac{V_n}{2}\left(1+\cos\left(n\omega-\gamma\right)\right) + \sum_{i=1}^{N}\sum_{j=i+1}^{N}\left(4\varepsilon_{ij}\left[\left(\frac{\sigma_{ij}}{r_{ij}}\right)^{12}-\left(\frac{\sigma_{ij}}{r_{ij}}\right)^{6}\right] + \frac{q_i q_j}{4\pi\varepsilon_0 r_{ij}}\right)$$

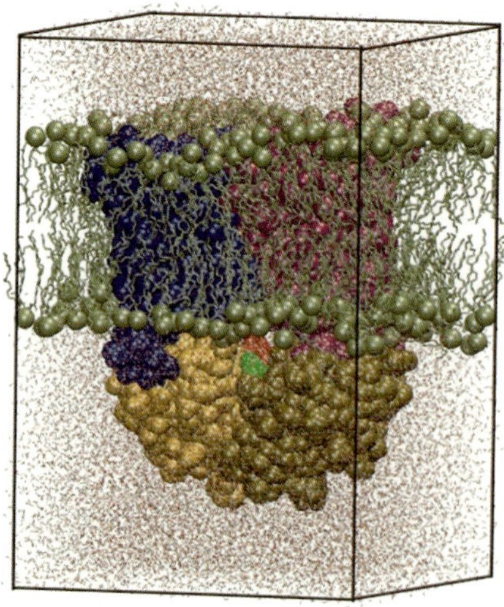

This potential function contains harmonic terms for bonds and angles, a cosine expansion for torsion angles, and Lennard–Jones and Coulomb interactions for nonbonded interactions. The constants k_i are harmonic force constants, l_i, is the current, $l_{i,0}$ the reference bond length, θ_i the current, $\theta_{i,0}$ the reference angle, V_n, n, and γ are the barrier height, multiplicity, and off-set from the origin for the cosine function used to describe dihedral angles (rotations around a central bond), ε and σ are Lennard–Jones parameters (a different pair for each possible combination of two different atom types), q_i and q_j are (partial) atomic charges, and r_{ij} is the distance between two atoms, and ε_0 (different from ε in the Lennard–Jones potential) is the dielectric constant of the medium. By using this potential function, the forces (the derivative of the potential with respect to position) on all atoms in the system of interest are calculated and used to solve classical equations of motions to generate a trajectory of all atoms in time. An example of a system studied with MD is shown in Fig. 3, which is a simulation snapshot of the ATP-Binding Cassette (ABC) transporter BtuCD [76] in a realistic environment consisting of lipids, water, and ions.

Fig. 3 An orthographic view of the periodic box for the ATP-bound BtuCD simulation, showing water (red and white), lipid with phosphorus atoms enlarged, and the protein. The transporter consists of two transmembrane domains (blue and purple) and two nucleotide-binding domains (orange and ochre). The two docked MgATP molecules are partially visible (green and red). Illustration courtesy of E. Oloo. See Ref. [119] for more details. Figures rendered with VMD.

The primary result of the simulation is a trajectory of all atoms in time, from which specific details of the system can be analyzed. This is an exciting concept, because atoms can be followed as they move in real time on a time scale of up to about 100 ns, although longer simulations have also been reported. In principle, any property that depends on coordinates, velocities, or forces can be calculated, given sufficient simulation time. No assumptions are required about the nature of the solvent, there is no need to choose dielectric boundaries because all atoms are explicitly present, and in principle all interactions (water–ions, water–protein, water–lipid, lipid–protein, etc.) are incorporated.

The main limitations of MD are its computational cost, the limited time and length scale that can be treated in a simulation, and technical limitations such as the accuracy of current empirical force fields.

4.2 Free Energy Calculations

Several useful extensions of the basic simulation method make it possible to calculate free energy differences from simulations [77, 78]. If a reaction coordinate can be identified, processes that are orders of magnitude slower can be studied than would be possible by direct simulation. Such a reaction coordinate could be a concerted conformational change, or a pathway for ion permeation in ion channels. In umbrella sampling, a biasing potential is used to restrict a simulated system to sample phase space within a specified region (called a *window*). By placing windows along the reaction coordinate, a free energy profile (also called *potential of mean force*) can be generated that will quantitatively describe why one region of space is more favorable than another.

The relative free energies of different side-chain mutations in a protein can be studied using free energy perturbation by slowly changing one side chain into another within the computer (also known as *computational alchemy*). This makes it possible to investigate in the computer the effect of mutations on the stability of a protein or on, for example, the binding affinity of a ligand. The theory behind this type of calculation is well developed [79], and the algorithms have been implemented in many MD software packages [80–83].

4.3 Brownian Dynamics Calculations

In BD simulations [84, 85], the trajectories of individual particles (ions, molecules) are calculated using the Langevin equation:

$$m_i \frac{d\mathbf{v}_i}{dt} = -\gamma_i \mathbf{v}_i + \mathbf{F}_R + \mathbf{F}_i$$

where m_i and $\mathbf{v}_i$ are the mass and velocity, respectively, of atom i. Water molecules are not included explicitly, but are present implicitly in the form of a friction coefficient $\gamma = kT/D_i m_i$ (where k is the Boltzmann constant, T is temperature, and D is the diffusion coefficient) and a stochastic force $\mathbf{F}_R$ arising from random collisions of water molecules with ions, obeying the fluctuation–dissipation theorem. $\mathbf{F}_i$ is the force due to other particles in the system as well as external sources, such as an applied electric field. When the friction is large and the motions are overdamped, the inertial term $m_i d\mathbf{v}/dt$ may be neglected, and the simplified form

$$\mathbf{v}_i = \frac{D_i}{kT}\mathbf{F}_i + \mathbf{F}_R$$

may be used. This is the approximation made in BD.

BD simulations require only a few input parameters: in its simplest form, only the diffusion coefficients of the different species of ions and the charge on the ions are needed, although the model can be refined. To study a protein such as an ion channel, the residues of the protein can be modeled as a set of partial charges, and some form of interaction potential between the mobile ions and the protein must be specified (see above).

The result of BD simulations is a large set of trajectories for diffusion particles such as ions, proteins, or ligands. By averaging over these trajectories, macroscopic properties such as the conductance of ion channels or association rate of protein–protein or protein–ligand encounters can be calculated. In addition, the simulations yield molecular details of the paths through space for the diffusing particles.

Although BD simulations are conceptually simple, replacing solvent by a continuum description and ignoring internal conformational changes in proteins are significant assumptions that require careful consideration. When BD simulations are valid they represent a very powerful method to study biological processes on a time scale that is much longer than can be reached by MD.

5
Example Applications

The current literature is very extensive, and hundreds of reports have been made based on each of the methods described above. For purposes of illustration, some examples of biochemical problems that have been addressed by modeling and simulation methods are provided in the following subsections.

5.1
Acetylcholinesterase

Acetylcholinesterase (AChE) is an extremely rapidly acting enzyme that hydrolyzes the neurotransmitter acetylcholine into choline (Ch) and acetic acid in order to terminate signaling in cholinergic synapses of the nervous system. AChE has been investigated in great detail using a number of computational methods, including MD simulations, continuum electrostatics calculations, and BD simulations, all of which have contributed to explaining the mechanism of action of this enzyme, and details of its inhibition.

The first crystal structure of AChE was reported in 1991, to reveal a narrow, 20 Å-long channel that connected the surface of the enzyme with the active site [86]. One of the first puzzles posed by the crystal structure of AChE was the question of how the substrate, acetylcholine, could gain access to the active site since, in the crystal structure, there was no unobstructed pathway to the active site. This original observation for AChE from the fish *Torpedo californica* was subsequently reiterated in mouse and human AChE, with several groups having applied MD simulations to resolve this issue. The aim was to explain the role of protein structural fluctuations in allowing access to the active site, and how these motions were related to the rapid kinetics of the enzyme. These motions appeared quite complex and spatially extensive, which suggested possible modes of regulation of the activity of the enzyme. Although such a mechanism has been observed in other proteins, including hemoglobin and myoglobin, the very fast reaction rates of AChE were difficult to reconcile with major structural rearrangements.

Computer simulations described two possible states for the channel – open and closed [87–89]. The closed state, in which the gorge is so narrow that the ligand cannot access the active site, is the favored state of AChE. In particular, the channel is lined by aromatic residues, the side-chain rotations of which act as a gate to control the passage of acetylcholine (as substrate); in addition, the reaction rate of AChE is limited by the diffusion of acetylcholine. In this regard, BD simulations represent a suitable method to study the encounter between acetylcholine and AChE. The acetylcholine diffuses under the influence of intermolecular interactions between itself and the protein, and of a random force that mimics the effect of the solvent. The net effect is that the protein "guides" the acetylcholine to its active site via an electrostatic mechanism, such that the rate at which acetylcholine reaches the active site is enhanced beyond the pure diffusional limit. In fact, negatively charged residues are located on the external surface of the enzyme, so as to create an electric field that serves as the main force for capturing the positively charged substrate.

A similar mechanism has since been identified for several other fast-reacting proteins, whereby the substrate can be funneled into a channel, aided by π–π stacking interactions, and with a pool of hydrophobic residues lining the gorge of the enzyme. At the same time, additional secondary channels open and close multiple times, which allows solvent to move in and out during translocation of the ligand. All of these steps have been successfully targeted by MD simulations, thus confirming that the pool of aromatic residues facilitates passage of the substrate into the channel and promotes the open form of the enzyme [89, 90].

AChE is the target of the neurotoxin fasciculin, a three-fingered peptide isolated from snake venoms. Subsequent BD of the peptide and AChE provided an insight into the binding kinetics of fasciculin and the structure of the resultant complex. In a series of BD simulations, Elcock *et al.* examined the effect of mutations in the protein on the association rate constants [91]. The electrostatic interaction between AChE and fasciculin that promotes association is illustrated in Fig. 4. In its simplest form, the BD simulations reproduced the correct order of rate constants for different mutants, although the absolute values were too large by about a factor of 30. In a more accurate treatment of the continuum electrostatics problem, this discrepancy can be reduced greatly. Interestingly, these calculations make it possible to distinguish between kinetic and thermodynamic effects and suggest that, in the case of fasciculin–AChE-binding,

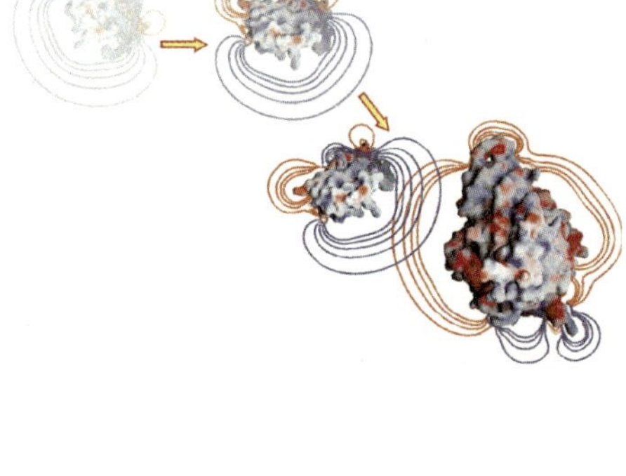

Fig. 4 Schematic illustration of the association of fasciculin (at left) with acetylcholinesterase. The blue and red contour lines indicate regions of positive and negative electrostatic potential, respectively. Illustration courtesy of D. Sept and A. Elcock. See Ref. [91] for more details.

mutations can separately affect the rate of association and the binding constant.

During the past 20 years, AChE has been the target of several drug development efforts, with particular interest being shown regarding its potential role in Alzheimer's disease (AD), a neurodegenerative disorder that is one of the most common forms of dementia. Whilst the majority of AD cases occur sporadically, a small percentage is related to gene mutations (familial AD). Nonetheless, both AD forms are characterized by common features, such as a reduced synthesis of acetylcholine due to the loss of cholinergic neurons and amyloid-β fibrillization. AChE mediates a large number of noncatalytic functions, including the initiation of amyloid aggregation [92], and it is for this reason, together with its catalytic function, that AChE has been identified as suitable target for the treatment of AD. The availability of several crystal structures of this enzyme in complex with different compounds has allowed the application of several molecular modeling approaches aimed at identifying new inhibitors effective against AD, such as ligand-based methods, structure-based techniques, or a combination of these. These methods, and the MD techniques cited above, have allowed research groups to acquire a detailed understanding of the mechanism of action of this enzyme and its dynamics, with subsequent promising results in drug discovery.

To date, in the treatment of AD, four acetylcholinesterase inhibitors (AChEIs) – tacrine, donepezil, rivastigmine, and galantamine – have achieved commercial status. Donepezil was first approved in 1996 by the US Food and Drug Administration (FDA) as an inhibitor of acetylcholine hydrolysis; subsequently, in 1999, the crystal structure of its complex with *T. californica* AChE was resolved [93], and a dual-binding mode of the drug developed. Donepezil is able to bind simultaneously to the catalytic site and peripheral sites of AChE, thus reducing not only the hydrolysis of acetylcholine but also amyloid-β aggregation. These findings led to the rational design of families of dual-binding-site AChEIs as "hit" compounds for new drug candidates. These included, in 2003, a new compound – AP2238 – that had been designed specifically to interact with the active and peripheral sites of AChE, and also with the aromatic residues lining the gorge of the enzyme [94]. Subsequently, with AP2238, an *in vitro* amyloid-β anti-aggregation effect of a dual-binding-site AChEI was reported for the first time. More recently, a new compound that has been rationally designed to serve as a dual-binding-site inhibitor, NP-0361, has entered clinical trials [95].

5.2 Water Transport through Aquaporins

Water can be transported across membranes via several mechanisms, including diffusion in lipid bilayers, osmosis in channels, and cotransport in cotransporters and uniporters [96]. The best-characterized route is that mediated by the aquaporins, a group of ubiquitous membrane proteins that help, crucially, to maintain the osmotic balance in cells. Since their discovery in red blood cells, the aquaporins have been identified in all kingdoms of life. Their sequence is highly conserved in plants and animals, and at least 11 isoforms have been characterized in humans. They have been divided into two classes, depending on whether they translocate only water molecules (aquaporins), or also small neutral solutes such as glycerol (aquaglyceroporins). The road to

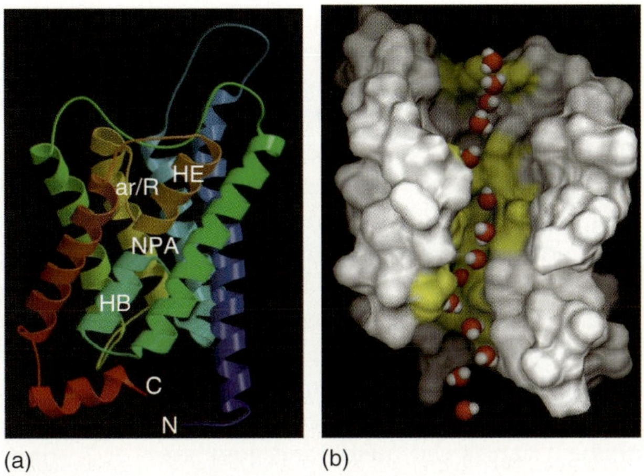

Fig. 5 (a) Ribbon diagram of the aquaporin monomer and (b) a spacefilling representation illustrating a water file through the channel pore. (a) Starting at the N terminus of the monomer, there are two transmembrane helices, followed by the coil-NPA-helix motif. A third transmembrane helix completes the first half of the protein. The second half of the monomer has two transmembrane helices, followed by the coil-NPA-helix motif, followed by another transmembrane helix. The path of the water channel is through the core of the protein, following the path of the coil motifs which meet at the NPA signature; (b) The pathway is more clearly illustrated. Here, the protein is rendered as a molecular surface in white, with the front surface of the protein cut away, so that the pore can be clearly seen in the middle of the protein. Surfaces colored yellow are those that interact most strongly with passing water molecules. The waterfile displayed is an overlay of a number of snapshots from the 10 ns simulation. The dipole inversion of water at the NPA motif is clearly illustrated here. Illustration courtesy of B. de Groot. See Refs [104] and [12] for more details.

the high-resolution structures of these proteins, and to an understanding in atomic detail of how they function, represents an excellent example of the link between experiment, modeling, and simulation [97].

The high sequence identity among aquaporins has indicated a common architecture that was later confirmed by structural data. In early 2000, De Groot and coworkers used 4.5 Å-resolution data (which was insufficient to identify the fold of the protein directly) to predict the fold of aquaporins, based on the constraints of helix packing, atomic force microscopy (AFM) data, and primary sequences [98]. From 1440 possible folds, it was possible to identify only a maximum of eight. Later that year, a 3.8 Å resolution structure of aquaporin became available, using cryoelectron microscopy [99], as well as a 2.2 Å-resolution X-ray crystal structure for a homolog of aquaporin, GlpF, the glycerol transport facilitator [100]. Within the membrane, aquaporins are arranged in tetramers. The fold of each monomer (see Fig. 5a) consists of six highly conserved transmembrane helices, with the N-terminal and C-terminal tails located on the cytoplasmic side of the membrane. The helices are connected by loops that differ among species; the loops connecting the second and third helices, and the fifth and sixth helices, are predominantly hydrophobic in nature, and demonstrate the characteristic conserved coil-NPA (Asn-Pro-Ala)-helix motif. On either side of the membrane, the protein pore is shaped by these two reentrant loops, where the NPA residues meet in the middle of the channel, thus conferring an "hourglass" arrangement to each monomer. The partial helices of the NPA motif are labeled HB and HE. The aromatic-rich constriction region (ar/R = aromatic/arginine; see Fig. 5a) is thought to contribute to the aquaporin's specificity. In order to preserve the electrochemical gradient across the membrane, ions are excluded from the pore by an electrostatic barrier, with a maximum value at the NPA motif [101].

A direct simulation of the aquaporin structure obtained using cryoelectron microscopy demonstrated a discontinuous water file through the channel, and obvious defects in the structure, particularly at the NPA motifs, based on an instability observed through the MD simulations [102, 103]. Further homology modeling using the newly available structures yielded an improved stability in the core of the channel [104]. Simulations of the high-resolution GlpF structure [105], and the improved homology model of aquaporin-1, each demonstrated continuous water files diffusing through the water channel (see Fig. 5b), at rates comparable to those observed experimentally.

During late 2001, a 2.2 Å-resolution X-ray crystal structure was obtained of human aquaporin-1 [106]. With confidence in the 3-D structure, the subsequent simulation studies were able to focus on measuring the physical properties of the channel using classical simulation methods. Attempts were also made to pinpoint the origin of the aquaporin's ability to transport water at high diffusion rates, while restricting proton transport. As the basis of proton transport is better addressed by quantum mechanical simulations, combinations of classical dynamics methods with semi-empirical methods have been employed by several groups to determine the basis for proton exclusion by these channels.

One particularly exciting point regarding the aquaporins as a case study, is that the biological function of these proteins – to serve as a water channel across the cell membrane – occurs at a rate which can be studied using state-of-the-art MD simulations. The time scale of water translocation mediated by aquaporins is of the order of nanoseconds, which is very fast compared to the mechanisms employed by most transporters, but easily accessible with computer simulations. Consequently, by conducting a series of simulations the deficiencies in the low-resolution crystal structures were identified not only on the basis of structural instability but also because the observed behavior *in silico* did not agree with that measured experimentally. Moreover, these observations helped to highlight the significance of certain structural motifs in the protein (namely, the reentrant loops with the NPA motif). The evolution of the present understanding of aquaporin structure through experiments, homology modeling, and simulations, has led to similar methods being applied to other aquaporins with as-yet unsolved structures [107].

5.3 ATP Binding Cassette Transporters

Unfortunately, many human proteins of major medical interest are difficult to obtain in sufficiently large quantities for structural studies, the only exceptions having been proteins that occur in large amounts (e.g., bovine lens protein rhodopsin [108] or aquaporin-1 from blood; see above). Over the past 10 years, an emerging theme in membrane protein structural biology has been the use of bacterial proteins to study the human homologs, with examples in the field of potassium channels [109] and ABC transporters [110]. The ABC transporters are modular mechanical machines that couple the hydrolysis of ATP with the transport of a wide variety of substrates, ranging from small ions to macromolecules, across membranes. Mutations in the genes encoding many of the 48 ABC transporters of human cells are associated with several diseases, such as cystic fibrosis. The increased expression of certain ABC transporters is a major cause of resistance to peptide antibiotics, antifungals, herbicides, anticancer drugs, and other cytotoxic agents. Interestingly, ABC transporters probably form the largest group of homologous proteins, and exist in all species.

From a structural point of view, these membrane proteins have cytosolic and membrane-spanning domains that undergo large conformational changes in order to allow translocation of the substrates. In particular, the common architecture consists of four domains, two transmembrane domains (TMDs), and two cytosolic nucleotide-binding domains (NBDs). The TMDs are organized in several transmembrane alpha-helices that form the translocation pathway for the substrates, and are characterized by considerable sequence variability. In contrast, the two NBDs are highly conserved, and contain the characteristic motifs found in all the ATP-binding proteins important for the coordination of ATP. After ATP binding, the NBDs dimerize, forming one functional unit that can couple the energy from ATP hydrolysis with the transport of the substrate.

Several transporters have been characterized structurally using X-ray crystallography, and have provided important advances in the knowledge of their mechanism(s) of action. In 2002, Locher *et al.* reported the first high-resolution (3.2 Å) crystal structure of a bacterial transporter, the vitamin B_{12} importer from *Escherichia coli* [76] whilst, in 2006, the full-length crystal structure of the bacterial exporter Sav1866 from *Staphylococcus aureus* was described at a resolution of 3.0 Å [111]. The latter transporter is a homodimer responsible for the extrusion of several drugs, including vinblastine and doxorubicin, that are used in cancer therapy. Sav1866 shares a considerable homology with several human multidrug transporters, and can be used as a template to study human homologs that are highly involved in drug resistance, especially in tumors. The conformation of the transporter in the crystal structure corresponds to a nucleotide-bound state (two ADP molecules are present in the binding pockets), with the NBDs tightly coupled, and the TMDs arranged to form a cavity that is open towards the extracellular side of the membrane (Fig. 6a). This conformation, which is described as an "outward-facing conformation," corresponds to a final step of the transport cycle, when the substrate leaves the cavity by diffusion as a consequence of its low affinity.

Recently, the crystal structure was determined of mouse P-glycoprotein, an ABC transporter that exports a variety of compounds and toxic agents [112]. P-glycoprotein is also known as a *multidrug resistance transporter*, as it extrudes chemotherapeutic drugs from cancer cells, and is closely related to the bacterial transporter Sav1866. The conformation of the transporter in this new crystal structure represents a possible initial step of the

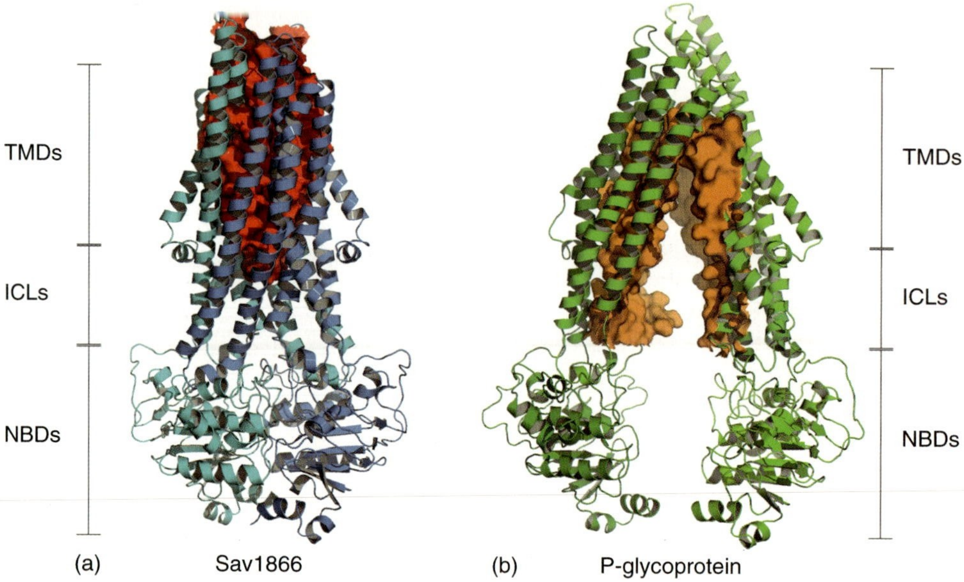

Fig. 6 (a) Crystal structure of the bacterial transporter Sav1866 (PDB code 2HYD). Sav1866 is a homodimer, and the two chains are represented in blue and cyan, respectively. The surface of the cavity, open toward the extracellular side of the membrane, is in red; (b) Crystal structure of the mouse P-glycoprotein (PDB code 3G5U). P-glycoprotein consists of one long polypeptide chain, folded into two similar subunits (colored dark and light green in the image). The surface of the cavity of the transporter, open toward the cytosol, is shown in orange. TMDs, transmembrane domains; NBDs, nucleotide-binding domains; ICLs, intracellular loops. Image created and rendered with Pymol 0.99 (The PyMOL Molecular Graphics System, Schrodinger, LLC).

transport cycle, with a 30 Å separation between the two NBDs (Fig. 6b). This conformation is described as an "inward-facing conformation," and may be the conformation with a high affinity for the substrates. In addition, two other crystal structures of P-glycoprotein, in complex with cyclic inhibitors, have provided structural insights into the binding pocket of the ligands.

Currently, crystal structures are available for other full-length transporters, such as MsbA [113], Hl1470/1 [114], $ModB_2C_2$ [115], molybdate/tungstate transporter [116], methionine transporter [117], and the maltose transporter MalFGK2 [118]. In addition, several structures of NBDs have been resolved. Despite this impressive progress in structural studies, several key questions regarding the basic mechanism of action of the ABC transporters remain unanswered. It is to be hoped that simulations and homology modeling can be used to progress towards an understanding of the dynamics of ABC transporters, and to translate these bacterial structures to human homologs, such as P-glycoprotein or the transporter associated with antigen processing (TAP).

Previously, MD simulation studies have been used to investigate the "real-time" dynamics of the vitamin B_{12} importer BtuCD from *E. coli*, based on the single snapshot captured in the crystal structure [119]. An example of a simulation model that incorporates BtuCD is shown in Fig. 3, where the crystal structure is incorporated into a realistic environment of phospholipids, water, and ions, and simulated for 20 ns, both in the absence and presence of MgATP in the two binding sites. The results showed that the docking of ATP to the catalytic pockets progressively draws the two cytoplasmic nucleotide-binding cassettes towards each other. In turn, movement of the cassettes into closer opposition induces a conformational rearrangement of the α-helices in the TMD. As a consequence, the shape of the translocation pathway is changed in such a manner that could aid the vectorial movement of vitamin B_{12}. Taken together, these results suggest that ATP binding may indeed represent the power stroke in the catalytic mechanism. Moreover, the occlusion of ATP at one catalytic site is mechanically coupled to opening of the nucleotide-binding pocket at the second site. This may indicate that asymmetric behavior at the two catalytic pockets forms the structural basis by which the transporter is able to alternate ATP hydrolysis from one site to the other. While this suggestion remains to be tested, this study represents an example of the use of MD simulations that build on a single crystal structure, to obtain more information concerning the full process in which the protein is involved.

TAP is the only ABC transporter with a specific role in the immune system [120]. Typically, antigenic peptides, derived from the degradation of cellular proteins, are presented at the cell surface to the T lymphocytes, to monitor events occurring inside the cell. In order to reach the cell surface, the peptides are translocated by TAP from the cytosol to the endoplasmic reticulum (ER) lumen, where they are loaded into nascent major histocompatibility complex (MHC) class I molecules. Several proteins are involved in the formation of MHC class I–peptide complexes, and the mature complexes are subsequently released from the ER to target the cell surface. TAP is a heterodimeric transporter that consists of two half-transporters, TAP1 (748 amino acids) and TAP2 (686 amino acids). Besides the TMD (which is formed by six transmembrane helices) and NBD, TAP1 and TAP2 have an additional N-terminal domain that mediates the interactions between TAP and other proteins. However, several aspects of its mechanism of action, including how peptide binding is related to ATP binding and hydrolysis, remain unclear [121].

Nonetheless, a comparison with other known ABC transporters led to the elucidation of some important features. In particular, from a functional point of view, two of the best-characterized ABC transporters are P-glycoprotein and the cystic fibrosis transmembrane conductance regulator (CFTR) [122]. The function of CFTR, which acts as both an ion channel and ion channel regulator, is critical for the transportation of fluid and electrolytes across epithelial cells. In the cases of both P-glycoprotein and CFTR, several residues have been identified as being crucial to their activity and substrate specificity. For example, in CFTR point mutations for Phe508, Tyr109, Pro208, and Glu92 can induce channel closure, while the mutation of His117, Phe334, Phe337, and Arg347 can affect chloride flow across the channel [123–125]. Alternatively, the cavity of P-glycoprotein is highly hydrophobic and lined by a large number of phenylalanine residues, the mutation of some of which (e.g., Phe72, Phe335, Phe978) directly affects both ligand binding and drug transport activities [126, 127]. Sequence alignments and comparisons between CFTR, P-glycoprotein, and TAP have allowed the identification of some residues of TAP1, including Glu263 and Phe265, that impair peptide transport [128].

Important structural insights for this antigenic transporter became available via recognition of the crystal structures of the monomeric and homodimeric TAP1-NBD [129, 130]. These structures explained the role of specific NBD mutations in affecting the binding and/or hydrolysis of ATP, together with peptide translocation. However, details of the full-length crystal structure for this transporter are required in order to completely characterize its mechanism of action. A homology model of TAP was proposed by Oancea *et al.* to elucidate some of the functional data referring to the peptide binding pocket and the crosstalk region between the TMDs and NBDs [131]. TAP is a member of the ABC subfamily B, together with the multidrug resistance protein and the bacterial transporter Sav1866 that was used as a template to build the TAP homology model.

Previous studies have identified multiple regions of TAP as being directly involved in binding of the peptide [132]. In particular, intracellular loop (ICL) 2 connects transmembrane helices 4 and 5, as well as a set of residues belonging to transmembrane helix 6. Functional data are also available with regards to the role of the ICL between the transmembrane helices 2 and 3 (named ICL 1), and which is mainly involved in the crosstalk between TMDs and NBDs. Consequently, ICLs 1 and 2 – both of which are located at the interface between TMDs and NBDs – show two different functions in terms of their peptide binding and translocation. The homology model of TAP supplied a structural context for these biochemical data, explaining the functional difference of the two ICL on the basis of specific interactions with highly conserved motifs of the NBDs. This model may be useful when designing and interpreting further experimental studies. In turn, the functional data may be used directly to improve the model, driving an iterative process that may advance the present understanding of medically important ABC transporters.

6
Perspectives

Today, molecular modeling methods have developed into standard techniques to address biochemical problems at the level of biomolecular structure. Several factors have contributed to this success, most notably the development of computers, which have become vastly more powerful during the past decades. Initially, simulations were restricted to very simple systems on time scales and at levels of detail that typically were inadequate to answer specific questions about proteins. However, with a 10-fold increase being provided in computing speed every five years, the solutions for a wide range of problems are now accessible directly via computer simulation, an example being water transport in the aquaporins. Today, even the fastest-folding proteins can be investigated directly by simulations, on the same time scale as the fastest experiments, thus providing unprecedented progress in the understanding of protein folding. In the same vein, computer graphics hardware has experienced a similarly rapid increase in speed, enabling complex surface representations to be created of large biomolecular complexes, together with interactive 3-D graphics.

A second factor has been the improvement in simulation and modeling software, which has in turn led to the development of software that is vastly more user-friendly than that available only a few years ago, and which can be operated on commodity hardware rather than exotic (and very expensive) workstations.

A third factor is the improvement in the algorithms and parameters that are used to describe the interactions between atoms, or to predict the structure of proteins. Today, with increasing accuracy and/or more efficient algorithms, molecular modeling is able to answer ever more-detailed and complicated questions on this subject.

Finally, the massive advances in experimental studies at the level of biomolecular structure have been fed back into computer modeling and, as a consequence, the ever-expanding database of high-resolution protein structures represents a key resource for the further development of modeling methods. Advanced single-molecule experiments may provide a direct link with simulations that in the past was very rare, while the increasing number of experimental techniques available can now be linked directly to the results of simulations. An example of such an advance is the data obtained from NMR spectroscopy studies, which can provide both structural and dynamical data on time scales that are comparable to those of simulations.

Because molecular modeling is such a broad area, it is difficult to indicate exactly where most progress can be expected in the future. Clearly, technical advances in hardware and software engineering, in addition to algorithm and parameter developments, will continue to increase the number of problems that can be addressed via molecular modeling. These advances will also facilitate the incorporation of computationally intense calculations such as quantum mechanics (QM) and QM/MM (molecular mechanics) methods into accessible programs that are more user-friendly to the molecular modeler. The use of Grid technology, distributed computing, and massively parallel commodity-based Beowulf clusters will bring central processor unit-intensive calculations within the reach of an increasing number of research teams, thereby allowing the study of larger and more complex systems. Yet, it is to be expected that the most important areas of growth will be a stronger link between modeling and experiments, the routine incorporation of modeling and simulation in otherwise mostly experimental studies, and improved methods to study protein–protein interactions and chemical reactions.

Acknowledgments

D.P.T. is a Scientist of the Alberta Heritage Foundation for Medical Research; all studies conducted in his laboratory were supported by the Canadian Institutes of Health Research.

References

1 Petsko, G., Ringe, D. (2004) *Protein Structure and Function*, New Science Press.
2 Pauling, L., Corey, R.B., Branson, H.R. (1951) The structure of proteins – 2 hydrogen-bonded helical configurations of the polypeptide chain. *Proc. Natl Acad. Sci. USA*, **37**, 205–211.
3 Vaughan, C.K., Buckle, A.M., Fersht, A.R. (1999) Structural response to mutation at a protein-protein interface. *J. Mol. Biol.*, **286**, 1487–1506.

4 Bourne, P., Weissig, H. (2003) *Structural Bioinformatics*, Wiley-Liss, Hoboken, New Jersey.

5 Lesk, A. (2001) *Introduction to Protein Architecture: The Structural Biology of Proteins*, Oxford University Press, New York.

6 Greenfield, N.J. (2007) Using circular dichroism spectra to estimate protein secondary structure. *Nat. Protoc.*, **1**, 2876–2890.

7 Barth, A., Zscherp, C. (2002) What vibrations tell us about proteins. *Q. Rev. Biophys.*, **35**, 369–430.

8 Wishart, D.S., Sykes, B.D., Richards, F.M. (1991) Relationship between nuclear magnetic resonance chemical shift and protein secondary structure. *J. Mol. Biol.*, **222**, 311–333.

9 Rupp, B. (2009) *Modern Biomolecular Crystallography*, Taylor and Francis, London.

10 Cavanagh, J. (2007) *Protein NMR Spectroscopy: Principles and Practice*, Academic Press, San Diego.

11 Lacapère, J. (2010) *Membrane Protein Structure Determination: Methods and Protocols*, Humana Press, Totowa, NJ.

12 de Groot, B.L., Engel, A., Grubmuller, H. (2003) The structure of the aquaporin-1 water channel: a comparison between cryo-electron microscopy and X-ray crystallography. *J. Mol. Biol.*, **325**, 485–493.

13 Leach, A.R. (2001) *Molecular Modelling: Principles and Applications*, Prentice-Hall, Essex, UK.

14 Schlick, T. (2010) *Molecular Modeling and Simulation: An Interdisciplinary Guide*, Springer, New York.

15 Rost, B. (2003) *Prediction in 1D: Secondary Structure, Membrane Helices, and Accessibility*, John Wiley & Sons, Inc., Hoboken, NJ.

16 Levinthal, C. (1969) How to Fold Graciously, in: *Mossbauer Spectroscopy in Biological Systems: Proceedings of a Meeting held at Allerton House, Monticello, Illinois*, University of Illinois Press, Champaign, IL, pp. 22–24.

17 Chan, H.S. (2005) Simplified Models of Protein Folding, in: Yip, S. (Ed.) *Handbook of Materials Modeling*, Springer, Netherlands, pp. 1823–1836.

18 Moult, J., Pedersen, J.T., Judson, R., Fidelis, K. (1995) A large-scale experiment to assess protein structure prediction methods. *Proteins Struct. Funct. Bioinf.*, **23**, ii–iv.

19 Bujnicki, J.M., Elofsson, A., Fischer, D., Rychlewski, L. (2001) Structure prediction meta server. *Bioinformatics*, **17**, 750–751.

20 Janin, J., Henrick, K., Moult, J., Eyck, L.T., Sternberg, M.J., Vajda, S., Vakser, I., Wodak, S.J. (2003) CAPRI: A critical assessment of predicted interactions. *Proteins*, **52**, 2–9.

21 Bonneau, R., Baker, D. (2001) Ab initio protein structure prediction: Progress and prospects. *Annu. Rev. Biophys. Biomol. Struct.*, **30**, 173–189.

22 Murzin, A.G., Brenner, S.E., Hubbard, T., Chothia, C. (1995) SCOP – a structural classification of proteins database for the investigation of sequences and structures. *J. Mol. Biol.*, **247**, 536–540.

23 Suhrer, S.J., Wiederstein, M., Sippl, M.J. (2007) QSCOP-SCOP quantified by structural relationships. *Bioinformatics*, **23**, 513–514.

24 Orengo, C.A., Michie, A.D., Jones, S., Jones, D.T., Swindells, M.B., Thornton, J.M. (1997) CATH – a hierarchic classification of protein domain structures. *Structure*, **5**, 1093–1108.

25 Michie, A.D., Orengo, C.A., Thornton, J.M. (1996) Analysis of domain structural class using an automated class assignment protocol. *J. Mol. Biol.*, **262**, 168–185.

26 Taylor, W.R., Flores, T.P., Orengo, C.A. (1994) Multiple protein-structure alignment. *Protein Sci.*, **3**, 1858–1870.

27 Holm, L., Ouzounis, C., Sander, C., Tuparev, G., Vriend, G. (1992) A database of protein-structure families with common folding motifs. *Protein Sci.*, **1**, 1691–1698.

28 Holm, L., Sander, C. (1996) Mapping the protein universe. *Science*, **273**, 595–602.

29 Altschul, S.F., Madden, T.L., Schaffer, A.A., Zhang, J.H., Zhang, Z., Miller, W., Lipman, D.J. (1997) Gapped BLAST and PSI-BLAST: a new generation of protein database search programs. *Nucleic Acids Res.*, **25**, 3389–3402.

30 Bowie, J., Luthy, R., Eisenberg, D. (1991) A method to identify protein sequences that fold into a known three-dimensional structure. *Science*, **253**, 164–170.

31 Sippl, M.J. (1990) Calculation of conformational ensembles from potentials of mean force – an approach to the knowledge-based prediction of local structures in globular-proteins. *J. Mol. Biol.*, **213**, 859–883.

32 Sippl, M.J. (1993) Boltzmann principle, knowledge-based mean fields and protein-folding – an approach to the computational determination of protein structures. *J. Comput. Aided Mol. Des.*, **7**, 473–501.

33 Sippl, M.J. (1995) Knowledge-based potentials for proteins. *Curr. Opin. Struct. Biol.*, **5**, 229–235.

34 Sippl, M.J., Weitckus, S. (1992) Detection of native-like models for amino-acid-sequences of unknown 3-Dimensional structure in a data-base of known protein conformations. *Proteins Struct. Funct. Genet.*, **13**, 258–271.

35 Sippl, M.J. (1993) Recognition of errors in 3-Dimensional structures of proteins. *Proteins Struct. Funct. Genet.*, **17**, 355–362.

36 Jones, D.T., Taylor, W.R., Thornton, J.M. (1992) A new approach to protein fold recognition. *Nature*, **358**, 86–89.

37 Taylor, W.R., Orengo, C.A. (1989) Protein-structure alignment. *J. Mol. Biol.*, **208**, 1–22.

38 Orengo, C.A., Taylor, W.R. (1990) A rapid method of protein-structure alignment. *J. Theor. Biol.*, **147**, 517–551.

39 Jones, D. (1998) THREADER: Protein Sequence Threading by Double Dynamic Programming, in: Steven, D.B.S., Salzberg, L., Simon, K. (Eds) *New Comprehensive Biochemistry*, Chapter 13, Elsevier, pp. 285–311.

40 Godzik, A. (2005) *Fold Recognition Methods*, in: Bourne, P.E. and Weissig, H. (Eds) Structural Bioinformatics, Vol. **44**, John Wiley & Sons, Inc., Hoboken, NJ.

41 Rychlewski, L., Fischer, D., Elofsson, A. (2003) LiveBench-6: large-scale automated evaluation of protein structure prediction servers. *Proteins-Struct. Funct. Genet.*, **53**, 542–547.

42 Tramontano, A. (2007) *Introduction to Bioinformatics*, Chapman & Hall (Taylor & Francis Group), London.

43 Abagyan, R.A., Batalov, S. (1997) Do aligned sequences share the same fold? *J. Mol. Biol.*, **273**, 355–368.

44 Rost, B. (1999) Twilight zone of protein sequence alignments. *Protein Eng.*, **12**, 85–94.

45 Krieger, E., Nabuurs, S.B., Vriend, G. (2003) *Homology Modeling*, John Wiley & Sons, Inc., Hoboken, NJ.

46 Sali, A., Blundell, T.L. (1993) Comparative protein modeling by satisfaction of spatial restraints. *J. Mol. Biol.*, **234**, 779–815.

47 Fiser, A.S., Sali, A. (2003) *MODELLER: generation and refinement of homology-based protein structure models*, in: Carter, C.W. Jr, Sweet, R.M. (Eds) Methods in Enzymology, Vol. **374**, *Macromolecular Crystallography Part D*, Academic Press, San Diego, pp. 461–491.

48 Vriend, G. (1990) What if – a molecular modeling and drug design program. *J. Mol. Graphics*, **8**, 52.

49 Guex, N., Peitsch, M.C. (1997) SWISS-MODEL and the Swiss-Pdb viewer: an environment for comparative protein modeling. *Electrophoresis*, **18**, 2714–2723.

50 Tate, J. (2003) *Molecular Visualization*, John Wiley & Sons, Inc., Hoboken, NJ.

51 Sayle, R.A., Milnerwhite, E.J. (1995) Rasmol – biomolecular graphics for all. *Trends Biochem. Sci.*, **20**, 374–376.

52 Humphrey, W., Dalke, A., Schulten, K. (1996) VMD: visual molecular dynamics. *J. Mol. Graphics*, **14**, 33–38.

53 Davis, M.E., McCammon, J.A. (1990) Electrostatics in biomolecular structure and dynamics. *Chem. Rev.*, **90**, 509–521.

54 Honig, B., Nicholls, A. (1995) Classical electrostatics in biology and chemistry. *Science*, **268**, 1144–1149.

55 Warshel, A., Aqvist, J. (1991) Microscopic Simulations of Chemical Processes in Proteins and the Role of Electrostatic Free Energy, in: Beveridge, D.L., Lavery, R. (Eds) Theoretical Biochemistry and Molecular Biophysics, Vol. **2**, *Proteins*, Adenine Press, Schenectady, New York, pp. 257–274.

56 Lee, L.P., Tidor, B. (2001) Barstar is electrostatically optimized for tight binding to barnase. *Nat. Struct. Biol.*, **8**, 73–76.

57 Gilson, M.K., Sharp, K.A., Honig, B.H. (1988) Calculating the electrostatic potential of molecules in solution: method and error assessment. *J. Comput. Chem.*, **9**, 327–335.

58 Davis, M.E., Madura, J.D., Luty, B.A., McCammon, J.A. (1991) Electrostatics and diffusion of molecules in solution: simulations with the University of Houston Brownian dynamics program. *Comput. Phys. Commun.*, **62**, 187–197.

59 Baker, N.A., Sept, D., Joseph, S., Holst, M.J., McCammon, J.A. (2001) Electrostatics of nanosystems: application to microtubules and the ribosome. *Proc. Natl Acad. Sci. USA*, **98**, 10037–10041.

60 Lengauer, T., Rarey, M. (1996) Computational methods for biomolecular docking. *Curr. Opin. Struct. Biol.*, **6**, 402–406.

61 Jorgensen, W.L. (2004) The many roles of computation in drug discovery. *Science*, **303**, 1813–1818.

62 Taylor, R.D., Jewsbury, P.J., Essex, J.W. (2002) A review of protein-small molecule docking methods. *J. Comput.-Aided Mol. Des.*, **16**, 151–166.

63 Walters, W.P., Stahl, M.T., Murcko, M.A. (1998) Virtual screening – an overview. *Drug Discovery Today*, **3**, 160–178.

64 Alvarez, J., Shoichet, B. (2005) *Virtual Screening in Drug Discovery*, Taylor & Francis, Boca Raton, Florida.

65 Schneider, G., Fechner, U. (2005) Computer-based de novo design of drug-like molecules. *Nat. Rev. Drug Discovery*, **4**, 649–663.

66 Halperin, I., Ma, B., Wolfson, H., Nussinov, R. (2002) Principles of docking: an overview of search algorithms and a guide to scoring functions. *Proteins Struct. Funct. Bioinf.*, **47**, 409–443.

67 Charifson, P.S., Corkery, J.J., Murcko, M.A., Walters, W.P. (1999) Consensus scoring: a method for obtaining improved hit rates from docking databases of three-dimensional structures into proteins. *J. Med. Chem.*, **42**, 5100–5109.

68 Vajda, S., Camacho, C.J. (2004) Protein-protein docking: is the glass half-full or half-empty? *Trends Biotechnol.*, **22**, 110–116.

69 Smith, G.R., Sternberg, M.J.E. (2002) Prediction of protein-protein interactions by docking methods. *Curr. Opin. Struct. Biol.*, **12**, 28–35.

70 Janin, J. (1995) Protein–protein recognition. *Prog. Biophys. Mol. Biol.*, **64**, 145–166.

71 Elcock, A.H., Sept, D., McCammon, J.A. (2001) Computer simulation of protein–protein interactions. *J. Phys. Chem. B*, **105**, 1504–1518.

72 Dominguez, C., Boelens, R., Bonvin, A.M.J.J. (2003) HADDOCK: a protein-protein docking approach based on biochemical or biophysical information. *J. Am. Chem. Soc.*, **125**, 1731–1737.

73 Karplus, M., McCammon, J.A. (2002) Molecular dynamics simulations of biomolecules. *Nat. Struct. Biol.*, **9**, 646–652.

74 Vangunsteren, W.F., Berendsen, H.J.C. (1990) Computer-simulation of molecular-dynamics – methodology, applications, and perspectives in chemistry. *Angew. Chem. Int. Ed. Engl.*, **29**, 992–1023.

75 Frenkel, D., Smit, B. (2002) *Understanding Molecular Simulation: from Algorithms to Applications*, Academic Press, New York.

76 Locher, K.P., Lee, A.T., Rees, D.C. (2002) The *E. coli* BtuCD structure: a framework for ABC transporter architecture and mechanism. *Science*, **296**, 1091–1098.

77 Kollman, P. (1993) Free-energy calculations – applications to chemical and biochemical phenomena. *Chem. Rev.*, **93**, 2395–2417.

78 Chipot, C., Pohorille, A. (2007) *Free Energy Calculations: Theory and Applications in Chemistry and Biology*, Springer, Berlin.

79 Straatsma, T.P., McCammon, J.A. (1992) Computational alchemy. *Annu. Rev. Phys. Chem.*, **43**, 407–435.

80 Van Der Spoel, D., Lindahl, E., Hess, B., Groenhof, G., Mark, A.E., Berendsen, H.J.C. (2005) Gromacs: Fast, flexible, and free. *J. Comput. Chem.*, **26**, 1701–1718.

81 Phillips, J.C., Braun, R., Wang, W., Gumbart, J., Tajkhorshid, E., Villa, E., Chipot, C., Skeel, R.D., Kalé, L., Schulten, K. (2005) Scalable molecular dynamics with NAMD. *J. Comput. Chem.*, **26**, 1781–1802.

82 Case, D.A., Cheatham, T.E., Darden, T., Gohlke, H., Luo, R., Merz, K.M., Onufriev, A., Simmerling, C., Wang, B., Woods, R.J. (2005) The Amber biomolecular simulation programs. *J. Comput. Chem.*, **26**, 1668–1688.

83 Brooks, B.R., Bruccoleri, R.E., Olafson, B.D., States, D.J., Swaminathan, S., Karplus, M. (1983) CHARMM: a program for macromolecular energy, minimization, and dynamics calculations. *J. Comput. Chem.*, **4**, 187–217.

84 van Gunsteren, W.F., Berendsen, H.J.C. (1982) Algorithms for Brownian dynamics. *Mol. Phys. Int. J. Interface Chem. Phys.*, **45**, 637–647.

85 Ermak, D.L., McCammon, J.A. (1978) Brownian dynamics with hydrodynamic interactions. *J. Chem. Phys.*, **69**, 1352–1360.

86 Sussman, J.L., Harel, M., Frolow, F., Oefner, C., Goldman, A., Toker, L., Silman, I. (1991) Atomic structure of acetylcholinesterase from *Torpedo californica* – a prototypic

acetylcholine-binding protein. *Science*, **253**, 872–879.

87 Shen, T.Y., Tai, K.H., Henchman, R.H., McCammon, J.A. (2002) Molecular dynamics of acetylcholinesterase. *Acc. Chem. Res.*, **35**, 332–340.

88 Zhou, H.X., Wlodek, S.T., McCammon, J.A. (1998) Conformation gating as a mechanism for enzyme specificity. *Proc. Natl Acad. Sci. USA*, **95**, 9280–9283.

89 Bui, J.M., Henchman, R.H., McCammon, J.A. (2003) The dynamics of ligand barrier crossing inside the acetylcholinesterase gorge. *Biophys. J.*, **85**, 2267–2272.

90 Branduardi, D., Gervasio, F.L., Cavalli, A., Recanatini, M., Parrinello, M. (2005) The role of the peripheral anionic site and cation-pi interactions in the ligand penetration of the human AChE gorge. *J. Am. Chem. Soc.*, **127**, 9147–9155.

91 Elcock, A.H., Gabdoulline, R.R., Wade, R.C., McCammon, J.A. (1999) Computer simulation of protein–protein association kinetics: acetylcholinesterase-fasciculin. *J. Mol. Biol.*, **291**, 149–162.

92 Soreq, H., Seidman, S. (2001) Acetylcholinesterase – new roles for an old actor. *Nat. Rev. Neurosci.*, **2**, 294–670.

93 Kryger, G., Silman, I., Sussman, J.L. (1999) Structure of acetylcholinesterase complexed with E2020 (Aricept®): implications for the design of new anti-Alzheimer drugs. *Structure*, **7**, 297–307.

94 Piazzi, L., Rampa, A., Bisi, A., Gobbi, S., Belluti, F., Cavalli, A., Bartolini, M., Andrisano, V., Valenti, P., Recanatini, M. (2003) 3-(4-{[benzyl(methyl)amino]methyl}-phenyl)-6,7-dimethoxy-2H-2-chromenone (AP2238) inhibits both acetylcholinesterase and acetylcholinesterase-induced beta-amyloid aggregation: a dual function lead for Alzheimer's disease therapy. *J. Med. Chem.*, **46**, 2279–2282.

95 Camps, P., Formosa, X., Galdeano, C., Go'mez, T., Munñoz-Torrero, D., Scarpellini, M., Viayna, E., Badia, A., Clos, M.V., Camins, A., Pallàs, M., Bartolini, M., Mancini, F., Andrisano, V., Estelrich, J., Lizondo, M., Bidon-Chanal, A., Luque, F.J. (2008) Novel donepezil-based inhibitors of acetyl- and butyrylcholinesterase and acetylcholinesterase-induced β-amyloid aggregation. *J. Med. Chem.*, **51**, 3588–3598.

96 Preston, G.M., Carroll, T.P., Guggino, W.B., Agre, P. (1992) Appearance of water channels in *Xenopus* oocytes expressing red cell CHIP28 protein. *Science*, **256**, 385–387.

97 Agre, P. (2004) Aquaporin water channels (Nobel lecture). *Angew. Chem. Int. Ed.*, **43**, 4278–4290.

98 de Groot, B.L., Heymann, J.B., Engel, A., Mitsuoka, K., Fujiyoshi, Y., Grubmüller, H. (2000) The fold of human aquaporin 1. *J. Mol. Biol.*, **300**, 987–994.

99 Murata, K., Mitsuoka, K., Hirai, T., Walz, T., Agre, P., Heymann, J.B., Engel, A., Fujiyoshi, Y. (2000) Structural determinants of water permeation through aquaporin-1. *Nature*, **407**, 599–605.

100 Fu, D., Libson, A., Miercke, L.J.W., Weitzman, C., Nollert, P., Krucinski, J., Stroud, R.M. (2000) Structure of a glycerol-conducting channel and the basis for its selectivity. *Science*, **290**, 481–486.

101 Zeidel, M.L., Nielsen, S., Smith, B.L., Ambudkar, S.V., Maunsbach, A.B., Agre, P. (1994) Ultrastructure, pharmacological inhibition, and transport selectivity of aquaporin channel-forming integral protein in proteoliposomes. *Biochemistry*, **33**, 1606–1615.

102 Zhu, F., Tajkhorshid, E., Schulten, K. (2001) Molecular dynamics study of aquaporin-1 water channel in a lipid bilayer. *FEBS Lett.*, **504**, 212–218.

103 Jensen, M.O., Tajkhorshid, E., Schulten, K. (2003) Electrostatic tuning of permeation and selectivity in aquaporin water channels. *Biophys. J.*, **85**, 2884–2899.

104 de Groot, B.L., Engel, A., Grubmüller, H. (2001) A refined structure of human aquaporin-1. *FEBS Lett.*, **504**, 206–211.

105 Jensen, M.Ø., Tajkhorshid, E., Schulten, K. (2001) The mechanism of glycerol conduction in aquaglyceroporins. *Structure*, **9**, 1083–1093.

106 Sui, H., Han, B.-G., Lee, J.K., Walian, P., Jap, B.K. (2001) Structural basis of water-specific transport through the AQP1 water channel. *Nature*, **414**, 872–878.

107 Law, R.J., Sansom, M.S.P. (2004) Homology modelling and molecular dynamics simulations: comparative studies of human aquaporin-1. *Eur. Biophys. J.*, **33**, 477–489.

108 Palczewski, K., Kumasaka, T., Hori, T., Behnke, C.A., Motoshima, H., Fox, B.A., Trong, I.L., Teller, D.C., Okada, T. Stenkamp, R.E., Yamamoto, M., Miyano,

M. (2000) Crystal structure of rhodopsin: a G protein-coupled receptor. *Science*, **289**, 739–745.

109 MacKinnon, R. (2004) Potassium channels and the atomic basis of selective ion conduction (Nobel lecture). *Angew. Chem. Int. Ed.*, **43**, 4265–4277.

110 Holland, I. (2003) *ABC Proteins: From Bacteria to Man*, Academic Press, Amsterdam.

111 Dawson, R.J.P., Locher, K.P. (2006) Structure of a bacterial multidrug ABC transporter. *Nature*, **443**, 180–185.

112 Aller, S.G., Yu, J., Ward, A., Weng, Y., Chittaboina, S., Zhuo, R.P., Harrell, P.M., Trinh, Y.T., Zhang, Q.H., Urbatsch, I.L., Chang, G. (2009) Structure of p-glycoprotein reveals a molecular basis for poly-specific drug binding. *Science*, **323**, 1718–1722.

113 Ward, A., Reyes, C.L., Yu, J., Roth, C.B., Chang, G. (2007) Flexibility in the ABC transporter MsbA: alternating access with a twist. *Proc. Natl Acad. Sci. USA*, **104**, 19005–19010.

114 Pinkett, H.W., Lee, A.T., Lum, P., Locher, K.P., Rees, D.C. (2007) An inward-facing conformation of a putative metal-chelate-type ABC transporter. *Science*, **315**, 373–377.

115 Hollenstein, K., Frei, D.C., Locher, K.P. (2007) Structure of an ABC transporter in complex with its binding protein. *Nature*, **446**, 213–216.

116 Gerber, S., Comellas-Bigler, M., Goetz, B.A., Locher, K.P. (2008) Structural basis of trans-inhibition in a molybdate/tungstate ABC transporter. *Science*, **321**, 246–250.

117 Kadaba, N.S., Kaiser, J.T., Johnson, E., Lee, A., Rees, D.C. (2008) The high-affinity *E. coli* methionine ABC transporter: structure and allosteric regulation. *Science*, **321**, 250–253.

118 Oldham, M.L., Khare, D., Quiocho, F.A., Davidson, A.L., Chen, J. (2007) Crystal structure of a catalytic intermediate of the maltose transporter. *Nature*, **450**, 515–521.

119 Oloo, E.O., Tieleman, D.P. (2004) Conformational transitions induced by the binding of MgATP to the vitamin B-12 ATP-binding cassette (ABC) transporter BtuCD. *J. Biol. Chem.*, **279**, 45013–45019.

120 Schöz, C., Tampé, R. (2005) The intracellular antigen transport machinery TAP in adaptive immunity and virus escape mechanisms. *J. Bioenerg. Biomembr.*, **37**, 509–515.

121 Procko, E., O'Mara, M.L., Bennett, W.F.D., Tieleman, D.P., Gaudet, R. (2009) The mechanism of ABC transporters: general lessons from structural and functional studies of an antigenic peptide transporter. *FASEB J.*, **23**, 1287–1302.

122 Gadsby, D.C., Vergani, P., Csanady, L. (2006) The ABC protein turned chloride channel whose failure causes cystic fibrosis. *Nature*, **440**, 477–483.

123 Nunes, V., Chillon, M., Dork, T., Tummler, B., Casals, T., Estivill, X. (1993) A new missense mutation (E92k) in the 1st transmembrane domain of the Cftr gene causes a benign cystic-fibrosis phenotype. *Hum. Mol. Genet.*, **2**, 79–80.

124 Sheppard, D.N., Rich, D.P., Ostedgaard, L.S., Gregory, R.J., Smith, A.E., Welsh, M.J. (1993) Mutations in Cftr associated with mild-disease-form Cl^- channels with altered pore properties. *Nature*, **362**, 160–164.

125 Schaedel, C., Kristofferson, A.C., Holmberg, L., Kornfalt, R. (1994) A novel cystic-fibrosis mutation, Y109c, in the first transmembrane domain of Cftr. *Hum. Mol. Genet.*, **3**, 1001–1002.

126 Loo, T.W., Clarke, D.M. (1993) Functional consequences of phenylalanine mutations in the predicted transmembrane domain of P-glycoprotein. *J. Biol. Chem.*, **268**, 19965–19972.

127 Beaudet, L., Urbatsch, I.L., Gros, P. (1998) Mutations in the nucleotide-binding sites of P-glycoprotein that affect substrate specificity modulate substrate-induced adenosine triphosphatase activity. *Biochemistry*, **37**, 9073–9082.

128 Ritz, U., Drexler, I., Sutter, D., Abele, R., Huber, C., Seliger, B. (2003) Impaired transporter associated with antigen processing (TAP) function attributable to a single amino acid alteration in the peptide TAP subunit TAP1. *J. Immunol.*, **170**, 941–946.

129 Gaudet, R., Wiley, D.C. (2001) Structure of the ABC ATPase domain of human TAP1, the transporter associated with antigen processing. *EMBO J.*, **20**, 4964–4972.

130 Procko, E., Ferrin-O'Connell, I., Ng, S.L., Gaudet, R. (2006) Distinct structural and functional properties of the ATPase sites in an asymmetric ABC transporter. *Mol. Cell*, **24**, 51–62.

131 Oancea, G., O'Mara, M.L., Bennett, W.F.D., Tieleman, D.P., Abele, R., Tampe, R. (2009) Structural arrangement of the transmission interface in the antigen ABC transport complex TAP. *Proc. Natl Acad. Sci. USA*, **106**, 5551–5556.

132 Nijenhuis, M., Hammerling, G.J. (1996) Multiple regions of the transporter associated with antigen processing (TAP) contribute to its peptide binding site. *J. Immunol.*, **157**, 5467–5477.

16
System Models for Inference on Mechanisms of Neuronal Dynamics

Klaas E. Stephan[1,2] and Karl J. Friston[2]
[1] University of Zurich, Laboratory for Social and Neural Systems Research, Department of Economics, Bluemlisalpstr. 10, 8006 Zurich, Switzerland
[2] University College London, Wellcome Trust Centre for Neuroimaging, Institute of Neurology, 12 Queen Square, London WC1N 3BG, UK

Systems Biology: Advances in Molecular Biology and Medicine, First Edition. Edited by Robert A. Meyers.

Keywords

Synaptic plasticity
The change in structural and functional transmission properties of synapses.

Neuromodulation
Modulatory effects on neuronal activity or synaptic transmission.

Dopamine
A neurotransmitter.

Acetylcholine
A neurotransmitter.

Dynamic causal modeling (DCM)
A Bayesian system identification technique that is used for inference on neuronal systems.

Granger causality
A time series model of causality.

Psychophysiological interactions
A regression model for studying contextual changes in connectivity.

Structural equation modeling
A regression-like model of effective connectivity.

Bayesian model selection
A model comparison which follows the principles of probability theory.

Understanding processes that result from the interaction of multiple elements requires mathematical models of system dynamics. This increasingly important theme has led to the revitalization of "systems biology," with a focus on molecular questions. However, system models have been used for much longer in computational neuroscience and neuroimaging to infer on causal mechanisms of neuronal dynamics, for example, effective connectivity. In this chapter, system models are reviewed for inferring effective connectivity from brain activity data obtained by functional magnetic resonance imaging or electroencephalography. Following an outline of general systems theory, a generic framework is provided for formalizing the description of systems; this guides the subsequent description of various models of effective connectivity. Attention is focused particularly on dynamic causal modeling (DCM), a Bayesian system identification approach which distinguishes between neuronal state equations and biophysical forward models, as well as methods for Bayesian model selection.

1 General System Theory and Systems Biology

Most scientific disciplines share one central goal, namely to understand *systems* – that is, ensembles of interacting elements. Today, this statement sounds almost trivial, yet in biology at least, the importance of the systems concept has been established only relatively recently. A key figure here was Ludwig von Bertalanffy, a biologist and philosopher, who wrote a series of seminal articles during the first half of the twentieth century in which he argued that complex biological phenomena, such as metabolic processes, could only be understood properly through a mathematical description of how system behavior emerged from its dynamics; that is, the temporal evolution of interactions among its constituent elements. In demonstrating the existence of *system isomorphisms* – that is, the general mathematical descriptions that explain the dynamic behavior of very different types of system at different scales and across fields as diverse as physics, biology, economy, and sociology – von Bertalanffy introduced a very general framework that became known as *general system theory* (see the collection of essays in Ref. [1]). By the 1940s, the systems concept had experienced a breakthrough in biology, which in turn led to the rise of *cybernetics*, "...the science of control and communication in the animal and the machine" [2, 3]. Today, biology uses the systems concept to address questions at all levels, from the molecular level to whole organisms and populations. Nowadays, the systems concept is omnipresent in biology, such that about a decade ago a special subfield was called into existence, namely "systems biology," which (as understood by most today) focuses mainly on systems phenomena at the genetic, molecular, and cell levels. Although this has often been perceived as a new development, it merely recapitulates many of the questions that were addressed by other fields over decades, following von Bertalanffy's initial studies of metabolism. In particular, computational neuroscience and neuroimaging have developed many approaches for understanding the key properties and mechanisms of neuronal systems. As will be discussed below, most of these approaches concern various forms of connectivity among neurons or neuronal populations. In this chapter, some of these approaches are reviewed, with the hope that those readers with a molecular or genetic background will recognize certain analogies that may inspire the way in which connectivity at the levels of proteins, genes, or cells is considered.

So, what exactly is a "system," and why is the systems concept so useful for framing scientific questions? Informally, a system can be defined as a set of elements which interact with each other in a spatially and temporally specific fashion. Before attempting a more formal definition of a system (see below), it should be remembered that one of the classic scientific methods is first to "analyze" a given phenomenon; that is, to break it down into atomic units and processes that can be investigated independently of each other. This approach is appealing because it reduces a complex problem to a set of simpler problems, each of which can be addressed under conditions that can be controlled more easily for potentially confounding influences. For example, if there was need to understand the physiological properties of a single neuron, the best approach might be to isolate that neuron from its environment (for example, allow it to grow in a dish) and then to map its responses to currents applied to various parts of its dendritic tree. Unfortunately, this analytic approach cannot fully predict the neuron's behavior if it were to form part of a neural system in the brain, and would then interact with other neurons. Rather, as part of a system, the response of an individual neuron to a particular synaptic input (or injected current) u_1 would depend on the spatial and temporal distribution of further inputs $u_2 \ldots \; u_n$ that its dendritic tree received from other neurons. If these additional inputs were to occur sufficiently close in time and space to u_1, they would affect the magnitude of the postsynaptic potential elicited by u_1, either linearly (by spatiotemporal summation) or nonlinearly (e.g., by changing the opening probability of voltage-gated channels) [4]. In other words, the connectivity between neurons in the system would mediate effects that could not be predicted by studying a single neuron.

Similar scenarios can be described for any other scientific field, such as biochemistry. For example, after having studied a set of different biochemical processes in isolation, it might not be possible to predict their collective dynamics. Here, the problem would be (as above) that the different processes may interact; typically, one process may change the substrate/product ratio of another process, or the efficacy of an enzyme that is relevant to a particular process may change due to the presence of allosteric (in)activators produced by a second process, or due to dynamic changes in gene expression mediated by a third process.

In summary, the general problem of analytical procedures in science is that they are blind to predicting the consequences that might arise from interactions between the elements of a system. Consequently, analytical procedures must be complemented with a theoretical framework that takes into account both the connectivity between the elements and also any external perturbations, in order to achieve a mechanistic explanation of the dynamics of the system as a whole. This framework is provided by *general system theory*.

2
A General Form for System Models

Why, then, is it useful at all to strive for formal mathematical definitions of systems? First (as described below), it allows the precise description of structure, function, and structure–function relationships (SFRs). Second, it allows a system's behavior to be predicted for situations where that system has not previously been observed (see Ref. [5] for an impressive collection of examples from biology). Third, it represents the only means to fully understand how a system works and, in particular, how a system's function could be restored if were to be rendered dysfunctional, for example, by disease [6].

In this section, deterministic differential equations with time-invariant parameters will be used as a mathematical framework. Of course, these are not the only possible mathematical representation of dynamic systems (see Ref. [7] for alternatives; later sections of this chapter will also provide details of discrete difference equations and stochastic differential equations, respectively). The underlying concept, however, is quite universal: a *system* is defined by ... a set of elements with n time-variant properties altogether that interact with each other. Each time-variant property $x_i (1 \leq i \leq n)$ is referred to as a *state variable*, and the n-vector $x(t)$ of all state variables in the system is called the *state vector* (or simply *state*) of the system at time t:

$$x(t) = \begin{bmatrix} x_1(t) \\ \vdots \\ x_n(t) \end{bmatrix} \tag{1}$$

Taking an ensemble of interacting neurons as an example, the system elements would correspond to the individual neurons, each of which is represented by one or several state variables. These state variables could refer to various neurophysiological properties, for example, postsynaptic potentials and the status of ion channels. This touches on an important distinction: in *system construction* or *simulation* (e.g., in engineering), the relevant state variables and their mutual dependencies are usually known, whereas in *system identification* (e.g., when trying to understand a biological system) they are not known. This means that a *model* of the system is always required that represents the current hypothesis regarding the structure of the system and how its function emerges from that structure (i.e., the SFR).

The crucial point here is that the state variables interact with each other; that is, the evolution of each state variable depends on at least one other state variable. This mutual functional dependence between the state variables of the system can be expressed, for example, by a set of ordinary differential equations that operate on the state vector:

$$\frac{dx}{dt} = \begin{bmatrix} f_1(x_1, \ldots, x_n) \\ \vdots \\ f_n(x_1, \ldots, x_n) \end{bmatrix} = F(x) \tag{2}$$

However, this description is not sufficient. First, the specific form of the dependencies f_i must be specified – that is, the nature of the causal relations between state variables. This requires a set of parameters θ which determine the form and strength of influences between state variables. In neural systems, these parameters usually correspond to time constants or strengths of the connections between the system elements. Second, in the case of non-autonomous systems (which are greatest interest to biology) the input into the system must be considered (e.g., sensory information entering the brain). Here, the set of all m known inputs is represented by the m-vector function $u(t)$. Accordingly, extending Eq. (2) leads to a general state equation for non-autonomous deterministic systems:

$$\frac{dx}{dt} = F(x, u, \theta) \tag{3}$$

where θ is the parameter vector of the system. Such a model provides a causal description of how system dynamics results from system structure, because it describes: (i) when and where external inputs enter the system; and (ii) how the state changes, induced by these inputs, evolve in time depending on the system's structure. As explained below in more detail (see Sect. 3), Eq. (3) therefore provides a general form for models of *effective connectivity* in neural systems – that is, the causal influences that neural units exert over one another [8].

At this point, two main assumptions have been made implicitly to simplify the exposition. First, it has been assumed that all processes in the system are deterministic and occur instantaneously. Random components (noise) and delays could be accounted for by using stochastic differential equations and delay differential equations, respectively [random differential equations employed when introducing stochastic dynamic causal models (DCMs) will be described later in this chapter]. Second, it is has been assumed that at least some of the inputs entering the system are known. This is a tenable assumption in neuroimaging, because the inputs are experimentally controlled variables, for example, changes in stimuli or instructions. It should be noted that the use of time-invariant dependencies f_i and parameters θ does not restrict the generality of the framework discussed here. Although the mathematical form of f_i per se is static, the use of time-varying inputs u allows for dynamic changes in which components of f_i are "activated." For example, by using box-car functions that are multiplied with the different terms of a polynomial function, it is possible to induce changes from linear to nonlinear behavior (and *vice versa*) over time. In addition, as there is no principled distinction between states and time-invariant parameters, the estimation of time-varying parameters can be treated as a state estimation problem.

On the basis of the general system description provided by Eq. (3), it can now be stated accurately that, given a particular system model, it is possible to define clearly structure, function, and SFRs (see Ref. [9] for a more detailed discussion):

- *Structure* is defined by the time-invariant components of the system; that is, the nature of θ (e.g., absence/presence and strength of connections) and the mathematical form of the state variable dependencies f_i.
- *Function* refers to those time-variant components of the system model that are conditional on its structure, that is, $x(t)$, but not $u(t)$.
- The SFR is represented by F: integrating F in time determines the temporal evolution of the system state x from time $t = 0$ up to a time point τ, given an initial state $x(0)$:

$$x(\tau) = x(0) + \int_0^{\tau} F(x, u, \theta) dt \qquad (4)$$

In other words, given a particular temporal sequence of inputs $u(t)$, Eq. (4) provides a complete description of how the dynamics of the system (i.e., the trajectory of its state vector x in time) results from its structure and initial state.

3
System Models for Inferring Functional Integration and Effective Connectivity from Neuroimaging Data

Modern cognitive neuroscience has adopted an explicit system perspective. A commonly accepted view is that the elements (brain regions) that constitute a given neuronal system are computationally specialized, but that the exact nature of their individual computations depends on context – for example, the inputs from other regions. The aggregate behavior of the system depends on these context-dependent interactions between the system components [10]. An equivalent perspective is provided by the twin concepts of functional specialization and functional integration [11]. *Functional specialization* assumes a local specialization for certain aspects of information processing, but allows for the possibility that this specialization is anatomically segregated across different cortical areas. The majority of current functional neuroimaging experiments have adopted this view, and interpret the areas that are activated by a certain task component as the elements of a distributed system. However, this explanation is incomplete as long as no insight is provided into how the locally specialized computations are bound together by context-dependent interactions among these areas – that is, the *functional integration* within the system. This functional integration within distributed neural systems can be characterized in two ways: functional connectivity; and effective connectivity.

Functional connectivity refers to statistical dependencies between regional time series [8, 12]. Thus, analyses of functional connectivity do not incorporate any knowledge or assumptions about the structure and the SFR of the system of interest. Depending on the amount of knowledge for the system under investigation, this may be a strength or a weakness. If the system is largely unknown, then functional connectivity approaches are useful because they can be used in an exploratory fashion, either by computing functional connectivity maps with reference to a particular seed region [13, 14], or by using a variety of multivariate techniques that find sets of voxels, the time series of which represent distinct (orthogonal or independent) components of the covariance structure of the data [15–17]. The information from these analyses can then be used to generate hypotheses about the system. Conversely, given sufficient information about the system structure and a specific hypothesis about the SFR of the system, models of effective connectivity are more powerful. In this chapter, attention is focused exclusively on models of effective connectivity.

Effective connectivity has been defined by various authors, but in complementary ways. For example, it has been defined by some as the causal influences that neural units exert over another [8], while others have proposed that " ... effective connectivity should be understood as the experiment- and time-dependent, simplest possible circuit diagram that would replicate the observed timing relationships between the recorded neurons" [18]. Both definitions emphasize that determining effective connectivity requires a causal model of the interactions between the elements of the neural system of interest. Such a causal model must take into account the external inputs that perturb the system, and the anatomical connections by which neural units influence each other. In other words, any such model is related to the general system model as formalized by Eq. (3).

While the equations presented in the previous section are fairly general, in order to illustrate how the concept of effective connectivity emerges naturally from system models, the special case of a linear dynamic system should be discussed. Although most natural phenomena are of a nonlinear nature, linear models play an important role in systems science because: (i) they are analytically tractable; and (ii) given sufficiently long observation periods and non-negligible external input, their dynamics are largely independent of the initial state [5]. Therefore, nonlinear systems are usually investigated in restricted subspaces of interest, using linear models as local approximations. The following model of n interacting brain regions is a simple linear case of Eq. (3) which uses a single state variable per region and m external inputs:

$$\begin{bmatrix} \frac{dx_1}{dt} \\ \vdots \\ \frac{dx_n}{dt} \end{bmatrix} = \begin{bmatrix} a_{11} & \cdots & a_{1n} \\ \vdots & \ddots & \vdots \\ a_{n1} & \cdots & a_{nn} \end{bmatrix} \begin{bmatrix} x_1 \\ \vdots \\ x_n \end{bmatrix} + \begin{bmatrix} c_{11} & \cdots & c_{1m} \\ \vdots & \ddots & \vdots \\ c_{n1} & \cdots & c_{nm} \end{bmatrix} \begin{bmatrix} u_1 \\ \vdots \\ u_m \end{bmatrix} \tag{5}$$

In this model, the change of any given element depends on the state of the other system elements, and also on external inputs which affect it either directly or indirectly. This system model can be written in compact matrix form as:

$$F(x) = \frac{dx}{dt} = Ax + Cu \tag{6}$$

where the non-zero values of A and C represent the parameters of the system (i.e., θ in Eq. (3)). The state of the system at time point τ can be obtained by integration (compare Eq. (4)):

$$x(\tau) = e^{A\tau}x(0) + \int_0^\tau e^{A(\tau-t)}\, Cu(t)\, dt \tag{7}$$

where e^{At} is the matrix exponential [5]. In this model, the system's behavior has two separable components: endogenous dynamics (parameterized by matrix A); and dynamics enforced by external inputs (parameterized by matrix C). The first term of Eq. (6) says that the change of the state variable x_i is a linear mixture of all state variables in the system, weighted by the parameters a_{ij}. By defining a particular parameter a_{ij} to be zero, it is possible to disallow for a direct effect of x_j on x_i. Conversely, any non-zero parameter a_{ij} represents a causal influence of the dynamics of x_j on that of x_i. The binarized parameter matrix $\tilde{A}$

$$\tilde{A} = \chi(A) = \begin{bmatrix} \chi(a_{11}) & \cdots & \chi(a_{1n}) \\ \vdots & \ddots & \vdots \\ \chi(a_{n1}) & \cdots & \chi(a_{nn}) \end{bmatrix},$$

$$\chi(a) = \begin{cases} 1 \text{ if } a \neq 0 \\ 0 \text{ if } a = 0 \end{cases} \tag{8}$$

represents the *structural connectivity* of the system model. Knowledge of structural connectivity is important for the specification of system models as it constrains (but does not determine) the effective connectivity [19, 20] and can also significantly improve the evidence of these models [21]. The definition of the structural connectivity is often guided by anatomical investigations in primates [22] or, more recently, and at a coarser spatial scale, by diffusion-weighted imaging (DWI) studies in humans (see Ref. [23] for caveats concerning the interpretation of DWI results in terms of connectivity).

In Eq. (6), the values of A represent the endogenous influences of system elements over each other, and thus correspond to the *effective connectivity* within the system. In contrast, the values of the matrix C represent the magnitude of the direct effects that external (e.g., sensory) inputs have on system elements. A and C represent the system parameters (θ) that must be estimated when fitting the model in Eq. (6) to measured data. Linear models of the type described by Eq. (6) have found widespread application in various scientific disciplines [1]. In Sect. 6, it will be seen that DCM (see Ref. [24]) extends the above formulation by bilinear terms that model context-dependencies of endogenous connection strengths.

The framework outlined so far is concerned with dynamic systems in continuous time, and thus uses differential equations. The same basic ideas, however, can also be applied to dynamic systems in discrete time (using difference equations), as well as to "static" systems where the system is assumed to be at equilibrium at each point of observation. The latter perspective, which is useful for regression-like equations, is used by classic system models for functional neuroimaging data, for example, psycho-physiological interactions (PPIs; [25]), structural equation modeling (SEM; [26, 27]), or multivariate autoregressive/vector autoregressive models (MAR/VAR; [28–30]). These approaches will be described in the following sections, before attention is focused on DCMs. In general, the models below will be explained in the context of their application to functional magnetic resonance imaging (fMRI) data. It should be noted, however, that all such models can be applied (at least in principle) to any other neuroimaging or electrophysiological data modality.

4 Regression-Based Models of Effective Connectivity

4.1 Psycho-Physiological Interaction (PPI)

The PPI represents one of the simplest models available to assess functional interactions in neuroimaging data [25]. Given a chosen reference time series y_0 (obtained from a reference voxel or region), PPI can be used to compute whole-brain connectivity maps of this reference voxel with all other voxels y_i in the brain, according to the regression-like equation:

$$y_i = ay_0 + b(y_0 \times u) + cu + X\beta + e \qquad (9)$$

where a is the strength of the intrinsic (context-independent) connectivity between y_0 and y_{i}. The bilinear term $y_0 \times u$ represents the interaction between physiological activity y_0 and a psychological variable u which can be construed as a contextual input into the system, modulating the connectivity between y_0 and y_{i} (here, $\times$ represents the Hadamard product, i.e., element-by-element multiplication). The third term describes the strength c by which the input u determines activity in y_{i} directly, independent of y_0. Finally, β are parameters for effects of no interest X (such as confounds), and e is a *Gaussian error* term.

Notwithstanding the fact that this is a nondynamic model, Eq. (9) contains the basic components of system descriptions as outlined by Eq. (3), and there is some similarity between its form and that of the state equation of DCM (Eq. (12); see below). However, since only pair-wise interactions are considered (i.e., separately between the reference voxel and all other brain voxels), this model is severely limited in its capacity to represent neural systems. This has also been highlighted in the initial description of PPIs [25]. Although PPIs are not a true system model, they have a useful role in exploring the functional interactions of a chosen region across the whole brain (e.g., Refs [31, 32]). This exploratory nature bears some similarity to the analyses of functional connectivity, except that the PPIs model the contextual modulation of connectivity, and this modulation has a directional character: testing for a PPI from y_0 to y_i is not identical to testing for a PPI from y_i to y_0. This is because regressing $y_0 \times u$ on y_i is not equivalent to regressing $y_i \times u$ on y_0.

4.2 Structural Equation Modeling (SEM)

Although SEM has been an established statistical technique in the social sciences for several decades, it was only introduced to neuroimaging during the early 1990s by McIntosh *et al.* [33]. It is a multivariate, hypothesis-driven technique that is based on a structural model which represents the hypothesis about the causal relations between several variables (for methodological details, see Refs [26, 27, 34, 35]). In the present context, these variables correspond to neuroimaging or electrophysiological measurements, such as BOLD (blood oxygen level-dependent) or electroencephalography (EEG) channel time series $y_1 \ldots y_n$ of n brain regions, and the hypothetical causal relations are based on anatomically plausible connections between the regions. The strength of each connection $y_i \rightarrow y_j$ is specified by a so-called "path coefficient" which, by analogy to a partial regression coefficient, indicates how the variance of y_j depends on the variance of y_i if all other influences on y_j are held constant.

The statistical model of SEM as it is typically used for neuroimaging data[1] can be summarized by the equation:

$$y = Ay + u \tag{10}$$

1) When applied to questions in the social sciences, SEM is typically used to infer relations between unobserved or "latent" variables. In this case, Eq. (11) must be combined with a model linking latent variables to observations.

where y is a $n \times s$ matrix of n area-specific time series with s scans each, A is a $n \times n$ matrix of path coefficients (with zeros for non-existent connections), and u is a $n \times s$ matrix of zero mean Gaussian error terms, which are driving the modeled system ("innovations"; see Eq. (11)). Parameter estimation is achieved by minimizing the difference between the observed and the modeled covariance matrix Σ of the areas [36]. For any given set of parameters, Σ can be computed by transforming Eq. (10):

$$\begin{aligned} y &= (I - A)^{-1} u \\ \Sigma &= yy^T \\ &= (I - A)^{-1} uu^T (I - A)^{-1^T} \end{aligned} \tag{11}$$

where I is the identity matrix and T denotes the transpose operator (Fig. 1). These equations illustrate that the model on which SEM rests has some similarities to the general equation for non-autonomous linear systems in Eq. (6) (except that SEM is a static model and the inputs to the modeled system are random noise). Moreover, the first line of Eq. (11) can be understood as a generative model of how system function results from the system's connectional structure: the measured time series y results by applying a function of the inter-regional connectivity matrix, that is, $(I - A)^{-1}$, to Gaussian innovations u.

In the case of fMRI or EEG/magnetoencephalography (MEG), the path coefficients of a SEM (i.e., the parameters in A) describe the effective connectivity of the system across the entire experimental session. What one would often prefer to know, however, is how the coupling between certain regions changes as a function of experimentally controlled context, for example, differences in coupling between two different tasks. Notably, SEM does not account for

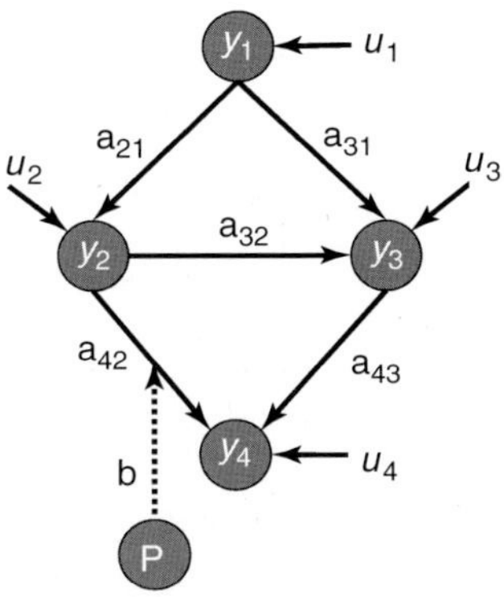

$$y_1 = u_1$$
$$y_2 = a_{21}y_1 + u_2$$
$$y_3 = a_{31}y_1 + a_{32}y_2 + u_3$$
$$y_4 = a_{43}y_3 + a_{42}y_2 + b(P \times y_2) + u_4$$

Fig. 1 Example of a structural equation model with four regions and a bilinear term (P), allowing for changes in effective connectivity across conditions.

temporal order: if all regional time series were permuted in the same fashion, the estimated parameters would not change. In case of blocked designs, this makes it possible to proceed as if one were dealing with positron emission tomography (PET) data, that is, to partition the time series into condition-specific sub-series and fit separate SEMs to them. These SEMs can then be compared statistically to test for condition-specific differences in effective connectivity (for examples, see Refs. [37, 38]). An alternative approach is to augment the model with bilinear terms (cf. Eq. (9) and Fig. 1) which represent the modulation of a given connection by experimentally controlled variables (e.g., Refs. [27, 39]). In this case, only a single SEM is fitted to the entire time series.

One limitation of SEM is that one is restricted to use structural models of relatively low complexity with regard to the number of connections that are being tested. This is because the model is fitted to replicate the observed covariance matrix; therefore, only a limited number of data points are available, leaving not enough degrees of freedom for testing complex models with many parameters [35]. For such cases, heuristics have been suggested that use multiple fitting steps in which different parameters are held constant while changing others (see Ref. [26] for an example). Recent developments with regard to SEM have focused on introducing different estimation principles and accounting for temporal structure [40, 41]. These developments bring SEM close to VAR models which are discussed in the next section.

5 Vector Autoregressive Models and Granger Causality

In contrast to classical SEM, autoregressive models explicitly address the temporal aspect of causality in time series. They take into account the causal dependence of the present based on the past: each data point of a regional time series is explained as a linear combination of past data points from the same region. Models which extend this approach to n brain regions have been termed *multivariate* or *vector autoregressive* models (MAR, VAR); these terms can be used interchangeably.

VAR models explain the n-vector of regional signals at time t (y_t) as a linear

$$\begin{bmatrix} y_t^{(1)} \\ y_t^{(2)} \\ y_t^{(3)} \end{bmatrix} = \sum_{i=1}^{p=1} \begin{bmatrix} y_{t-1}^{(1)} \\ y_{t-1}^{(2)} \\ y_{t-1}^{(3)} \end{bmatrix} \begin{bmatrix} a_{11} & 0 & 0 \\ a_{21} & a_{22} & 0 \\ 0 & a_{23} & a_{33} \end{bmatrix} + \begin{bmatrix} e_t^{(1)} \\ e_t^{(2)} \\ e_t^{(3)} \end{bmatrix}$$

$$= \begin{bmatrix} a_{11}y_{t-1}^{(1)} \\ a_{22}y_{t-1}^{(2)} + a_{21}y_{t-1}^{(1)} \\ a_{33}y_{t-1}^{(3)} + a_{32}y_{t-1}^{(2)} \end{bmatrix} + \begin{bmatrix} e_t^{(1)} \\ e_t^{(2)} \\ e_t^{(3)} \end{bmatrix}$$

Fig. 2 A simple system consisting of three hierarchically connected regions, whose dynamics is described by a first-order vector autoregressive model.

combination of p past data vectors whose contributions are weighted by the parameter matrices A_i:

$$y_t = \sum_{i=1}^{p} y_{t-i}A_i + u_t \tag{12}$$

VAR models thus represent directed influences among a set of regions whose causal interactions are inferred via their mutual predictability from past time points. Figure 2 shows a concrete example of a first-order VAR model, describing a simple system consisting of three hierarchically connected regions. Although VAR has been an established statistical technique in time series analysis for many decades, specific implementations for neuroimaging were suggested only relatively recently [28] suggested a VAR implementation that allowed for the inclusion of bilinear variables representing modulatory effects of contextual variables on connections and used a Bayesian parameter estimation scheme specifically developed for VAR models [42]. This Bayesian scheme also determined the optimal model order, that is, the number of past time points (p in Eq. (12)) to be considered by the model. Another VAR approach, based on the idea of "Granger causality" [43], was proposed by Goebel *et al.* [29] and Roebroeck *et al.* [30]. Given two regional time-series y_1 and y_2, this framework considers y_1 to be caused by y_2 if its dynamics can be predicted better using past values from y_1 and y_2 as opposed to using past values of y_1 alone. This notion of causality thus rests on "temporal precedence" of one regional BOLD signal relative to another. It can be tested efficiently using the F statistic, corresponding to a nested model comparison. For fMRI, this approach has been implemented in terms of Granger causality mapping (GCM; [30]) which computes maps of voxels that are identified as sources or targets of directed influences for a selected region-of-interest.

6
The Importance of Biophysical Forward Models

An important limitation of the models of effective connectivity described above is that they operate at the level of the measured signals. This is a serious problem because the causal architecture of the system that we would like to identify is expressed at the level of neuronal dynamics which is not directly observed using non-invasive techniques. For example, in EEG and MEG data there is a difference between signals measured at each electrode and the underlying neuronal activity: changes in neural activity in different brain regions lead to changes in electric potentials that superimpose linearly. The scalp electrodes therefore record a mixture, with unknown weightings, of potentials generated by a number of different sources. Even more problematic is the case of fMRI data, where PPI, SEM, and VAR are fitted to the measured time series which result from a hemodynamic convolution of the underlying neural activity. Since these classical models do not include the forward model linking neuronal activity to the measured hemodynamic data, analyses of inter-regional connectivity performed at the level of hemodynamic responses can be problematic. For example, different brain regions can exhibit marked differences in neurovascular coupling, and these differences, expressed in terms of different response latencies and waveforms, may lead to false inference about connectivity. It has recently been demonstrated that this is not just a theoretical concern but is practically relevant: [44] performed simultaneous fMRI measurements and intra-cortical recordings in the brains of epileptic rats, demonstrating that the source of the epileptic discharges could only be identified if the model of effective connectivity explicitly accounted for the hemodynamic convolution. Specifically, they showed that a DCM (with a hemodynamic forward model) enabled the correct identification of the epileptic source, whereas a VAR model based on the concept of Granger causality did not, unless it was augmented by a hemodynamic forward model.

These results by David *et al.* [44] triggered a methodological discussion about the relative strengths and weaknesses of fMRI analyses based on Granger causality and DCM, respectively; for example, see Refs. [45–47]. More importantly, they highlighted that inference about connectivity between neural units requires models that combine two things: (i) a parsimonious but neurobiologically plausible model of neural population dynamics and (ii) a biophysically plausible forward model that describes the transformation from neural activity to the measured signal. Such models make it possible to fit jointly the parameters of the neural and of the forward model and provide a mechanistic explanation for the observed time series. In principle, any of the models described above could be combined with a modality-specific forward model, and indeed, VAR models have previously been combined with linear forward models to explain EEG data [48]. So far, however, DCM is the only approach where the marriage between models of neural dynamics and biophysical forward models is a mandatory component (see Fig. 1).

7
Dynamic Causal Modeling (DCM) for fMRI

DCM is a generic framework for Bayesian identification of mechanisms governing neuronal dynamics, where the latter cannot be observed directly but are only expressed indirectly via more macroscopic measures, such as fMRI, EEG, MEG, or local field potentials (LFPs) (Fig. 3). The general idea of DCM is identical for all data modalities (Fig. 4), but the mathematical details of its implementation vary greatly. Here, we begin discussing DCM by describing its original implementation for fMRI data, in terms of a model resting on bilinear deterministic differential equations. Extensions to nonlinear and

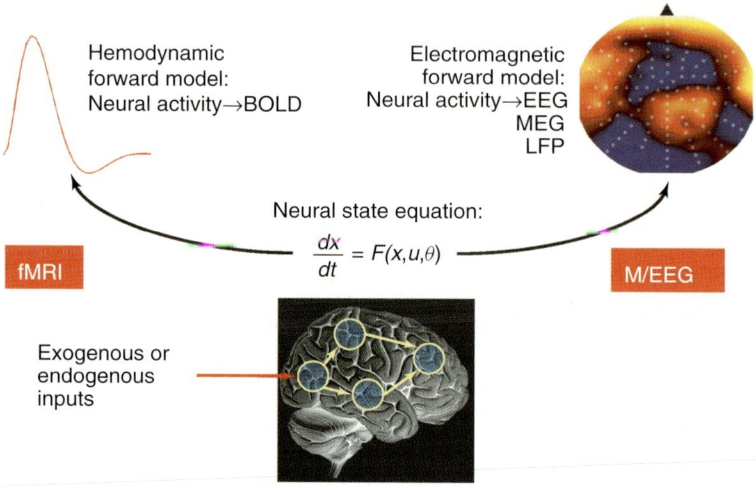

Fig. 3 A summary of the general concept of dynamic causal modeling (DCM). A model of (hidden) neuronal processes is combined with a modality-specific biophysical forward model that predicts the expected measurement for a given neuronal state. Bayesian inversion is then used to infer on the parameters governing the hidden neuronal dynamics.

stochastic equations as well as to other measurement types than fMRI are described in later sections below.

The classical version of DCM for fMRI rests on a simple deterministic model of neuronal population dynamics in a system of n interacting brain regions. In its original form Ref. [24], it models the change of a neural state vector x in time, with each region in the system being represented by a single state variable, using the following bilinear differential equation:

$$\frac{dx}{dt} = F\left(x, u, \theta^{n}\right) = \left(A + \sum_{j=1}^{m} u_j B^{(j)}\right) x + Cu \qquad (13)$$

This bilinear neural state equation results from a Taylor expansion to any nonlinear dynamic system (for details, see Refs. [24, 49]). More precisely, this bilinear form is the simplest low-order approximation that accounts both for endogenous and exogenous causes of system dynamics. Note that it follows the general form for deterministic system models introduced by Eq. (8): the modeled state changes are a function of the current system state, the inputs u and some parameters $\theta^{(n)} = \{A, B, C\}$ that define the nature of the interactions among brain regions at a neuronal level. The neural state variables represent a summary index of neural population dynamics in the respective regions. The neural dynamics are driven by experimentally controlled external inputs that can enter the model in two different ways: they can elicit responses through direct influences on specific regions (e.g., evoked responses in early sensory cortices; the C matrix) or they can modulate the coupling among regions (e.g., during learning or attention; the B matrices). Note that Eq. (13) does not account for conduction delays in either inputs or inter-regional influences. This is not necessary because, due to the large

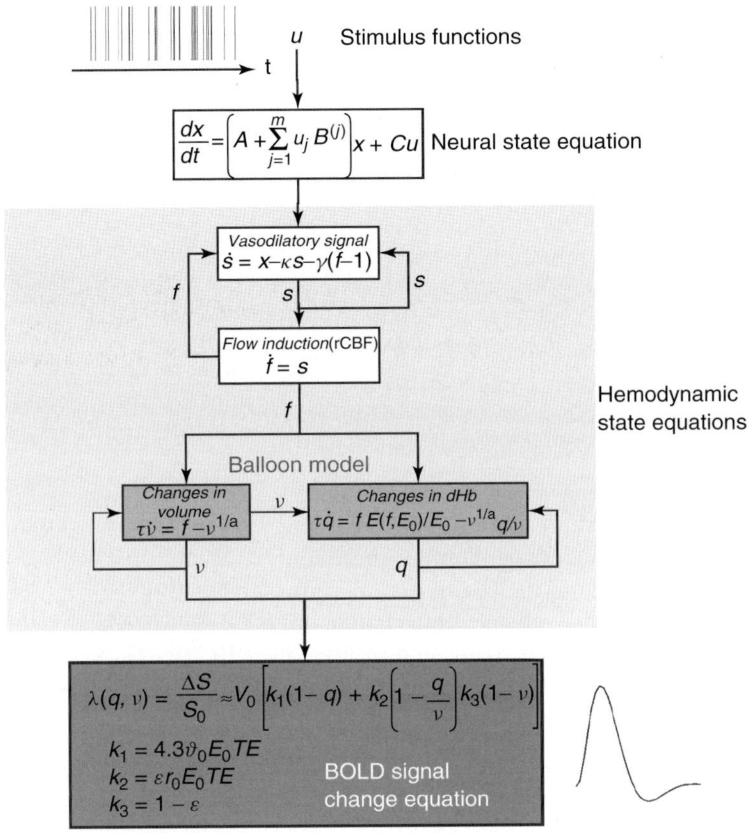

Fig. 4 Schematic summary of the neuronal and hemodynamic models included in classical bilinear DCM. Experimentally controlled input functions u evoke neural responses x, modeled by a bilinear differential state equation, which trigger a hemodynamic cascade, modeled by four state equations with five parameters. These hemodynamic parameters comprise the rate constant of the vasodilatory signal decay (κ), the rate constant for auto-regulatory feedback by blood flow (γ), transit time (τ), Grubb's vessel stiffness exponent (α), and capillary resting net oxygen extraction (E_0). The so-called "Balloon model" consists of the two equations describing the dynamics of blood volume (ν) and deoxyhemoglobin content (q) (light gray boxes). Integrating the state equations for a given set of inputs and parameters produces predicted time-series for ν and q which enter a BOLD signal equation λ (dark gray box) to give a predicted BOLD response. Reproduced with permission from Ref. [53]; © 2007, Elsevier.

regional variability in hemodynamic response latencies, fMRI data do not possess enough temporal information to enable estimation of inter-regional axonal conduction delays which are typically in the order of 10–20 ms (note that the differential latencies of the hemodynamic response are accommodated by region-specific biophysical parameters in the hemodynamic model described below). This was verified by Stephan *et al.* [49] who showed in simulations that DCM parameter estimates were not affected by introducing artificial delays of up to ±1 s. In contrast, conduction delays are an important part of DCM for electrophysiological measures, such as event-related potentials [50], which are described below.

Given the bilinear state equation in Eq. (13), the neural parameters $\theta^{(n)}$ can be expressed as partial derivatives of F:

$$\begin{aligned} A &= \left.\frac{\partial F}{\partial x}\right|_{u=0} \\ B^{(j)} &= \frac{\partial^2 F}{\partial x \partial u_j} \\ C &= \left.\frac{\partial F}{\partial u}\right|_{x=0} \end{aligned} \tag{14}$$

As can be seen from these equations, the matrix A represents the endogenous (fixed or context-independent) connectivity among the regions in the absence of input, the matrices $B^{(j)}$ encode the change in connectivity induced by the j-th input u_j, and C embodies the strength of direct (driving) inputs on neuronal activity. In most instances, the parameters of primary interest are the modulatory ones (i.e., the matrices $B^{(j)}$) since they encode how experimentally controlled manipulations change the connection strengths in the system.

DCM for fMRI combines this model of neural dynamics with an experimentally validated hemodynamic model that describes the transformation of neuronal activity into a BOLD response. This so-called "Balloon model" was initially formulated by Buxton *et al.* [51] and later extended and refined by Friston *et al.* [52] and Stephan *et al.* [53]. Briefly, it consists of a set of differential equations that describe the relations between different hemodynamic state variables, using a set of parameters $\theta^{(h)}$. More specifically, changes in neural activity elicit a vasodilatory signal that leads to increases in blood flow and subsequently to changes in blood volume and deoxyhemoglobin content. The predicted BOLD signal is a non-linear function of blood volume and deoxyhemoglobin content. This hemodynamic model is summarized by Fig. 4 and described in detail by Stephan *et al.* [53].

Together, the neuronal and hemodynamic state equations furnish a deterministic DCM. For any given combination of parameters $\theta = \{\theta^{(n)}, \theta^{(h)}\}$ and inputs u, the measured BOLD response y is modeled as the predicted BOLD signal $h(x, \theta)$ plus a linear mixture of confounds $X\beta$ (e.g., signal drift) and Gaussian observation error ε:

$$y = h(x, u, \theta) + X\beta + \varepsilon \tag{15}$$

The combined neural and hemodynamic parameter set θ is estimated from the measured BOLD data, using a fully Bayesian approach with empirical priors for the hemodynamic parameters and conservative zero-mean shrinkage priors for the coupling parameters.[2] Details of the parameter estimation scheme, which rests on a fixed-form variational Bayesian algorithm, using a Laplace (Gaussian) approximation to the true posterior, can be found in [54, 55].

Once the parameters of a DCM have been estimated from measured BOLD data, the posterior distributions of the parameter estimates can be used to test hypotheses about connection strengths. Due to the Laplace approximation, the posterior distributions are defined by their maximum a posteriori (MAP) estimate

2) A shrinkage prior to DCM is characterized by a mean of zero and high precision (small variance). Consequently, the posterior estimates are "shrunk" towards the prior expectation of zero, unless the data contain strong evidence to the contrary. This means that the posterior expectation of parameter estimates in DCMs that are fitted to random data is zero.

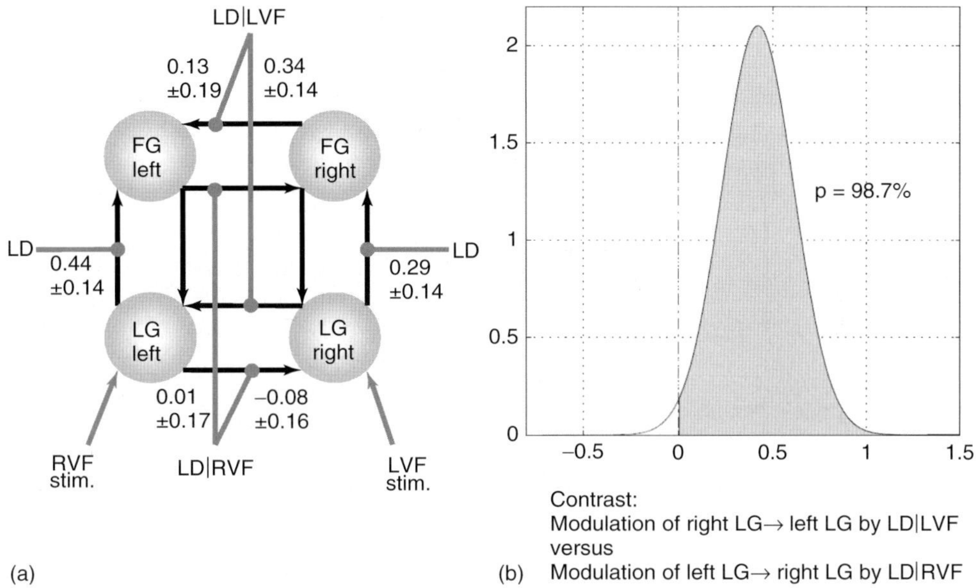

Fig. 5 An example of a single subject DCM that was used to study asymmetries in interhemispheric connections during a letter decision task. LG, lingual gyrus; FG, fusiform gyrus; LDs, letter decisions; LD|VF = letter decisions conditional on the visual field (VF) of stimulus presentation. (a) The values denote the maximum a posteriori (MAP) estimates of the parameters (±square root of the posterior variances; units: 1/s = Hz). For clarity, only the parameters of interest, that is, the modulatory parameters of inter- and intra-hemispheric connections, are shown; (b) Asymmetry of callosal connections with regard to contextual modulation. The plots show the probability (98.7%) that the modulation of the right LG → left LG connection is stronger than the modulation of the left LG → right LG connection. Figure adapted, with permission, from Figs 5 and 6 in Ref. [56].

and their posterior covariance. Usually, the hypotheses to be tested concern context-dependent changes in coupling (i.e., the matrices $B^{(j)}$ in Eq. (13)). An example, originally reported by Stephan *et al.* [56], where DCM was applied to fMRI data from a single subject is given by Fig. 5. Here, the hypothesis was tested that in the ventral stream of the visual system the strength of right-to-left inter-hemispheric connections would increase during a letter decision task, but only when the word stimuli were presented in the left visual field and were thus initially received by the right hemisphere, necessitating transfer of stimulus information to the specialized left hemisphere. This hypothesis was tested by constructing a four-area model of ventral stream areas, comprising the lingual and fusiform gyri in both hemispheres (Fig. 5a), and comparing the modulatory influences of task, conditional on the visual field of stimulus presentation, for inter-hemispheric connections in both directions. This comparison, based on the MAP estimates and the posterior covariances of the modulatory parameters, indicated that for this particular subject and for the connections between left and right lingual gyrus the hypothesized asymmetry in interhemispheric transfer existed with a probability of 98.7% (Fig. 5b). Other examples of single-subject analyses can be found in [49, 57, 58].

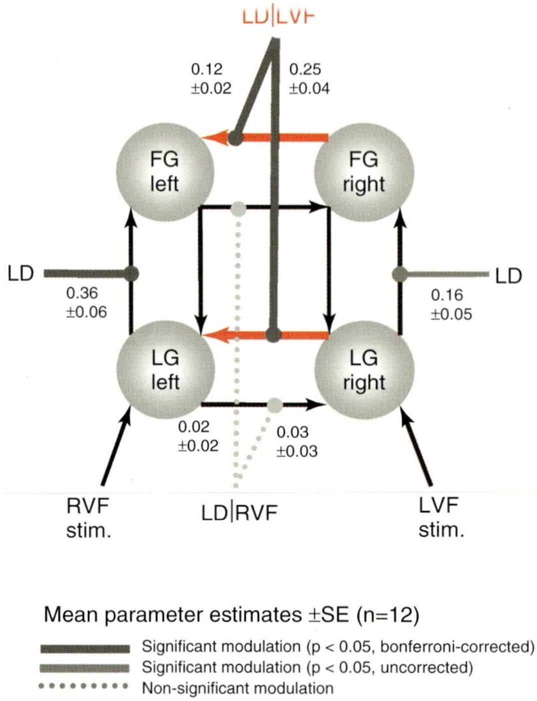

Fig. 6 The same model as in Fig. 5, using the same nomenclature, but now presenting the results of a group analysis of 12 subjects. This model indicates that interhemispheric connections are modulated by a letter decision (LD) task, but conditional on the visual field of stimulation (LD|VF). Furthermore, there is a highly significant asymmetry in the strength of interhemispheric modulations. This indicates asymmetric information transfer from the non-dominant right to the dominant left hemisphere (red arrows). The average modulatory parameter estimates (±standard errors) are shown alongside the modulatory inputs. Dark gray arrows: significant effects surviving Bonferroni's correction; light gray, significant effects not surviving Bonferroni's correction; dotted lines, nonsignificant effects. For clarity, this figure shows only the modulatory parameters; the values of all other parameters can be found in Ref. [60]. Figure adapted, with permission, from Fig. 3 in Ref. [60].

For statistical inference about parameters at the group level, various options exist [59]. One commonly used approach, corresponding to a random effects analysis, is to enter the conditional estimates of interest into a classical second-level analysis, for example, a *t*-test on the MAP estimates of a particular parameter across subjects. As an example, Fig. 6 shows a group analysis (across 12 subjects) using the model of inter-hemispheric integration in the visual system shown in Fig. 5. One can see that the asymmetry in inter-hemispheric connection strengths found for the single subject in Fig. 5 also holds for the group level.

An alternative approach is to use Bayesian statistics at the group level as well. This can be done by computing, for a given parameter, one joint posterior density across all subjects, treating the posterior of one subject as the prior for the next. This approach can be more sensitive; its disadvantage, however, is that it corresponds to a fixed effects analysis and thus does not allow for inference beyond the particular group studied.

8
Bayesian Model Selection (BMS)

Model comparison and selection is central to the scientific process, in that it allows one to evaluate different hypotheses about the way data are caused [58, 61]. Nearly all scientific reporting rests upon some form of model comparison, which represents a probabilistic statement about the belief in one hypothesis relative to some other(s), given some observations. In other words: given some observed data, which of several alternative models is optimal? The decision cannot be made solely by comparing the relative fit of competing models. One also needs to account for differences in complexity; that is, the number of free parameters and the degree of their inter-dependency. This is important because as model complexity increases, fit increases monotonically, but at some point the model will start fitting noise that is specific to the particular data (i.e., "over-fitting") and thus becomes less generalizable across multiple realizations of the same underlying generative process. Therefore, the question "What is the optimal model?" can be reformulated as "What is the model that represents the best balance between fit and complexity?" This is the model that maximizes the model evidence:[3)]

$$p(y|m) = \int p(y|\theta, m)p(\theta|m)\, d\theta \tag{16}$$

3) Model comparison based on the evidence is appropriate if all models have identical a priori probabilities. If this is the case, the model evidence p(y—m) is identical to the posterior probability of the model, p(m—y).

Here, the numbers of free parameters (as well as the functional form of the generative model that determines their interdependencies) are subsumed by the integration. Unfortunately, this integral cannot usually be solved analytically; therefore an approximation to the model evidence is used. This is usually a lower bound on the log evidence [55]:

$$F = \log p(y|m) - KL\left[q(\theta), p(\theta|y, m)\right] \tag{17}$$

Because of its relation to variational calculus and statistical physics, this free energy bound F is often referred to as the "*negative free energy*" or "*variational free energy*" [55, 62, 63]. Here, KL denotes the Kullback-Leibler divergence between the approximating posterior density $q(\theta)$ and the true posterior $p(\theta|y, m)$ [55]. After convergence of the estimation scheme, the KL term is minimized and $F \approx \ln p(y|m)$. Rewriting Eq. (17) as

$$F = \langle \log p(y|\theta, m)\rangle_q - KL(q(\theta), p(\theta)) \tag{18}$$

adds a useful perspective. The two terms in Eq. (18) can be understood as encoding the two opposing requirements of a good model: it explains the data and is as simple as possible (i.e., uses minimal number of parameters that deviate minimally from their priors).

To quantify the relative goodness of two models m_i and m_j, the differences in their log evidences can be transformed into a *Bayes factor* (BF):

$$BF_{ij} = \frac{p(y|m_i)}{p(y|m_j)} \approx \exp(F_i - F_j) \tag{19}$$

The group Bayes factor (GBF) for any given model m_i, relative to m_j, is equivalent to the product of the subject-specific Bayes factors for a given model comparison

$$GBF_{ij} = \exp\left(\sum_k F_i - F_j\right) = \prod_k B_{ij}^k \quad (20)$$

where k indexes subjects [53]. It rests on the assumption that model evidences are independent across subjects; this is tenable if the subjects are independent samples from the population. Furthermore, given a flat prior on the model, the product of model evidences is equivalent to multiplying the posterior probabilities of all models.

The GBF suffers from two major problems. First, it corresponds to a fixed effects analysis, making it vulnerable to the existence of outlier subjects. A second problem is that the GBF does not represent a formal evaluation of the density on model space itself and rests upon a slightly odd generative model for group data: the GBF encodes the relative probability that the data were generated by one model relative to another, assuming the data were generated by the same model for all subjects. What we really want, however, is the density from which models are sampled to generate subject-specific data. A hierarchical model

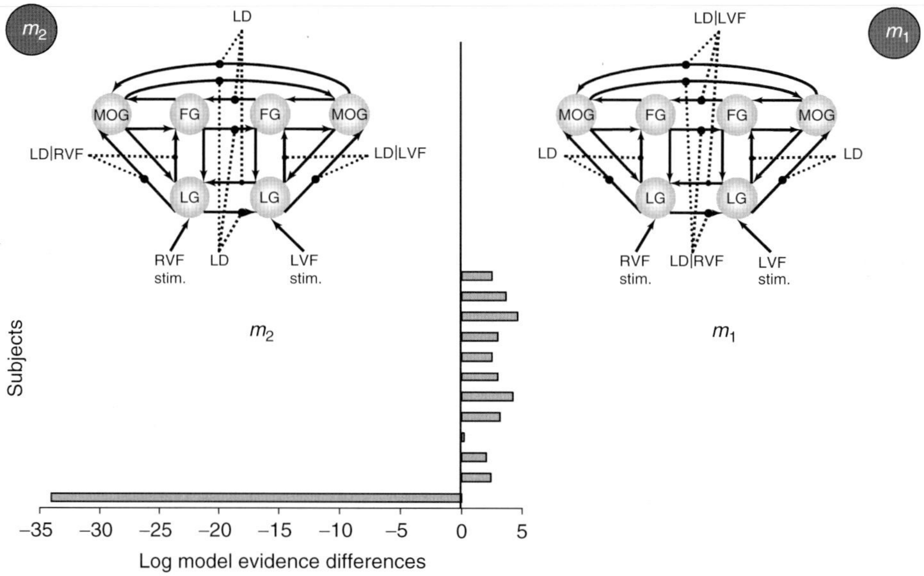

Fig. 7 Comparison of DCMs describing alternative mechanisms of inter-hemispheric integration in terms of context-dependent modulation of connections. Two variants of a six-area model of the ventral visual stream, comprising the lingual gyrus (LG), middle occipital gyrus (MOG) and fusiform gyrus (FG) in both hemispheres were compared. In the first model, m_1, inter-hemispheric connections were modulated by a letter decision (LD) task, but conditional on the visual field of stimulus presentation (LD|VF); intra-hemispheric connections were modulated by LD alone. In the second model, m_2, these modulations were reversed: inter-hemispheric connections were modulated by LD and intra-hemispheric connections were modulated by LD|VF alone. The distribution of log-evidence differences across the 12 subjects is shown in the center of the figure. Although m_1 was superior in 11 of the 12 subjects, a single outlier was so extreme that model comparison based on the group Bayes factor (GBF) would have favored m_2 (GBF = 15 in favor of m_2). Figure adapted, with permission, from Fig. 5 in Ref. [64].

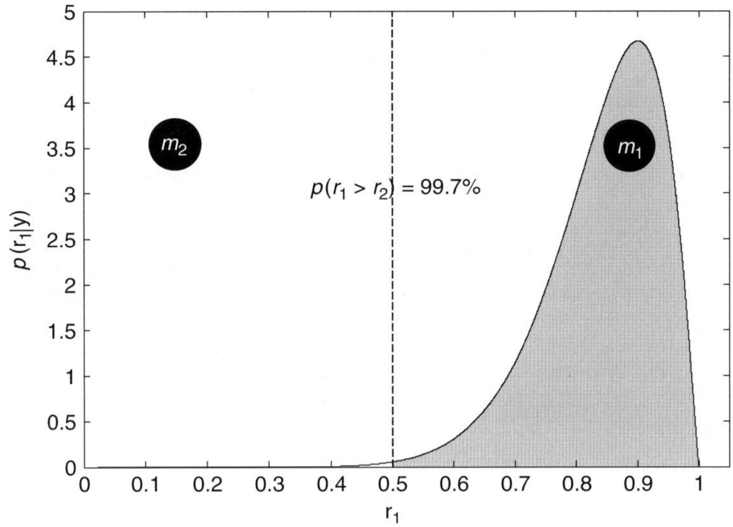

Fig. 8 Application of Bayesian random effects model selection [64] to the models shown in Fig. 3. The Dirichlet density describing the probability of model m_1 in Fig. 7 given the measured data across the group. The shaded area represents the exceedance probability $\varphi_1 = p(r_1 > 0.5|y;\alpha)$ of m_1 being a more likely model than the alternative model m_2 (compare Fig. 3). In contrast to the conventional fixed effects indices, such as the GBF or inference based on frequentist statistics, the variational random effects Bayesian model selection method was not affected by the strong outlier subject shown by Fig. 7: the exceedance probability in favor of m_1 was $\phi_1 = 99.7\%$. Figure adapted, with permission, from Fig. 6 in Ref. [64].

that provides such estimates of model probabilities using a variational Bayesian approach was introduced by Stephan *et al.* [64]. This model treats models as random effects and allows one to compute their expected posterior probabilities as well as the "exceedance probability" that a given model has a higher probability of having generated the observed data across the group than any other model considered. Figures 7 and 8 show an example where two models of inter-hemispheric integration in the visual system [60] are compared. A fixed effects analysis, using the GBF, would prefer model m_2 even though it has higher evidence in only one subject. In contrast, the random effects approach based on the hierarchical variational model clearly prefers model m_2, resulting in an exceedance probability of >99% for model m_1 (Fig. 8).

Bayesian Model Selection (BMS) can be used for any model of effective connectivity. Indeed, special cases of it are frequently used for approaches other than DCM, for example, nested model comparisons (in SEM or Granger causality based on VAR models) or limit approximations to the model evidence, such as Akaike Information Criterion (AIC) or Bayesian Information Criterion (BIC) [34, 65]. In practice, however, it has most often been used in the context of DCM where it plays a central role. Generally, the search for the best model, amongst those deemed plausible a priori, precedes (and is often more important than) inference on the parameters of the model selected (for an overview of procedures, see Ref. [66]. Many studies have used BMS to adjudicate among competing DCMs for fMRI (e.g., Refs. [60, 67–70]). This approach, to search for a single best model (from a set of alternatives) and then proceed to inference on its parameters, is pursued most often. However, alternatives to this single-model approach exist. For example, one can partition model space and make inferences about model families [64, 71]. Alternatively, one can use Bayesian model averaging, where the parameter estimates of each model considered are weighted by the posterior probability of the model [71].

9
Nonlinear, Two-State and Stochastic DCMs for fMRI

Since its first description by Friston *et al.* [24], DCM for fMRI has been extended in several ways. For example, an extension of the forward model enables precise sampling from predicted responses, regardless of the repetition time (TR) of fMRI data acquisition [72]. Another development was to represent each region in the model by two state variables and distinguish between population activity of excitatory and inhibitory neurons [73]. This has the potential advantage of explicitly distinguishing between intrinsic (within-region) connections that can be excitatory or inhibitory, depending on subpopulation, and long-range (inter-regional) connectivity which, within the cortex, is always glutamatergic.

Another extension concerned the inclusion of nonlinear modulatory effects [49]. This enables one to represent activity-dependent changes in effective connectivity that correspond to multiplicative interactions among synaptic inputs. This form of neuronal gain modulation is an important phenomenon in neuronal circuits and arises from short-term synaptic plasticity, synaptic gating, or neuromodulation (c.f. Ref. [74]). To enable a realistic representation of how neuronal populations modulate the gains of other populations, one needs to model nonlinear interactions amongst the n states of a given DCM. For this purpose, one can use a second-order Taylor series [49], resulting in the following nonlinear differential equation:

$$f(x,u) = \frac{dx}{dt} = \left(A + \sum_{i=1}^{m} u_i B^{(i)} + \sum_{j=1}^{n} x_j D^{(j)} \right) x + Cu \qquad (21)$$

Here, the $D^{(j)}$ matrices encode which of the n regions gate which connections in the system. Specifically, any non-zero entry $D^{(j)}_{kl}$ indicates that responses of region k to inputs from region l depend on activity in region j. Figure 9 shows a simple example, with synthetic data generated by a nonlinear DCM. This illustrates the sort of dynamics, both at the neuronal and hemodynamic level which this sort of model exhibits.

The ability to model how connection strengths are gated by the activity of one or several neuronal populations is critical for various applications, for example, for marrying computational learning models with DCM [75, 76], but also for mechanistic accounts of the effects of attention. Stephan *et al.* [49] applied nonlinear DCM to a single-subject data set from a blocked fMRI study of attention to visual motion [27]. They inverted and compared four different models, each of which embodied a different explanation for the empirical finding that V5 responses increased

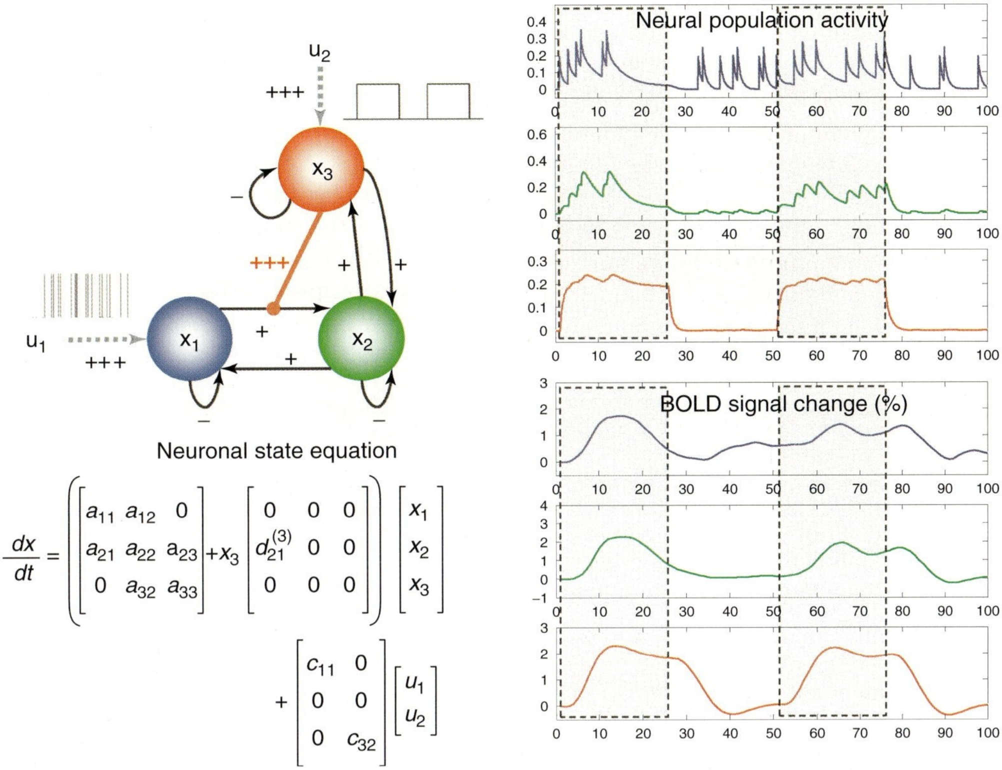

Fig. 9 An example of neuronal and hemodynamic dynamics modeled by nonlinear DCMs. The right-hand panel shows synthetic neuronal and BOLD time-series that were generated using the nonlinear DCM shown in the left-hand panel. In this model, neuronal population activity x_1 (blue) is driven by irregularly spaced random events. Activity in x_2 (green) is driven through a connection from x_1; critically, the strength of this connection depends on activity in a third population, x_3 (red), which receives a connection from x_2 but also receives a direct input from a box-car input. The effect of nonlinear modulation can be seen easily: responses of x_2 to x_1 become negligible when x_3 activity is low. Conversely, x_2 responds vigorously to x_1 inputs when the $x_1 \rightarrow x_2$ connection is gated by x_3 activity. Strengths of connections are indicated by symbols (−: negative; +: weakly positive; +++: strongly positive). Figure reproduced, with permission, from Ref. [49]; © 2008, Elsevier.

during attention to motion, compared to unattended motion. The most likely model was one in which the gain of the V1 → V5 connection depended on the activity in the posterior parietal cortex (PPC), a region on which attention was modeled to exert a direct effect (this could result, for example, from cholinergic inputs from the brainstem). Examining the posterior density of the modulatory parameter in this model indicates that a positive nonlinear gating of the V1 → V5 connection by attention can be assumed to exist with 99.1% confidence (see Fig. 10), conditional on the present model.

The most recent development in DCM for fMRI concerns stochastic models based on random differential equations. Stochastic DCMs have been made possible by advances in the development

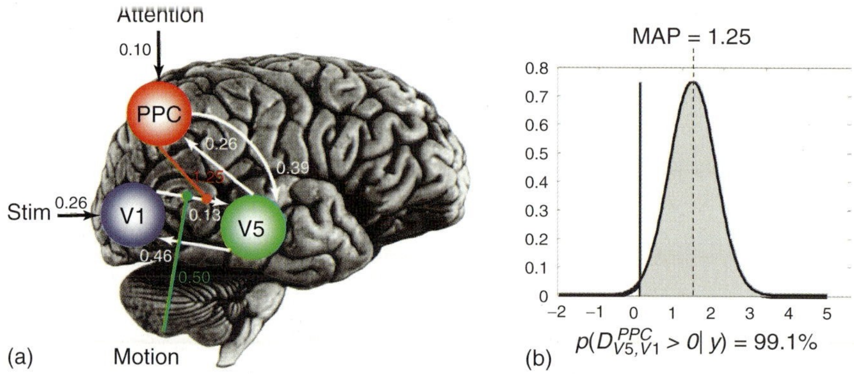

Fig. 10 Application of nonlinear DCM to single subject fMRI data from an attention to motion paradigm by Buchel *et al.* [27]. (a) Maximum a posteriori estimates of all parameters. PPC, posterior parietal cortex; (b) Posterior density of the estimate for the nonlinear modulation parameter for the V1 → V5 connection. It is 99.1% confident (conditional on the present model) that the true parameter value is larger than zero or, in other words, that there is an increase in gain of V5 responses to V1 inputs that is due to PPC-mediated gating (i.e., interactions in the synaptic inputs from V1 and PPC to V5). Figure reproduced, with permission, from Ref. [49]; © 2008, Elsevier.

of computationally efficient Bayesian inversion techniques for triple-estimation of state, parameters and hyperparameters [77–80]. In contrast to deterministic DCMs, stochastic DCMs for fMRI represent neuronal and hemodynamic states s in generalized coordinates of motion: $\tilde{s} = [s, s', s'', \ldots]^T$. This has several fundamental advantages [79]; most importantly, this scheme can accommodate temporal correlations in random fluctuations on the hidden states, which are typical for biological systems (e.g., 1/f spectra). This circumvents the need to make Markovian or Wiener assumptions about the noise. In stochastic DCMs for fMRI the state equation of bilinear deterministic DCMs (Eq. 13) is extended to account for stochastic effects at the neuronal level [81]:

$$\begin{aligned} \frac{dx}{dt} &= (A + \sum_j u_j B^{(j)})x + Cv + \omega^{(x)} \\ v &= u + \omega^{(v)} \end{aligned} \tag{22}$$

Here, the random state fluctuations $\omega^{(x)}$ have an unknown precision and smoothness that are represented by hyperparameters. For example, under Gaussian assumptions about the state noise, the hyperparameters comprise log-precisions π and smoothness σ encoding correlations among the generalized motion of states $\tilde{\omega} = [\omega, \omega', \omega'', \ldots]^T$. Notably, the random differential equation above makes the inputs $u(t) \in \{0, 1\}$ priors on hidden neuronal causes $v(t) \in \mathcal{R}$, because the fluctuations induce uncertainty about how inputs influence neuronal activity. Crucially, when state noise has very low amplitude, Eq. (22) reduces to the ordinary differential equation used in deterministic DCMs for fMRI (c.f. Eq. 13).

Although stochastic DCM for fMRI is still a developing technique, initial work by Li *et al.* [81] and Friston *et al.* [46] has demonstrated the feasibility of inverting stochastic causal models, given typical fMRI data, their robustness and plausible estimates of state noise. Generally, the ability to model endogenous or random fluctuations on hidden neuronal (and other physiological) states greatly expands the range of dynamics that can be modeled with DCM and provides a novel and potentially more plausible perspective on how regional fMRI signals are generated. Moreover, in contrast to deterministic DCMs, stochastic DCMs can also be applied to "resting state" fMRI data where neuronal dynamics is driven by endogenous inputs.

10
DCM for Electrophysiological Data

Electrophysiological data obtained noninvasively, for example, with EEG or MEG, or invasively, for example, with LFP recordings, can be analyzed in numerous ways in both time and frequency domains. Over the past few years, implementations of DCM have been proposed for various electrophysiological data features, including event-related potentials (ERPs; [50]), induced responses [82], spatially distributed responses [83], steady-state responses expressed via cross-spectral densities [84], or phase coupling [85]. The mathematical formulation of the state equations varies considerably across these DCMs. As a concrete example, DCM are discussed for ERPs, as measured with EEG or MEG. This model was the first DCM for electrophysiological responses. Following its introduction by David *et al.* [50], it was refined in subsequent papers [86, 87] and also forms the basis for DCM of steady-state responses expressed in the spectral domain [84].

DCM for ERPs was developed as a biologically plausible model of how event-related responses result from the dynamics in coupled neural ensembles. It assembles a network of coupled cortical sources consisting of neural masses (representing cortical macrocolumns) that are connected according to the connectivity rules of hierarchical sensory systems. These rules distinguish connections with different laminar patterns of origin and termination, that is, forward, backward, and lateral connections [88]. These inter-areal corticocortical connections are excitatory, using glutamate as neurotransmitter, and arise from pyramidal cells.

Each region or source is modeled as a microcircuit in which three neuronal subpopulations are assigned to different layers and linked via intrinsic (intra-areal) connections. Within this model, excitatory interneurons (spiny stellate cells) are found in layer 4 and receive forward connections (from hierarchically lower areas or thalamic nuclei). Excitatory pyramidal cells and inhibitory interneurons are both considered to occupy infra- and supragranular layers and receive backward and lateral inputs.

The structure of intrinsic connections and neural state equations as used in the most recent version of this model [84] are summarized in Fig. 11. To perturb the system and model event-related responses, the network receives inputs u from subcortical (e.g., thalamic) structures to the spiny stellate cells in layer 4. These inputs are modeled as two parameterized components, a gamma density function (representing an event-related burst of input that is delayed and dispersed by subcortical synapses and axonal conduction delay) and a discrete cosine set (representing fluctuations in input over peristimulus time). The influence of this input on each source is controlled by a parameter vector C. Overall, the DCM is specified in terms of the state equations f shown in Fig. 11

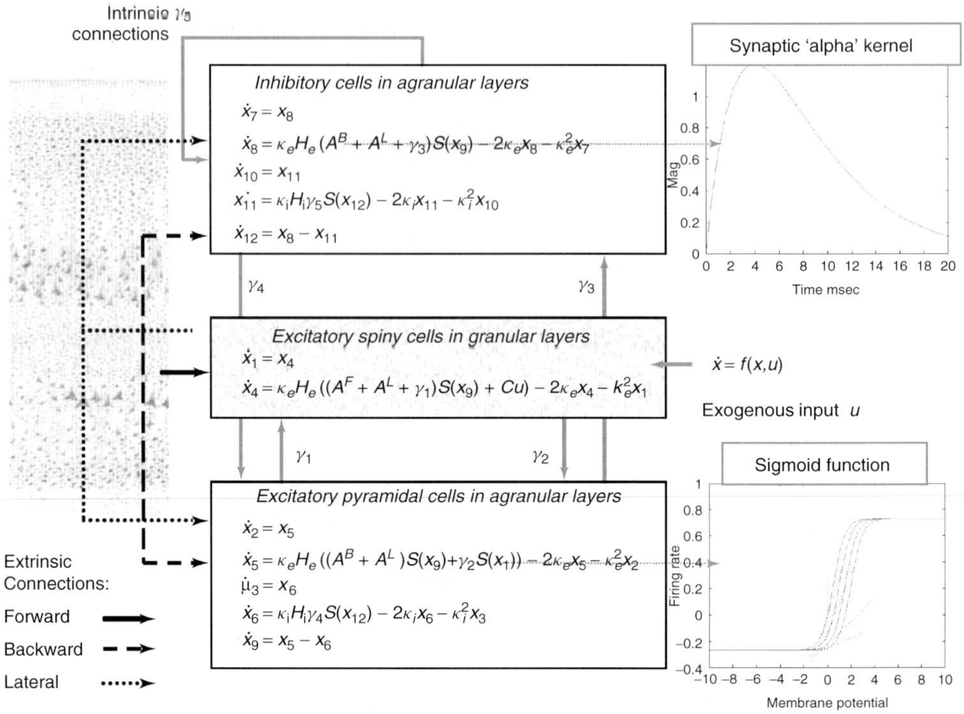

Fig. 11 Schematic summary of the neural model in DCM for event-related responses (ERPs). This schema shows the state equations describing the dynamics of a microcircuit representing an individual region (source). Each region contains three subpopulations (pyramidal, spiny stellate, and inhibitory interneurons) that are linked by intrinsic connections and have been assigned to supragranular, granular, and infragranular cortical layers. Different regions are coupled through extrinsic (long-range) excitatory connections as described in the main text. Adapted, with permission, from Fig. 1 in Ref. [84].

and a linear output equation

$$\begin{aligned} \frac{dx}{dt} &= f(x, u, \theta) \\ y &= Lx_0 + \varepsilon \end{aligned} \tag{23}$$

where x_0 represents the transmembrane potential of pyramidal cells and L is a lead field matrix coupling electrical sources to the EEG channels. Observation noise comprises white and pink noise components [84]. In comparison to DCM for fMRI, the forward model is a simple linearity as opposed to the nonlinear hemodynamic model in DCM for fMRI. In contrast, the state equations of DCM for ERPs are much more complex and realistic (compare Figs 4 and 11). As an example, the state equation for the inhibitory subpopulation is

$$\begin{aligned} \dot{x}_7 &= x_8 \\ \dot{x}_8 &= \frac{H_e}{\tau_e}((A^B + A^L + \gamma_3 I) S(x_0)) \\ &\quad - \frac{2x_8}{\tau_e} - \frac{x_7}{\tau_e^2} \end{aligned} \tag{24}$$

The parameter matrices A^F, A^B, A^L encode forward, backward, and lateral connections respectively. Within each subpopulation, the dynamics of neural states are determined by two operators. The first transforms the average density of presynaptic inputs into the average postsynaptic membrane potential. This is modeled by a linear transformation with excitatory (e) and inhibitory (i) kernels parameterized by $H_{e,i}$ and $\tau_{e,i}$. $H_{e,i}$ control the maximum postsynaptic potential and $\tau_{e,i}$ represent lumped rate constants (i.e., lumped across dendritic spines and the dendritic tree). The second operator S transforms the average potential of each subpopulation into an average firing rate. This is assumed to be instantaneous and is a sigmoid function. Intra-areal interactions among the subpopulations depend on constants $\gamma_{1\ldots4}$ which control the strength of intrinsic connections and reflect the total number of synapses expressed by each subpopulation. In Eq. (24), the top line expresses the rate of change of voltage as a function of current. The second line specifies how current changes as a function of voltage, current, and presynaptic input from extrinsic and intrinsic sources. For simplification, our description here has omitted the fact that in DCM for ERPs all intra- and inter-areal connections have conduction delays. This is implemented by delay differential equations [50].

Due to their relatively fine-grained representation of biophysical mechanisms and synaptic physiology, DCMs for electrophysiological hold great promise for unraveling the mechanisms which govern the dynamics of neural circuits. This includes processes such as synaptic plasticity and neuromodulation which play a key role in pathophysiological processes (see next section). Figure 12 shows an example from a mismatch negativity paradigm in which the presentation of unexpected tones within a series of predictable auditory stimuli induces short-term plasticity in connections between primary auditory cortex and the superior temporal gyrus [89]. One of the most recent developments concerns the construction of DCMs which, in contrast to the neural mass model formulation in Fig. 11, consist of conductance-based state equations modeling the contributions of specific ion channels to postsynaptic currents [90, 91]. Provided they are successfully validated (see next section), this new class of models may enable even more detailed inference of low-level processes from macroscopic brain activity data in the future and could play an important role in using DCMs for diagnostic purposes.

11 Future Applications and Challenges for Inferring Effective Connectivity from Neuroimaging and Electrophysiological Data

Since the first application of effective connectivity analyses to fMRI data [92], there has been an exponential increase in papers dealing with this topic. Similarly, the use of effective connectivity analyses of EEG and MEG data has greatly increased over the past few years. It is expected that several application domains for models of effective connectivity will prove to be particularly exciting and fruitful in the near future.

The first domain is the integration of neurophysiologically interpretable models of effective connectivity with models describing computational aspects of brain function. This is critical for making inferences on short-term synaptic plasticity from neuroimaging data. According to

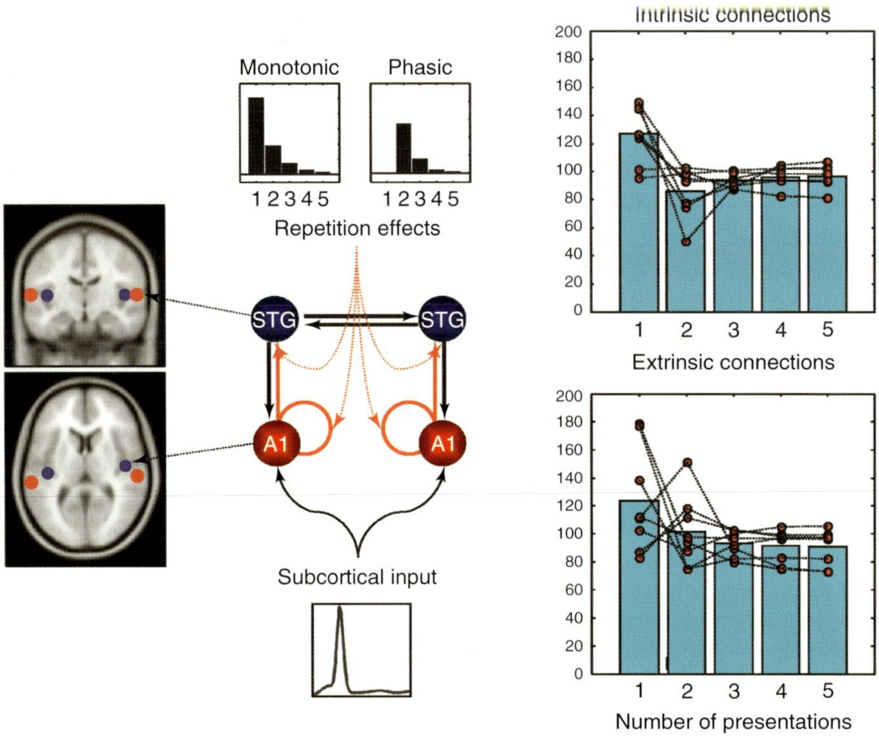

Fig. 12 DCM for event-related responses (ERPs) applied to MEG data from a roving mismatch negativity paradigm [89]. This model comprises four sources (primary auditory cortex, A1, and superior temporal gyrus, STG) in both hemispheres that are connected with forward (red), backward (black), and intrinsic (red) connections. The panel on the left shows the source locations. The DCM receives (parameterized) subcortical input at the A1 sources, which elicit transient perturbations in the remaining sources. Plasticity is modeled by changes in intrinsic and forward connections (red) that are a mixture of monotonic (upper left) and phasic (upper right) repetition-specific effects. Adapted, with permission, from Figs 3 and 4 in Ref. [89].

various current learning theories (e.g., Refs [93–96]), trial-wise prediction errors play a key role in determining the deployment of short-term synaptic plasticity in neuronal circuits implementing the learning process. Simply speaking, the need to reconfigure neuronal circuits during learning should be inversely proportional to how well those neuronal circuits are capable of predicting their inputs, for example, sensory stimuli. Using this notion to obtain indices of synaptic plasticity from fMRI (or other) data requires embedding computational models of learning (such as reinforcement or Bayesian learning models, respectively) into DCMs, such that the modulation of connectivity becomes a function of trial-wise prediction error quantities from the computational models. A first demonstration of how this could be achieved was provided recently by den Ouden *et al.* [75, 76], who studied trial-wise changes in effective connectivity, as an expression of prediction error-induced synaptic plasticity, during both implicit and explicit associative learning.

The second domain concerns model validation. This is crucial for any modeling approach and comprises at least two areas: (i) robustness of the statistical techniques (e.g., robustness of the inversion scheme, assumptions about the noise); and (ii) validity of the biophysical interpretability of model parameters (cf. Ref. [97] for a review of these issues for DCM). While some issues can be investigated using simulation studies [64, 66, 98–100], many questions of validation can only be addressed by means of well-controlled animal studies, where specific effects are induced or specific mechanisms are manipulated and where there is a subsequent need to test whether the model can infer the known effects or manipulations. Investigating the role of physiological confounds for effective connectivity models for fMRI, such as the inter-regional variability of the BOLD response as discussed above, or the influence of neuromodulatory transmitters on vascular responses, also requires animal studies with multimodal measurements. While some studies of this type exist and have provided encouraging and useful results [44, 101], much more systematic investigations of this type are required in the future.

Finally, provided that their validation is successful, a highly important goal is the development of models with clinical utility, for example, as diagnostic tools. Although models of effective connectivity have already been applied to clinical populations [102–106], the critical challenge for the future will be to develop models for which parameter estimates have sufficient sensitivity and specificity to delineate subgroups of patients characterized by different pathophysiological mechanisms. Practical applications of this type will be facilitated by the ongoing development of "generative embedding" techniques that combine DCMs with multivariate classification methods. In brief, this approach uses DCMs as a principled dimensionality reduction device for projecting the high-dimensional data onto a low-dimensional space (i.e., the model parameter estimates), which then serve as features for subsequent classification. This approach has been used successfully for predictions concerning stimuli (trial-by-trial; [107]) or diagnostic status (subject-by-subject) [108]. It is envisaged that this technique will be of considerable practical utility, for example, when dealing with psychiatric spectrum diseases such as schizophrenia, where attempts may be made to establish DCMs that can represent, in conjunction with pharmacological challenges and learning paradigms, specific abnormalities in the regulation of synaptic plasticity by neuromodulatory transmitters such as dopamine or acetylcholine [109].

References

1 Von Bertalanffy, L. (1969) *General System Theory*, George Braziller, New York.

2 Wiener, N. (1948) *Cybernetics*, John Wiley & Sons, Inc., New York.

3 Ashby, W.R. (1956) *An Introduction to Cybernetics*, Chapman & Hall, London.

4 Magee, J.C., Johnston, D. (2005) Plasticity of dendritic function. *Curr. Opin. Neurobiol.*, **15**, 334–342.

5 Bossel, H. (1992) *Modellbildung und Simulation*, Vieweg, Braunschweig.

6 Payne, B.R., Lomber, S.G. (2001) Reconstructing functional systems after lesions of cerebral cortex. *Nat. Rev. Neurosci.*, **2**, 911–919.

7 Bar-Yam, Y. (1997) *Dynamics of Complex Systems*, Addison-Wesley, Reading.

8 Friston, K.J. (1994) Functional and effective connectivity in neuroimaging: a synthesis. *Hum. Brain Mapp.*, **2**, 56–78.

9 Stephan, K.E. (2001) On the role of general system theory for functional neuroimaging. *J. Anat.*, **205**, 443–470.
10 McIntosh, A.R. (2000) Towards a network theory of cognition. *Neural. Netw.*, **13**, 861–870.
11 Friston, K.J. (2002) Beyond phrenology: what can neuroimaging tell us about distributed circuitry? *Annu. Rev. Neurosci.*, **25**, 221–250.
12 Gerstein, G.L., Perkel, D.H. (1969) Simultaneously recorded trains of action potentials: analysis and functional interpretation. *Science*, **164**, 828–830.
13 Horwitz, B., Rumsey, J.M., Donohue, B.C. (1998) Functional connectivity of the angular gyrus in normal reading and dyslexia. *Proc. Natl Acad. Sci. USA*, **95**, 8939–8944.
14 McIntosh, A.R., Rajah, M.N., Lobaugh, N.J. (2003) Functional connectivity of the medial temporal lobe relates to learning and awareness. *J. Neurosci.*, **23**, 6520–6528.
15 Beckmann, C.F., DeLuca, M., Devlin, J.T., Smith, S.M. (2005) Investigations into resting-state connectivity using independent component analysis. *Philos. Trans. R. Soc. Lond., B, Biol. Sci.*, **360**, 1001–1013.
16 Friston, K.J., Büchel, C. (2004) Functional Connectivity: Eigenimages and Multivariate Analyses, in: Frackowiack, R.S. (Ed.) *Human Brain Function*, 2nd edition, Elsevier, New York, pp. 999–1018.
17 McIntosh, A.R., Lobaugh, N.J. (2004) Partial least squares analysis of neuroimaging data: applications and advances. *NeuroImage*, **23** (Suppl. 1), S250–S263.
18 Aertsen, A., Preissl, H. (1991) Dynamics of Activity and Connectivity in Physiological Neuronal Networks, in: Schuster, H.G. (Ed.) *Non Linear Dynamics and Neuronal Networks*, VCH Publishers, New York, pp. 281–302.
19 Honey, C.J., Kotter, R., Breakspear, M., Sporns, O. (2007) Network structure of cerebral cortex shapes functional connectivity on multiple time scales. *Proc. Natl Acad. Sci. USA*, **104**, 10240–10245.
20 Sporns, O., Tononi, G., Edelman, G.M. (2000) Theoretical neuroanatomy: relating anatomical and functional connectivity in graphs and cortical connection matrices. *Cereb. Cortex*, **10**, 127–141.
21 Stephan, K.E., Tittgemeyer, M., Knosche, T.R., Moran, R.J., Friston, K.J. (2009) Tractography-based priors for dynamic causal models. *NeuroImage*, **47**, 1628–1638.
22 Stephan, K.E., Kamper, L., Bozkurt, A., Burns, G.A., Young, M.P., Kötter, R. (2001) Advanced database methodology for the Collation of Connectivity data on the Macaque brain (CoCoMac). *Philos. Trans. R. Soc. Lond., B, Biol. Sci.*, **356**, 1159–1186.
23 Jones, D.K. (2010) Challenges and limitations of quantifying brain connectivity in vivo with diffusion MRI. *Imaging Med.*, **2**, 341–355.
24 Friston, K.J., Harrison, L., Penny, W. (2003) Dynamic causal modelling. *NeuroImage*, **19**, 1273–1302.
25 Friston, K.J., Buechel, C., Fink, G.R., Morris, J., Rolls, E., Dolan, R.J. (1997) Psychophysiological and modulatory interactions in neuroimaging. *NeuroImage*, **6**, 218–229.
26 McIntosh, A.R., Gonzales-Lima, F. (1994) Structural equation modelling and its application to network analysis in functional brain imaging. *Brain Mapp.*, **2**, 2–22.
27 Buchel, C., Friston, K.J. (1997) Modulation of connectivity in visual pathways by attention: cortical interactions evaluated with structural equation modelling and fMRI. *Cereb. Cortex*, **7**, 768–778.
28 Harrison, L., Penny, W.D., Friston, K. (2003) Multivariate autoregressive modeling of fMRI time series. *NeuroImage*, **19**, 1477–1491.
29 Goebel, R., Roebroeck, A., Kim, D.S., Formisano, E. (2003) Investigating directed cortical interactions in time-resolved fMRI data using vector autoregressive modeling and Granger causality mapping. *Magn. Reson. Imaging*, **21**, 1251–1261.
30 Roebroeck, A., Formisano, E., Goebel, R. (2005) Mapping directed influence over the brain using Granger causality and fMRI. *NeuroImage*, **25**, 230–242.
31 Macaluso, E., Frith, C.D., Driver, J. (2000) Modulation of human visual cortex by crossmodal spatial attention. *Science*, **289**, 1206–1208.
32 Stephan, K.E., Marshall, J.C., Friston, K.J., Rowe, J.B., Ritzl, A., Zilles, K., Fink, G.R. (2003) Lateralized cognitive processes and lateralized task control in the human brain. *Science*, **301**, 384–386.

33 McIntosh, A.R., Cooper, R.M., Gonzalez-Lima, F. (1991) Metabolic activation of the rat visual system by patterned light and footshock. *Neurosci. Lett.*, **133**, 311–314.

34 Bullmore, E., Horwitz, B., Honey, G., Brammer, M., Williams, S., Sharma, T. (2000) How good is good enough in path analysis of fMRI data? *NeuroImage*, **11**, 289–301.

35 Penny, W.D., Stephan, K.E., Mechelli, A., Friston, K.J. (2004) Modelling functional integration: a comparison of structural equation and dynamic causal models. *NeuroImage*, **23** (Suppl. 1), S264–S274.

36 Bollen, K.A. (1989) *Structural Equations with Latent Variables*, John Wiley & Sons, Inc., New York.

37 Buchel, C., Coull, J.T., Friston, K.J. (1999) The predictive value of changes in effective connectivity for human learning. *Science*, **283**, 1538–1541.

38 Honey, G.D., Fu, C.H., Kim, J., Brammer, M.J., Croudace, T.J., Suckling, J., Pich, E.M., Williams, S.C., Bullmore, E.T. (2002) Effects of verbal working memory load on corticocortical connectivity modeled by path analysis of functional magnetic resonance imaging data. *NeuroImage*, **17**, 573–582.

39 Rowe, J., Stephan, K.E., Friston, K., Frackowiak, R., Lees, A., Passingham, R. (2002) Attention to action in Parkinson's disease: impaired effective connectivity among frontal cortical regions. *Brain*, **125**, 276–289.

40 Hyvarinen, A., Zhang, K., Shimizu, S., Hoyer, P.O. (2010) Estimation of a structural vector autoregression model using non-Gaussianity. *J. Mach. Learn. Res.*, **11**, 1709–1731.

41 Kim, J., Zhu, W., Chang, L., Bentler, P.M., Ernst, T. (2007) Unified structural equation modeling approach for the analysis of multisubject, multivariate functional MRI data. *Hum. Brain Mapp.*, **28**, 85–93.

42 Penny, W., Roberts, S. (2002) Bayesian multivariate autoregressive models with structured priors. *IEE Proc. Vis. Image Signal Process.*, **149**, 33–41.

43 Granger, C.W.J. (1969) Investigating causal relations by econometric models and cross-spectral methods. *Econometrica*, **37**, 414.

44 David, O., Guillemain, I., Saillet, S., Reyt, S., Deransart, C., Segebarth, C., Depaulis, A. (2008) Identifying neural drivers with functional MRI: an electrophysiological validation. *PLoS Biol.*, **6**, 2683–2697.

45 Roebroeck, A., Formisano, E., Goebel, R. (2011) The identification of networks in the brain using fMRI: model selection, causality and deconvolution. *NeuroImage*, **58**, 296–302.

46 Friston, K.J., Li, B., Daunizeau, J., Stephan, K.E. (2010) Network discovery with DCM. *NeuroImage*, **56**, 1202–1221.

47 David, O. (2011) fMRI connectivity, meaning and empiricism. Comments on: Roebroeck, A., Formisano, E., Goebel, R. The identification of interacting networks in the brain using fMRI: model selection, causality and deconvolution. *NeuroImage*, **58**, 306–309; author reply 310–311.

48 Yamashita, O., Galka, A., Ozaki, T., Biscay, R., Valdes-Sosa, P. (2004) Recursive penalized least squares solution for dynamical inverse problems of EEG generation. *Hum. Brain Mapp.*, **21**, 221–235.

49 Stephan, K.E., Kasper, L., Harrison, L.M., Daunizeau, J., den Ouden, H.E., Breakspear, M., Friston, K.J. (2008) Nonlinear dynamic causal models for fMRI. *NeuroImage*, **42**, 649–662.

50 David, O., Kiebel, S.J., Harrison, L.M., Mattout, J., Kilner, J.M., Friston, K.J. (2006) Dynamic causal modeling of evoked responses in EEG and MEG. *NeuroImage*, **30**, 1255–1272.

51 Buxton, R.B., Wong, E.C., Frank, L.R. (1998) Dynamics of blood flow and oxygenation changes during brain activation: the balloon model. *Magn. Reson. Med.*, **39**, 855–864.

52 Friston, K.J., Mechelli, A., Turner, R., Price, C.J. (2000) Nonlinear responses in fMRI: the Balloon model, Volterra kernels, and other hemodynamics. *NeuroImage*, **12**, 466–477.

53 Stephan, K.E., Weiskopf, N., Drysdale, P.M., Robinson, P.A., Friston, K.J. (2007) Comparing hemodynamic models with DCM. *NeuroImage*, **38**, 387–401.

54 Friston, K.J. (2002) Bayesian estimation of dynamical systems: an application to fMRI. *NeuroImage*, **16**, 513–530.

55 Friston, K., Mattout, J., Trujillo-Barreto, N., Ashburner, J., Penny, W. (2007) Variational

free energy and the Laplace approximation. *NeuroImage*, **34**, 220–234.

56 Stephan, K.E., Penny, W.D., Marshall, J.C., Fink, G.R., Friston, K.J. (2005) Investigating the functional role of callosal connections with dynamic causal models. *Ann. N. Y. Acad. Sci.*, **1064**, 16–36.

57 Mechelli, A., Price, C.J., Noppeney, U., Friston, K.J. (2003) A dynamic causal modeling study on category effects: bottom-up or top-down mediation? *J. Cogn. Neurosci.*, **15**, 925–934.

58 Penny, W.D., Stephan, K.E., Mechelli, A., Friston, K.J. (2004) Comparing dynamic causal models. *NeuroImage*, **22**, 1157–1172.

59 Kasess, C.H., Stephan, K.E., Weissenbacher, A., Pezawas, L., Moser, E., Windischberger, C. (2010) Multi-subject analyses with dynamic causal modeling. *NeuroImage*, **49**, 3065–3074.

60 Stephan, K.E., Marshall, J.C., Penny, W.D., Friston, K.J., Fink, G.R. (2007) Interhemispheric integration of visual processing during task-driven lateralization. *J. Neurosci.*, **27**, 3512–3522.

61 Pitt, M.A., Myung, I.J. (2002) When a good fit can be bad. *Trends Cogn. Sci.*, **6**, 421–425.

62 Mac Kay, D.J.C. (2003) *Information Theory, Inference, and Learning Algorithms*, Cambridge University Press, Cambridge.

63 Neal, R.M., Hinton, G.E. (1998) A View of the EM Algorithm that Justifies Incremental Sparse and Other Variants, in: Jordan, M.I. (Ed.) *Learning in Graphical Models*, Kluwer Academic Publishers, Dordrecht, pp. 355–368.

64 Stephan, K.E., Penny, W.D., Daunizeau, J., Moran, R.J., Friston, K.J. (2009) Bayesian model selection for group studies. *NeuroImage*, **46**, 1004–1017.

65 Ramsey, J.D., Hanson, S.J., Hanson, C., Halchenko, Y.O., Poldrack, R.A., Glymour, C. (2010) Six problems for causal inference from fMRI. *NeuroImage*, **49**, 1545–1558.

66 Stephan, K.E., Penny, W.D., Moran, R.J., den Ouden, H.E., Daunizeau, J., Friston, K.J. (2010) Ten simple rules for dynamic causal modeling. *NeuroImage*, **49**, 3099–3109.

67 Grol, M.J., Majdandzic, J., Stephan, K.E., Verhagen, L., Dijkerman, H.C., Bekkering, H., Verstraten, F.A., Toni, I. (2007) Parieto-frontal connectivity during visually guided grasping. *J. Neurosci.*, **27**, 11877–11887.

68 Kumar, S., Stephan, K.E., Warren, J.D., Friston, K.J., Griffiths, T.D. (2007) Hierarchical processing of auditory objects in humans. *PLoS Comput. Biol.*, **3**, e100.

69 Leff, A.P., Schofield, T.M., Stephan, K.E., Crinion, J.T., Friston, K.J., Price, C.J. (2008) The cortical dynamics of intelligible speech. *J. Neurosci.*, **28**, 13209–13215.

70 Summerfield, C., Koechlin, E. (2008) A neural representation of prior information during perceptual inference. *Neuron*, **59**, 336–347.

71 Penny, W.D., Stephan, K.E., Daunizeau, J., Rosa, M.J., Friston, K.J., Schofield, T.M., Leff, A.P. (2010) Comparing families of dynamic causal models. *PLoS Comput. Biol.*, **6**, e1000709.

72 Kiebel, S.J., David, O., Friston, K.J. (2006) Dynamic causal modelling of evoked responses in EEG/MEG with lead field parameterization. *NeuroImage*, **30**, 1273–1284.

73 Marreiros, A.C., Kiebel, S.J., Friston, K.J. (2008) Dynamic causal modelling for fMRI: a two-state model. *NeuroImage*, **39**, 269–278.

74 Salinas, E., Sejnowski, T.J. (2001) Gain modulation in the central nervous system: where behavior, neurophysiology, and computation meet. *Neuroscientist*, **7**, 430–440.

75 den Ouden, H.E., Friston, K.J., Daw, N.D., McIntosh, A.R., Stephan, K.E. (2009) A dual role for prediction error in associative learning. *Cereb. Cortex*, **19**, 1175–1185.

76 den Ouden, H.E., Daunizeau, J., Roiser, J., Friston, K.J., Stephan, K.E. (2010) Striatal prediction error modulates cortical coupling. *J. Neurosci.*, **30**, 3210–3219.

77 Friston, K.J., Trujillo-Barreto, N., Daunizeau, J. (2008) DEM: a variational treatment of dynamic systems. *NeuroImage*, **41**, 849–885.

78 Friston, K.J. (2008) Variational filtering. *NeuroImage*, **41**, 747–766.

79 Friston, K., Stephan, K.E., Li, B., Daunizeau, J. (2010) Generalised filtering. *Math. Probl. Eng.*, **621**, 670.

80 Daunizeau, J., Friston, K.J., Kiebel, S.J. (2009) Variational Bayesian identification and prediction of stochastic nonlinear

dynamic causal models. *Physica D*, **238**, 2089–2118.

81 Li, B., Daunizeau, J., Stephan, K.E., Penny W., Hu, D., Friston, K. (2011) Generalised filtering and stochastic DCM for fMRI. *NeuroImage*, **58**, 442–457.

82 Chen, C.C., Kiebel, S.J., Friston, K.J. (2008) Dynamic causal modelling of induced responses. *NeuroImage*, **41**, 1293–1312.

83 Daunizeau, J., Kiebel, S.J., Friston, K.J. (2009) Dynamic causal modelling of distributed electromagnetic responses. *NeuroImage*, **47**, 590–601.

84 Moran, R.J., Stephan, K.E., Seidenbecher, T., Pape, H.C., Dolan, R.J., Friston, K.J. (2009) Dynamic causal models of steady-state responses. *NeuroImage*, **44**, 796–811.

85 Penny, W.D., Litvak, V., Fuentemilla, L., Duzel, E., Friston, K. (2009) Dynamic Causal Models for phase coupling. *J. Neurosci. Methods*, **183**, 19–30.

86 Kiebel, S.J., Garrido, M.I., Friston, K.J. (2007) Dynamic causal modelling of evoked responses: the role of intrinsic connections. *NeuroImage*, **36**, 332–345.

87 Moran, R.J., Stephan, K.E., Kiebel, S.J., Rombach, N., O'Connor, W.T., Murphy, K.J., Reilly, R.B., Friston, K.J. (2008) Bayesian estimation of synaptic physiology from the spectral responses of neural masses. *NeuroImage*, **42**, 272–284.

88 Felleman, D.J., Van Essen, D.C. (1991) Distributed hierarchical processing in the primate cerebral cortex. *Cereb. Cortex*, **1**, 1–47.

89 Garrido, M.I., Kilner, J.M., Kiebel, S.J., Stephan, K.E., Baldeweg, T., Friston, K.J. (2009) Repetition suppression and plasticity in the human brain. *NeuroImage*, **48**, 269–279.

90 Marreiros, A.C., Kiebel, S.J., Friston, K.J. (2010) A dynamic causal model study of neuronal population dynamics. *NeuroImage*, **51**, 91–101.

91 Moran, R.J., Stephan, K.E., Dolan, R.J., Friston, K.J. (2011) Consistent spectral predictors for dynamic causal models of steady-state responses. *NeuroImage*, **55**, 1694–1708.

92 Friston, K.J., Ungerleider, L.G., Jezzard, P., Turner, R. (1995) Characterizing modulatory interactions between areas V1 and V2 in human cortex: a new treatment of functional MRI data. *Hum. Brain Mapp.*, **2**, 211–224.

93 Montague, P.R., Dayan, P., Sejnowski, T.J. (1996) A framework for mesencephalic dopamine systems based on predictive Hebbian learning. *J. Neurosci.*, **16**, 1936–1947.

94 Rao, R.P., Ballard, D.H. (1999) Predictive coding in the visual cortex: a functional interpretation of some extra-classical receptive-field effects. *Nat. Neurosci.*, **2**, 79–87.

95 Schultz, W., Dickinson, A. (2000) Neuronal coding of prediction errors. *Annu. Rev. Neurosci.*, **23**, 473–500.

96 Friston, K. (2009) The free-energy principle: a rough guide to the brain? *Trends Cogn. Sci.*, **13**, 293–301.

97 Daunizeau, J., David, O., Stephan, K.E. (2010) Dynamic causal modelling: a critical review of the biophysical and statistical foundations. *NeuroImage*, **58**, 312–322.

98 Lee, L., Friston, K., Horwitz, B. (2006) Large-scale neural models and dynamic causal modelling. *NeuroImage*, **30**, 1243–1254.

99 Marrelec, G., Kim, J., Doyon, J., Horwitz, B. (2009) Large-scale neural model validation of partial correlation analysis for effective connectivity investigation in functional MRI. *Hum. Brain Mapp.*, **30**, 941–950.

100 Smith, S.M., Miller, K.L., Salimi-Khorshidi, G., Webster, M., Beckmann, C.F., Nichols, T.E., Ramsey, J.D., Woolrich, M.W. (2011) Network modelling methods for FMRI. *NeuroImage*, **54**, 875–891.

101 Reyt, S., Picq, C., Sinniger, V., Clarencon, D., Bonaz, B., David, O. (2010) Dynamic Causal Modelling and physiological confounds: a functional MRI study of vagus nerve stimulation. *NeuroImage*, **52**, 1456–1464.

102 Allen, P., Stephan, K.E., Mechelli, A., Day, F., Ward, N., Dalton, J., Williams, S.C., McGuire, P. (2010) Cingulate activity and fronto-temporal connectivity in people with prodromal signs of psychosis. *NeuroImage*, **49**, 947–955.

103 Grefkes, C., Nowak, D.A., Eickhoff, S.B., Dafotakis, M., Kust, J., Karbe, H., Fink, G.R. (2008) Cortical connectivity after subcortical stroke assessed with functional magnetic resonance imaging. *Ann. Neurol.*, **63**, 236–246.

104 Rowe, J.B., Hughes, L.E., Barker, R.A., Owen, A.M. (2010) Dynamic causal modelling of effective connectivity from fMRI: are results reproducible and sensitive to Parkinson's disease and its treatment? *NeuroImage*, **52**, 1015–1026.

105 Seminowicz, D.A., Mayberg, H.S., McIntosh, A.R., Goldapple, K., Kennedy, S., Segal, Z., Rafi-Tari, S. (2004) Limbic-frontal circuitry in major depression: a path modeling meta-analysis. *NeuroImage*, **22**, 409–418.

106 Sonty, S.P., Mesulam, M.M., Weintraub, S., Johnson, N.A., Parrish, T.B., Gitelman, D.R. (2007) Altered effective connectivity within the language network in primary progressive aphasia. *J. Neurosci.*, **27**, 1334–1345.

107 Brodersen, K.H., Haiss, F., Ong, C.S., Jung, F., Tittgemeyer, M., Buhmann, J.M., Weber, B., Stephan, K.E. (2011) Model-based feature construction for multivariate decoding. *NeuroImage*, **56**, 601–615.

108 Brodersen, K.H., Schofield, T.M., Leff, A.P., Ong, C.S., Lomakina, E.I., Buhmann, J.M., Stephan, K.E. (2011) Generative embedding for model-based classification of fMRI data. *PLoS Comput. Biol.*, **7**, e1002079.

109 Stephan, K.E., Friston, K.J., Frith, C.D. (2009) Dysconnection in schizophrenia: from abnormal synaptic plasticity to failures of self-monitoring. *Schizophr. Bull.*, **35**, 509–527.

17
Systems Biology of the Liver

Ian David Lockhart Bogle, Rajiv Jalan, Elizabeth Shephard, Robert Seymour, Antony Finkelstein, Thomas Sumner, and Anne Warner
University College London, CoMPLEX – Centre for Mathematics and Physics in the Life Sciences and Experimental Biology, Gower Street, London WC1E 6BT, UK

Systems Biology: Advances in Molecular Biology and Medicine, First Edition. Edited by Robert A. Meyers.

Keywords

Modeling
Developing sets of mathematical equations that can predict the behavior of physiological functions with predictions obtained through a computational implementation of those equations.

Glucose homeostasis
The metabolic processes that regulate the level of glucose in the bloodstream. This involves the conversion of glucose into glycogen by the action of glucagon and the conversion of glycogen back into glucose by the action of insulin.

Hepatocytes
The cells that constitute the main parenchymal cells of the liver.

Hepatic vein
The vein that drains blood from the liver into the systemic circulation.

Hepatic artery
The artery that supplies oxygenated blood to the liver.

Portal vein
The vein that drains blood from the spleen, stomach, and intestines, and then supplies the same blood to the liver.

Glycogenolysis
The processes involved in the breakdown of glycogen to glucose.

Gluconeogenesis
The synthesis of glucose from amino acids and other organic molecules.

Glycogenesis
The formation of glycogen from glucose.

Nonalcoholic fatty liver disease (NAFLD)
A disease of the liver where fat is deposited in the liver due to metabolic disturbances; the condition is not due to alcohol consumption.

Model management
The process of developing, connecting, executing, and archiving a set of computational models.

Composite model
A model of a system consisting of a complex set of phenomena composed of a number of communicating computational models.

This chapter explores the potential and progress of using systems biology to provide an understanding of functioning of the liver, the main organ of metabolism regulation in the body. The range of studies for modeling the metabolic, genetic, and physiological systems that affect the main functions of the liver, although currently at an early stage, are discussed. In particular, the modeling of a key liver function – glucose homeostasis – is used to exemplify the progress and potential of systems biology in the liver. Many models used to predict liver behavior are capable of predicting quantitative behavior to within an order of magnitude accuracy; however, for use in patient analysis a greater fidelity of the models is required to provide an even greater level of accuracy. Many challenges remain to understand and predict other complex interacting functions of the liver, beyond glucose homeostasis. Systems biology provides an opportunity to unravel the complex interacting phenomena involved in liver behavior that otherwise would be impossible. A key opportunity here is to employ systems biology for clinical purposes, with individualized medicines, the aim being to develop an *in silico* model such that the potential hepatoxicity of drugs could be predicted in individuals, allowing the development of safer treatments.

1 Background

The analysis of liver function presents a major challenge, due to the inherent complexity of liver biology and its connected systems. In this chapter, the progress that has been made towards the systems biology of the liver is documented, including the construction of integrated system models of liver function through an iterative cycle of model building, data generation, and experimental validation. Typically, integrated system models will aid in a better understanding of normal function and life-long health and well-being and, in the longer term, to the analysis of disease and the development of therapeutic strategies. Yet, despite much progress having been made, the ongoing studies – and, indeed, the systems biology approach that they exemplify – are in their infancy.

1.1 Liver Function and Structure

The liver is one of the most complex organs of the body, the complexity arising because the liver fulfills a number of key functions in the physiology of an organism. For example, the liver stores molecules for release on demand by metabolism, it synthesizes proteins from amino acid components absorbed by the gut, and it also plays a key role in the detoxification of the blood. In addition, the liver generates bile salts which are stored in the gallbladder and later released into the gut to assist with digestive processes.

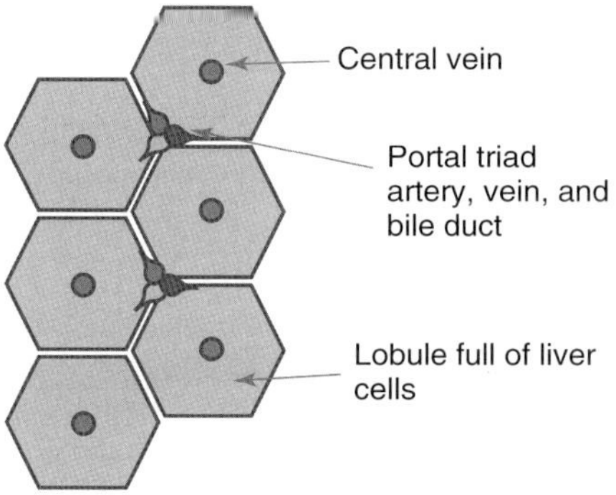

Fig. 1 Schematic showing the structural organization of the liver.

The liver has a lobular structure, with each lobule being divided into acini arranged around a central hepatic vein. Ingested foods absorbed by the gut are transported by the blood into the hepatic portal vein, in which they are then carried directly into the liver (see Fig. 1). The blood enters the liver via the hepatic artery and the portal vein; blood from each of these vessels then flows around each cord of *hepatocytes* – the cells in the liver which carry out the main functions. The cords of hepatocyte cells are usually referred to as the *liver plates*.

The liver is a central element of the chemical system in the body, as it undertakes several functions of carbohydrate metabolism, including: the synthesis of glucose from amino acids and other organic molecules (*gluconeogenesis*); the breakdown of glycogen into glucose (*glycogenolysis*); the formation of glycogen from glucose (*glycogenesis*); and amino acid synthesis. It is through these processes that the liver plays a key role in the control of glucose homeostasis – the control of glucose levels in the bloodstream.

The liver also plays a role in lipid metabolism, including the synthesis of cholesterol and the production of triglycerides or fats (*lipogenesis*), and of a range of coagulation factors. Typically, the liver breaks down insulin, hemoglobin, ammonia, toxic substances, and most medicinal products (via drug metabolism) such that, ultimately, these materials are excreted either into the bile or into the urine. The liver also converts a range of other proteins for storage, and produces certain proteins for immunological effects and for the regulation of blood pressure.

1.2 Liver Disease

Diseases of the liver occur as a result of malfunction of the above-described processes:

- *Diabetes* is the result of an inability of the pancreas to produce insulin, thereby disabling glucose regulation in which the liver plays an important part.
- *Nonalcoholic fatty liver disease* (NAFLD) is manifested by a build-up of fatty deposits in the liver that is not related to alcohol intake, but rather results from insulin resistance when insulin becomes less effective at regulating the glucose levels.
- Some *drugs* may irreversibly damage the hepatocytes, such that they become unable to fulfill some or all of the functions described above.
- *Hepatitis* is caused by a viral infection that interferes with the metabolism of hepatocytes in various ways.
- *Cirrhosis* is a consequence of chronic liver disease, causing scarring of the liver tissue that is formed during regeneration of the liver cells.
- *Cancer* is also an important cause of the loss of liver function.

In order to explore the behavior of key functions involving the liver, it is necessary also to consider interactions with other organs, though this represents a hugely complex task. Yet, such complexity can be managed by using models which take account of the interactions between the constituent parts – an approach which is at the heart of systems biology. In this respect, the regulation of glucose is an excellent example of a way in which systems biology can assist in the understanding and treatment of liver disease, since the process involves key chemical and metabolic processes, interactions with other organs, and is directly relevant to common liver diseases, such as diabetes.

1.3 Liver Systems Biology

Such difficult integrated problems can be tackled through computational model building, which allows the results obtained from small-scale experiments to be integrated to generate a computational model of the functioning organ. In the past, this approach has been applied successfully to the vertebrate heart, where the Noble/Hunter model [1] has provided a sufficiently faithful representation of cardiac function for the American Food and Drug Administration to have authorized it as an early "test bed" for drug development. In the same way, a computational model of the liver might not only improve the understanding of liver biology but, in the longer term, also help with the analysis, understanding and, therefore, the treatment of liver disease. In fact, together with models of associated organs, it may become possible to analyze liver behavior within the context of a functioning animal.

For the experimental analysis of system behavior, it is first essential to separate liver function into manageable components, so that methods appropriate to each component can be applied. A model allows these components then to be reintegrated, while retaining the ability to manipulate individual elements of the system and to observe the consequences on overall liver function. Today, analysis is no longer limited by what is experimentally feasible – it is possible to obtain estimates of parameters that are difficult to measure, while the consequences of alterations in specific functions can be observed in a model and then related back to liver behavior. In turn, this may lead to novel approaches for analyzing the individual functions of the liver.

System models of the liver would be useful in a number of ways. Initially, they would help in exploring and understanding any system-wide toxicity effects, they would be valuable for drug discovery and for optimizing drug delivery and, eventually, they would be useful in the diagnosis and treatment of liver disease.

There is, of course, a long way to go before comprehensive models are available for these purposes. Nonetheless, the aim of this chapter is to discuss the current progress and future challenges relating to the systems biology of the liver. For this purpose, the biochemistry and physiology of liver function will be discussed in detail, with particular reference to disease states. Various attempts to model the whole liver system, with relevant discussions, are then outlined, in addition to aspects of modeling liver physiology as a key building block for building a system model. A description is then provided of a system model of liver function, notably in relation to the regulation of glucose (a key function of the liver), which has been assembled using components from a range of sources. At least potentially, this approach may become widespread for developing a wide range of physiological system models. Finally, an overview is provided of future studies targeted at the creation of comprehensive systems biology models of the liver and associated systems. In this case, the aim is not only to improve the present understanding of liver function, but also to identify clinical applications, with the main goal of individualized medicines.

2
Liver Physiology, Biochemistry, and Disease

Today, humankind is exposed constantly to "foreign" chemicals, including dietary components, environmental pollutants, and therapeutic drugs. Following their oral ingestion, these chemicals are transported from the intestine to the liver via the bloodstream. Subsequently, the survival of the organism depends on the efficacy of the liver's battery of detoxifying enzymes to remove these chemicals (which typically are small hydrophobic molecules that can easily enter the cells) from the body, thus preventing their build-up to potentially toxic concentrations.

In time, the liver has evolved an elegant system for converting these chemicals into more hydrophilic molecules that can then exit the body via the bile, faeces, or urine. The group of enzymes that detoxify the body by helping to remove these chemicals are collectively known as "drug-metabolizing enzymes" (DMEs), this name being coined because many drugs are substrates for these enzymes.

Drug metabolism (detoxification) involves two phases: Phase 1, and Phase 2. The key Phase 1 enzymes include the cytochrome P450 monooxygenases (CYPs) and the flavin-containing monooxygenases (FMOs), both of which incorporate an atom of molecular oxygen into the substrate. This incorporation of oxygen prepares the foreign chemical for further metabolic treatment by the Phase 2 enzymes, which then add a chemical group – such as a sulfate, small peptide, or sugar – so that the foreign chemical is made more water-soluble and thus is more easily excreted from the body. Because of the vast array of signaling systems, and the interaction of a large number of interacting metabolic pathways, drug development is hugely confounded by hepatotoxicity that can be either predictable or unpredictable.

According to the British Society for the Study of Liver Disease, hepatic disease is the tenth most common cause of death worldwide (www.basl.org.uk). Although acute liver failure is a rare condition, chronic liver disease and cirrhosis are much more common. The most important cause of acute liver failure in the developed world (including the UK and USA) is paracetamol (acetaminophen) overdose, whereas in the developing world and in many parts of Europe viral hepatitis is the main cause. Worldwide, the main causes of cirrhosis are alcohol abuse, infection with hepatitis viruses (types B and C), although the most recent epidemic is due to fatty liver.

Approximately 10% of people in Western Europe and the USA have NAFLD, and today the incidence of the condition is increasing rapidly. Typically, NAFLD will progress to cirrhosis, culminating in liver failure and hepatocellular carcinoma. An underlying NAFLD is also associated with a reduced quality of life on both psychological and physical domains, and predisposes patients to exacerbated liver injury following stresses such as ischemia, infection, and excessive drug intake. The pathogenesis of NAFLD and its consequences is complex, as it involves widespread disturbances in insulin–glucagon-mediated glucose homeostasis, and engages *interorgan*, *intercellular*, and *intracellular* factors. Those organs that are critically involved are the liver, pancreas, muscle, and fat, and it is likely that in order to develop an understanding of NAFLD the *interorgan* trafficking of insulin, glucagon, and glucose must first be addressed. Widespread alterations in the expression of various *intercellular* signaling molecules (such as connexins) are regulated by the hormones insulin and glucagon, and impact on the ability of the cells to communicate effectively, thereby producing an environment for the adverse effects of xenobiotics and progression to cirrhosis and hepatocellular cancer [2]. Today, NAFLD is commonly observed in association with diabetes and obesity, both of which are directly associated with an altered insulin sensitivity that may impact on *intracellular* insulin–glucagon signaling pathways.

Given the complexity of the pathophysiological mechanisms involved in the development of NAFLD and its downstream effects, current biological approaches have proven to be largely ineffective for relating events at the interorgan, intercellular, and intercellular levels. In this respect, the systems biology approach is particularly exciting, as it may help to reveal the pathophysiological mechanisms involved by focusing attention on these seemingly disparate mechanisms and interactions. Since the changes in glucose homeostasis that occur in conditions such as diabetes, obesity, and the metabolic syndrome account for the majority of cases of NAFLD, the development of an in-silico model glucose metabolism might have the potential to define unexpected therapeutic targets, drug interactions, and nutritional effects, and hence allow individually tailored therapies. Moreover, as the main mechanism underlying the pathogenesis of NAFLD in these situations is insulin resistance, a systems biology approach may provide novel solutions for the better management of diabetes and other insulin-resistance-related syndromes.

Having discussed in detail the benefits of a systems biology approach to the problems associated with NAFLD, it is clear that these arguments apply equally to the other liver diseases that have been detailed above and which cause complex metabolic effects in systems that involve the liver and associated organs. Moreover, as many such systems involve glucose homeostasis, with its central role in liver function, this might serve as an ideal starting point to develop systems biology approaches for the liver.

3
Modeling of Genetic and Metabolic Phenomena in the Liver

Although, currently, systems biology approaches to the exploration of liver behavior are at an early stage, attempts have been made previously at the comprehensive genomic analysis and modeling of important sections of the hepatic system. Unfortunately, however, this has often been conducted in association with other organs since, for certain functions, it is impossible to consider the liver in isolation. Additional efforts have involved data gathering to explore system-wide behavior, while others have been targeted at particular disease states. The current progress of the subject will be reviewed in the following subsections.

Efforts to exploit proteomic and metabolomic information have been the most common to date. For example, Yan *et al.* [3] employed a system-based proteomic analysis to explore interferon response in human liver cells, while Andersson *et al.* [4] explored the understanding of elevated serum alanine transaminase levels in a clinical trial with ximelagtran. Subsequently, Achur *et al.* [5] explored the effects of circulating cytokines as biomarkers of alcohol abuse and alcoholism, while Wu *et al.* [6] used a modeling approach to explore the dynamics of insulin receptor substrate–protein kinase R (IRS–PKR) signaling in liver cells.

As other chapters have testified, the genomics revolution – along with modeling capabilities – has imparted considerable impetus to systems biology. Recently, a group from China published the details of a comprehensive database of the human liver proteome, Liverbase [7] (available at http://liverbase.hupo.org.cn), which contains 6788 proteins and 11 205 significantly expressed genes, and allows systematic searches to be conducted within specific functional or metabolic pathways. Subsequently, a large German research project, Hepatosys, has been working towards developing a comprehensive validated map of systems transduction pathways in hepatocytes [8]. Clearly, these projects will add significantly to the efforts in using systems biology to explore liver behavior and disease.

Because of the complexity of any problem relating to human disease, modeling efforts have sought to focus attention on the factors associated with specific behaviors or disease states. Ierapetritou *et al.* [9] explored the behavior of xenobiotic metabolism in the liver, while Li and Chan [10] employed a systems biology approach to explore metabolic pathways for identifying toxicity pathways induced by free fatty acids and tumor necrosis factor (TNF)-α in human blastoma cells, by integrating gene expression, metabolite profile, toxicity measurements and a priori knowledge to identify gene targets for modulating liver toxicity. Subsequently, Eveillard *et al.* [11] explored the metabolic effects of di-(2-ethylhexyl)-phthalate by comparing *in silico* results with data from adult mice exposed to the compound. This analysis showed which pathways were disrupted, and led to the conclusion that other phthalates could also produce adverse impacts. Schnackenberg *et al.* [12] explored the acute effects of valproic acid.

Although current toxicity studies are likely to be of increasing importance, they require exploration beyond protein and metabolism behavior, since distributed effects exist which may vary across the liver plate and lobule [13]. The majority of current studies assume a homogeneity of hepatocytes ([14–17]).

Although, today, hepatologists think "... in terms of hypothesis-driven clinical research using Oslerian principles and now a systems biology approach" [18], there is an increasing recognition that more should be done to support this approach. In particular, it should be possible to easily explore any system-wide effects of networks and perturbations to individual genes, for example. Yet, whilst the present review highlights the range of investigations that have been conducted, much remains to be done to bring these studies together to enable a true system-wide exploration of the complex genetic and metabolic aspects of liver function.

4 Modeling Liver Physiology

The investigations described so far have explored the genetic and metabolic effects in the liver, and their response to various stimuli. As noted above, the liver consists of a number of interacting cell and tissue types, whereby the hepatocytes lie in plates in the liver lobule and communicate by activating connexin proteins through gap junctions which are difficult to explore experimentally.

Any experimental analysis of liver function is complicated by the range of physiological processes that need to be analyzed, typically on an individual basis. Thus, whilst some functions are best analyzed in cultured hepatocytes, others lend themselves to the use of liver tissue slices; the latter approach is especially true for analyses carried out to assess the variation in functions across the liver plate. The integration of a range of results obtained using different techniques can benefit from a formal model-based framework, however, ultimately to provide a scaffold on which a complete picture of the physiology of the liver can be built.

The simplest form of linked model is that of two cells separated by a gap junction. The gap junctions are constructed from so-called connexin proteins; these are membrane-embedded proteins that contain both intracellular and extracellular loops. In this case, six connexin molecules are linked together to form a connexon; the connexons then link across the intercellular space to form gap junctions (for excellent descriptions of gap junctions, see Refs [19] and [20].

Gap junctions range in size from small plaques containing few connexons, to large plaques that are several microns in diameter and incorporate several hundred connexin channels. The gap junctions found in the adult liver tend to be large, so that much of the cell membrane surrounding each hepatocyte forms a gap junction. In the liver, the gap junctions are constructed from connexin 32, which shows little sensitivity to any variation in physiological conditions. Typically, connexin 32 gap junctions are neither pH- nor calcium-sensitive, and show very little change in response to intracellular variations in second messenger molecules such as cyclic AMP and cyclic GMP. When the system is well coupled, gap junctional communication is affected only minimally by the properties of the cells that form the coupled system. However, if the level of cell–cell coupling falls, then the intercellular communication pathway will begin to have a significant effect on the properties of the system (this issue is discussed in depth in Ref. [21]).

The behavior of gap junctions between hepatocytes is complex, the critical factor being the way in which different connexins interact with each other, and whether the junctions contain more than one connexin. The likelihood is that, in diseased hepatocytes, there will be a potential for nonlinear behavior because connexins 43, 32, and 26 are all expressed. Potential complexities are most likely to arise in multiple connexin-expressing diseased livers, a situation which is best tackled by using a systems biology approach, given the difficulties of measurement and the complexity involved.

Various examples exist of models being used to explore the physiological effects of liver injury. For example, Subramanian *et al.* [22] developed a set of ordinary differential equation models in Matlab to explore drug-induced liver injury. In this case, the model embraced energy metabolism, glutathione metabolism, fatty acid metabolism, bile salt metabolism and transport, canalicular contraction, and calcium release from the endoplasmic reticulum (ER). It also created predictions of the specific output effects of exposure to four drugs, namely buthionine sulfoximine, daphnetoxin, perhexilline, and amiodarone. Lagoa *et al.* [23] later explored the role of initial trauma in response to injury and hemorrhage.

Several impediments have been encountered in progressing towards the development of systems approaches to the analysis of liver function:

- A huge body of data have been reported concerning liver function, much of which has not been quantified.
- Where such data have been quantified, the details of a wide variety of models have been reported on individual phenomena, and on some parts of the liver system. Unfortunately, the combination of these details with others is often impossible without conducting a major rewrite of the model.
- Much data are available on basic biological behavior (though rarely as raw data). There is also a huge untapped resource of patient-related information that is often maintained in clinical databases.
- In many cases, modeling efforts have not been focused on defining a purpose for the model. Whilst it is useful to confirm biological hypotheses for disease-related problems, the models may include much that is irrelevant, leading to confusion, and to incorrect behaviors and predictions.

In the past, there has been a need to develop systems which allow a greater degree of flexibility in the modeling process, both with regards to the sources of models and the ability to replace parts that have been superseded or have proved to be inadequate. Recently, a system of this type has been developed at University College, London, and used to model a liver system phenomenon. This is described briefly in the next section, prior to details of the glucose homeostasis model itself being presented.

5 A Composite Systems Biology Model to Predict Glucose Homeostasis

Glucose homeostasis is a central chemical regulation that is critical for an organism to maintain a healthy status; it is, therefore, a useful starting point to develop a systems biology approach to liver function. Both, Liu and Tang [16] and Liu *et al.* [15] have modeled glucose homeostasis, focusing specifically on the dynamics of the insulin signaling pathway, along with the action and kinetics of key enzymes involved in the process of glucose mobilization by glycogen phosphorylase and glycogen synthase in the liver. Together, these authors constructed a model that consisted of approximately 30 differential equations with 70 parameters that were taken from the literature and obtained from fitting to high-level data (also available from the literature). This model performed well under certain circumstances and, notably, provided a qualitative match to experimentally acquired insulin production data.

However, because of the complexity of biological systems it is difficult to obtain all of the information and models from one source. One approach which circumvented this problem was to use a system that allowed models from multiple sources to be linked and solved together. Such a

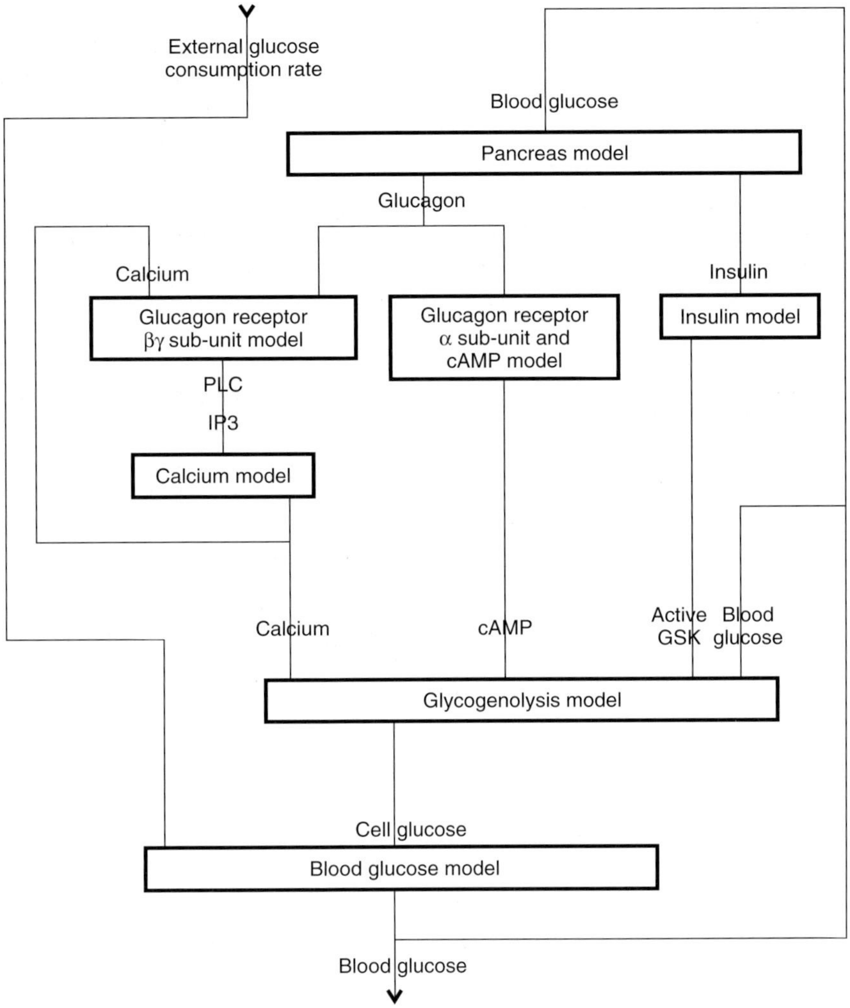

Fig. 2 Structure of the glucose homeostasis model.

model management system was described by Hetherington *et al.* [14], Margoninski *et al.* [24], and Saffrey *et al.* [25], and has been used to develop a systems biology model of glucose homeostasis, one of the key functions of the liver and associated organs. This system enabled the incremental development of a single, gradually expanding model, with each refinement being made only when the existing elements had been well validated. The system made it possible to integrate models of different mathematical forms, independently of the solution technique that might be employed for individual components. This approach is described here to demonstrate its value, and is exemplified by the development of a "composite" model of glucose homeostasis.

The composite model brings together a series of sub-models, some taken from or based on established models, and others developed *ab initio*, of aspects of the physiology of glucose homeostasis. The current choice of component models was determined by a judgment of the necessary physiological and biochemical phenomena required to describe the basic features of the glucose homeostasis system, and the key elements of its control. Some sub-models represent the physiology at a greater level of detail than others.

5.1 Component Models

Seven component models describe the major elements of glucose homeostasis with an overarching feedback cycle: *liver* → *blood* → *pancreas* → *liver*. This cycle is moderated via the blood by the dietary intake of glucose and the active consumption of glucose in the organism's activity. The model aims to mimic quantitatively the behavior of glucose regulation following external stimuli to the liver in both a healthy state and with diabetes (which is a common cause of NAFLD). The interchange between glucose and glycogen is influenced by four factors: glucagon via calcium; glucagon via cAMP; insulin; and directly by cellular glucose levels. These are the key factors for a central model necessary for the computation to represent competing enzyme phosphorylation equilibria. These nested feedback structures, and the inputs and outputs of each component model, are represented in Fig. 2.

Five of the seven models represent aspects of hepatocyte physiology:

- Activation of G-proteins by glucagon (glucagon receptor model).
- Cytoplasmic calcium concentration changes (calcium model).
- The cyclic AMP pathway (cAMP model).
- Inactivation of glycogen synthase kinase 3 (GSK3) by insulin (insulin model).
- The synthesis and breakdown of glycogen by the liver (glycogenolysis model).

The models treat the whole body of hepatocytes across the liver as a single cellular component, and the parameters reflect this. The two remaining components are:

- A simple model of the production of the hormones glucagon and insulin by the pancreas (pancreas model).
- A model mediating between pancreas and liver representing blood transport (blood transport model).

Each component model is described briefly below, and each was validated separately, allowing independent verification of the component's behavior in a context which did not depend on its embedding within the larger system. The detailed mathematical descriptions of each sub-model and its parameters have been reported in Ref. [14], and the validation of each model in Ref. [26]. A summary of the models is provided in Table 1.

5.2 Glucagon Receptor Model (A)

This module represents the activation of a G-protein-coupled receptor by a hormone stimulus. The G-protein-coupled receptor is an important element of many signaling pathways, and this module can therefore be used as a component of several signaling systems. Receptor activation of the G-protein causes GDP → GTP exchange, which results in conformational changes in the G-protein and separation

Tab. 1 Details of component models of the glucose homeostasis composite model.

Model name	*Biology*	*Source*	*Computational language*
Glucagon receptor model (A)	Glucagon receptor βγ sub-units, desensitization, PLC	[27, 28]	Mathematica
Calcium model (B)	IP3, calcium oscillations	[29]	XPPAUT
cAMP model (C)	Glucagon receptor α sub-unit, cAMP, PKA	*Ab initio*	Mathematica
Insulin model (D)	Insulin receptor, PIP, GSK	*Ab initio*	XPPAUT
Blood model (E)	Metabolic demand, trans-membrane glucose transport	*Ab initio*	XPPAUT
Glycogenolysis model (F)	Control of glycogen phosphorylase and synthase	[30–33]	XPPAUT
Pancreas model (G)	Glucagon and insulin production	*Ab initio*	Mathematica

GSK, glycogen synthase kinase; IP_3, Inositol triphosphate; PIP, Phosphatidylinositol-(3,4,5)-triphosphate; PLC, Phospholipase C.

of the Gα subunit from Gβγ subunits. Both, GTP-bound Gα and Gβγ are able to regulate effectors [34, 35].

Activation of the α-subunit is considered under the cAMP model. The glucagon receptor model describes activation of the βγ sub-units of the G-protein, and is based on the studies of Nauroschat and an der Heiden [36] and Riccobene *et al.* [27]. The known effect that calcium increases receptor inactivation, as modeled in Kummer *et al.* [28], has been added. The processes modeled are: ligand–receptor binding and dissociation; receptor sequestration and desequestration and its dependence on receptor phosphorylation state; receptor phosphorylation and its dependence on active G-protein and ligand-binding; G-protein activation and inactivation and its dependence on calcium and phospholipase C; and the production of phospholipase C by active G-protein.

5.3 Calcium Model (B)

This describes the calcium signaling pathway activated by inositol-1,4,5-triphosphate (IP_3), taken from Höfer [29]. It is a simplification of Höfer's model, in which all switch-like response curves are unified as a Hill function with a common coefficient, as discussed in Ref. [14]. The model includes IP_3-dependent calcium entry through the cell membrane, calcium- and IP_3-dependent release of calcium by the ER, and the ER and membrane calcium pumps.

5.4 cAMP Model (C)

This model, developed *ab initio*, represents activation of the G-protein βγ subunit of the glucagon receptor model (A), which has different dynamics from the βγ subunit associated with the calcium pathway described above. The key modeled processes here are the activation of the receptor, the production of cAMP by the active receptor complex, the elevated rate of cAMP production by activated receptor, the activation of protein kinase A (PKA), and the potential nuclear localization of PKA, which will become important when the current composite model is extended to cover the transcriptional changes that result from the hormonal stimulation of hepatocytes. Like the calcium component, the cAMP model is based largely on Hill-function response dynamics. The receptor–ligand binding dynamics are assumed to be fast, so that the ligand-bound receptor is not a separate quantity, but rather is obtained from the total receptor population.

5.5 Insulin Model (D)

The current composite model uses a very simple, *ab initio* model for the insulin receptor and associated pathway. Insulin increases the amount of active (α-form) glycogen synthase, resulting in an increased conversion of glucose to glycogen, in part by inactivating GSK. The model describes the inactivation of GSK as a threshold response to insulin with a relaxation time delay.

5.6 Blood Model (E)

The blood model was developed *ab initio* to describe the movements of glucose between the blood, the liver, and the pancreas. Blood glucose is transported into the hepatocyte, and there converted into glucose-6-phosphate. The membrane glucose transporter (GLUT) is passive, but selective, transporting only glucose, and not glucose-6-phosphate. As a result, the transport appears to be active, with glucose apparently pumped into the cell with a slower transport time constant than might be expected for a strictly passive GLUT transporter. In this and the glycogenolysis models (model F; see below), a single intracellular glucose pseudoconcentration variable represents the effective total glucose and glucose-6-phosphate. This illusory-pump model fits the data for a perfusive radiolabeling experiment in pigs conducted by Munk *et al.* [37], in which it was found necessary to consider separately a forward clearance rate into cells (the effective pumping), and a reverse rate out of the cells. The apparent active pumping of glucose actually represents the glucose to glucose-6-phosphate equilibrium; the GLUT pump itself is known to be an *effective rapid passive transporter*. The radiolabeling methods of Munk *et al.* [37] could not distinguish between the intracellular glucose and the glucose held in the cell as glycogen or in the phosphorylated form; consequently, an apparent increase in cellular glucose pseudoconcentration above the concentration in the surrounding medium will reflect glucose held in other forms. The advective effects of blood circulation are neglected; the blood is treated as a single effective volume, in contact with liver and pancreas, and mediates communication between the two via blood glucose concentration variables.

An important input to this model, which is also an input to the composite model, represents feeding, exercise, and the basic metabolic drain of the (unmodeled) rest of the body. This exogenous rate of input/output of blood glucose is denoted by M. In this case, M is defined as 0, indicating a zero net glucose supply/demand.

In this way, $M = 0$ can be considered to be a situation where a person is given a glucose drip that exactly matches the constant metabolic demand of their resting state. An increase in M represents an increase in blood glucose as, for example, after feeding, while a fall in M represents a fall in blood glucose as, for example, during fasting. In this way, it is possible to avoid representing the rest of the body as a separate component.

The model involves a quantity that represents the effective volume ratio of liver cytoplasm to blood volume, so that a unit concentration fall of cellular glucose in the liver produces v_f units of blood glucose concentration increase. The transmembrane glucose transport parameters are based on the values in Munk *et al.* [37], where the steady-state ratio of cellular pseudoconcentration to blood concentration is somewhat over unity, and the timescale of response is of order 1 min. The selection of appropriate values is explained in Ref. [26].

5.7 Glycogenolysis Model (F)

Many factors control glycogen breakdown and synthesis. This model incorporates direct control by cellular glucose and glucose-6-phosphate, based on the model in Ref. [30], control by calcium ions, based on the model in Ref. [31], control by cAMP via PKA, and by insulin via GSK, based on the representations in Refs [32] and [33]. The model includes a direct control of glucose by glucose, in addition to pancreatic-hormone control. In order to reconcile these four competing inputs, it would be possible (at least in principle) to model the full enzyme kinetic scheme consisting of glycogen phosphorylase and synthase, their phosphatases and kinases, and the additional phosphatases and kinases involved. This phosphorylation cycle cascade is, however, extremely complicated and not sufficiently well-characterized experimentally for detailed dynamic modeling to be possible. Instead, a simplified, pragmatic approach has been chosen. The overall decision system, whereby the protein cycles that compute the behavior in terms of the cell's second messengers, rather than being modeled in detail, is represented using fuzzy logic. A Boolean logical expression, based on each of the four inputs, determines whether glycogen build-up and breakdown are on or off. This Boolean expression is extended to the domain where the four inputs are not themselves Boolean variables, but rather, intermediate degrees of cellular activity in the unit interval 0–1. The model is completed by defining a simple saturating function to represent the saturating kinetics of the enzymes, and specifying response times for these activities to the inputs.

5.8 Pancreas Model (G)

To complete the glucohomeostatic feedback cycle, a pancreas model is needed that is responsible for the production of the hormones glucagon and insulin. In order to retain a focus on hepatic physiology, a particularly simple pancreas model was developed *ab initio* in which the release of glucagon or insulin followed time-delayed threshold responses. This is modeled using a logarithmic response to the ratio of blood glucose to a fixed reference level, defined so that there is a zone within which the rate of glucagon or insulin release is small, but which rises steeply as

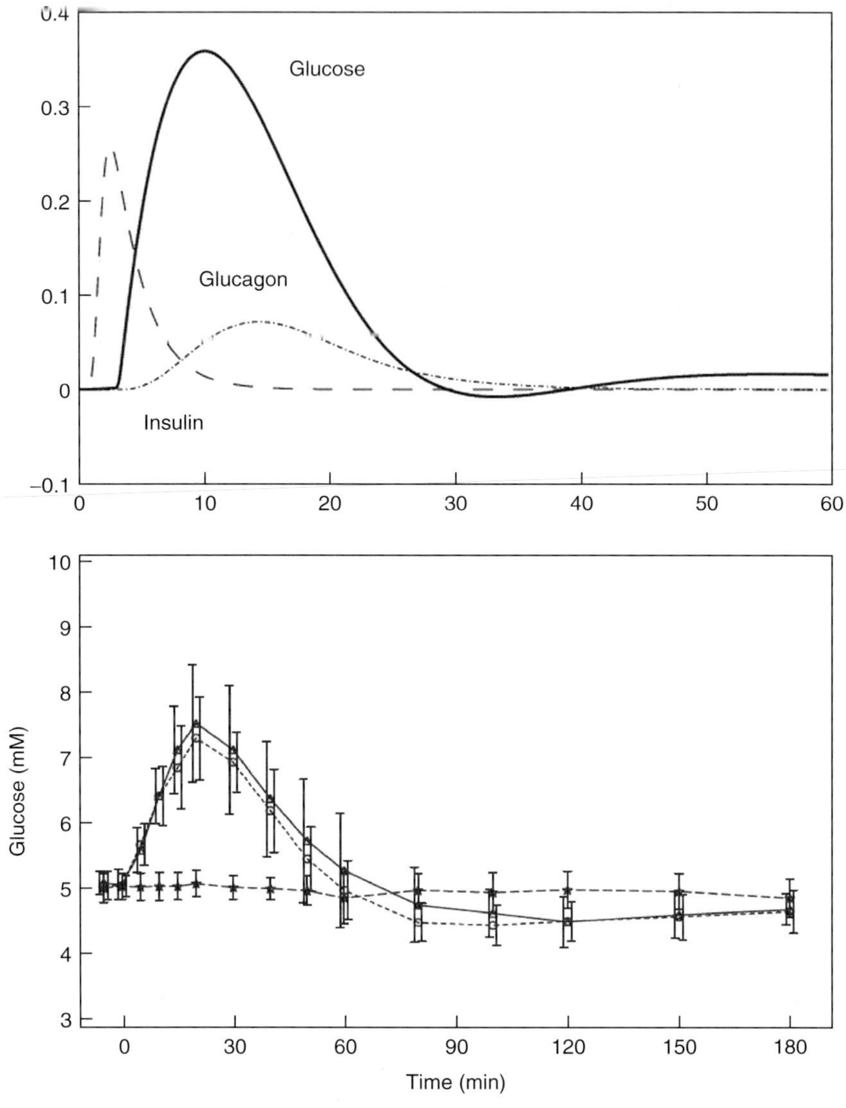

Fig. 3 Comparison of the model with a glucagon challenge, showing qualitative similarity with experimentally observed glucose transient. See text for details. Lower panel reproduced with permission from Ref. [38].

the perimeter of this zone is approached. An additional factor not modeled directly is the time scale of the circulation between pancreas and liver, which is represented using only the pancreatic and liver response times.

5.9 Composite Glucose Homeostasis Model Results

The composite model composed of these seven models has been shown to display expected qualitative behavior in response to a range of glucose challenges. Validating models in systems biology have the added requirement of validating each component, as well as a complete system model. Although both of these stages have been described in detail elsewhere [14, 26], the main results will be summarized here to demonstrate the validity of the approach and to exemplify the state of the art in systems biology of the liver.

The model was used to explore the behavior following a glucagon challenge, as proposed by Lockton and Poucher [38]. The virtual experiment was undertaken once the model was fully constructed; the results from the model are shown in Fig. 3. In this study, fasting subjects were challenged with a bolus of glucagon, after which the rise in their blood glucose level (taken as the consequence of glycogenolysis) was monitored. The model produced a transient that matched the patient data for both glucose and insulin, and which was also noted to have resulted from the challenge. Although the time scale of the model was about twice as fast as the data, there was evidence that there might be substantial variation in time scales between individuals. Thus, the model predicted qualitatively correct behavior, and was within the margin of error for predicting approximate quantitative behavior.

The predictions from the model following a range of glucose inputs are shown in Fig. 4. In this case, Fig. 4a shows

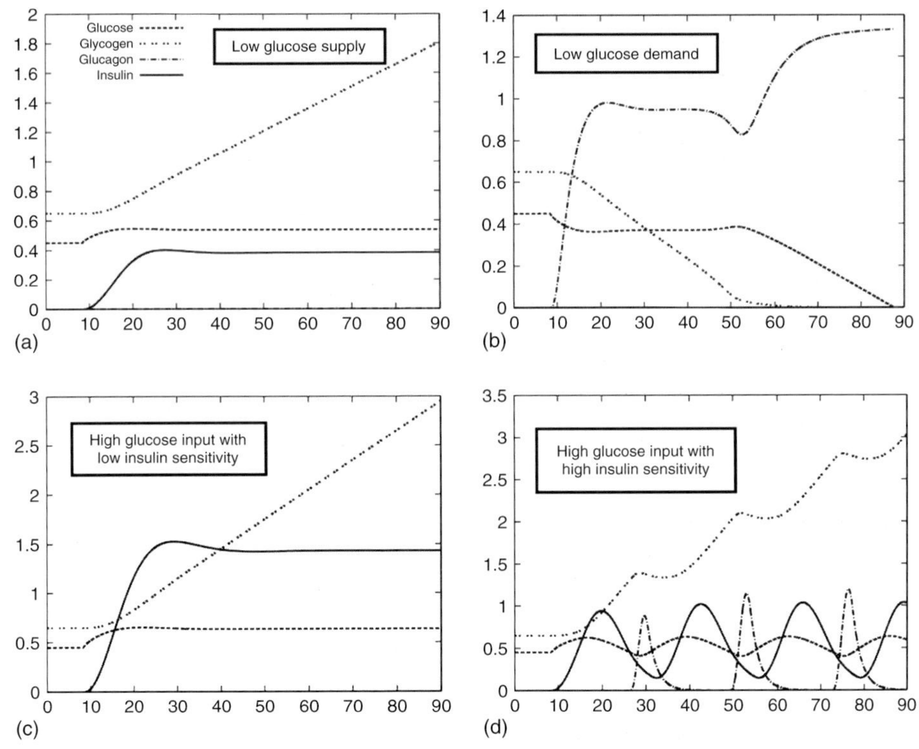

Fig. 4 (a–d) Model responses to various glucose stimuli for differing characteristics. See text for details.

the consequence of a small glucose input increase, which results in a raised glucose concentration and a steadily rising glycogen level, while Fig. 4b shows the results for decreased glucose input. In Fig. 4c, a larger increase in glucose input shows that a normal glucose homeostasis can still be maintained by the direct control mechanism. This is because a rise in cellular glucose concentration stimulates glycogen synthesis, despite the reduced sensitivity to insulin that prevents a fully hormone-driven response. Finally, Fig. 4d shows that if the GSK resistance to insulin is lower than normal, then all of the parameters oscillate. A more detailed discussion of these results may be found in Ref. [26]. The model matched naturally occurring ultradian oscillations [14]. For large, sustained glucose inputs, the model shows that blood glucose would increase without limit over time if there were no way of eliminating or utilizing the excess. This would normally be taken up by fat metabolism pathways, but these were not included in the model.

5.10 Sensitivity Studies

Sensitivity analysis techniques can be applied to composite models of this type to determine which parts of the system most strongly affect behavior, and which parameters are most influential in obtaining accurate predictions. A number of global sensitivity analysis methods are available. In this context, the Morris method [39] has been found to provide valuable insights into the model behavior, while avoiding the high computational costs of more rigorous techniques such as that of Sobol. The Morris method is a screening approach in which each parameter is varied individually around a fixed value, and its effects on the model output are then calculated. The process is repeated at multiple points in the parameter space to capture the effects of nonlinearities and interactions between parameters.

Summary results from the sensitivity analysis study are shown in Fig 5. Here,

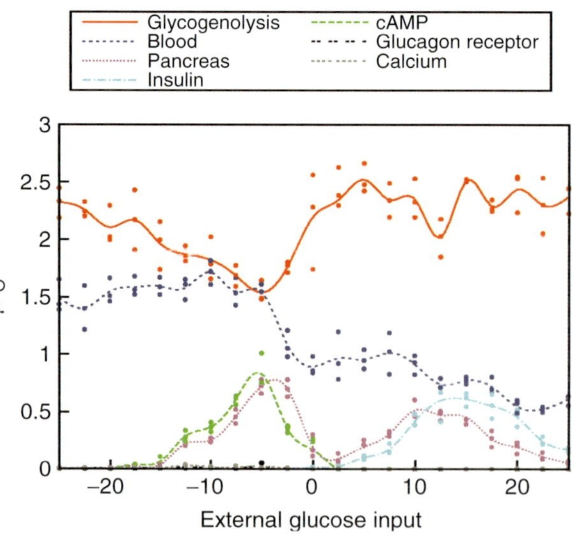

Fig. 5 Sensitivity analysis results for composite glucose homeostasis model showing the overall Morris measure as a function of external glucose input.

the blood and glycogenolysis models were found to be the two most important across all inputs (a higher value of the Morris measure μ_o^* for both positive and negative changes to glucose input). The high sensitivity values for the glycogenolysis model imply that the synthesis and breakdown of glycogen are the most important parts of this system. This result is consistent with the expected behavior of the system; the amount of glucose being stored or released by the liver is crucial to determining the blood glucose level. A parametric analysis shows that, for this model, the threshold value at which glucose affects the activity of glycogen phosphorylase and glycogen synthase is the most important parameter. This suggests that the autoregulation of hepatic glucose production by glucose is an important part of the control of blood glucose levels. The glucagon receptor and calcium models are unimportant for all external input values, while the importance of the pancreas, insulin and cAMP models depend on the sign of the glucose input. Further details of the sensitivity analysis of the liver model can be found in Ref. [40].

Given the complexity of biological systems, analyses of this type will be crucial in enabling research groups and other users to identify weaknesses in the models, to identify the parameters for which accuracy of the data are most important, and to provide general confidence in the models.

6 Conclusions and Future Developments

Although, currently, the development of systems biology models for liver systems is in its infancy, significant progress can be made in this area by using only simple models. In order to more accurately quantify behavior, however, more complex metabolic and physiological models are required. Necessarily, the specific areas which need to be developed will be dependent on the purpose for which the model is to be used. Whilst it is not realistic to obtain many accurate measurements, for reasons of cost, time, and/or feasibility, sensitivity analysis will help to inform which measurements are crucial for accurate predictions. This will be particularly important when moving towards the use of systems biology for individualized medicine.

Systems biology approaches employ the iteration of model development and experimental exploration to develop an understanding of a complex biological system. Based on the above discussions, a number of aspects of the glucose homeostasis model emerge as obvious candidates for improvement as a result of the modeling process, and further experimental exploration will be required:

1. Since the glycogenolysis model has been identified as the most important, the development or identification of a more mechanistic model of the regulation of glycogen phosphorylase and glycogen synthase would allow the role of these processes in the control of blood glucose to be investigated in greater detail.
2. The pancreas model is particularly simple, and may not accurately reflect the dynamic response.
3. Certain drugs [e.g., paracetamol (acetaminophen)] affect hepatocytes close to the hepatic artery, but not those closer to the hepatic vein, yet current models do not allow for variation across

the liver plate. This effect, which is known as *zonation*, requires both modeling and experimental investigation to obtain a suitable model for incorporation into a system model. If this can be achieved it could have a significant impact on understanding the toxicity of certain drugs.

4. Metabolism is an integrated process. Glucose is taken up by the gut and consumed by all cells (the major consumers being the heart and brain). As in the liver, glucose is both stored and released by the kidney, stored but not released by skeletal muscle, and significantly modulated by adipose tissue. A comprehensive, system-level model of glucose metabolism is needed that will represent, with variable degrees of precision, all of these processes, but which is centered around liver function. The model should include the effects on the liver of inter-organ movement of solutes. Such a model would allow the re-examination of results obtained with simpler models, and permit the development and testing of models that reflect different physiological states. In the longer term, a basic system model that reproduces normal glucose homeostasis satisfactorily would underpin any future analysis of disease processes by incorporating the alterations in insulin sensitivity that accompany diabetes and NAFLD.
5. In the longer term, the ability to identify the key measurements of an individual's characteristics and symptoms would be required in order to use a composite model to predict individual responses to stimuli.

In this way, it will be possible to build fuller systems biology models of the liver. Today, new models and data are being developed in many locations worldwide; consequently, the use of model management will be important to facilitate the easy incorporation of models and data from these many sources, and to facilitate their validation and, if necessary, correction.

Whilst, clearly, the goal must be to derive accurate models, complete accuracy will never be possible because of the complexity of biological systems; rather, it is more important to ensure that a model is "fit for purpose." However, it is important to ensure that such purpose is clear at an early stage of the modeling process. For clinical investigations, the disease being studied will provide very strong indications as to which phenomena should be included in a model (or a composite model). In this case, sensitivity studies and matching with data will help to indicate which models are most important, and which parameters need to be measured accurately.

It is likely that future systems biology investigations will focus on model development for specific disease targets, or assist in the testing of basic biological hypotheses. The development of a generic system should allow the formulation and solution of all such problems by employing mathematical model solvers that will lead to progress in the development of physiological models for both biological and clinical use. This will, in turn, greatly facilitate the development of systems biology of the liver and its associated systems, leading hopefully to a greater understanding of system-wide phenomena and, in time, to the diagnosis and treatment of liver disease.

References

1 Noble, D. (2002) Modelling the heart: insights, failures and progress. *BioEssays*, **24**, 1155–1163.

2 Vinken, M., Henkens, T., De Rop, E., Fraczek, J., Vanhaecke, T., Rogiers, V. (2008) Biology and pathobiology of gap junctional channels in hepatocytes. *Hepatology*, **47** (3), 1077–1088.

3 Yan, W., Lee, H., Yi, E.C., Reiss, D., Shannon, P., Kwieciszewski, B.K., Coito, C., Li, X.J., Keller, A., Eng, J., Galitski, T., Goodlett, D.R., Aebersold, R., Katze, M.G. (2004) System based proteomic analysis of the interferon response in human liver cells. *Genome Biol.*, **5** (8), R54.

4 Andersson, U., Lindberg, J., Wang, S., Balasubramanian, R., Marcusson-Ståhl, M., Hannula, M., Zeng, C., Juhasz, P.J., Kolmert, J., Bäckström, J., Nord, L., Nilsson, K., Martin, S., Glinghammar, B., Cederbrant, K., Schuppe-Koistinen, I. (2009) A systems biology approach to understanding elevated serum alanine transaminase levels in a clinical trial with ximelagatran. *Biomarkers*, **14** (8), 572–586.

5 Achur, R.N., Freeman, W.M., Vrana, K.E. (2010) Circulating cytokines as biomarkers of alcohol abuse and alcoholism. *J. Neuroimmune. Pharmacol.*, **5** (1), 83–91.

6 Wu, M., Yang, X., Chan, C. (2009) A dynamic analysis of IRS-PKR signaling in liver cells: a discrete modeling approach. *PLoS ONE*, **4** (12), e8040.

7 Sun, A., Jiang, Y., Wang, X., Liu, Q., Zhong, F., He, Q., Guan, W., Li, H., Sun, Y., Shi, L., Yu, H., Yang, D., Xu, Y., Song, Y., Tong, W., Li, D., Lin, C., Hao, Y., Geng, C., Yun, D., Zhang, X., Yuan, X., Chen, P., Zhu, Y., Li, Y., Liang, S., Zhao, X., Liu, S., He, F. (2010) Liverbase: a comprehensive view of human liver biology. *J. Proteome Res.*, **9** (1), 50–58.

8 Klingmuller, U., Bauer, A., Bohl, S., Nickel, P.J., Breitkopf, K., Dooley, S., Zellmer, S., Kern, C., Merfort, I., Sparna, T., Donauer, J., Walz, G., Geyer, M., Kreutz, C., Hermes, M., Gotschel, F., Hecht, A., Walter, D., Egger, L., Neubert, K., Borner, C., Brulport, M., Schormann, W., Sauer, C., Baumann, F., Preiss, R., Macnelly, S., Godoy, P., Wiercinska, E., Ciuclan, L., Edelmann, J., Zeilinger, K., Heinrich, M., Zanger, U.M., Gebhardt, R., Maiwald, T., Heinrich, R., Timmer, J., Weizsacker, F.V., Hengstler, J.G. (2006) Primary mouse hepatocytes for systems biology approaches: a standardized in vitro system for modelling of signal transduction pathways. *Syst. Biol.*, **153**, 433–447.

9 Ierapetritou, M.G., Georgopoulos, P.G., Roth, C.M., Androulakis, L.P. (2009) Tissue-level modeling of xenobiotic metabolism in liver: an emerging tool for enabling clinical translational research. *Clin. Transl. Sci.*, **2** (3), 228–237.

10 Li, Z., Chan, C. (2009) Systems biology for identifying liver toxicity pathways. *BMC Proc.*, **3** (Suppl. 2), S2.

11 Eveillard, A., Lasserre, F., de Tayrac, M., Polizzi, A., Claus, S., Canlet, C., Mselli-Lakhal, L., Gotardi, G., Paris, A., Guillou, H., Martin, P.G., Pineau, T. (2009) Identification of potential mechanisms of toxicity after di-(2-ethylhexyl)-phthalate (DEHP) adult exposure in the liver using a systems biology approach. *Toxicol. Appl. Pharmacol.*, **236** (3), 282–292.

12 Schnackenberg, L.K., Jones, R.C., Thyparambil, S., Taylor, J.T., Han, T., Tong, W., Hansen, D.K., Fuscoe, J.C., Edmondson, R.D., Beger, R.D., Dragan, Y.P. (2006) An integrated study of acute effects of valproic acid in the liver using metabolomics, proteomics, and transcriptomics platforms. *OMICS*, **10** (1), 1–14.

13 Jungermann, K., Katz, N. (1989) Functional specialization of different hepatocyte populations. *Physiol. Rev.*, **69**, 708–764.

14 Hetherington, J., Bogle, I.D.L., Saffrey, P., Margoninski, O., Li, L., Varela Rey, M., Yamaji, S., Baigent, S., Ashmore, J., Page, K., Seymour, R.M., Finkelstein, A., Warner, A. (2007) Addressing the challenges of multiscale model management in systems biology. *Comput. Chem. Eng.*, **31**, 962–979.

15 Liu, W., Hsin, C.C., Tang, F. (2009) A molecular mathematical model of glucose mobilization and uptake. *Math. Biosci.*, **221** (2), 121–129.

16 Liu, W., Tang, F. (2008) Modeling a simplified regulatory system of blood glucose at molecular levels. *J. Theor. Biol.*, **252**, 608.

17 Ohno, H., Naito, Y., Nakajima, H., Tomita, M. (2008) Construction of a biological tissue

model based on a single cell model: a computer simulation of metabolic heterogeneity in the liver lobule. *Artif. Life*, **14** (1), 3–28.

18 Mishra, L. (2008) The 21st century hepatologist and a systems biology based approach to liver diseases. *Hepatology*, **48** (6), 1731–1733.

19 Harris, A.L. (2001) Emerging issues of connexin channels: biophysics fills the gap. *Q. Rev. Biophys.*, **34** (3), 325–472.

20 Segretaina, D., Falk, M.M. (2004) Regulation of connexin biosynthesis, assembly, gap junction formation, and removal. *Biochim. Biophys. Acta*, **1662**, 3–21.

21 Baigent, S., Stark, J., Warner, A. (2001) Convergent dynamics of two cells coupled by a nonlinear gap junction. *Nonlinear Anal.*, **47**, 257–268.

22 Subramanian, K., Raghavan, S., Rajan Bhat, A., Das, S., Bajpai Dikshit, J., Kumar, R., Narasimha, M.K., Nalini, R., Radhakrishnan, R., Raghunathan, S. (2008) A systems biology based integrative framework to enhance the predictivity of in vitro methods for drug-induced liver injury. *Expert Opin. Drug Safety*, **7** (6), 647–662.

23 Lagoa, C.E., Bartels, J., Baratt, A., Tseng, G., Clermont, G., Fink, M.P., Billiar, T.R., Vodovotz, Y. (2006) The role of initial trauma in the host's response to injury and hemorrhage: insights from a correlation of mathematical simulations and hepatic transcriptomic analysis. *Shock*, **26** (6), 592–600.

24 Margoninski, O., Saffrey, P., Hetherington, J., Li, L., Finkelstein, A., Warner, A. (2006) A specification language and a framework for the execution of composite models in systems biology. *LNCS Trans. Comput. Syst. Biol. VII.*, **4230**, 163–124.

25 Saffrey, P., Margoninski, O., Hetherington, J., Varela-Rey, M., Yamaji, S., Finkelstein, A., Bogle, D., Warner, A. (2007) End-to-end information management for systems biology. *LNCS Trans. Comput. Syst. Biol VIII*, **4780**, 77–91.

26 Hetherington, J., Sumner, T., Seymour, R.M., Li, L., Varela Rey, M., Yamaji, S., Saffrey, P., Margoninski, O., Bogle, I.D.L., Finkelstein, A., Warner, A. (2010) A composite computational model of liver glucose homeostasis. *Proc. Roy. Soc. Interface*, published online June 15, 2011. doi: 10.1098/rsif.2011.0141.

27 Riccobene, T.A., Omann, G.M., Linderman, J.J. (1999) Modeling activation and desensitization of G-protein-coupled receptors provides insight into ligand efficiency. *J. Theor. Biol.*, **200**, 207–222.

28 Kummer, U., Olsen, L., Dixon, C., Green, A., Bornberg, E. (2000) Switching from simple to complex oscillations in calcium signaling. *Biophys. J*, **79**, 1188–1195.

29 Höfer, T. (1999) Model of intercellular calcium oscillations in hepatocytes: synchronization of heterogeneous cells. *Biophys. J.*, **77**, 1244–1256.

30 Cardenas, M.L., Goldbeter, A. (1996) The glucose-induced switch between glycogen phosphorylase and glycogen synthase in the liver: outlines of a theoretical approach. *J. Theor. Biol.*, **182**, 421–426.

31 Gall, D., Baus, E., Dupont, G. (2000) Activation of the liver glycogen phosphorylase by Ca^{2+} oscillations: a theoretical study. *J. Theor. Biol.*, **207**, 445–454.

32 Bollen, M., Keppens, S., Stalmans, W. (1998) Specific features of glycogen metabolism in the liver. *Biochem. J.*, **336**, 13–31.

33 King, M.W. (2005) The Medical Biochemistry Page. http://web.indstate.edu/thcme/mwking/. Peer reviewed webpage, via MedEdPortal (accessed 29 October 2010).

34 Grishina, G., Berlot, C.H. (1997) Identification of common and distinct residues involved in the interaction of alphai2 and alphas with adenylyl cyclase. *J. Biol. Chem.*, **272**, 20619–20626.

35 Ford, C.E., Skiba, N.P., Bae, H., Daaka, Y., Reuveny, E., Shekter, L.R., Rosal, R., Weng, G., Yang, C.S., Iyengar, R., Miller, R.J., Jan, L.Y., Lefkowitz, R.J., Hamm, H.E. (1998) Molecular basis for interactions of G protein betagamma subunits with effectors. *Science*, **280**, 1271–1274.

36 Nauroschat, J., an der Heiden, U. (1997) A theoretical approach to G-protein modulation of cellular responsiveness. *J. Math. Biol.*, **35**, 609–627.

37 Munk, O.I., Bass, L., Roelsgaard, K., Bender, D., Hansen, S.B., Kelding, S. (2001) Liver kinetics of glucose analogs measured in pigs by pet: Importance of dual-input blood sampling. *J. Nucl. Med.*, **42**, 439–441.

38 Lockton, J.A., Poucher, S.M. (2007) Single dose glucagon (0.5 mg iv bolus) administration in healthy human volunteers is a robust

model for assessment of glycogenolysis: characterisation of the glucose excursion after glucagon challenge. *J. Pharmacol. Toxicol. Methods*, **55**, 86–90.

39 Morris, M.D. (1991) Factorial sampling plans for preliminary computational experiments. *Technometrics*, **33** (2), 161–174.

40 Sumner, T. (2010) Sensitivity analysis in systems biology modelling and its application to a multiscale model of blood glucose homeostasis. PhD Thesis, University College London.

Part IV
Systems Biology in Medicine and Disease

18
Inferring Networks for Diseases

Mikael Benson[1] *and Michael A. Langston*[2]
[1]*University of Linköping, The Center for Individualized Medicine, Linköping, 58185, Sweden*
[2]*University of Tennessee, Department of Electrical Engineering and Computer Science, 203 Claxton Complex, 1122 Volunteer Boulevard, Knoxville, TN 37996-3450, USA*

Systems Biology: Advances in Molecular Biology and Medicine, First Edition. Edited by Robert A. Meyers.

Keywords

Hub
A node with high connectivity.

Connectivity (or degree)
The number of links a node has to other nodes in a network.

Modularity
A group of physically or functionally linked nodes (or molecules) that work together to achieve a distinct function.

Diseasome
A bipartite network formed by interlinked human diseases and disease genes..

Genome-wide association studies
Studies in which a dense array of genetic markers which captures a substantial proportion of common variation in genome sequence, is typed in a set of DNA samples that are informative for a trait of interest. The aim is to map susceptibility effects through the detection of associations between genotype frequency and trait status.

1

An important focus of medical research is to identify individual disease genes that are potential diagnostic and/or therapeutic targets. The most common diseases, such as allergies and diabetes, are complex in the sense that they are caused by altered interactions between tens, hundreds, or even thousands, of genes. Whilst the majority of those genes each have small effects, their combined effects may be large. In addition, environmental factors play major roles in complex diseases. Moreover, different combinations of genes and environmental factors may give rise to diseases with similar manifestations. An important clinical implication here is that patients who appear to have the same disease may not respond equally to the same medication; indeed, some may even deteriorate or experience serious adverse side effects. If it were possible to determine the combinations of genes and environmental factors that give rise to different diseases, this information could be used to "personalize" medication. It is also possible that an improved understanding of these factors could lead to diagnostic and/or therapeutic strategies for the prevention of disease. Network-based analysis may provide a theoretical and methodological framework to harness the complexity of common diseases whereby, instead of focusing on individual genes (or their products), the scale could be changed to network modules of interacting and functionally related genes. Such modules could then be used to determine the main pathogenic mechanisms in complex diseases, and how those mechanisms vary between patients. This information might, in turn, be used to identify combinations of genes (or their products) suitable as diagnostic and/or therapeutic targets.

1 Common Diseases Are Complex

Although clinical medicine has made significant advances during the past few decades, for most common diseases the same problems remain: (i) there is a limited understanding of the causal mechanism(s); (ii) in general, it is not possible to predict and prevent disease; (iii) often, there are no phenotypic differences or diagnostic markers to identify such groups; and (iv) many common drugs have either limited or no effects in large groups of patients. Consequently, the pharmaceutical industry continues to struggle with the ever-increasing and staggeringly high costs of drug development.

All of these problems are interlinked and depend ultimately on the complexity of common diseases, which may exist at different levels. At the genetic (or epigenetic) level, thousands of genes (or their products) may be involved in complex diseases [1, 2] (for simplicity, "genes or their products" will henceforth be referred to simply as "genes"). Although a small portion of these genes have major effects, and may therefore be selected as therapeutic and/or diagnostic candidates, the large proportion of disease-associated genes will each have minor effects, although their combined effects may be very large. Thus, the identification alone of large-effect genes may not suffice for diagnostic and therapeutic purposes. The disease-causing effects may be either *direct*, perhaps by altering the composition of the encoded protein, or *indirect*, by affecting gene expression through alterations in the regulatory regions of the genome. In addition, epigenetic changes such as DNA methylation, histone modifications, or noncoding RNAs can cause disease. Taken together, changes such these may involve several molecular species including DNA,

RNA, proteins, and metabolites that may amount in total to hundreds of thousands of different molecules and their variants. In addition to such complexity, these molecules may be variably affected in different cell types, and in different compartments of the body. Moreover, considerable inter-individual variations can occur in complex diseases, depending on the different combinations of genes in the same pathways, causing the same disease phenotypes.

Another important cause of disease is that of environmental factors, acting either alone or in combination with genetic factors. An example of this is the allergies that occur typically during the pollen season, when approximately 20–30% of the population in the Western world develops symptoms of rhinitis or asthma. Yet, one hundred years ago such symptoms were rare! Although allergy has a considerable heredity, the major increase in its prevalence indicates a strong environmental component. Major environmental influences have also become evident in the etiologies of common diseases such as obesity, metabolic disorders, and cancer.

If it were possible to fully understand the multilayered complexity that underlies common diseases, this information could ideally be used to predict, prevent, or treat disease in an individualized manner. However, large-scale measurements, analytical techniques, as well as a theoretical framework would be required to organize the available information. During the past decade, genome-wide techniques, high-performance computing (HPC) and bioinformatics have each made significant advances, while networks have provided a theoretical framework by which to analyze the data that has become available. Although the challenges of integrating and applying these advances are clearly daunting, some preliminary studies have already yielded clinically relevant results. One such example has been the identification of combinations of diagnostic serum proteins for the prediction of Alzheimer's disease [3]. Those diagnostic proteins that can be used to personalize medication could be regarded as the "low-hanging fruit" in the context of clinical applications of network-based analysis. Ideally, clinicians would routinely profile proteins in the patient's saliva, blood, or urine for purposes such as personalized medication. Whilst the same principles could also lead to improved drug development, it is difficult to assess if – and when – network-based analysis would reach the clinic. On the one hand, there are the huge challenges outlined above, while on the other hand network-based analysis has been applied to clinical research for only the past four years. Previously, such analytical principles would have been developed on model organisms and experimental data acquired for at least than six or seven years. In an attempt to illustrate the rapid development in this field, the dates of publication for some of the "landmark" studies have been provided in the text.

In this chapter, attention will be focused on how network-based analysis can be applied to interpret high-throughput data from the studies of complex diseases, on the problems encountered, and on the current state of the art as well as future perspectives.

2
Network-Based Analysis: A Brief Summary

During the past few decades, many scientific disciplines have been faced with the same problem, namely the interpretation of increasingly huge datasets. Indeed, this problem simply reflects the abundance of complex systems in Nature and in everyday life. The problem was first addressed in a themed issue of the journal *Science* in 1999 [4], when scientists from a diversity of fields that included biomedicine, economics, sociology, and ecology described their respective complex systems, and realized that a common denominator might exist among them. One vital realization was that a predictive understanding of the systems could not be achieved by the detailed study of individual components; rather, the combination of many components would be required. Although at the time there was no common theoretical framework to describe and analyze these complex systems, it was only four years later, in *Science*, that the proposal was made to use these networks for the creation of just such a framework [5].

Briefly, networks are graphical models in which the nodes represent interactors, and the edges interactions (Fig. 1). In a landmark study in 1999, Barabasi *et al.* showed that networks in highly diverse complex systems – whether molecules in cells, movie actors, or computers on the world-wide web – would display a common property; namely, that the number of connections per node followed a power-law distribution [6]. Although at the time this property was referred to as *scale-freeness*, the viability of such a set-up was later questioned. Nonetheless, although the distribution may not always follow a power-law, many networks in Nature appear to be characterized by a few nodes being highly connected (termed "hubs"), while the vast majority of the nodes would have few connections. Intuitively, it seems likely that hubs would be more important to the network, as can be inferred from everyday experiences of social networks; invariably, a "hub" individual within a group will have a greater impact than the other individuals. The importance of hubs was later demonstrated in an analysis of yeast cell proteins, whereby the deletion of hub proteins were much more likely to affect fitness than would the non-hub proteins [7].

Modularity is another common property that has been identified in a wide range of networks, including social groups, ecosystems, computers on the internet, and gene interaction networks. As will be discussed below, both connectivity and modularity have great relevance for networks in disease.

3 Networks in Diseases

In another study conducted in 2006, Lim *et al.* showed that 18 of 23 known disease genes in ataxia were interconnected

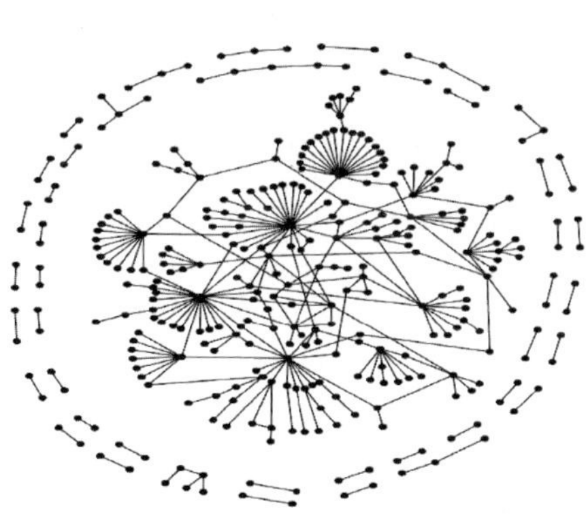

Fig. 1 Networks consist of nodes and edges, which represent interactors and interactions, respectively. In Nature, networks tend to be characterized by a few nodes having multiple interactions, while the majority of nodes have few interactions. Another common property is modularity. Functionally related and interacting nodes form modules.

in the gene interaction network [8]. The importance of this point lay in the fact that the gene interaction network could be used as a template to organize the human disease genes that had been identified in genomic high-throughput studies. In fact, this point had been suggested following a pioneering study carried out in 2001, when Jenssen and colleagues showed that cancer genes tended to form small modules of interacting and functionally related genes [9]. This study was based on an automated text-mining of all abstracts from MedLine, to construct a network of co-cited genes. The underlying assumptions were that: (i) co-citation reflected biologically meaningful relationships; and (ii) the disease genes were interconnected. These assumptions were validated by using the network as a template to organize differentially expressed genes from public gene expression microarray data derived from cancer studies. The authors found that small modules of interacting genes in the network were significantly enriched for differentially expressed genes.

The principle of mapping disease-associated genes onto a network model of gene interaction was later applied in several studies that employed increasingly refined network models. This approach has been facilitated by the availability of an increasing wealth of genomic high-throughput data in public repositories. In 2004, Segal *et al.* performed a meta-analysis of gene expression microarray, using the results of many thousands of experiments available in such a repository [10]. This resulted in the identification of a set of modules of co-expressed and functionally related genes, where the modules represented different functions such as cell proliferation and metabolism. The importance of this study lay in the fact that, instead of hundreds or thousands of differentially expressed genes resulting from a gene expression microarray study, the scale would be changed to only a few modules. Nonetheless, these modules should provide an overview of the main cellular functions that were involved in the disease, and could be further dissected to identify pathways and individual key regulatory genes.

An additional important aspect is that modules can be used to identify novel disease genes; that is, *known disease genes can be used to trace novel genes* by applying different strategies. For example, from a study of breast cancer conducted in 2007, Pujana *et al.* used four known disease genes to identify co-expressed genes in a microarray study. This resulted in the identification of four modules of co-expressed genes that were then analyzed to identify a novel candidate gene; this was subsequently validated by functional and genetic studies [11]. The clinical relevance of modules was demonstrated by a study of Alzheimer's disease, in 2007, whereby the authors identified a module of proteins that interacted with proteins known to be associated with the disease. By using that module, 18 signaling proteins were identified in blood plasma that could be used to classify blinded samples from Alzheimer's disease and control subjects, with close to 90% accuracy. Moreover, it was possible to predict which patients with mild cognitive impairment would later develop Alzheimer's disease [3]. Most importantly, if the analytical principles behind the aforementioned studies of cancer and Alzheimer's disease are generally applicable, this could have a major impact on clinical research aimed at predicting, preventing, and treating disease.

In a commentary in *The New England Journal of Medicine* from 2007, A.L. Barabasi suggested that disease modules could be expanded from genes and mRNAs to include multiple other molecular layers, such as metabolites and enzymes [12]. Another important suggestion was based on recent findings from studies of obesity whereby, although a specific gene could explain a significant portion of the risk for developing obesity, social networks played a much greater role. In fact, in a large population-based study, the risk was found to be strongly associated with social contact with other obese individuals [13]. In a similar way, familial or sexual networks may have significant roles in the spread of infectious diseases, and this led to a suggestion that diseases could be explained by multiple network layers that included not only molecules but also environmental factors. From a clinical perspective, this suggestion may be seen as a formal and structured counterpart to the principles behind the diagnostic work-up procedures carried out by clinicians on an everyday basis. As discussed later in Section X, this multilayered, network-based perspective may have significant clinical implications for the future.

3.1
Diseasome-Wide Studies Indicate that the Same Network Principles may be Generally Applicable to Complex Diseases

Although medical research often focuses on individual diseases, most human diseases are not independent of each other; rather, they show significant comorbidity. For example, obesity is linked to diabetes, cardiovascular diseases, and asthma, which in turn suggests that the network modules that underlie each of these diseases are also interlinked. In 2007, Goh *et al.* proposed a conceptually different approach by constructing a network model of human diseases and disease genes, termed the "diseasome" [14]. Briefly, Goh and colleagues used information from the Online Mendelian Inheritance in Man (OMIM) database to construct a two-layered network that consisted of two different sets of nodes: one set corresponded to all known genetic disorders, whilst the other set corresponded to all known disease genes.

A disorder and a gene were considered to be connected by a link if mutations in that gene were implicated in that disorder. The diseasome was based on OMIM data from 1284 disorders and 1777 disease genes; subsequently, the authors showed that genes associated with similar disorders demonstrated both a higher likelihood of physical interactions between their products, and a higher expression profiling similarity for their transcripts. This supported the existence of distinct disease-specific functional modules. In other words, phenotypically similar diseases were associated with interacting and functionally related genes. As mutations in hub genes are most likely to affect cellular fitness, a likely hypothesis would be that disease genes would serve as hubs. However, this proved not to be the case; instead a study conducted in 2008 showed that disease genes tended to have intermediate connectivities, situated between essential and non-disease genes [15], although the authors did concede that these studies may be confounded by both knowledge- and selection bias. In 2009, however, Barrenäs *et al.* analyzed genes that had been identified by genome-wide association studies (GWAS) from several complex diseases, and also demonstrated modularity [16] (Fig. 2). The advantage of this approach was that, unlike the data from the OMIM database, the disease genes identified by GWAS are not

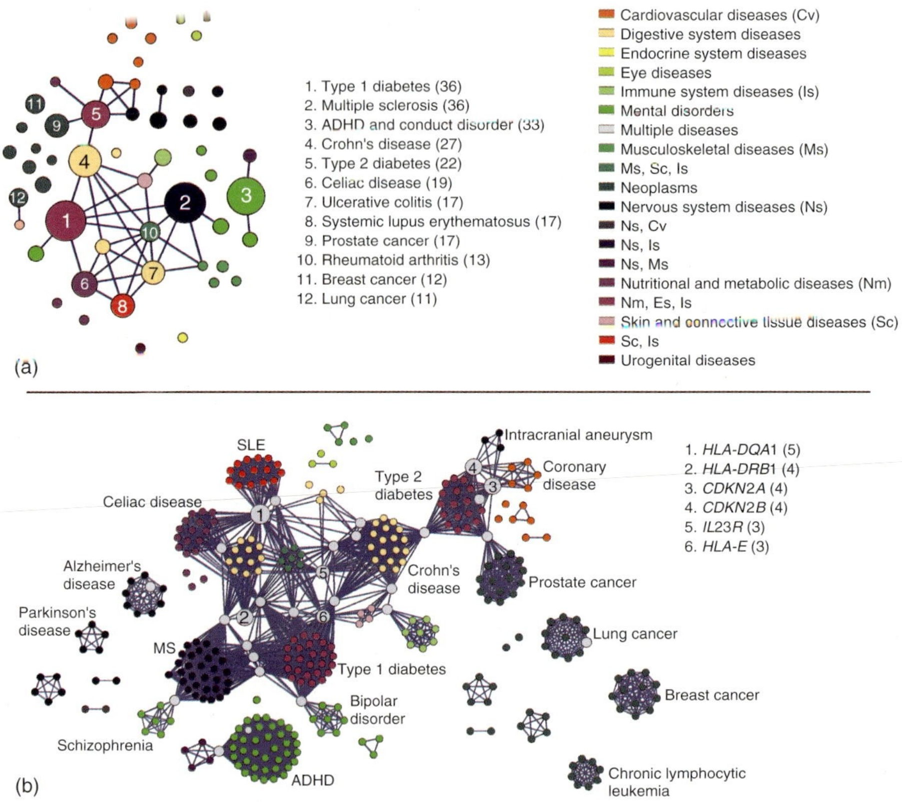

Fig. 2 Networks of complex diseases and complex disease genes. (a) Complex disease network (CDN). Each node is a complex disease studied in GWAS, with the link representing a sharing of disease genes. The color of the nodes corresponds to the disease class, as identified using Medical Subject Headings (MeSH) terms (shown at right). Notably, complex diseases are difficult to define using a single MeSH term. The node size refers to the number of associated genes identified. Diseases with the highest number of associated genes identified through GWAS are listed on the right-hand side, where the numbers in parentheses indicate the number of associated genes; (b) Complex Disease Gene Network (CGN). Each node represents a gene, and connections between two genes represent their association with the same disease. The node size refers to the number of diseases with which a gene is associated. Genes associated with many diseases are listed at the top right-hand side; the number of diseases with which they are implicated is shown in parentheses. The nodes (highlighted in gray) in the lung cancer and Alzheimer's disease gene clusters are singular associations in idiopathic pulmonary fibrosis and narcolepsy, respectively. ADHD, attention deficit hyperactive disorder; SLE, systemic lupus erythematosus; MS, multiple sclerosis.

selected based on prior knowledge. Ultimately, these studies showed that human disease genes have common topological properties, an important implication being that the same network principles may be generally applicable to complex diseases. This, in turn, would have implications for studies both on the diseasome and on an individual disease scale, as discussed in the following sections.

3.2
Diseasome Studies to Gain Systems Level Understanding of Complex Diseases and Disease Genes

Diseasome studies have shown that there is a considerable overlap between diseases and disease genes; moreover, there is also an increasing recognition that phenotypically similar diseases may be caused by similar genes, pathways, or modules. Conversely, it is also possible that some phenotypically dissimilar diseases may share disease genes; for example, asthma and eczema may both be associated with increased IgE levels.

Currently applied diagnostic criteria are the result of a balance between an improved understanding of pathogenic mechanisms, and diagnostic classifications that were originally based on phenotypic observations made by physicians during thousands of years, mostly without any understanding of the underlying pathogenic mechanisms. It is possible that, in the near future, improved phenotypic and genotypic methods and studies will lead to more precise and, in some cases, an altered classification of human diseases. Network-based analyses on the diseasome-scale may provide a framework for such studies and reclassification [17, 18]

It is also possible that the elucidation of disease gene properties on a diseasome-scale may result in completely novel questions and perspectives that may help to understand pathogenic mechanisms, and also have implications for the identification of diagnostic and therapeutic targets. Approximately 15% of human genes are known to be disease genes; in fact, observations arising from diseasome-wide studies suggest that several genes are shared by different diseases. So, what are the network and functional properties of these genes? Are they more connected with or likely to regulate other disease genes? If so, could they serve as diagnostic and therapeutic targets?

Several examples of drugs exist for which the original indications have either been changed, or have increased because of pleiotropic effects, and this has led to the suggestion that shared disease genes should be examined as therapeutic targets for multiple diseases. However, shared genes also lead to many important questions: Why are such genes shared by different diseases? And could they reveal information relating to the environmental factors that have resulted in the increase of complex diseases during past decades?

3.3
Top-Down Diseasome Studies to Identify Novel Candidate Genes in Individual Diseases

The identification of candidate genes in individual complex diseases requires combinations of multiple genomic, experimental, and bioinformatic methods but, unfortunately, this approach is complex, time-consuming, and costly. Clearly, it would be very useful to create a complementary strategy that would simplify the search and narrow down the number of candidates for the experimental and clinical validation stages. One possibility for this might be to use the increasingly large data sources in the public domain; for example, repositories exist with disease genes that include GWAS and gene expression microarray data acquired from hundreds of thousands of patients. Indeed, such repositories have already been utilized to study the properties of human disease genes [14–16, 19–21]. It is proposed that inferences from such studies might be used to define and prioritize candidate genes, the "top-down, diseasome-based strategy" for which is outlined in Fig. 3, where the diseasome is the "top" starting point, and the individual diseases are the final "down" points:

1. The diseasome-scale is based on constructing a cellular network model based on known human gene interactions. Known disease genes from the literature and GWAS data in the public domain are then mapped on to that model, and the resulting genes analyzed to determine the system level properties of human disease genes. But what are their topological, functional, regulatory, and pharmaceutical properties? It is hypothesized that such properties may be used to identify novel genes.
2. On the disease class scale, these properties are analyzed in a more detailed map of specific disease classes, for example, immune system or oncological diseases. First, modules are defined using gene expression microarray data in the public domain, as described previously [10]. Next, the hypothesis is tested that the systems properties of disease genes (as identified in step 1) can be used to locate candidate genes in the modules. Such tests can be performed by using disease genes defined from the literature, as well as GWAS genes in public repositories.
3. On the disease-scale, the systems properties are used to identify candidate genes that can then be tested by applying detailed functional studies in individual diseases.
4. On the clinical scale, candidate diagnostic and therapeutic targets are tested in both animal and clinical studies.

The advantage of this strategy is that, instead of identifying hundreds or thousands of candidates by gene expression microarrays, the number of candidates available for costly and time-consuming functional studies may be narrowed down. Notably, this first step exploits the huge datasets and repositories available within the public domain.

Whilst the above two sections have been based on a diseasome perspective, it may also be possible to employ a network-based analysis of individual diseases in order to extrapolate diseasome-level principles. This proposal is based on the assumption that the same network-based analytical principles are likely to be generally applicable to complex diseases. From a practical point of view, this has one important implication: *That the analytical principles of network-based analysis should be applied to and developed on the most tractable complex diseases, and subsequently applied to other complex diseases.* This implication will serve as the focal point of the following sections.

4
Disease-Scale Network-Based Analyses of a Model Disease: Seasonal Allergic Rhinitis

As discussed in Sect. 1, the large number of genes, cells, and tissues that are involved in complex diseases present a formidable research challenge that is compounded by additional problems of disease heterogeneity and practical difficulties of a clinical nature. The condition of stroke is an extreme example of such variation, with its many different clinical presentations and etiologies, and its ability to affect multiple cells with effects that differ even between patients. In stroke, many of these variables are only partially

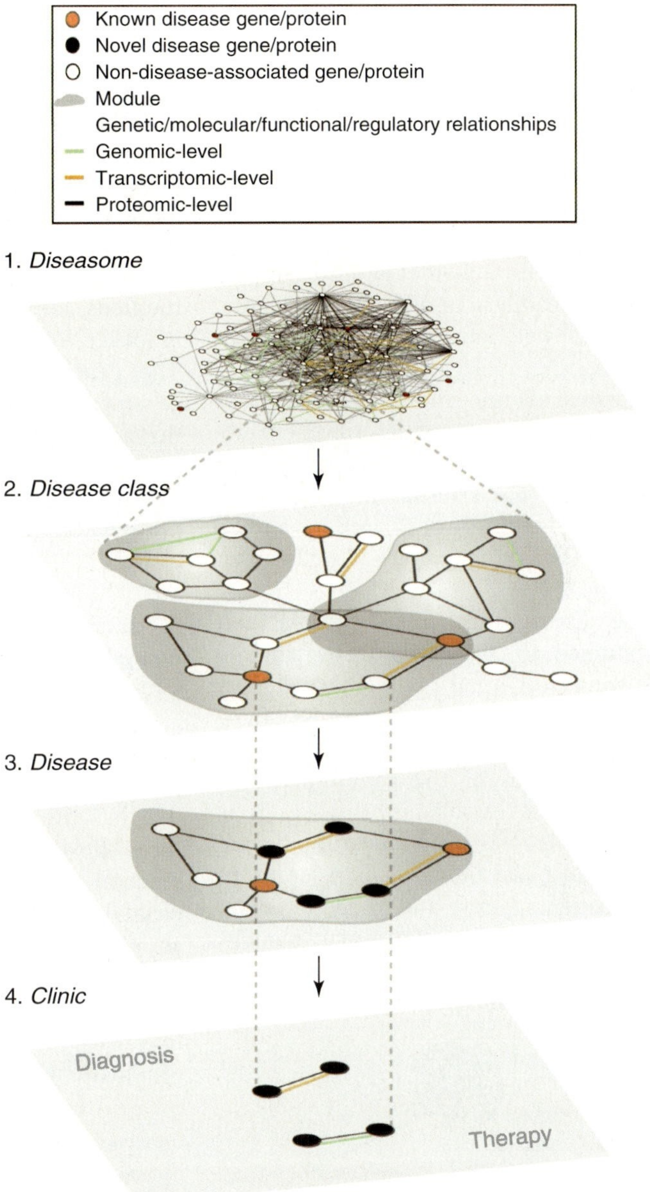

Fig. 3 The top-down, diseasome-based strategy to study human diseases. 1. A model of the diseasome is constructed and analyzed to define modules of disease genes. 2. Modules of a specific disease class are analyzed to identify novel genes among the interactors of known disease genes. 3. A module with novel genes is studied in detail in a specific disease. 4. The novel genes are tested as diagnostic and therapeutic targets in clinical studies. Image provided by Miguel Pujana.

known, the location of the disease makes it almost impossible to obtain human study samples, while the many unknown factors make it difficult to develop reliable cellular or animal models.

Ideally, a disease model would be characterized by each the following factors being identified: the external cause; the main disease cell (or cells); and a read-out. In addition, the cells and other clinical samples would be readily accessible from patients, along with well-defined *in vitro* and *in vivo* models.

Seasonal allergic rhinitis (SAR) demonstrates all of the above-described characteristics, while details of the main disease mechanism, the external cause (pollen), the cells involved (Th2 cells), and a read-out (Th2 cytokines) are all well known [22] (Fig. 4).

The disease mechanism of SAR can be studied in Th2 cells from patients following an allergen- or diluent challenge. This means that the disease process can be analyzed in paired samples that represent different stages of the condition, from pre-disease to established disease. It is also possible to analyze the entire disease process in nasal fluids or biopsies taken from patients before and after the pollen season, as well as in animal models. The disease phenotype is unambiguous (runny, itchy nose during the pollen season, and positive tests for allergy against pollen), while the disease severity can be readily assessed by a simple physical examination.

Thus, SAR may represent an optimal model to apply and develop the methods of a network-based analysis of complex diseases. The following sections will be based on the authors' experiences of studies with SAR to develop network-based methods aimed at identifying markers in order to personalize the medication. The general relevance of the methods is supported by diseasome-scale studies, the details of which are also presented. (These research investigations have been funded by the European Commission; http://www.multimod-project.eu/.)

4.1 A Tractable Strategy for Network-Based Analysis to Identify Diagnostic Markers in Clinical Studies

Following a challenge with pollen extract, patients with SAR develop a local reaction that has the cellular and molecular features shown schematically in Fig. 4. This reaction can be studied in nasal fluids, nasal fluid cells and nasal biopsies, as well as other tissues such as the skin. Analyses of such compartments have revealed that thousands of genes are involved in this reaction, a functional understanding of which can be achieved by sorting the

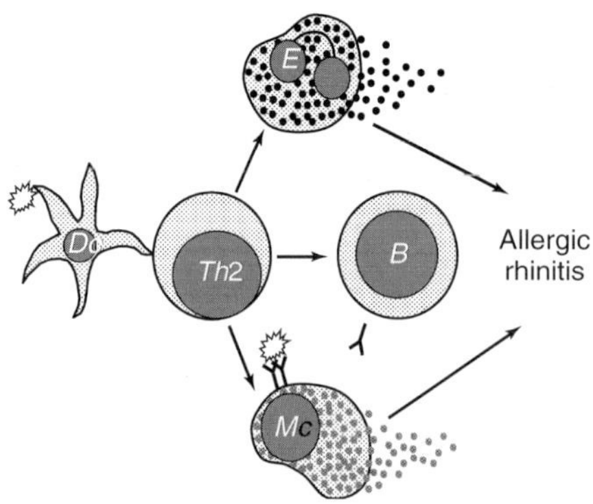

Fig. 4 A simplified overview of the pathogenesis of seasonal allergic rhinitis. Pollen (white jagged circle to the left) is taken up by a dendritic cell (Dc), which activates Th2 cells, which in turn activate eosinophils (E), B cells (B), and mast cells (Mc) to release different inflammatory proteins, including the Th2 cytokines interleukin (IL)-4, IL-5, and IL-13.

genes into functional categories using gene ontology (GO) terms, gene set enrichment analysis (GSEA), or different pathway analysis tools. In the case of clinical studies aimed at identifying diagnostic markers, a network-based analytical strategy was found to be the most tractable method (Fig. 5). This strategy can be outlined as follows

1. Disease-associated genes are identified and organized into putative protein–protein interaction (PPI) networks.
2. Networks are dissected to identify modules of genes with distinct biological functions.
3. Modules are further decomposed to elucidate putative pathways and individual genes with key probable regulatory functions.
4. The mRNA modules are expanded to include other layers, ranging from DNA to protein. This is achieved by adding data from complementary high-throughput experiments, the ultimate aim being to obtain multilayer

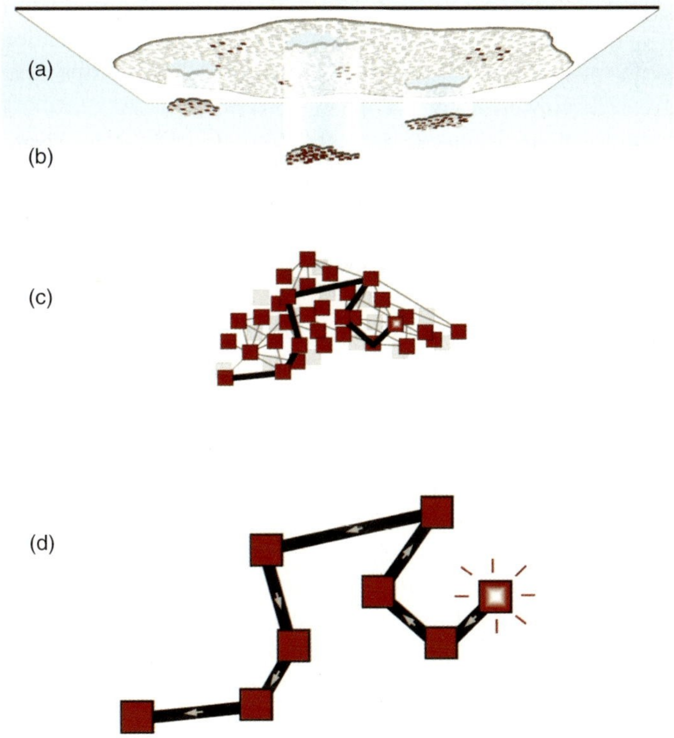

Fig. 5 Network-based analysis of DNA microarray data. (a) Genes identified by DNA microarray analysis of allergic disease (red) are mapped on to an interaction network formed by all human genes (gray); (b) Modules of interacting genes that represent disease-associated biological functions are identified; (c) A module is dissected to find a pathway; (d) The pathway is analyzed to find a putative disease-causing gene (in this case exemplified by an up-stream gene).

modules (MLMs) that include information concerning all layers and regulatory elements.
5. Protein markers for individual variations are extracted from the MLMs.
6. The markers are then tested diagnostically to generate a personalized medication for each patient.

The first stages of the strategy were undertaken in studies of allergen-challenged Th2 cells and skin, which resulted in an identification of the modules, pathways, and individual genes. The corresponding proteins, some of which were potential diagnostic markers, could then be measured in both the cell supernatant and nasal fluid. Specifically, three pathways were found to be enriched for the differentially expressed genes in allergen-challenged skin, namely those induced by interleukin-4 (IL-4), transforming growth factor-beta (TGFβ), and chemokine ligand (CCL4). Whilst the first two of the three pathways had been described previously in allergy, the third pathway, involving CCL4, was novel. CCL4 is a chemokine that is released from Th2 cells and is known to attract eosinophils (the important effector cells in allergy). This finding suggested that the allergen-challenge to the skin had resulted in Th2 cells releasing CCL4, which in turn attracted the eosinophils. In support, a strong correlation was found between the expression levels of CCL4 and a gene that is specifically expressed in eosinophils, while CCL4 protein levels were also found to be significantly higher in nasal fluids from patients *during* the pollen season compared to pre-season. These findings demonstrate how network-based analysis can be used to structure and dissect complex data, in order to identify potential diagnostic markers. The analyses were performed using a commercial bioinformatic tool, known as ingenuity pathway analysis (IPA), which is very tractable for clinical studies. Unfortunately, however, the research strategy was confounded by two problems:

- The analytical tool was based on PPI defined by the manual curation of biomedical literature; this implied a knowledge-bias, such that the PPI might not be generally relevant to different cells and tissues.
- The disease-associated genes were defined by differential expression in gene expression microarrays.

Differential expression does not provide any information regarding inter-gene relationships, the alteration of which may be very important for the disease. For example, whilst the Th2 cells release the cytokine IL-4, which induces allergic inflammation, the Th2 cells are in turn inhibited by Th1 cells through the Th1 cytokine interferon-γ (IFN-γ). Thus, in theory an individual with high levels of IL-4 and IFN-γ could be healthy, whereas another with high IL-4 levels and low IFN-γ levels could be ill. Such interdependencies – which are common in complex diseases – would not be detected by differential expression. The means by which these problems were overcome are outlined in the next two sections.

4.2 Combining Different Sources of Biological Information to Identify Disease Modules and Genes

Recently, high-throughput methods to analyze single nucleotide polymorphisms (SNPs), proteins, and different regulatory elements such as microRNAs and DNA methylation, have been developed.

Because the layers and elements are interdependent, an analysis of dependencies can be used for the step-wise development and cross-validation of disease modules. The concept here is similar to solving a Rubik's cube, in that layers and elements that match each other are used to build modules of increasing complexity and accuracy. As an example, variations in gene expression may depend on SNPs (see Fig. 2); however, this association can only be validated by determining whether individuals with and without the SNPs have the expected differences in mRNA and protein expression. As another example, if gene expression microarrays have led to the identification of a disease-associated transcription factor (TF), then the targets of that TF can be identified using other methods such as chromatin immunoprecipitation chips (ChIP-chips) or chromatin immunoprecipitation-sequencing (ChIP-seq). It is proposed that such interdependencies could be used to build an MLM of increasing detail and accuracy, the main advantages being that the MLMs can be analyzed to identify more accurately disease pathways and key regulatory genes (as described in Fig. 1). In addition, they may contribute to an understanding of *why* those pathways and genes are involved. For example, the increased activity of a pathway could be due to a specific TF, which in turn could depend on a SNP that increases the expression of that TF. From a clinical perspective, the tracing of expression changes to the SNP may contribute to an understanding of how, and why, pathogenic mechanisms vary between patients. This information might be used for the hypothesis-based selection of protein markers for individualized medications.

4.3 Identification and Validation of Pathways Using Gene Expression Microarrays and SNP Analysis

As discussed above, gene expression is partially dependent on SNPs. Consequently, a disease-associated pathway may be identified by gene expression microarray analysis, and validated by the SNP analysis of genes or regulatory regions related to that pathway. According to the disease model, the Th2 cell plays a key role in allergy, and analyses of gene expression microarray data from allergen-challenged Th2 cells have supported this proposal. The T-cell receptor (TCR) signaling pathway was significantly enriched for genes identified by the gene expression microarray analysis. The next stage was to analyze SNPs in genes belonging to that pathway in an independent material of patients with SAR and healthy controls. This resulted in the identification of SNPs that differed significantly between patients and controls. These findings supported the suggestion that a network-based analysis could be used to identify modules and pathways that can, in turn, be validated by SNPs. On a genome-wide scale, however, this combination of differing methods and many sources of biological data present considerable computational and bioinformatic challenges.

5 Graph Theoretical Tools for Network-Based Analysis

Graph theory offers a level of abstraction that is suitable for dealing with the extreme computational challenges presented by high-throughput biological data. In this section, the terminology will be used that is generally applied to theoretical contexts: vertices are typically employed to represent genes, gene products, and other relevant entities; while edges are used to denote communication, correlation, or other forms of association. This approach can produce genome-scale graphs, in which the research groups seek relatively small, dense subgraphs in an effort to identify putative disease genes and modules.

Consider, for example, the analysis of gene coexpression data, for which differential analysis has proven to be a useful approach. A frequent first step is the relatively simple task of *differential expression*. It is commonplace to identify statistically significant, differentially expressed genes using a variety of tools, such as a mixed model ANOVA [23]. However, differential expression can of course deal only with individual genes (vertices). Nonetheless, given sufficient data the research group can build high-quality correlation matrices, threshold them to obtain correlation graphs, and then also study *differential correlation* [24, 25]. By this is meant a significant relationship (set of edges) found in one or more conditions, but not in others. With differential correlations, a variety of central questions can often be studied that may include determining relationships that exist between genes prior to a stimulus (but not afterwards and vice versa), deciding how relationships differ by tissue type, and identifying the putative factors that activate and suppress various networks. While differential expression is found at the vertex level, and differential correlation is reflected at the edge level, it may be helpful go a step further with *differential topology* [24, 25]. In this way, it is possible to identify changes that are best seen at the subgraph (i.e., network) level. For example, a dense subgraph that appears in disease, but not in health, may highlight a credible network of genes for further investigation.

One frequent goal is the elucidation of molecular response mechanisms in the form of genetic regulatory networks. In order to perform the requisite analysis, it is necessary to develop novel graph algorithms to generate highly distilled gene sets, to produce scalable implementations for cutting-edge HPC platforms, to use these implementations to extract gene sets suggestive of co-regulation, and to perform genomic data mining to highlight the most promising gene sets for detailed scrutiny [24, 26, 27]. In this respect, a wealth of clustering methods has been proposed (see Refs [28–32]), the most common being *hierarchical*, in which all genes begin in their own clusters and are eventually merged into one; or *centroid*, in which genes are organized into a predefined number of clusters by iterative adjustments based on similarity [33]. The clusters that these methods produce are typically disjoint, which places an artificial limitation on the biology under study in that many genes play important roles in multiple, but distinct, pathways [34]. Some recently developed clustering techniques – such as those employing factor analysis [35] – do not require exclusive cluster membership for single genes but, unfortunately, these tend to produce biologically uninterpretable factors without the incorporation of prior biological information [36]. Relevance networks [34, 37, 38] have been proposed as a means to overcome the limitations of traditional clustering methods. Without an algorithmic means to extract the aggregate relationships between multiple genes, however, the most interesting relationships (those with tight connections between multiple genes) remain embedded within the vast sea of correlations.

It has been necessary, therefore, to develop more powerful tools to extract subsets of coordinately regulated genes from large aggregates of gene expression data, and *graph theory* has offered unique advantages to this problem. Novel algorithms have been based on decades of basic research, and constitute a class of tools that can help to elucidate relationships in highly complex data structures, such as matrices of correlations across thousands of genes. In this respect, graph algorithms offer a means to extract meaningful aggregates of genes from within the relevance network framework. Weighted graphs are produced from this type of data; these consist of vertices representing genes and edges, the weights of which are indicative of the correlation between each pair of vertices (genes). Given a suitable threshold, t, edges with weights less than t are discarded, while edges with weights of at least t are retained [39, 40]; this produces an unweighted graph, G, the structural properties of which are of great interest. Similar representations have already been shown to be suitable for probing and determining the structure of biological networks, including the extraction of evolutionarily conserved modules of co-expressed genes [41–43].

The challenge – once the graph is created – is not to study the graph in its entirety but rather to extract its embedded subgraphs, or small, tightly connected regions of the graph that represent subsets of genes with strong correlations between every pair of its members and are thus likely to represent biologically significant interactions. In the most extreme case, in which a subgraph contains all possible edges between its vertices, this structure is termed a *clique*. Here, each and every pair of vertices is joined by an edge, from which can be inferred some form of co-regulation among the corresponding genes. Clique is widely known for its application in a variety of combinatorial settings, a great number of which are relevant to computational molecular biology [44]. It is particularly useful in microarray analysis, because it addresses the previously noted shortcomings of traditional clustering algorithms. Cliques need not be disjoint; a vertex can reside in more than one clique, just as a gene can operate in more than one regulatory network. In terms of gene expression, it has been said that clique represents the most trusted potential for identifying a set of interacting genes [34]. This general methodology is placed in the broader context of clustering for microarray analysis in Fig. 6.

The inputs to clique are an undirected graph G with n vertices, and a parameter $k \leq n$. The question asked is whether G contains a clique of size k, that is, a subgraph isomorphic to K_k. Subgraph isomorphism – and clique in particular – is NP-complete. from which it follows that there is no known algorithm for deciding clique that runs in time polynomial in the size of the input. Of course, clique could be solved by generating and checking all candidate solutions, but this brute force approach requires $O(n^k)$ time and is thus prohibitively slow, even for problem instances of only modest size. The temptation might be to try and solve clique approximately, rather than exactly, but clique is so difficult that guaranteeing solutions even to within only n^{ε} cannot be accomplished within polynomial time for any $\varepsilon > 0$, unless $P = NP$ [45]. It must be noted here that the use of thresholding and clique is distinct from a superficially similar approach employed in "signature algorithms" (including those described in Refs [46] and [47]), because with signature algorithms there is no notion of

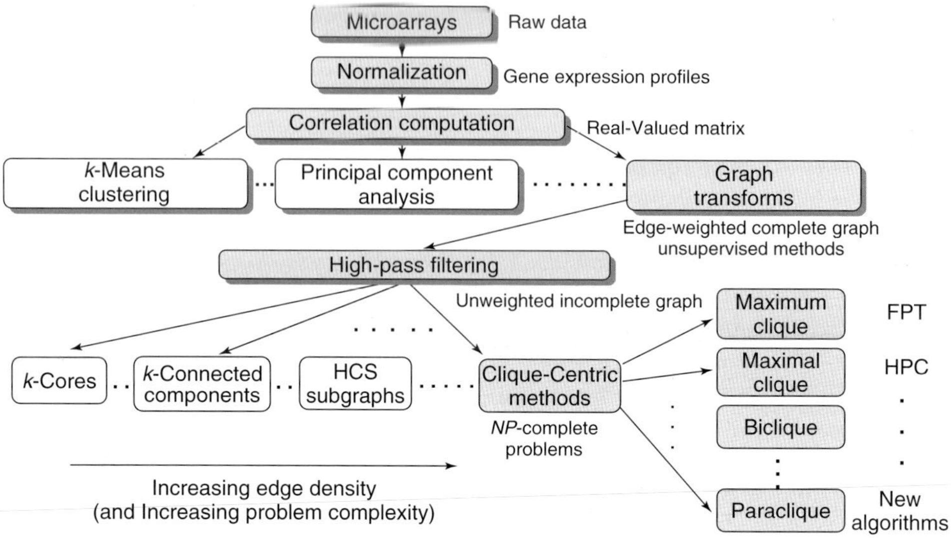

Fig. 6 An overview of microarray data analysis. An approach using graph algorithms is highlighted in gray. Raw array data are obtained and normalized. A variety of correlation computations may be employed to compute a gene–gene matrix, which is viewed as a weighted, complete graph. Thresholding then produces a corresponding unweighted, incomplete graph. Depending on the experimental set-up, a number of clique-centric algorithms may be used to extract dense subgraphs. *NP*-completeness demands powerful strategies. These include fixed-parameter tractability (FPT), high-performance computing (HPC), and innovative new algorithms to decompose complex relationships and ameliorate the effects of noise.

extreme edge density. Instead, modules derived with signature algorithms depend heavily on the seeds chosen, the number of iterations employed, and other factors. Likewise, the time required to run these algorithms depends on seeds, iterations, and convergence criteria. Another interesting approach is presented in Ref. [48]. This differs from graph theoretical methods, in that partial correlations are used without thresholding, and clusters are identified by genes sharing partial correlations that are stronger within the cluster than surrounding correlations. While the idea of partial correlations is intriguing, there is a danger that true relationships may be statistically adjusted away. For example, if three genes are correlated, then partial correlations between any two, adjusted for the third, could be zero, and that cluster would disappear.

A powerful, emergent recipe for clique finding is based on fixed parameter tractability (FPT), the roots of which can be traced at least as far back as the studies described in Refs [49, 50]. A problem is FPT if it has an algorithm that runs in $O(f(k)n^c)$ time, where n is the problem size, k is the input parameter, and c is a constant that is independent of both n and k. Clique is not FPT, however, unless the W hierarchy collapses [51]. (The W hierarchy, whose lowest level is FPT, can be viewed as a fixed-parameter analog of the polynomial hierarchy, whose lowest level is *P*.) Such a collapse is widely viewed as an exceedingly unlikely event, roughly on a par with the likelihood of the collapse of the polynomial hierarchy [52]. Thus, attention is focused instead on clique's complementary dual, the *vertex cover* problem. Consider G', the complement of G: G' has the same vertex set as G, but edges present in G are absent in G', and vice versa. The question now asked is whether G contains a set C of k vertices that covers every edge in G, where an edge is said to be covered if either or both of its endpoints are in C. Like clique, vertex cover is NP-complete; unlike clique, however, vertex cover is FPT. The crucial observation here is this: a vertex cover of size k in G turns out to be exactly the complement of a clique of size $n - k$ in G'. Thus, a minimum vertex cover in G is sought, and the search is continued to find the desired maximum clique in G'. Currently, the fastest known vertex cover algorithm runs in $O(1.2759^k k^{1.5} + kn)$ time [53]; contrast this with $O(n^k)$. The requisite exponential growth (assuming $P \neq NP$) is therefore reduced to a mere *additive* term. Recent investigations on this subject have been described in Refs [54–56].

Computational difficulties in network analysis arise when the problem of interest is intractable *and* exact solutions are preferred. For example, the computation of gene–gene and gene–phenotype relations frequently calls for the extraction of maximal cliques. (A maximal clique is one to which no more vertices can be added. It need not be the largest clique in the graph.) In the worst case, a graph of size n can have as many as $3^{n/3}$ maximal cliques [57], which can be exceedingly memory-intensive, even on massively parallel shared-memory supercomputers [58]. In addition to complex coding issues, it is necessary to argue successfully for access time, only to find that supercomputers are often in more or less prototype mode, with a relatively short mean time to failure and a plethora of upgrade and reliability issues. Moreover, it is also essential to deal with so-called "disruptive technologies," including distributed memory and multi-core processors. Special approaches and considerable expertise in both algorithms and implementations is generally required to obtain satisfactory performance on supercomputer platforms. One such handy approach is depicted in Fig. 7.

The integration of multiple genome-scale data sets brings with it huge algorithmic challenges for modern systems biology. In such an environment, bipartite graphs have proven useful in representing relationships across heterogeneous data types, with the interpretation of these relationships accomplished through an enumeration of maximal *bicliques* (see Fig. 8) (note that maximal is a local measure of optimality). No new vertex (and thus no new edge) can be added to a maximal biclique without sacrificing the biclique property. Maximum, on the other hand, is a global measure of optimality. No biclique can contain more vertices than a vertex-maximum biclique. No biclique can contain more edges than an edge-maximum biclique. As molecular response networks occur in many sizes, attention is usually focused on computing and enumerating maximal bicliques. Again, this task can be exceedingly memory-intensive, even on massively parallel shared-memory supercomputers. Fortunately, recent progress [59] has reduced run times by as much as 2 orders of magnitude, bringing genome-scale solutions now within reach.

Noise is a major concern. In observing that clique is an ideal cluster definition, every vertex (in this case every transcript) in a clique must be highly correlated with all others in that clique, by definition. Microarray data, on the other hand – at

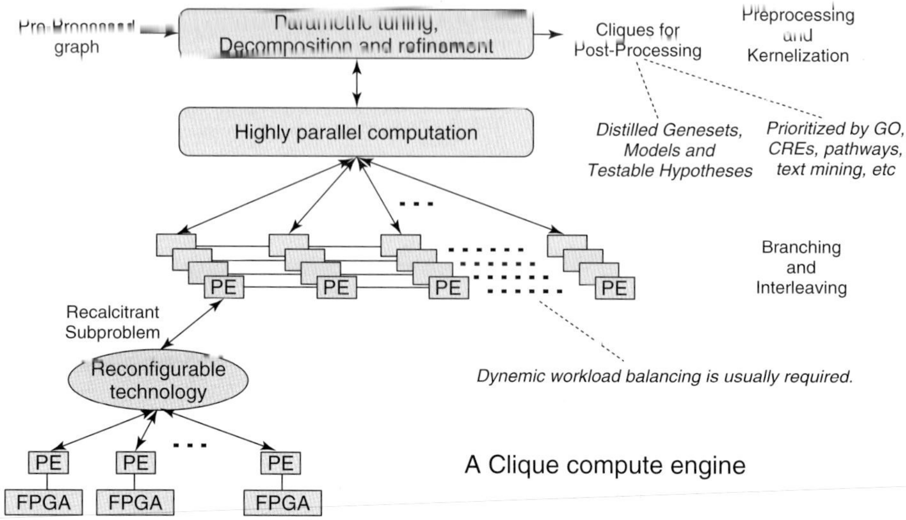

Fig. 7 A clique compute engine. Raw data are transformed into a finite and simple, but generally enormous, graph. Parametric tuning, preprocessing, and kernelization produce a compute core aka kernel, the size of which is linear in the chosen parameter (clique size). Exhaustive tree-structured branching computations are performed on this *NP*-complete core, with each processing element (PE) assigned a subtree. Especially difficult subproblems require dynamic load balancing and, in extreme cases, hardware acceleration in the form of field programmable gate arrays (FPGAs).

least as generated under current technologies – are inherently noisy. It seems highly unlikely that noise alone could cause an entire clique's worth of correlation coefficients to be excessively high, whereas common clustering methods including KNN and K-Means can allow transcripts into a group that are related due to noise. Thus, in contrast to many other popular approaches, clique does not seem to be plagued with false positives. On the other hand, if even one coefficient is incorrectly found to be too low, then the clique is lost, equaling a false negative. Several edges may even be missing from the largest subgraphs. The result is that a typical clique analysis may yield exceedingly large numbers of modest-sized, highly-overlapping cliques [60]. To aggregate these data, the need is to solve something akin to a dense-*k*-subgraph [61], which is as difficult as clique, even on graphs of maximum degree 3. To accomplish this, the research teams have devised novel methods such as *k* clique communities [62] and paraclique [63]; the former operates bottom-up, and the latter top-down. A paraclique, for example, is a clique augmented with vertices in a highly controlled manner to maintain density. (Algorithmic details are available in Refs [25, 63], and extensive empirical testing on real data in Refs [64, 65].)

6
An Example of a Combinatorial, Network-Based Study to Identify a Disease Module in SAR

The disease model in Fig. 4 is a simplification, in that several other cells are involved in SAR. In order to test the

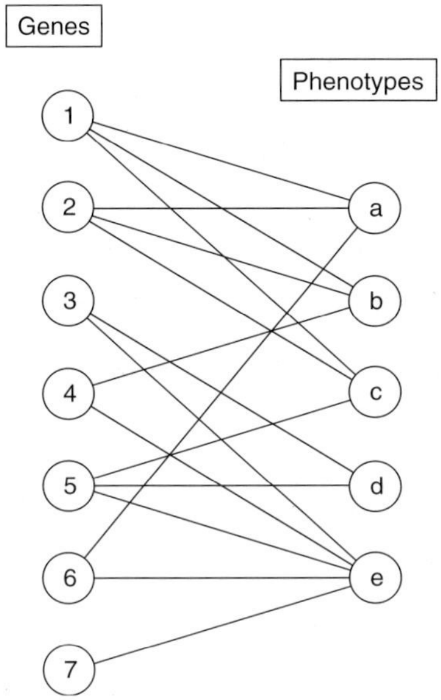

Fig. 8 A bipartite graph illustrates associations between disparate data types. In this example, genes are correlated with phenotypes. Vertex sets {1,2,a,b,c} and {3,4,5,6,7,e} are two of the graph's maximal bicliques. The former is also edge-maximum. With its six edges, no other biclique has more, and the latter is vertex-maximum. With its six vertices, no other biclique has more.

model, and the relevance of the Th2 cell, a clique-based analysis was performed of the gene expression microarray data from cells and tissues from patients with different forms of allergic disease (SAR, asthma, and eczema) [66]; this led to an identification of the most differentially correlated genes in the material. Pathway analysis using the IPA tool showed that TCR signaling was the pathway to be most enriched for differentially correlated genes, thus supporting the relevance of the disease model in Fig. 4. In order to define a module of genes that included the TCR signaling pathway, an examination was made to determine if this pathway was coregulated by the same TF, or by combinations of TFs. This was achieved by searching for combinations of transcription factor binding site s (TFBSs) that were conserved in different species and shared by the genes in the pathway. The rationale behind this was that functionally related genes are often coregulated by the same TFs. The search was performed using a commercial bioinformatic tool (Genomatix), but no shared combinations could be identified. Rather, a key gene in the pathway (using a knock-out mouse model) was identified, after which the bioinformatic tool was used to seek a TF that regulated that gene. Next, the targets of that TF were identified, and those that were differentially expressed in allergen-challenged CD4+ cells were selected; these genes were then arranged into a module, as described in Fig. 5. Previously, it has been shown that highly connected disease genes may be more relevant to the disease. Indeed, the most well-connected genes included several known disease genes as well as one novel gene, which were validated experimentally. The results of this study showed how a combinatorial, module-based approach could be used to identify and prioritize between disease genes, and the strategy was subsequently reviewed and applied to a study of atherosclerosis [67, 68]. Whilst this concept supported the proposal that the same network-based principles are generally applicable to complex diseases, one problem with both studies was that the modules were identified using sequence-based predictions of genes coregulated by the same TFs. Consequently, as such a strategy may be confounded by the low specificity of such predictions, improved methods for target predictions are called for.

6.1 Combining Gene Expression Microarrays, ChIP-Chip, and siRNA-Mediated Knockdowns to Identify a Module in SAR

In order to identify more accurately a module of coregulated genes in SAR, a gene expression microarray analysis of allergen-challenged CD4+ cells was first used to identify a highly regulated TF, the known functions of which supported its pathogenic relevance. Next, a ChIP-chip analysis was performed to identify the targets of that TF, and this resulted in the identification of almost 2000 targets (such a large number is not uncommon in ChIP-chip studies). The number was reduced by searching for actively regulated genes by combining the data from the gene expression microarrays with RNA polymerase II ChIP-chips (these genes formed part of several modules). Based on the disease model in Fig. 4, attention was focused on a module that was involved in regulating the release of Th2 cytokines. However, further analysis of that module revealed that the original TF regulated a secondary TF which, in turn, activated the release of Th2 cytokines. This analysis was based on performing siRNA-mediated knockdowns of the TFs in combination of gene expression microarrays before and after knockdown. Although complex and resource-demanding, this type of analysis may be generally applicable to identify modules in complex diseases. It should also be noted that the analysis may be simplified by the increasing availability of ChIP-chip data in the public domain, which in turn may result in the development of improved bioinformatic tools to identify modules of genes coregulated by the same TF.

6.2 Network-Based Analysis of Clinical Samples to Identify Diagnostic Markers

As previously discussed, the identification of diagnostic markers is complicated by the involvement of multiple cells and tissues, many of which may not be accessible to clinical studies. Although, in SAR, the nasal fluid proteins may contain diagnostic proteins, alterations in the concentrations of such proteins will depend on complex changes in the local secretion and transudation of nasal fluid from the blood. Furthermore, those proteins may be derived from nasal fluid cells and nasal mucosal cells, as well as from the blood. In order to identify possible diagnostic proteins for response to treatment with cortisone, a series of gene expression microarrays of the nasal fluid cells and the nasal mucosa were performed under different conditions, in addition to a proteomic analysis of the nasal fluids. This resulted in the identification of several pathways that overlapped only partially. The next stage was to select proteins encoded by the genes in those pathways, which were highly and differentially expressed. About half of these proteins were found to differ significantly from those in the nasal fluids of another cohort of patients before and after treatment with cortisone. In contrast, a network-based analysis of allergen-challenged $CD4^+$ cells and skin led to the identification of putative diagnostic proteins in the nasal fluids. The clinical relevance of these findings may lie in the identification of diagnostic proteins, perhaps relying on the network-based high-throughput analysis of several different cells and tissues, and that combinations of proteins representing different pathways and compartments are required for diagnostic purposes. It is also possible that the strategy could be applied to candidate modules of disease genes, rather than to individual candidate genes. Such modules might be useful for diagnostic purposes, as has been demonstrated in a study of liver cancer (Segal liver cancer), where the gene expression microarrays signatures of liver cancer were correlated to disease severity as determined using computed tomography [69].

7
Problems in Applying Network-Based Analyses to Clinical Research

Studies of different complex diseases have already demonstrated the feasibility of applying network-based analyses for concrete clinical purposes, such as identifying markers for personalized medication [3, 66, 70, 71]. Yet, relatively few clinical studies have been conducted to date in which such approaches have been applied. The reasons for this situation continue to be discussed extensively among systems biologists, with one important factor proposed to be the relatively short time span taken to develop genomic high-throughput technologies as tools for network-based analysis. A second factor is that the analysis of multiple genes may be seen as a paradigmatic shift from the detailed analyses of individual genes that previously has played such a dominant part in medical research. Furthermore, network-based analysis in clinical research requires large multidisciplinary collaborations that may include the services of clinicians and biomedical research teams, as well as experts in genomics, statistics, bioinformatics, and computer science [72]. Unfortunately, many academic organizations are simply not structured to conduct such research, and the investigators are generally not trained across several disciplines – though this would greatly facilitate such collaborations. Nonetheless, today several centers continue to focus on systems biology, which will undoubtedly contribute to the translation of network-based analyses for clinical research.

8
Future Perspectives

Although still in their infancy, recent advances in genomic high-throughput technologies, the increasing wealth of genomic data in public repositories, and the emergence of tools for network-based analysis should have significant implications for clinical research in the near future. On the disease-scale, network-based analysis has already resulted in the identification of protein markers to predict the occurrence of Alzheimer's disease [3], as well as candidate markers in several other diseases, including allergy [71]. By integrating high-throughput data from different genomic layers, such as DNA, mRNA, microRNAs, methylation, metabolites, and proteins, disease-associated MLMs may contribute to an understanding of disease mechanisms, and why such mechanisms may vary between individuals. This may, in turn, contribute to improved diagnostic and therapeutic targets for preventive and personalized treatments. Ideally, this will ultimately lead to physicians routinely measuring a set of disease markers in saliva, blood, or urine in order to optimize a patient's treatment. From a pharmaceutical perspective, an improved understanding of disease mechanisms – and how they vary between individuals – may improve the chances of developing effective drugs that would more accurately target subgroups of patients who would respond better to treatment. Conversely, an increased understanding of the mechanisms underlying a lack of response, or even a negative response, to a pharmaceutical agent may lead to the creation of novel drugs.

Another important future challenge might be to better understand the time scale of diseases, which may span from just a susceptibility to a life-threatening situation, and include stages of great heterogeneity. Such a scale can vary widely among different diseases; for example, cancer and obesity may develop over decades, whilst meningitis can develop from a light condition to a life-threatening disease within a few hours. In chronic diseases, multilayer network-based analyses of different data derived from long-term, population-based studies may lead to the creation of diagnostic and therapeutic tools that could specifically target different stages of the diseases. It is also possible that the same principles might later be applied to combined diagnostic and therapeutic strategies to create real-time interventions in acute diseases. This would require the development of methods for the continuous measurement and analysis of diagnostic markers, as well as the administration of therapeutic molecules.

On the diseasome-scale, meta-analyses of public repositories containing genomic high-throughput data, as well as phenotypic disease information, are likely to contribute to an understanding of how diseases and groups of diseases relate to underlying disease mechanisms, as well as to environmental factors. It is possible that this will unravel shared disease mechanisms and, in turn, lead to improved disease classifications based on combined functional and phenotypic observations.

9 Conclusions

The network-based analysis of genomic high-throughput data and related datasets may have far-reaching diagnostic and therapeutic implications. Although significant challenges remain, clinical studies have already been conducted illustrating the feasibility of identifying diagnostic markers in complex diseases. Such studies are likely to generate interest in the application of network-based analysis and contribute to the dissemination of such analyses.

Abbreviations

ChIP-chip	Chromatin immunoprecipitation promoter tiling arrays
ChIP-seq	Chromatin immunoprecipitation sequencing
CGN	Complex disease gene network
CNV	Copy number variants
GEO	Gene expression Omnibus
GO	Gene ontology
GSEA	Gene set enrichment analysis
GWAS	Genome-wide association study
IFN	Interferon
IL	Interleukin
IPA	Ingenuity pathway analysis
MLM	Multi-layer modules
OMIM	Online Mendelian inheritance in man
PPI	Protein–protein interaction
SAR	Seasonal allergic rhinitis
siRNA	Short interfering RNA
SNP	Single nucleotide polymorphism
TF	Transcription factor
TFBS	Transcription factor binding site.

References

1 Petretto, E., Liu, E.T., Aitman, T.J. (2007) A gene harvest revealing the archeology and complexity of human disease. *Nat. Genet.*, **39** (11), 1299–1301.

2 Plomin, R., Haworth, C.M., Davis, O.S. (2009) Common disorders are quantitative traits. *Nat. Rev. Genet.*, **10** (12), 872–878.

3 Ray, S., Britschgi, M., Herbert, C., Takeda-Uchimura, Y., Boxer, A., Blennow, K., Friedman, L.F., Galasko, D.R., Jutel, M., Karydas, A., Kaye, J.A., Leszek, J., Miller, B.L., Minthon, L., Quinn, J.F., Rabinovici, G.D., Robinson, W.H., Sabbagh, M.N., So, Y.T., Sparks, D.L., Tabaton, M., Tinklenberg, J., Yesavage, J.A., Tibshirani, R., Wyss-Coray, T. (2007) Classification and prediction of clinical Alzheimer's diagnosis based on plasma signaling proteins. *Nat. Med.*, **13** (11), 1359–1362.

4 Special Issue (1999) Complex systems. *Science*, **284** (5411), 1–212.

5 (2003) Networks in biology. *Science*, **301** (5641), 1797–1980.

6 Barabasi, A.L., Albert, R. (1999) Emergence of scaling in random networks. *Science*, **286** (5439), 509–512.

7 Jeong, H., Mason, S.P., Barabási, A.L., Oltvai, Z.N. (2001) Lethality and centrality in protein networks. *Nature*, **411** (6833), 41–42.

8 Lim, J., Hao, T., Shaw, C., Patel, A.J., Szabó, G., Rual, J.F., Fisk, C.J., Li, N., Smolyar, A., Hill, D.E., Barabási, A.L., Vidal, M., Zoghbi, H.Y. (2006) A protein-protein interaction network for human inherited ataxias and disorders of Purkinje cell degeneration. *Cell*, **125** (4), 801–814.

9 Jenssen, T.K., Laegreid, A., Komorowski, J., Hovig, E. (2001) A literature network of human genes for high-throughput analysis of gene expression. *Nat. Genet.*, **28** (1), 21–28.

10 Segal, E., Friedman, N., Koller, D., Regev, A. (2004) A module map showing conditional activity of expression modules in cancer. *Nat. Genet.*, **36** (10), 1090–1098.

11 Pujana, M.A., Han, J.D., Starita, L.M., Stevens, K.N., Tewari, M., Ahn, J.S., Rennert, G., Moreno, V., Kirchhoff, T., Gold, B., Assmann, V., Elshamy, W.M., Rual, J.F., Levine, D., Rozek, L.S., Gelman, R.S., Gunsalus, K.C., Greenberg, R.A., Sobhian, B., Bertin, N., Venkatesan, K., Ayivi-Guedehoussou, N., Solé, X., Hernández, P., Lázaro, C., Nathanson, K.L., Weber, B.L., Cusick, M.E., Hill, D.E., Offit, K., Livingston, D.M., Gruber, S.B., Parvin, J.D., Vidal, M. (2007) Network modeling links breast cancer susceptibility and centrosome dysfunction. *Nat. Genet.*, **39** (11), 1338–1349.

12 Barabási, A.L. (2007) Network medicine–from obesity to the "diseasome". *N. Engl. J. Med.*, **357** (4), 404–407.

13 Christakis, N.A., Fowler, J.H. (2007) The spread of obesity in a large social network over 32 years. *N. Engl. J. Med.*, **357** (4), 370–379.

14 Goh, K.I., Cusick, M.E., Valle, D., Childs, B., Vidal, M., Barabási, A.L. (2007) The human disease network. *Proc. Natl Acad. Sci. USA*, **104** (21), 8685–8690.

15 Feldman, I., Rzhetsky, A., Vitkup, D. (2008) Network properties of genes harboring inherited disease mutations. *Proc. Natl Acad. Sci. USA*, **105** (11), 4323–4328.

16 Barrenas, F., Chavali, S., Holme, P., Mobini, R., Benson, M. (2009) Network properties of complex human disease genes identified through genome-wide association studies. *PLoS ONE*, **4** (11), e8090.

17 Loscalzo, J., Kohane, I., Barabasi, A.L. (2007) Human disease classification in the postgenomic era: a complex systems approach to human pathobiology. *Mol. Syst. Biol.*, **3**, 124.

18 Hidalgo, C.A., Blumm, N., Barabási, A.L., Christakis, N.A. (2009) A dynamic network approach for the study of human phenotypes. *PLoS Comput. Biol.*, **5** (4), e1000353.

19 Chavali, S., Barrenas, F., Kanduri, K., Benson, M. (2010) Network properties of human disease genes with pleiotropic effects. *BMC Syst. Biol.*, **4**, 78.

20 Wachi, S., Yoneda, K., Wu, R. (2005) Interactome-transcriptome analysis reveals the high centrality of genes differentially expressed in lung cancer tissues. *Bioinformatics*, **21** (23), 4205–4208.

21 Jonsson, P.F., Bates, P.A. (2006) Global topological features of cancer proteins in the human interactome. *Bioinformatics*, **22** (18), 2291–2297.

22 Benson, M., Carlsson, L., Guillot, G., Jernås, M., Langston, M.A., Rudemo, M., Andersson, B. (2006) A network-based analysis of allergen-challenged CD4+ T cells

from patients with allergic rhinitis. *Genes Immun.*, **7** (6), 514–521.

23 Kerr, M.K., Martin, M., Churchill, G.A. (2000) Analysis of variance for gene expression microarray data. *J. Comput. Biol.*, **7** (6), 819–837.

24 Voy, B.H., Scharff, J.A., Perkins, A.D., Saxton, A.M., Borate, B., Chesler, E.J., Branstetter, L.K., Langston, M.A. (2006) Extracting gene networks for low-dose radiation using graph theoretical algorithms. *PLoS Comput. Biol.*, **2** (7), e89.

25 Langston, M.A., Perkins, A.D., Saxton, A.M., Scharff, J.A., Voy, B.H. (2008) Innovative computational methods for transcriptomic data analysis: a case study in the use of FPT for practical algorithm design and implementation. *Comput. J.*, **51**, 26–38.

26 Baldwin, N.E., Chesler, E.J., Kirov, S., Langston, M.A., Snoddy, J.R., Williams, R.W., Zhang, B. (2005) Computational, integrative, and comparative methods for the elucidation of genetic coexpression networks. *J. Biomed. Biotechnol.*, **2005** (2), 172–180.

27 Chesler, E.J., Lu, L., Shou, S., Qu, Y., Gu, J., Wang, J., Wang, J., Hsu, H.C., Mountz, J.D., Baldwin, N.E., Langston, M.A., Threadgill, D.W., Manly, K.F., Williams, R.W. (2005) Complex trait analysis of gene expression uncovers polygenic and pleiotropic networks that modulate nervous system function. *Nat. Genet.*, **37** (3), 233–242.

28 Bellaachia, A., Portnoy, D., Chen, Y., Elkahloun, A.G. (2002) E-cast: a data mining algorithm for gene expression data. Proceedings, Workshop on Data Mining in Bioinformatics, Edmonton, Alberta, Canada, pp. 49–54.

29 Ben-Dor, A., Bruhn, L., Friedman, N., Nachman, I., Schummer, M., Yakhini, Z. (2000) Tissue classification with gene expression profiles. *J. Comput. Biol.*, **7** (3-4), 559–583.

30 Ben-Dor, A., Shamir, R., Yakhini, Z. (1999) Clustering gene expression patterns. *J. Comput. Biol.*, **6** (3-4), 281–297.

31 Hansen, P., Jaumard, B. (1997) Cluster Analysis and Mathematical Programming. *Math. Program.*, **79**, 191–215.

32 Hartuv, E., Schmitt, A., Lange, J., Meier-Ewert, S., Lehrachs, H., Shamir, R. (1999) An algorithm for clustering cDNAs for gene expression analysis. Proceedings, RECOMB, Lyon, France, pp. 188–197.

33 Slonim, D.K. (2002) From patterns to pathways: gene expression data analysis comes of age. *Nat. Genet.*, **32** (Suppl.), 502–508.

34 Butte, A.J., Tamayo, P., Slonim, D., Golub, T.R., Kohane, I.S. (2000) Discovering functional relationships between RNA expression and chemotherapeutic susceptibility using relevance networks. *Proc. Natl Acad. Sci. USA*, **97** (22), 12182–12186.

35 Alter, O., Brown, P.O., Botstein, D. (2000) Singular value decomposition for genome-wide expression data processing and modeling. *Proc. Natl Acad. Sci. USA*, **97** (18), 10101–10106.

36 Girolami, M., Breitling, R. (2004) Biologically valid linear factor models of gene expression. *Bioinformatics*, **20** (17), 3021–3033.

37 Patti, M.E., Butte, A.J., Crunkhorn, S., Cusi, K., Berria, R., Kashyap, S., Miyazaki, Y., Kohane, I., Costello, M., Saccone, R., Landaker, E.J., Goldfine, A.B., Mun, E., DeFronzo, R., Finlayson, J., Kahn, C.R., Mandarino, L.J. (2003) Coordinated reduction of genes of oxidative metabolism in humans with insulin resistance and diabetes: potential role of PGC1 and NRF1. *Proc. Natl Acad. Sci. USA*, **100** (14), 8466–8471.

38 Allocco, D.J., Kohane, I.S., Butte, A.J. (2004) Quantifying the relationship between co-expression, co-regulation and gene function. *BMC Bioinf.*, **5**, 18.

39 Borate, B.R., Chesler, E.J., Langston, M.A., Saxton, A.M., Voy, B.H. (2009) Comparison of threshold selection methods for microarray gene co-expression matrices. *BMC Res. Notes*, **2**, 240.

40 Perkins, A.D., Langston, M.A. (2009) Threshold selection in gene co-expression networks using spectral graph theory techniques. *BMC Bioinf.*, **10** (Suppl. 11), S4.

41 Alon, U. (2003) Biological networks: the tinkerer as an engineer. *Science*, **301**, 1866–1867.

42 Barabási, A.L., Oltvai, Z.N. (2004) Network biology: understanding the cell's functional organization. *Nat. Rev. Genet.*, **5** (2), 101–113.

43 Oltvai, Z.N., Barabási, A.L. (2002) Systems biology. Life's complexity pyramid. *Science*, **298** (5594), 763–764.

44 Setubal, J.C., Meidanis, J. (Eds) (1997) *Introduction to Computational Molecular Biology*, PWS Publishing Company, Boston.

45 Feige, U., Goldwasser, S., Lovasz, L., Safra, S., Szegedy M. (1991) Approximating the maximum clique is almost NP-complete. Proceedings, IEEE Symposium on the Foundations of Computer Science, San Juan, Puerto Rico, pp. 2–12.

46 Ihmels, J., Friedlander, G., Bergmann, S., Sarig, O., Ziv, Y., Barkai, N. (2002) Revealing modular organization in the yeast transcriptional network. *Nat. Genet.*, **31** (4), 370–377.

47 Kloster, M., Tang, C., Wingreen, N.S. (2005) Finding regulatory modules through large-scale gene-expression data analysis. *Bioinformatics*, **21** (7), 1172–1179.

48 Magwene, P.M., Kim, J. (2004) Estimating genomic coexpression networks using first-order conditional independence. *Genome Biol.*, **5** (12), R100.

49 Fellows, M.R., Langston, M.A. (1988) Nonconstructive tools for proving polynomial-time decidability. *J. ACM*, **35**, 727–739.

50 Fellows, M.R., Langston, M.A. (1994) On search, decision and the efficiency of polynomial-time algorithms. *J. Comput. Syst. Sci.*, **49**, 769–779.

51 Downey, R.G., Fellows, M.R. (Eds) (1999) *Parameterized Complexity*, Springer, New York.

52 Garey, M.R., Johnson, D.S. (Eds) (1990) *Computers and Intractability; a Guide to the Theory of Np-Completeness*, W. H. Freeman and Company.

53 Chandran, L.S., Grandoni, F. (2005) Refined memorisation for vertex cover. *Inf. Process. Lett.*, **93** (3), 123–131.

54 Abu-Khzam, F.N., Langston, M.A., Shanbhag, P., Symons, C.T. (2006) Scalable parallel algorithms for FPT problems. *Algorithmica*, **45**, 269–284.

55 Abu-Khzam, F.N., Langston, M.A., Suters, W.H. (2005) Effective vertex cover kernelization: a tale of two algorithms. Proceedings, ACS/IEEE International Conference on Computer Systems and Applications, Cairo, Egypt, doi:

56 Abu-Khzam, F.N., Collins, R.L., Fellows, M.R., Langston, M.A., Suters, W.H., Symons C.T. (2004) Kernelization algorithms for the vertex cover problem: theory and experiments. Proceedings, Workshop on Algorithm Engineering and Experiments (ALENEX), New Orleans, Louisiana, USA, pp. 62–69.

57 Tomita, E., Tanaka, A., Takahashi, H. (2006) The worst-case time complexity for generating all maximal cliques and computational experiments. *Theor. Comput. Sci.*, **363**, 28–42.

58 Zhang, Y., Abu-Khzam, F.N., Baldwin, N.E., Chesler, E.J., Langston, M.A., Samatova, N.F. (2005) Genome-scale computational approaches to memory-intensive applications in systems biology. Proceedings, Supercomputing, Seattle, Washington, USA, doi:

59 Zhang, Y., Chesler, E.J., Langston, M.A. (2008) On finding bicliques in bipartite graphs: a novel algorithm with application to the integration of diverse biological data types. Proceedings, Hawai'i International Conference on System Sciences, Waikoloa Village, Big Island, Hawaii, USA, doi:

60 Langston, M.A., Lan, L., Peng, X., Baldwin, N.E., Symons, C.T., Zhang, B., Snoddy, J.R. (2005) A Combinatorial Approach to the Analysis of Differential Gene Expression Data: The Use of Graph Algorithms for Disease Prediction and Screening, in: Johnson, K.F., Lin, S.M. (Eds) *Methods of Microarray Data Analysis IV, Papers from CAMDA '03*, Kluwer Academic Publishers, Boston, pp. 223–238.

61 Feige, U., Peleg, D., Kortsarz, G. (2001) The dense K-subgraph problem. *Algorithmica*, **29**, 410–421.

62 Palla, G., Derényi, I., Farkas, I., Vicsek, T. (2005) Uncovering the overlapping community structure of complex networks in nature and society. *Nature*, **435** (7043), 814–818.

63 Chesler, E.J., Langston, M.A. (2006) Combinatorial Genetic Regulatory Network Analysis Tools for High Throughput Transcriptomic Data, in: Eskin, E. (Ed.) *Systems Biology and Regulatory Genomics*, Vol. **4023**, Springer, pp. 150–165.

64 Eblen, J.D., Gerling, I.C., Saxton, A.M., Wu, J., Snoddy, J.R., Langston, M.A. (2009) Graph Algorithms for Integrated Biological Analysis, with Applications to Type 1 Diabetes Data, in: Chaovalitwongse, W.A. (Ed.) *Clustering Challenges in Biological Networks*, World Scientific, pp. 207–222.

65 Jay, J.J., Eblen, J.D., Zhang, Y., Benson, M., Perkins, A.D., Saxton, A.M., Voy, B.H.,

Chesler, E.J., Langston, M.A. (2011) A systematic comparison of genome scale clustering algorithms. *Lecture Notes in Bioinformatics*, **6674**, 416–427.

66 Mobini, R., Andersson, B.A., Erjefält, J., Hahn-Zoric, M., Langston, M.A., Perkins, A.D., Cardell, L.O., Benson, M. (2009) A module-based analytical strategy to identify novel disease-associated genes shows an inhibitory role for interleukin 7 receptor in allergic inflammation. *BMC Syst. Biol.*, **3**, 19.

67 Tegnér, J. (2009) Can modular analysis identify disease-associated candidate genes for therapeutics? *J. Biol.*, **8** (5), 48.

68 Hägg, S., Skogsberg, J., Lundström, J., Noori, P., Nilsson, R., Zhong, H., Maleki, S., Shang, M.M., Brinne, B., Bradshaw, M., Bajic, V.B., Samnegård, A., Silveira, A., Kaplan, L.M., Gigante, B., Leander, K., de Faire, U., Rosfors, S., Lockowandt, U., Liska, J., Konrad, P., Takolander, R., Franco-Cereceda, A., Schadt, E.E., Ivert, T., Hamsten, A., Tegnér, J., Björkegren, J. (2009) Multi-organ expression profiling uncovers a gene module in coronary artery disease involving transendothelial migration of leukocytes and LIM domain binding 2: The Stockholm Atherosclerosis Gene Expression (STAGE) study. *PLoS Genet.*, **5** (12), e1000754.

69 Segal, E., Sirlin, C.B., Ooi, C., Adler, A.S., Gollub, J., Chen, X., Chan, B.K., Matcuk, G.R., Barry, C.T., Chang, H.Y., Kuo, M.D. (2007) Decoding global gene expression programs in liver cancer by noninvasive imaging. *Nat. Biotechnol.*, **25** (6), 675–680.

70 Wang, H., Barrenäs, F., Bruhn, S., Mobini, R., Benson, M. (2009) Increased IFN-gamma activity in seasonal allergic rhinitis is decreased by corticosteroid treatment. *J. Allergy Clin. Immunol.*, **124** (6), 1360–1362.

71 Wang, H., Chavali, S., Mobini, R., Muraro, A., Barbon, F., Boldrin, D., Åberg, N., Benson, M. (2011) A pathway-based approach to find novel markers of local glucocorticoid treatment in intermittent allergic rhinitis. *Allergy*, **66** (1), 132–140.

72 Clermont, G., Auffray, C., Moreau, Y., Rocke, D.M., Dalevi, D., Dubhashi, D., Marshall, D.R., Raasch, P., Dehne, F., Provero, P., Tegner, J., Aronow, B.J., Langston, M.A., Benson, M. (2009) Bridging the gap between systems biology and medicine. *Genome Med.*, **1** (9), 88.

19
Personalized Medicine (Predictive and Preventive)

Alette M. Wessels[1], Robert R. Bies[1], and John Urquhart[2,3]
[1]Indiana University School of Medicine, Department of Clinical Pharmacology, Department of Medicine, 1001 W. 10th Street, W.D. Myers Bldg W7125, Indianapolis, IN 46202, USA
[2]AARDEX Group, S.A., Ave de la Gare, 29, 1950 Sion, Switzerland
[3]University of California San Francisco Medical Center, Department of Biomedical Engineering and Therapeutic Sciences, 505 Parnassus Ave, San Francisco, CA 94122, USA

Systems Biology: Advances in Molecular Biology and Medicine, First Edition. Edited by Robert A. Meyers.

Keywords

Personalized medicine; history; and changing definition
Medical practice is inherently personalized. Its long history has a consistent theme of listening to and questioning patients, careful examination of patients with available means, diagnosis according to the prevailing classification of disease, and prescription of treatment with available means. The twentieth century has seen a rise in pharmaceutical medicine and great advances in surgery, increasingly refined definitions and classification of disease, and greatly expanded means for examining patients. These changes have refocused the meaning of "personalized medicine" to: the achievement of low-variability responses to the administration of appropriately prescribed drugs.

Pharmacogenomics
The study of the influence of genetic variations on drug responses in patients that is accomplished by correlating gene expression or single nucleotide polymorphisms (SNPs) with a drug's efficacy and/or toxicity. The biochemical mechanisms involved in pharmacokinetics and pharmacodynamics appear to have many pharmacogenomic underpinnings.

Dosage form kinetics
The kinetics of drug release from formulated drugs.

Patient adherence
The likelihood of a patient to maintain his/her acceptance of a prescribed drug dosing regimen.

Pharmacokinetics
The relationship between drug release kinetics from an administered dosage form and the ensuing time course of drug concentration in plasma.

Pharmacodynamics
The relationship between the time course of drug concentration in plasma and drug action.

Although personalized medicine has a long history, today it implies low-variability responses to appropriately prescribed drugs. Achieving large reductions in variability requires reductions to be made in each of the many sources of variance in drug response. The four principal sources are: (i) the kinetics of drug release from formulated drug; (ii) the patients' variable adherence to prescribed dosing regimens; (iii) pharmacokinetics; and (iv) pharmacodynamics. Each source contributes nonlinearly to the overall variability in the relationship between the administered dose and the time courses of the drug's actions: the overall variance in drug response is the square root of the sums of the squares of each source's respective variances. This nonlinearity is a system property that arises from statistical principles; thus, large reductions in an individual source of variability will cause only a modest reduction in the overall variability in drug response. To achieve a substantial reduction in overall variability requires substantial reductions in each principal source of variance. Pharmacogenomics underlies the biochemical mechanisms involved in both pharmacokinetics and pharmacodynamics, and advances in this area will most likely lead to reductions in variance arising from these two sources. This alone will not achieve the goal of making major reductions in the variability of responses to drug treatment, however. Rather, to achieve that goal – seen as the capstone of personalized medicine – will require comparable reductions to be made in all four major sources of variance in drug response.

1
Some History and Some Fundamentals

The term "personalized medicine," when taken at face value, signifies a clinical process that has a history reaching back beyond the time of Hippocrates (460–377 BCE), who was a physician, scholar, and author in the Periclean age of ancient Greece. The "clinical process" is the questioning and examination of a patient by an observant physician, giving rise to the physician's judgment about the nature of the patient's problems, interpreted against the prevailing classification of diseases, leading to a recommendation for treatment, chosen from available means. Of course, in ancient Greece medicine was a primitive art, with crude conceptions of disease, limited means of observation, and few effective methods of treatment.

However, by some 1500 years later, medicine had acquired a set of principles of diagnosis and treatments, codified in the *Canon of Medicine,* written by Abd Allah Ibn Sina, who was known in the West by his Latinized name, Avicenna (c. 980–1037) [1]. Ibn Sina's *Canon of*

Medicine was the universal textbook of medicine and surgery until the mid seventeenth century, in both Europe and the Middle East.

1.1 On Medicaments

Until the end of the nineteenth century, medicaments were limited to poorly standardized botanicals and animal products. For example, Ibn Sina recommended that surgeons use the following to arrest bleeding: rabbit-hair, cobweb, egg white, and the cautery (*Canon of Medicine*, p. 535). The transition to chemistry-based pharmaceuticals began with the acetylation of salicylic acid in 1853 and its subsequent development, at the laboratories of Bayer during the 1890s, into a pharmaceutical product that became known as *aspirin*. Perhaps most surprising, the period between the discovery of aspirin and its first bedside use in 1899 involved a 46-year delay.

1.2 On Microorganisms

The decade of the 1880s witnessed an unprecedented amount of intensive research into, and the discovery of, many of the microorganisms that played causative roles in not only the fermentation of food products but also animal and human diseases, including tuberculosis, syphilis, pneumonia, typhoid fever, cholera, erysipelas and other streptococcal infections, gangrene, and anthrax. The immediate practical consequences of this were the adoption of antiseptic methods in surgery, in addition to hygienic methods in sanitation and food preparation, based on the germ-theory of disease.

Although, not all diseases have bacterial or viral origins, bacterial infections had always been recognized as the predominant causes of death in infants, children, mothers in childbirth, and adults of all ages. Subsequently, the recognition and wide acceptance of the germ-theory of disease set the stage for remarkable changes in these mortality patterns, first from the expanded application of hygienic methods, followed by the advent of the sulfa drugs during the mid-1930s, and penicillin and other antibiotics during the late 1940s and beyond. Concomitant improvements were also made in the area of nutrition.

During the first decade of the twentieth century, the 1908 Nobel Laureate in Physiology or Medicine, Paul Ehrlich, screened hundreds of synthetic organic compounds for their therapeutic action, while working at the pharmaceutical company Casella in Frankfurt. Ultimately, he arrived at an arsenical compound known as *arsphenamine* which bore the number 606, indicating its position among agents screened against *Treponema pallidum*, the causative agent of syphilis, which had been pandemic in western Europe at least since the fifteenth century. Unfortunately, arsphenamine proved to be a far-from-ideal chemotherapeutic agent, and drastic reductions in the prevalence of syphilis, and in the incidence of new cases thereof, were forced to await the discovery and industrial development of penicillin during and after World War II.

1.3 Visualizing the Body's Interior

In 1895, Wilhelm Röntgen earned the first Nobel Prize in Physics by his discovery of a then-new form of radiation, which he named "X-rays," and which enabled visualization of the interior structures of the body. This discovery, combined with several subsequent inventions, led to the start of a century of development of means for the noninvasive visualization of the interior of the human body, with increasing sensitivity and resolution. The extensive clinical utility of X-ray examinations in medical care stimulated the search for other means of minimally or noninvasive means for directly visualizing the body's interior, with its often uniquely personalized variations. More recently, ultrasound and fiber-optics have greatly expanded the range of what can be seen and treated, an example being the entire interior of the human gastrointestinal tract, using fiber-optic endoscopy. Concomitantly, the range of applications of X-rays was also greatly expanded as the sensitivity of X-ray imaging was increased and means were developed to place contrast media at local sites (e.g., the coronary arteries) to define with great precision the sites and sizes of obstructive lesions.

1.4 The Beginnings of Biotechnology-Based Pharmaceuticals

The discovery and subsequent development during the 1920s of a commercial means for preparing insulin (this was the subject of a controversial Nobel Prize in Medicine or Physiology in 1923) led to a transformation in the treatment of diabetes mellitus (DM). Suddenly, DM was transformed from being a certain death sentence to a manageable condition, despite extensive complications that included a high incidence of premature death, blindness, neuropathy, renal failure, myocardial infarction, and limb loss from premature peripheral arterial disease. Since that time, substantial progress has been made in the prevention of these complications, however, in part by increasing the efficacy of personalized metabolic management, based on patient-administered tests to determine metabolic status and to make corresponding adjustments of insulin dosage and modifications to the patients' diets.

1.5 The Beginnings of Systems Analyses in Physiology

The earliest applications of systems theory occurred almost simultaneously, and somewhat bizarrely, at the Bell Laboratories as an automatic volume control for amplitude modulated (AM) radio, and in investigations by the German physiologist, Eberhard Koch, of the self-regulation of blood pressure via the carotid sinus reflex [2]. The earliest recorded uses of systems analyses in physiology and biomedical engineering were during World War II, and these were continued post war using computers that originally had been applied to areas such as the disencryption of coded messages, the physiological regulation of ventilation for high-altitude flying, and the control of ballistics. Although the modeling and simulation of pharmacometrics and clinical trials of new drugs were commenced during the 1970s, however, the subject emerged as a discrete discipline only during the 1990s.

1.6 The Beginnings of Anti-Microbial Pharmaceuticals

It was during the 1930s that chemists at Bayer, in Germany, first discovered the antibacterial effects of an agent known as *prontosil*. Strangely, prontosil lacked an antibacterial action *in vitro* but in humans was strikingly active against streptococcal infections in particular. Very soon, it was shown by a French research group that prontosil was in fact a *prodrug* that was metabolized in the body to form the active metabolite, sulfanilamide, together with another inactive molecule. As a consequence, sulfanilamide became the first widely used antimicrobial product.

During World War II, massive efforts were made to adapt laboratory-scale fermentations to an industrial scale, in order to produce large amounts of penicillin, which was deemed to be the first of the antibiotics. Indeed, the many successful uses of penicillin led to its discoverer, Alexander Fleming, and its two industrial developers, Howard Florey and Ernst Chain, being awarded the Nobel Prize in Physiology and Medicine in 1945. Subsequently, in the aftermath of World War II, many successful research investigations were focused on identifying new antibiotics from natural sources, and this led to the discovery and development of manufacturing methods for streptomycin, the tetracyclines, erythromycin, and other antibiotics. Penicillin provided a close approximation to Ehrlich's original concept of a so-called "magic bullet" that could kill bacteria in patients, without adversely affecting their cells, tissues, or organs. Although, in some cases, patients may develop an allergy to penicillin which, in extreme cases may be fatal, penicillin has nonetheless withstood the "test of time" in its claim to be otherwise harmless.

1.7 The Beginnings of Large-Scale Public Support for Biomedical Research

The public support for biomedical research and product development during World War II led to the creation of large national research institutions capable of both performing and financing biomedical research at the highest levels. Such public policies were adopted not only in the US, with the National Institutes of Health and the National Science Foundation, but also in Canada, UK, France, Germany, Netherlands, Sweden, and other western European countries. One early scientific triumph that had been heavily supported by public monies was the discovery of the structure of DNA by Watson, Crick, Wilkins, and Franklin in 1953. This success, which was recognized in 1962 by the award of the Nobel Prize in Physiology and Medicine, set the stage for a vast amount of research that ultimately would result an ability to identify the molecular bases for human diseases and to derive their novel treatments.

1.8 The Rise and Partial Fall of Coronary Artery Disease as a Leading Cause of Death

Myocardial infarction was not recognized as a clinical entity until 1912 [3], at a time when cardiovascular diseases were considered among the leading causes of death, but were not yet at the forefront. By about 1920, however, cardiovascular disease (which included myocardial infarction and ischemic heart disease) had begun to emerge as the leading cause of death in the US, Canada, and western Europe. Subsequently, by the end of World War II there was a steady, year-on-year growth in the proportion of deaths attributed to cardiovascular disease, with myocardial infarction as a leading contributor.

The National Heart Institute, as one of the US National Institutes of Health, was formed during the late 1940s, and had an initial annual budget of US$ 500 000 – about equal to that year's budget for agricultural research on a disease afflicting Long Island potatoes! The situation changed very rapidly, however, as a strong national consensus was formed to support the research needed to launch an attack on cardiovascular and other leading diseases. The number of deaths due to cardiovascular disease peaked shortly after about 1970, and then began a long and steady decline of approximately 2% per year. Whilst the specific reasons for this decline were unclear, it was most likely due to a mixture of changes in "lifestyles" (e.g., diet and exercise), an increasingly improved means of treating hypertension, the development of a series of lipid-modifying drugs, with the introduction in 1987 of the first among a series of lipid-modifying drugs termed "*statins*," and an emerging consensus regarding the target levels of the various fractions of cholesterol at which the patients and their physicians should aim. The basic discoveries that led to the creation of the "statins" were the subject of the 1985 Nobel Prize in Physiology and Medicine.

Another outstanding line of progress has been the development of an expanding array of surgical approaches to relieve obstructions to the coronary arterial blood flow. These techniques have included coronary artery bypass surgery and coronary angioplasty, in addition to an expanding array of radiographic methods capable of precisely defining the location and extent of atheromatous lesions in various branches of the coronary arterial tree. Following their identification, such lesions could then either be bypassed by vascular grafting or attacked directly, using angioplastic therapies.

Today, it is important to celebrate – under the rubric of "personalized medicine" – the ability to visualize, quantify and, to attack with great precision the highly individualized obstructive lesions in the coronary arterial tree of each patient.

1.9 The Beginnings of Organized Surveillance for Drug-Related Hazards

During the first decade of the twentieth century a series of catastrophic adverse effects with some new vaccines and medicines led to a major expansion in the regulations pertaining to the research and development of new drugs and medical devices. Subsequently, these regulations have stimulated the widening use and development of statistical methods for evaluating the efficacy and safety of new drugs and devices. Such statistical methods are based on the randomized assignment of patients who meet certain criteria for inclusion into or exclusion from other groups of people to be studied, whereby one group receives the test agent and the other group receives either a dummy (placebo) treatment or an existing treatment, the aim being to ascertain which is the more effective approach and the most free from harm. The purpose, of course, is to compare like with like in the differential responses of the two groups to experimental versus established treatments. The result can be considered only a rough degree of personalization, expressed as the averaged attributes of groups of like-defined patients. Whilst this is, in a sense, a step away from "personalization," it has proven its mettle in elevating considerations of drug efficacy from the realm of experts' opinions to an objective proof of causal relations between defined treatments and reliably measured biomedical effects.

The optimal statistical evaluation of strictly personalized methods of treatment remains an open question, however. For example, while comparing like with like is fundamental, it is much more difficult to judge the limits of "like" within the diversity, for example, of distribution throughout the coronary arterial tree of obstructive lesions of various sizes, in various locations, to cite but one clinically important, fine-grained resolution of disease. Whilst the evaluation is much more straightforward when a categorical genetic factor can be identified among the patients' responses to a given treatment, the challenge of comparing like with like becomes more difficult when the outcome of genetic testing is not categorical, but rather is a series of probabilities of future events that define subgroups of patients.

1.10
And So?

The foregoing approximation of the transition from Ibn Sina's eleventh century *Canon of Medicine* to today's vast libraries of basic and applied medical sciences can be characterized as a sometimes erratic, often long-interrupted progress from successive scientific discoveries towards an ever-more precise personalized characterization of disease.

Since about 1840, attention has begun to be focused on gaining an increasingly precise classification of disease, based on the earlier findings of William Cullen of Edinburgh, but driven in a new direction by the first medical statistician, William Farr, who wrote:

> *The advantages of a uniform statistical nomenclature, however imperfect, are so obvious, that it is surprising no attention has been paid to its enforcement in Bills of Mortality. Each disease has, in many instances, been denoted by three or four terms, and each term has been applied to as many different diseases: vague, inconvenient names have been employed, or complications have been registered instead of primary diseases. The nomenclature is of as much importance in this department of inquiry as weights and measures in the physical sciences, and should be settled without delay. [4]*

In other words, the first – or perhaps even the zeroth – step in any scientific effort is the establishment of a sound classification of, or taxonomy for, the elements of the field of interest. This point is not a quibble over semantics, but rather expresses the need to attain a sharply defined platform for rational discussion, which in turn will define what to measure and analyze, leading to eventual comparisons of like with like.

2
Where Are We Today?

In today's usage, the term "*personalized medicine*" is something of a code-word for a pharmacogenomics-based individualization of the treatment of disease. Perforce, it is not a satisfactory term, because code-words with meanings that are known to some, but hidden from many, are not satisfactory scientific terms. Thus, there is a taxonomic problem in the area of applying pharmacogenomic discoveries to medical practice. With regards to "personalization," the ways in which physicians have in the past examined and diagnosed patients have been described above. This has involved the physicians having the best tools at their disposal at the time, such that they could recommend a disease-specific treatment which, at the time, was the most effective and least hazardous available (as judged by the physician), given the methods of diagnosis, the concepts of disease, and the treatments available.

Whereas, a millennium ago the treatment for bleeding may have been rabbit hair, spider webs, the white of an egg and cautery, today it may be highly purified thrombin, applied locally, in addition to a microscopically (and perhaps also endoscopically) guided electro-cautery for the ligation of small blood vessels. Progress in this direction may be slow, however, as illustrated by the fact that Ibn Sina's recommendation at about 1000 CE to use spider web to facilitate the coagulation of blood was echoed in Sir William Osler's textbook on medical practice in 1892 [5]; in other words, there was no change in over 900 years.

During the twentieth century, however, not only did the pace of discovery accelerate, but the lag-times between discoveries and their implementation into medical care were also shortened. The diagnoses gained in precision, as evidenced by the steady increase in the size of successive volumes of the *International Classification of Disease* (ICD). As importantly, methods available for quantifying the hazards of various treatments have also gained in precision, such that great caution has been injected into the adoption of new treatments. The statistical problem of group characteristics versus "personalized" characteristics was captured by a contemporary news item concerning problems in professional fees paid for oncologic treatments: "The new fee is meant to encourage doctors to follow standard treatments rather than opting too often for individualized and unproven courses of therapy, which can include the most expensive drug combinations" [6]. There stands exposed the tension between group-based versus personalized approaches.

At the same time, another source of tension has begun to emerge between what is usually referred to as "*customary practice*" in the region or community in which a physician practices, and what may be rapidly evolving changes in contemporary standards of evidence-based medical practice. When a patient's clinical course turns out badly, the physician is usually shielded from charges of malpractice if his or her care has been consistent with "customary practice" in the region. However, a physician who has sought to apply a contemporary standard of evidence-based practice may find his or her liability ambiguous with respect to charges of malpractice. Doubtless, this ambiguity has contributed to the estimated average 17-year lag period in the adoption of new research findings into clinical practice [7], although it also recalls the words of Hippocrates:

Life is short,
the art long,
opportunity fleeting,
experiment fallible,
judgment difficult.

Or, as W.S. Gilbert wrote, some 2500 years later:

Things are seldom what they
seem,/Skim milk masquerades
as cream.

In prose, it can be said that a judicious interval of time is advisable between the publication of peer-reviewed, new experimental results and their implementation in clinical practice. In judging how long this interval should be, however, the devil lies in the details, which often revolve around the possible fallibility of experiment. Thus, it is a neat point of case-by-case judgment as to how long to wait, or how soon to begin, when implementing new findings in medical practice. Decisions are tilted one way or the other as certain areas of clinical investigation gain and lose status as fads, either gaining or losing exceptional economic support for the conduct of research.

Recently, the US National Institutes of Health and the FDA published a perspective regarding their intentions and incentives to incorporate concepts of personalized medicine (i.e., pharmacogenomics) into decision-making, during drug discovery and development, and in clinical practice [8]. At this point, the concept of personalized medicine is mean as a code-word for pharmacogenomics, despite its history in terms and aspects of the individualization of therapy. A sometimes used alternative term here is *"tailored therapeutics,"* which implies " ... giving the right treatment to the right person at the right time," and ensuring that these specifications are in fact met in the "real world." ("Real world" is another contemporary code-phrase that alludes to potential differences between the circumstances of carefully designed, controlled, and disciplined clinical trials, compared to actually happens when, in routine medical care, and in their own ways, physicians prescribe and patients execute the instructions that accompany the prescription.) In effect, it alludes to the fallibility of experiments (i.e., clinical trials) the results of which may be scientifically sound but, for indeterminate reasons, do not predict the results obtained in routine medical care (i.e., the "real world"). Varying degrees of patients' adherence to treatment regimens (as discussed below) may be a leading source of variance in drug actions, and differ substantively between randomized trials, controlled trials, and clinical practice.

3
Summing Up

Thus, two major questions and two paradigmatic problems are faced:

- *Question 1*: How well does "tailored therapeutics," as currently practiced or envisioned, succeed in reducing variability in patients' responses to such treatment?
- *Question 2*: What is needed to close the gap between the answer to the first question and the desired goal?
- *Paradigm problem 1*: How can the prevailing statistical methods of analyzing groups of patients' drug responses be reconciled with the goal of personalizing or "tailoring" prescribed treatment?
- *Paradigm problem 2*: What are the inevitably added costs of tailored or customized therapies?

In pursuing the sartorial metaphor for "tailoring" therapeutics to individual patients' characteristics, it is common knowledge that the close-fitting of an off-the-rack man's suit adds only modestly to the cost of the garment. In contrast, what the British call "bespoke" tailoring – that is, to construct a suit from "the ground up," individualized to a man's specific measurements at various points from toe to head – roughly trebles the cost of the garment. Although rough, it is a cautionary metaphor for personalized medicine, in an era in which technological advances permit unprecedentedly fine distinctions in "personalization," but not without added costs to do so.

4
Examples of Recent Advances in Personalized Medicine Based on Pharmacogenomic Findings

With these issues in mind, attention can now be turned to some noteworthy examples of recent advances in personalized medicine, based on pharmacogenomic findings. In this case, pharmacogenomics can be defined as: "The study of the influence of genetic variations on drug responses in patients that is accomplished by correlating gene expression or single nucleotide polymorphisms (SNPs) with a drug's efficacy and/or toxicity."

4.1
Chronic Myeloid Leukemia

Until relatively recently, the diagnosis of chronic myeloid leukemia (CML) was a *de facto* sentence of death. Following its recognition as a distinct disease in 1840, a distinctive chromosomal feature of CML, known as the *Philadelphia chromosome*, was first discovered and described in 1960 by Peter Nowell and David Hungerford, each working at research institutions in Philadelphia. In 1973, Janet D. Rowley at the University of Chicago identified the mechanism by which the Philadelphia chromosome arises as a translocation, creating the unique fusion gene which is termed *BCR-ABL* and which is now believed to be the principal cause of the chronic phase of CML, by virtue of its excessive tyrosine kinase activity. The rationale for treating CML with the Novartis compound *imatinib* (Gleevec®) has been the desire to inhibit this tyrosine kinase, and thus to reduce the abnormally high rate of production of abnormal leukocytes. The treatment of patients with imatinib leads to three successive stages of response, the first of which is termed a "complete cytogenic response" (CCyR), and which corresponds to a 2-log (100-fold) reduction in transcript levels of the abnormal *BCR-ABL* gene. After five years of treatment with imatinib, about half of all treated patients will have achieved a 3-log (1000-fold) reduction in transcript levels, which brings the patient to the second of the three levels of response, the "major molecular response (MMR). This is deemed not only clinically beneficial, but is also an important therapeutic milepost. Continuing treatment brings about 10% of patients to an approximately 4-log (10 000-fold) reduction in transcript levels, by which point the levels become undetectable; this is deemed a "complete molecular response" (CMR). A persistent CMR may be the equivalent of a cure, and allows treatment with imatinib to be discontinued [9]. Imatinib was developed during the late 1990s by a multidisciplinary research group that included biochemist Nicholas Lydon (at Ciba-Geigy, a precursor company of today's Novartis), oncologist Brian Druker (Oregon Health and Science University), Carlo Gambacorti-Passerini (at the University of Milano Bicocca, Italy), John Goldman (at the Royal Hammersmith Hospital in London, UK), and Charles Sawyers (of the Memorial Sloan-Kettering Cancer Center, New York).

Whether the heterogeneous responses are due to heterogeneity in molecular mechanisms, or to the patients' variable adherence to the imatinib dosing regimen (one 400 mg dose per day), is not known. It is clear, however, from studies conducted by David Marin and associates [9], at the Royal Hammersmith Hospital, London, that the overt failures in treatment with imatinib have been (at least in the original series of patients studied by them) uniformly attributable to a poor adherence to the prescribed once-daily dose of imatinib. Whilst this finding was surprising, the occurrence of less-than-optimal dosing in the treatment of various tumors had been described in several studies reported since about 1990 [10]. Thus, "even" in the treatment of various malignancies, a substantial minority of patients fail to take their medicines, either as prescribed or at all [11].

At the time of writing this chapter, only a few months have elapsed since Marin's first report [9], and undoubtedly more results will be forthcoming. The critical questions here are whether frequently recurring lapses in dosing play a major or minor role in failures of prescribed imatinib treatment, and whether assumptions about "drug resistance" can instead be explained by clinically unrecognized poor- or non-adherence. Perhaps this question should be rephrased as: What are the sources of variability in the response to prescribed treatment with imatinib and, by extension, subsequently developed tyrosine kinase inhibitors? (This question will be returned to later in the chapter.)

Two points in the CML story warrant attention. The first is to note the time-line, notably that a period of 120 years elapsed from the first identification of CML as a distinct disease, until an identification of the Philadelphia chromosome. A further 12 years then passed until the underlying genetic mechanism was identified regarding the abnormal gene, *BCR-ABL*, and another 18 years until the therapeutic role (i.e., the pharmacological inhibition of tyrosine kinase) was first demonstrated. Subsequently, another nine years elapsed until a demonstration of the evident role of variable patient adherence to the recommended dosing regimen for imatinib. Not yet fully evaluated has been the impact of temporary gaps in the patients' exposure to imatinib that usually prevail among many patients prescribed chronic-use medicines [10, 11]. The question remains: Do such temporary gaps in exposure to imatinib or other tyrosine kinase inhibitors predispose to the emergence of tumor cells that have acquired resistance to imatinib and/or other tyrosine kinase inhibitors? Albeit with different nouns, the analogous question has been frequently posed in the treatment of infectious diseases (bacterial, viral, parasitic) with antimicrobial agents. where suspicions abound that an intermittency of treatment is one source of emergent drug-resistant microorganisms, though hard evidence on the point remains sparse. Thus, further studies must be carried out, as is usually the case, because " ... experiment is fallible, and decision is difficult." The second point is that there was a minimal lag between the first demonstration of the remarkably positive effect of treating CML with imatinib and its implementation in routine treatment of the disease. The American drug regulatory agency, the FDA, has provided an expeditious review of imatinib and its revolutionary labeling (http://en.wikipedia.org/wiki/imatinib).

Meanwhile, the US National Cancer Institute has made an important point regarding the unique properties of imatinib and, by inference, other tyrosine kinase inhibitors, as follows:

"Imatinib (Gleevec®) represents a new class of cancer drugs and a new way of thinking about cancer. These molecularly targeted drugs are different because they target abnormal proteins that are fundamental to the cancer itself. Most current cancer therapies lack specificity, killing both cancer and normal cells. This

is one reason why many people who undergo chemotherapy experience unwanted side effects from their medications. But Gleevec and other drugs in development are designed to zero in on specific cancer-causing molecules, eliminating cancer cells while avoiding serious damage to other, non-cancerous cells. In the case of Gleevec, the drug is targeted at the bcr-abl protein in CML cells. Gleevec also affects other messenger systems in a cell, which may contribute to its toxicity [12]."

It is Ehrlich's "magic bullet" in another guise.

4.2
Use of Warfarin in Long-Term Anticoagulation, to Prevent Arterial Embolization in Patients with Chronic Atrial Fibrillation

The achievement of an optimal response to warfarin therapy, the most widely prescribed oral anticoagulant drug, involves navigating a narrow path between uncontrolled bleeding due to an excessive anticoagulant effect of warfarin on the one hand, and a high risk due to an insufficient anticoagulant effect of warfarin, of embolization of clot (thrombus) from a fibrillating left atrium or its appendage, on the other hand,. The narrowness of this path arises from warfarin's narrow therapeutic range, which complicates the effort to minimize the risk of intravascular thrombosis.

The dose of warfarin can be optimized by starting low and carefully making small stepwise, upward adjustments, until the desired level of anticoagulant effect is reached. This is determined with a blood test to monitor the International Normalized Ratio (INR), the normal values of which range between 0.8 and 1.2 (as a ratio, the INR is dimensionless.) The usually accepted range of INR values for the prevention of intra-atrial or intravascular thrombus or embolus formation in atrial fibrillation is 2.0–3.0. An INR value below 2.0 is associated with an increased risk of thrombus/embolus formation, while a level above 3.0 is associated with an increased risk of uncontrolled bleeding from one or more sites in or on the body. Common sites of thrombosis formation are the deep veins of the lower legs and a fibrillating left atrium, especially in the left atrial appendage, which has the shape of a cul-de-sac. When normal atrial contractions occur, the left atrium and its appendage contract with each beat of the heart and empty most of their contained blood. In atrial fibrillation, however, the atrium and its appendage cease to contract in a coordinated manner, such that blood tends to pool in what is usually an enlarged atrium and appendage. The pooling of blood predisposes to clotting, and the resulting clot may break loose into the stream of incoming blood, enter the left ventricle, and then be pumped out into the aorta by the ensuing beat of the heart. From the aorta, it is a matter of chance as to whether the clot – which is now termed an *embolus* because it is a clot (thrombus) in motion – will pass into one of the carotid arteries and thence into one of the cerebral arteries, occluding it, and creating an infarction of the region of the brain normally served by the suddenly embolus-blocked vessel. The clinical result – a "*stroke*" – is characterized by loss of the functions that the blood-deprived region of the brain had previously maintained. It is also possible for an embolus to be carried by the aortic bloodstream instead to other major arteries, blocking the blood flow to, for example, the gastrointestinal tract, the kidneys, or the legs. Thus, the aim of anticoagulation is to avoid such organ damage by preventing blood clot formation.

The need to optimize the dose of warfarin arises from the need to have "enough" anticoagulation, but not "too much." When the anticoagulation effect is excessive, bleeding may occur from the nose, the gastrointestinal tract, the urinary bladder, various lesions of the skin, or anywhere else within or on the body where mechanical or chemical trauma can occur and initiate bleeding that does not stop until an out-of-range INR level is brought down to a satisfactory level (i.e., 2.0–3.0). Keeping the INR level within that range reduces the risk of either intravascular or intracardiac thrombosis and associated complications by 50–80%.

Warfarin's actions are a complicated story that starts with its inhibition of the vitamin K-dependent synthesis of biologically active forms of the calcium-dependent clotting factors II, VII, IX, and X, as well as the regulatory factors proteins C, S, and Z. Variations in two genes, the vitamin K epoxide reductase (VKOR) complex 1 (*VKORC1*) and the cytochrome P450, polypeptide 9 (*CYP2C9*) have been associated with changes in warfarin's dose–time-effect relationship. These changes result in longer times needed to achieve a stable INR level at a given daily dose of warfarin, and variations in the magnitude of the INR change elicited by a given change in warfarin's daily dose.

Thus, unlike other situations in personalized medicine where, for example, an abnormal gene results in an abnormally high rate of production of a key metabolite (as in CML), or a critical gene is missing such that a normal or therapeutically desirable mechanism is inoperative, the story relating to the pharmacogenomics of warfarin action is a changing gradation of warfarin's dose-dependent and time-dependent actions. Thus, the results of the recently introduced pharmacogenomics test do not relieve the prescriber of a need to titrate the warfarin dose to an optimal level. Instead, it puts the prescriber on notice that the corridor between too-much warfarin and too-little may in certain patients be narrower than usual and that, after a change in warfarin dosage, more time may be required in order for a steady INR level to be reached.

The difficulty that the prescriber faces, in attempting to evaluate the results of dose-titration, is the high incidence of intermittent omissions of warfarin doses by the large fraction of patients who are variably nonadherent. Stephen Kimmel and colleagues at University of Pennsylvania have evaluated the prevalence of nonadherence using the objective method of electronic medication event monitoring to assess patient adherence with the warfarin regimen [13, 14]. Their data have revealed that approximately 23% of the 155 patients in their study missed doses on 10–20% of treatment days; about 25% of the155 patients missed doses on 20–50% of treatment days; and a further 13% missed doses on 50–100% of treatment days. Meanwhile, the patients' self-reports of their dosing usually overestimated their actual dosing (as has been found in many other disease conditions). Consequently, the physician who is attempting to identify an optimal dose of warfarin is titrating in the face of a sometimes substantial, but usually opaque, degree of underdosing, due to poor adherence to the prescribed dosing regimen. Such recurrent underdosing tends to lead the physician to overestimate the optimal dose. The remedy to this problem would be to compile the patient's warfarin dosing history, and to take those data into account when judging what might be an optimal daily dose of warfarin to prescribe.

At the time of writing this chapter, warfarin – which has used as a pharmaceutical agent since the early 1950s – is about to be made obsolete by the recent approval and imminent entry into the US market of a novel oral anticoagulant, *dabigatran*, which has been developed by the German pharmaceutical company, Boehringer-Ingelheim. Dabigatran is the prodrug of a direct thrombin inhibitor and, as such, does not require repeated INR testing; neither is it subject to the food–drug interactions that vex the management of warfarin-based anticoagulation. Dabigatran has certain drug–drug interactions, but these are less widely-ranging than those of warfarin. Dabigatran was identified during the early 1980s as a potential candidate to replace warfarin, and approved for marketing by the European Medicines Agency in the spring of 2008, and a few months later in Canada. Presumably, postmarketing surveillance in these early adopting countries has raised the level of experience with dabigatran greatly beyond the numbers of patients studied in preregistration studies. Thus, there would appear to be a clear pathway for American prescribers to abandon the use of warfarin.

Another aspect of the timeline for progress in the oral anticoagulant field will be the time-course of the decline in numbers of warfarin-treated patients in the US and other major countries, as the use of dabigatran rises. The time-dependent rate of switching from warfarin to dabigatran is another facet of the timeline for the adoption of this medical innovation; though the pricing of dabigatrin will doubtless influence the rates at which these changes occur.

Dabigatrin is the first of four orally active anticoagulants that are, at the time of writing, heading towards the pharmaceutical marketplace, and which can be expected to be involved in a high level of competition. This will most likely have the effect of stimulating the design and execution of comparative clinical studies for the other three new agents that are able to gain approval for marketing in the US.

It should be noted that, in this and other subjects in this chapter, emphasis on the US market is not provincialism but rather a reflection of the fact that the US pharmaceutical market is not only the largest in terms of sales, but also the only major national market to have unregulated pricing. Thus, the US market is a disproportionately large contributor to the profits for pharmaceutical agents.

4.3 Von Gierke's Disease

Another disease of genetic origin, von Gierke's disease, occurs when the body lacks the enzyme glucose-6-phosphatase (which effects the release of glucose from glycogen), as well as the genetic mechanisms that govern its formation. As a consequence, infants with von Gierke's disease relentlessly accumulate glycogen in the kidneys and liver, although the use of special diets and infusions of glucose can prolong survival. Other one-gene/one-enzyme diseases follow a similar course, though some have begun to be treated by the exogenous administration of the missing enzyme. These treatments represent another aspect of technological advances that embody the principle of personalized medicine.

A key point implicit in the example of von Gierke's disease is that the congenital absence of a genetic mechanism necessary for the production of a crucial enzyme can have catastrophic consequences that tax the metabolic understanding and ingenuity of physicians seeking therapeutic means to bypass the critical block.

4.4 Estrogen-Sensitivity in Cancers of the Breast

Two important, recent advances in the understanding and treatment of breast cancer have been the discoveries of: (i) the human epidermal growth factor receptor (HER-2); and (ii) the ability to ascertain the estrogen-sensitivity of individual patients' breast cancers.

About one-fourth of breast cancers have the HER-2 receptor, which indicates a desirability for an early-stage treatment with the monoclonal antibody, trastuzumab. The correct treatment of patients who meet these criteria – having both the HER-2 receptor and early-stage breast cancer – can extend the patient's life and delay any reactivation of the disease. As with the hyperactive tyrosine kinase in CML, this too relates to an abnormally amplified expression of genetic function, the normalization of which achieves a substantial suppression of the cancer. However, while many patients are spared a recurrence of disease, many others are not; hence, while unprecedented in the magnitude of beneficial effect, the aforesaid advances constitute a partial victory. The earlier findings were confirmed in two major studies which were conducted concurrently, with the results being reported in 2005 [15, 16].

The timeline of these studies was as follows. In 1987, Slamon, Ullrich, and others reported their discovery of the amplification of the *HER-2* gene in about 30% of patients' with estrogen-sensitive breast cancer. Such a discovery led to the initiation of investigations at Genentech (a pioneering biotechnology company), to identify a means of reversing the effects of the amplified *HER-2* gene. In 1998, Genentech's monoclonal antibody to *HER-2*, known as *trastuzumab* (Herceptin®), was approved by FDA on the basis of positive results from a limited amount of clinical testing. In 2000, several major, placebo-controlled trials were begun to assess, in patients for whom treatment with the monoclonal antibody was indicated, the monoclonal antibody's impact on disease progression. The trial results became available in 2005. Thus, the total time that elapsed between the 1987 discovery of the amplified *HER-2* gene and the successful development and FDA approval of the monoclonal antibody, trastuzumab, and the striking results of two major trials of unequivocal size and scope in 2005, was 18 years. Yet, this 18-year time-line was remarkably short for a new treatment of a chronic disease to proceed from laboratory to practice, and most likely approaches a limit that is inescapable, without radical changes in the current methods of project management, clinical trials designs, and the analysis of trial results. New drug development is often fraught with unpredictable delays or setbacks, which trastuzumab somehow appears to have largely or completely escaped.

Another notable – if less dramatic – finding has been the development of a battery of tests for genes involved in breast cancer, termed a "*multiparameter gene expression analysis.*" In newly diagnosed patients with node-negative, estrogen receptor-positive breast cancer, the Oncotype DX® assay (Genomic Health Inc.) can be used to predict the risk of recurrence in patients treated with the estrogen-blocking prodrug, tamoxifen. The assay procedure is used to identify a subset of patients who are most likely to achieve the greatest therapeutic benefit from tamoxifen treatment alone, and who thus may not require the addition of chemotherapy to their treatment regimen. The various forms of chemotherapy aim at achieving a narrow zone between killing the cancer cells and sparing normal cells. In practice, chemotherapy incurs risks of a variety of complications arising from suboptimal achievement of the aforesaid objective.

One of the many difficulties in interpreting the results of multiple tests arises from changes in available therapeutic choices. For example, in 2002 the aromatase inhibitors entered the market for the adjuvant treatment of breast cancer in post-menopausal women, with patent protection, premium pricing, and strong promotion. These aromatase inhibitors competed with generic tamoxifen which, by then, was without patent protection, premium pricing, or strong promotion. Moreover, tamoxifen is a prodrug, giving rise to two metabolites that each possess estrogen-blocking activity, albeit with different potencies. Thus, both scientific and economic factors were present that could influence the decision-making process.

The American Society for Clinical Oncology, which reviews and makes recommendations for the constantly changing field of oncology, recently summarized its stance with the following statement:

It is important to emphasize that guidelines and technology assessments cannot always account for individual variation among patients. They are not intended to supplant physician judgment with respect to particular patients or special clinical situations, and cannot be considered inclusive of all proper methods of care or exclusive of other treatments reasonably directed at obtaining the same result [17].

In part, of course, a "legal hand" can be seen in the writing of this statement, although it also echoes the theme with which this chapter began, namely that the principle of personalized medicine is persistent throughout the history of medicine.

5
Analysis of Variability in Drug Response: Sources, Surprises, and Solutions

Another perspective on personalized medicine focuses on the extent of variability in patients' responses to prescribed medicines. An obvious implication of either of the terms "*personalized medicine*" or "*tailored therapeutics*" is that the prescribing physician is able to prescribe so as to achieve the desired level of drug response in the treated patient. The achievement of this goal begins with an analysis of the known sources of variability in drug response, understanding how they interact and cumulate, and finding ways to minimize variance in the relation between dosing instructions and drug actions. In so doing, the first stage is to consider the pathway that is followed by the drug molecules as they make their way from:

- their formulation in oral dosage forms (tablets or capsules) stored in a package;
- to the patient who ingests (or not), the dosage form(s) at times that may differ from what is called for by the prescription, or not at all;
- from the ingested dosage form to distribution within the body, establishing a more or less predictable time course of drug concentration in plasma, and being eliminated from the body by various means;
- to react in a concentration-dependent manner with drug receptors on or within cells, and by various means, to exert predictable actions, both desired and undesired.

The four steps in the above-described pathway, from prescribed dosage to drug response, are shown schematically in Fig. 1, where the colored rectangles depict four processes, respectively:

- Drug formulation
- Patient adherence
- Pharmacokinetics
- Pharmacodynamics.

5.1 The Harter–Peck Model

A model is now considered (as formulated originally by two FDA physician-scientists, John Harter and Carl Peck) of the multicomponent system which relates drug dose and responses [18]. As the measure of variance in drug response and the processes by which a drug response is generated, Harter and Peck chose the statistical parameter, "coefficient of variation" (CV): this is a value's standard deviation expressed as a percentage of its mean value. The Harter–Peck model achieves several objectives: First, it describes the principal components from which variance arises in drug response, plus it allows a simulation of the consequences of major reductions of variance arising from each principal component. The framework of the model (see Fig. 1) shows the flow of increasing variance proceeding from left to right, as each of the four processes contribute their share of variance. The four processes and their colors in Fig. 1 are, from left to right:

- Drug formulation (blue), CV = ±20%.
- Patient adherence (red), CV = ±50%.
- Pharmacokinetics (gray), CV = ±50%.
- Pharmacodynamics (orange), CV = ±30%.

The rationales for these choices of variance is discussed below.

In Fig. 1, the horizontal arrow at the extreme left, labeled "prescription," specifies the name of the drug, the size of individual doses, and either the frequency at which the prescribed dose is to be given, or the time interval that should elapse between successive doses. Of course, dosing frequency and dosing interval are approximately interconvertible; for example, "one dose twice daily" and "one dose taken every 12 h" have approximately the same meaning, although "every 12 h" is more precise than "twice daily." Potential misunderstandings lurk in hastily given dosing instructions. For example, an occasional patient will misinterpret the word "daily" as meaning that dosing should occur within each interval of daylight. That misinterpretation creates no problem with a single daily dose, but it can create curious errors in adherence if the instructions call for one dose twice daily, or more often.

Note at the extreme right of Fig. 1 the word "response"; this is depicted not as a single value but rather as a range of values spanning the range

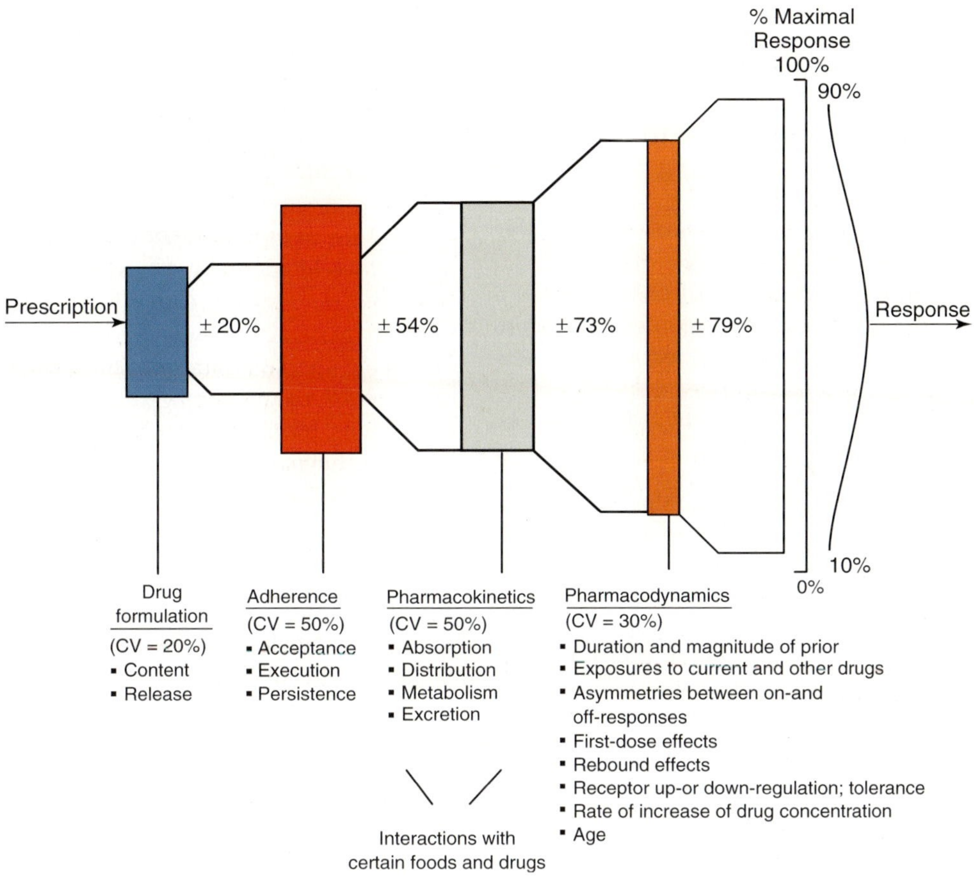

Fig. 1 Sources of variability in drug response. Adapted from Ref. [18].

between about 10% and 90% of the maximal response – obviously, this is a widely variable response to a given dose.

Moving from left to right, the blue process is virtual, because it represents oral dosage forms (tablets or capsules) formulated with a specified amount drug (± a single digit CV). Moreover, the drug is formulated for release with specified kinetics during a certain time period after ingestion. *In vivo* rates of drug release are generally much more variable than is the drug content, which gives rise to the model's assigned value of a 20% CV for the formulated drug. The process is characterized as virtual because dosage forms do not function while sitting in their package. The dosage form only commences to release its contained drug after it has been ingested which, of course, is conditional upon adherence, the process depicted by the red rectangle in Fig. 1. The adherence process represents whether and when the patient ingests the dosage form, dose by dose.

5.1.1 Notes on Patient Adherence

The horizontal arrow at the left extremity of Fig. 1, labeled "Prescription," represents the prescriber's intention, which specifies a time series of prescribed amounts of drug (doses) to be ingested, at specified intervals or frequency. The prescription defines a standard against which the patient's actual dose-taking is compared. A series of questions thus arise:

- Did the patient start taking the medicine? If so, when?
- How well or how poorly does the patient execute the prescription's instruction to ingest the prescribed dose at the prescribed intervals?
- When does the patient's dosing completely stop?

Quantitative answers to these questions define the patient's external exposure to the drug in question and, when compared to the prescribed dosing regimen, the patient's adherence. As the CV of 50% implies, patient adherence is highly variable both between and within individual patients, and is prevalent in all fields of ambulatory medical care, seemingly irrespective of severity or prognosis of disease, nature and degree of symptoms, and the availability of medicines that, if taken correctly, can manage or cure the disease in question (www.iadherence.org) [10, 19, 20].

Variable adherence thus prevails as the "rule" rather than as the "exception" whenever patients are responsible for taking prescribed medicines. Variable adherence can also be regarded as yet another dimension of personalized medicine, in the sense that patients tend to have individualized patterns of recurrent deviations from the dosing regimens they have been prescribed.

5.1.2 Notes on the Origins of Adherence Research

The field of adherence research has been catalyzed by two important advances within the past three decades. One of these is the growth and development of the field of pharmacoepidemiology, which has taught the importance of ascertaining soundly quantified exposure of patients to prescription drugs. Pharmacoepidemiology has also brought new methods for assessing drug actions during "real-world" use, as opposed to highly structured, randomized, controlled clinical trials. In general, controlled clinical trials include special efforts made to assure the patients' continuation with the trial medications, relative to what occurs in routine clinical practice – that is, "real-world" use. Although pharmacoepidemiology has mainly been focused on the detection and quantification of unexpected hazards from clinical uses of various drugs, its methods also have a role in the detection and quantification of unexpected benefits from the clinical use of various drugs.

The second of the two advances in adherence research has been the development and expanding applications of electronic methods for compiling drug dosing histories – mainly by electronic time-stamping of entries into the drug package, or of dose removal from the drug package. These methods, which are known as "*electronic medication-event monitoring*," have brought data with high temporal resolution to adherence research, which previously had depended on low temporal resolution methods. The latter included the questioning of patients at scheduled medical visits, or tracking patients' aggregate use or non-use of prescribed medicines through counting untaken doses at times of prescription refills, or determining intervals between refills. Gathering data at the usually one- to three-month intervals between refills means the acquisition of data acquisition a few times a year, which gives a very low temporal resolution of the patients' dose-taking behavior. Gathering data by questioning patients is also complicated by an incomplete recall of when doses might have been omitted, plus the reluctance of some patients to admit to deviations from their physicians' dosing instructions.

Five main findings characterize recent research on patient adherence:

- The most frequently occurring errors involve underdosing, with about four out of five errors being delays or the omissions of scheduled doses. Some patients infrequently take an extra dose when uncertainty arises about whether or not the scheduled dose had recently been taken.
- The biggest single discrepancy between prescribers' intentions and patients' actual dosing is the patients' early, complete discontinuation of dosing of medicines prescribed for long-term use. Most of these complete discontinuations escape clinical detection, despite their occurrence in major chronic diseases, such as hypertension, lipid disorders, osteoporosis, HIV infection, post-transplant immunosuppression, and various anticancer treatments [10]. A large minority or a modest majority of patients completely discontinue dosing on their own initiative during the first year of treatment with medicines intended for indefinitely long use; complete discontinuation of dosing continues to occur in subsequent years up to about year 5, by which time only about 10% of an original group of patients will still be engaged with the prescribed dosing regimen.
- Dose omissions sometimes occur sequentially; these are called *drug "holidays"* when three or more days' doses are consecutively omitted.
- These and other temporal patterns of deviations from prescribed dosing regimens are substantially similar across major diseases, more or less independent of prognosis, or prospects for good outcomes from correct dosing, [10, 11].
- The most important point is that the clinical recognition of poor adherence is poor, aptly described as being no better than a "coin-toss." Thus, without adequate methods for compiling dosing history data, adherence problems are incompletely detected and are usually ignored.

Although a number of other aspects of the patients' variable taking of prescribed medicines have been recognized, the foregoing are the most consequential. Just how consequential depends on the drug, the disease and its severity, and other concomitant diseases that the patient might be suffering from.

With respect to a complete cessation of dosing, those patients who do not take an aptly prescribed medicine will lose its benefits and, of course, also its hazards, and revert to findings among the recipients of placebo in controlled trials. The consequences of occasional lapses in dosing are more difficult to ascertain, because the patients' exposure to drug is only partial in the aggregate, but may follow very different temporal patterns of exposure and ultimately have the same aggregate amount of drug taken [10].

With anti-infective drugs for infectious diseases, a broad consensus among experts supports the view that the intermittent dosing of antibiotics and other anti-infective agents predisposes to the emergence of drug-resistant microorganisms. This same consensus prevails in the treatment of parasitic diseases, notably malaria. The field of oncology is just now awakening to the prevalence of poor adherence to anticancer drug regimens meant for patients to self-administer. As with "drug-resistant hypertension," about half of which is due to clinically unrecognized nonadherence [21], "drug resistance" to certain antineoplastic agents in oncology may turn out to have a large component of clinically unrecognized poor or nonadherence.

Proceeding to the right in Fig. 1, the process colored light gray involves the conversion of the ingested dose of drug and its kinetics of drug release from the dosage form, into a time-series of concentrations of drug in the blood. The key sequential events in that process are summarized in the following paragraphs.

5.1.3 Notes on Pharmacokinetics

The swallowed dose is carried, by one or more traveling waves of esophageal contraction, to the stomach, where gastric contractions mix the dosage form with gastric secretions; this causes the dosage form to begin dissolving and to be broken into small pieces. In due course, the dissolved drug and/or fragments of the original dosage form are passed onwards into the duodenum, where the acidity of the stomach juices are neutralized by the alkaline duodenal and pancreatic secretions. Most of the absorption of drug into the bloodstream occurs in the duodenum and the upper part of the small intestine, with the result that the majority of the ingested drug enters the bloodstream within several hours after ingestion of the dosage form. The motility patterns of the small intestine usually carry water and solutes toward the colon by a somewhat erratically to-and-fro motion, reaching the colon about 3 h after exiting the stomach. The absorption of some drugs may continue within the colon, but not of others.

Once in the bloodstream, the drug is carried throughout the body, reaching tissues via transcapillary diffusion and/or ultrafiltration. Concurrently, the absorbed drug may undergo more or less extensive metabolic conversions into inactive metabolites, which are excreted in the urine and/or the feces. Any unmetabolized drug may also be subject to excretion in the urine or the feces.

Irrespective of the details of drug metabolism and excretion, the general truism is: What enters the body, whether by ingestion or injection, departs the body, whether chemically intact or as a variety of metabolites derived from the original compound. Thus, if adherence is maintained, there is a continuous turnover of drug within the body, with concentrations rising or falling depending on the direction and magnitude of differences between the rate of drug entry into, and the rate of drug exit from, the bloodstream.

5.2 Notes on Pharmacokinetic Variability

The multiplicity of events that an ingested drug might encounter during the ADME (absorption distribution, metabolism, and excretion) processes are subject to variability arising from the multiple variables that prevail. Each of these may have a certain influence on the ADME processes, including diet, nutritional status, exercise, external temperature, hydration of the body, the presence of disease(s) and their respective severities, and drug treatments underway. These factors will account in part for the high degree of variability in the relationship between the ingested dose and the ensuing time-course of drug concentration in the blood, or in the plasma.

A substantial part of the variability in drug concentrations arising from pharmacokinetics occurs as a consequence of the longstanding practice of formulating drugs so that they are relatively quickly released from their dosage form, and thus made available for absorption during a relatively short period of time. The result is that drug concentrations in blood rise rapidly to a peak that usually occurs early during the interval between successive doses. This usually creates a large fluctuation within each time-interval between successive doses in the time-course of drug concentration in the blood. The net effect of this longstanding, "default" approach to drug formulation is to create the automotive equivalent of a very bumpy road when, in many therapeutic situations, the rational objective is the pharmacokinetic counterpart of a "smooth ride"; that is, minimal fluctuations in concentration within the intervals between successively taken doses.

In effect, then, one of the largest sources of variability in drug concentration – and thus drug action – derives from this default mode of drug formulation. There are, to be sure, certain infrequently occurring circumstances in which the optimal temporal pattern of a drug's concentration in blood or plasma calls for a short, sharp bolus of drug, leading to an early, high peak in concentration, followed by a gradual decline that continues until the next dose is taken.

A set of mechanisms in the wall of the gastrointestinal tract prevents the absorption of many substances that enter via the subject's diet. These mechanisms create a gauntlet of enzymes through which many ingested substances must pass, during which these substances are converted into metabolites that may or may not have pharmacological activity. The enzymes in question are collectively referred to as *mixed-function oxidases*, and have acquired an array of encoded letters and numbers. These oxidative enzymes are subject to wide within- and between-patient variations in their levels of activity, depending on prior exposure to various foods or drugs. Thus, they sometimes become a substantial source of variability in determining how much drug is absorbed, unchanged, into the bloodstream. In fact, a number of drugs have such an extensive preabsorption metabolism that only single-digit percentages of ingested drug are absorbed, unchanged, into the bloodstream. Drug metabolism that occurs during the process of drug absorption is often referred to collectively as "*first-pass effects*" (this is a major topic that is described and discussed in detail elsewhere in these volumes).

The net result of all of the foregoing aspects of pharmacokinetics is that, issuing from the corresponding process in Fig. 1, is a time-course of drug concentration in blood or plasma that reflects an aggregate of upstream variations in: variable patient adherence; the "default" design of dosage forms that impact the body with short, sharp boluse of drug, erratic kinetics of drug release from the ingested dosage form, variable first-pass metabolism, and other variations in the ADME processes.

The result is that the collective processes that comprise "pharmacokinetics" are a major source of variability in drug response; indeed, sometimes they are the leading source. Typically, pharmacokinetics will vies with patient adherence as the largest source of variability in drug response.

If variability is considered as a pollutant being added to a river at various points along its flowstream, then the further downstream, the higher will be the levels of pollution – unless there is some type of intermediate process that acts to reduce the pollution. Just how effective one of the nonlinear aspects of pharmacodynamics (the orange rectangle in Fig. 1) may be to block the downstream propagation of variability is considered in the next subsection.

5.2.1 Notes on Pharmacodynamic Variability

The majority of the body's tissues contain receptors with which the molecules of a drug can interact and, through that interaction, initiate a chain of cellular events that lead to the occurrence of the drug's actions. Most drugs have more than a single action (hence the plural "actions"). As a general rule, the magnitudes of each of a drug's actions depends on the concentration of that drug in the blood or plasma, which in turn makes drug concentrations a natural focus of interest. However, the relationships between the concentration of drug in the blood or plasma and the drug actions are nonlinear, in several senses. The most likely nonlinearity is that, in the lower range of concentrations, as the blood levels rise or fall, the drug action will rise or fall relatively steeply. However, at an intermediate range of concentration the drug concentration curve will bend and flatten, and its slope will fall toward zero; as a consequence the drug action will plateau, with little or no changes in drug action despite large changes in drug concentrations in the blood or plasma. Indeed, such plateauing of the concentration-dependence of drug actions acts to block any variability that might arise from the three upstream processes of pharmacokinetics, adherence, and drug formulation. This relative blockade of the propagation of variability is only one facet of pharmacodynamics; other facets are listed in the lower right-hand corner of Fig. 1.

Thus, pharmacodynamics is also a rich source of variability. Of course, not all of the listed types or sources of pharmacodynamic variability will usually come into play simultaneously, and there will be certain redundancies in the list shown in Fig. 1. For example, a drug that has a prominent first-dose effect will most likely not be tolerated (except in very low doses) by patients who have never before been treated with that drug (so-called "drug-naïve" patients). Thus, the prescriber must start treatment with a dose that is sufficiently low to be tolerated by the patient, and then continue with that dose for a few days, before increasing it to a somewhat higher level. Such a process of stepwise increases in doses every few days would be repeated until the patient reaches their so-called "therapeutic" dose.

The need for this pattern of treatment is most common with drugs that prevent cardiac arrhythmias. With such agents, the dose titration is usually carried out in-hospital with continuous monitoring of the patient's electrocardiogram and other cardiovascular variables, to ensure that the upward steps in dosing are not proceeding too rapidly, or with too-high steps. Most antiarrhythmic drugs have latent pro-arrhythmic effects and tend, if the dose is escalated too rapidly, to induce the very types of arrhythmia that the drug is being prescribed to prevent. The currently available antiarrhythmic drugs do provide some protection against disabling or sometimes fatal disturbances in cardiac rhythm, though only at therapeutic doses that are attainable only after the patient has run the gauntlet of dose-titration and its risk of pro-arrhythmic effects.

It might be supposed that, when the dose-titration has been successfully concluded, all should be well; however, this ideal state is always subject to variability in the patient's adherence to the prescribed dosing regimen. One form of such variability that recurs in some patients is the multiday drug "holiday," when a patient omits several days' doses before resuming dosing. These *drug "holidays"* are defined as three or more days of consecutively omitted doses, and it is well demonstrated that the likelihood of longer holidays diminishes, though holidays as long as 10 days have been observed [20].

One important, yet unanswered, question here is: How long a drug holiday can a patient incur without having to be re-titrated? If the answer were to be seven days, then a situation might be encountered where the patient's failure to take the drug would potentially to lead to a fatal complication, if the seven-or-more-day holiday-taker were to resume dosing at full-strength, unbeknown to anyone except the patient. Fatal arrhythmias occur in some, but not all, patients treated with anti-arrhythmic drugs that also have pro-arrhythmic effects. That some patients, but not others, have fatal arrhythmias is an extreme manifestation of variability in drug response.

Another holiday-related form of drug toxicity occurs with the most commonly prescribed type of β-adrenergic receptor antagonist ("beta blockers"), which has the effect, during long-term dosing, of provoking an increase in the sensitivity of the beta-adrenergic receptors. This change, which is a silent form of physiological adaptation to long-term beta-receptor blockade, is referred to as "receptor upregulation." When the drug is taken on a regular basis, such an increase in receptor sensitivity is masked by the drug's receptor-blocking action. However, when a patient begins a drug holiday the drug will be cleared from the blood within 24–48 h, and the blocking effect on beta-receptors will disappear concurrently. The receptors generally persist in an upregulated state for a fortnight or more, during which time they decline gradually to their pretreatment state. During the time that the receptors are still upregulated but unblocked, the patient will be exposed to markedly exaggerated effects of any endogenously secreted β-agonists, such as epinephrine (adrenaline) and norepinephrine (noradrenaline). These effects include palpitations, feelings of impending doom, coronary arterial spasms, the appearance of signs and symptoms of cardiac ischemia, and/or sudden death. Collectively, these symptoms are known as the "*rebound effects*" of beta-receptor antagonists, and represent a pharmacodynamic contribution to the variability observed in the dose–response relationship.

With yet other drugs – notably, organic nitrates such as nitroglycerine and nitrates of isosorbide – continuous dosing leads to a downregulation of nitrate receptors and a loss of the drugs' vasodilatory and other effects. Thus, a nitrate drug that is effective today may not be effective tomorrow, and will continue not to be so until a "nitrate holiday" is interposed for a day or two to restore responsivity to the particular organic nitrate being used. If continuous dosing is resumed after the "nitrate holiday," the loss of response recurs; this so-called "nitrate tolerance" represents another form of variability in drug response.

Evidence has also been provided (see Ref. [2]) that at least two drugs and one steroid hormone exert actions which are related not only to their actual plasma concentrations but also to the rate of increase of such concentrations. The best-known example of this phenomenon is nifedipine [2], whereby a hazardous side effect of *reflex tachycardia* is triggered if the drug's plasma concentration is increased at a rapid rate. However, when nifedipine was administered in an osmotic pump formulation that released the drug at a constant rate throughout an 18 h period, the drug showed a typical blood pressure-lowering effect but did not trigger the reflex tachycardia seen in over one-third of patients prescribed a conventional, quick-release form of nifedipine (i.e., the "default" formulation). In effect, any drug with a pharmacodynamic profile that causes a major change in responsiveness when the plasma concentration is increased will appear to show an erratic relationship between the concentration of the drug and its effects. Such confusion can be resolved, however, by employing a correctly designed, model-based simulation that links such sensitivity not only to the rate of increase in concentration but also to absolute drug concentrations.

Typically – but not always – the concentration of a drug in blood is the main driver of its actions, especially if the rate of change in drug concentration has an influence on the patient's response to that drug. And therein lies the importance in determining a drug's pharmacodynamic profile, to provide details of any aspect of drug exposure that might influence the patient's response to that drug.

One complicating factor when analyzing a drug's pharmacodynamics (the orange rectangle in Fig. 1) is the great diversity of actions that different drugs may manifest. Typical examples include blood pressure-lowering, the suppression of gastric acid secretion and of retroviral replication, the reversal of erectile dysfunction, the suppression of skin inflammation and of disturbances of cardiac rhythm, the relief of urinary tract outflow obstruction, of pain, and of urinary urgency, the lessening of insomnia, the killing of infecting bacteria, the reduction of nasal mucus formation, and the alleviation of mental depression.

Each of these diverse actions is measurable, by one means or another, with some needing to be followed from minute to minute, second to second, or even at millisecond intervals, to obtain a full perspective on the dynamics of drug action. An example would be the measurement of blood pressure before, during, and after treatment with one or more antihypertensive drugs. In contrast, the actions of certain other drugs can be adequately quantified and followed with a much less frequent sampling of data. It is a matter of scientific judgment to decide on the necessary frequency of measurements in the analysis of drug actions. For example, some measurements can be gathered automatically and noninvasively, while others entail interference with the patient's usual activities, such as the withdrawal of serial blood samples, the taking of photographs, and the administration of a questionnaire. Yet, even a very brief questionnaire may have limits as to how frequently it can be readministered without becoming objectionable to the patient.

Typically, a single dose of most analgesics will result in a pain-relieving action that rises to a peak and then declines such that, within a period of 4–6 h, the analgesic action has faded to a point where another dose is required if the analgesia is to be continued. The assessment of analgesic action is clearly subjective, and may be made by a variety of methods that include direct questioning, perhaps asking a patient to indicate their present level of pain on a linear analog scale that extends from no pain at all to the worst pain ever experienced.

In summary, the processes referred to as "*pharmacodynamics*" represent the drug's interactions with its receptors, the mechanisms set in motion by linkage of the drug molecules to those receptors, with a set of chemical and, in some instances, physical mechanisms that link the drug-bound receptor to intracellular chemical "machinery" that generates the sought-for effect of the drug. Often, this results in a highly nonlinear relationship between drug concentration and drug action, with many surprises arising from mechanisms that lie within the orange rectangle in Fig. 1. One of the key reasons for assigning a lower CV (30%) to pharmacodynamics than to pharmacokinetics (50%) is the plateauing of the concentration–effect relationship that occurs with most drugs (see above).

Some of the nonlinear aspects of pharmacodynamics may create major stumbling blocks in clinical drug development. One pressing problem for drug developers is the time required to probe a new drug's pharmacodynamics sufficiently well to define a recommended dosage regimen that will withstand the marketplace pressures to reduce the recommended dose, with the aim of lowering treatment costs. In the past, post-marketing dose reductions of at least 50% have been made for about one in four drugs [22, 23]. Recently, it has been confirmed that these post-marketing dose reductions are not only continuing to occur (R. Crombag and C.C. Peck, personal communication), but that they are now implemented faster than ever before. In fact, such dose reductions may threaten the commercial viability of the one-fourth (approximately) of new drugs for which the dose – and thus the revenue – is reduced.

In previous decades, pharmacokinetic considerations have dominated the definition of what seemed to be an optimal dosing regimen. Experience with drugs such as omeprazole and clopidogrel have shown that pharmacodynamic – rather than pharmacokinetic – factors may dominate the identification of an optimal dosing regimens. For example, omeprazole has a pharmacokinetic half-life of less than 1 h, and initially was considered by its developers to be impractical for oral use, as such a short half-life would surely require dosing at impracticably short intervals of a few hours. In reality, however, the drug's pharmacodynamics involve its binding to, and complete inactivation of, the enzyme in the gastric mucosa that is directly involved in gastric acid secretion. Thus, the duration of omeprazole's ability to block gastric acid secretion depended not on its plasma half-life, but rather on the time required for the synthesis of a new, unblocked gastric mucosal enzyme – which is several days. Subsequently, omeprazole went on to become (in its time) the largest-selling drug in the history of the pharmaceutical industry. Unfortunately, this and other "false starts" that occurred during the premarket development of omeprazole not only took time but had also significantly reduced the drug's patent life by the time it eventually entered the market place.

6 How Variance Aggregates in a Multicomponent System

At this point, a systems principle is encountered that governs the way in which variance from the four sources discussed above determines the overall variability in the relationship between the prescribed dosing regimen and the time-course of drug response. The overall, or *aggregated*, variance in the prescribed dose–response relationship is the square root of the sum of the squares of the variance in each component. This principle has had many "real-world" applications, perhaps the most "real" of which was an adoption of the teachings of the late W. Edwards Deming by various Japanese industries, as they were rebuilt after World War II [24]. During the War, Deming had worked in "operations research," a discipline whose name has since fallen from use but which laid the foundations for the post-war development of statistical quality control methods, systems analysis, and control theory. It may seem strange that a method employed in industrial quality control is carried over into physiology and medicine, but biological processes are not exempt from either physical laws or statistical principles.

Initially, the estimates that Harter and Peck made were for an anti-asthmatic drug – theophylline – which, during the 1980s, acquired two reputations: (i) it was the best-studied of then widely used drugs, from the perspective of quantitative clinical pharmacology, which later became known as *pharmacometrics*; and (ii) it was an especially difficult drug to use in clinical practice because of the unusually high variability of its dose–effect relationship. The variability was such that, if the usual starting dose were to be prescribed to a patient of normal body size, and a medical examination and history uncomplicated by anything other than the patient's susceptibility to acute attacks of asthma, then the response that could be expected would range from virtually nil to near-maximal. The wide-ranging possibility of drug response from a recommended dose given to an "ideal" patient is indicated by the long, curved bracket at the right-hand edge of Fig. 1.

The aggregation of variance is also illustrated in Fig. 1, as the sequence of four numbers preceded by a ±sign. Not shown in Fig. 1 are the respective CV values, which were discussed previously and which Harter and Peck used as an estimate of the variance contributed by each of the four elements:

	Drug formulation	*Adherence*	*Pharmacokinetics*	*Pharmacodynamics*
CV	20%	50%	50%	30%
Square	400	2500	2500	900
		Σ squares: 6300		
		Square root: 79%		

This high level of variability, a CV of 79%, approximates the clinical picture.

An informative analogy to illustrate the meaning of such high variability would be, rather than considering drug actions, to consider the responses of an elevator in a 20-storey building to commands from its control panel. When the button for floor 10 is pressed, the elevator goes instead to floor 18; the next time, pressing floor 10 sends the elevator to floor 7, and the time after that to floor 16. After only a few such misadventures the elevator would be shut down and the elevator maintenance service would be urgently requested to return the elevator to its expected level of performance.

Although the numbers in Fig. 1 are oversimplified, as with any model, they allow simulations to be run in order to determine what might be accomplished by various, rather large reductions in the variability of these four subsystems of the simulated system.

To begin, suppose that, through incisive pharmacogenomic research, testing, and individualization of treatment, the CV of the pharmacokinetic process can be successfully reduced from 50% to 10%; this is an 80% reduction, and a technological triumph. The ensuing change in the sum of the squares, however, is: 400 + 2500 + 100 + 900 = 3900, the square root of which is 62.4%; this is a disappointingly small reduction of approximately 20% in the system's variability, despite wiping out the vast majority of the original variance in the pharmacokinetics.

Suppose then that similar methods are applied and the variance due to pharmacodynamics is successfully reduced from 30% to 3% – another technological triumph. The associated change in the square root of the sum of the squares will then be: 400 + 2500 + 100 + 9 = 55%, which is another modest decrease in overall variability.

In effect, with these two exercises, about as much variance reduction has been achieved as might reasonably be expected, based on quite remarkably large improvements in the two processes upon which the vast majority of attention is focused in clinical pharmacology and pharmacogenomics.

Following the logic of this elementary systems analysis, it can be seen that, if the variation in the dose–response relationship is to be brought down successfully to the upper end of the manageable range, then both drug formulation and patient adherence must be attacked.

This can be achieved by incorporating the best advances in membrane technology, as adapted to drug formulation. One valid system by which this might be achieved is the *ALZET mini-osmotic pump*, a multipurpose drug-delivery system that has been designed for subcutaneous or intraperitoneal implantation in laboratory animals. As this system has about a 10% CV in the *in vivo* rate of delivery of its contents, and this value as considered to be attainable for a drug formulation, then the new array of numbers will be: 100 + 2500 + 100 + 9, which sums to 2709, the square root of which is 52%. Thus, another very substantial reduction in variance has yielded only a very small reduction in the variability of the system's performance.

The remaining process that is in need of improvement is *patient adherence*. In this case, the adept use is supposed of electronic monitoring methods that are integrated into a system not only to contact patients when doses are late, but also to provide a timely remotivation when the patients' dosing histories are changed in ways that might portend an imminent discontinuation of dosing. Currently, data relating to improvements in adherence associated with various approaches to medication management are too sparse, and too early in respect to their learning curves, to make a confident estimate of how much could reasonably be expected to be achieved. Nonetheless, a conservative estimate should be made that the CV of adherence could be cut in half. The result of achieving such a cautious estimate would be:

$$100 + 625 + 100 + 9 = 834,$$

the square root of which is 29%.

If it were possible to reduce the CV of adherence to one-third of its original level, then the result would be:

$$100 + 279 + 100 + 9 = 488,$$

the square root of which is 22%.

It should be noted that it makes no difference which sequence is used when attacking the different sources of variance.

Deming proposed that the attack on sources of variance should be prioritized by rank-ordering them by measured or estimated magnitude, and then to re-rank them according to the prevailing technical capabilities for reducing their variability. Over time, technical capabilities are likely to improve to a degree that makes formerly intractable processes improvable. Deming also suggested that it was economically foolish to push for the lowest possible levels of variance, which would exceed the expectations of the customers.

7 Some Provisos

Not all drugs are as variable as theophylline which, although exceptional in its high level of variability, has been rendered obsolete by the development and introduction during the past 25 years of a number of anti-asthmatic medicines that seem to have far less variability. Of course, variable adherence, because of its poor recognition by clinical means, has often gone unrecognized as a major contributor to variability, and has mistakenly been ascribed to other elements in the system that collectively determine drug response and variability.

There is no doubt that certain absolutes can enter the picture portrayed above. One such absolute is the categorical but closeted non-acceptance by some patients of a prescribed medicine, to a point where the drug is never taken by the patient, despite its having been prescribed. Another absolute is a categorical absence of a particular enzyme that is a necessary condition for a response to a prescribed treatment. Yet another is an exceptional barrier to drug absorption, or an exceptionally high rate of drug metabolism, such that standard doses have little or no effect. In these instances, there is no evident response to the prescribed medicine, although that view is seen through a lens made foggy by placebo effects and, often, also by a lack of knowledge concerning adherence. Although clearly an important variable, adherence is not yet generally considered to be deserving of measurement in clinical situations where the reasons for drug nonresponsiveness are ambiguous. In light of the foregoing analyses, and in recognition of the numerous costly clinical alternatives that hang in the balance, it might reasonably be considered that the measurement of parameters of adherence would be justified by ruling in or ruling out nonadherence as the cause of a poor or nonresponse to drug treatment.

8 An Overview

Improvements beyond reducing variability in drug response by over 70%, as illustrated in the final numerical exercise, would most likely incur major costs for diminishing results. The final exercise, then, is an approximation of the limit of variance-reduction in drug response – "the 22% solution." Getting there entails multiple approaches, in each of the four processes designated in Fig. 1. The achievement of low-variability response to drug treatment – which is the capstone of personalized medicine – is therefore a systems problem that must be approached as such. While forecasting great triumphs in pharmacogenomic research and their successful applications to clinical medicine forms part of the solution, other sources of variance – most notably the patients' variable adherence and dosage form design – must also to be curtailed substantially in order to rid drug responses of most of their prevailing variability. Low-variability in drug responses is, indeed, a hallmark of personalized medicine.

References

1 Abu Ali al-Husayn ibn Abd Allah ibn Sina, A. (1999) The Canon of Medicine, in: Bakhtiar L. (Ed.) *Great Books of the Islamic World*, KAZI Publications, Inc., Chicago, p. 710.

2 Urquhart, J. (2002) Simulation for Designing Clinical Trials: A Pharmacokinetic-Pharmacodynamic Modeling Perspective, in: Kimko, H.C., Duffull, S.B. (Eds) *History-Informed Perspectives on the Modeling and Simulation of Therapeutic Drug Actions*, Marcel Dekker, New York, pp. 245–269.

3 Braunwald, E. (1997) Shattuck lecture. Cardiovascular medicine at the turn of the millennium: triumphs, concerns, and opportunities. *N. Engl. J. Med.*, **337** (19), 1360–1369.

4 First Annual Report (1839) Registrar General of England and Wales. London.

5 Osler, W. (1892) *Principles and Practice of Medicine*, D. Appleton and Company, New York.

6 Abelson, R. (2010) Insurers test new cancer pay systems. *New York Times*, October 20, 2010, p. B1.

7 Orszag, P. (2010) Malpractice methodology. *New York Times*, October 20, 2010, p. B1.

8 Hamburg, M.A., Collins, F.S. (2010) The path to personalized medicine. *N. Engl. J. Med.*, **363** (4), 301–304.

9 Marin, D., Bazeos, A., Mahon, F.X., Eliasson, L., Milojkovic, D., Bua, M., Apperley, J.F., Szydlo, R., Desai, R., Kozlowski, K., Paliompeis, C., Latham, V., Foroni, L., Molimard, M., Reid, A., Rezvani, K., de Lavallade, H., Guallar, C., Goldman, J., Khorashad, J.S. (2010) Adherence is the critical factor for achieving molecular responses in patients with chronic myeloid leukemia who achieve complete cytogenetic responses on imatinib. *J. Clin. Oncol.*, **28** (14), 2381–2388.

10 Blaschke T, Osterberg L, Vrijens B, Urquhart J. (2012) Adherence to medications: insights arising from studies on the unreliable link between prescribed and actual drug dosing histories. *Ann Rev Pharmacol Toxicol*, **52** 275–301.

11 Urquhart, J. (1997) The electronic medication event monitor. Lessons for pharmacotherapy. *Clin. Pharmacokinet.*, **32** (5), 345–356.

12 Mendonca, L.S., Moreira, J.N., de Lima, M.C., Simoes, S. (2010) Co-encapsulation of anti-BCR-ABL siRNA and imatinib mesylate in transferrin receptor-targeted sterically stabilized liposomes for chronic myeloid leukemia treatment. *Biotechnol. Bioeng.*, **107** (5), 884–893.

13 Parker, C.S., Chen, Z., Price, M., Gross, R., Metlay, J.P., Christie, J.D., Brensinger, C.M., Newcomb, C.W., Samaha, F.F., Kimmel, S.E. (2007) Adherence to warfarin assessed by electronic pill caps, clinician assessment, and patient reports: results from the IN-RANGE study. *J. Gen. Intern. Med.*, **22** (9), 1254–1259.

14 Kimmel, S.E., Chen, Z., Price, M., Parker, C.S., Metlay, J.P., Christie, J.D., Brensinger, C.M., Newcomb, C.W., Samaha, F.F., Gross, R. (2007) The influence of patient adherence on anticoagulation control with warfarin: results from the International Normalized Ratio Adherence and Genetics (IN-RANGE) Study. *Arch. Intern. Med.*, **167** (3), 229–235.

15 Piccart-Gebhart, M.J., Procter, M., Leyland-Jones, B., Goldhirsch, A., Untch, M., Smith, I., Gianni, L., Baselga, J., Bell, R., Jackisch, C., Cameron, D., Dowsett, M., Barrios, C.H., Steger, G., Huang, C.S., Andersson, M., Inbar, M., Lichinitser, M., Lang, I., Nitz, U., Iwata, H., Thomssen, C., Lohrisch, C., Suter, T.M., Ruschoff, J., Suto, T., Greatorex, V., Ward, C., Straehle, C., McFadden, E., Dolci, M.S., Gelber, R.D. (2005) Trastuzumab after adjuvant chemotherapy in HER2-positive breast cancer. *N. Engl. J. Med.*, **353** (16), 1659–1672.

16 Romond, E.H., Perez, E.A., Bryant, J., Suman, V.J., Geyer Jr, C.E., Davidson, N.E., Tan-Chiu, E., Martino, S., Paik, S., Kaufman, P.A., Swain, S.M., Pisansky, T.M., Fehrenbacher, L., Kutteh, L.A., Vogel, V.G., Visscher, D.W., Yothers, G., Jenkins, R.B., Brown, A.M., Dakhil, S.R., Mamounas, E.P., Lingle, W.L., Klein, P.M., Ingle, J.N., Wolmark, N. (2005) Trastuzumab plus adjuvant chemotherapy for operable HER2-positive breast cancer. *N. Engl. J. Med.*, **353** (16), 1673–1684.

17 Harris, L., Fritsche, H., Mennel, R., Norton, L., Ravdin, P., Taube, S., Somerfield, M.R., Hayes, D.F., Bast, R.C. Jr. (2007) American Society of Clinical Oncology 2007 update of recommendations for the use of tumor markers in breast cancer. *J. Clin. Oncol.*, **25** (33), 5287–5312.
18 Harter, J.G., Peck, C.C. (1991) Chronobiology. Suggestions for integrating it into drug development. *Ann. N. Y. Acad. Sci.*, **618**, 563–571.
19 Urquhart, J. (1998) Pharmacodynamics of variable patient compliance: implications for pharmaceutical value. *Adv. Drug. Deliv. Rev.*, **33** (3), 207–219.
20 Vrijens, B., Vincze, G., Kristanto, P., Urquhart, J., Burnier, M. (2008) Adherence to prescribed antihypertensive drug treatments: longitudinal study of electronically compiled dosing histories. *Br. Med. J.*, **336** (7653), 1114–1117.
21 Burnier, M., Schneider, M.P., Chiolero, A., Stubi, C.L., Brunner, H.R. (2001) Electronic compliance monitoring in resistant hypertension: the basis for rational therapeutic decisions. *J. Hypertens.*, **19** (2), 335–341.
22 Cross, J., Lee, H., Westelinck, A., Nelson, J., Grudzinskas, C., Peck, C. (2002) Postmarketing drug dosage changes of 499 FDA-approved new molecular entities, 1980-1999. *Pharmacoepidemiol. Drug Saf.*, **11** (6), 439–446.
23 Heerdink, E.R., Urquhart, J., Leufkens, H.G. (2002) Changes in prescribed drug doses after market introduction. *Pharmacoepidemiol. Drug Saf.*, **11** (6), 447–453.
24 Deming, W.E. (1982) *Out of the Crisis*, MIT Center for Advanced Engineering Study, Cambridge.

Part V
Systems Biology of Organisms

Systems Biology: Advances in Molecular Biology and Medicine, First Edition. Edited by Robert A. Meyers.

20
Microbiomes

Ramana Madupu[1], Yu-Hui Rogers[1], Doug Rusch[2], Jason Miller[1], Konstantinos Krampis[1], and Karen E. Nelson[1]
[1]J. Craig Venter Institute, 9704 Medical Center Drive, Rockville, MD 20850, USA
[2]Center for Genomics and Bioinformatics at Indiana University, Bloomington, Jordan Hall A302D 1001 E. 3rd Street, Bloomington, IN 47405-3700, USA

Systems Biology: Advances in Molecular Biology and Medicine, First Edition. Edited by Robert A. Meyers.

Keywords

Microbiome
The totality of microbes and the collective genomes of all the organisms and environmental interactions, within a defined environment. A defined environment might be the gut of a human being, or an animal. Thus, the microbiome usually includes microbiota and their complete genetic elements.

Microbiota
The collection of microbes that colonize a particular body site or a host.

Metagenomics
Metagenomics is a culture-independent study of microbial communities contained in an environmental sample. Metagenomics describes the functional and sequence-based analysis of genetic material from a mixed community of organisms contained in an environmental sample.

Personalized medicine
The customization of an individual patient's preventive and therapeutic care, based on genetic or other information.

Metagenomics, also referred to as *environmental* or *community genomics*, has brought about radical changes in the ability to analyze complex microbial communities by direct sampling of their natural habitat. Metagenomics has truly revolutionized biology and medicine, and changed the way in which genomics is studied. To date, many metagenomic studies have been undertaken, with samples from diverse habitats including the oceans, soil, air, human, and animal hosts having been subject to metagenomic examinations. Currently, huge national and international projects, aimed at elucidating the biogeography of microbial communities living within and on the human body, are well underway. The analysis of human microbiome data has brought about a paradigm shift in the present understanding of the role of resident microorganisms in human health and disease, and brings nontraditional areas such as gut ecology to the forefront of personalized medicine. In parallel, rapid technological advances in DNA sequencing methods have reduced the time and costs associated with sequencing while at the same time significantly increasing the data output. As genome sequencing becomes cheaper, it is being applied to sequence complex metagenomes, and large-scale 16S ribosomal DNA sequencing has become far more routine. Today, metagenomics is proving to be a powerful tool, considerably enhancing the present understanding of the extent and role of microbial diversity in their natural habitats, and in many ecologically important environments, with far greater implications on human health and disease. An overview of the current literature, together with details of projects and the state-of-the-art in microbiome studies, are presented in this chapter.

1 Introduction

Traditionally, genome-sequencing projects have relied on the availability of cultivated isolates to determine the complete genetic complement of an organism by sequencing the base pairs of its DNA. Inherently, however, this approach is significantly limited in its inability to recover genetic information from uncultivated microorganisms. *Metagenomics* is the study of microorganisms recovered from an environment in a culture-independent fashion, typically with the goal of simply measuring the population structure, genetic diversity, and ecological relationships of the microbes in a sample, but increasingly with the broader goal of building a systems biological model of the community. The study of single microbial species at the genomic level, and the concurrent development of tools to sequence, assemble, and analyze these species, have accelerated the field of metagenomics. By using the power of genomic analysis, the emerging field of metagenomics provides novel approaches for analyzing genetic information, to reveal important characteristics of microbial communities within a sample. The assemblage of microorganisms that reside within or on human and animal hosts has been termed the *indigenous microbiota*. The term *microbiome*, as originally introduced by Joshua Lederberg, essentially refers to the community of microbes, their genetic elements (genomes), and environmental interactions in an "ecological niche" such as the human gut or a soil sample.

The true capabilities of new sequencing technologies and progress in the field were exemplified by the first large-scale environmental metagenomic study, conducted by Craig Venter and his colleagues at the J. Craig Venter Institute (JCVI), and reported in 2004 [1]. The results included details of the sequencing and analysis of samples collected from the Sargasso Sea by using whole-genome shotgun (WGS) sequencing, coupled with high-performance computing developed to sequence the human genome. Through this study, the team reported the discovery of more than 1.2 million new genes. This initial study evolved to become JCVI's Global Ocean Sampling [2] initiative (as described in more detail below) which to date has generated millions of sequences derived from sites in the oceans, worldwide. To date, the Global Ocean Sampling (GOS) expedition to study marine microbial diversity from a complex environment remains the largest protein dataset [1, 2], and has also resulted in a number of spin-off studies that have been focused on the datasets that were initially generated [3–5].

Other large-scale metagenomic projects, such as the National Institutes of Health (NIH) -funded Human Microbiome Project (HMP), which was initiated to study human-associated microbial communities, have continued to evolve along with smaller single-investigator-driven projects focused on animals that include ruminants [6], canines [7], poultry [8], and nonhuman primates [9], as well as various insect species [10, 11]. The main results of these investigations have been the generation a tremendous amount of data, together with an acceleration of computational tool development aimed at interpreting genome sequence data, and at interpreting species diversity, as well as the provision of laboratory tools to identify, culture, and sequence novel species. Today, studies such as the HMP continue to make inroads into understanding the larger picture of human health, in wellness and disease, in addition to the environmental impact of microbial inhabitants on disease, immune status, and human physiology.

2
History of Microbial Diversity Studies

In the past, 16S rDNA gene analysis has been used widely to identify microbial species and to interrogate the diversity of microbes in a range of environments. The value in using the 16S rDNA gene as a phylogenetic marker derives from its being a necessary component of the ribosomal apparatus of microbial cells, and hence being present in all prokaryotic genomes. It also has a relatively slow rate of evolution coupled with highly conserved motifs, which are ideal for primer design. Several faster-evolving (variable) segments within the 16S rDNA gene sequence provide useful phylogenetic information, and are compatible with some of the next-generation sequencing technologies. Lastly, there is little evidence that the 16S gene moves by lateral gene transfer; thus, it tracks closely with the protein-coding genes that define a particular microbial lineage.

The 16S rDNA gene can, however, be less than perfect, as its copy number varies from species to species and even between strains, with copy numbers ranging anywhere from one to a dozen copies per cell. As such, 16S rDNA gene sequencing is an unreliable indicator of the actual abundance of the species being investigated. Intraspecies differences in ribosomal operons have also been described [12–15], again leading to the possible misinterpretation of results obtained from surveys. Furthermore, while 16S has largely been successful at classifying organisms down to the species level, it evolves too slowly to capture the diversity within species. Finally, the primer design and amplification processes that are used to obtain the 16S rDNA gene sequences are also known to introduce biases and errors in the process, including the formation of chimeric products, biased rates of amplification, and a failure to amplify some targets due to primer mismatch.

Regardless of these limitations, 16S rDNA sequencing continues to produce a wealth of survey data. The latest update from the Ribosomal Database Project [16], (http://rdp.cme.msu.edu/; RDP Release 10, Update 26 :: 28 March 2011) includes a total of 1,613,063 16S rRNAs from species that inhabit many different environments. Other public curated collections of 16S rRNA sequences that are available as online resources include GreenGenes [17], SILVA [18], and EZ-Taxon [19]. In the absence of better avenues for detailed comparative purposes, 16S gene-based approaches for characterizing microbial diversity have become popular tools for microbial ecology. Moving past a heavy reliance on Sanger-based full-length 16S rDNA gene sequencing where the approximate 1500 base-pair sequence was the goal, newer sequencing technologies have made it possible to generate hundreds of thousands of 16S genes from a single environment, allowing the community to be studied in unprecedented detail.

The newer generation of sequencing technologies generate shorter read-lengths, and as such have been used primarily in large surveys and for the inference of microbial diversity in ecological samples [20]. Most of the studies that employ the use of next-generation sequencing technologies tend to focus on one or more variable regions for inference of the population diversity [21]. A multiple-primer sequencing strategy on the Illumina sequencing platform to obtain longer 16S sequences needed for taxonomic analysis of mixed populations of bacteria was recently described [22].

The large genomic sequence data sets in multiple data formats present novel computational challenges for the exploration of metagenomic sequence data and classification of 16S rRNA sequences. Many new bioinformatic tools geared towards high-throughput analyses have been deployed to assess microbial phylogeny and the diversity of complex flora [23, 24]. The next-generation sequencing technologies offer the ability to sample genomes of organisms that occur at low abundances in communities, at a much lower cost. This has resulted in a high discovery rate of previously uncultured and novel organisms.

3 Microbiomes

3.1 Oceans

The oceans are teeming with microscopic organisms that have been characterized only to a limited extent, but whose actions are crucial to maintaining the biosphere. At the same time, the all-encompassing term "ocean" belies the complex and varied habitats that make up marine ecosystems serving to downplay the diverse communities that exist within it. The GOS expedition, which started in 2003 with a pilot project in the Sargasso Sea, is one of the first and largest environmental genomic studies. The approach was to collect hundreds of liters of surface seawater and to filter this into different size fractions dominated by eukaryotes (3.0 μm), eukaryotes and bacteria (0.8 μm), bacteria (0.1 μm), and viruses. Initially, the sequencing costs were the limiting factor, and consequently most attention was placed on the 0.1 μm bacterial fraction, where the gene density was expected to be greatest and the potential for confounding factors such as introns or repetitive elements would be minimized. Marine microbial communities are thought to be fairly complex, and to contain thousands of distinct taxa. Given that so little was known about these microbial communities, however, the goal was to circumnavigate the globe and to examine the genomic content of these communities, explore the diversity present in these systems, and attempt to gain a better understanding of how microbes, communities, and the environment interact.

The initial pilot project, conducted in the Sargasso Sea, generated over 1.7 million Sanger sequencing reads, from which over 1.2 million peptide sequences were identified. Many of these peptides provided a tremendous expansion in known protein families, as exemplified by the proteorhodopsin gene family. The proteorhodopsins function as a fast, light-driven proton pump, and are thought to provide an important (if limited) advantage to heterotrophic organisms in nutrient-poor environments, such as those found in the open ocean. Previous studies, including polymerase chain reaction (PCR)-based surveys had characterized a total of 67 different proteorhodopsins from a handful of different clades. In the Sargasso Sea pilot study, 782 additional proteorhodopsins were identified that belonged to a variety of unsampled clades, further confirming the widespread distribution of this gene family in marine ecosystems. The expansion of the proteorhodopsins was typical of many characterized protein families. Approximately 65% of the new peptides matched previously characterized orphan hypothetical genes, thus identifying entirely new protein families. The growth in existing protein families, and the identification of new families, was both exciting and troubling, however, the major impediment when handling these large volumes of data being a lack of data mining tools to detect metagenomically isolated genes and to develop better models for describing the function and role of gene families. Nonetheless, the Sargasso Sea pilot study ultimately demonstrated the power of metagenomics for identifying novel genes and linking them to their environment.

The GOS expedition began officially in 2004, starting with a transect down the eastern seaboard, through the Gulf Stream and the Gulf of Mexico, across the Panama Canal, and ending just outside of the Galapagos Islands. Phase I included 41 different locations covering a number of different biomes and spanning a wide range of environmental conditions. In a series of reports, the Phase I dataset was analyzed either from a community composition perspective in which tools such as assembly, fragment recruitment, and comparative metagenomics were applied [25], or from a protein-content perspective, where techniques for clustering the peptide data were developed to explore the growth of protein families and the diversity within protein families [26]. The overarching result of these studies was that the diversity within the marine environment was tremendous. In fact, diversity within the taxonomic classification of species was so extensive that no two cells would be expected to be the same in terms of gene content. Within protein families, the breadth of the analysis led to the discovery of protein families in the Eubacteria that previously had only been associated with eukaryotes or the Archaea. Phase I also included studies of the viral fractions examining the co-occurrence of phage and host, along with the types of genes associated with viral populations.

Currently, the GOS expedition continues to collect and analyze samples. As of 2011, the GOS includes hundreds of samples from the circumnavigation, along with additional samples from the west coast of North America, from the Antarctic Ocean and from Europe, including the Baltic and Mediterranean Seas. Recently, the Moore foundation sponsored the analysis of a diverse set of 137 cultivated marine microbes isolated from across the globe. Many of these genomes were not well represented in the GOS, either because they are found in different size fractions or because they were not abundant in marine surface waters. Those that were abundant constitute the "planktonic" fraction of the microbial community, and show adaptations to living in low-nutrient environments. Improvements in metagenomic assembly and binning are helping to identify genomes for many of the uncharacterized and uncultivated clades in the ocean [27].

3.2 Air

The atmosphere is a complex and very dynamic environment that, to date, has been poorly characterized in terms of its biological content. In a typical day, an average adult breathes in over 10,000 liters of air while simultaneously exposing themselves to the array of organisms that use the air as a mechanism for their dispersal. Air plays host to a diverse array of organisms, including pollen granules, resilient fungal and microbial spores, aerosolized bacteria, and viruses, which are either free-floating or attached to small dust particles. That these microbial entities have an impact on human health is well accepted, yet much concerning the diversity, distribution, and activity of these airborne organisms is unclear. For example, do indoor and outdoor environments differ in their biological composition? Have microbes adapted to live in climate-controlled environments? Do different indoor environments host different airborne communities? Does this result in a greater exposure of humans to organisms possessing virulence factors and pathogenicity islands? In the longer term, it would be beneficial to know how the bio-aerial community can be effectively monitored, and to determine effective methods of intervention that would both minimize exposure to potentially harmful organisms and increase exposure to protective organisms.

The challenge when studying the aerial microbial community is that the atmospheric load is approximately 10 bacteria per liter, while that of seawater may be in excess of 10^9. At issue here is the need for sufficient genetic material to perform sequencing and other molecular techniques, ideally without having to resort to amplifying the DNA, which can – and will – introduce biases. Furthermore, such small quantities of biological material make it easy to introduce contaminants that can swamp any real signal. Air samples also contain a variety of nonbiological materials, including small dirt, metal, and dust particles that must be removed but which initially are either attached to or mixed with the microorganisms. Finally, many organisms are contained in pollen granules and spores that are extremely resistant to lysis. Given all of these issues, it is especially important for research groups to design their experiments and controls very carefully, in order to ensure that the data acquired by these steps are both reliable and unbiased.

Given all of these challenges, it is not surprising that only a small number of indoor or outdoor air studies have been conducted. In the largest air metagenomic study carried out to date [28], samples of indoor air from two shopping centers in Singapore were compared with nearby soil and water samples. This data showed both air samples to be less diverse, and distinct from the other environmental samples.

3.3 Host-Associated Microbiomes

Host–microbiome studies have provided insights into the interactions between species and their mutualistic inhabitants. In theory, they should also allow a distinction to be made between microbes that are triggers for disease conditions, neutral organisms that are "along for the ride," and microbes that are beneficial. Many of these studies have focused on the microbial diversity associated with the intestinal community. For example, a little-known fact is that some of the earliest metagenomic studies were actually focused on the microbial diversity of the rumen [6]. In this case, Bryan White and colleagues at the University of Illinois, along with colleagues at the JCVI in Rockville, MD, USA, were funded by the US Department of Agriculture (USDA) to study the impact of diet on changes in the microbial populations of the rumen in cattle [6].

More recently, Kim *et al.* [29] tallied the microbial diversity estimates from all available rumen studies, based on a meta-analysis of all available curated 16S rRNA gene sequences deposited in the RDP database. As of November 2010, totals of 13,478 bacterial and 3,516 archaeal rRNA sequences had been described, that could be assigned to 5,271 species-level operational taxonomic units

(OTUs) that represented 19 existing phyla. The Firmicutes, Bacteroidetes, and Proteobacteria were found to be the most predominant phyla. In total, 180 existing genera were represented, with nearly all of the archaeal sequences being assigned to the phylum Euryarchaeota and representing 12 existing archaeal genera. The survey results suggested that there is a significant amount of microbial diversity yet to be described at the metagenomic level, and that this will hopefully be achieved with the new advances in sequencing technology.

Recent studies [7] to characterize the phylogeny and functional diversity of the canine gastrointestinal microbiome were the first of their type. In this case, fecal material from healthy adult dogs fed a low-fiber control diet (K9C), or a diet containing 7.5% beet pulp (K9BP), was used for 454 pyrosequencing. The dominant phyla included the Bacteroidetes/Chlorobi group and Firmicutes (which comprised about 35% of all sequences), followed by Proteobacteria (13–15%) and Fusobacteria (7–8%). The archaeal species identified in this study were not influenced by diet, and represented approximately 1% of all sequences. Less than 0.4% of the sequences were of viral origin, and more than 99% of these were associated with bacteriophages. Interestingly, the clustering of several gastrointestinal metagenomes demonstrated similarity between dogs, humans, and mice.

White and colleagues at the JCVI have also recently reported their findings on nonhuman primates. A comparative analysis of the microbial populations derived from the gastrointestinal microbiomes of nonhuman primate species may possibly provide a better understanding of the evolution of nonhuman primate diversity. Based on an analysis of fecal samples from three different wild nonhuman primate species, namely black-and-white colobus (*Colubus guereza*), red colobus (*Piliocolobus tephrosceles*), and red-tailed guenon (*Cercopithecus ascanius*), the Firmicutes were shown to comprise the vast majority of the phyla, while the microbial community composition was more similar within the same species than among different species. Moreover, when compared to humans, the microbial diversity was distinct and related to the host phylogeny. Diet was also seen to be a possible influence on community structure.

3.4 Human Body

In humans, the microflora represents an integral part of the genetic landscape and evolution. It has been estimated that the number of microbial cells resident in the human body outnumbers those of the host by a factor of at least 10 : 1, where complex and diverse species have been identified in the different body sites. The results of previous microbiome studies have also shown that microbial communities vary within and between human hosts. For example, the analysis of fecal microbial communities of obese and lean twins have shown that, despite a shared "core microbiome" at the gene level within family members, phylum-level changes (along with a host of other factors) influence the overall metabolic profile of an individual [30]. Another recently conducted gut microbiome study highlighted the influence of diet in shaping the structure and functional capabilities of the gut microbiota [31]. Evidence from this and other studies has suggested that bacterial lineages are unique to the host, and a "core microbiome" is largely defined through ecological interactions within hosts, as well as host variability in terms of dietary nutrients, genotype, illness, antibiotic use, and colonization history [32].

Currently, several microbiome studies are ongoing designed to examine any differences between the microbiomes of healthy patients and those of patients suffering from disease, under the umbrella of the International Human Microbiome Consortium (IHMC) (www.human-microbiome.org) and the Human Microbiome Project (HMP). The goal of the IHMC is to create a common set of guiding principles and policies, and to generate shared data resources that would enable research groups to conduct comprehensive studies on the human microbiome. Currently, several groups worldwide are participating in human microbiome research, conducting studies on various aspects of the microbiome in healthy adults and its effects on disease states. Members of the IHMC include agencies from Canada, United States, European Commission, Japan, and Australia. The NIH initiated the HMP project as an initiative of the NIH Roadmap for Biomedical Research (http://nihroadmap.nih.gov), with the goal of comprehensively characterizing the microbial flora within and on the surface of the human body, and to understand the role of the microbiome in human health and disease. The objectives, timeline, and complete details of implementation of the project have been described in the maker report [33].

The METAHIT consortium (http://www.metahit.eu/), which is an active participant in the IHMC, reported the first large-scale metagenomic study of gut microbiome of 124 human subjects of European origin [34]. For this, the authors analyzed fecal samples using Illumina sequencing technology, and reported the characterization of 3.3 million nonredundant microbial genes, which were largely shared among the individuals. The microbial gene catalog generated through this project was approximately 150-fold larger than the human gene complement. Most recently, the MetaHIT consortium reported an additional analysis of previously published data sets from 22 newly sequenced fecal metagenomes of individuals originating from four countries across continents. In this case, the data were classified into three clusters, mostly driven by species composition, and referred to as *distinct enterotypes*, which suggested the existence of a limited number of well-balanced, host–microbial symbiotic states [35].

During recent years, through the HMP and other similar projects around the globe [36], access to the human microbiome and an understanding of its influence on health and disease has become of utmost importance. Today, the HMP has generated metagenomic sequence data from 15 to 18 body sites of 300 "normal" individuals, many of whom have become repeat sample donors. In parallel to the metagenomic work, culture-based, nonculture-based, and single-cell approaches have allowed access to the genomes of reference species that will provide insight into the genetic diversity of these microorganisms, as well as acting as a scaffolding for the metagenomic datasets [37].

The first large-scale report from the HMP consortium described 178 reference genomes that were generated as a result of much sequencing and annotation effort [33]. From these genomes, approximately 547,968 polypeptides that were greater than 100 amino acids in length were identified, of which 29,987 were unique; that is, they had not been identified previously when the datasets were compared against all publicly available data. Although, at first sight, this might appear to be a large dataset, it is known that the genomes included in this initial study – despite being the largest conglomeration of sequenced genomes in a single publication – represent but a subset of the species associated with the human body, based on the interrogation of available 16S surveys of the human body. The continued data-mining of these and other datasets that are being generated by the HMP and international consortia (see Ref. [35]), as well as by other smaller groups, is anticipated to reveal additional significant details of gene clusters, antibiotic markers, plasmids, and phages.

3.5 Virome Studies

Although, most of the initial microbiome research (especially gut microbe studies) were focused on exploring the eubacterial community, archaea, viruses, fungi, and other microbes have also been frequently detected in environmental samples [38]. The discovery of RNA viruses in healthy human fecal samples [39] led to a general acceptance that the normal range of microbes in the human gut includes a wide variety of viruses, including enteric viruses, bacteriophages, and many known and unknown human viruses. Advances in sequencing technologies and the subsequent development of novel bioinformatic methods have allowed clinical research groups to increasingly adapt genomics-oriented approaches for translational medicine. Today, metagenomic sequencing has opened new avenues for the discovery of novel viral species from clinical specimens, and also provides opportunities for understanding the etiology of unexplained illnesses. Deep-sequencing analyses of viruses, aided by metagenomics-based tools, have allowed an in-depth understanding of host–pathogen interactions to predict the existence of emerging and re-emerging virus-borne infectious diseases [40], and to examine the role of common viral infections interacting with the host microbiome in the context of complex diseases [41].

4 Single-Cell Genomics

Single-cell genomics is a relatively new, but very powerful, technique that can be used to examine the DNA and (potentially) also the RNA content of one cell, in simultaneous fashion. It has been estimated that more than 99% of the microbes that exist on Earth currently elude standard cultivation. Hence, the major benefit of the single-cell technique is that it provides an ability to isolate one cell at a time from a complex environmental sample, to amplify its genomic contents and, eventually, to sequence that genome. At its core, single-cell genomics promises to "digitize" microbial ecology by allowing research groups to examine entire communities at their most granular level – that of the individual cell. Today, individual cells can be isolated relatively easily by dilution, or preferably by cell sorting using flow cytometry. The latter technique is of great potential, since in theory it can be used in conjunction with various stains for the selective isolation of cells of interest from a heterogeneous population. In this case, the cells of interest might either be actively dividing or carrying out specific metabolic processes. After isolation, the cells are lysed and then undergo multiple displacement amplification (MDA) to provide a billion-fold amplification of the DNA to an abundance that will support sequencing; the MDA is accomplished using the phiX polymerase that is both fast and accurate. After amplification the individual cells are tested for the presence of the 16S or 18S rRNA gene; this not only allows the research team to pick and choose which cells to take for further analysis, but also permits the generation of large banks of amplified single cells, and the selection of targets of interest while avoiding potentially redundant sequencing.

The use of single-cell genomes overcomes one of the greatest drawbacks of shotgun metagenomic techniques – notably, that genes and pathways can be identified but often cannot be explicitly linked one to another, principally because of the diversity within microbial communities and populations. This lack of context means that it is very difficult to determine the overall biological capabilities of uncultivated organisms and, by extension, to build accurate models of entire communities. The sequencing of single cells explicitly determines this context, and promises to provide a much better view of the diversity and capabilities of organisms within a community.

Nonetheless, single-cell genomics is a young field in which many technical challenges have still to be overcome. In order to minimize damage to the DNA, the cell lysis must be performed very gently; unfortunately, however, many cells require rather harsh techniques to ensure an efficient lysis, and therefore this technique may not be amenable to some classes of microorganisms. As the initial MDA is very sensitive to the starting conditions, the genome may very often be amplified unevenly, perhaps many thousand-fold. Whereas, normalization and low-cost sequencing can partially compensate for such an uneven coverage, any individually amplified genome will typically contain between 10% and 90% of the complete genome. This problem necessitates the sequencing of multiple cells, and generally reduces the efficiency of the approach. Finally, due to the extreme sensitivity of MDA, even the smallest levels of contamination may become problematic, and every batch of enzymes and new molecular techniques must be carefully tested to ensure purity. The development of protocols to address these issues, and to identify only the best, most evenly amplified contaminant-free genomes, is key to the future success of single-cell genomics.

5
Sequence Technologies and Tools

The rapid advancement of sequencing technologies during recent years has resulted in a number of production-ready "Next Generation" (NextGen) sequencing instruments. These NextGen platforms have largely replaced the capillary-based Sanger instruments, and have become the primary tools for producing sequence data for the majority of applications in the field of biology and genomics, including the study of microbiomes. The NextGen instruments allow a large amount of sequence data to be generated in a massively parallel manner, and thereby significantly increase the experimental output to the gigabasepair (Gbp) level per experiment, and at a significantly lower cost compared to Sanger-based technologies. Examples of these instruments include the Roche/454 GS FLX (Genome Sequencer), Illumina/Solexa GA (Genome Analyzer), and ABI SOLiD (Supported Oligonucleotide Ligation and Detection) platforms. Rapid and significant technological advances continue to be made to these platforms in terms of the quantity and the quality of the sequence data generated, and their utility for a wider set of experimental design and application.

Furthermore, several other approaches are currently in development that promise to increase the speed and reduce the costs of sequencing even further. As a consequence, the generation of genome sequence data is no longer proprietary to large genome centers, as smaller research laboratories will have access to NextGen sequencing machines. The interpretation and extraction of useful biological data from the large amount of NextGen sequence data generated from microbiome samples remains a major challenge, however, due to the complexity of the microbiome samples and the lack of standards. Today, approaches and computational tools for analyzing the NextGen microbiome sequence data are still in the process of being defined and developed, pushing costs away from the data generation stages into the data analysis.

5.1 454 (Roche) GS Sequencing Technology

The first of the NextGen platforms to become commercially available, in 2005, was the 454 GS system, a novel and highly parallel system based on pyrosequencing chemistry [42]. In this system, sample preparation involves the fragmentation of genomic DNA, followed by the ligation of adaptor sequences and clonal amplification of the target DNA on micron-sized beads, using an emulsion-based PCR (emPCR) method [43]. The sample preparation process is much more simple than that of Sanger sequencing, which contributes to not only an improved cost-effectiveness but also a significant improvement in throughput. The sequencing-by-synthesis reactions are performed directly on the template-carrying beads, which are preloaded into a microfabricated glass plate containing 1.6 million reactor wells, each of volume 1 pl. In order to determine the DNA sequence, the four nucleotides are delivered sequentially through the plate, while a charge-coupled device (CCD) camera detects those wells from which light is emitted as nucleotide incorporation occurs.

Currently, a single 454 GS FLX machine run, using the Titanium reagents, produces over 400 Mb of sequence data with reads of approximately 375 bp in length from over 1.2 million distinct sequencing-by-synthesis reactions, which take about 7 h to complete. Since its entry to the marketplace, the 454 pyrosequencing technology has been adapted to a wide variety of microbiome applications, including shotgun, 16S rRNA gene, and transcriptomic sequencing, so as to generate over 300 published reports.

5.1.1 Illumina/Solexa GA Sequencing Technology

The Illumina GA IIx and HiSeq 2000 systems have the highest throughput, are the most widely used NextGen platforms available, and employ the massively parallel sequencing-by-synthesis approach. Sample preparation process in this system involves the fragmentation of genomic DNA, followed by the ligation of adaptor sequences. DNA template amplification is achieved by the solid-phase bridge PCR method, which allows amplification on a solid substrate in a free-flowing cell by virtue of a localized amplification of genome fragments; these are referred to as *clusters* that have been adaptor-ligated [44]. The resulting PCR continues to amplify in a localized fashion to form a discrete cluster of a single molecular species; the latter is then used for a primer-mediated base incorporation using fluorescently labeled terminator bases and a modified polymerase. The use of fluorescent dye terminator bases results in a marked reduction of insertion and deletion sequencing errors as compared to the 454 technology. However, the use of a modified polymerase and nucleotides may result in base incorporation errors, thus increasing the number of substitution bases found. The sequence detection process, via the imaging of polymerase-incorporated fluorescently tagged bases, results in over 600 billion bases (HiSeq 2000) of sequences per instrument run, with read lengths of up to 150 bp. Due to their large capacity and low sequence production costs, the Illumina GA IIx and HiSeq have become the most widely used platforms for several microbiome applications, including shotgun and transcriptomic sequencing sufficient to generate almost 20 published reports.

5.1.2 ABI SOLiD Sequencing Technology

The ABI SOLiD analyzer utilizes a hybridization–ligation-based approach for sequence detection, and combines the optimality of PCR on bead with an open interface for the detection of a large number of sequences. Much like the 454 process, single adaptor-ligated genomic fragments are amplified by emulsion PCR on beads, which are then deposited onto a glass slide. The bases are detected by successive cycles of primer hybridization and detection probe ligation; this is followed by cleavage of part of the detection probe to release fluorescence in one of four colors that can be used to detect the dinucleotide bases incorporated into the detection probe. The system has a throughput of more than 50 billion bases per run, at a read length of 35–75 bp. The ABI SOLiD platform has also been adapted to a number of microbiome applications including metagenomic sequencing and microRNA expression analysis.

5.2 Assembly

Assembly is a standard bioinformatics treatment for shotgun sequence data. Whereas, the shotgun method [45] enables high-throughput sequencing by sampling DNA in random fashion, the shotgun assembly method [46] reconstructs genomic sequences from the random samples. The assembled sequences are longer than the individual reads, and they can be more accurate as they represent the consensus of redundant reads. Thus, assembly enriches both sequence length and accuracy, and these enrichments act to improve the yield from any subsequent sequence analysis, including gene recognition and pathway detection.

The results of an assembly are presented as *contigs* and *scaffolds*. Contigs are multiple sequence alignments of reads, and have a consensus sequence, whereas scaffolds (or super-contigs) are collections of contigs. Linear scaffolds, which are the most common type, define the relative order and orientation of contigs, plus the gap sizes between contigs. The gap sizes are estimated from the expected read pair separation distances. Whereas, contigs summarize one or more underlying reads at every position, scaffolds include gaps to summarize any missing or unresolved sequence.

In the GOS expedition, water was collected from the Atlantic and Pacific Oceans and, after size filtration for bacterial cells, DNA samples were sequenced on Sanger sequencing platforms at the JCVI. The data were subject to many analyses, including assembly, with one of the assembly strategies employing the Celera Assembler

[47]. The software in this case was parameterized to be very aggressive, accepting even nucleotide sequence identity as low as 84% as an indication of common origin of two reads. In fact, the software required modification to operate at such low thresholds. With a species abundance too low, there was little hope of recovering individual genomes; instead, sequences were constructed that might represent diverse strains, whole species, or even larger clades.

The assembly strategy generated several benefits; notably, it connected many novel GOS proteins (which otherwise might have been impossible to assign to taxa) to 16S or *rec*A taxonomic markers [2]. Some genes were linked to others, including a type-II glutamine synthetase; at that time, this gene was presumed to have a type-I variant in prokaryotes only, and a type-II variant in eukaryotes only. However, assembly linked the type-II variant convincingly to other prokaryotic genes; otherwise, the gene might have been regarded as eukaryotic contamination. Finally, the assembled contigs were used to nucleate gene-specific clusters in a downstream bioinformatics analysis of the reads [48].

Unfortunately, assembly involves risk; even on single-genome data sets assembly algorithms can omit repeated genome segments [49] or combine unrelated sequences that happen to share a repeated segment. These are also problems of nonassembly and mis-assembly, or rather false-negative assembly and false-positive assembly. These problems are exacerbated by the added complexity found in metagenomics data sets. Bork and colleagues [50] noted that microbiome community complexity induced a low coverage of individual species, and that even individual species present polymorphism between the genomes. Likewise, Pop [51] noted that variable abundance and polymorphism in metagenomics data challenged the traditional assembly software, while Kunin and colleagues [52] reported that genomic repeats would lead to assembly software to generate chimeric sequences from metagenomics data. Rusch and coworkers found likely chimera in the GOS assemblies [2], and later reconstructed putative whole genomes that assembly had missed. For complex microbiomes in particular, assembly-free analysis can reduce the exposure to risk. Assembly-free analysis methods include clustering reads according to their similarity with reference sequences [48], fragment recruitment to known sequences [53], phylogenetic profiling [54, 55], *ab initio* gene prediction, and gene recognition by homology [56].

It may be illuminating to challenge assembly software with data sets for which the underlying genomes are known. As simulation rarely captures all of the nuances of life, these challenges may define minimal requirements. MetaSim [57] is useful software for simulating reads given reference sequences, a relative-abundance model of the community structure, and error models for particular sequencers. An early test of simulated metagenomics Sanger reads showed that, for the three assemblers tested, only the longest contigs were reliable [58]. Recently, the Human Genome Project challenged various assemblers with real reads from mock communities that had been generated by the combination of bacteria of which the genomes had been sequenced, including some closely related genomes. One even-abundance and one staggered-abundance community were sequenced using Illumina technology. When six assemblers were tested, including ABySS [59], CABOG [60], CLC (http://clcbio.com), Newbler (http://454.com), SOAP de novo, and Velvet [61, 62], the packages each performed similarly. Across all assemblies, most contigs that were 300 bp or longer had full-length alignments to a reference: 85% in even-abundance and 80% in staggered-abundance, average per assembly. There were between zero and 10 chimera per assembly, defined as having more than one nonoverlapping alignment to a reference that was 1000 bp, or longer; however, none of the assemblers even came close to capturing one complete genome. The maximum contig size per assembly was, on average, 272 kbp in even-abundance and 148 kbp in staggered-abundance; hence, it would appear that staggering the genome abundance levels increased the assembly difficulty. It should be noted that all of the assembly software tested had been developed for single-genomes, and not for metagenomics.

Currently, metagenomics-specific assembly software is either lacking or immature. Since assembly software development is a lengthy process, most large-scale microbiome projects have re-used existing software that was designed for single-genome applications only. The MetaHIT Illumina reads collected from human stool samples were assembled with the SOAP *de novo* software, which had been developed for single-genome data [63, 64]. The GOS Sanger reads collected from ocean water were assembled using a modified version of the Celera Assembler software [1, 2], which had also been developed for single-genome data [48]. Microbiome assembly should become more sensitive and accurate as computer scientists develop algorithms and software that is specific to the task.

In the near future, microbiome assembly will be challenged by larger data sets as sequencing efficiency continues to improve. Assembly software developers may respond to this situation by incorporating more sophisticated distributed processing models, or by relying on the increasing availability of powerful, multi-core, high-RAM computer servers. Software may also grow to exploit the increasing relevance of data beyond the shotgun reads themselves. This includes reference genome sequences, transcript sequence data, and single-cell genome sequence.

5.3 Data Analysis

5.3.1 Fragment Recruitment

Today, whole-genome and metagenome studies provide a clearer picture of the conservation and diversity found within and across microbial species. Complete reference genomes are powerful tools for exploring the metabolic and functional capabilities of an organism, and provide a snapshot of the diversity that is present in any single isolate. As multiple strains of bacteria from the same species have been sequenced it has become clear how different even very closely related organisms can be [65]. By measuring the "genomic fluidity" of a population, it is increasingly possible to estimate the gene content associated with a particular clade of microbes [66]. Because metagenomic studies do not require cultivation they can reveal the true extent of the diversity in a particular population and environment. The technique of fragment recruitment is a powerful approach for measuring both the abundance and the diversity of organisms in a community relative to known reference sequences [25]. Moreover, when applied to metagenomic assemblies, it may be used as a tool for analyzing the accuracy of the assembly and to bin scaffolds into genomic units.

The approach to fragment recruitment is technically simple: the sequencing reads are aligned at low stringency to complete and draft genomes using a favorite sequence alignment tool (typically BLASTN). A simple filter removes any poorly aligned reads, after which the remaining alignments are plotted to show the position (on the reference sequence) and the identity of the alignment (number of correctly aligned bases divided by the length of the read). The value of these simple percentage identity plots [67] can be further enhanced by coloring individual reads, based on various forms of metadata. This additional information is typically environmental metadata depicting the sample and/or environment characteristics associated with a specific read. Alternatively, if mate-paired reads were to be recruited, the additional information could be in the form of structural metadata describing the relative placement and orientation of the paired reads. Structural information can indicate whether paired reads are closer or further away than expected, and indicate a possible deletion or insertion relative to the reference. Although the fragment recruitment concept is straightforward, these plots contain a depth of information that belies their simple origins.

Recruitment plots can be both an attractive and elegant means of organizing metagenomic information. Generally, recruitment provides useful information only when the metagenome contains reads from the same species or genus as one or more of the references genomes. Given a suitable reference, the recruited reads will be more or less evenly distributed along the length of the reference, and the density (number of reads per kilobyte) [66–68] will be indicative of the abundance of that genus or species in the environment. In general, identity indicates how closely related – and therefore how representative – a genome is of the environmental organisms. Typically, there are gaps in the recruitment where tens or hundreds of kilobases of the reference will have few, if any, recruited reads; these correspond to hypervariable segments that appear to be strain-specific and contain largely uncharacterized proteins [25, 68]. In some cases there will be segments of the genome that recruit reads from only a subset of samples, thus highlighting any potential geographically or environmentally adaptive genes. The plots can be become very complicated as the abundance of related organisms goes up, and as the number of related strains in the community increases; moreover, artifacts may emerge due to the recruitment process. Virtually every microbial genome will have some recruitment, not because they are present in a particular metagenomic sample but rather because conserved genes and motifs will align to even very distantly related genes from other organisms.

The application of fragment recruitment to assemblies derived from a metagenome can be used to validate scaffolds or to identify chimeric joins that can, in theory, be broken to improve the assembly. Valid scaffolds will have the same general distribution and abundance of reads along their lengths; furthermore, there should be large numbers of "good mate pairs" that are the correct distance apart and properly oriented, distributed along the length of the scaffold. The presence of high-identity good mates almost ensures that at least one clone from the environment will support that specific scaffold layout. Unfortunately, however, the converse is not true – poorly mated, high-identity reads do not invalidate a specific scaffold layout, but rather suggest that there may be other valid layouts (variants) in the environment.

The Advanced Recruitment Viewer, which is available through the JCVI website, provides a mechanism to view precomputed recruitment plots for GOS, HMP, and other metagenomic studies. It also provides a number of basic capabilities for assessing abundance, examining recruitment in the context of existing annotation, viewing structural metadata, and exporting recruited reads of interest.

5.4 Data Management

Data generation and management during the sequencing of microbiomes can be divided into three distinct stages depending on the datatypes, the methods used for generating each type, and the required storage space. The first stage involves collection of the raw instrument data; these are image files containing light-intensity values emitted by the fluorescent nucleotides added during each sequencing cycle, and captured by the image sensor of the instrument (for a detailed review of sequencing chemistry, see Ref. [69]). At the second stage, data are obtained from the quantification and quality check of the intensity values, and then converted to the corresponding nucleotides in order to acquire sequence reads, using algorithms specific to each sequencing technology. The manufacturer usually implements these algorithms in software packages supplied along with the sequencing instrument. The third and final stage is of the greatest scientific value, and is where meaningful sequence data are extracted from the generated data. This involves running assembly algorithms [70] with the sequence reads as input in order to reconstruct the original genome, followed by an analysis through bioinformatic pipelines for gene calling, functional annotation, and the discovery of other interesting biological features on the genome.

These stages are interdependent, and two opposing trends are observed as sequencing data are processed and moved from the first to the third stage. One trend is to increase the scientific value of the data, while the other is a reduction in size and storage requirements. More specifically, the raw image files have no value for scientific analysis after quantification and conversion to sequence reads, except if used as reference for a quality check of the reads which, nonetheless, would be impractical given their data size. On the other hand, sequence reads can be used as input to assembly algorithms, in order to reconstruct the complete genome sequence of an organism; alternatively, they may be mapped to annotated genomes for polymorphism discovery and the quantification of gene expression in RNAseq experiments. As can be seen from the data relating to various sequencing instruments in Table 1, moving from stage one to stage two results in a data reduction of between a 22-fold (Illumina II) and threefold (SOLiD 3) for both fragment and paired-end libraries. Finally, assembled metagenomes have the most value as they can be used to discover new full-length genes, and can be used for comparative and evolutionary analysis with cultivated genomes.

Since the introduction of the ABI SOLiD instrument in 2007, sequencing technologies have continued to move in a direction where throughput is increasing while the cost per sequenced DNA base is decreasing; in fact, the costs have fallen by half

Tab. 1 Storage requirements at the three stages of data analysis, for the different sequencing technologies.

Data units: gigabytes (GB)		***Roche/454***			***Illumina/Solexa***			***ABI/SOLID***		
		GS20	***FLX***	***Ti***	***I***	***II***	***IIx***	***1***	***2***	***3***
Stage one images	Fragment	10	10	30	500	1100	2800	1800	2500	1900
	Paired-end	[a]	10	30	1000	2200	5600	3600	5000	3800
Stage two sequence	Fragment	0.5	1	4	30	50	[b]	100	140	600
	Paired-end	[a]	1	4	60	100	[b]	200	280	1200

[a]This instrument cannot generate paired-end libraries.
[b]Data are not available.
The first row of numbers at each stage refers to fragment data, while the second row is for paired-end library data [71]. This instrument cannot generate paired-end libraries.
Source: Genome Center at Washington University, Dooling 2010, http://www.politigenomics.com/next-generation-sequencing-informatics.

at about every five months. Furthermore, from the information provided in Table 1 it is apparent that data production on the order of terabytes per run (1 TB = 1000 GB) is becoming the norm for next-generation sequencing; hence, it would be possible to overwhelm the storage capacity of most laboratories with only a few sequencing runs. Consequently, significant costs for data storage must be considered in addition to the expense of acquiring the sequencing instrument. Recently, prices per gigabyte have been following a downward trend, similar to the cost per sequenced DNA base, but this has occurred at a reduced rate, being halved about every 14 months [72]. Despite the price drop in sequencing, which is expected to lead to the "US$ 1000 personal genome," the investment required for storage systems and the overall costs for bioinformatic infrastructure may increase this figure by many orders of magnitude.

The storage price per gigabyte also depends on the technology used. At one end of the spectrum are consumer-grade external hard drives that cost approximately US$ 150 per TB (first quarter of year 2011), whereas at the opposite end are highly scalable clustered storage systems with automatic replication, archiving, and disaster recovery, but costing hundreds of thousands of dollars. Many intermediate cost tiers exist between these two extremes, however, with systems having different combinations of capacity, performance, and data reliability characteristics. A key guiding principle when designing storage infrastructures is the scientific value placed on each of the three data stages of next-generation sequencing, which in turn determines the investment amounts in the different storage tiers. More specifically, an ideal configuration would be composed of large amounts of cheap storage attached to the sequencing instrument; these would hold the bulk of the stage one image data on a temporary basis, until the data are processed and quantified into sequence reads. The second component of the system would hold the stage two sequence data; this would consist of high-speed disks because the performance is critical and this is where most of the computation, including assembly, mapping, and annotation, would take place. The second component must also be made disaster-proof by using replicated disk systems, with the sequence data and analysis results (stage three data) safeguarded against hardware failures, as these will be used for scientific publications and, in most cases, released to the public following completion of the project. A third and final component is that of archival storage, which is used for the long-term preservation of almost all data generated from sequencing projects. This entails two key characteristics: first, to provide large volumes of storage at very low cost; and second, to include a disaster-proof status. The first characteristic is achieved by using disk drives or even magnetic tape – that is, not highly networked in order to achieve a low cost. The second characteristic can be achieved through the off-site replication of data, or by using the services of commercial vendors who offer the preservation of magnetic tapes in disaster-proof vaults. Another option for archival storage has, until recently, been NCBI's Short Read Archive (SRA, http://www.ncbi.nlm.nih.gov/sra), which is offered as a free service to research groups.

Given the significant costs that can result from storage systems, data management plans should be drawn up at the early stages of the experiment design and the submission of proposals involving the creation of large amounts of next-generation sequencing data. The length of time required for completion, and the extent of the data analysis to be performed as part of the funded project, represent major factors in setting a data retention policy and planning for storage costs. This can also help to determine whether, and for how long, the images, sequence reads and sequence assemblies should be retained for the particular goals of a specific sequencing project. As a rough guiding principle, stage one raw image data may be kept for up to two weeks until the quality checks, base calling, and conversion to sequence reads are complete. In contrast, stage two sequence reads, and their stage three derivatives (assembly, annotation, etc.), must be retained anywhere from one year until the results are published, or even up to a decade, if the research institute acts as a resource center for the scientific community. For example, the original plan for the 1000 Genomes Project (http://www.1000genomes.org) was for research groups submitting sequenced genomes to include all generated files. The average storage requirement for both the images and sequence reads was estimated at about 50 bytes per sequenced base. However, later in the project the participating members decided to submit only reads along with some metadata regarding their quality, which reduced the requirement to a single byte per base.

An alternative option to investing in informatics infrastructure, would be to obtain storage capacity from a "cloud computing" service. Cloud services provide research groups with the ability to store data and to perform bioinformatic analysis on an essentially unlimited pool of Virtual Machine (VM) servers, without the groups owning or maintaining any computer hardware. The charge model used by cloud service providers is similar to that used by utilities such as electricity, and customers are billed based on the amounts of computational resources consumed. This system may operate better for smaller research laboratories, as it avoids the need to invest in computer hardware and data center infrastructure, the cost of which cannot be justified for just a few experiments. For example, at the Amazon Simple Storage Service (S3, http://aws.amazon. com/s3), prices range down from US$ 0.14 per stored GB per month (as of the first quarter of 2011). This price applies to the first 1,000 GB of storage used, after which there is a tiered charge that allows the cost to be reduced as the usage increases. For example, an additional 49,000 GB would cost US$ 0.125 per GB per month.

An important concern when exchanging sequence data among collaborating research centers, using either cloud-based storage or downloading data from public databases, is the data transfer bottleneck that occurs across the Internet. According to results reported by the Amazon cloud service, when using an average Internet connection 600 GB of data would take approximately one week for transfer, while using a high-speed line would require about two days. One other option offered by the Amazon service for them physically to ship the disk drives to the company's local branch (http://aws.amazon.com/importexport), and have the data copied directly to the cloud servers; this would cost US$ 80 for disk drives containing up to 4 TB of data. Despite the data transfer bottleneck, data storage using a cloud service has the advantage that large-scale sequencing datasets can be easily exchanged among collaborators on a worldwide basis. Inherent in the design of Amazon S3 service is a replication of data across several physical storage locations for disaster prevention; this is currently available on US East and West regions, European Union (Ireland), and Asia Pacific (Singapore). For the data upload, the research team can choose the closest region to minimize data transfer latency over the Internet. The cloud service can then automatically replicate the data to different locations as part of the disaster prevention policy, thus allowing collaborating research groups to download the data from their closest region.

6 Future Perspectives

Currently, an evolving knowledge of the microbiome continues to provide a greater understanding of the vital role of baseline commensal microbes in the various defined environments of the human body. The focus of current microbiome studies is to assess the totality of microbes and their genomes to determine the role of host–microbe and microbe–microbe interactions. Consequently, it will be possible to expand the present knowledge of the microbiome's influences in shaping the host ecosystem and immunity, while evaluating their importance in the occurrence of disease and the maintenance of a healthy homeostasis. This knowledge could, potentially, have far-reaching implications in the diagnosis of disease risk and disease pathogenesis, and also in the design of microbiome-focused therapeutics and "personalized medicine."

Acknowledgments

The authors thank Eddy Navarro for IT help and advice on metagenomic data management at JCVI.

References

1 Venter, J.C., Remington, K., Heidelberg, J.F., Halpern, A.L., Rusch, D., Eisen, J.A., Wu, D., Paulsen, I., Nelson, K.E., Nelson, W., Fouts, D.E., Levy, S., Knap, A.H., Lomas,

M.W., Nealson, K., White, O., Peterson, J., Hoffman, J., Parsons, R., Baden-Tillson, H., Pfannkoch, C., Rogers, Y.H., Smith, H.O. (2004) Environmental genome shotgun sequencing of the Sargasso Sea. *Science*, **304** (5667), 66–74.

2 Rusch, D.B., Halpern, A.L., Sutton, G., Heidelberg, K.B., Williamson, S., Yooseph, S., Wu, D., Eisen, J.A., Hoffman, J.M., Remington, K., Beeson, K., Tran, B., Smith, H., Baden-Tillson, H., Stewart, C., Thorpe, J., Freeman, J., Andrews-Pfannkoch, C., Venter, J.E., Li, K., Kravitz, S., Heidelberg, J.F., Utterback, T., Rogers, Y.H., Falcón, L.I., Souza, V., Bonilla-Rosso, G., Eguiarte, L.E., Karl, D.M., Sathyendranath, S., Platt, T., Bermingham, E., Gallardo, V., Tamayo-Castillo, G., Ferrari, M.R., Strausberg, R.L., Nealson, K., Friedman, R., Frazier, M., Venter, J.C. (2007) The Sorcerer II global ocean sampling expedition: northwest Atlantic through eastern tropical Pacific. *PLoS Biol.*, **5** (3), e77.

3 Sharon, I., Alperovitch, A., Rohwer, F., Haynes, M., Glaser, F., Atamna-Ismaeel, N., Pinter, R.Y., Partensky, F., Koonin, E.V., Wolf, Y.I., Nelson, N., Béjà, O. (2009) Photosystem I gene cassettes are present in marine virus genomes. *Nature*, **461** (7261), 258–262.

4 Comeau, A.M., Arbiol, C., Krisch, H.M. (2010) Gene network visualization and quantitative synteny analysis of more than 300 marine T4-like phage scaffolds from the GOS metagenome. *Mol. Biol. Evol.*, **27** (8), 1935–1944.

5 Sorokin, V.A., Gelfand, M.S., Artamonova, I.I. (2010) Evolutionary dynamics of clustered irregularly interspaced short palindromic repeat systems in the ocean metagenome. *Appl. Environ. Microbiol.*, **76** (7), 2136–2144.

6 Brulc, J.M., Antonopoulos, D.A., Miller, M.E., Wilson, M.K., Yannarell, A.C., Dinsdale, E.A., Edwards, R.E., Frank, E.D., Emerson, J.B., Wacklin, P., Coutinho, P.M., Henrissat, B., Nelson, K.E., White, B.A. (2009) Gene-centric metagenomics of the fiber-adherent bovine rumen microbiome reveals forage specific glycoside hydrolases. *Proc. Natl Acad. Sci. USA*, **106** (6), 1948–1953.

7 Swanson, K.S., Dowd, S.E., Suchodolski, J.S., Middelbos, I.S., Vester, B.M., Barry, K.A., Nelson, K.E., Torralba, M., Henrissat, B., Coutinho, P.M., Cann, I.K., White, B.A., Fahey, G.C. Jr. (2010) Phylogenetic and gene-centric metagenomics of the canine intestinal microbiome reveals similarities with humans and mice. *ISME J.*, **5** (4), 639–649.

8 Qu, A., Brulc, J.M., Wilson, M.K., Law, B.F., Theoret, J.R., Joens, L.A., Konkel, M.E., Angly, F., Dinsdale, E.A., Edwards, R.A., Nelson, K.E., White, B.A. (2008) Comparative metagenomics reveals host specific metavirulomes and horizontal gene transfer elements in the chicken cecum microbiome. *PLoS ONE*, **3** (8), e2945.

9 Yildirim, S., Yeoman, C.J., Sipos, M., Torralba, M., Wilson, B.A., Goldberg, T.L., Stumpf, R.M., Leigh, S.R., White, B.A., Nelson, K.E. (2010) Characterization of the fecal microbiome from non-human wild primates reveals species specific microbial communities. *PLoS ONE*, **5** (11), e13963.

10 Allen, H.K., Cloud-Hansen, K.A., Wolinski, J.M., Guan, C., Greene, S., Lu, S., Boeyink, M., Broderick, N.A., Raffa, K.F., Handelsman, J. (2009) Resident microbiota of the gypsy moth midgut harbors antibiotic resistance determinants. *DNA Cell Biol.*, **28** (3), 109–117.

11 Suen, G., Scott, J.J., Aylward, F.O., Adams, S.M., Tringe, S.G., Pinto-Tomás, A.A., Foster, C.E., Pauly, M., Weimer, P.J., Barry, K.W., Goodwin, L.A., Bouffard, P., Li, L., Osterberger, J., Harkins, T.T., Slater, S.C., Donohue, T.J., Currie, C.R. (2010) An insect herbivore microbiome with high plant biomass-degrading capacity. *PLoS Genet.*, **6** (9), e1001129.

12 Mylvaganam, S., Dennis, P.P. (1992) Sequence heterogeneity between the two genes encoding 16S rRNA from the halophilic archaebacterium *Haloarcula marismortui*. *Genetics*, **130** (3), 399–410.

13 Lopez-Lopez, A., Benlloch, S., Bonfá, M., Rodríguez-Valera, F., Mira, A. (2007) Intragenomic 16S rDNA divergence in *Haloarcula marismortui* is an adaptation to different temperatures. *J. Mol. Evol.*, **65** (6), 687–696.

14 Pei, A.Y., Oberdorf, W.E., Nossa, C.W., Agarwal, A., Chokshi, P., Gerz, E.A., Jin, Z., Lee, P., Yang, L., Poles, M., Brown, S.M., Sotero, S., Desantis, T., Brodie, E., Nelson, K., Pei, Z. (2010) Diversity of 16S rRNA genes within individual prokaryotic

genomes. *Appl. Environ. Microbiol.*, **76** (12), 3886–3897.

15 Ray, A.E., Connon, S.A., Sheridan, P.P., Gilbreath, J., Shields, M., Newby, D.T., Fujita, Y., Magnuson, T.S. (2010) Intragenomic heterogeneity of the 16S rRNA gene in strain UFO1 caused by a 100-bp insertion in helix 6. *FEMS Microbiol. Ecol.*, **72** (3), 343–353.

16 Cole, J.R., Wang, Q., Cardenas, E., Fish, J., Chai, B., Farris, R.J., Kulam-Syed-Mohideen, A.S., McGarrell, D.M., Marsh, T., Garrity, G.M., Tiedje, J.M. (2009) The Ribosomal database project: improved alignments and new tools for rRNA analysis. *Nucleic Acids Res.*, **37**, D141–D145.

17 DeSantis, T.Z., Hugenholtz, P., Larsen, N., Rojas, M., Brodie, E.L., Keller, K., Huber, T., Dalevi, D., Hu, P., Andersen, G.L. (2006) Greengenes, a chimera-checked 16S rRNA gene database and workbench compatible with ARB. *Appl. Environ. Microbiol.*, **72** (7), 5069–5072.

18 Pruesse, E., Quast, C., Knittel, K., Fuchs, B.M., Ludwig, W., Peplies, J., Glöckner, F.O. (2007) SILVA: a comprehensive online resource for quality checked and aligned ribosomal RNA sequence data compatible with ARB. *Nucleic Acids Res.*, **35** (21), 7188–7196.

19 Chun, J., Lee, J.H., Jung, Y., Kim, M., Kim, S., Kim, B.K., Lim, Y.W. (2007) EzTaxon: a web-based tool for the identification of prokaryotes based on 16S ribosomal RNA gene sequences. *Int. J. Syst. Evol. Microbiol.*, **57** (10), 2259–2261.

20 Unno, T., Jang, J., Han, D., Kim, J.H., Sadowsky, M.J., Kim, O.S., Chun, J., Hur, H.G. (2010) Use of barcoded pyrosequencing and shared OTUs to determine sources of fecal bacteria in watersheds. *Environ. Sci. Technol.*, **44** (20), 7777–7782.

21 Thompson, F.L., Bruce, T., Gonzalez, A., Cardoso, A., Clementino, M., Costagliola, M., Hozbor, C., Otero, E., Piccini, C., Peressutti, S., Schmieder, R., Edwards, R., Smith, M., Takiyama, L.R., Vieira, R., Paranhos, R., Artigas, L.F. (2011) Coastal bacterioplankton community diversity along a latitudinal gradient in Latin America by means of V6 tag pyrosequencing. *Arch. Microbiol.*, **193** (2), 105–114.

22 Ram, J.L., Karim, A. S., Sendler, E.D., Kato, I. (2011) Strategy for microbiome analysis using 16S rRNA gene sequence analysis on the Illumina sequencing platform. *Syst. Biol. Reprod. Med.*, **57** (3), 162–170.

23 Petrosino, J.F., Highlander, S, Luna, R.A., Gibbs, R.A., Versalovic, J. (2009) Metagenomic pyrosequencing and microbial identification. *Clin. Chem.*, **55** (5), 856–866.

24 Haas, B.J. (2011) Chimeric 16S rRNA sequence formation and detection in Sanger and 454-pyrosequenced PCR amplicons. *Genome Res.*, **21** (3), 494–504.

25 Rusch, D.B., Halpern, A.L., Sutton, G., Heidelberg, K.B., Williamson, S., Yooseph, S., Wu, D., Eisen, J.A., Hoffman, J.M., Remington, K., Beeson, K., Tran, B., Smith, H., Baden-Tillson, H., Stewart, C., Thorpe, J., Freeman, J., Andrews-Pfannkoch, C., Venter, J.E., Li, K., Kravitz, S., Heidelberg, J.F., Utterback, T., Rogers, Y.H., Falcón, L.I., Souza, V., Bonilla-Rosso, G., Eguiarte, L.E., Karl, D.M., Sathyendranath, S., Platt, T., Bermingham, E., Gallardo, V., Tamayo-Castillo, G., Ferrari, M.R., Strausberg, R.L., Nealson, K., Friedman, R., Frazier, M., Venter, J.C. (2007) The Sorcerer II global ocean sampling expedition: northwest Atlantic through eastern tropical Pacific. *PLoS Biol.*, **5** (3), e77.

26 Yooseph, S., Sutton, G., Rusch, D.B., Halpern, A.L., Williamson, S.J., Remington, K., Eisen, J.A., Heidelberg, K.B., Manning, G., Li, W., Jaroszewski, L., Cieplak, P., Miller, C.S., Li, H., Mashiyama, S.T., Joachimiak, M.P., van Belle, C., Chandonia, J.M., Soergel, D.A., Zhai, Y., Natarajan, K., Lee, S., Raphael, B.J., Bafna, V., Friedman, R., Brenner, S.E., Godzik, A., Eisenberg, D., Dixon, J.E., Taylor, S.S., Strausberg, R.L., Frazier, M., Venter, J.C. (2007) The Sorcerer II global ocean sampling expedition: expanding the universe of protein families. *PLoS Biol.*, **5** (3), e16.

27 Rusch, D.B., Martiny, A.C., Dupont, C.L., Halpern, A.L., Venter, J.C. (2010) Characterization of *Prochlorococcus* clades from iron-depleted oceanic regions. *Proc. Natl Acad. Sci. USA*, **107** (37), 16184–16189.

28 Tringe, S.G., Zhang, T, Liu, X., Yu, Y., Lee, W.H., Yap, J., Yao, F., Suan, S.T., Ing, S.K., Haynes, M., Rohwer, F., Wei, C.L., Tan, P., Bristow, J., Rubin, E.M., Ruan, Y. (2008) The airborne metagenome in an indoor urban environment. *PLoS ONE*, **3** (4), e1862.

29 Kim, M., Morrison, M., Yu, Z. (2011) Status of the phylogenetic diversity census of ruminal microbiomes. *FEMS Microbiol. Ecol.*, **76** (1), 49–63.

30 Turnbaugh, P.J., Hamady, M., Yatsunenko, T., Cantarel, B.L., Duncan, A., Ley, R.E., Sogin, M.L., Jones, W.J., Roe, B.A., Affourtit, J.P., Egholm, M., Henrissat, B., Heath, A.C., Knight, R., Gordon, J.I. (2009) A core gut microbiome in obese and lean twins. *Nature*, **457** (7228), 480–484.

31 Muegge, B.D., Kuczynski, J., Knights, D., Clemente, J.C., González, A., Fontana, L., Henrissat, B., Knight, R., Gordon, J.I. (2011) Diet drives convergence in gut microbiome functions across mammalian phylogeny and within humans. *Science*, **332** (6032), 970–974.

32 Arora, T., Sharma, R. (2011) Fermentation potential of the gut microbiome: implications for energy homeostasis and weight management. *Nutr. Rev.*, **69** (2), 99–106.

33 Human Microbiome Jumpstart Reference Strains Consortium, Nelson, K.E., Weinstock, G.M., Highlander, S.K., Worley, K.C., Creasy, H.H., Wortman, J.R., Rusch, D.B., Mitreva, M., Sodergren, E., Chinwalla, A.T., Feldgarden, M., Gevers, D., Haas, B.J., Madupu, R., Ward, D.V., Birren, B.W., Gibbs, R.A., Methe, B., Petrosino, J.F., Strausberg, R.L., Sutton, G.G., White, O.R., Wilson, R.K., Durkin, S., Giglio, M.G., Gujja, S., Howarth, C., Kodira, C.D., Kyrpides, N., Mehta, T., Muzny, D.M., Pearson, M., Pepin, K., Pati, A., Qin, X., Yandava, C., Zeng, Q., Zhang, L., Berlin, A.M., Chen, L., Hepburn, T.A., Johnson, J., McCorrison, J., Miller, J., Minx, P., Nusbaum, C., Russ, C., Sykes, S.M., Tomlinson, C.M., Young, S., Warren, W.C., Badger, J., Crabtree, J., Markowitz, V.M., Orvis, J., Cree, A., Ferriera, S., Fulton, L.L., Fulton, R.S., Gillis, M., Hemphill, L.D., Joshi, V., Kovar, C., Torralba, M., Wetterstrand, K.A., Abouellleil, A., Wollam, A.M., Buhay, C.J., Ding, Y., Dugan, S., FitzGerald, M.G., Holder, M., Hostetler, J., Clifton, S.W., Allen-Vercoe, E., Earl, A.M., Farmer, C.N., Liolios, K., Surette, M.G., Xu, Q., Pohl, C., Wilczek-Boney, K., Zhu, D. (2010) A catalog of reference genomes from the human microbiome. *Science*, **328** (5981), 994–999.

34 Qin, J., Li, R., Raes, J., Arumugam, M., Burgdorf, K.S., Manichanh, C., Nielsen, T., Pons, N., Levenez, F., Yamada, T., Mende, D.R., Li, J., Xu, J., Li, S., Li, D., Cao, J., Wang, B., Liang, H., Zheng, H., Xie, Y., Tap, J., Lepage, P., Bertalan, M., Batto, J.M., Hansen, T., Le Paslier, D., Linneberg, A., Nielsen, H.B., Pelletier, E., Renault, P., Sicheritz-Ponten, T., Turner, K., Zhu, H., Yu, C., Li, S., Jian, M., Zhou, Y., Li, Y., Zhang, X., Li, S., Qin, N., Yang, H., Wang, J., Brunak, S., Doré, J., Guarner, F., Kristiansen, K., Pedersen, O., Parkhill, J., Weissenbach, J., MetaHIT Consortium, Bork, P., Ehrlich, S.D., Wang, J. (2010) A human gut microbial gene catalogue established by metagenomic sequencing. *Nature*, **464** (7285), 59–65.

35 Arumugam, M., Raes, J., Pelletier, E., Le Paslier, D., Yamada, T., Mende, D.R., Fernandes, G.R., Tap, J., Bruls, T., Batto, J.M., Bertalan, M., Borruel, N., Casellas, F., Fernandez, L., Gautier, L., Hansen, T., Hattori, M., Hayashi, T., Kleerebezem, M., Kurokawa, K., Leclerc, M., Levenez, F., Manichanh, C., Nielsen, H.B., Nielsen, T., Pons, N., Poulain, J., Qin, J., Sicheritz-Ponten, T., Tims, S., Torrents, D., Ugarte, E., Zoetendal, E.G., Wang, J., Guarner, F., Pedersen O., de Vos, W.M., Brunak, S., Doré, J., MetaHIT Consortium, Antolín, M., Artiguenave, F., Blottiere, H.M., Almeida, M., Brechot, C., Cara, C., Chervaux, C., Cultrone, A., Delorme, C., Denariaz, G., Dervyn, R., Foerstner, K.U., Friss, C., van de Guchte, M., Guedon, E., Haimet, F., Huber, W., van Hylckama-Vlieg, J., Jamet, A., Juste, C., Kaci, G., Knol, J., Lakhdari, O., Layec, S., Le Roux, K., Maguin, E., Mérieux, A., Melo Minardi, R., M'rini, C., Muller, J., Oozeer, R., Parkhill, J., Renault, P., Rescigno, M., Sanchez, N., Sunagawa, S., Torrejon, A., Turner, K., Vandemeulebrouck, G., Varela, E., Winogradsky, Y., Zeller, G., Weissenbach, J., Ehrlich, S.D., Bork, P. (2011) Enterotypes of the human gut microbiome. *Nature*, **473** (7346), 174–180.

36 NIH HMP Working Group, Peterson, J., Garges, S., Giovanni, M., McInnes, P., Wang, L., Schloss, J.A., Bonazzi, V., McEwen, J.E., Wetterstrand, K.A., Deal, C., Baker, C.C., Di Francesco, V., Howcroft, T.K., Karp, R.W., Lunsford, R.D.,

Wellington, C.R., Belachew, T., Wright, M., Giblin, C., David, H., Mills, M., Salomon, R., Mullins, C., Akolkar, B., Begg, L., Davis, C., Grandison, L., Humble, M., Khalsa, J., Little, A.R., Peavy, H., Pontzer, C., Portnoy, M., Sayre, M.H., Starke-Reed, P., Zakhari, S., Read, J., Watson, B., Guyer, M. (2009) The NIH human microbiome project. *Genome Res.*, **19** (12), 2317–2323.

37 Yeoman, C.J., Yildirim, S., Thomas, S.M., Durkin, A.S., Torralba, M., Sutton, G., Buhay, C.J., Ding, Y., Dugan-Rocha, S.P., Muzny, D.M., Qin, X., Gibbs, R.A., Leigh, S.R., Stumpf, R., White, B.A., Highlander, S.K., Nelson, K.E., Wilson, B.A. (2010) Comparative genomics of *Gardnerella vaginalis* strains reveals substantial differences in metabolic and virulence potential. *PLoS ONE*, **5** (8), e12411.

38 Mai, V., Draganov, P.V. (2009) Recent advances and remaining gaps in our knowledge of associations between gut microbiota and human health. *World J. Gastroenterol.*, **15** (1), 81–85.

39 Zhang, T., Breitbart, M., Lee, W.H., Run, J.-Q., Wei, C.L., Ling Soh, S.W., Hibberd, M.L., Liu, E.T., Rohwer, F., Ruan, Y. (2006) RNA viral community in human feces: prevalence of plant pathogenic viruses. *PLoS Biol.*, **4** (1), e3.

40 Nakamura, S., Yang, C.S., Sakon, N., Ueda, M., Tougan, T., Yamashita, A., Goto, N., Takahashi, K., Yasunaga, T., Ikuta, K., Mizutani, T., Okamoto, Y., Tagami, M., Morita, R., Maeda, N., Kawai, J., Hayashizaki, Y., Nagai, Y., Horii, T., Iida, T., Nakaya, T. (2009) Direct metagenomic detection of viral pathogens in nasal and fecal specimens using an unbiased high-throughput sequencing approach. *PLoS ONE*, **4** (1), e4219.

41 Foxman, E.F., Iwasaki, A. (2011) Genome-virome interactions: examining the role of common viral infections in complex disease. *Nat. Rev. Microbiol.*, **9** (4), 254–264.

42 Nyren, P., Pettersson, B., Uhlen, M. (1993) Solid phase DNA minisequencing by an enzymatic luminometric inorganic pyrophosphate detection assay. *Anal. Biochem.*, **208** (1), 171–175.

43 Dressman, D., Yan, H., Traverso, G., Kinzler, K.W., Vogelstein, B. (2003) Transforming single DNA molecules into fluorescent magnetic particles for detection and enumeration of genetic variations. *Proc. Natl Acad. Sci. USA*, **100**, 8817–8822.

44 Bentley, D.R. (2006) Whole genome re-sequencing. *Curr. Opin. Genet. Dev.*, **16**, 545–552.

45 Sanger, F., Coulson, A.R., Barrell, B.G., Smith, A.J., Roe, B.A. (1980) Cloning in single-stranded bacteriophage as an aid to rapid DNA sequencing. *J. Mol. Biol.*, **143** (2), 161–178.

46 Staden, R. (1979) A strategy of DNA sequencing employing computer programs. *Nucleic Acids Res.*, **6** (7), 2601–2610.

47 Myers, E.W., Sutton, G.G., Delcher, A.L., Dew, I.M., Fasulo, D.P., Flanigan, M.J., Kravitz, S.A., Mobarry, C.M., Reinert, K.H., Remington, K.A., Anson, E.L., Bolanos, R.A., Chou, H.H., Jordan, C.M., Halpern, A.L., Lonardi, S., Beasley, E.M., Brandon, R.C., Chen, L., Dunn, P.J., Lai, Z., Liang, Y., Nusskern, D.R., Zhan, M., Zhang, Q., Zheng, X., Rubin, G.M., Adams, M.D., Venter, J.C. (2000) A whole-genome assembly of *Drosophila*. *Science*, **287** (5461), 2196–2204.

48 Yooseph, S., Sutton, G., Rusch, D.B., Halpern, A.L., Williamson, S.J., Remington, K., Eisen, J.A., Heidelberg, K.B., Manning, G., Li, W., Jaroszewski, L., Cieplak, P., Miller, C.S., Li, H., Mashiyama, S.T., Joachimiak, M.P., van Belle, C., Chandonia, J.M., Soergel, D.A., Zhai, Y., Natarajan, K., Lee, S., Raphael, B.J., Bafna, V., Friedman, R., Brenner, S.E., Godzik, A., Eisenberg, D., Dixon, J.E., Taylor, S.S., Strausberg, R.L., Frazier, M., Venter, J.C. (2007) The Sorcerer II global ocean sampling expedition: expanding the universe of protein families. *PLoS Biol.*, **5** (3), e16.

49 Alkan, C., Sajjadian, S., Eichler, E.E. (2011) Limitations of next-generation genome sequence assembly. *Nat. Methods*, **8** (1), 61–65.

50 Raes, J., Foerstner, K.U., Bork, P. (2007) Get the most out of your metagenome: computational analysis of environmental sequence data. *Curr. Opin. Microbiol.*, **10** (5), 490–498.

51 Pop, M. (2009) Genome assembly reborn: recent computational challenges. *Brief. Bioinform.*, **10** (4), 354–366.

52 Kunin, V., Copeland, A., Lapidus, A., Mavromatis, K., Hugenholtz, P. (2008) A bioinformatician's guide to metagenomics. *Microbiol. Mol. Biol. Rev.*, **72** (4), 557–578.

53 Inskeep, W.P., Rusch, D.B., Jay, Z.J., Herrgard, M.J., Kozubal, M.A., Richardson, T.H., Macur, R.E., Hamamura, N., Jennings, R., Fouke, B.W., Reysenbach, A.L., Roberto, F., Young, M., Schwartz, A., Boyd, E.S., Badger, J.H., Mathur, E.J., Ortmann, A.C., Bateson, M., Geesey, G., Frazier, M. (2010) Metagenomes from high-temperature chemotrophic systems reveal geochemical controls on microbial community structure and function. *PLoS ONE*, **5** (3), e9773.

54 Huson, D.H., Auch, A.F., Qi, J., Schuster, S.C. (2007) MEGAN analysis of metagenomic data. *Genome Res.*, **17** (3), 377–386.

55 McHardy, A.C., Martín, H.G., Tsirigos, A., Hugenholtz, P., Rigoutsos, I. (2007) Accurate phylogenetic classification of variable-length DNA fragments. *Nat. Methods*, **4** (1), 63–72.

56 Meyer, F., Paarmann, D., D'Souza, M., Olson, R., Glass, E.M., Kubal, M., Paczian, T., Rodriguez, A., Stevens, R., Wilke, A., Wilkening, J., Edwards, R.A. (2008) The metagenomics RAST server – a public resource for the automatic phylogenetic and functional analysis of metagenomes. *BMC Bioinformatics*, **9**, 386.

57 Richter, D.C., Ott, F., Auch, A.F., Schmid, R., Huson, D.H. (2008) MetaSim: a sequencing simulator for genomics and metagenomics. *PLoS ONE*, **3** (10), e3373.

58 Mavromatis, K., Ivanova, N., Barry, K., Shapiro, H., Goltsman, E., McHardy, A.C., Rigoutsos, I., Salamov, A., Korzeniewski, F., Land, M., Lapidus, A., Grigoriev, I., Richardson, P., Hugenholtz, P., Kyrpides, N.C. (2007) Use of simulated data sets to evaluate the fidelity of metagenomic processing methods. *Nat. Methods*, **4** (6), 495–500.

59 Simpson, J.T., Wong, K., Jackman, S.D., Schein, J.E., Jones, S.J., Birol, I. (2009) ABySS: A parallel assembler for short read sequence data. *Genome Res.*, **19** (6), 1117–1123.

60 Miller, J.R., Delcher, A.L., Koren, S., Venter, E., Walenz, B.P., Brownley, A., Johnson, J., Li, K., Mobarry, C., Sutton, G. (2008) Aggressive assembly of pyrosequencing reads with mates. *Bioinformatics*, **24** (24), 2818–2824.

61 Zerbino, D.R., Birney, E. (2008) Velvet: algorithms for *de novo* short read assembly using de Bruijn graphs. *Genome Res.*, **18** (5), 821–829.

62 Zerbino, D.R., McEwen, G.K., Margulies, E.H., Birney, E. (2009) Pebble and rock band: heuristic resolution of repeats and scaffolding in the velvet short-read de novo assembler. *PLoS ONE*, **4** (12), e8407.

63 Li, R., Fan, W., Tian, G., Zhu, H., He, L., Cai, J., Huang, Q., Cai, Q., Li, B., Bai, Y., Zhang, Z., Zhang, Y., Wang, W., Li, J., Wei, F., Li, H., Jian, M., Li, J., Zhang, Z., Nielsen, R., Li, D., Gu, W., Yang, Z., Xuan, Z., Ryder, O.A., Leung, F.C., Zhou, Y., Cao, J., Sun, X., Fu, Y., Fang, X., Guo, X., Wang, B., Hou, R., Shen, F., Mu, B., Ni, P., Lin, R., Qian, W., Wang, G., Yu, C., Nie, W., Wang, J., Wu, Z., Liang, H., Min, J., Wu, Q., Cheng, S., Ruan, J., Wang, M., Shi, Z., Wen, M., Liu, B., Ren, X., Zheng, H., Dong, D., Cook, K., Shan, G., Zhang, H., Kosiol, C., Xie, X., Lu, Z., Zheng, H., Li, Y., Steiner, C.C., Lam, T.T., Lin, S., Zhang, Q., Li, G., Tian, J., Gong, T., Liu, H., Zhang, D., Fang, L., Ye, C., Zhang, J., Hu, W., Xu, A., Ren, Y., Zhang, G., Bruford, M.W., Li, Q., Ma, L., Guo, Y., An, N., Hu, Y., Zheng, Y., Shi, Y., Li, Z., Liu, Q., Chen, Y., Zhao, J., Qu, N., Zhao, S., Tian, F., Wang, X., Wang, H., Xu, L., Liu, X., Vinar, T., Wang, Y., Lam, T.W., Yiu, S.M., Liu, S., Zhang, H., Li, D., Huang, Y., Wang, X., Yang, G., Jiang, Z., Wang, J., Qin, N., Li, L., Li, J., Bolund, L., Kristiansen, K., Wong, G.K., Olson, M., Zhang, X., Li, S., Yang, H., Wang, J., Wang, J. (2010) The sequence and de novo assembly of the giant panda genome. *Nature*, **463** (7279), 311–317.

64 Li, R., Zhu, H., Ruan, J., Qian, W., Fang, X., Shi, Z., Li, Y., Li, S., Shan, G., Kristiansen, K., Li, S., Yang, H., Wang, J., Wang, J. (2010) De novo assembly of human genomes with massively parallel short read sequencing. *Genome Res.*, **20** (2), 265–272.

65 Medini, D., Donati, C., Tettelin, H., Masignani, V., Rappuoli, R. (2005) The microbial pan-genome. *Curr. Opin. Genet. Dev.*, **6**, 589–594.

66 Kislyuk, A.O., Haegeman, B., Bergman, N.H., Weitz, J.S. (2011) Genomic fluidity:

an integrative view of gene diversity within microbial populations. *BMC Genomics*, **12**, 32.

67 Schwartz, S., Zang, Z., Frazer, K.A., Smit, A., Riemer, C., Bouck, J., Gibbs, R., Hardison, R., Miller, W. (2000) PipMaker – a web server for aligning two genomic DNA sequences. *Genome Res.*, **10**, 577–586.

68 Coleman, M.L., M.B.S., Martiny, A.C., Steglich, C., Barry, K., Delong, E.F., Chisholm, S.W. (2006) Genomic islands and the ecology and evolution of *Prochlorococcus. Science*, **311**, 1768–1770.

69 Metzker, M.L. (2010) Sequencing technologies – the next generation. *Nat. Rev. Genet.*, **11** (1), 31–46.

70 Miller, J.R., Koren, S., Sutton, G. (2010) Assembly algorithms for next-generation sequencing data. *Genomics*, **95** (6), 315–327.

71 Wu, D., Hartman, A., Ward, N., Eisen, J.A. (2008) An automated phylogenetic tree-based small subunit rRNA taxonomy and alignment pipeline. *PLoS ONE*, **3** (7), e2566.

72 Baker, M. (2010) Next-generation sequencing: adjusting to data overload. *Nat. Methods*, **7**, 495–499.

21
Synthetic Biology: Implications and Uses

Sanjay Vashee, Mikkel A. Algire, Michael G. Montague, and Michele S. Garfinkel
J. Craig Venter Institute, Synthetic Biology, 9704 Medical Center Drive, Rockville, MD 20850, USA

Systems Biology: Advances in Molecular Biology and Medicine, First Edition. Edited by Robert A. Meyers.

Keywords

Synthetic biology
An effort to construct biological systems, which may include entire biosynthetic pathways, synthetic organelles and cellular structures, and whole organisms, that have medical, industrial, and scientific applications. This is achieved via the application of engineering principles, such as hierarchical design, modular reusable parts, the isolation of unrelated functions, and standard interfaces.

Synthetic cell
A cell that is controlled solely by a genome that was assembled from chemically synthesized pieces of DNA.

DNA assembly
The building of larger DNA fragments from smaller DNA fragments.

Circuits
A collection of various modular component parts that responds to an input signal that is then relayed to produce an output signal.

Compartmentalization
The spatial sequestering of substrates, intermediates, products, enzymes, and activities.

Synthetic biology is an effort to construct and engineer biological systems, ranging from individual genetic elements, to biosynthetic pathways, to whole organisms. The results of these engineering efforts can be of great value to human interests such as medicine and industry. In this chapter, advances in DNA assembly technologies are reviewed, and how these advanced DNA assembly technologies, in conjunction with the application of engineering principles such as modular parts, have facilitated the rational engineering of organisms to obtain desired functions or to understand complex cellular behavior, are highlighted. The recent creation of a synthetic cell is also described. Finally, the societal concerns posed by synthetic biology are discussed.

1 Introduction

The field of Synthetic Biology can be considered more as an engineering discipline, and less as an empirical science. Efforts to create artificial life systems, both in biochemical systems [1] and in software environments [2], may also be considered as Synthetic Biology, though these are beyond the scope of this chapter. Synthetic Biology is viewed as the effort to construct and engineer biological systems of value to human interests. Such efforts can range in scopes far larger than the traditional genetic engineering of genes, to include the engineering of entire biosynthetic pathways complete with the regulation of the genes in that pathway [3, 4], synthetic organelles and cellular structures [5], whole organisms [6–9], and even ecosystems [10–13]. Synthetic Biology has the ambition to apply classical engineering principles such as hierarchical design, modular reusable parts, the isolation of unrelated functions, and standard interfaces. The empirical fields that correlate to Synthetic Biology are Systems Biology, Genetics, and Molecular Biology.

Synthetic Biology is not a new field, but rather extends back into prehistory. For example, it has been determined that the process of engineering maize – a highly optimized domestic agricultural crop plant – from the wild grass teosinte began over 9000 years ago [14]. The method used by the pre-Columbian cultivators of teosinte was simple artificial selection which, as such, is very slow. However, with the discovery of laws of inheritance and natural selection [15–17], and the suggestion that DNA was the chemical medium of inheritance [18], the scene was set to engineer a living system in a far more direct and rapid manner. A prominent example of this is the Dupont *Escherichia coli* strain used for the production of 1,3 propanediol, in which case an entire biosynthetic pathway has been added to *E. coli*, and the metabolism of the bacterium substantially altered to allow for a majority of the carbon feedstock (glycerol) to be converted into the economically valuable chemical 1,3 propanediol [6, 8, 9]. This feat, which was begun prior to the development of most of the Synthetic Biology techniques reviewed in this chapter, took many years and substantial investment to achieve. Yet, with recent advances in the field, such bioengineering projects will become faster to develop, easier to operate, and also much more ambitious.

During recent years, Synthetic Biology has progressed in a manner which is very different from those of other engineering disciplines. This is because, unlike architecture or software engineering, there is already a reservoir of highly sophisticated and complex functional parts to be found in Nature, and consequently most efforts in Synthetic Biology have been focused on harnessing that natural resource base. In general, two basic approaches have been undertaken to achieve this feat. The first approach has been to engineer natural organisms so as to incorporate recombinant pathways and other such desirable attributes. This method has the advantage of not requiring the capability to build – nor require an understanding of – massive biological systems such as genomes and metabolisms. However, it does have the disadvantage of being undefined; that is, whilst certain genes of the organism might be the result of human intervention, most of the genome remains wild-type, and is neither subject to human control nor necessarily operating within the limits of human knowledge.

The alternate approach is to use functional components, originally "mined" from Nature, such as promoters, aptamers, protein–protein interaction domains, terminators, or ribosome-binding sites. These functional components can be cataloged and then used to compose larger defined constructions of genes, pathways, and even whole genomes. Because all the components of a synthetic biological system that are constructed in such an approach have precisely defined properties, a high degree of predictive control over the final product is afforded. Indeed, this more defined approach has become synonymous with advances in Synthetic Biology.

Historically, one of the main limitations in following this defined approach relates to the knowledge of these natural systems that serve as a source of parts. As biology has been characterized, both new components for synthetic biology – and also new tools to utilize those components – have become available. In turn, a new Synthetic Biology capability has driven greater advances in the understanding of biology. This has been most obvious in the development of tools for the synthesis and manipulation of DNA, the first of which were developed via the discovery of restriction endonucleases, DNA ligases, and the creation of recombinant DNA molecules [19–24]. These tools allowed the development of the recombinant DNA cloning and expression techniques that ultimately made possible the exploitation of enzymes with desired activities and properties. Notably, the development of the polymerase chain reaction (PCR) greatly increased the ability to amplify and manipulate DNA [25–27]. The subsequent combination of recombinant DNA cloning techniques with the PCR allowed a much greater exploitation of natural biological components, such as thermostable DNA polymerases, and this in turn made the PCR more robust and practical. Today, the PCR has become an indispensable tool for biology. Thus, knowledge of natural systems has led to an improved technology for exploiting those systems, which has in turn provided an improved knowledge of biological systems in a double feedback loop, thus improving both scientific understanding and technological capabilities. As shown in Fig. 1, as knowledge of the fundamental principles of biology have continued to grow, it has been possible to take a more defined engineering approach.

In the past, advances in synthetic biology have been bounded by the capacity to assemble and modify DNA, as well as knowledge of biological parts and circuits that such DNA might encode. Correspondingly, these two areas are directly addressed in the following sections, with details of synthetic biological pathways, synthetic genomes, synthetic organelles, and even synthetic organisms provided as

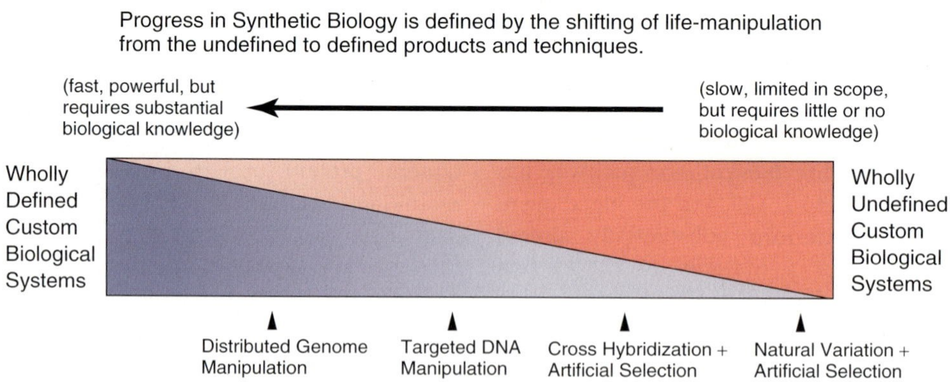

Fig. 1 Increasingly defined biological engineering. A schematic of how biological engineering has emphasized a more defined and rational design as it has advanced, based on a greater knowledge of natural biological systems.

examples of the capabilities that Synthetic Biology has already begun to deliver. These new capabilities create, in turn, new societal challenges, and these are also discussed.

2
DNA Assembly and Modification

As discussed above, one way to view Synthetic Biology is as an engineering discipline aimed at manipulating cellular systems to produce a *de novo*-designed function that does not exist in the natural organism. As with other engineering fields, however, Synthetic Biology is dependent on the tools and techniques available.

Organisms carry out a variety of reactions aimed at their self-sustenance and self-replication, with such reactions being carried out by the proteins and RNAs encoded in the organism's genome. As the sequence of the DNA directs production of the proteins and RNAs of an organism, control of the cellular DNA therefore allows for an ability to direct the functions of a cell. Based on this principle, many of the basic tools utilized in Synthetic Biology are aimed at producing defined sequences of DNA molecules, and easily manipulating the DNA content of an organism. The DNA molecules necessary for Synthetic Biology purposes can vary greatly in length, from individual DNA parts, genes and plasmids (containing tens to thousands of base pairs) to biosynthetic pathways and genetic circuits (thousands to millions of base pairs) to synthesizing whole genomes (viral and bacterial).

For many years, gene cloning and DNA assembly were dominated by the use of restriction endonucleases and DNA ligases [19–24]. While some well-designed restriction enzyme-based methods are still commonly in use [28, 29], these methods are gradually being superseded by the development of very rapid, more robust and less limited DNA assembly techniques. For example, although BioBricks were originally designed to be assembled with a restriction enzyme/ligation method [30], more recently an *in vitro* homologous recombination was adapted to increase the flexibility and speed of BioBrick assembly [31].

The starting materials for DNA sequence construction may include chemically synthesized DNA oligonucleotides (oligos), natural DNA fragments, PCR products, or a combination of all three sources. Defined short single-stranded oligos have been commercially available as a commodity for many years, usually for use as primers in PCR, mutagenesis, and sequencing reactions. Since chemically synthesized DNA oligos are of a user-defined sequence, this allows for a nucleotide level control of gene sequences and even entire genome sequences – a firm requirement when designing new functions in organisms. The idea of synthesizing genes from DNA oligos is not new; previously, oligos have been used to synthesize genes such as the alanine tRNA from yeast [32] and the human leukocyte interferon gene [33]. Likewise, the gene encoding a mammalian hormone, somatostatin, was synthesized and expressed in *E. coli* [34]. It is only recently that the lower costs of oligonucleotide synthesis and DNA sequencing have been combined to allow the development of more cost-effective and rapid methods for assembling groups of oligos into synthetic pieces of DNA or genes [35–39]. Today, several commercial gene synthesis companies exist that are able to produce custom genes/DNA at an accessible cost per base pair, although such costs can quickly become prohibitive if numerous different DNAs are required. The cost prohibition of large-scale gene synthesis can, in part, be overcome by utilizing natural DNA fragments and PCR products in the DNA assembly for those sections of DNA that do not need to be created synthetically.

Typically, the process of constructing DNA is hierarchical (Figs 2 and 3). Briefly, groups of smaller DNAs (single- or double-stranded) are mixed and assembled into larger DNA pieces. Figure 2 shows double-stranded DNAs being assembled into a larger construct, but the process can begin with single-stranded oligos as the substrates. These larger pieces (subassemblies) are then grouped and assembled. These steps can be repeated until the final full-length DNA construct is obtained, whether it is a gene or genome. The DNA pieces to be assembled must have homologous overlapping ends, the overlaps being important because the DNA assembly techniques utilize homologous recombination. For example, if three pieces of DNA (A, B, and C) are to be assembled into a single DNA molecule, then one end of piece A must have an overlap with piece B, and the other end of B must overlap piece C. This configuration will result in a linear DNA molecule, A–B–C. In order to generate a circle from these pieces, the end of C must overlap piece A. The assembly of DNA into a circle is most often achieved with a DNA piece which contains sequences that enable the final construct to be cloned into a desired host (e.g., *E. coli, Saccharomyces cerevisiae* or *Bacillus subtilis*) [37, 38, 40–42]. The DNA homologous recombination reaction can be carried out completely *in vitro* in one step, either with an enzyme mix [37] or by using the PCR [35]. The reaction can also be performed *in vivo* by utilizing the natural homologous recombination activity of an organism, such as *S. cerevisiae* (yeast) and *B. subtilis* [7, 40–44]. Occasionally, a method will include an *in vitro* step to perform a partial reaction (DNA chewback/DNA annealing/DNA extension), and an *in vivo* step to complete the reaction (DNA repair) [38, 45, 46].

The various *in vitro* homologous recombination methods used to assemble double-stranded DNA or single-stranded oligos share the same general mechanism (Fig. 2). The nucleotides are first removed from one strand of the overlapping ends of the adjacent double-stranded DNAs, thus creating single-stranded ends of the DNA (Step 1). This process is analogous to a restriction enzyme digestion creating complementary sticky ends of DNA, except that the single stranded ends are typically 20–60 nucleotides long. Depending on the method used, the nucleotides can be removed by applying an exonuclease activity from either the 5′ or 3′ ends of the DNA [37, 38]. The creation of single-stranded overhangs (Step 1) is not necessary if oligos are used as the starting material for DNA assembly, because the oligos are already single-stranded. The single-stranded DNA overhangs are complementary to each other on adjacent molecules, and thus are able to anneal (Step 2). If the ends of the molecule being constructed are complementary, then the final construct will be circular (this is the normal method used when DNA is being assembled into a cloning or expression vector). The final step to the reaction is repair of the DNA. In the *in vitro* reaction, a DNA polymerase is used to fill the gaps, while a DNA ligase seals the nicks so as to create the larger assembled DNA molecule. The DNA repair activity of *E. coli* can be utilized to complete the reaction after the annealing step

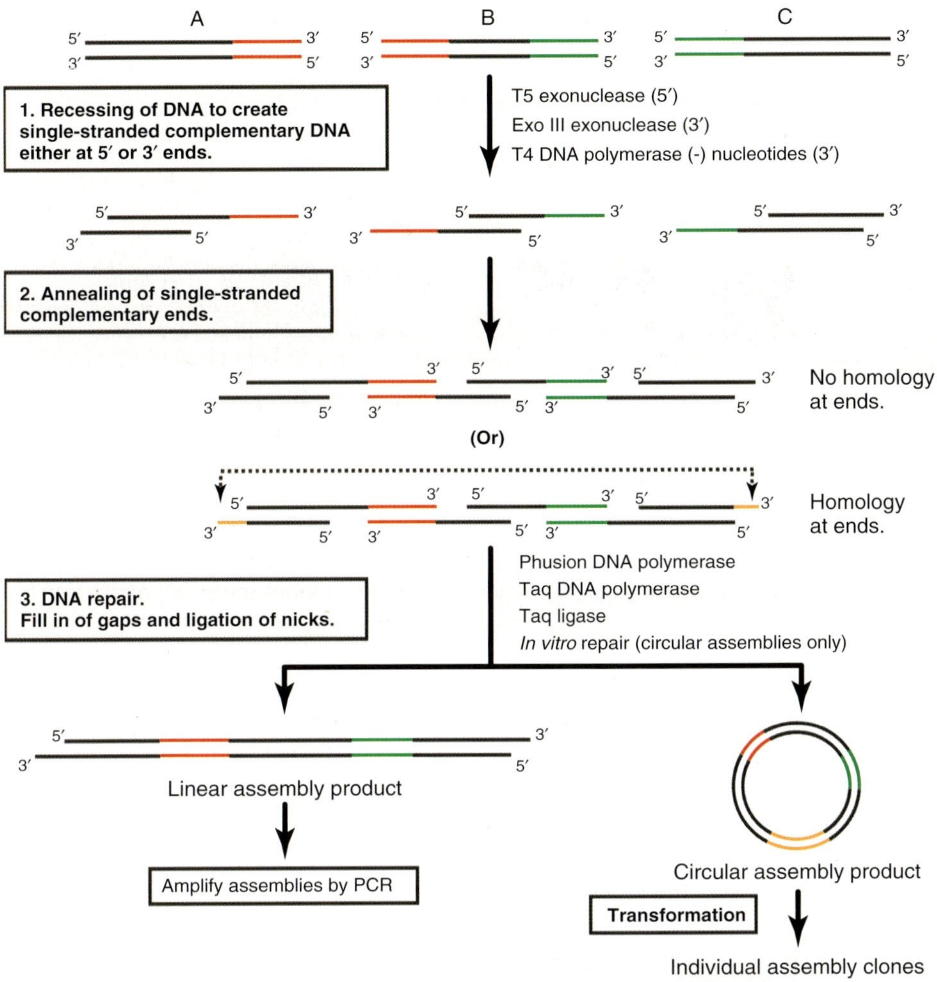

Fig. 2 Schematic depicting *in vitro* homologous recombination DNA assembly. First, nucleotides are removed from either the 5′ or the 3′ ends of the DNA pieces (5′ removal depicted). This step can be performed by several enzymes. The newly exposed single-stranded homologous ends (red, green, or yellow regions) on the adjacent pieces are complementary, and can anneal. Providing homology at the ends of the DNA pieces will result in the assembly of a circular DNA molecule. Following the annealing step, the DNA is repaired by filling in the gaps with a DNA polymerase and sealing any nicks with DNA ligase. A linear assembly product can be amplified by using the PCR and used in further assembly reactions. The circular assembly products are transformed into the appropriate host in order to isolated individual clones with the final assembled molecule. The assemblies can also be repaired *in vivo* after transformation by the native activities of *E. coli*.

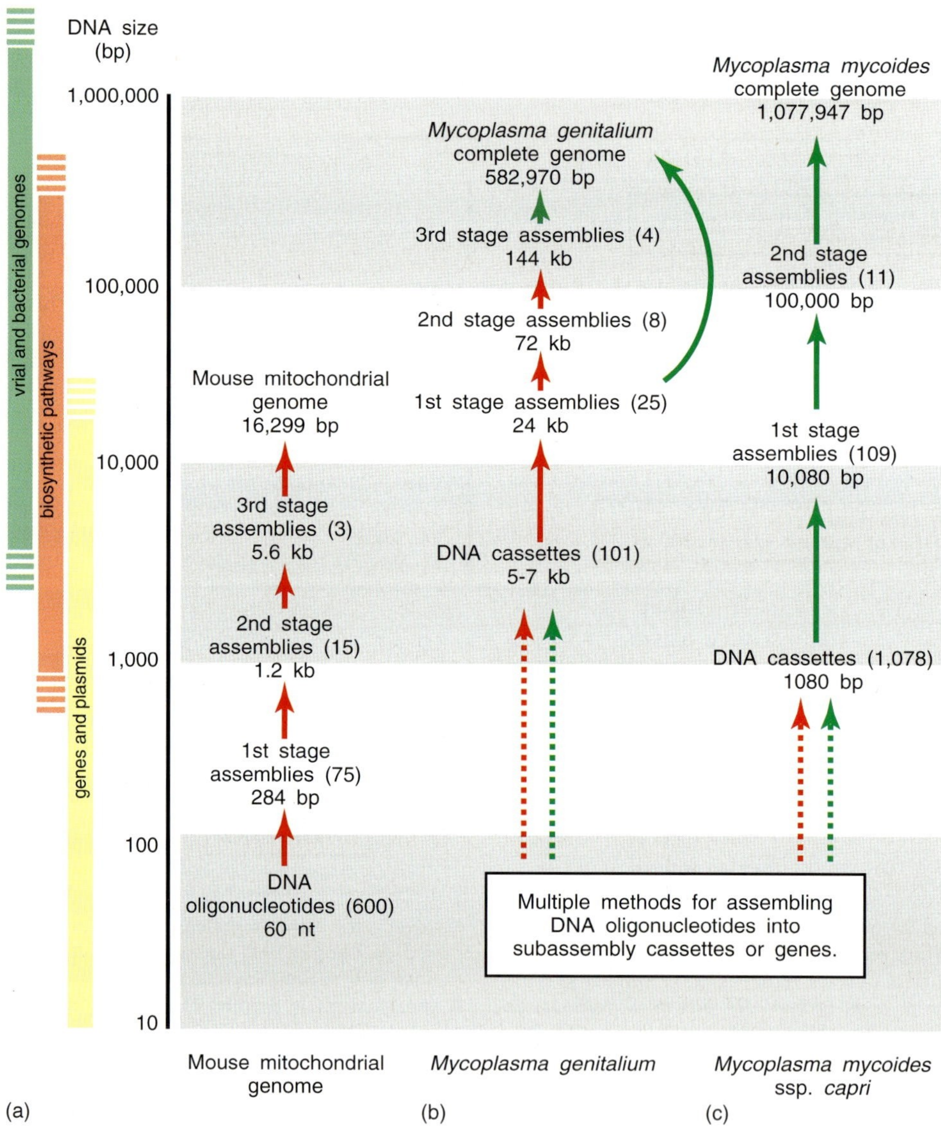

vrial and bacterial genomes
biosynthetic pathways
genes and plasmids
DNA size (bp)
1,000,000
100,000
10,000
1,000
100
10
Mouse mitochondrial genome 16,299 bp
3rd stage assemblies (3) 5.6 kb
2nd stage assemblies (15) 1.2 kb
1st stage assemblies (75) 284 bp
DNA oligonucleotides (600) 60 nt
Mycoplasma genitalium complete genome 582,970 bp
3rd stage assemblies (4) 144 kb
2nd stage assemblies (8) 72 kb
1st stage assemblies (25) 24 kb
DNA cassettes (101) 5-7 kb
Mycoplasma mycoides complete genome 1,077,947 bp
2nd stage assemblies (11) 100,000 bp
1st stage assemblies (109) 10,080 bp
DNA cassettes (1,078) 1080 bp
Multiple methods for assembling DNA oligonucleotides into subassembly cassettes or genes.
Mouse mitochondrial genome
Mycoplasma genitalium
Mycoplasma mycoides ssp. capri
(a)
(b)
(c)

[38] and after other enzymatic steps that remove errors [36]. The final assembly can be amplified by using the PCR and, if necessary, used in another round of assembly to generate even larger DNA molecules. If the final assembly is circular. it can be transformed into the appropriate host in order to generate individual assembly clones.

The construction of DNA from oligos can also be performed *in vivo* using the recombination activity of an organism. The yeast *S. cerevisiae* has robust homologous recombination activity, as well as an ability to take up multiple pieces of double- or single stranded DNA [47–51]. Previously, Gibson and colleagues have shown that yeast can assemble at least 38 overlapping single-stranded 60-mer oligos directly into a plasmid, thus forming a 1170 bp DNA insert. Alternatively, fewer – but longer – oligonucleotides (up to 200 nt in length) that overlap by as little as 20 bp can also be used to assemble 1100 bp assemblies directly into the desired plasmid [43]. The ability of yeast to support at least 2 Mb of cloned DNA also makes yeast a good host for assembling large constructs.

At this point, a discussion of gene synthesis and DNA assembly methods, within the context of the synthesis of three different genomes, will be used to highlight and demonstrate a variety of DNA assembly methods that have been used to synthesize DNA sequences, starting from single-stranded oligos, and the hierarchical assembly of the DNA subfragments into complete genomes. This is not meant to be a comprehensive list of all available methods and techniques; rather, the intention is to demonstrate the flexibility of recently used methods in DNA assembly. Whilst the discussion will be within the framework of whole genome synthesis, these techniques can also be used to synthesize and/or assemble any DNA of interest, from a few base pairs in length to over a million.

Gibson and colleagues have synthesized three genomes using both *in vitro* and *in vivo* assembly techniques [7, 40, 41]. Figure 3 shows a schematic flow of the hierarchical synthesis of the mouse mitochondrial genome (16 299 bp) (this has also been assembled, using a different method, by Itaya *et al.*; see below),

Fig. 3 Synthesis of mitochondrial and bacterial genomes. The hierarchical assembly of three genomes is depicted with the sizes of the intermediate subassemblies and final products on a logarithmic scale. The red arrows represent *in vitro* assembly, and the green arrows *in vivo* assembly in yeast. Values in parentheses indicate the number of pieces at that stage in the assembly. The colored bars on the left represent the several different DNA molecule classes that can be produced, and their relative sizes. (a) The mouse mitochondrial genome was synthesized starting from 60 nucleotide-long oligonucleotides in four stages. All of the assembly steps were performed *in vitro*; (b) The *Mycoplasma genitalium* genome was assembled from 5 to 7 kb cassettes purchased from a custom DNA synthesis company. Both, *in vitro* and *in vivo* DNA assembly techniques were utilized in the genome construction. The final stage of the assembly was performed *in vivo*, using yeast. The ability of yeast to take up multiple DNA pieces can eliminate several rounds of construction, as demonstrated by the 25-piece *in vivo* assembly of the genome; (c) The *Mycoplasma mycoides* ssp. *capri* genome was assembled from over one thousand 1080 bp purchased DNA cassettes. The three assembly steps for this genome were all performed *in vivo*, with yeast as the host. The dotted red and green lines represent the availability of several *in vitro* and *in vivo* DNA synthesis methods that can be applied to produce the DNA cassettes for use in hierarchal DNA assemblies.

the *Mycoplasma genitalium* genome (582 970 bp), and the *Mycoplasma mycoides* ssp. *capri* genome (1 077 947 bp). The basis of these DNA assembly methods is the DNA homologous recombination of overlapping DNA fragments.

The mouse mitochondrial genome was synthesized starting from 600 overlapping 60 nt-long single-stranded DNAs (60-mers), using an *in vitro* one-step assembly method (as discussed above) and PCR [52]. The first stage of genome construction produced 284 bp subassemblies by assembling groups of eight oligos directly into pUC19 (Fig. 3a). Assembling the oligos directly into a cloning vector allowed the individual assemblies to be cloned, isolated, and sequence-verified before continuing with the construction process. Following sequence verification, the correct 284 bp first stage assemblies were amplified by PCR in order to generate more material. Following amplification, the first-stage assemblies were pooled into overlapping groups of five, and again assembled *in vitro* to produce the 1.2 kb second stage assembly intermediates. The second stage assemblies were not propagated in a host organism, but rather were PCR-amplified immediately following the *in vitro* assembly reaction. These 15 PCR products were pooled into groups of five and then joined to form the 5.6 kb third stage assembly intermediates. The assembly products were amplified by PCR (as before), and these three PCR products were then assembled to form the complete synthetic mouse mitochondrial genome. The final assembly reaction with the three PCR products also contained a bacterial artificial chromosome (BAC), so that the finished mitochondrial genome could be cloned into *E. coli*. This case demonstrates not only the ability to perform large-scale DNA construction almost entirely *in vitro*, but also that the process is amenable to automation.

In 2008, the first synthetic bacterial genome was synthesized at the J. Craig Venter Institute [40]. The process to assemble the synthetic 582 970 bp *Mycoplasma genitalium* genome was hierarchical – similar to the mouse mitochondrial genome assembly (although the *M. genitalium* genome is about 35-fold larger). The synthetic *M. genitalium* genome was assembled from 101 synthetic DNA cassettes each of about 5–7 kb in length (Fig. 3b). In this case, the cassettes were synthesized from oligos by several different gene synthesis companies, and verified by sequencing. The cassettes overlapped their adjacent neighbors by an average of about 80 bp. In order to allow the formation of the circular genome, cassette 1 overlapped cassette 101.

The main challenge in the synthesis of the *M. genitalium* genome was the assembly and cloning of synthetic DNA molecules larger than those previously known. In the first stage, sets of four adjacent cassettes were assembled by *in vitro* recombination into a BAC vector to form circularized recombinant plasmids with about 24 kb inserts that were then released by restriction enzyme-mediated digestion in preparation for the next assembly stage. The 25 first-stage assemblies were taken three at a time to form the 72 kb second-stage assemblies, again by *in vitro* recombination. In the third stage, the 72 kb second-stage assemblies were taken two at a time to produce four third-stage assemblies, each of approximately one-quarter-genome (144 kb) in size. The first three stages of assembly were performed by *in vitro* recombination and cloned into *E. coli* in order to generate more DNA for the subsequent rounds of assembly. The final stage of the *M. genitalium* genome assembly was carried out *in vivo* by utilizing the homologous recombination activity of *S. cerevisiae*. The last stage of the genome assembly consisted of six overlapping pieces of DNA to generate the complete *M. genitalium* genome (one yeast vector, two fragments of quarter 3, and quarters 1, 2, and 4). The final step was performed *in vivo* because limitations of the cloning host and *in vitro* assembly reaction became apparent. It is possible that larger assemblies (280–580 kb) are not stable in *E. coli*, but it is also possible that the circularization of large DNA molecules may be inefficient during the *in vitro* recombination reaction, and/or that handling large DNA molecules in solution leads to breakage of the DNA before transformation.

Subsequently, the powerful ability of yeast to take up and assemble multiple large fragments of DNA was demonstrated by taking the 25 first-stage assemblies and assembling them in one step by using yeast [41]. This proved to be significant because it allows for fewer assembly steps, and thus greatly reduces the time required to construct large DNAs.

By leveraging the DNA uptake and recombination capability of yeast, a three-stage hierarchical strategy was designed to assemble the 1 077 947 bp *M. mycoides* ssp. *capri* genome. In this case, the assembly steps were performed entirely *in vivo* by transformation and homologous recombination in yeast, following the initial DNA cassette constructions (Figs 3c and 4) [7]. This differs from the strategy used to construct the mouse mitochondrial and *M. genitalium* genomes, which used in part an *in vitro* homologous recombination reaction. The cassettes designed to assemble into the complete genome were generally each of 1080 bp, with 80 bp overlaps to adjacent cassettes. As with *M. genitalium*, the 1078 cassettes (each 1080 bp long) were all produced commercially by the assembly of chemically synthesized oligos. To assist in the assembly process, DNA cassettes and assembly intermediates were designed to recombine in the presence of vector elements to allow for growth and selection in yeast. During the first stage of assembly, groups of 10 of the 1080 bp DNA cassettes and a vector were recombined in yeast to produce circular subassembly plasmids; these were then transferred to *E. coli* in order to easily generate the quantities of subassembly DNA required for the second-stage assembly step.

For the second-stage assemblies, 10 of the 10 kb assemblies were pooled and their respective cloning vectors transformed into yeast to produce 100 kb assembly intermediates. Circular plasmid DNA was extracted from yeast in order to proceed to the final assembly stage. In the final stage, 11 of the second-stage assemblies (100 kb each) were pooled, and the yeast transformation procedure was repeated a final time to produce the circular *M. mycoides* ssp. *capri* genome.

Recently, various alternatives to using yeast to clone and assemble large DNAs *in vivo* have been described. For example, Itaya and colleagues also constructed the complete recombinant mouse mitochondrion (16.3 kb) and rice chloroplast (134.5 kb) genomes from their small contiguous DNA pieces. In these cases, the starting DNAs for the assembly were derived via a PCR and assembled in *B. subtilis* [53]. The latter bacterium has a very large capacity to uptake and assemble DNA, as demonstrated by cloning of the 3.5 Mb genome of the photosynthetic bacterium *Synechocystis* into the *B. subtilis* genome [42, 54]. This was accomplished

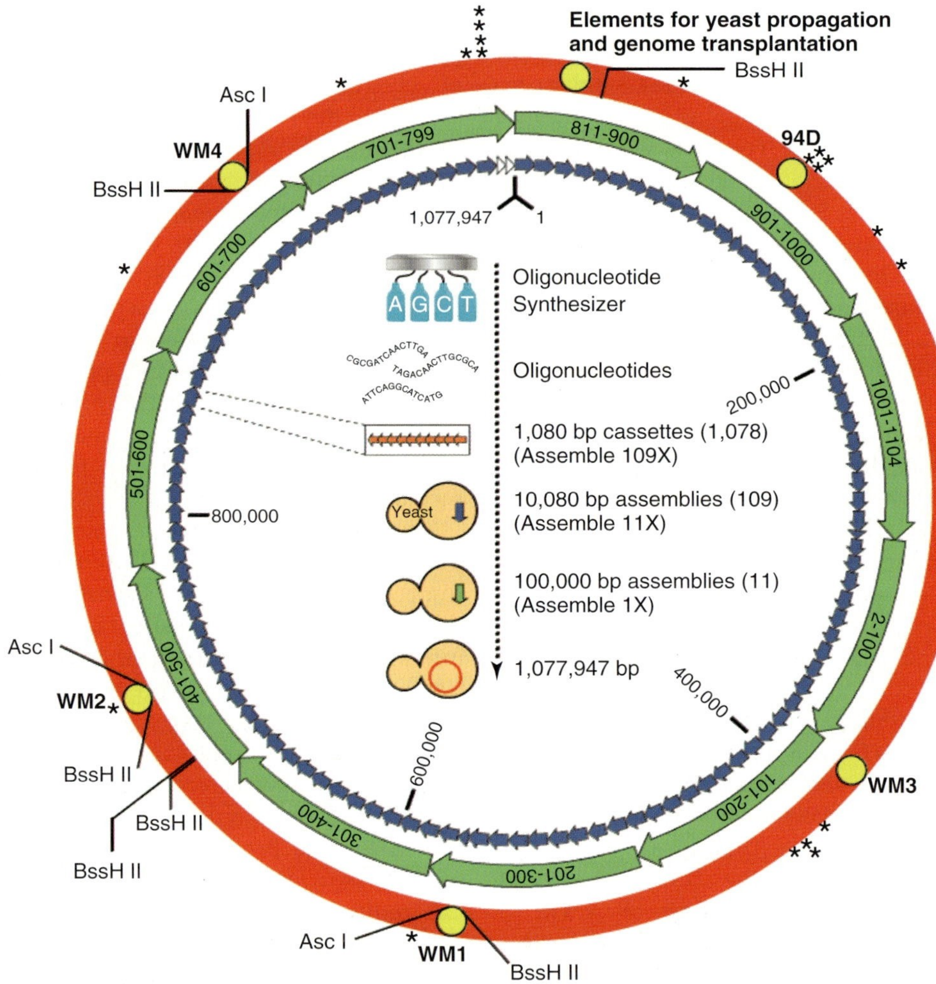

Fig. 4 The assembly of a synthetic *M. mycoides* ssp. *capri* genome in yeast. A synthetic *M. mycoides* genome was assembled from 1078 overlapping DNA cassettes in three steps. In the first step, 1080-bp cassettes (orange arrows), produced from overlapping synthetic oligonucleotides, were recombined in sets of 10 to produce 109 approximately 10 kb assemblies (blue arrows). These were then recombined in sets of 10 to produce 11 approximately 100 kb assemblies (green arrows). In the final stage of assembly, these 11 fragments were recombined into the complete genome (red circle). With the exception of two constructs that were enzymatically pieced together *in vitro* (white arrows), assemblies were carried out by *in vivo* homologous recombination in yeast. Major variations from the natural genome are shown as yellow circles. These include four watermarked regions (WM1 to WM4), a 4 kb region that was intentionally deleted (94D), and elements for growth in yeast and genome transplantation. In addition, there are 20 locations with nucleotide polymorphisms (asterisks). Coordinates of the genome are relative to the first nucleotide of the natural *M. mycoides* ssp. *capri* sequence. The designed sequence is 1 077 947 bp.

by progressively assembling and editing contiguous DNA regions that cover the entire *Synechocystis* genome. It is important to note that the *Synechocystis* genome was not a circular free molecule (as are the other genomes), but rather was incorporated as two pieces into the *B. subtilis* genome.

Several general features have been identified that might be desirable in a DNA synthesis method. Although, ideally, the method should have a low cost per base pair of DNA synthesized, the present cost of DNA synthesis is dominated by the price of the starting oligos. Oligonucleotides obtained from DNA microarrays can greatly reduce the cost of gene synthesis (estimated to be as much as an order of magnitude), because thousands of oligos can be synthesized on a single chip on a small scale. Gene synthesis from DNA microarrays can be hampered by the small quantity and complex mixture of the oligos obtained [55–57]. Regardless of the source of oligos, the cost of DNA synthesis can be greatly impacted by the amount of sequencing required to find a correct clone, and this leads to the importance of accuracy in gene synthesis. Errors in gene synthesis are unavoidable because the process of creating the starting DNA oligos is not perfect, and some fraction of the oligos will inevitably contain errors. Whilst the starting oligos can be purified by using different methods in order to minimize the number of error-containing oligos, such processes are expensive and time- consuming, and also are not compatible with high-throughput gene synthesis. The sequencing of multiple clones is sometimes sufficient to identify the correct synthesized DNA, depending on the efficiency of the assembly method used and the size of the DNA construct. Several error-correction methods have been developed, however, in attempts to reduce the amount of sequencing required [36, 58, 59].

The goals of Synthetic Biology often require DNA to be manipulated from the nucleotide to the genome level. Although the methods available to generate synthetic DNAs from genes to genomes have been discussed, in many cases a genome may need to be modified by inserting, deleting, or replacing a gene or sequence. These modifications may be necessary in only a few places, either individually or simultaneously in several noncontiguous places. Consequently, several DNA manipulation techniques have been developed to allow these types of change.

Recombineering uses the activity of lambda phage enzymes to catalyze highly efficient homologous recombination *in vivo* [60]. The lambda Red system allows an easy and rapid modification of the genome of a compatible host. However, because only short homologous overlaps are required for recombination with the lambda Red system, it is possible to use PCR to easily generate the modifying DNA by incorporating the homologous overlaps into the primer design. Recombineering can be used to insert, delete or replace a gene or sequence. Moreover, if multiple modifications are needed in the same organism the process can be repeated sequentially, although the time required to make more than a few changes can become significant. This system is available for *E. coli* [61, 62], *Pseudomonas* [63], and *Salmonella* [64], while a similar system based on a different phage has been developed for *Mycobacterium tuberculosis* and *M. smegmatis* [65]. Based on the studies with *M. tuberculosis*, it is not unreasonable to speculate that a similar recombineering system can be implemented in many more organisms by exploiting their native phages.

By using the same lambda recombination proteins, Wang and colleagues developed the technique of multiplex automated genome engineering (MAGE) for the large-scale programming and evolution of *E. coli* cells [66]. MAGE employs a mixture of single-stranded oligos to simultaneously target many locations on the chromosome for modification either in a single cell, or across a population of cells. Selection markers are not necessary because the process is efficient, iterative, and cumulative. The highest efficiencies of MAGE are observed when small changes are being made to the genome (a few base pairs), but the efficiency is much reduced when larger changes such as insertions (>20 bp) or deletions (>1000 bp) are attempted. The MAGE process is able rapidly to produce combinatorial genomic diversity. Indeed, the power of MAGE has been demonstrated by tuning the translation of 20 endogenous genes and optimizing the production of lycopene in *E. coli*. Warner *et al.* have combined a molecular barcode technology with recombineering to develop trackable multiplex recombineering (TRMR) [67], such that thousands of specific genetic modifications can be produced simultaneously, by recombineering. In this case, each modification is associated with a molecular barcode, and the barcode sequences and microarrays can then be used to quantify the allele frequency in the population; this, in turn, allows mapping of the genetic modifications that affect a trait of interest. This technique may be useful when engineering a trait for which there is limited genetic knowledge.

A method for the rapid engineering of multiple genetic changes in yeast was developed by Suzuki and colleagues [68]. This method, referred to as "*Green Monster*," employs an inducible green fluorescent protein (GFP) reporter gene to create individual deletions in separate yeast strains. The deletions in the individual strains are then combined by repeated rounds of mating, meiosis, and flow cytometry-based enrichment. In each sexual cycle, progeny bearing an ever-increasing number of altered loci are enriched on the basis of gene dosage. If it could be adapted to extrachromosomal DNAs, the Green Monster might represent a valuable technique for altering large DNAs or bacterial genomes which are cloned in *S. cerevisiae*. Although the method has been demonstrated using *S. cerevisiae*, it should be possible to extend the technology to bacteria with a mating equivalent such as bacterial conjugation (e.g., *E. coli*).

In some cases, when the expression from a high gene dosage is desired, the use of plasmids may not be a feasible option. In this case, copies of the gene of interest could be inserted into the genome by using some of the above-described methods, but this may be both labor- and time-intensive if a high gene dosage is required. In an attempt to overcome this problem, Tyo and colleagues used a plasmid-free, high-gene copy expression system termed chemically inducible chromosomal evolution (CIChE) to evolve an *E. coli* chromosome with about 40 copies of a recombinant pathway [69]. This was achieved by creating a cassette that contained the genes of interest, along with a gene encoding antibiotic resistance for chloramphenicol. When the strain was then grown in increasing concentrations of chloramphenicol, the selective pressure of the increasing antibiotic concentration resulted in duplications of the cassette and, therefore, also of the antibiotic resistance gene by *recA*-dependent homologous recombination. When the desired cassette copy number was reached, *recA* could be deleted to prevent any homologous recombination that could alter the cassette copy number.

As shown above, many good methods have been devised for constructing and assembling either synthetic or natural DNA. However, several factors must be considered and balanced when deciding on a DNA assembly strategy; these should include the cost of the method, the time required, the individual's experience with the different host organisms, and whether there is a need for vast combinatorial numbers. Fortunately, the different assembly methods are diverse and robust, so that a large range of requirements can be met. In general, it appears that yeast may be the preferred host when constructing large DNAs, although several of the methods available to generate genes or smaller (ca. 1 kb) subassemblies for the construction of larger DNAs are equally suitable, depending on the user's preferences.

3
Modular Parts and Circuits

One core component of Synthetic Biology is to apply engineering-based approaches of modularization, rationalization, and modeling to control cellular behavior, in order to obtain desired functions or to understand complex biological systems [70, 71]. As a result, an increasing number of synthetic biologists have begun to apply electrical circuit analogies to biological pathways as a means of designing and generating synthetic genetic devices which can then be placed into cells to control their behavior [72]. The first examples of genetic circuits – the toggle switch and the repressilator – were demonstrated about a decade ago [73, 74], since when there has been an ever-expanding number of reports of various types of biological circuit, including additional genetic switches [75, 76], other oscillators [77–80] and memory networks [81, 82], as well as other electronic-inspired genetic devices [70] such as pulse generators [83], logic gates [75, 84], filters [85], and communications modules [86, 87]. Today, these devices have begun to be used for practical applications in biosensing, therapeutics, and in the generation of important industrial products. As the scope of this chapter is very broad, it is only possible here to discuss these synthetic genetic devices and their uses with limited representative examples. A number of excellent recent reviews will provide more detailed information on these synthetic gene networks [70, 71, 88–90].

The basic design of circuits is the assembly of various modular component parts that respond to an input signal that is then relayed to produce an output signal. For biological circuits, the component parts have their origins in the vast amount of basic research in all aspects of the biological functions of organisms. The combined effort of many research groups has led to great understanding of the various pathways that organisms use to respond to environmental signals, such as light or quorum sensing. Accordingly, many of the components of these pathways – such as signaling proteins and transcription factors, as well as the promoters they control – have been identified from a wide variety of organisms [71, 89, 91]. In order to create some biological devices, synthetic biologists have taken components from one organism and placed them in a different model organism. For example, Danino *et al.* synthesized synchronized genetic oscillators by placing elements of the quorum sensing machineries of *Vibrio fischeri* and *Bacillus thurigensis* into the model organism, *E. coli* [77]. Given the difference in GC (guanine–cytosine) content, codon usage, transcription and translation processes between the various organisms, it was not initially clear whether the devices created from the different organism-derived component parts would function as intended. However, to overcome this difficulty several methods have been adapted.

First, web applications such as GeneDesign [92] have been developed that enable synthetic biologists to design the component parts according to the specifications of the host organism (see Table 1). In addition, as discussed above, the recent advances in gene synthesis, as well as the decrease in the cost of oligos and the proliferation of gene synthesis companies, have allowed synthetic biologists simply to purchase the component parts.

Second, Knight, Rettberg, Endy and others have founded the Registry of Standard Parts (Table 1) to provide a framework for synthetic biologists to devise biological circuits, pathways, and other genetically encoded systems. The idea here is to emulate the engineering principles involved in the construction of such things as electronic devices [93]. The Registry of Standard Parts provides a catalog of a wide variety of biological parts, such as transcription promoters and terminators, ribosome binding sites and regulatory proteins, as well as a variety of chassis in which the parts can function. This facilitates the ability of the synthetic biologists to tailor their biological devices by allowing them to choose from a variety of parts whilst, at the same time, providing a wealth of information on how these parts can function. The Registry also allows for interfacing with other web applications such as modeling tools. As an example, users can input information from the Registry into SynBioSS designer to generate kinetic models for the selected biological constructs, and provide a picture of how these constructs would influence the behavior of the whole [94].

With these advances in place, synthetic biologists are today beginning to use DNA assembly techniques to piece together modular parts into controlled biological pathways as well as biological circuits in a wide spectrum of applications, such as biosensing, the production of therapeutics and biofuels, as well as understanding complex cellular behavior and even ecosystems. For example, synthetic biosensors can be used to detect various environmental signals and then to prompt cells to enter a programmed behavior [70]. In one study, Kobayashi *et al.* generated a genetic device that could detect DNA damage and, through a designed activation of the SOS pathway, program *E. coli* cells to enter a biofilm state [95]. The ability was also demonstrated of designed bacteria to produce invasin from *Yersinia pseudotuberculosis,* upon the detection of an hypoxic environment of tumor cells, which in turn allowed the bacteria to invade the tumor cells [96]. In a particularly innovative experiment, Looger *et al.* computationally redesigned protein–ligand specificities to construct receptors that could bind trinitrotoluene or L-lactate. These receptors were then incorporated into synthetic bacterial signal transduction pathways to regulate gene expression in response to extracellular trinitrotoluene or L-lactate [97]. The results of these studies confirm the promise that genetic biosensors of this type may be useful for detecting any desired environment or signal, and to allow the host organism to respond in a targeted manner.

Biological circuits have also shown their possible value in the antibacterial therapeutic field. For example, bacteriophages

Tab. 1 Resources for synthetic biology.

Resource	URL	Comments
Synthetic Biology Resources	http://www.istl.org/10-spring/internet1.html	Provides links to various resources on the internet, including synthetic biology associations, centers of research, ethics, training and educational resources, and journals
Synthetic Biology.net	http://www.syntheticbiology.net/index.aspx	Portal for professionals in synthetic biology providing information on news, events, products, suppliers, etc., regarding synthetic biology
Synthetic Biology Project	http://www.synbioproject.org/	Established as an initiative of the Foresight and Governance Program of Woodrow Wilson International Center for Scholars to foster informed public and policy discourse concerning the advancement of Synthetic Biology
BioBricks Foundation	http://bbf.openwetware.org/	Encourages the development and responsible use of technologies based on BioBrick™ standard DNA parts to allow synthetic biologists to program living organisms in the same way a computer scientist can program a computer (see below)
Registry of Standard Biological Parts	http://partsregistry.org/Main_Page	A collection of genetic parts that can be mixed and matched to build synthetic biology devices and systems
SynBERC Synthetic Biology Engineering Research Center	http://www.synberc.org/	Mission is to develop technologies to build biological components and assemble them into integrated systems to perform designed tasks, train engineers for biology, and educate the public on Synthetic Biology
BIOFAB	http://www.biofab.org/	Biological design–build facility that aims to produce useful collections of standard biological parts available to academic and commercial users
JBEI Registry	https://public.jbeir.org/	Also aims to provide standard DNA parts for Synthetic Biology
GeneDesign	http://www.genedesign.org/	Set of web applications that provides public access to a nucleotide manipulation pipeline for Synthetic Biology
SynBioSS Designer	http://synbioss.sourceforge.net/	Software suite for the generation, storage, retrieval, and quantitative simulation of synthetic biological networks

were engineered to suppress the SOS pathway of bacteria and enhance the killing of antibiotic-resistant bacteria, "persister cells," and biofilm cells [98]. Another area where the ability to easily design and place modular parts into a designed biological pathway is proving useful is in the control of metabolic flux for the production of industrially important materials or chemicals. Recently, the Stephanopoulos group has reported an ability to increase the titer of taxadiene (an intermediate of the potent anticancer drug, Taxol) in an engineered *E. coli* strain [99]. This effect was accomplished by first partitioning the taxadiene metabolic pathway into two modules – a native upstream methylerythritol-phosphate (MEP) pathway forming isopentenyl pyrophosphate, and a yew tree-based heterologous downstream terpenoid-forming pathway – and then varying the module's expression simultaneously to obtain an improved balanced pathway.

Genetic devices are also proving their worth in helping to understand the underlying basic principles of coordinated complex cell behavior. As examples, two relatively recent studies have highlighted efforts to emulate pattern formation that is important for development in higher eukaryotes [85, 100]. The latter study used an *in vitro* approach with DNA-coated paramagnetic beads fixed by magnets in an artificial chamber to form artificial transcription–translation networks that generate simple patterns [100]. In the former study, Basu *et al.* engineered two genetically distinct populations of bacteria (acyl-homoserine lactone senders and receivers) and manually overlaid them in different configurations to produce different patterns [85]. More recently, instead of using two distinct bacterial populations, Tabor *et al.* genetically engineered an isogenic community of *E. coli* cells to sense light, to communicate to identify light-dark edges, and produce an image [101]. Similar biological circuits to those discussed above have also been utilized to construct synthetic ecosystems or biofilms to model and better understand microbial communities [10–13]. While relatively simple genetic devices have been used thus far, it is expected that a thorough characterization of their performance, together with improved predictive mathematical tools, will allow for the design and construction of more elaborate circuits to program cells and cellular communities for functions that mimic those of natural systems.

4 Spatial Regulation

Many of the parts and circuits detailed above involve the regulation of genes or gene-products. For most of the time, when gene regulation is referred to, it is assumed to be a temporal regulation – that is, the increase or decrease of expression of a gene that was, at a previous *time*, in the other state. It is, however, also possible to regulate genes and gene products in space. The ability to compartmentalize or localize enzymatic activities has long been recognized as a powerful means of optimizing catalysis. In Nature, this can take the form of macromolecular complexes that actively channel substrates from one enzymatic active site of a biosynthetic pathway to the next [102–105], organelles and compartments that can physically separate certain enzymes and substrates from the rest of the cellular processes [106, 107], or it may simply be the result of reduced diffusion due to co-localization [108, 109]. Just as compartmentalization and co-localization can take many forms, so too they offer many advantages.

These include the mitigation of toxicity of intermediates of a biosynthetic pathway, the protection of intermediates from diffusion or degradation, the elimination of unproductive side reactions as a consequence of the other biological activities of the host organism, or improving activity by driving kinetics (for an excellent review of both natural and engineered systems of spatial control over cellular processes, see Ref. [108]). In the past, organic chemists have been inspired by these natural systems and have created a wide variety of biomimetic catalysts that incorporate such features as cyclodextrin covalent linkers that allow many enzyme molecules to be tethered to one another, leading to improved kinetics [110]. However, whilst inspired by biology, and often using components derived from living organisms, such systems are not wholly biological and rarely operate in the biological context from which their component enzymes are derived. Consequently, such cell-free systems, which lack the self-replication capability of living cells, often suffer from problems of enzyme stability and purification. An excellent review of biochemical constructs that incorporate compartmentalization and co-localization based around such diverse methods as covalent and noncovalent linkers, of micelles made from both synthetic polymers and lipids, and of vesicles and viral particles used as nanoreactors, has been prepared by Vriezema *et al.* [111].

Here, examples of engineered, cellular compartmentalization strategies will be discussed, with emphasis placed on the design principles that they embody and use.

4.1 Co-Localization

Dueber *et al.* were able to use the synthetic biology principles of modular parts and co-localization to create a heterologous synthetic protein complex in *E. coli* of three enzymes: acetoacetyl-CoA thiolase (A to B); hydroxy-methylglutaryl-CoA synthase (HMGS); and hydroxymethylglutaryl-CoA reductase (HMGR) [3]. These three genes, derived from yeast, compose a pathway that produces mevalonate from acetyl-CoA. Mevalonate is a precursor for the production of chemicals in the industrially and medically valuable large isoprenoid family [112]. However, due to very different levels of activity, these enzymes – even if expressed at optimal levels – result in a build-up of the toxic intermediate hydroxymethylglutaryl-CoA (HMG-CoA). To overcome this, Dueber *et al.* organized the three-enzyme pathway into a synthetic complex of enzymes on a "scaffold" by using the well-characterized signal processing protein–protein interaction domains of the metazoan cells [mouse SRC Homology 3 (SH3) and Post Synaptic Density protein, *Drosophila* Disc Large Tumor Suppressor, and Zonula Occludens-1 protein (PDZ) domains and the rat GTPase protein Binding Domain (GBD domain] with their corresponding ligands. Each ligand is a small tag-like sequence that can be added to either end of a protein, and which binds specifically to its corresponding domain. The SH3, PDZ and GBD domains were then expressed as a fusion protein that would recruit the three ligand-tagged enzymes of the mevalonate pathway into a single complex or "scaffold." The main aspects of the scaffold optimization and design involved the order of the binding domains (and, consequently, of the pathway enzymes with their ligands), and whether the ligand sequence was fused to the N or C terminus of each enzyme. Another aspect to be optimized was the number of each enzyme that was recruited to the scaffold. Typically, a scaffold might be designed to have three PDZ domains, but only one SH3 and one GBD domain; this caused three copies of one of the pathway enzymes to be recruited to the complex, but only one each of the other enzymes. When these optimizations were made, the result was a dramatic 77-fold increase in the yield of mevalonate due to a reduction in the metabolic load on the cell and toxicity associated with HMG-CoA build up. The general nature of the method was also proved by its application to the pathway for glucaric acid synthesis [3].

The principle of synthetic complexes and co-localization can be used to improve not only synthesis pathways but also degradative pathways. For example, the fungal cellulase Cel6A has been docked onto a bacterial mini-cellulosome to achieve a greater cellulose degradation and, as with the mevalonate biosynthetic pathway discussed above, issues of geometry and organization of the final synthetic scaffold strongly affected the efficacy of the result [113, 114]. As the technology of scaffold design improves, the development may begin of synthetic pathways that resemble tryptophan synthesis [102], polyketide synthesis [104] or carbamoyl phosphate synthetase [105], where the enzymes are not merely co-located but actually channel the substrate down a tunnel.

4.2 Compartmentalization

Instead of recruiting enzymes to a scaffold and achieving a degree of control over the diffusion of substrates from enzyme to enzyme of a pathway, it might be advantageous to create a separate cellular compartment to physically encapsulate specific enzymes and substrates. An impressive example of this has been seen in the investigations of Parsons *et al.* [5], who characterized the genes of a natural bacterial organelle, which they called a bacterial microcompartment (BMC). BMCs are polyhedral protein shells that are associated with specific biosynthetic pathways. The most well-studied BMC is the carboxysome, which is associated with the fixation of carbon in Cyanobacteria [115]. Parsons *et al.* focused their attention on the *pdu* operon which contains a number of genes, some of which are implicated in the construction of a BMC, while others are associated with *Salmonella enterica* serovar. *typhimurium* LT2's pathway for the conversion of 1,2-propanediol into propionaldehyde, 1-propanol, and propionyl-CoA. These enzymes, and the reactions they catalyze, are localized within the BMC that the *pdu* operon encodes [106]. Based initially on the sequence similarity to carboxysome proteins, and later on the results of experiments where certain proteins were subtracted, Parsons and coworkers identified a set of five genes (*PduA*, *PduB*, *PduJ*, *PduK*, and *PduN*) which produced six proteins (*PduB* produces two versions of the protein) that are necessary and sufficient to produce empty BMC compartments in *E. coli*, similar to those produced by the native *pdu* operon in *S. enterica* serovar. *typhimurium* LT2. Although, not necessary, PduU was shown to help regulate the size of the BMC to the approximately 100 nm dimensions it has natively. If the full set of required genes for BMC construction was not present, then large intracellular structures such as sheets, filaments, or hexagonal lattices were observed. To prevent these structures from forming, the wild-type order and orientation of the remaining *Pdu* genes needed to be maintained. Lastly, by using the N-terminal region of a gene in the *Pdu* operon, *PduV*, it was possible to create a tag that localized GFP into the BMC.

Although Parsons *et al.* failed to demonstrate any enzymatic activity inside the resulting BMC, this was achieved inside a compartment that was similar in some ways, but prepared from a cowpea chlorotic mottle virus capsid, albeit with the very simple single enzyme system of horseradish peroxidase [116]. It was also shown possible to recruit compartments that are already present in a cell, such as the periplasmic space [117].

In keeping with the modular approach central to most Synthetic Biology methods, Parsons *et al.* and Deuber *et al.* devised systems by which tags could be fused to enzymes, allowing for their localization and thus the spatial regulation.

5
The Synthetic Cell

The organisms that have been engineered thus far for industrial or other purposes have had the advantage of being easily manipulated genetically, because of high transformation efficiencies and good recombination activities. Unfortunately, many industrially relevant organisms do not possess these characteristics and therefore, cannot easily be engineered; for these and other intractable organisms, novel engineering methods are necessary. One approach, as adopted by the research team at the J. Craig Venter Institute, has been to build a minimal cell that contains only essential genes, the functions of which have been characterized, in an attempt to understand the basic principles of life. Such a cell may also provide a base into which various pathways can be placed to synthesize industrially important products, but in a more energy efficient manner than is possible with the presently engineered organisms.

During their efforts to build a minimal cell, the J. Craig Venter Institute group recently reported the creation of a bacterial cell that could be controlled by a chemically synthesized genome [7]. The research group described in detail the design, synthesis, and assembly of the 1.08 mega-base pair *M. mycoides* JCVI-syn1.0 genome (see Fig. 4), starting from digitized genome sequence information, and its transplantation into a recipient cell to create new *M. mycoides* cells that are controlled only by the synthetic chromosome. To distinguish the synthetic genome from the natural genome, the researchers placed "watermark" sequences and included other designed gene deletions and polymorphisms in the synthetic genome (see Fig. 4). Even though the cytoplasm of the recipient cell is not synthetic, the research team referred to the cells produced after the transplantation process as "synthetic cells," because they are controlled solely by a genome that was assembled from chemically synthesized pieces of DNA. These synthetic cells have expected phenotypic properties, although the JCVI-syn 1.0 transplants grew slightly faster than a control strain.

These studies, using several *Mycoplasma* species, were the culmination of extensive efforts over a number of years that led to the development of several novel technologies (as summarized in Fig. 5). First, as discussed above, the team developed a strategy of assembling viral-sized pieces to produce large DNA molecules that allowed them to assemble bacterial genomes in *S. cerevisiae* [40, 41, 43]. Second, the team established additional methods to clone whole bacterial genomes as centromeric plasmids in yeast [118]. Third, they developed methods to transplant the genome of one bacterial species, *M. mycoides* ssp.

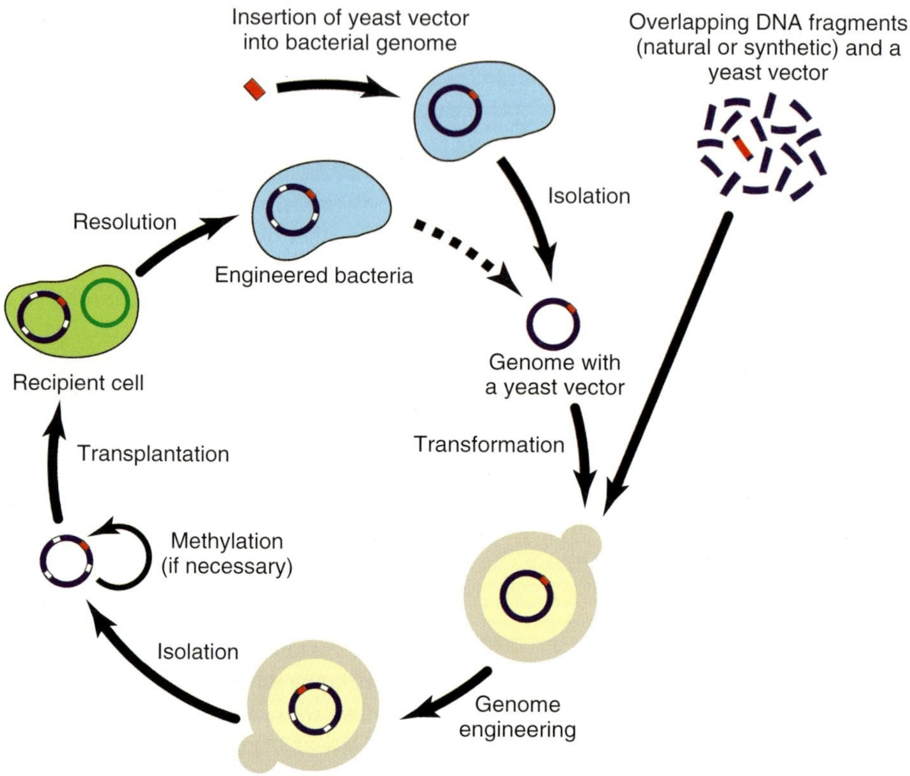

Fig. 5 Moving a bacterial genome into yeast, engineering the genome, and its re-installation into a bacterium, by genome transplantation. A yeast vector is inserted into a bacterial genome by transformation, and the genome is then cloned into yeast. Alternatively, a bacterial genome is cloned by transforming overlapping DNA fragments (natural or synthetic) together with a yeast vector into yeast and allowing the host's homologous recombination system to assemble an intact genome. After cloning, the repertoire of yeast genetic methods is used to create insertions, deletions, rearrangements or any combination of modifications in the bacterial genome. This engineered genome is then isolated and transplanted into a recipient cell to generate an engineered bacterium. Prior to transplantation, it may be necessary to methylate the donor DNA in order to protect it from the recipient cell's restriction system(s). This cycle can be repeated starting from the newly engineered genome (dashed arrow).

capri, into a different recipient bacterial cell species, *M. capricolum* ssp. *capricolum*, to obtain cells of the donor *M. mycoides* ssp. *capri* [119]. These studies were extended when the group was able to transplant the genome of *M. mycoides* ssp. *capri*, which this time was isolated from yeast as a centromeric plasmid, into recipient *M. capricolum* ssp. *capricolum* cells and produced viable *M. mycoides* ssp. *capri* cells [120]. The team also genetically altered the *M. mycoides* ssp. *capri* genome in yeast by using the host's powerful genetic tools and newly developed tools [121] to produce a new strain of *M. mycoides* ssp. *capri* that would not have been possible with the tools currently available for these bacterial species.

Taken together, the Venter Institute research team has developed a series of technologies that enabled them to clone whole bacterial genomes, whether from natural sources or synthetic pieces, to manipulate them, and to transplant them to produce viable bacterial cells (see Fig. 5). It will be interesting to see in the future whether this technology can be utilized for other intractable organisms.

6
Societal Challenges Posed by Synthetic Biology

Along with the potential of significant benefit, all new technologies raise societal concerns. With respect to biotechnology in general, these can be described as concerns about bioterrorism, laboratory safety, harm to the environment, the distribution of benefits, and ethical and religious concerns [122–125].

Synthetic Biology itself – both at the level of research and in the application of such research to the development of new products – also raises a variety of societal concerns, some of which are identical to those raised by all biotechnology, though some may be unique. These new or unique concerns may be especially important for the governance of the new technology. To its credit, the community of research workers that identify as synthetic biologists recognized at a very early stage that these societal concerns were both real and legitimate. A good number of the synthetic biologists have worked with a variety of policymaking and social science research communities, to ensure that the studies being carried out would be performed in a safe and ethical manner. Given much interaction between and among these various communities, concerns that are unique – or that have been recognized as of great importance, even if not unique – have been well analyzed, and while the policy problems have in no way been solved the challenges have been very well articulated.

The first sets of societal concerns that were dealt with in detail by these communities were those of biosecurity and biosafety. There is a constellation of ethical issues (discussed below) that are of no less importance than security and safety. However, it was clear very early on that if Synthetic Biology were to result in hazards that could not be mitigated for the research teams or for society as a whole, then there would need to be a moratorium on such studies. Consequently, these societal concerns regarding security and safety were analyzed first by a variety of policy researchers. Hence, the safety and security analyses remain, for now, somewhat more advanced than the ethics analyses.

In particular as much of the recent Synthetic Biology studies have been conducted in a post "9–11" environment, concerns relating to biosecurity were at the forefront of virtually all policy analyses. Specifically, the ability to synthesize genomes means that, at least in some cases, access to pathogens can no longer be physically limited as long as the sequences are publicly available. For now, the concerns are about increasing the ease with which viruses, such as 1918 influenza, smallpox and Ebola, can be obtained [125]. Additionally, Synthetic Biology may eventually provide a relatively straightforward way to construct pathogens with increased virulence, by allowing those with nefarious intent to add a variety of pathogenesis factors directly to a viral genome or bacterial chromosome.

In order to deal with these potential malicious applications, both the United States Government [126] and a consortium of companies providing synthetic DNA [127] have released guidelines for the screening both of synthetic DNA orders, and of the customers placing these orders. Questions as to whether these guidelines should be legally binding, the full range of what the companies should be screening for, and whether orders for smaller pieces of DNA (oligonucleotides) should also be screened are all currently under discussion.

Current biosafety or laboratory safety concerns are mostly focused on the speed and scale that Synthetic Biology brings to research, and some concerns about workers in the field who have not been trained as microbiologists. At institutes with a formal approach to dealing with biosafety – such as universities, research foundations and scientific companies – institutional biosafety committees and other institutional groups will likely be taking up questions in Synthetic Biology research in the same way that they do all other biological/biotechnological research. In an earlier report [125], policy researchers at the JCVI (Michele Garfinkel and Robert Friedman), MIT (Drew Endy), and the Center for Strategic and International Studies (Gerald Epstein) have described several options for such bodies for educating themselves, and their institutional research teams, about what steps must be taken to ensure that the research is safe. A more generic set of concerns about the research teams, and how to mitigate any possible biosafety dangers, was discussed even earlier in a report from the National Academy of Sciences [128]. Yet another set of people interested in synthetic biology has also raised concern, namely the "do it yourself" (DIY) community [129]. Whilst for the moment, little is being done by the DIY community that is clearly "synthetic biology," there has been much discussion among the group eventually to employ those technologies. In anticipation of this possibility, the Presidential Commission for the Study of Bioethical Issues, which is in the process of completing a report on and recommendations for synthetic biology, has addressed the issues of DIY specifically [130].

Although the safety and security concerns may not be "solved," it appears at least that the great majority of issues have been laid out, and at least for the moment there seem to be no risks that would lead to the conclusion that the research should be banned, or even severely restricted. (This also appears to be the conclusion of the Presidential Bioethical Commission, although its current recommendations are only in draft form.) Thus, the policy- and social science research communities have turned at least some of their attention to other, broader societal challenges brought about by the potential use of Synthetic Biology technologies.

Concerns about harm to the environment from accidental or planned releases of engineered microbes date to discussions at the Asilomar meetings during the mid-1970s. The two critical concerns are that an engineered microbe will grow out of control if released accidentally or as part of a planned release, and that DNA from an engineered organism may be transferred to a related organism. These concerns have been dealt with over time via guidance and regulation dealing with the containment of genetically modified organisms and rules for testing these organisms for release into the environment. Several US Government agencies are currently reviewing several sets of regulations and guidance to understand whether they are sufficient to deal with the use of many new microbes in open environments. For example, the *NIH Guidelines* for working with recombinant DNA are currently being reviewed to assure that guidance which was written to deal specifically with recombinant DNA applies equally to synthetic DNA [131].

In addition to the potential harm to people or to the environment (as discussed above), there are societal challenges beyond the physical. The distribution of benefits and risks is a very long-standing concern, and one which surfaces for virtually every new technology. Ownership as defined by intellectual property rights, the concentration of knowledge, and resources in a small number of firms or institutes – and how and whether these resources should be shared – are issues that have become particularly acute around research groups (both academic and commercial) who wish to develop products. Interestingly, the Synthetic Biology community, in addition to civil society organizations, has placed this issue at the forefront of many of its own discussions [132]. These discussions could well lead to a better understanding of distribution concerns and possible solutions generally.

Finally, hubris – sometimes called *"playing God"* – might be the major nonphysical concern in Synthetic Biology, even if it is not fully unique in this case. In brief, concerns about hubris are focused on a key issue: Are there actions that human beings simply should not take? In the case of constructing a synthetic cell, these questions arise for many communities, from religious traditions to policymakers. Is constructing a synthetic cell creating life? If it is, is it hubris? If it is not creating life, then what would define creating life? In either case, is this hubris? And how might conducting such experiments as constructing a synthetic cell change how human beings think of themselves, both individually and with respect to other organisms and the environment in general? There are long and thoughtful writings – both fiction and nonfiction, and both recent and deep in history – about hubristic pursuits that will not be reviewed here. However, it should be noted that these questions are being studied in detail with respect to Synthetic Biology by philosophers, ethicists and theologians [133].

7
Concluding Remarks

Today, the capability is available to create any arbitrary DNA sequence, and to express that sequence in a wide variety of living systems. By using modular parts with standardized structures, this DNA assembly capability has allowed the engineering of pathways, organelles, organisms, tissues, and even ecosystems. Unlike previous genetic engineering methods and selective breeding methods, these engineering projects have the potential to be cheap, fast, and easy. As these capabilities are further refined and extended, an era can be anticipated in which biological engineering will impact every industry and activity of humankind.

Acknowledgments

The authors thank Dr. Carole Lartigue and Dr. Chuck Merryman, and also the Synthetic Biology members of the J. Craig Venter Institute (JCVI) for their stimulating discussions. Apologies are offered to any Synthetic Biology colleagues whose work has not been cited in this chapter. Funding for the research conducted at the JCVI was provided by Synthetic Genomics, Inc. and by the Office of Science (BER), U.S. Department of Energy, Grant No. **DE-FC02-02ER63453**.

References

1 Mansy, S.S., Szostak, J.W. (2009) Reconstructing the emergence of cellular life through the synthesis of model protocells. *Cold Spring Harbor Symp. Quant. Biol.*, **74**, 47–54.

2 Ray, T.S. (1992) An Approach to the Synthesis of Life, in: Langton, C.G. (Ed.) *Artificial Life II*, Addison-Wesley Publishing Company, Inc., pp. 371–408.

3 Dueber, J.E., Wu, G.C., Malmirchegini, G.R., Moon, T.S., Petzold, C.J., Ullal, A.V., Prather, K.L., Keasling, J.D. (2009) Synthetic protein scaffolds provide modular control over metabolic flux. *Nat. Biotechnol.*, **27** (8), 753–759.

4 Ro, D.K., Paradise, E.M., Ouellet, M., Fisher, K.J., Newman, K.L., Ndungu, J.M., Ho, K.A., Eachus, R.A., Ham, T.S., Kirby, J., Chang, M.C., Withers, S.T., Shiba, Y., Sarpong, R., Keasling, J.D. (2006) Production of the antimalarial drug precursor artemisinic acid in engineered yeast. *Nature*, **440** (7086), 940–943.

5 Parsons, J.B., Frank, S., Bhella, D., Liang, M., Prentice, M.B., Mulvihill, D.P., Warren, M.J. (2010) Synthesis of empty bacterial microcompartments, directed organelle protein incorporation, and evidence of filament-associated organelle movement. *Mol. Cell*, **38** (2), 305–315.

6 Biebl, H., Menzel, K., Zeng, A.P., Deckwer, W.D. (1999) Microbial production of 1,3-propanediol. *Appl. Microbiol. Biotechnol.*, **52** (3), 289–297.

7 Gibson, D.G., Glass, J.I., Lartigue, C., Noskov, V.N., Chuang, R.Y., Algire, M.A., Benders, G.A., Montague, M.G., Ma, L., Moodie, M.M., Merryman, C., Vashee, S., Krishnakumar, R., Assad-Garcia, N., Andrews-Pfannkoch, C., Denisova, E.A., Young, L., Qi, Z.Q., Segall-Shapiro, T.H., Calvey, C.H., Parmar, P.P., Hutchison, C.A., III, Smith, H.O., Venter, J.C. (2010) Creation of a bacterial cell controlled by a chemically synthesized genome. *Science*, **329**, 52–56.

8 Nakamura, C.E., Whited, G.M. (2003) Metabolic engineering for the microbial production of 1,3-propanediol. *Curr. Opin. Biotechnol.*, **14** (5), 454–459.

9 Zeng, A.P., Biebl, H. (2002) Bulk chemicals from biotechnology: the case of 1,3-propanediol production and the new trends. *Adv. Biochem. Eng. Biotechnol.*, **74**, 239–259.

10 Balagadde, F.K., Song, H., Ozaki, J., Collins, C.H., Barnet, M., Arnold, F.H., Quake, S.R., You, L. (2008) A synthetic *Escherichia coli* predator-prey ecosystem. *Mol. Syst. Biol.*, **4**, 187.

11 Brenner, K., Karig, D.K., Weiss, R., Arnold, F.H. (2007) Engineered bidirectional communication mediates a consensus in a microbial biofilm consortium. *Proc. Natl Acad. Sci. USA*, **104** (44), 17300–17304.

12 Hu, B., Du, J., Zou, R.Y., Yuan, Y.J. (2010) An environment-sensitive synthetic microbial ecosystem. *PLoS One*, **5** (5), e10619.

13 Weber, W., Daoud-El Baba, M., Fussenegger, M. (2007) Synthetic ecosystems based on airborne inter- and intrakingdom communication. *Proc. Natl Acad. Sci. USA*, **104** (25), 10435–10440.

14 Matsuoka, Y., Vigouroux, Y., Goodman, M.M., Sanchez, G.J., Buckler, E., Doebley, J. (2002) A single domestication for maize shown by multilocus microsatellite genotyping. *Proc. Natl Acad. Sci. USA*, **99** (9), 6080–6084.

15 Darwin, C. (1859) *On the Origin of Species by Means of Natural Selection, or the Preservation of Favoured Races in the Struggle for Life*, D. Appleton, New York.

16 Fisher, R. (1936) Has Mendel's work been rediscovered? *Ann. Sci.*, **1** (2), 115–137.

17 Mendel, G. (1866) Versuche über Pflanzen-Hybriden, Verhandlungen des naturforschenden Verein Brunn. Abh. IV.

18 Avery, O.T., Macleod, C.M., McCarty, M. (1944) Studies on the chemical nature of the substance inducing transformation of pneumococcal types: induction of transformation by a deoxyribonucleic acid fraction isolated from *Pneumococcus* type Iii. *J. Exp. Med.*, **79** (2), 137–158.

19 Weiss, B., Richardson, C.C. (1967) Enzymatic breakage and joining of deoxyribonucleic acid, I. Repair of single-strand breaks in DNA by an enzyme system from *Escherichia coli* infected with T4 bacteriophage. *Proc. Natl Acad. Sci. USA*, **57** (4), 1021–1028.

20 Smith, H.O., Wilcox, K.W. (1970) A restriction enzyme from *Hemophilus influenzae*. I.

Purification and general properties. *J. Mol. Biol.*, **51** (2), 379–391.

21 Danna, K., Nathans, D. (1971) Specific cleavage of simian virus 40 DNA by restriction endonuclease of *Hemophilus influenzae*. *Proc. Natl Acad. Sci. USA*, **68** (12), 2913–2917.

22 Jackson, D.A., Symons, R.H., Berg, P. (1972) Biochemical method for inserting new genetic information into DNA of Simian Virus 40: circular SV40 DNA molecules containing lambda phage genes and the galactose operon of *Escherichia coli*. *Proc. Natl Acad. Sci. USA*, **69** (10), 2904–2909.

23 Cohen, S.N., Chang, A.C., Boyer, H.W., Helling, R.B. (1973) Construction of biologically functional bacterial plasmids in vitro. *Proc. Natl Acad. Sci. USA*, **70** (11), 3240–3244.

24 Lobban, P.E., Kaiser, A.D. (1973) Enzymatic end-to end joining of DNA molecules. *J. Mol. Biol.*, **78** (3), 453–471.

25 Saiki, R.K., Gelfand, D.H., Stoffel, S., Scharf, S.J., Higuchi, R., Horn, G.T., Mullis, K.B., Erlich, H.A. (1988) Primer-directed enzymatic amplification of DNA with a thermostable DNA polymerase. *Science*, **239** (4839), 487–491.

26 Mullis, K.B., Faloona, F.A. (1987) Specific synthesis of DNA in vitro via a polymerase-catalyzed chain reaction. *Methods Enzymol.*, **155**, 335–350.

27 Saiki, R.K., Scharf, S., Faloona, F., Mullis, K.B., Horn, G.T., Erlich, H.A., Arnheim, N. (1985) Enzymatic amplification of beta-globin genomic sequences and restriction site analysis for diagnosis of sickle cell anemia. *Science*, **230** (4732), 1350–1354.

28 Engler, C., Gruetzner, R., Kandzia, R., Marillonnet, S. (2009) Golden gate shuffling: a one-pot DNA shuffling method based on type IIs restriction enzymes. *PLoS One*, **4** (5), e5553.

29 Engler, C., Kandzia, R., Marillonnet, S. (2008) A one pot, one step, precision cloning method with high throughput capability. *PLoS One*, **3** (11), e3647.

30 Shetty, R.P., Endy, D., Knight, T.F. Jr (2008) Engineering BioBrick vectors from BioBrick parts. *J. Biol. Eng.*, **2**, 5.

31 Sleight, S.C., Bartley, B.A., Lieviant, J.A., Sauro, H.M. (2010) In-fusion biobrick assembly and re-engineering. *Nucleic Acids Res.*, **38** (8), 2624–2636.

32 Khorana, H.G., Agarwal, K.L., Buchi, H., Caruthers, M.H., Gupta, N.K., Kleppe, K., Kumar, A., Otsuka, E., RajBhandary, U.L., Van de Sande, J.H., Sgaramella, V., Terao, T., Weber, H., Yamada, T. (1972) Studies on polynucleotides. 103. Total synthesis of the structural gene for an alanine transfer ribonucleic acid from yeast. *J. Mol. Biol.*, **72** (2), 209–217.

33 Edge, M.D., Green, A.R., Heathcliffe, G.R., Meacock, P.A., Schuch, W., Scanlon, D.B., Atkinson, T.C., Newton, C.R., Markham, A.F. (1981) Total synthesis of a human leukocyte interferon gene. *Nature*, **292** (5825), 756–762.

34 Itakura, K., Hirose, T., Crea, R., Riggs, A.D., Heyneker, H.L., Bolivar, F., Boyer, H.W. (1977) Expression in *Escherichia coli* of a chemically synthesized gene for the hormone somatostatin. *Science*, **198** (4321), 1056–1063.

35 Stemmer, W.P., Crameri, A., Ha, K.D., Brennan, T.M., Heyneker, H.L. (1995) Single-step assembly of a gene and entire plasmid from large numbers of oligodeoxyribonucleotides. *Gene*, **164** (1), 49–53.

36 Bang, D., Church, G.M. (2008) Gene synthesis by circular assembly amplification. *Nat. Methods*, **5** (1), 37–39.

37 Gibson, D.G., Young, L., Chuang, R.Y., Venter, J.C., Hutchison, C.A., III, Smith, H.O. (2009) Enzymatic assembly of DNA molecules up to several hundred kilobases. *Nat. Methods*, **6** (5), 343–345.

38 Li, M.Z., Elledge, S.J. (2007) Harnessing homologous recombination in vitro to generate recombinant DNA via SLIC. *Nat. Methods*, **4** (3), 251–256.

39 Smith, H.O., Hutchison, C.A., III, Pfannkoch, C., Venter, J.C. (2003) Generating a synthetic genome by whole genome assembly: phiX174 bacteriophage from synthetic oligonucleotides. *Proc. Natl Acad. Sci. USA*, **100** (26), 15440–15445.

40 Gibson, D.G., Benders, G.A., Andrews-Pfannkoch, C., Denisova, E.A., Baden-Tillson, H., Zaveri, J., Stockwell, T.B., Brownley, A., Thomas, D.W., Algire, M.A., Merryman, C., Young, L., Noskov, V.N., Glass, J.I., Venter, J.C., Hutchison, C.A., III, Smith, H.O. (2008) Complete

chemical synthesis, assembly, and cloning of a *Mycoplasma genitalium* genome. *Science*, **319** (5867), 1215–1220.

41 Gibson, D.G., Benders, G.A., Axelrod, K.C., Zaveri, J., Algire, M.A., Moodie, M., Montague, M.G., Venter, J.C., Smith, H.O., Hutchison, C.A., III (2008) One-step assembly in yeast of 25 overlapping DNA fragments to form a complete synthetic *Mycoplasma genitalium* genome. *Proc. Natl Acad. Sci. USA*, **105** (51), 20404–20409.

42 Tsuge, K., Matsui, K., Itaya, M. (2003) One step assembly of multiple DNA fragments with a designed order and orientation in *Bacillus subtilis* plasmid. *Nucleic Acids Res.*, **31** (21), e133.

43 Gibson, D.G. (2009) Synthesis of DNA fragments in yeast by one-step assembly of overlapping oligonucleotides. *Nucleic Acids Res.*, **37** (20), 6984–6990.

44 Shao, Z., Zhao, H. (2009) DNA assembler, an in vivo genetic method for rapid construction of biochemical pathways. *Nucleic Acids Res.*, **37** (2), e16.

45 Quan, J., Tian, J. (2009) Circular polymerase extension cloning of complex gene libraries and pathways. *PLoS One*, **4** (7), e6441.

46 Zhu, B., Cai, G., Hall, E.O., Freeman, G.J. (2007) In-fusion assembly: seamless engineering of multidomain fusion proteins, modular vectors, and mutations. *Biotechniques*, **43** (3), 354–359.

47 Ma, H., Kunes, S., Schatz, P.J., Botstein, D. (1987) Plasmid construction by homologous recombination in yeast. *Gene*, **58** (2–3), 201–216.

48 Moerschell, R.P., Tsunasawa, S., Sherman, F. (1988) Transformation of yeast with synthetic oligonucleotides. *Proc. Natl Acad. Sci. USA*, **85** (2), 524–528.

49 Orr-Weaver, T.L., Szostak, J.W., Rothstein, R.J. (1981) Yeast transformation: a model system for the study of recombination. *Proc. Natl Acad. Sci. USA*, **78** (10), 6354–6358.

50 Raymond, C.K., Pownder, T.A., Sexson, S.L. (1999) General method for plasmid construction using homologous recombination. *Biotechniques*, **26** (1), 134–138, 140–141.

51 Raymond, C.K., Sims, E.H., Olson, M.V. (2002) Linker-mediated recombinational subcloning of large DNA fragments using yeast. *Genome Res.*, **12** (1), 190–197.

52 Gibson, D.G., Smith, H.O., Hutchison, C.A., III, Venter, J.C., Merryman, C. (2010) Chemical synthesis of the mouse mitochondrial genome. *Nat. Methods*, **7**, 901–903.

53 Itaya, M., Fujita, K., Kuroki, A., Tsuge, K. (2008) Bottom-up genome assembly using the Bacillus subtilis genome vector. *Nat. Methods*, **5** (1), 41–43.

54 Itaya, M., Tsuge, K., Koizumi, M., Fujita, K. (2005) Combining two genomes in one cell: stable cloning of the *Synechocystis* PCC6803 genome in the *Bacillus subtilis* 168 genome. *Proc. Natl Acad. Sci. USA*, **102** (44), 15971–15976.

55 Richmond, K.E., Li, M.H., Rodesch, M.J., Patel, M., Lowe, A.M., Kim, C., Chu, L.L., Venkataramaian, N., Flickinger, S.F., Kaysen, J., Belshaw, P.J., Sussman, M.R., Cerrina, F. (2004) Amplification and assembly of chip-eluted DNA (AACED): a method for high-throughput gene synthesis. *Nucleic Acids Res.*, **32** (17), 5011–5018.

56 Tian, J., Gong, H., Sheng, N., Zhou, X., Gulari, E., Gao, X., Church, G. (2004) Accurate multiplex gene synthesis from programmable DNA microchips. *Nature*, **432** (7020), 1050–1054.

57 Zhou, X., Cai, S., Hong, A., You, Q., Yu, P., Sheng, N., Srivannavit, O., Muranjan, S., Rouillard, J.M., Xia, Y., Zhang, X., Xiang, Q., Ganesh, R., Zhu, Q., Matejko, A., Gulari, E., Gao, X. (2004) Microfluidic PicoArray synthesis of oligodeoxynucleotides and simultaneous assembling of multiple DNA sequences. *Nucleic Acids Res.*, **32** (18), 5409–5417.

58 Carr, P.A., Park, J.S., Lee, Y.J., Yu, T., Zhang, S., Jacobson, J.M. (2004) Protein-mediated error correction for de novo DNA synthesis. *Nucleic Acids Res.*, **32** (20), e162.

59 Binkowski, B.F., Richmond, K.E., Kaysen, J., Sussman, M.R., Belshaw, P.J. (2005) Correcting errors in synthetic DNA through consensus shuffling. *Nucleic Acids Res.*, **33** (6), e55.

60 Sawitzke, J., Thomason, L., Costantino, N., Bubunenko, M., Datta, S. (2007) Recombineering: in vivo genetic engineering in *E. coli*, *S. enterica*, and beyond. *Methods Enzymol.*, **421**, 171–199.

61 Herring, C.D., Glasner, J.D., Blattner, F.R. (2003) Gene replacement without selection: regulated suppression of amber mutations in *Escherichia coli*. *Gene*, **311**, 153–163.

62 Datsenko, K.A., Wanner, B.L. (2000) One-step inactivation of chromosomal genes in *Escherichia coli* K-12 using PCR products. *Proc. Natl Acad. Sci. USA*, **97** (12), 6640–6645.

63 Lesic, B., Rahme, L.G. (2008) Use of the lambda Red recombinase system to rapidly generate mutants in *Pseudomonas aeruginosa*. *BMC Mol. Biol.*, **9**, 20.

64 Karlinsey, J.E. (2007) Lambda-Red genetic engineering in *Salmonella enterica* serovar. *typhimurium*. *Methods Enzymol.*, **421**, 199–209.

65 van Kessel, J.C., Hatfull, G.F. (2008) Mycobacterial recombineering. *Methods Mol. Biol.*, **435**, 203–215.

66 Wang, H.H., Isaacs, F.J., Carr, P.A., Sun, Z.Z., Xu, G., Forest, C.R., Church, G.M. (2009) Programming cells by multiplex genome engineering and accelerated evolution. *Nature*, **460** (7257), 894–898.

67 Warner, J.R., Reeder, P.J., Karimpour-Fard, A., Woodruff, L.B., Gill, R.T. (2010) Rapid profiling of a microbial genome using mixtures of barcoded oligonucleotides. *Nat. Biotechnol.*, **28** (8), 856–862.

68 Suzuki, Y., St Onge, R.P., Mani, R., King, O.D., Heilbut, A., Labunskyy, V.M., Chen, W., Pham, L., Zhang, L.V., Tong, A.H., Nislow, C., Giaever, G., Gladyshev, V.N., Vidal, M., Schow, P., Lehar, J., Roth, F.P. (2010) Knocking out multi-gene redundancies via cycles of sexual assortment and fluorescence selection. *Nat. Methods*, **8**, 159–164.

69 Tyo, K.E., Ajikumar, P.K., Stephanopoulos, G. (2009) Stabilized gene duplication enables long-term selection-free heterologous pathway expression. *Nat. Biotechnol.*, **27** (8), 760–765.

70 Khalil, A.S., Collins, J.J. (2010) Synthetic biology: applications come of age. *Nat. Rev. Genet.*, **11** (5), 367–379.

71 Mukherji, S., van Oudenaarden, A. (2009) Synthetic biology: understanding biological design from synthetic circuits. *Nat. Rev. Genet.*, **10** (12), 859–871.

72 McAdams, H.H., Arkin, A. (2000) Towards a circuit engineering discipline. *Curr. Biol.*, **10** (8), R318–R320.

73 Elowitz, M.B., Leibler, S. (2000) A synthetic oscillatory network of transcriptional regulators. *Nature*, **403** (6767), 335–338.

74 Gardner, T.S., Cantor, C.R., Collins, J.J. (2000) Construction of a genetic toggle switch in *Escherichia coli*. *Nature*, **403** (6767), 339–342.

75 Dueber, J.E., Yeh, B.J., Chak, K., Lim, W.A. (2003) Reprogramming control of an allosteric signaling switch through modular recombination. *Science*, **301** (5641), 1904–1908.

76 Kramer, B.P., Fussenegger, M. (2005) Hysteresis in a synthetic mammalian gene network. *Proc. Natl Acad. Sci. USA*, **102** (27), 9517–9522.

77 Danino, T., Mondragon-Palomino, O., Tsimring, L., Hasty, J. (2010) A synchronized quorum of genetic clocks. *Nature*, **463** (7279), 326–330.

78 Fung, E., Wong, W.W., Suen, J.K., Bulter, T., Lee, S.G., Liao, J.C. (2005) A synthetic gene-metabolic oscillator. *Nature*, **435** (7038), 118–122.

79 Stricker, J., Cookson, S., Bennett, M.R., Mather, W.H., Tsimring, L.S., Hasty, J. (2008) A fast, robust and tunable synthetic gene oscillator. *Nature*, **456** (7221), 516–519.

80 Tigges, M., Marquez-Lago, T.T., Stelling, J., Fussenegger, M. (2009) A tunable synthetic mammalian oscillator. *Nature*, **457** (7227), 309–312.

81 Ajo-Franklin, C.M., Drubin, D.A., Eskin, J.A., Gee, E.P., Landgraf, D., Phillips, I., Silver, P.A. (2007) Rational design of memory in eukaryotic cells. *Genes Dev.*, **21** (18), 2271–2276.

82 Friedland, A.E., Lu, T.K., Wang, X., Shi, D., Church, G., Collins, J.J. (2009) Synthetic gene networks that count. *Science*, **324** (5931), 1199–1202.

83 Basu, S., Mehreja, R., Thiberge, S., Chen, M.T., Weiss, R. (2004) Spatiotemporal control of gene expression with pulse-generating networks. *Proc. Natl Acad. Sci. USA*, **101** (17), 6355–6360.

84 Anderson, J.C., Voigt, C.A., Arkin, A.P. (2007) Environmental signal integration by a modular AND gate. *Mol. Syst. Biol.*, **3**, 133.

85 Basu, S., Gerchman, Y., Collins, C.H., Arnold, F.H., Weiss, R. (2005) A synthetic multicellular system for programmed

pattern formation. *Nature*, **434** (7037), 1130–1134.

86 Bulter, T., Lee, S.G., Wong, W.W., Fung, E., Connor, M.R., Liao, J.C. (2004) Design of artificial cell-cell communication using gene and metabolic networks. *Proc. Natl Acad. Sci. USA*, **101** (8), 2299–2304.

87 You, L., Cox, R.S., III, Weiss, R., Arnold, F.H. (2004) Programmed population control by cell-cell communication and regulated killing. *Nature*, **428** (6985), 868–871.

88 Aubel, D., Fussenegger, M. (2010) Mammalian synthetic biology–from tools to therapies. *BioEssays*, **32** (4), 332–345.

89 Grunberg, R., Serrano, L. (2010) Strategies for protein synthetic biology. *Nucleic Acids Res.*, **38** (8), 2663–2675.

90 Haynes, K.A., Silver, P.A. (2009) Eukaryotic systems broaden the scope of synthetic biology. *J. Cell Biol.*, **187** (5), 589–596.

91 Choudhary, S., Schmidt-Dannert, C. (2010) Applications of quorum sensing in biotechnology. *Appl. Microbiol. Biotechnol.*, **86** (5), 1267–1279.

92 Richardson, S.M., Nunley, P.W., Yarrington, R.M., Boeke, J.D., Bader, J.S. (2010) GeneDesign 3.0 is an updated synthetic biology toolkit. *Nucleic Acids Res.*, **38** (8), 2603–2606.

93 Endy, D. (2005) Foundations for engineering biology. *Nature*, **438** (7067), 449–453.

94 Weeding, E., Houle, J., Kaznessis, Y.N. (2010) SynBioSS designer: a web-based tool for the automated generation of kinetic models for synthetic biological constructs. *Brief. Bioinform.*, **11** (4), 394–402.

95 Kobayashi, H., Kaern, M., Araki, M., Chung, K., Gardner, T.S., Cantor, C.R., Collins, J.J. (2004) Programmable cells: interfacing natural and engineered gene networks. *Proc. Natl Acad. Sci. USA*, **101** (22), 8414–8419.

96 Anderson, J.C., Clarke, E.J., Arkin, A.P., Voigt, C.A. (2006) Environmentally controlled invasion of cancer cells by engineered bacteria. *J. Mol. Biol.*, **355** (4), 619–627.

97 Looger, L.L., Dwyer, M.A., Smith, J.J., Hellinga, H.W. (2003) Computational design of receptor and sensor proteins with novel functions. *Nature*, **423** (6936), 185–190.

98 Lu, T.K., Collins, J.J. (2009) Engineered bacteriophage targeting gene networks as adjuvants for antibiotic therapy. *Proc. Natl Acad. Sci. USA*, **106** (12), 4629–4634.

99 Ajikumar, P.K., Xiao, W.-H., Tyo, K.E.J., Wang, Y., Simeon, F., Leonard, E., Mucha, O., Phon, T.H., Pfeifer, B., Stephanopoulos, G. (2010) Isoprenoid pathway optimization for taxol precursor overproduction in *Escherichia coli*. *Science*, **330** (6000), 70–74.

100 Isalan, M., Lemerle, C., Serrano, L. (2005) Engineering gene networks to emulate *Drosophila* embryonic pattern formation. *PLoS Biol.*, **3** (3), e64.

101 Tabor, J.J., Salis, H.M., Simpson, Z.B., Chevalier, A.A., Levskaya, A., Marcotte, E.M., Voigt, C.A., Ellington, A.D. (2009) A synthetic genetic edge detection program. *Cell*, **137** (7), 1272–1281.

102 Hyde, C.C., Ahmed, S.A., Padlan, E.A., Miles, E.W., Davies, D.R. (1988) Three-dimensional structure of the tryptophan synthase alpha 2 beta 2 multienzyme complex from *Salmonella typhimurium*. *J. Biol. Chem.*, **263** (33), 17857–17871.

103 Jorgensen, K., Rasmussen, A.V., Morant, M., Nielsen, A.H., Bjarnholt, N., Zagrobelny, M., Bak, S., Moller, B.L. (2005) Metabolon formation and metabolic channeling in the biosynthesis of plant natural products. *Curr. Opin. Plant Biol.*, **8** (3), 280–291.

104 Pfeifer, B.A., Khosla, C. (2001) Biosynthesis of polyketides in heterologous hosts. *Microbiol. Mol. Biol. Rev.*, **65** (1), 106–118.

105 Thoden, J.B., Holden, H.M., Wesenberg, G., Raushel, F.M., Rayment, I. (1997) Structure of carbamoyl phosphate synthetase: a journey of 96 A from substrate to product. *Biochemistry*, **36** (21), 6305–6316.

106 Bobik, T.A., Havemann, G.D., Busch, R.J., Williams, D.S., Aldrich, H.C. (1999) The propanediol utilization (pdu) operon of *Salmonella enterica* serovar. *typhimurium* LT2 includes genes necessary for formation of polyhedral organelles involved in coenzyme B(12)-dependent 1, 2-propanediol degradation. *J. Bacteriol.*, **181** (19), 5967–5975.

107 Kofoid, E., Rappleye, C., Stojiljkovic, I., Roth, J. (1999) The 17-gene ethanolamine (eut) operon of *Salmonella typhimurium*

encodes five homologues of carboxysome shell proteins. *J. Bacteriol.*, **181** (17), 5317–5329.

108 Conrado, R.J., Mansell, T.J., Varner, J.D., DeLisa, M.P. (2007) Stochastic reaction-diffusion simulation of enzyme compartmentalization reveals improved catalytic efficiency for a synthetic metabolic pathway. *Metab. Eng.*, **9** (4), 355–363.

109 Pettersson, H., Pettersson, G. (2001) Kinetics of the coupled reaction catalysed by a fusion protein of beta-galactosidase and galactose dehydrogenase. *Biochim. Biophys. Acta*, **1549** (2), 155–160.

110 Wenz, G. (1994) Cyclodextrins as building blocks for supramolecular structures and functional units. *Angew. Chem. Int. Ed. Engl.*, **33** (8), 803–822.

111 Vriezema, D.M., Comellas Aragones, M., Elemans, J.A., Cornelissen, J.J., Rowan, A.E., Nolte, R.J. (2005) Self-assembled nanoreactors. *Chem. Rev.*, **105** (4), 1445–1489.

112 Martin, V.J., Pitera, D.J., Withers, S.T., Newman, J.D., Keasling, J.D. (2003) Engineering a mevalonate pathway in *Escherichia coli* for production of terpenoids. *Nat. Biotechnol.*, **21** (7), 796–802.

113 Mingardon, F., Chanal, A., Lopez-Contreras, A.M., Dray, C., Bayer, E.A., Fierobe, H.P. (2007) Incorporation of fungal cellulases in bacterial minicellulosomes yields viable, synergistically acting cellulolytic complexes. *Appl. Environ. Microbiol.*, **73** (12), 3822–3832.

114 Mingardon, F., Chanal, A., Tardif, C., Bayer, E.A., Fierobe, H.P. (2007) Exploration of new geometries in cellulosome-like chimeras. *Appl. Environ. Microbiol.*, **73** (22), 7138–7149.

115 Iancu, C.V., Ding, H.J., Morris, D.M., Dias, D.P., Gonzales, A.D., Martino, A., Jensen, G.J. (2007) The structure of isolated *Synechococcus* strain WH8102 carboxysomes as revealed by electron cryotomography. *J. Mol. Biol.*, **372** (3), 764–773.

116 Comellas-Aragones, M., Engelkamp, H., Claessen, V.I., Sommerdijk, N.A., Rowan, A.E., Christianen, P.C., Maan, J.C., Verduin, B.J., Cornelissen, J.J., Nolte, R.J. (2007) A virus-based single-enzyme nanoreactor. *Nat. Nanotechnol.*, **2** (10), 635–639.

117 Yang, C., Freudl, R., Qiao, C. (2009) Export of methyl parathion hydrolase to the periplasm by the twin-arginine translocation pathway in *Escherichia coli*. *J. Agric. Food Chem.*, **57** (19), 8901–8905.

118 Benders, G.A., Noskov, V.N., Denisova, E.A., Lartigue, C., Gibson, D.G., Assad-Garcia, N., Chuang, R.Y., Carrera, W., Moodie, M., Algire, M.A., Phan, Q., Alperovich, N., Vashee, S., Merryman, C., Venter, J.C., Smith, H.O., Glass, J.I., Hutchison, C.A., III (2010) Cloning whole bacterial genomes in yeast. *Nucleic Acids Res.*, **38** (8), 2558–2569.

119 Lartigue, C., Glass, J.I., Alperovich, N., Pieper, R., Parmar, P.P., Hutchison, C.A., III, Smith, H.O., Venter, J.C. (2007) Genome transplantation in bacteria: changing one species to another. *Science*, **317** (5838), 632–638.

120 Lartigue, C., Vashee, S., Algire, M.A., Chuang, R.Y., Benders, G.A., Ma, L., Noskov, V.N., Denisova, E.A., Gibson, D.G., Assad-Garcia, N., Alperovich, N., Thomas, D.W., Merryman, C., Hutchison, C.A., III, Smith, H.O., Venter, J.C., Glass, J.I. (2009) Creating bacterial strains from genomes that have been cloned and engineered in yeast. *Science*, **325** (5948), 1693–1696.

121 Noskov, V.N., Segall-Shapiro, T.H., Chuang, R.Y. (2010) Tandem repeat coupled with endonuclease cleavage (TREC): a seamless modification tool for genome engineering in yeast. *Nucleic Acids Res.*, **38** (8), 2570–2576.

122 Cho, M.K., Magnus, D., Caplan, A.L., McGee, D. (1999) Policy forum: genetics. Ethical considerations in synthesizing a minimal genome. *Science*, **286** (5447), 2087, 2089–2090.

123 Cho, M.K., Relman, D.A. (2010) Genetic technologies. Synthetic 'life,' ethics, national security, and public discourse. *Science*, **329** (5987), 38–39.

124 De Vriend, H.C. (2006) *Constructing Life – Early Social Reflections on the Emerging Field of Synthetic Biology*, Working Document 97, Rathenau Institute, The Hague.

125 Garfinkel, M.S., Endy, D., Epstein, G.L., Friedman, R.M. (2007) Synthetic genomics: Options for governance. *Biosecur. Bioterror.*, **5** (4), 359–362.

126 Anonymous (2010) *Screening Framework Guidance for Providers of Synthetic Double-stranded DNA*, Department of Health and Human Services. Available at: http://www.phe.gov/Preparedness/legal/guidance/syndna/Documents/syndna-guidance.pdf (accessed 7 April 2011).

127 International Gene Synthesis Consortium (2009) Harmonized Screening Protocol: Gene Sequence & Customer Screening to Promote Biosecurity. Available at: http://www.genesynthesisconsortium.org/Gene_Synthesis_Consortium/Harmonized_Screening_Protocol.html (accessed 7 April 2011).

128 Committee on Research Standards and Practices to Prevent the Destructive Application of Biotechnology (2004) in: National Research Council of the National Academies (Ed.) *Biotechnology Research in An Age of Bioterrorism*, The National Academies Press, Washington, DC.

129 DIYbio (2010) An Institution for the Amateur Biologist. Available at: http://diybio.org/blog (accessed 7 April 2011).

130 The Presidential Commission for the Study of Bioethical Issues (2010) Agenda for September 13–14. Available at: http://www.bioethics.gov/meetings/091310/ (accessed 7 April 2011).

131 Corrigan-Curay, J. (2009) *NIH Guidelines for Research Involving Recombinant DNA: Biosafety and Synthetic Nucleic Acids*, Department of Health and Human Services. Available at: http://oba.od. nih.gov/biosecurity/meetings/200912/CorriganCuray_NIH%20Guidelines%20Update.pdf (accessed 7 April 2011).

132 Anonymous (2007) Presentations on Safety, Security, and Societal Issues, in Synthetic Biology 3.0. Available at: http://www.syntheticbiology3.ethz.ch/monday.htm (accessed 7 April 2011).

133 Kaebnick, G., Murray, T.H., Parens, E. (2009) Ethical Issues in Synthetic Biology. Available at: http://www.thehastingscenter.org/Research/Detail.aspx?id=1548 (accessed 7 April 2011).

22
Plant Systems Biology

Sonia Osorio and Alisdair R. Fernie
Max Planck Institut für Molekulare Pflanzenphysiologie, Am Mühlenberg 1, 14476 Potsdam-Golm, Germany

Keywords

Systems biology
The combination of molecular and system-level studies, applying a synergistic approach involving modeling, theory, and experiment.

Functional genomics
The field of molecular biology that uses many possible approaches (i.e., genome, transcriptome, proteome, and metabolome) to help understand the properties and function of the entirety of an organism's genes and gene products.

Transcriptomics
The quantification of the transcriptome, the complete set of transcripts in a cell, and their abundance, for a specific developmental stage or physiological condition.

Systems Biology: Advances in Molecular Biology and Medicine, First Edition. Edited by Robert A. Meyers.

Proteomics
A term coined to encompass a field that attempts to understand the expression, function, and regulation of the entire set of proteins, or "proteome" encoded by an organism.

Metabolomics
The quantitative complement of low-molecular-weight metabolites present in a cell under a given set of physiological conditions.

Systems biology aims to integrate the diverse disciplines of biology, chemistry, statistics, and engineering into a cohesive science, and its current challenge is to identify the regulation networks that characterize a living organism. Systems biology combines experimental analyses and creates mathematical models to provide an understanding of how metabolic behavior emerges. Metabolomics has clearly demonstrated its value in elucidating the interaction of the biological system's genome with its environments.

1 Introduction

Organisms are composed of multiple individual components – that is, cells and their corresponding chemical complements, genes, proteins, and metabolites – which are all governed by an intricate network of interactions. This network is not static; rather, the various components evolve and adapt dynamically to internal and environmental changes. The study of this complex system as a single entity is a challenge that has been traditionally addressed by studying the different components of the system in isolation. Although such approaches have produced a significant amount of knowledge and understanding, they are limited in their ability to predict the effects of any alterations in single or multiple components on the dynamics of the whole system.

The discipline of systems biology attempts to shift the way in which an organism is perceived to address the complexity of living systems.

Systems biology can be defined as " . . . a study that aims to understand and the living cell as complete system" [1, 2], and considered as an integration across multiple high-output platform, systems, and disciplines [3, 4].

During most of the twentieth century, biochemistry was tackled in a reductionist manner whereby a cell would be reduced to its component parts which, in turn, would be examined in isolation. In this case, the inherent assumption was that the various parts would function in isolation as they did in the cell. Although much biological knowledge was accrued via such approaches, several problems became apparent when applying the reductionist philosophy. The most obvious problems were the metabolic complexity, and the fact that biological processes are rarely controlled by a single molecular entity. In fact, control was seen to be generally shared across the different entities rather than to be exerted at a single point [5]. In this way, the switch from the reductionist philosophy to a more holistic approach has raised several new

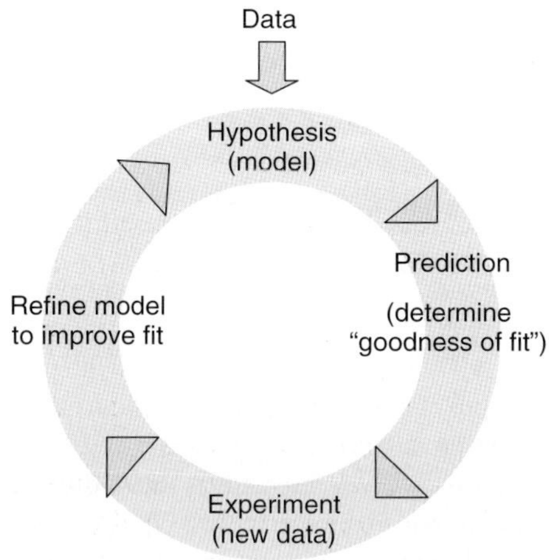

Fig. 1 Systems biology research process.

challenges, including functional genomics where attention is focused specifically on the biochemical complement of the cells and tissues.

The recent years have witnessed an explosion both in developments in technology and in experimental techniques that can be applied to interrogate, simultaneously, many elements of a system. The emergence of high-throughput genomic tools that allow the cataloguing of changes in transcripts, proteins, and metabolites levels in parallel (omics) has made necessary the integration of these parameters. Such studies are designed to identify the network analysis of metabolism that connects the different elements of the system, the main challenge being to integrate the different disciplines of biology, chemistry, statistics, and engineering into a cohesive science.

Systems biology is hypothesis-driven, global, quantitative, iterative, integrative, and dynamic. Its practice begins with the acquisition of global sets of biological data from as many hierarchical levels of information as possible (i.e., DNA sequences, RNA expression, protein, or lipid abundance). Systems biology was initiated due to a need to summarize into useful mathematical models the increasing quantities of high-throughput experimental data that were being acquired at the cell and subcellular levels from the genome, transcriptome, proteome, and metabolome. A simplified process of a systems biological research project includes the following stages (see Fig. 1):

1. If possible, an initial model of the molecular interactions governing pathway function, drawn from previous genetic and biochemical research of the system of interest, is constructed based on biological evidence that can be used for computational analysis.
2. Each pathway component is perturbed via a series of genetic (e.g., gene deletions or overexpressions) or environmental (e.g., changes in growth conditions or temperature) manipulations. The corresponding global cellular response to each perturbation can then be detected and quantified, using technologies for large-scale measurements.
3. The data generated in these experiments are then compared against the predictions and, if necessary, the model is refined to improve its fit with the observations.
4. New hypotheses are formulated to explain any observations not predicted by the model.

The next phase would be to design additional perturbation experiments capable of testing these hypotheses, and then to iteratively repeat steps 2, 3, and 4.

Within this framework, it is necessary to develop analytical tools and strategies that allow relevant biological processes to be described, and information to be extracted from parallel analyses carried out with different profiling platforms.

2 Network Analysis

The network analysis of living systems is an essential component of contemporary systems biology. It is targeted at the assemblage of mutual dependences between interacting systems elements into an integrated view of whole-system functioning. The behavior of biological systems can be deduced from their mathematical models, which are then iteratively refined through a hypothesis-driven process of system perturbation and data integration. Cycles of this process will result in more accurate models; ultimately, these models will explain the systems-level properties of the biological system of interest. Once the model is sufficiently accurate and detailed, it will allow two tasks to be accomplished that were never previously possible: (i) to predict the behavior of the system, given any perturbation; and (ii) to redesign or perturb the molecular network in order to create new emergent properties.

One important concept in system biology is that of the network. According to *Webster's Dictionary*, a network is an intricately connected system of things or people. A type of a biological network is defined by what these "things" are (nodes, vertices, etc.), what the nature of their connections (edges) is and, ideally, why these "things" are connected. In what are currently termed *metabolic networks* or *biochemical reaction networks*, the vertices are represented by metabolites (substances), while the metabolic reaction are represented by directed edges, which interconnect substrates and the products of these reactions. A metabolic network describes the potential pathways that may be used by a cell to accomplish metabolic processes and to use pathway information and data from multiple measurement platforms, tissues, and species to elucidate the mechanistic underpinning of complex phenotypes.

3 Plant Systems Biology

In plants, the initiative to develop systems biology has facilitated the development of functional genomics as a scientific discipline. Similarly, plant systems biology was defined [6] using computational modeling approaches to predict a plant cell (ome) from an underlying genomic understanding. Systems biology also exemplifies the proposition that the simulation of biological systems is a broad field. The goal of functional genomics is to understand the relationship between an organism's genome and its phenotype. The term *functional genomics* is often used broadly to refer to the many possible approaches (i.e., genome, transcriptome, proteome, and metabolome) to understanding the properties and function of the entirety of an organism's genes and gene products. In this part of the chapter, attention will be focused on the growing role of transcriptomics, proteomics, and metabolomics in plant system biology.

Plants are endowed with remarkable flexibility in gene expression that modulates their metabolism and development in response to a wide variety of environmental cues. *Transcriptomics* is the quantification of the transcriptome, the complete set of transcripts in a cell, and their abundance, for a specific developmental stage or physiological condition [7]. It provides the ability to identify quantitative and qualitative differences in gene expression when comparing multiple mRNA populations, and has had a significant impact in nearly all facets of biological science (a huge amount of data is stored in publically accessible databases). Microarrays have long been the platform of choice for many transcriptomic studies, although sequence-based approaches and other related techniques are becoming increasingly attractive with the development of next-generation sequencing (NGS) technology.

The gene expression microarray is a very powerful tool for exploring the expression level of large numbers of transcripts in a single experiment. Currently available commercial microarrays can be used to track the expression levels of 60 000 or more transcripts, and the results of many thousands of microarray studies have since been reported.

Microarray experiments can be used for the identification of target genes for transcriptional regulators, as well as the characterization of complex expression patterns to reveal unexpected dependencies within transcriptional networks.

Today, new sequencing technologies are dramatically accelerating progress in forward genetics. Typically, the identification of genetic variants has become easier with the advent of NGS methods [8], in addition to a major impact on genome-wide association studies (GWAS) [8–11]. The application of NGS methods to the cloning of major-effect genetic variants in crop species is straightforward. Similarly, this will accelerate the discovery of those genes responsible for quantitative trait loci [11].

In order to obtain a better understanding of the molecular basis of an observed phenotype in a plant, it is essential to know which genes are deregulated. This is of particular importance for the functional analysis of transcriptional regulators, since in order to understand the biological function of a transcriptional regulator itself it is often necessary to recognize the genes that this factor regulates. One classical approach towards identifying target genes regulated by a transcriptional factor is to compare the transcriptional profile of a mutant for that transcription factor (TF) with that of the corresponding wild-type. The challenge here is to distinguish these primary target genes from genes that are deregulated in response to the deregulated primary targets. Subsequently, when potential primary target genes are validated using immunoprecipitation with specific antibodies directed against the TF, the DNA of the target locus should become enriched in the precipitate. In this way, the TF should become directly associated with the locus of its target gene.

The technique of mRNA immunopurification has been utilized to achieve a better resolution of the gene networks associated with development and stress responses in plants. This approach was used to profile the total cellular mRNAs from protoplasts of root cell types from *Arabidopsis* that had been exposed to nitrogen or perturbed by iron and sodium chloride, to demonstrate that the environment plays a role in decisions regarding cell identity [12, 13]. A similar study was recently carried out on a specific cell population of *Arabidopsis* seedlings in which the mRNA in the ribosome complexes was profiled. When the seedlings were exposed to a brief period of hypoxia, a pronounced reprioritization of mRNA enrichment in the cell-specific translatomes occurred, including a ubiquitous rise in 49 mRNAs encoding the TFs, signaling proteins, anaerobic metabolism enzymes and uncharacterized proteins, as well as TF family member mRNAs that were markedly reconfigured by hypoxia at both global and cell-specific levels [14].

Whilst transcriptomics provides a useful overview of global gene expression, *proteomics* is often used as a complementary technique to provide a comprehensive insight into the protein profile of an organism. The term proteomics was coined to encompass a field that attempts to understand the expression, function, and regulation of the entire set of proteins, or "proteome," encoded by an organism [15, 16]. The high-throughput quantification of proteins is essentially important for all systems biology approaches, and provides complementary information on steady-state gene expression and perturbation-induced systems responses. Quantitative information on proteins is necessary to infer regulatory events that take place between the expression of a gene and the metabolite that is synthesized by the gene product. It has been shown in different biological systems that, with any rapid metabolism perturbation (i.e., stress conditions), no correlations appear to exist between the transcript, protein, and metabolite levels, which in turn suggests that regulation occurs at different nodes in the network.

The quantitative analysis of protein expression is, therefore, an important tool for the examination of complex biological systems. Despite its importance, quantitative proteomics remains a challenging task because of the high dynamic range of protein concentrations in the cell, and the variation in the physical properties of proteins.

The sequencing of complete genomes has advanced the present understanding of biological systems, mostly by enabling a broad range of technologies for the analysis of gene functions, and also by providing information regarding the theoretical protein-coding capacity of organism. During recent years, mass spectrometry (MS), using instruments with increased detection sensitivities, together with protein and peptide fractionation technologies and data analysis tools, have facilitated the cataloguing of proteomes to acquire information concerning the functional properties and activities of the genome [17, 18]. Until now, two-dimensional gel electrophoresis (2-DGE), coupled to MS, has been the most widely used method for dissecting proteomes. However, gel-free proteomics, such as multidimensional protein identification technology (MudPIT), are now beginning to emerge as high-throughput techniques for the elucidation of proteomes.

For a better understanding of the regulatory network of gene expression at the protein level of biological systems, is important to have access to catalogs (databases) of expressed proteins. In this vein, an intensive analysis has been performed in *Arabidopsis* to assemble a high-density proteome map [19], whereby 13 029 proteins have been identified that include 10 902 distinct proteins identified from plant organs (including roots, cotyledons, juvenile leaves, flower buds, open flowers, carpels, siliques, and seeds), while 8698 proteins have been identified from undifferentiated cultured cells. Together, these proteins represent assignments for almost 50% of all predicted *Arabidopsis* gene models. This study has provided quantitative protein information that can be used to establish correlations between transcript and protein accumulation in different organs, and also to identify organ-specific protein biomarkers to facilitate targeted quantitative proteomics surveys [19]. The study has recently been expanded to cover the early maturation stages of *Arabidopsis* seed development [16].

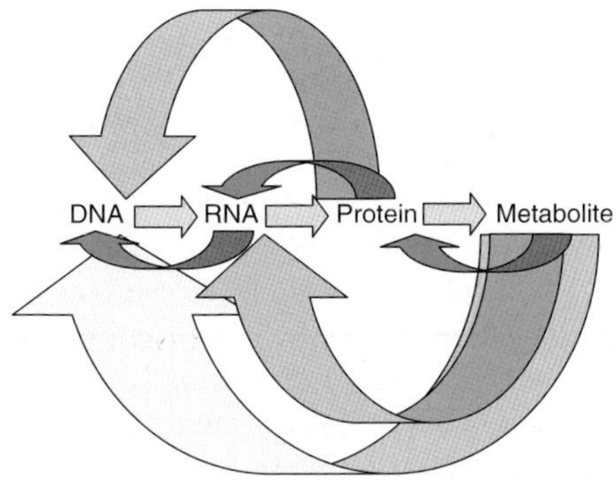

Fig. 2 General schematic of the omics organization, where the flow is from genes to transcripts to proteins to metabolites to phenotype. The dogmatic genes are shown to encode proteins, thus altering the influence of the metabolite.

Cell proliferation is the main driving force for plant growth. Although the basic underlying mechanisms of cell division are conserved among all eukaryotes, the *Arabidopsis* genome contains a collection of cell-cycle regulatory genes that is intriguingly large when compared to other eukaryotes. By using a targeted proteomics approach, it has been possible to successfully map a first draft of the basic cell cycle complex machinery of *Arabidopsis*, providing many new insights into plant cell division [20]. In this study, over 100 new candidate cell cycle proteins provided a comprehensive view on heterodimeric cyclin-dependent kinase (CDK)–cyclin complexes. This strongly underscored the functional diversification among the large family of cyclins, and reflected the pivotal role of cell-cycle regulation in the developmental plasticity of plants. In this study, the cell-cycle interactome was subdivided according to the functional classes of the baits and revealed a complex core cell cycle machinery.

Following the trend set by other post-genomic approaches, recent years have witnessed a veritable explosion both in the interest afforded to metabolite profiling, and the range of metabolites that can be measured using these techniques [21–25].

Metabolomics is the latest technique among a crop of tools that are being applied in the name of functional genomics (i.e., transcriptomics and proteomics). The *metabolome*, in turn, is defined as the " ... quantitative complement of low-molecular-weight metabolites present in a cell under a given set of physiological conditions" [26].

Metabolomics is often – and indeed should be – viewed as a complementary technique to other functional genomics approaches. First, the metabolome is directly related to the phenotype; metabolic fluxes are regulated not only by gene expression, but also by post-transcriptional and post-translational events. Or, put more simply, the metabolome more directly influences the phenotype than do either transcripts or proteins (Fig. 2). Second, although it should be noted that such a linear view is overly simplistic and greatly masks the complexity inherent in cellular regulation, changes in the metabolome are often amplified relative to changes in the transcriptome or proteome. An analysis of the complex network of plant metabolites requires analytical methods that are capable of recording information relating to as many metabolites as possible.

Metabolic profiling – the parallel detection of the levels of multiple metabolites in a single extract – is a relatively new scientific field. However, during the past few years the rapid development of high-throughput metabolic profiling has facilitated the analysis of chemically diverse metabolites in widely divergent genetic populations.

Initial system-based approaches that used data from metabolite profiling involved comprehensive correlation analyses between all of the metabolites profiled in potato tubers. The results of these studies confirmed that most metabolites were minimally correlated, yet some were tightly regulated and others nonlinearly regulated [27]. Recently, more sophisticated metabolic-network analyses have been undertaken by plotting networks in which the metabolites are represented by nodes that are interconnected by lines representing the correlative behavior that intertwines them [28]. These studies have allowed the actions of a small theoretical network on the strength of correlations between the metabolites that constitute these networks.

An even more comprehensive approach is that provided by integration of the data provided by measuring metabolites, proteins, and/or mRNA from the same sample, and then assessing the connectivity across the different molecular entities. In this vein, metabolite–transcript correlation from large data sets collected throughout development and ripening in tomato fruits has revealed different aspects of the regulation of metabolism during fruit ripening [29]. It was apparent that transcript abundance was less strictly coordinated by the functional group than by metabolite abundance, which suggested that post-translational mechanisms dominated the metabolic regulation. Nevertheless, there were some correlations between specific transcripts and metabolites, while several novel associations were identified that could provide potential targets for the manipulation of fruit compositional traits. Finally, there was a strong relationship between ripening-associated transcripts and specific metabolite groups, such as tricarboxylic acid cycle organic acids and sugar phosphates, thus underlining the importance of the respective metabolic pathways during fruit development. A similar analysis was performed on the development and postharvest drying (withering) of grapevine berries, in which transcriptomic, proteomic, and metabolomic data were integrated [30]. In this case, transcripts and proteins were identified that became modulated during withering, in addition to specific classes of metabolites that accumulated at the same time; the latter were used to select subdata sets of variables. It was also possible to identify variables representing selected molecular processes that occurred specifically during berry withering.

The use of metabolomics to assign gene function has also been carried out in several transcriptomes studies, with profiling being used in conjunction with knockout mutagenesis to identify the specific function of *PAL1* and *PAL2* genes of phenylpropanoid metabolism [31], the Myb-like TF, PAP1 [32], the flavonol arabinosyltransferase (UGT78D3) and UDP-rhamnose (RHM) in the flavonoid pathway [33], and the UDP-glucose pyrophosphorylase3 (UGP3) in sulfolipid synthesis [34].

Of all the available genomics tools, metabolite profiling offers arguably the best combinations of practical performance and cost per sample. The expression of almost every gene from yeast and *Escherichia coli* genomes in *Arabidopsis*, and the subsequent metabolite profiling with gas chromatography–mass spectrometry (GC–MS) and liquid chromatography–mass spectrometry (LC–MS), has recently been achieved [35]. This approach was deliberately nonbiased with respect to both the choice of genes and metabolites measured, because of two main objectives: (i) to explore gene function at the level of protein activity; and (ii) to explore the consequences of introducing new proteins into the metabolic network of *Arabidopsis*.

As a technique, GC–MS is versatile and well suited to the nontargeted metabolite profiling of volatile and thermally stable polar and nonpolar metabolites. Classes of metabolites routinely measured with GC–MS include sugars, amino acids, phosphorylated metabolites, organic acids, lipids, and amines. Typically, most of these compounds are nonvolatile and so require comparatively complex derivatization stages to be included during their sample preparation, to induce volatility and thermal stability.

In contrast, sample preparation for LC–MS is simple, and the technique caters well for compounds that are not well suited to GC–MS (large, thermally labile, etc.). Unfortunately, LC is technically demanding and requires the use of large amounts of MS-grade solvents that are expensive not only to purchase but also to dispose of. The technical reproducibility of LC is also not trivial, particularly compared to GC, and retention time drift may also be problematic. There is also the issue that no one column is suitable for a wide range of analyte classes. Secondary metabolites are typically unsuited to GC–MS as they are often too large, particularly after derivatization.

The approach of focusing on individual genes can be easily extended to explore the phenotypic relevance of genome regions. When, recently, the GC–MS profiling of the leaf and mature fruit of breeding populations of wild tomato species was analyzed [36], changes in metabolite content were identified that were potentially important with respect to both stress responses and nutritional benefits. Subsequently, metabolite profiling was applied to an identification of the pericarp metabolite quantitative trait loci (QTL) [37, 38], using a set of well-characterized tomato introgression lines (ILs) of the wild species, *Solanum pennellii* [38]. This resulted in the identification of 889 QTL that were stable over two independent harvests and governed the accumulation of 74 metabolites, including important primary metabolites (e.g., sugars, organic acids, essential amino acids), intermediate metabolites, and vitamins. Furthermore, many of the metabolites for which QTL were detected were important nutritionally, agronomically, or organoleptically; consequently, the incorporation of genetic material from wild species represents an attractive alternative to transgenic approaches for crop improvement. In follow-up studies, the heritability of these traits was investigated by analyzing an additional year's harvest and evaluating the metabolite profiles of heterozygous introgression lines (ILHs) [39]. The study results revealed that most of the metabolic QTL (174 of 332) were dominantly inherited, with relatively high proportions of either additively (61 of 332) or recessively (80 of 332) inherited QTL, while a negligible number of QTL displayed an overdominant inheritance. Interestingly, the mode of inheritance differed quantitatively between diverse classes of compounds (i.e., between sugars and acids), whereas several metabolite pairs belonging to the same pathway displayed a similar mode of inheritance at the same chromosomal loci. However, an evaluation of the association between morphological and metabolic traits in the ILHs revealed that this correlation was far less prominent than in the ILs. Thus, the possibility to uncouple an enhanced metabolite content from any penalties with respect to plant performance and fecundity, and the redevelopment of hybrid genetic material, could prove to be an important advance in the use of genomics-driven breeding approaches. Currently, stress responses in plants in their own right are also beginning to be evaluated using multiple genomics tools, with recent investigations having revealed important information on responses to sulfur starvation, nitrogen, and low-temperature stresses [24, 40].

To recapitulate, the integration of metabolite profiling with other genomics tools is beginning to prove very effective in gene-functional annotation, and in the identification of candidate genes for biotechnology and/or breeding strategies. Hence, systems biology will facilitate the development of functional genomics as a scientific discipline by generating novel hypotheses through quantitative modeling. Plant systems biology is a combination of molecular- and system-level studies that applies a synergistic approach involving modeling, theory, and experiment. Yet, today the scientific community is collecting new and detailed information much faster than it can be assimilated, understood, integrated, and used. This situation has arisen at least in part because the resources provided by modeling at higher hierarchical scales are not being tapped into.

Modeling aspects emerge because of the need for instruments capable of summarizing these increasing quantities of experimental data. Models can play roles not only in data synthesis and prediction, but also in heuristics and system design [41], and may often help to identify gaps in the present knowledge. However, the bridging of a gap between a large body of experimental data and knowledge and a potentially useful mathematical model is not trivial.

In the plant sciences, much effort has been devoted to modeling the growth of crops and plants in relation to the environmental conditions.

In agriculture, the first application to be developed was the process-based model (PBM), which simulates physiological processes and provides an account of metabolism and crop growth in term of mass variables per unit area of land. In contrast, in a functional–structural plant model (FSPM) the plants are treated as a set of interconnected elementary units, and the development over the time or structure of plants, as governed by physiological processes that depend on environmental factors, is explicitly described [42, 43]. In the FSPM, the organs of an individual plant are the primary objects of simulation, though different and interacting plants can be included in one simulation. In this case of the FSPM, two types have been identified: (i) a dynamic model, which generates the state of the system at time step $t + \Delta t$ from the state at time t and the rate of change during the time step Δt; and (ii) a static structural model, that describes a three-dimensional structure at a particular time *in silico*. Unfortunately, these models lack the provision to calculate the change in structure over time [44].

4 Concluding Remarks

Initially, integrated data analysis was introduced as the intermediate level of a systems biology approach to analyze different "omics" datasets; that is, genome-wide measurements of transcript, protein levels or protein–protein interactions, and metabolite levels, the aim being to assess a coherent understanding of biological function. Each of the techniques described above has shown a remarkable value for comprehensive phenotyping strategies and for functional genomics. Yet, in the future, transcriptomics will be led by the development of NGS technologies (provided that data analysis can keep pace) while proteomics will, at least for the immediate future, continue to be dominated by 2-DGE. The accessibility to technology for gel-free proteomics will surely also be improved, leading to its increased use. One significant advantage of gel-free proteomics is its quantitative power through the use of isobaric tags for relative and absolute quantification (iTRAQ), spectral counting, and other techniques. In particular, the field of metabolomics should be improved with the increased resolution of mass spectrometry, while the next generation of instruments should also greatly aid in compound identification, leading to the presence of fewer and fewer "unknowns." Nuclear magnetic resonance techniques are also becoming increasingly sensitive. Yet, whilst each of these platforms holds much future potential, it is their use in combination that continues to create the most excitement, and the simultaneous measurement of transcripts, proteins and metabolites from the same sample will surely promote the study of systems biology in the future. This should lead, ultimately, to the mapping of cellular protein functions to discrete positions in both space and time, and the mathematical modeling of complex biological processes.

To date, the integrative genomics approaches taken have yielded rich descriptive data networks. The challenge remains, however, to elucidate the mechanisms that underlie the behavior of these networks.

References

1 Kitano, H. (2002) Systems biology: a brief overview. *Science*, **295**, 1662–1664.

2 Barabasi, A.L., Oltvai, Z.N. (2004) Network biology: understanding the cell's functional organization. *Nat. Rev. Genet.*, **5**, 101–113.

3 McCulloch, A.D., Paternostro, G. (2005) Cardiac systems biology. *Ann. N. Y. Acad. Sci.*, **1047**, 283–295.

4 Sweetlove, L.J., Last, R.L., Fernie, A.R. (2003) Predictive metabolic engineering: a goal for systems biology. *Plant Physiol.*, **132**, 420–425.

5 ter Kuile, B.H., Westerhoff, H.V. (2001) Transcriptome meets metabolome: hierarchical and metabolic regulation of the glycolytic pathway. *FEBS Lett.*, **500**, 169–171.

6 Minorsky, P.V. (2003) Achieving the in silico plant. Systems biology and the future of plant biological research. *Plant Physiol.*, **132**, 404–409.

7 Wang, Z., Gerstein, M., Snyder, M. (2009) RNA-Seq: a revolutionary tool for transcriptomics. *Nat. Rev. Genet.*, **10**, 57–63.

8 Metzker, M.L. (2010) Sequencing technologies – the next generation. *Nat. Rev. Genet.*, **11**, 31–46.

9 Nordborg, M., Weigel, D. (2008) Next-generation genetics in plants. *Nature*, **456**, 720–723.

10 Bergelson, J., Roux, F. (2010) Towards identifying genes underlying ecologically relevant traits in *Arabidopsis thaliana*. *Nat. Rev. Genet.*, **11**, 867–879.

11 Schneeberger, K., Weigel, D. (2011) Fast-forward genetics enabled by new sequencing technologies. *Trends Plant Sci.*, **16**, 282–288.

12 Dinneny, J.R., Long, T.A., Wang, J.Y., Jung, J.W., Mace, D., Pointer, S., Barron, C., Brady, S.M., Schiefelbein, J., Benfey, P.N. (2008) Cell identity mediates the response of *Arabidopsis* roots to abiotic stress. *Science*, **320**, 942–945.

13 Gifford, M.L., Dean, A., Gutierrez, R.A., Coruzzi, G.M., Birnbaum, K.D. (2008) Cell-specific nitrogen responses mediate developmental plasticity. *Proc. Natl Acad. Sci. USA*, **105**, 803–808.

14 Mustroph, A., Zanetti, M.E., Jang, C.J., Holtan, H.E., Repetti, P.P., Galbraith, D.W., Girke, T., Bailey-Serres, J. (2009) Profiling translatomes of discrete cell populations resolves altered cellular priorities during

hypoxia in *Arabidopsis*. *Proc. Natl Acad. Sci. USA*, **106**, 18843–18848.

15 Wasinger, V.C., Cordwell, S.J., Cerpa-Poljak, A., Yan, J.X., Gooley, A.A., Wilkins, M.R., Duncan, M.W., Harris, R., Williams, K.L., Humphery-Smith, I. (1995) Progress with gene-product mapping of the Mollicutes: *Mycoplasma genitalium*. *Electrophoresis*, **16**, 1090–1094.

16 Zhu, H., Bilgin, M., Snyder, M. (2003) Proteomics. *Annu. Rev. Biochem.*, **72**, 783–812.

17 Cox, J., Mann, M. (2007) Is proteomics the new genomics? *Cell*, **130**, 395–398.

18 Eriksson, J., Fenyo, D. (2007) Improving the success rate of proteome analysis by modeling protein-abundance distributions and experimental designs. *Nat. Biotechnol.*, **25**, 651–655.

19 Baerenfaller, K., Grossmann, J., Grobei, M.A., Hull, R., Hirsch-Hoffmann, M., Yalovsky, S., Zimmermann, P., Grossniklaus, U., Gruissem, W., Baginsky, S. (2008) Genome-scale proteomics reveals *Arabidopsis thaliana* gene models and proteome dynamics. *Science*, **320**, 938–941.

20 Van Leene, J., Hollunder, J., Eeckhout, D., Persiau, G., Van De Slijke, E., Stals, H., Van Isterdael, G., Verkest, A., Neirynck, S., Buffel, Y., De Bodt, S., Maere, S., Laukens, K., Pharazyn, A., Ferreira, P.C., Eloy, N., Renne, C., Meyer, C., Faure, J.D., Steinbrenner, J., Beynon, J., Larkin, J.C., Van de Peer, Y., Hilson, P., Kuiper, M., De Veylder, L., Van Onckelen, H., Inze, D., Witters, E., De Jaeger, G. (2010) Targeted interactomics reveals a complex core cell cycle machinery in *Arabidopsis thaliana*. *Mol. Syst. Biol.*, **6**, 397.

21 Fiehn, O., Kopka, J., Dormann, P., Altmann, T., Trethewey, R.N., Willmitzer, L. (2000) Metabolite profiling for plant functional genomics. *Nat. Biotechnol.*, **18**, 1157–1161.

22 Sumner, L.W., Mendes, P., Dixon, R.A. (2003) Plant metabolomics: large-scale phytochemistry in the functional genomics era. *Phytochemistry*, **62**, 817–836.

23 Hall, R.D., Vos, C.H.R., Verhoeven, H.A., Bino, R.J. (2005) Metabolomics for the assessment of functional diversity and quality traits in plants, in: Vaidyanathan, S., Harrigan, G.G., Goodacre, R. (Eds) *Metabolome Analyses: Strategies for Systems Biology*, Springer, pp. 31–44.

24 Saito, K., Dixon, R.A., Willmitzer, L. (2006) *Plant Metabolomics*, Springer, Heidelberg.

25 Hall, R.D., Brouwer, I.D., Fitzgerald, M.A. (2008) Plant metabolomics and its potential application for human nutrition. *Physiol. Plant*, **132**, 162–175.

26 Oliver, S.G., Winson, M.K., Kell, D.B., Baganz, F. (1998) Systematic functional analysis of the yeast genome. *Trends Biotechnol.*, **16**, 373–378.

27 Roessner, U., Luedemann, A., Brust, D., Fiehn, O., Linke, T., Willmitzer, L., Fernie, A. (2001) Metabolic profiling allows comprehensive phenotyping of genetically or environmentally modified plant systems. *Plant Cell*, **13**, 11–29.

28 Weckwerth, W., Loureiro, M.E., Wenzel, K., Fiehn, O. (2004) Differential metabolic networks unravel the effects of silent plant phenotypes. *Proc. Natl Acad. Sci. USA*, **101**, 7809–7814.

29 Carrari, F., Baxter, C., Usadel, B., Urbanczyk-Wochniak, E., Zanor, M.I., Nunes-Nesi, A., Nikiforova, V., Centero, D., Ratzka, A., Pauly, M., Sweetlove, L.J., Fernie, A.R. (2006) Integrated analysis of metabolite and transcript levels reveals the metabolic shifts that underlie tomato fruit development and highlight regulatory aspects of metabolic network behavior. *Plant Physiol.*, **142**, 1380–1396.

30 Zamboni, A., Di Carli, M., Guzzo, F., Stocchero, M., Zenoni, S., Ferrarini, A., Tononi, P., Toffali, K., Desiderio, A., Lilley, K.S., Enrico Pe, M., Benvenuto, E., Delledonne, M., Pezzotti, M. (2010) Identification of putative stage-specific grapevine berry biomarkers and omics data integration into networks. *Plant Physiol.*, **154**, 1439–1459.

31 Rohde, A., Morreel, K., Ralph, J., Goeminne, G., Hostyn, V., De Rycke, R., Kushnir, S., Van Doorsselaere, J., Joseleau, J.P., Vuylsteke, M., Van Driessche, G., Van Beeumen, J., Messens, E., Boerjan, W. (2004) Molecular phenotyping of the pal1 and pal2 mutants of *Arabidopsis thaliana* reveals far-reaching consequences on phenylpropanoid, amino acid, and carbohydrate metabolism. *Plant Cell*, **16**, 2749–2771.

32 Tohge, T., Nishiyama, Y., Hirai, M.Y., Yano, M., Nakajima, J., Awazuhara, M., Inoue, E., Takahashi, H., Goodenowe, D.B.,

Kitayama, M., Noji, M., Yamazaki, M., Saito, K. (2005) Functional genomics by integrated analysis of metabolome and transcriptome of *Arabidopsis* plants over-expressing an MYB transcription factor. *Plant J.*, **42**, 218–235.

33 Yonekura-Sakakibara, K., Tohge, T., Matsuda, F., Nakabayashi, R., Takayama, H., Niida, R., Watanabe-Takahashi, A., Inoue, E., Saito, K. (2008) Comprehensive flavonol profiling and transcriptome coexpression analysis leading to decoding gene-metabolite correlations in *Arabidopsis*. *Plant Cell*, **20**, 2160–2176.

34 Okazaki, Y., Shimojima, M., Sawada, Y., Toyooka, K., Narisawa, T., Mochida, K., Tanaka, H., Matsuda, F., Hirai, A., Hirai, M.Y., Ohta, H., Saito, K. (2009) A chloroplastic UDP-glucose pyrophosphorylase from *Arabidopsis* is the committed enzyme for the first step of sulfolipid biosynthesis. *Plant Cell*, **21**, 892–909.

35 Fernie, A.R., Trethewey, R.N., Krotzky, A.J., Willmitzer, L. (2004) Metabolite profiling: from diagnostics to systems biology. *Nat. Rev. Mol. Cell. Biol.*, **5**, 763–769.

36 Schauer, N., Zamir, D., Fernie, A.R. (2005) Metabolic profiling of leaves and fruit of wild species tomato: a survey of the *Solanum lycopersicum* complex. *J. Exp. Bot.*, **56**, 297–307.

37 Schauer, N., Semel, Y., Roessner, U., Gur, A., Balbo, I., Carrari, F., Pleban, T., Perez-Melis, A., Bruedigam, C., Kopka, J., Willmitzer, L., Zamir, D., Fernie, A.R. (2006) Comprehensive metabolic profiling and phenotyping of interspecific introgression lines for tomato improvement. *Nat. Biotechnol.*, **24**, 447–454.

38 Eshed, Y., Zamir, D. (1995) An introgression line population of *Lycopersicon pennellii* in the cultivated tomato enables the identification and fine mapping of yield-associated QTL. *Genetics*, **141**, 1147–1162.

39 Schauer, N., Semel, Y., Balbo, I., Steinfath, M., Repsilber, D., Selbig, J., Pleban, T., Zamir, D., Fernie, A.R. (2008) Mode of inheritance of primary metabolic traits in tomato. *Plant Cell*, **20**, 509–523.

40 Saito, K., Dixon, R.A., Willmitzer, L. (2006) in: Nagata, T. Lörz, H. Widholm, J.M. (Eds) *Plant Metabolomics*, Vol. **57**, Springer.

41 Yin, X., Struik, P.C. (2008) Applying modelling experiences from the past to shape crop systems biology: the need to converge crop physiology and functional genomics. *New Phytol.*, **179**, 629–642.

42 Vos, J., Marcelis, L.F.M., Evers, J.B. (2007) Functional-structural plant modelling in crop production: Adding a dimension, in: Vos, J., Marcelis, L.F.M., de Visser, P.H.B., Struik, P.C., Evers, J.B. (Eds) *Functional-Structural Plant Modelling in Crop Production*, Springer, pp. 1–12.

43 Hanan, J., Prusinkiewicz, P. (2008) Foreword: studying plants with functional-structural models. *Funct. Plant Biol.*, **35**, i–iii.

44 Vos, J., Evers, J.B., Buck-Sorlin, G.H., Andrieu, B., Chelle, M., Visser, P.H.B. (2010) Functional-structural plant modelling: a new versatile tool in crop science. *J. Exp. Bot.*, **61**, 2101–2115.

Index

Systems Biology: Advances in Molecular Biology and Medicine, First Edition. Edited by Robert A. Meyers.

d